Developments in Environmental Modelling, 25

# Models of the Ecological Hierarchy

## Developments in Environmental Modelling

1. ENERGY AND ECOLOGICAL MODELLING edited by W.J. Mitsch, R.W. Bossermann and J.M. Klopatek, 1981
2. WATER MANAGEMENT MODELS IN PRACTICE: A CASE STUDY OF THE ASWAN HIGH DAM by D. Whittington and G. Guariso, 1983
3. NUMERICAL ECOLOGY by L. Legendre and P. Legendre, 1983

4A. APPLICATION OF ECOLOGICAL MODELLING IN ENVIRONMENTAL MANAGEMENT PART A edited by S.E. Jørgensen, 1983

4B. APPLICATION OF ECOLOGICAL MODELLING IN ENVIRONMENTAL MANAGEMENT PART B edited by S.E. Jørgensen and W.J. Mitsch, 1983

5. ANALYSIS OF ECOLOGICAL SYSTEMS: STATE-OF-THE-ART IN ECOLOGICAL MODELLING edited by W.K. Lauenroth, G.V. Skogerboe and M. Flug, 1983
6. MODELLING THE FATE AND EFFECT OF TOXIC SUBSTANCES IN THE ENVIRONMENT edited by S.E. Jørgensen, 1984
7. MATHEMATICAL MODELS IN BIOLOGICAL WASTE WATER TREATMENT edited by S.E. Jørgensen and M.J. Gromiec, 1985
8. FRESHWATER ECOSYSTEMS: MODELLING AND SIMULATION by M. Straškraba and A.H. Gnauck, 1985
9. FUNDAMENTALS OF ECOLOGICAL MODELLING by S.E. Jørgensen, 1986
10. AGRICULTURAL NONPOINT SOURCE POLLUTION: MODEL SELECTION AND APPLICATION edited by A. Giorgini and F. Zingales, 1986
11. MATHEMATICAL MODELLING OF ENVIRONMENTAL AND ECOLOGICAL SYSTEMS edited by J.B. Shukia, T.G. Hallam and V. Capasso, 1987
12. WETLAND MODELLING edited by W.J. Mitsch, M. Straškraba and S.E. Jørgensen, 1988
13. ADVANCES IN ENVIRONMENTAL MODELLING edited by A. Marani, 1988
14. MATHEMATICAL SUBMODELS IN WATER QUALITY SYSTEMS edited by S.E. Jørgensen and M.J. Gromiec, 1989
15. ENVIRONMENTAL MODELS: EMISSIONS AND CONSEQUENCES edited by J. Fenhann, H. Larsen, G.A. Mackenzie and B. Rasmussen, 1990
16. MODELLING IN ECOTOXICOLOGY edited by S.E. Jørgensen, 1990
17. MODELLING IN ENVIRONMENTAL CHEMISTRY edited by S.E. Jørgensen, 1991
18. INTRODUCTION TO ENVIRONMENTAL MANAGEMENT edited by P.E. Hansen and S.E. Jørgensen, 1991
19. FUNDAMENTALS OF ECOLOGICAL MODELLING by S.E. Jørgensen, 1994
20. NUMERICAL ECOLOGY 2nd English edition by Pierre Legendre and Louis Legendre
21. FUNDAMENTALS OF ECOLOGICAL MODELLING, Third Edition by G. Bendoricchio and S.E. Jørgensen
22. ENVIRONMENTAL FORESIGHT AND MODELS A MANIFESTO edited by M.B. Beck
23. ENTROPY PRINCIPLE FOR THE DEVELOPMENT OF COMPLEX BIOTIC SYSTEMS: ORGANISMS, ECOSYSTEMS, THE EARTH by I. Aoki
24. NUMERICAL ECOLOGY 3rd English Edition by Pierre Legendre and Louis Legendre

Developments in Environmental Modelling, 25

# Models of the Ecological Hierarchy

## From Molecules to the Ecosphere

Ferenc Jordán
The Microsoft Research-University of Trento, Centre for Computational & Systems Biology (COSBI), Italy.

Sven Erik Jørgensen
Environmental Chemistry Section, Copenhagen University, Copenhagen Ø. Denmark.

ELSEVIER

AMSTERDAM • BOSTON • HEIDELBERG • LONDON
NEW YORK • OXFORD • PARIS • SAN DIEGO
SAN FRANCISCO • SINGAPORE • SYDNEY • TOKYO

Elsevier
Radarweg 29, PO Box 211, 1000 AE Amsterdam, The Netherlands
The Boulevard, Langford Lane, Kidlington, Oxford OX5 1GB, UK



**British Library Cataloguing in Publication Data**
A catalogue record for this book is available from the British Library

**Library of Congress Cataloging-in-Publication Data**
A catalog record for this book is available from the Library of Congress



Printed and bound by CPI Group (UK) Ltd, Croydon, CR0 4YY
Transferred to digital print 2012

ISBN: 978-0-444-59396-2

# Contents

## Part 2 Models of Populations



## Part 3 Models of Interacting Species, Communities



## Part 4 Models of Ecological Networks, Food Chains and Food Webs

## Part 6 Models of Ecosystems



## Part 7 Models of Landscapes

# Preface

This book is based on selected papers covering the presentations at the 7th European Conference on Ecological Modelling, organized by ISEM and hosted by The Microsoft Research—University of Trento Center for Computational and Systems Biology from 30 May to 2 June, 2011 in Riva del Garda, Italy. The key theme of the conference was "models of all levels of the ecological hierarchy". The idea behind the conference was to demonstrate that we do have models of all the levels of the ecological hierarchy—from molecules, genes and cells to ecosystems, landscapes and the ecosphere. Yet, we do not have many models that include several levels and their interactions simultaneously, but it is expected that we will see more and more multilevel models in the future, which would be important progress in ecological modeling, because the ecological hierarchical levels are interacting and all levels in principle influence all the other levels. It implies that only a model that considers these interactions in more or less detail will be able to capture properly the ecological responses to, for instance, human impacts. We need, however, more experience in the development of models that include the interactions between hierarchical levels: this book gives some first prescriptions on how to integrate more hierarchical levels and their interactions in a model. This is discussed in the Introduction, in the last section (Models of Interactions between Hierarchical Levels and Hierarchy Theory) and in the last chapter (Conclusive Remarks About Hierarchy Theory and Multilevel Models).

The book has an Introduction and 32 chapters including the Conclusions, divided into nine sections, focusing on different levels of the ecological hierarchy. The first section is called "Models of the Infra-Individual Level". It has five chapters that cover quantum chemical modeling in the molecular ecology, evolution before life, gene flow modeling, cell population dynamics and molecular interaction networks. The second section has three population dynamic models, while the third section looks at the level of communities and has two chapters. The next level, ecosystems, probably the most widely modeled hierarchical level, has three sections. One section covers networks, food chains and food webs. It has four chapters. The next ecosystem section is entitled "Models of Elements Cycling in Ecosystems" and has three illustrative chapters and the last section about ecosystems is called "Models of Ecosystems" with three chapters. The section "Models of Landscapes" has five chapters. Generally, we have seen an increasing number of papers in the scientific literature about models of landscapes, because of the increased computer power today that allows us to model entire landscapes, consisting of several ecosystems. The last hierarchical level, the ecosphere, has three chapters encompassing the global carbon cycle, the impact on quality of life for biosphere dynamics and the global energy flow and its impacts on the ecological footprint. As already mentioned, we need to develop models that are able to include the interactions between hierarchical levels, which is the focus of the last section together with hierarchy theory in general. Hierarchy theory is of course the prerequisite for a better model coverage of the interactions between hierarchical levels and to understand ecosystem reactions generally, because all the levels are significant for the ecosystem reactions. The last chapter attempts to summarize the conclusions that we can best draw from the 31 selected papers, and presents the state-of-the-art in hierarchy theory.

Ferenc Jordán, Sven Erik Jørgensen

March 2012, Trento and Copenhagen

# Contributors

**Alessandro Acquavita** (239)

ARPA FVG Friuli Venezia-Giulia Regional Environmental Protection Agency, Technical Scientific Management Department, Via Cairoli 14, Palmanova (UD), Italy

**Stefano Allesina** (297)

Department of Ecology & Evolution, Computation Institute, The University of Chicago, Chicago IL, USA

**Annalisa Appice** (35)

Dipartimento di Informatica, Università degli Studi di Bari "Aldo Moro," via Orabona 4, Bari, Italy

**Francisco Arreguín-Sánchez** (501)

Centro Interdisciplinario de Ciencias Marinas del IPN, Apartado, Postal 592, La Paz, 23000 Baja California Sur, Mexico

**Polina Artushenko** (3)

L.V. Kirensky Institute of Physics, SB RAS, Krasnoyarsk, 660036, Russia

**Niko Balkenhol** (391)

Department of Forest Zoology and Forest Conservation, University of Goettingen, Buesgenweg 3, 37077 Goettingen, Germany

**Yurii Baranchikov** (3)

V.N. Sukachev Institute of Forest, SB RAS, Krasnoyarsk, 660036, Russia

**Sergey I. Bartsev** (447)

Institute of biophysics SB RAS, Krasnoyarsk, 660036, Russia

**Pavel V. Belolipetsky** (447)

Institute of biophysics SB RAS, Krasnoyarsk, 660036, Russia; Institute of computational modeling SB RAS, Krasnoyarsk, 660036, Russia

**Maria Bezrukova** (331)

Institute of Physico-Chemical and Biological Problems in Soil Science of the Russian Academy of Sciences, 142290 Institutskaya ul., 2, Pushchino, Moscow Region, Russia

**Antonio Bodini** (297)

Department of Environmental Sciences, University of Parma, Viale Usberti 11/A, 43100 Parma, Italy

**Cristina Bondavalli** (297)

Department of Environmental Sciences, University of Parma, Viale Usberti 11/A, 43100 Parma, Italy

**Yann Bozec** (275)

Station biologique de Roscoff UMR CNRS-UPMC 7144-Equipe Chimie Marine, Place Gearges Teissier 29682 Roscoff, France

**Oksana Y. Buzhdygan** (181)

Department of Ecology and Biomonitoring, Chernivtsi National University, Chernivtsi 58012, Ukraine

**Donata M. Canu** (239)

OGS, The National Institute of Oceanography and Experimental Geophysics, Borgo Grotta Gigante 42c, Sgonico (TS), Italy

**Maria T. Carone** (431)

Envix_Lab, STAT Department, University of Molise, C.da Fonte Lappone, 86090, Pesche (IS), Italy; IMAA, Italian National Research Council—CNR, C.da S.ta Loja, 85050, Tito (PZ), Italy

**Maria L. Carranza** (431)

Envix_Lab, STAT Department, University of Molise, C.da Fonte Lappone, 86090, Pesche (IS), Italy

**Elisa Carraro** (259)

IRSA-CNR, Via del Mulino 19, 20861 Brugherio (MB), Italy

**Michelangelo Ceci** (35)

Dipartimento di Informatica, Università degli Studi di Bari "Aldo Moro," via Orabona 4, Bari, Italy

**Federica Ciocchetta** (517)

The Microsoft Research, University of Trento Centre for Computational and Systems Biology, Piazza Manifattura 1, 38068 Rovereto, Italy

**Marco Ciolli** (355)

Università di Trento, Dipartimento di Ingegneria Civile e Ambientale (DICA), via Mesiano 77, 38123 Trento, Italy

**José-Manuel Zaldívar Comenges** (51)

European Commission, Joint Research Centre, Institute for Health and Consumer Protection, Systems Toxicology Unit, Via E. Fermi 2749, Ispra (VA), Italy

**Diego Copetti** (259)

IRSA-CNR, Via del Mulino 19, 20861 Brugherio (MB), Italy

**Attila Csikász-Nagy** (73)

The Microsoft Research - University of Trento Centre for Computational and Systems Biology, Piazza Manifattura 1, 38068, Rovereto, Italy

**Brigolin Daniele** (165)

Department of Environmental Sciences, Informatics and Statistics, Università Ca' Foscari Venezia, Dorsoduro 2137, 30123 Venezia, Italy

**Hein de Baar** (275)

Royal Netherlands Institute for Sea Research (NIOZ), Department of Marine Chemistry and Geology, P.O. Box 59, NL-1790 AB Den Burg, Texel, Netherlands

**Marko Debeljak** (35)

Department of Knowledge Technologies, Jožef Stefan Institute, Jamova cesta, Ljubljana, Slovenia; Jožef Stefan International Postgraduate School, Jamova cesta, Ljubljana, Slovenia

**Andrey G. Degermendzhi** (447)

Institute of biophysics SB RAS, Krasnoyarsk, 660036, Russia

**Luca Delucchi** (391)

Fondazione Edmund Mach, Research and Innovation Centre, Department of Biodiversity and Molecular Ecology, Via E. Mach 1, 38010 S. Michele all'Adige (TN), Italy

**Masatoshi Denda** (115)

Water Environment Group River Restoration Team, Public Works Research Institute, 1-6 Minamihara, Tsukuba City, Ibaraki 305-8516, Japan

**Santiago R. Doyle** (533)

Instituto de Ciencias, Universidad Nacional de General Sarmiento, J. M. Gutiérrez 1150, Los Polvorines, Argentina; Conicet, Argentina

**Martin Drechsler** (413)

Helmholtz Centre for Environmental Research - UFZ, Department of Ecological Modelling, Permoserstr. 15, 04318 Leipzig, Germany

**Sašo Džeroski** (35)

Department of Knowledge Technologies, Jožef Stefan Institute, Jamova cesta, Ljubljana, Slovenia; Jožef Stefan International Postgraduate School, Jamova cesta, Ljubljana, Slovenia; Centre of Excellence for Integrated Approaches in Chemistry and Biology of Proteins

**Khalid Elkalay** (275)

SAEDD laboratory, Ecole Supérieure de Technologie d'Essaouira Km 9, Route d'Agadir, Essaouira Aljadida BP. 383, Essaouira, Morocco and EnSa laboratory (FP Safi), Morocco

**Chrisantha Fernando** (15)

School of Electronic Engineering and Computer Science (EECS), Queen Mary, University of London, Mile End Road, London E1 4NS, UK

**Fabrizio Ferretti** (355)

Unità di ricerca per la gestione dei sistemi forestali dell'Appennino, Consiglio per la Ricerca e la sperimentazione in Agricoltura (CRA), Isernia, Italy

**Tadesse Fetahi** (217)

Addis Ababa University, College of Natural Science, Department of Zoological Sciences, P. O. Box 1176, Addis Ababa, Ethiopia

**Michele Forlin** (517)

Centre for Integrative Biology, University of Trento, via delle Regole 101, 38123 Trento, Italy

**Efim Frisman** (149)

Institute for Complex Analysis of Regional Problems, Far Eastern Branch Russian Academy of Sciences, Sholom-Aleihem St., 4, Birobidzhan 679016, Russia

**Anne Ghisla** (391)

Fondazione Edmund Mach, Research and Innovation Centre, Department of Biodiversity and Molecular Ecology, Via E. Mach 1, 38010 S. Michele all'Adige (TN), Italy

**Pavel Grabarnik** (331)

Institute of Physico-Chemical and Biological Problems in Soil Science of the Russian Academy of Sciences, 142290 Institutskaya ul., 2, Pushchino, Moscow Region, Russia

**Nicolas Guyennon** (259)

IRSA-CNR, Via Salaria km 29,300 10 00015 Monterotondo St., Rome, Italy

**Florian Hartig** (413)

Helmholtz Centre for Environmental Research - UFZ, Department of Ecological Modelling, Permoserstr. 15, 04318 Leipzig, Germany

**Heidi C. Hauffe** (391)

Fondazione Edmund Mach, Research and Innovation Centre, Department of Biodiversity and Molecular Ecology, Via E. Mach 1, 38010 S. Michele all'Adige (TN), Italy

**Attila R. Imre** (391)

Has Centre for Energy Research, H-1525, POB 49, Budapest, Hungary

**Yuliya D. Ivanova** (459)

Institute of Biophysics SB RAS, Krasnoyarsk, 660036, Russia

**Imrich Jakab** (375)

Department of Ecology and Environmental Science, Faculty of Natural Sciences, Constantine the Philosopher University in Nitra, Nitra, Slovakia

**Karin Johst** (413)

Helmholtz Centre for Environmental Research - UFZ, Department of Ecological Modelling, Permoserstr. 15, 04318 Leipzig, Germany

**Ferenc Jordán** (391)

The Microsoft Research-University of Trento Centre for Computational and Systems Biology, Piazza Manifattura 1, 38068, Rovereto, Italy

**Sven Erik Jørgensen** (547)

Copenhagen University, University Park 2, 2100 Copenhagen Ø, Denmark

**Richard Judson** (51)

National Center for Computational Toxicology, US EPA, Research Triangle Park, NC, USA

**Caner Kazanci** (201)

Faculty of Engineering and Department of Mathematics, University of Georgia, Athens, GA 30602, USA

**Karima Khalil** (275)

SAEDD laboratory Ecole Supérieure de Technologie d'Essaouira Km 9, Route d'Agadir, Essaouira Aljadida BP. 383, Essaouira, Morocco and EnSa Laboratory (FP Safi), Morocco

**Yulia Khoraskina** (331)

Institute of Physico-Chemical and Biological Problems in Soil Science of the Russian Academy of Sciences, 142290 Institutskaya ul., 2, Pushchino, Moscow Region, Russia

**Christopher D. Knightes** (239)

United States Environmental Protection Agency, Office of Research and Development, National Exposure Research Laboratory, Ecosystems Research Division, Athens, GA, USA

**Wataru Koketsu** (115)

Oyo Corporation Ecology and Civil Engineering Research Institute, 275 Aza Ishibatake Oaza Nishikata, Miharu-machi, Tamura-gun, Fukushima 963-7722, Japan

**Alexey Kolobov** (131)

Institute for Complex Analysis of Regional Problems, Far Eastern Branch of the Russian Academy of Science, Sholom-Aleykhem St., 4, Birobidzhan 679016, Russia

**Alexander Komarov** (331)

Institute of Physico-Chemical and Biological Problems in Soil Science of the Russian Academy of Sciences, 142290 Institutskaya ul., 2, Pushchino, Moscow Region, Russia

**Tomas Lovato** (319)

Centro Euro-Mediterraneo sui Cambiamenti Climatici—CMCC, Viale A. Moro 44, I-40127 Bologna; Department of Environmental Sciences, Informatics and Statistics, Università Ca' Foscari Venezia, Dorsoduro 2137, I-30123 Venezia

**Anna Loy** (431)

Envix_Lab, STAT Department, University of Molise, C.da Fonte Lappone, 86090, Pesche (IS), Italy

**Qianqian Ma** (201)

Department of Biological and Agricultural Engineering, University of Georgia, Athens, GA 30602, USA

**Donato Malerba** (35)

Dipartimento di Informatica, Università degli Studi di Bari "Aldo Moro," via Orabona 4, Bari, Italy

**Emanuela C. Manfredi** (259)

IRSA-CNR, Via del Mulino 19, 20861 Brugherio (MB), Italy

**Giorgio Mattassi** (239)

ARPA FVG Friuli Venezia-Giulia Regional Environmental Protection Agency, Technical Scientific Management Department, Via Cairoli 14, Palmanova (UD), Italy

**Seyoum Mengistou** (217)

Addis Ababa University, College of Natural Science, Department of Zoological Sciences, P.O. Box 1176, Addis Ababa, Ethiopia

**Alexey Mikhailov** (331)

Institute of Physico-Chemical and Biological Problems in Soil Science of the Russian Academy of Sciences, 142290 Institutskaya ul., 2, Pushchino, Moscow Region, Russia

**Natalia Mikhailova** (331)

Institute of Mathematical Problems of Biology of the Russian Academy of Sciences, 142290 Institutskaya ul., 4, Pushchino, Moscow Region, Russia

**Junji Miwa** (115)

Water Environment Group River Restoration Team, Public Works Research Institute, 1-6 Minamihara, Tsukuba City, Ibaraki 305-8516, Japan

**Fernando R. Momo** (533)

Instituto De Ciencias, Universidad Nacional De General Sarmiento, J. M. Gutiérrez 1150, Los Polvorines, Argentina; Inedes, Universidad Nacional de Luján, Argentina

**Harini Nagendra** (391)

Ashoka Trust for Research in Ecology and the Environment, Royal Enclave, Srirampura, Jakkur P.O., Bangalore 560064, India; Center for the Study of Institutions, Population, and Environmental Change, Indiana University, 408 N. Indiana Avenue, Bloomington, IN 47408, USA

**David Neale** (391)

Department of Plant Sciences, University of California at Davis, Davis, CA, USA

**Markus Neteler** (391)

Fondazione Edmund Mach, Research and Innovation Centre, Department of Biodiversity and Molecular Ecology, Via E. Mach 1, 38010 S. Michele all'Adige (TN), Italy

**Galina Neverova** (149)

Institute for Complex Analysis of Regional Problems, Far Eastern Branch Russian Academy of Sciences, Sholom-Aleihem St., 4, Birobidzhan 679016, Russia

**Irvin W. Osborne-Lee** (469)

Texas Gulf Coast Environmental Data (TEXGED) Center, Chemical Engineering Department, Prairie View A&M University, Prairie View, TX, 77446, USA

**Sergey Ovchinnikov** (3)

L.V. Kirensky Institute of Physics, SB RAS, Krasnoyarsk, 660036, Russia; Siberian Federal University, Krasnoyarsk, 660041, Russia

**Tamara Ovchinnikova** (3)

V.N. Sukachev Institute of Forest, SB RAS, Krasnoyarsk, 660036, Russia

**Bernard C. Patten** (181)

Odum School of Ecology and Faculty of Engineering, University of Georgia, Athens, GA 30602, USA

**Giovanni Pecenik** (319)

Department of Environmental Sciences, Informatics and Statistics, Università Ca' Foscari Venezia, Dorsoduro 2137, I-30123 Venezia

**Peter Petluš** (375)

Department of Ecology and Environmental Science, Faculty of Natural Sciences, Constantine the Philosopher University in Nitra, Nitra, Slovakia

**Davide Prandi** (517)

Centre for Integrative Biology, University of Trento, via delle Regole 101, 38123 Trento, Italy

**Oksana Revutskaya** (149)

Institute for Complex Analysis of Regional Problems, Far Eastern Branch Russian Academy of Sciences, Sholom-Aleihem St., 4, Birobidzhan 679016, Russia

**Carlo Ricotta** (391)

Department of Environmental Biology, University of Rome "La Sapienza", Piazzale Aldo Moro 5, 00185 Rome, Italy

**Pastres Roberto** (165)

Department of Environmental Sciences, Informatics and Statistics, Università Ca' Foscari Venezia, Dorsoduro 2137, 30123 Venezia, Italy

**Duccio Rocchini** (391)

Fondazione Edmund Mach, Research and Innovation Centre, Department of Biodiversity and Molecular Ecology, Via E. Mach 1, 38010 S. Michele all'Adige (TN), Italy

**Piet Ruardij** (275)

Royal Netherlands Institute for Sea Research (NIOZ), Department of Marine Chemistry and Geology, P.O. Box 59, NL-1790 AB Den Burg, Texel, Netherlands

**Svitlana S. Rudenko** (181)

Department of Ecology and Biomonitoring, Chernivtsi National University, Chernivtsi 58012, Ukraine

**Franco Salerno** (259)

IRSA-CNR, Via del Mulino 19, 20861 Brugherio (MB), Italy

**Isabella Scroccaro** (239)

ARPA FVG Friuli Venezia-Giulia Regional Environmental Protection Agency, Technical Scientific Management Department, Via Cairoli 14, Palmanova (UD), Italy

**Safwat H. Shakir Hanna** (469)

Texas Gulf Coast Environmental Data (TEXGED) Center, Chemical Engineering Department, Prairie View A&M University, Prairie View, TX, 77446, USA

**Vladimir Shanin** (331)

Institute of Physico-Chemical and Biological Problems in Soil Science of the Russian Academy of Sciences, 142290 Institutskaya ul., 2, Pushchino, Moscow Region, Russia

**Anton L. Shchemel** (459)

Institute of biophysics SB RAS, Krasnoyarsk, 660036, Russia

**Tiziana Simoniello** (431)

IMAA, Italian National Research Council—CNR, C.da S.ta Loja, 85050, Tito (PZ), Italy

**Anders Skonhoft** (105)

Economic Department, NTNU, Trondheim, Norway

**Cosimo Solidoro** (239)

OGS, The National Institute of Oceanography and Experimental Geophysics, Borgo Grotta Giagante 42c, Sgonico (TS), Italy

**Daniela Stojanova** (35)

Department of Knowledge Technologies, Jožef Stefan Institute, Jamova cesta, Ljubljana, Slovenia; Jožef Stefan International Postgraduate School, Jamova cesta, Ljubljana, Slovenia

**Vladislav Sukhovol'sky** (3)

V.N. Sukachev Institute of Forest, SB RAS, Krasnoyarsk, 660036, Russia

**Gianni Tartari** (259)

IRSA-CNR, Via del Mulino 19, 20861 Brugherio (MB), Italy

**Clara Tattoni** (355)

Museo delle Scienze, Sezione Zoologia dei Vertebrati, Via Calepina 14, 38122 Trento, Italy

**Helmuth Thomas** (275)

Dalhousie University, Department of Oceanography, 1355 Oxford Street, Halifax, Nova Scotia, B3H 4J1, Canada

**Felix Tomilin** (3)

L.V. Kirensky Institute of Physics, SB RAS, Krasnoyarsk, 660036, Russia; Siberian Federal University, Krasnoyarsk, 660041, Russia

**José E. Ure** (533)

Instituto de Ciencias, Universidad Nacional de General Sarmiento, J. M. Gutiérrez 1150, Los Polvorines, Argentina

**Federico Vaggi** (73)

The Microsoft Research - University of Trento Centre for Computational and Systems Biology, Piazza Manifattura 1, 38068, Rovereto, Italy

**Lucia Valsecchi** (259)

IRSA-CNR, Via del Mulino 19, 20861 Brugherio (MB), Italy

**Claudio Varotto** (391)

Fondazione Edmund Mach, Research and Innovation Centre, Department of Biodiversity and Molecular Ecology, Via E. Mach 1, 38010 S. Michele all'Adige (TN), Italy

**Vera Vasas** (15)

Departament de Genètica i de Microbiologia, Grup de Biologia Evolutiva (GBE), Universitat Autònoma de Barcelona, 08193 Bellaterra, Barcelona, Spain

**Cristiano Vernesi** (391)

Fondazione Edmund Mach, Research and Innovation Centre, Department of Biodiversity and Molecular Ecology, Via E. Mach 1, 38010 S. Michele all'Adige (TN), Italy

**Gaetano Viviano** (259)

IRSA-CNR, Via del Mulino 19, 20861 Brugherio (MB), Italy

**Polina Volkova** (3)

V.N. Sukachev Institute of Forest, SB RAS, Krasnoyarsk, 660036, Russia

**Evgenii Vysotskii** (3)

Institute of Biophysics, SB RAS, Krasnoyarsk, 660036, Russia

**John Wambaugh** (51)

National Center for Computational Toxicology, US EPA, Research Triangle Park, NC, USA

**Martin Wegmann** (391)

University of Würzburg, Institute of Geography, Department of Remote Sensing, Remote Sensing and Biodiversity Research, Am Hubland, 97074 Würzburg, Germany

**Thomas Wohlgemuth** (391)

WSL Swiss Federal Institute for Forest, Snow and Landscape Research, Zürcherstrasse 111, CH-8903 Birmensdorf, Switzerland

**Ayalew Wondie** (217)

Department of Biology, Bahir Dar University, Ethiopia

**Efim Frisman** (131)

Institute for Complex Analysis of Regional Problems, Far Eastern Branch of the Russian Academy of Science, Sholom-Aleykhem St., 4, Birobidzhan 679016, Russia

**Oksana Zhdanova** (91)

Institute of Automation and Control Processes, Far Eastern Branch of the Russian Academy of Science, Vladivostok, Russia

# Introduction

Sven Erik Jørgensen*, Ferenc Jordán†

*COPENHAGEN UNIVERSITY, UNIVERSITY PARK 2, 2100 COPENHAGEN Ø, DENMARK, †THE MICROSOFT RESEARCH – UNIVERSITY OF TRENTO, CENTRE FOR COMPUTATIONAL AND SYSTEMS BIOLOGY, PIAZZA MANIFATTURA 1, 38068, ROVERETO, TN, ITALY

## 1. Hierarchy Theory

The ecological hierarchy is easy to observe. The biochemical processes take place in the cells, which have molecular components and structure to control the processes and protect the genome. In vertebrates, there are different types of cells, which are specialized to carry out the biochemical processes that are characteristic for different organs: the liver, the muscles, the kidneys, the heart and so on. The cells carrying out the processes that take place in the liver make up the liver and so on. It is proper and effective hierarchical solution that the cells that have certain biochemical functions are working together to ensure the functions of the organs. The next hierarchical level after the organs is the individuals of different species. The individuals of the same species are working together in populations that have numerous methods to ensure survival and growth for the individuals. The grazers form for instance a herd that makes it more

**Table 1** Relationship between Hierarchical Level, Openness (Area/Volume Ratio), and Approximate Values of the Four-Scale Hierarchical Properties, Presented by Simon (1973): Energy/Volume, Space Scale, Timescale, and Behavioral Frequency

| Hierarchical level | Openness[1, 3] (A/V, $m^{-1}$) | Energy[2] (kJ/$m^3$) | Space scale[1] (m) | Timescale[1] (s) | Dynamics[3] (g/$m^3$/s) |
|---|---|---|---|---|---|
| Molecules | $10^9$ | $10^9$ | $10^{-9}$ | $<10^{-3}$ | $10^4$–$10^6$ |
| Cells | $10^5$ | $10^5$ | $10^5$ | 10–$10^3$ | 1-$10^2$ |
| Organs | $10^2$ | $10^2$ | $10^{-2}$ | $10^4$–$10^6$ | $10^{-3}$–0.1 |
| Organisms | 1 | 1 | 1 | $10^6$–$10^8$ | $10^{-5}$–$10^{-3}$ |
| Populations | $10^{-2}$ | $10^{-2}$ | $10^2$ | $10^8$–$10^{10}$ | $10^{-7}$–$10^{-5}$ |
| Ecosystems | $10^{-4}$ | $10^{-4}$ | $10^4$ | $10^{10}$–$10^{12}$ | $10^{-9}$–$10^{-7}$ |
| Ecosphere | $10^{-7}$ | $10^{-7}$ | $2*10^7$ | $10^{13}$–$10^{14}$ | $10^{-11}$–$10^{-12}$ |

[1]Openness, spatial scale and timescale are inverse to hierarchical scale.

[2]Energy and matter exchange at each level depend on openness, measured as available exchange area relative to volume. Electromagnetic energy as solar photons comes in small packages (quanta, hν, where h is Planck's constant and ν is frequency), which makes only utilization at the molecular level possible. However, cross-scale interactive coupling makes energy usable at all hierarchical levels.

[3]Openness correlates with (and determines) the behavioral frequencies of hierarchical levels.

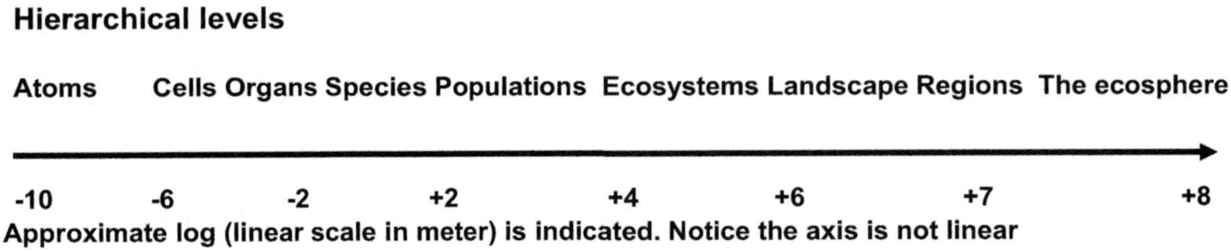


**FIGURE 1** The presented hierarchical levels are shown. Log to the linear scale in meter is indicated, but notice that the scale of the axis is not linear.

difficult for the predators to attack the individuals of the herd. On the other side, the predators hunt together to obtain by cooperation a higher probability for a successful hunting. Populations are also using communication among the individuals to increase the probability for survival. Populations are interacting in a network and make up together with the non-biological components of the environment, the ecosystem. The interactions in networks have a synergistic effect that is able to increase the utilization efficiency of matter, energy and information (Patten, 1991). Therefore, ecosystems have emerging properties. Landscapes are formed by interactions among several ecosystems and regions comprise of many landscapes. The entire living matter on the earth makes up the biosphere and the biosphere plus the non-biological components are denoted the ecosphere.

The entire hierarchy is illustrated in Figure 1 and the figure indicates the corresponding spatial scale, see also Table 1. The different hierarchical levels have not only different spatial scale but also different timescale and different openness, which is the surface area relatively to the volume. The openness determines the exchange with the environment and therefore also the energy density and the dynamics, as indicated in Table 1.

Higher hierarchical levels change therefore more slowly than lower levels due to the difference in spatial scale and the dynamics. It can be shown quantitatively (see Jørgensen, 2012 and Jørgensen and Nielsen, in press), how the variations in one level of the hierarchy are averaged and therefore result in less variations in the next, higher level. The variation is with other words damped when we go up through the hierarchical levels, which is a clear advantage by the hierarchical organization. Variations and disturbances of the lower levels have adynamic that may affect the lower levels in the hierarchy, but due to the hierarchical organization, the higher levels are much less affected because the variations caused by the disturbances are leveled off. Another clear advantage of the hierarchical organization is that a malfunction of one level can easily be eliminated by replacing a few components on the lower level. The function of an organ can for instance be improved by a renewal of a few cells by the organisms and an ecosystem may grow better by replacing one species with another one that is better fitted to new emergent conditions.

## 2. Interactions between the Hierarchical Levels

Figure 2 shows how the processes on one level of the hierarchy determine the conditions in the next level and how a level regulates and controls a lower level by feedbacks. A specific level consists of interacting and cooperative entities that are integrated components in a higher organizational level. The interaction of the entities in one level produce an integral activity of the whole, or expressed differently the dynamic of a lower level generates the behavior of the higher level. The variation of a whole level is significantly smaller than the sums of the variation of the parts, but thc degree of freedom of single processes is limited by the feedback regulation from the higher level. It is often necessary by development of ecological models/models of ecosystems to consider several hierarchical levels. Identification of the level of organization and selection of the needed complexity of the model are not trivial problems. It is possible to indicate as much as 19 hierarchical levels in living systems, but to include all of them in an ecological model is of course an impossible task, mainly due to lack of data and a general understanding of nature. Only the nine levels shown in Figure 1 are in most cases considered by development of ecological models and in most cases only one to three may be in some rare cases four levels will be included in an ecological model. Usually, it is not difficult to select the focal level, where the problem is or where the components of interest operate. The

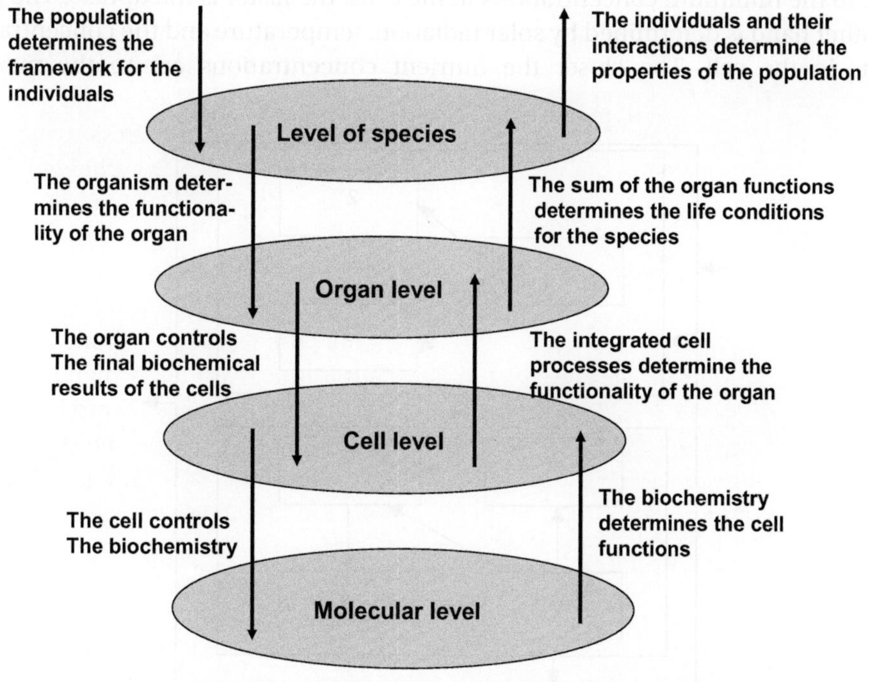


**FIGURE 2** A focal level has constraints from both lower and upper levels. The lower level determines to a high extent the processes and the upper level determines many of the constraints on the ecosystem.

level one step lower than the focal level may sometimes be of relevance for a good description of the processes. For instance, the primary production of an ecosystem is determined by the processes taken place in the individual plants. The level one step higher than the focal level determines many of the constraints, see Figure 2. It is therefore often difficult to understand a particular behavior of an ecosystem at a particular level, see Allen and Starr (1982), without also examining the behavior of the lower and the higher level.

## 3. Models with Two or More Hierarchical Levels

Due to the interaction between the hierarchical levels, we need of course models which include two or more hierarchical levels.

Figure 3 illustrates a model with three hierarchical levels, which might be needed if a multi-goal model is constructed. The first level could for instance be a hydrological model, the next level could be an eutrophication model and the third level can be a model of phytoplankton growth, considering the intracellular nutrient concentrations. Each sub-model will have its own conceptual diagram. The nutrients are taken up by phytoplankton in the third level model by a rate that is determined by the temperature, nutrient concentration in the cells and in the water. The closer the nutrient concentrations in the cells are to the minimum concentrations in the cells, the faster is the uptake. The growth on the other hand is determined by solar radiation, temperature and the concentration of nutrients in the cell. The closer the nutrient concentrations are to the maximum

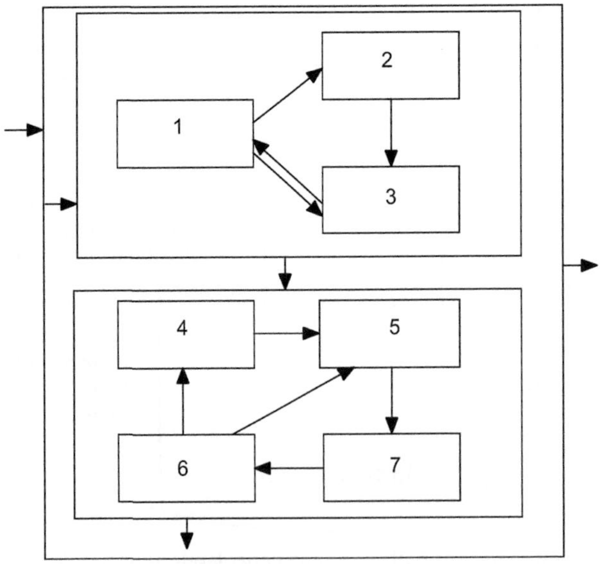


**FIGURE 3** Conceptualization of a model with three levels of hierarchical organization.

concentration, the faster is the growth. This description is provided according to phytoplankton physiology. Models that consider as well the distribution and effects of toxic substances, might often require three hierarchical levels: one for the hydrodynamics or aerodynamics to account for the distribution, one for the chemical and biochemical processes of the toxic substances in the environment and the third and last for the effect on the organism level. Ecological models have been constructed for all the hierarchical levels from the cells to the ecosphere, except organs. It is in the ecological literature possible to find models of species included their immediate environment, denoted habitats. There are also models between the two levels, populations and ecosystems, either focusing on the interactions of two or more species or on the entire community. Ecosystem models often include several of the abiological factors influencing the populations of the ecosystems.

It is the hope of the editors that this volume clearly demonstrates that we are today able to model all hierarchical levels. As it is frequently preferable to develop models that encompass more than one hierarchical levels, it is our assumption that this book will inspire more ecological modelers to develop multi-hierarchical models in the future. It will be interesting to see, whether such models will be able better to describe some of the processes that are significant for a better ecological understanding.

## 4. The Frequency of Disturbances Follow the Hierarchical Organization

Figure 4 shows how the recovery time after a catastrophic event decreases with the area damaged. It means that the higher levels in the hierarchy ecosystems, landscape, regions and the entire ecosphere require longer time to recover, but of course is not surprising when a larger area is damaged and the openness determining the dynamic is decreasing with the area.

Fortunately, the frequency of bigger catastrophic events is decreasing with the magnitude of the catastrophes. Figure 4 illustrates on the same graph the recovery time and the frequency, and it is shown that they are opposite—they, so to say, neutralize each other. The probability for a catastrophe that would damage large areas—ecosystems, landscape, regions and the entire ecosphere—is therefore decreasing with the hierarchical level. At the same time, as it was discussed above the damping effect on the disturbance is increasing with the hierarchical level. The hierarchical organization is therefore beneficial for the stability (survival or maintenance) of the larger systems of the hierarchy. At the same time, have the lower levels (see Table 1) a very high dynamics, which ensure a rapid renewal. The changes or the renewal is therefore a bottom-up effect. The higher levels, that control by feedbacks the lower levels, obtain by the damping effect a lower frequency iof disturbances. The hierarchical organization is at the same time important for the renewal and development of nature and for the conservations of particularly the larger systems in nature.

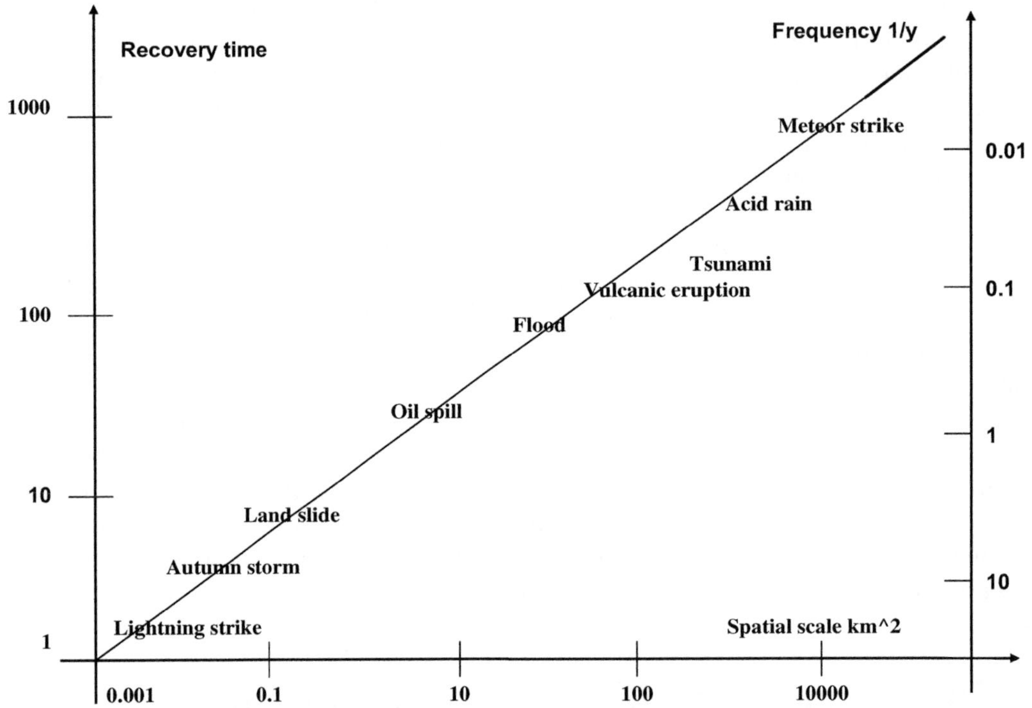


**FIGURE 4** The recovery time after a catastrophic event increases with the damaged area. The relationship is a straight line in a log–log diagram, which implies that the relationship is a power function. The inverse frequency of the damaged area is also a power function and follows the same relationship.

## 5. An Overview of the Models Presented in this Book

This book has 31 chapters that are divided into nine parts, covering the various hierarchical levels plus the interactions between hierarchical levels and hierarchy theory.

The first part covers models at the infra-individual levels. It contains very interesting papers, because all five papers show new application of ecological models and they are all based on very creative ideas. The five papers of the first part open for new possibilities in ecological modeling.

There are very few models focusing on the molecular and the cellular level of the ecological hierarchy, but the five first chapters show that such models have been developed. These hierarchical levels may be modeled more extensively in the future and the five chapters show actually that the models of molecular and cell processes have a potential and are clearly applicable. The first chapter presents a model of pheromone molecules produced by the gypsy moth and the Siberian silk moth. We do know in ecology that pheromone and other hormones can have an enormous and sometimes unexpected impact on ecological processes, for instance on population growth. It can therefore be foreseen that models in molecular ecology will play a bigger role in the future. The second paper shows an another interesting application of molecular ecological models, namely as tools to explain how chemical processes of complex organic molecules could lead to development of life. There is indeed a potential to use molecular models to test various hypothesis for the development of life, which is one of the core questions in biology and ecology. The third paper in this first part looks into gene flow models and models similar to the presented one may very well be used to model genetic pollution resulting from GMO, which would be a very important and very much needed application of ecological models. The fourth paper has developed an integrated model that couples a fate model with a cell growth model and a toxic effect model to analyze the dynamics of cell populations when subject to a toxicant in cell-based assay experiments. So, this paper points also toward new applications of ecological models: to test hypotheses that try to interpret assay experiments. The last and the fifth chapter in the first part of the book investigates molecular interaction networks responsible for various biological functions and uses network measures to identify the most important genes that control the biological process under investigation. These genes may play a major role in integrating the multiplicity of processes. A workflow where molecular interaction networks of genes associated with a given biological process are collected was developed.

Population dynamic is the second most applied model type. About 25% of all ecological models are population dynamic models, see Jørgensen (2008). Of the three models in this second part of the book, two models have age structure and one is using IBM, which most probably will be applied more in the future for the development of population dynamic models. A review of the last 4 years of the issues of the journal *Ecological Modeling* shows that it is a clear tendency. Population dynamic models are as all types of ecological models developing toward a higher complexity, which is demonstrated in the next part, where two papers are presented. They are showing how to include spatial competition and two age-structured interacting populations in our population dynamic models.

The populations in ecosystems are connected by food chains or rather food webs or ecological networks. These connections are decisive for a better understanding of ecosystems and how ecosystems work as systems. The fourth part contains four papers focusing on the connections of population and how it is possible to deal with the extra complications in ecological modeling. The theory of ecological networks, the hierarchical theory and the thermodynamic theory of ecosystems are fundamental to modern ecosystem theory. Two of the papers, Chapters 12 and 13, in this part are more theoretical in the sense that they deal with both the models of networks and with network theory applied to understand system-wide properties. The two chapters demonstrate therefore also the increasing use of ecological models to develop and improve ecosystem theory.

Many ecological models are based on element cycles in ecosystems, for instance eutrophication models that are very often based on cycling of nitrogen and phosphorus, and sometimes also silica and carbon (in aquatic ecosystems). The part entitled "Models of Element Cycling in Ecosystems" contains three papers dealing with cycling in ecosystems. Many pollution problems are best quantified by the use of cycling models, because all elements and many chemical compounds are in principle cycling in the ecosystems.

The part "Models of Ecosystems" contains three papers. Ecosystem models are of course extensively found in the modeling literature. Ecosystem models are frequently applied to develop ecosystem theory or test ecosystem theoretical hypotheses. Moreover, ecosystem models are applied to capture a holistic picture of ecosystems, including when an environmental management problem is the focus of the model development. Use of models for better managing ecosystems is compulsory. The three papers in this part cover

**(1)** cities as ecosystems,
**(2)** a lake model and
**(3)** a paper focusing on the use of a component-based framework for development of ecosystem models.

Recently, an increasing number of papers dealing with landscape models have been published in *Ecological Modeling*. Ecosystems are not closed system, but open to transfer of matter, energy and information. Consequently, ecosystems can hardly be properly modeled separately, but the important properties of openness have to be accounted for in the development of ecosystem models. Therefore, landscape models are the natural next step of model development after ecosystem models. It is on the one side important to cover the openness of ecosystems in ecosystem models, which of course is inevitable by the development of landscape models, but on the other side it is also important to model entire landscape, but it has indeed an environmental interest. Today, we have sufficient computer power to run models of several interacting ecosystems that form a landscape. Therefore, it is not surprising that we have seen more and more landscape models in the model literature and the experience of these models is that the interactions between two or more ecosystems are very important processes for ecosystems and for the development of the landscapes. They are indispensable in the planning of sustainable landscapes, as it is demonstrated in Chapters 22 and25, but also for quantifying the biodiversity in the landscape as presented in Chapters 23 and 24.

**(1)** the second last part, "Global models—models of the ecosphere" demonstrates clearly the application of these global models. The three papers in this part look into the global carbon cycling, which of course is the core in all the models of the greenhouse effect.
**(2)** the influence of global changes on the quality of life for the biosphere.
**(3)** the development of the ecosphere expressed by the use of ecological footprints.

The book has in 28 chapters which clearly show that it is possible to model all levels (except the levels of organs) of the ecological hierarchy, which is a very important observation for the future development of ecological models, as we do know that the interactions between two or more hierarchical levels are vital for a proper description of any of the ecological hierarchical levels. The obvious next questions are of course:

**(1)** how can we model these interactions?
**(2)** how can we develop multi-level ecological models?
**(3)** is our present hierarchical theory sufficient? An overview of the theory has shortly been presented in the first four sections of this chapter—but are additions to the existing theory needed?

The three questions will be discussed in the last chapter, where we will include the last part named "Models of Interactions between Hierarchical Levels and Hierarchy Theory" into the discussion of these crucial questions.

We believe that the variety of ecological models presented in the book will inspire young ecologists to broaden their knowledge also in a vertical direction of biological organization.

## References

Allen, T.F.H., Starr, T.B., 1982. Hierarchy, Perspectives for Ecological Complexity. The University of Chicago Press, Chicago and London, 310 pp.

Jørgensen, S.E., 2008. An Overview of the model types available for development of ecological models. Ecol. Model. 215, 3–9.

Jørgensen, S.E., 2012. Introduction to Systems Ecology. CRC, Baton Rouge. Florida, 320 pp.

Jørgensen, S.E. and Nielsen, S.N., The properties of the ecological hierarchy and their application as ecological indicators. Ecol. Indic, in press.

Patten, B.C., 1991. Network ecology: indirect determination of the life–environment relationship in ecosystems. In: Higashi, M., Burns, T.P. (Eds.), Theoretical Ecosystem Ecology: The Network Perspective. Cambridge University Press, London, pp. 288–351.

Simon, H.A., 1973. The organization of complex systems. In: Pattee, H.H. (Ed.), Hierarchy Theory – The Challenge of Complex Systems. Braziller, New York, pp. 1–27.

PART 1

# Models of the Infraindividual Level

# 1

# Quantum Chemical Modeling in the Molecular Ecology

Sergey Ovchinnikov*,†, Felix Tomilin*,†, Polina Artushenko*, Vladislav Sukhovol'sky**, Tamara Ovchinnikova**, Palina Volkova**, Yurii Baranchikov**, Evgenii Vysotskii$

*L.V. KIRENSKY INSTITUTE OF PHYSICS, SB RAS, KRASNOYARSK, 660036, RUSSIA, †SIBERIAN FEDERAL UNIVERSITY, KRASNOYARSK 660041, RUSSIA, **V.N. SUKACHEV INSTITUTE OF FOREST, SB RAS, KRASNOYARSK, 660036, RUSSIA, $INSTITUTE OF BIOPHYSICS, SB RAS, KRASNOYARSK, 660036, RUSSIA

## 1.1 Introduction

During their life, plants and animals synthesize and secrete into the environment volatile substances of different chemical natures. Flows of the molecules of these substances form "information fields" in the terrestrial ecosystems. It is possible for individuals in ecosystems to use these "fields" for searching habitats, food, sexual partners, etc. The air medium is a channel through which the information is transmitted. The study of the information flows and properties of the information channels in ecosystems opens up the possibilities for understanding the behavior of individuals in the populations and controls the ecological processes in the ecosystems. In the framework of the information theory, common tasks assess the reliability of the information transfer and the noise in the information channels. Reliability and noise level in the chemical communication channel are largely dependent on the characteristic lifetime of the molecules (the carriers of information) and on their stability to the modifying factors (electromagnetic radiation in the ultraviolet (UV), visible, and infrared ranges; air temperature and precipitation; and content of various substances in the air).

Convenient objects used to study the information flows in terrestrial ecosystems and to assess the impact of environmental factors are insects. The chemical composition of sex pheromones is known for several thousand species of insects (http://www.pherobase.com). However, the experimental study of the pheromones molecular stability to the effects of the modifying factors and the assessment of the impact of the environmental factors on the sexual behavior of insects are associated with technical difficulties. In addition, it is a time-consuming process that requires considerable financial resources.

Bioluminescence is the emission of visible light by an organism. This phenomenon is very widespread in the biosphere. However, the vast majority of bioluminescent

Models of the Ecological Hierarchy. DOI: http://dx.doi.org/10.1016/B978-0-444-59396-2.00001-8

organisms reside in the ocean; 80% of the more than 700 genera known to contain luminous species are marine (Shimomura, 2006). As a result of its prevalence, bioluminescence plays an important role in the ecology of the ocean. Its importance is clearer in the context of the essentially dark environment below 1000 m because a large number of organisms retain functional eyes to detect the bioluminescence at depths where sunlight never penetrates (Warrant and Locket, 2004). The bioluminescence can serve animals for the defense against predators, as an aid in locating food, for communication, and for camouflage, as well as for attracting a mate by means of species-specific spatial or temporal patterns of the light emission. Thus, in fact, the bioluminescence and the pheromone molecules implement similar functions in ecological systems.

Modern quantum chemistry provides a powerful tool to study complex molecules, their atomic and electronic structures in the ground and excited states, as well as unstable intermediates and transition states. The approach based on the density functional theory (DFT) incorporates the electron–electron Coulomb and exchange interactions beyond the well-known single-electron Hartree–Fock approximation (Schmidt et al., 1993). In this paper we propose to use quantum theoretical calculations for addressing the above-mentioned problems in molecular ecology and as an alternative to laboratory experiments. We consider two case studies. The first is represented by the pheromone molecules of different forest moths and their response to the substances present in the forest air and to the electromagnetic radiation. The second case study deals with the molecular mechanism of bioluminescence in $Ca^{2+}$-regulated photoproteins, which are mainly responsible for the luminescence of marine coelenterates.

## 1.2 Pheromone Molecules and Their Interaction With the Environment

The pheromones of the gypsy moth, *Limantria dispar* L., and the Siberian silk moth, *Dendrolimus sibiricus superans* Tschetv, have been studied. The gypsy moth pheromone has only one pheromone component, the disparlure, (7*R*,8*S*)-*cis*-7,8-epoxy-2-methyloctadecane, that belongs to the epoxy family (Bierl et al., 1970). The Siberian silk moth pheromone has two major components: the (*Z*,*E*)-5,7-dodecadienal (aldehyde) and the (*Z*,*E*)-5,7-dodecadien-1-ol (spirit) (Khrimian et al., 2002; Klun et al., 2000; Pletnev et al., 2000).

The quantum chemical computations of the electronic structure and analysis of the atomic–electronic structure for these compounds in the ground and excited states have been carried out. We used the configuration interaction method (configurations were constructed from 10 occupied and 6 unoccupied orbitals) of the semiempirical Hartree–Fock method PM3 (Parameterization Method 3) (Stewart, 1989) with the Open-Mopac2007 (http://openmopac.net) and the HyperChem 7.52 (http://www.hyper.com) programs. The most advanced method of density functional theory (DFT) B3LYP/

6-31(p,d) (Becke, 1993; Lee et al., 1988) has been incorporated in the GAMESS package (Schmidt et al., 1993).

For each pheromone molecule studied the following characteristics were calculated:

- The molecule total energy $E_{total}$.
- The electric dipole moment $D$, forming due to shifting of the electron cloud toward one of the atoms. This dipole moment shows how polar the molecule is.
- The oscillator strength (OS); this dimensionless quantity gives the information on the probability of the excitation, i.e., the intensity of transitions between the ground and excited states. The more the OS, the larger the intensity of the transition.
- The wavelength $\lambda$ and the absorption energy $E_{abs}$, which correspond to the absorption edge of molecule ($E = hc/\lambda$, where $E$ is the energy, $h$ is the Planck constant, and $c$ is the light velocity).

Three types of the conformers were chosen for each pheromone molecule: the linear structure (K1), the most curved structure (K2) and the armchair structure (K3). These conformers differ by the disposition of their carbon skeleton relative to the functional group of the molecule. In Fig. 1.1, conformers of the (7*R*,8*S*)-*cis*-7,8-epoxy-2-methyl-octadecane (disparlure) are presented as an example.

Pheromone molecules of the Siberian silk moth, *Dendrolimus superans*, have the $sp^2$-hybridized molecular orbital of carbon atoms, which form bonds in the same plane at an angle of 120°. This leads to the formation of the linear section in all the conformers and make structures K1 and K3 similar.

The calculated characteristics for the studied pheromone molecules are presented in Table 1.1.

As it is seen from Table 1.1, there is no large difference in the total energy value, $E_{total}$, for the different conformers of individual pheromone molecules. Maximal difference in the energy (6 kcal mol$^{-1}$) between the structure K2 and the rest of the conformers is found for the disparlure molecule. Similarity in the K1 and K3 structures for the pheromone molecules causes similarity in the conformers' properties. The K2 conformer's dipole moment is different from the dipole moments of the K1 and K3 structures by 0.1–0.4 D. Conformers' dipole moments range from 1.2 to 2.7 D. For the considered pheromone molecules, the dipole moment provides the ability to interact with polar molecules, for example, with water molecules, $H_2O$ (dipole moment 1.8 D), from the air.

In Table 1.1 the wavelengths, $\lambda$, and the absorption energies, $E_{abs}$, that correspond to the maximum of the pheromone molecules' electron transitions are listed. The absorbing power, $E_{abs}$, for the considered transitions depends on the quantity and relative position of the multiple bonds between the carbon atoms. These compounds absorb in the UV spectrum.

For the pheromone molecules of *D. sibiricus*, which have the conjugated double bonds, the absorbing wavelengths lie in the range of 260–265 nm. The other functional groups have less influence on the energy of the transition to the excited state.

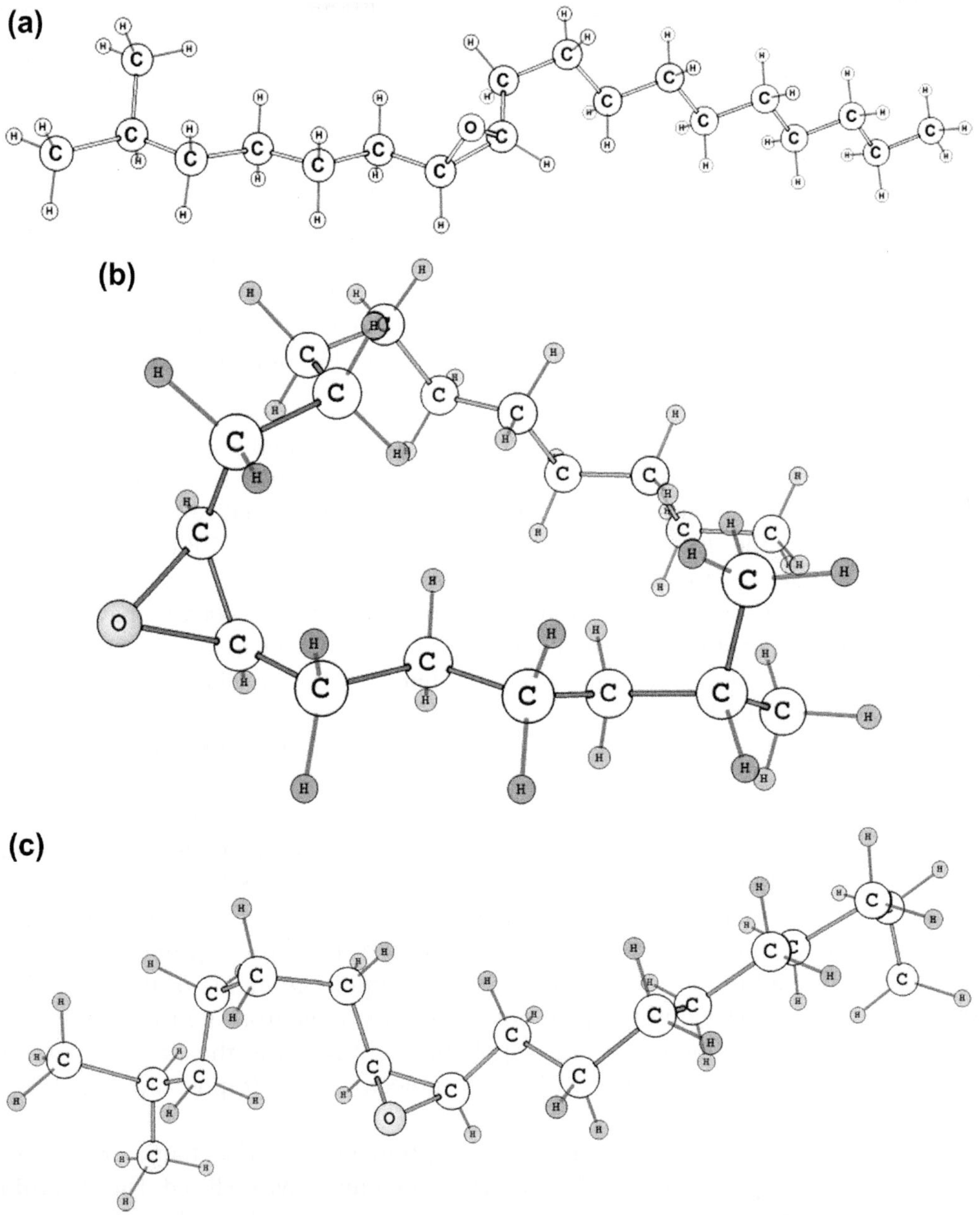

**FIGURE 1.1** Conformers of the (7*R*,8*S*)-*cis*-7,8-epoxy-2-methyloctadecane (disparlure). (a) Conformer K1, (b) conformer K2, and (c) conformer K3.

**Table 1.1** The Characteristics of the Studied Pheromone Molecules' Conformers

| Species | Pheromone molecule | Conformer | Molecule characteristics | | | | |
|---|---|---|---|---|---|---|---|
| | | | $E_{total}$ (kcal mol$^{-1}$) | $D$ (Debye) | $\lambda$ (nm) | $E_{abs}$ (eV) | OS |
| *L. dispar* | (7*R*,8*S*)-*cis*-7,8-Epoxy-2-methyloctadecane 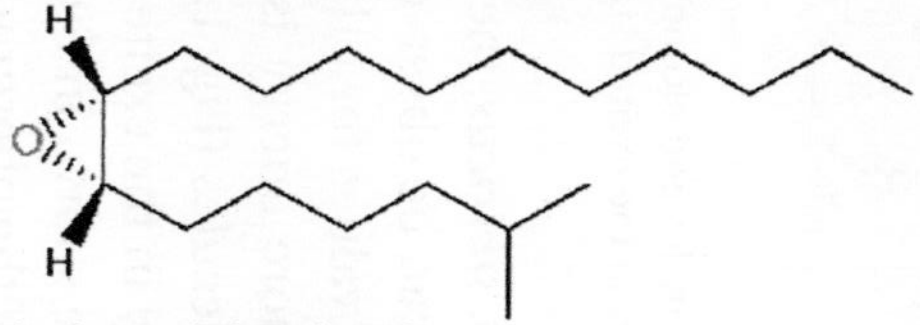 | K1 | −72,257 | 1.8 | 164 | 7.52 | 0.21 |
| | | K2 | −72,262 | 2.0 | 149 | 8.31 | 0.23 |
| | | K3 | −72,256 | 1.7 | 166 | 7.45 | 0.24 |
| *D. sibiri-cus* | (*Z*,*E*)-5,7-dodecadienal  | K1 | −46,681 | 2.3 | 259 | 4.77 | 0.92 |
| | | K2 | −46,681 | 2.7 | 262 | 4.71 | 0.94 |
| | | K3 | −46,681 | 2.5 | 261 | 4.74 | 0.95 |
| | (*Z*,*E*)-5,7-dodecadien-1-ol  | K1 | −47,402 | 1.5 | 262 | 4.71 | 0.91 |
| | | K2 | −47,402 | 1.6 | 262 | 4.73 | 0.90 |
| | | K3 | −47,403 | 1.5 | 263 | 4.70 | 0.90 |

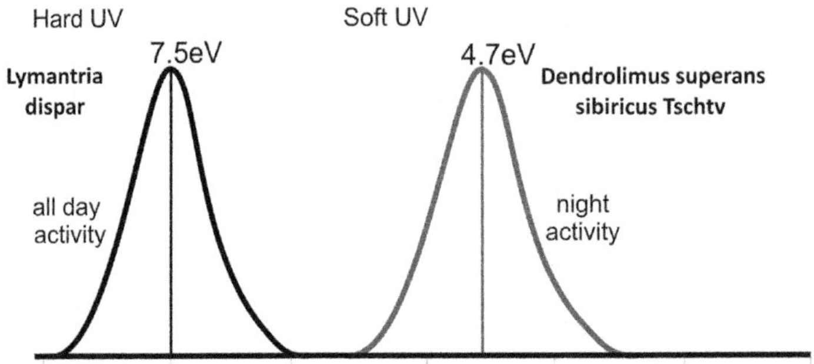

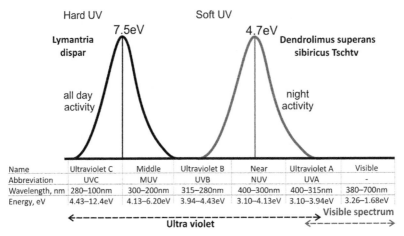

| Name | Ultraviolet C | Middle | Ultraviolet B | Near | Ultraviolet A | Visible |
|---|---|---|---|---|---|---|
| Abbreviation | UVC | MUV | UVB | NUV | UVA | - |
| Wavelength, nm | 280–100nm | 300–200nm | 315–280nm | 400–300nm | 400–315nm | 380–700nm |
| Energy, eV | 4.43–12.4eV | 4.13–6.20eV | 3.94–4.43eV | 3.10–4.13eV | 3.10–3.94eV | 3.26–1.68eV |

Ultra violet ←→ Visible spectrum

**FIGURE 1.2** The pheromone molecules' absorption edge.

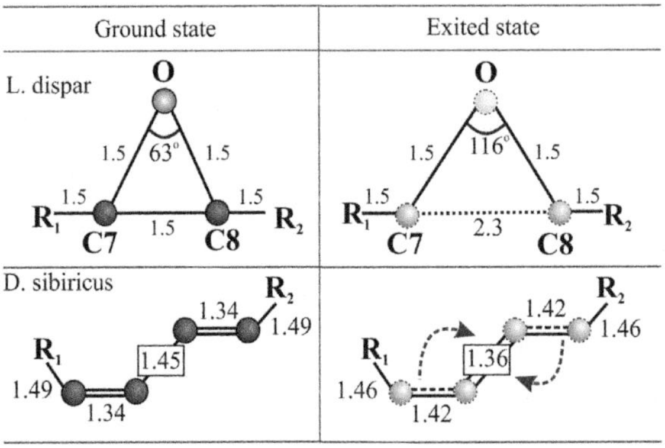


**FIGURE 1.3** Configurations of the diene group in aldehydes and alcohols of the pheromone components of the Siberian silk moth and the gypsy moth and its changes in the excited states. The bond length is in Angstroms.

Unlike the pheromone molecules of *D. sibiricus*, the pheromone molecule of *L. dispar* does not have multiple bonds, and as it can be observed from Table 1.1, the energy $E_{abs}$ for this molecule differs greatly from the values for other studied pheromone molecules (by approximately 2.5 eV). Therefore, more energy is needed to excite the disparlure molecule than to excite all the other molecules (Fig. 1.2).

Minimization of the molecule's energy in the excited state has revealed the change in the atomic geometry; the largest ones are shown in Fig. 1.3. These changes result in the sharp reduction of the activation energy of the pheromone chemical reaction with a water molecule. Notice that after reaction with water the pheromone molecule becomes deactivated. For example, the activation energy is equal to 290 kJ $mol^{-1}$ in the ground state ($H_2O$ + (*Z,E*)-5,7-dodecadien-1-ol) and no activation barrier in the excited state ($H_2O$ + *(*Z,E*)-5,7-dodecadien-1-ol) of the Siberian silk moth.

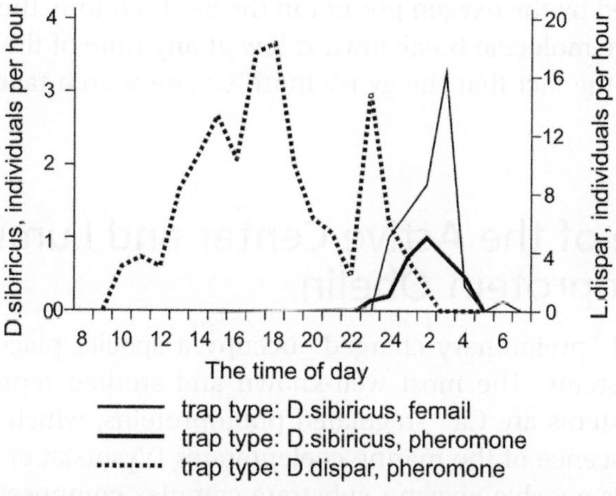


**FIGURE 1.4** Mate search activity during the day for the insect species resistant (gypsy moth) and nonresistant (Siberian silk moth) to the external factor pheromone molecules.

Thereby our calculations show that the solar radiation can change the configuration of the *D. sibiricus* pheromone molecules. During the morning hours, when the solar radiation is rather strong, the atom spacing and the bond angles can be changed in these pheromone molecules. This causes chemical reactivity of the molecule to increase and results in the deactivation of the *D. sibiricus* pheromone. In contrast, the gypsy moth pheromone (7*R*,8*S*)-*cis*-7,8-epoxy-2-methyloctadecane requires the high-energy UV radiation (e.g., lightning) to be deactivated.

We can compare our theoretical conclusions with the experimental data (Fig. 1.4) on the pheromone activity. The Siberian silk moth is active only during the night hours, whereas the gypsy moth is active in both periods.

Thus, the information losses in the pheromone communication channel of the Siberian silk moth are minimized during evening and night hours because the pheromone molecules do not get deactivated by light. The pheromone molecule components emitted by the female insects during the previous evening and night do not transmit information anymore because during that time the insects could have moved to a different location when the molecules were randomly carried around by turbulent airflows. However, during the following daylight hours these molecules can get deactivated, which will decrease the noise level in the pheromone channel by the next evening. Then influence of the external environmental factors can be seen as a mechanism assisting breakdown of the pheromone molecules that do not transmit information about the location of the female insect anymore.

The absorption band for the *L. dispar* pheromone molecules lies in the zone of the hard UV light (~200 nm), and for the *L. dispar* pheromone molecules to be affected by light, a greater energy has to be applied than that applied in the case of the Siberian silk moth pheromones. Usually the solar radiation in the zone of the hard UV is rather low because it is absorbed by the oxygen present in the air; therefore, the risk of information loss due to disparlure molecule breakdown is low at any time of the day. These conclusions correlate with the fact that the gypsy moth's mate search takes place throughout the day.

## 1.3 Structure of the Active Center and Luminescence in the Photoprotein Obelin

The systems named "preliminary charged" occupy a special place among numerous bioluminescence systems. The most well-known and studied representatives of such bioluminescence systems are $Ca^{2+}$-regulated photoproteins, which are mainly responsible for the luminescence of the marine coelenterates (Vysotskii et al., 2006). The photoprotein molecule is a stable enzyme–substrate complex composed of a monosubunit polypeptide and an oxygen-preactivated substrate, 2-hydroperoxycoelenterazine, which is noncovalently bound to the protein. Bioluminescence is initiated by the calcium ions and emerges due to the oxidative decarboxylation of the substrate bound to the protein. This causes formation of the reaction product, the coelenteramide (CLM), in an excited state. The transition of the CLM from the excited to the ground state is accompanied by light emission. The bioluminescence of the photoproteins is observed in the range of 465–495 nm and depends on the particular organism from which the photoprotein is isolated. Since the moment of discovery, the $Ca^{2+}$-regulated photoproteins have been studied intensively. The structures of the substrate and the reaction product have been determined, and the chemical mechanism of the bioluminescence reaction has been proposed. The spatial structures of several ligand-dependent conformational states of the photoproteins have been recently determined. However, the spatial structure of the protein provides information only about the static state of the protein molecule and the amino acids in the active center. It is still unknown what changes take place during the reaction. The state-of-the-art quantum chemical modeling allows this gap to be filled. In this section we used the quantum chemical methods to model the molecular mechanisms underlying the fluorescence of the $Ca^{2+}$-discharged photoprotein obelin from the hydroid *Obelia longissima.*

In a previous study, the fluorescence of the CLM and its analogs in solvents with various polarities was studied and it was found that the excited CLM can form five different ionic forms depending on the solvent (Shimomura and Teranishi, 2000; Mori et al., 2006). Theoretical models were constructed for all five CLM ionic forms using the quantum chemical semiempirical method PM3 (Becke, 1993; Stewart, 1989). Because, when modeling the absorption spectra and luminescence, it is also necessary to take into account the electron–electron Coulomb interaction, the calculations were conducted using the configuration interaction method (121 configurations). The calculated CLM emission wavelength in vacuum without any environment differs considerably from the experimental values. Therefore, the nearest amino acid environment of the cavity

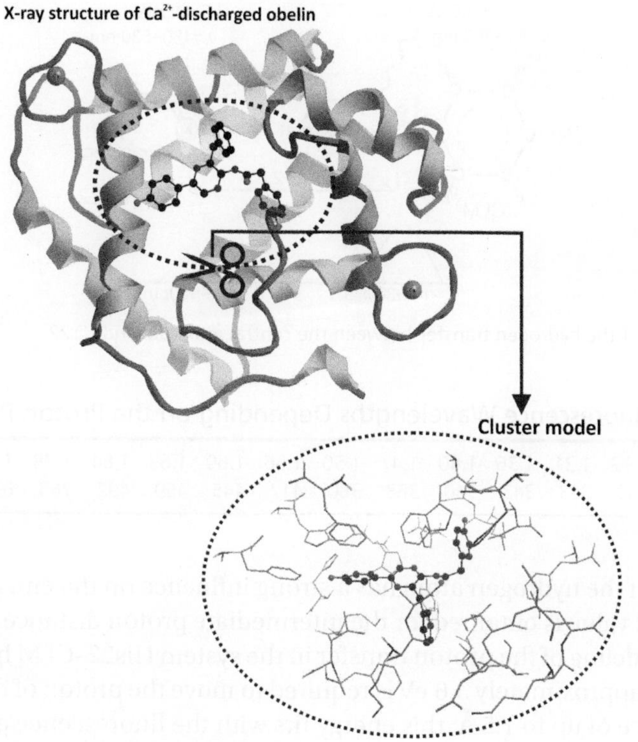


**FIGURE 1.5** The obelin and the model cluster for computations.

containing the CLM was included into the cluster model in the assumption that it could strongly influence the CLM excited state. The CLM and the amino acid environment do not interact chemically, and the effect of the environment is electrostatic. Thus, we computed the CLM molecule (about 50 atoms) and its nearest amino acid environment at a distance of about 4 Å from the CLM atoms, namely, His22, Gly143, Tyr190, Met171, Trp114, Phe72, Trp179, Phe28, Ala46, Phe88, Val118, Trp92, Ile50, Met25, Leu29, Gly115, Thr172, His175, Ile144, Asp49, Ser142, Lis53, Lis45, Cys51, Leu54, His64, Phe122, and $5H_2O$ (overall, about 500 atoms; Fig. 1.5).

To model the fluorescence process, it is necessary to test all possible CLM forms that could be the candidate emitters. Correspondingly, we have computed the absorption and fluorescence spectra for all ionic forms (Tomilin et al., 2008). However, all known static forms do not allow to reproduce the experimental spectra. Two static forms are related by the possible proton transfer from the CLM phenolic group to His22. We assume that the proton exchange may be a dynamical process.

Therefore, we modeled the position of the hydrogen atom between the CLM phenolic group and His22. For this purpose, the distance between the outermost hydrogen positions, 2.42 Å (Fig. 1.6), was divided into 30 equal parts to calculate the emission wavelength at each position (Table 1.2).

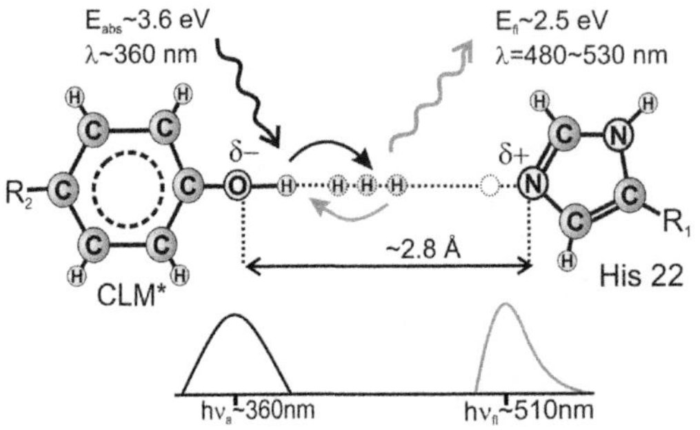


**FIGURE 1.6** Modeling of the hydrogen transfer between the central molecule and His22.

**Table 1.2** The Fluorescence Wavelengths Depending on the Proton Position

| CLM–H distance (Å) | 0.92 | 1.31 | 1.36 | 1.40 | 1.47 | 1.50 | 1.56 | 1.60 | 1.63 | 1.64 | 1.78 | 1.85 | 1.90 | 1.96 | 2.02 |
|---|---|---|---|---|---|---|---|---|---|---|---|---|---|---|---|
| λ (nm) | 343 | 343 | 343 | 340 | 363 | 380 | 417 | 445 | 469 | 492 | 763 | 633 | 620 | 615 | 610 |

The position of the hydrogen atom has a strong influence on the emission wavelength. The experimental value is obtained for the intermediate proton distance 1.60–1.64 Å from the CLM. The modeling of the proton transfer in the system His22–CLM has demonstrated that an energy of approximately 3.6 eV is required to move the proton of the CLM phenolic group to a distance of up to 1.5 Å; this energy fits with the fluorescence excitation energy. When the proton returns to the initial state, the energy of 2.3–2.5 eV is emitted. As the energy during the transition from the excited to the ground states is emitted spontaneously, the proton can return to the CLM from various distances; correspondingly, this broadens the fluorescence peak. In essence, this model reflects the formation of an ion-pair proton transfer complex (Mori et al., 2006; Shimomura and Teranishi, 2000).

## 1.4 Conclusions

The analysis on these two case studies showed that the existing quantum mechanics methods and computer software are quite suitable for calculating characteristics of such small molecules as lepidopterous pheromone molecules. A suggested approach allows estimating how resistant the pheromone molecules of some dangerous taiga pests are to the external environmental factors. The connection between the behavioral activity of an insect and the time of the day when the pheromone molecules are less influenced by the external factors is of considerable interest. Such a connection was found for *D. sibiricus* and *L. dispar*, and we believe that further juxtaposition of the pheromone molecules' properties, environmental conditions and sexual activity times can shed light on the evolutionary mechanisms underlying pheromone communication formation. It can also give clues on why the populations of certain insect species attend to occupy forest biomes, with certain biogeographic characteristics.

The methods applied to study the molecular mechanism of fluorescence have revealed that the fluorescence of the $Ca^{2+}$-discharged obelin (and possibly other $Ca^{2+}$-discharged photoproteins) is well modeled by moving the proton from the oxygen of the CLM phenolic group to the nitrogen of the amino acid His22.

## Acknowledgments

The authors would like to acknowledge the RFBR (grant no. 12-04-00119), FCP program "Kadry" GK P333, Government of Russian Federation's grant no. 11. G34.31.058, President of Russian Federation's grant NSh no. 1044.2012.2, Joint Super Computer Center (Moscow) (MBC-100K), and HPC IKIT of the Siberian Federal University (Krasnoyarsk) for financial support of this research.

## References

Becke, A.D., 1993. Density-functional thermochemistry. III. The role of exact exchange. J. Chem. Phys. 98, 5648–5652.

Bierl, B.A., Beroza, M., Collier, C.W., 1970. Potent sex attractant of the gypsy moth: its isolation, identification, and synthesis. Science 170, 87–89.

Khrimian, A., Klun, J.A., Hijji, Y., et al., 2002. Syntheses of (*Z,E*)-5,7-dodecadienol and (*E,Z*)-10,12-hexadecadienol, Lepidoptera pheromone components, via zinc reduction of enyne precursors. Test of pheromone efficacy against the Siberian moth. J. Agric. Food Chem. 50 (22), 6366–6370.

Klun, J.A., Baranchikov, Y.N., Mastro, V.C., Hijji, Yo, Nicholson, J., Ragenovich, I., Vshivkova, T.A., 2000. A sex attractant for the Siberian moth—*Dendrolimus superans sibiricus* (Lepidoptera: Lasiocampidae). J. Entomol. Sci. 36, 84–92.

Lee, C., Yang, W., Parr, R.G., 1988. Development of the Colle-Salvetti correlation-energy formula into a functional of the electron density. Phys. Rev. B 37, 785–789.

Mori, K., Maki, S., Niwa, H., Ikeda, H., Hirano, T., 2006. Real light emitter in the bioluminescence of the calcium-activated photoproteins aequorin and obelin: light emission from the singlet-excited state of coelenteramide phenolate anion in a contact ion pair. Tetrahedron 62, 6272–6288.

Pletnev, V.A., Ponomarev, V.L., Vendilo, N.V., Kurbatov, S.A., Lebedeva, K.V., 2000. Pheromone search of Siberian silk moth Dendrolimus superans sibiricus (Lepidoptera: Lasiocampidae). Agrochemistry 6, 67–72 (in Russian).

Schmidt, M.W., Baldridge, K.K., Boatz, J.A., Elbert, S.T., Gordon, M.S., Hensen, J.H., Koseki, S., Matsunaga, N., Nguen, K.A., Su, S., Windus, T.L., Dupius, M., Montgomery, J.A., 1993. General atomic and molecular electronic structure system. J. Comput. Chem. 14, 1347–1363.

Shimomura, O., Teranishi, K., 2000. Light-emitters involved in the luminescence of coelenterazine. Luminescence 15, 51–58.

Shimomura, O., 2006. Bioluminescence: Chemical Principles and Methods. World Scientific, New Jersey.

Stewart, J.J.P., 1989. Optimization of parameters for semiempirical methods I. Method. J. Comput. Chem. 10, 209–220.

Tomilin, F.N., Antipina, LYu, Vysotski, E.S., Ovchinnikov, S.G., Gitelzon II, , 2008. Fluorescence of Ca-discharged obeline: structure and molecular mechanism of emitter formation. Dokl. Biochem. Biophys. 422, 279.

Vysotskii, E.S., Markova, S.V., Frank, L.A., 2006. Calcium-regulated photoproteins of marine coelenterates. Mol. Biol. 40 (3), 404–417 (Moscow).

Warrant, E.J., Locket, N.A., 2004. Vision in the deep sea. Biol. Rev. Camb. Philos. Soc. 79 (3), 671–712.

# 2

# Evolution before Life

Vera Vasas*, Chrisantha Fernando[†]

*DEPARTAMENT DE GENÈTICA I DE MICROBIOLOGIA, GRUP DE BIOLOGIA EVOLUTIVA (GBE), UNIVERSITAT AUTÒNOMA DE BARCELONA, 08193 BELLATERRA, BARCELONA, SPAIN,
[†]SCHOOL OF ELECTRONIC ENGINEERING AND COMPUTER SCIENCE (EECS), QUEEN MARY, UNIVERSITY OF LONDON, MILE END ROAD, LONDON E1 4NS, UK

## 2.1 Introduction

If you want to make mice, the recipe is very easy; take a male and a female mouse, give them food, wait, and soon you will have more of them anyone would ever ask for. Try, however, to assemble a Frankenstein mouse in an organic chemistry laboratory, and you are up to certain disappointment. All life, as we know it, is autocatalytic; its creation requires already existing life. But how did it all begin?

Speaking in the most general terms, we expect that an autocatalytic chemical (Fig. 2.1) must have spontaneously appeared from the random chemistry of prebiotic Earth (Lifson, 1997). An autocatalyst can be any molecule or set of molecules that catalyze their production from environmentally available compounds, as $A$ in the following reaction:

$$x + A \rightarrow 2A \tag{1}$$

As opposed to heterocatalysis, as $H$ in the following reaction:

$$x + H \rightarrow y + H \tag{2}$$

The hypothetical autocatalyst grew to macroscopic levels while exhausting its reactants, and competition appeared among the slightly different autocatalysts that were produced by random chemical changes (Lifson, 1997). Thus, natural selection of chemical compounds, along with the continuous interaction between autocatalysts and their changing environment, kicked off the evolution leading to the wondrous diversity of current life.

A spontaneous synthesis of an autocatalyst initiates an explosion-like phenomenon. For a single catalyst molecule, with the velocity of one catalyzed reaction per microsecond, it would take more time than the estimated age of the universe to produce a mole of products—in contrast, for an autocatalyst with the same catalytic properties, this task would require 79 μs (Lifson, 1997). This exponential growth allows the increase of replicating molecules in a geometrical progression that makes the "survival of the fittest" possible (Szathmáry, 1991). As Eigen has shown in his seminal paper (Eigen, 1971), if autocatalysts (with certain specified properties) are reliably supplied with energy, their evolution becomes inevitable.

Models of the Ecological Hierarchy. DOI: http://dx.doi.org/10.1016/B978-0-444-59396-2.00002-X
ISSN 0167-8892, 

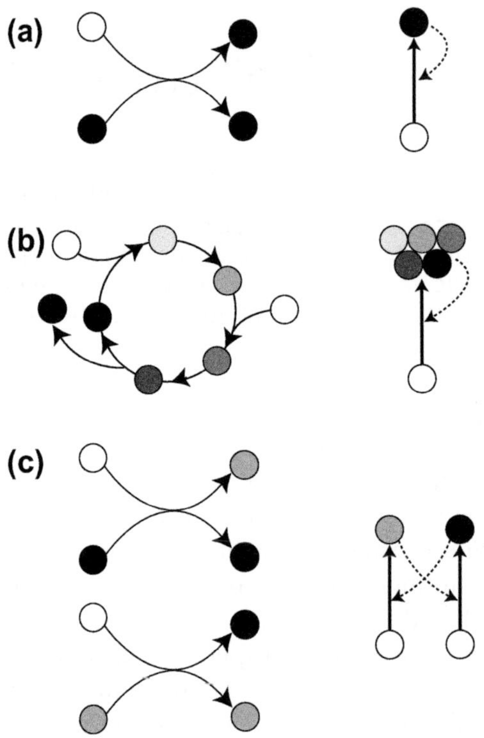


**FIGURE 2.1 Three basic motifs for direct and indirect autocatalysis.** White circles: food molecules available in the environment; grey circles: different nonfood molecules. Solid lines: reactions; dotted lines: catalytic activities. (a) A molecular autocatalyst is a molecule that catalyses its own formation from food. (b) An autocatalytic cycle comprises several stoichiometric reactions and can be regarded as one composite autocatalytic entity. Here, each turn of the cycle doubles its mass. (c) Indirect autocatalysis can arise from heterocatalysis if the catalytic dependencies form a closed circular path, called an autocatalytic loop.

However, the "certain specified properties" passage is important—not all autocatalysts can evolve. A fire is certainly autocatalytic, but it will always remain the very same fire; its properties entirely dependent on the material it is consuming. For natural selection to happen, we need a population of units that is capable of multiplication (one entity can give rise to many), variation (entities are not all alike, and some kinds are more likely to survive and multiply than others), and heredity (like begets like) (Fig. 2.2) (Maynard Smith, 1986). No matter how fascinating behavior the autocatalysts can show, they must pass on hereditary variation to be relevant for the origin of life. In other words, they must be informational replicators (Szathmáry, 2000). The aim of this chapter is to review the various theoretical models that have been proposed to explain the origin of evolvable systems and to provide a unified theory of chemical organizations that allow for their selectability. Doing so, we hope to highlight yet another possible use of network models.

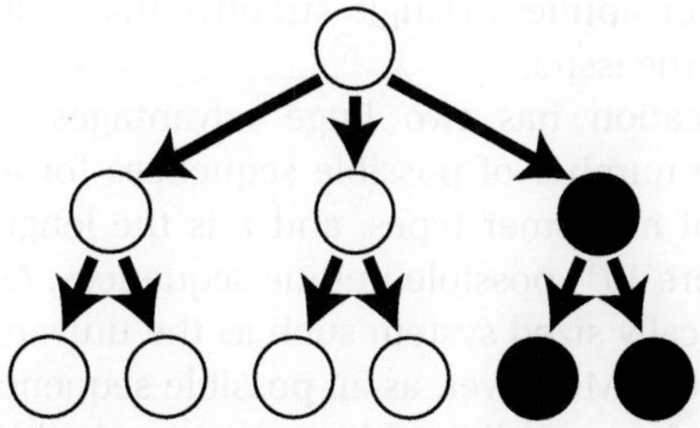

**FIGURE 2.2 Requirements for evolution.** Any population of units can evolve if the entities have the properties of multiplication (one entity can give rise to many), variation (entities are not all alike, and some kinds are more likely to survive and multiply than others), and heredity (like begets like).

## 2.2 Template-Based Replicators

All present life uses polynucleotide strings as the basis of storing and transmitting information. DNA and RNA replication is template based, meaning that the sequence of molecules on the parent strand is replicated (with mutation) to the child strand by a topographic mapping between the parent and child strand. Let us imagine a hypothetical polymer, held together with strong bonds, whose different building blocks can weakly bind identical monomers (Fig. 2.3). Due to the proximity of the monomers a new strand can form, and if it manages to separate from the original strand, we have an autocatalytic self-replicator—a molecule that catalyses its own formation from the food provided.

It is interesting to note that although such a self-copying mechanism sounds theoretically plausible, nature does not seem to use this mechanism [see Eigen (1971) for a possible explanation]. Instead, complimentary monomers form weak bonds in polynucleotide strings, resulting in + and − strands (Fig. 2.3). Here, both strands catalyze the formation of the other one, and autocatalysis arises as a two-member autocatalytic loop

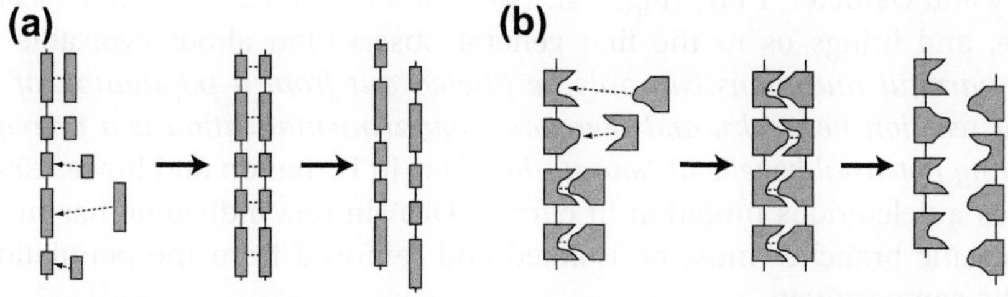


**FIGURE 2.3 Template replication.** The parent strand contains a sequence of molecules connected by strong bonds. Identical (a) or complementary (b) monomers attach specifically to their counterpart by weak bonds and due to their proximity are connected together into a new strand by strong bonds. The power of template replication is that all possible sequences replicate equally well.

(Fig. 2.1). This interpretation applies to single-stranded RNA, while in the case of DNA the double helix complicates the issue.

Template-based replication has two huge advantages over any other possible autocatalytic system. The number of possible sequences for a string of length $L$ is $n^L$, where $n$ is the number of monomer types, and $L$ is the length of the string. Even for $n = 2$ and $L = 100$, there are $10^{30}$ possible unique sequences. As this is more than could be realized in any realistically sized system such as the universe, variation is effectively unlimited (Szathmáry, 2000). Moreover, as all possible sequences use the same copying mechanism and are therefore equally stable and transmissible (information is storage based, Hogeweg, 1998), any mutation that might occur from a copying error is hereditary. In our terminology, whenever there is a mistake in either the $-$strand catalyzing the production of the $+$ strand or vice versa, a new, distinct autocatalytic $\pm$ collective will appear. If it replicates faster, it will take over the population; if not, it will quickly disappear.

There is, however, a major problem with increasing complexity in RNA. If, say, the probability of correct nonenzymatic pairing is $q = 0.99$ (rather the maximum value under optimal conditions), at polymer length $L = 100$, the probability of producing a correct copy is only $q^L = 0.366$ (Eigen, 1971). This means that reproducibility is essentially lost for polymers longer than a hundred monomers; however, a replicase enzyme that could possibly help the fidelity of copying would require a much longer code than this (Eigen's paradox, Eigen, 1971).

If long polymers have too high a mutational load, we can envision a set of distinct short replicators working together to the groups' benefit. In such a system, complexity could increase by the diversification of replicators, up to the point where enzymes could be produced. The crucial issue here is to couple the dynamics of distinct autocatalytic loops in such a way that diminishes the competition and the subsequent extinction of slower replicators.

Eigen (1971) has proposed the hypercycle for this coupling, where each polynucleotide helps the autocatalysis of the next polynucleotide (Fig. 2.4) in a cyclic design. However, it can be shown that if polynucleotides are allowed to mutate and produce proteins with different catalytic properties, shortcuts and parasites inevitably destroy the cycle (Szathmáry and Demeter, 1987) (Fig. 2.4). The problem of parasites is not specific to the hypercycle, and brings us to the first general observation about evolvable chemical networks: *harmful mutations can only be filtered out from a population of compartmentalized reaction networks, and therefore compartmentalization is a prerequisite for accumulating potential beneficial "adaptations"* (as in Fernando and Rowe, 2007). Much the same as a deleterious mutation in current DNA in one individual has no effect on others, parasitic branches must be isolated and removed from the population by the death of the compartment.

The compartmentalized hypercycle model, as expected, is now evolutionary stable (Zintharas et al., 2002); but then again, once compartments are available and link the destinies of replicating polynucleotides together, there is no need for linking them via

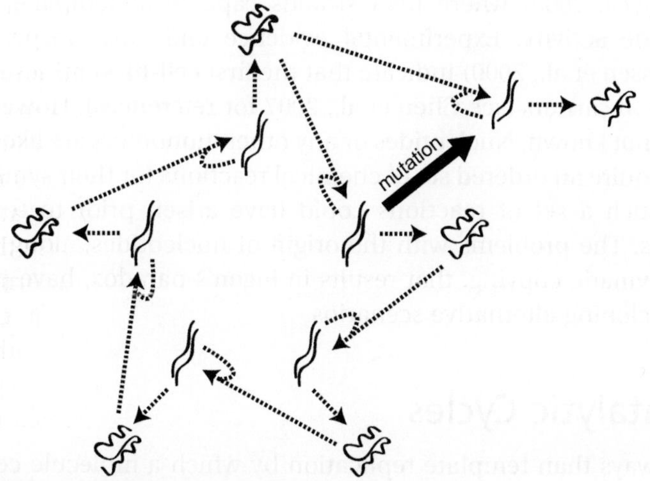


**FIGURE 2.4 The hypercycle.** Dotted lines indicate catalysis. The hypercycle consist of a set of polynucleotide strings (inside) that are translated by an unspecified mechanism to proteins (outside), arranged in a circular fashion where the product of each polynucleotide assists the template replication of the next one. The hypercycle is susceptible to parasitic mutations (grey) that benefit from its catalytic activities but does not code useful proteins.

a hypercycle. According to the stochastic corrector model (Grey et al., 1995; Szathmáry and Demeter, 1987), a simple package of competing replicators can be reliably maintained, given that each replicator is needed for the survival of the protocell. As compartments grow and divide, the daughter compartments can inherit any combination of distinct replicators, but selection at the level of the group removes the unfit combinations (Fig. 2.5). This is certainly a promising first step, and relevant experiments testing its experimental feasibility are on their way (Prof. Andrew Griffiths, personal communication).

Since the Nobel-prize winning finding that RNA can act as an enzyme (Guerrier-Takada et al., 1983; Kruger et al., 1982), it is usually agreed that there existed an RNA world

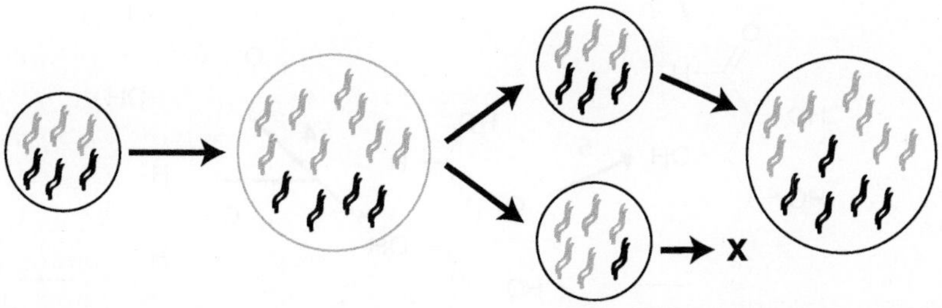


**FIGURE 2.5 The stochastic corrector model.** A collection of different template replicators (black and grey lines) can be reliably maintained if compartmentalized. Let us assume that the optimal ratio for protocell growth is 1:1. Although the grey type replicates faster (internal competition), daughter compartments with equal proportions will grow faster than those containing mostly one type (group selection).

(Gilbert, 1986; Joyce, 2002) where RNA strands capable of template replication also possessed catalytic activity. Experimental evidence and close inspection of current biochemistry (Nissen et al., 2000) indicate that the first cell-like entities might have used RNA as information carriers (see Chen et al., 2007 for references). However, the route to the RNA world is not known. Nucleotides or any other monomers are likely to be complex molecules that require an ordered set of chemical reactions for their synthesis, and we do not know how such a set of reactions could have arisen prior to template-encoded catalytic enzymes. The problems with the origin of nucleotides, along with the unreliability of nonenzymatic copying, that results in Eigen's paradox, have provided continuous fuel for developing alternative scenarios.

## 2.3 Autocatalytic Cycles

There are other ways than template replication by which a molecule could be autocatalytic. A ready example is the formose reaction (Breslow, 1959). Here, formaldehyde reacts with glycolaldehyde, and through a series of complex interconversions, two glycolaldehyde molecules are produced (Fig. 2.6). Such autocatalytic cycles behave the same way as molecular autocatalysts; the only difference is that here several reaction

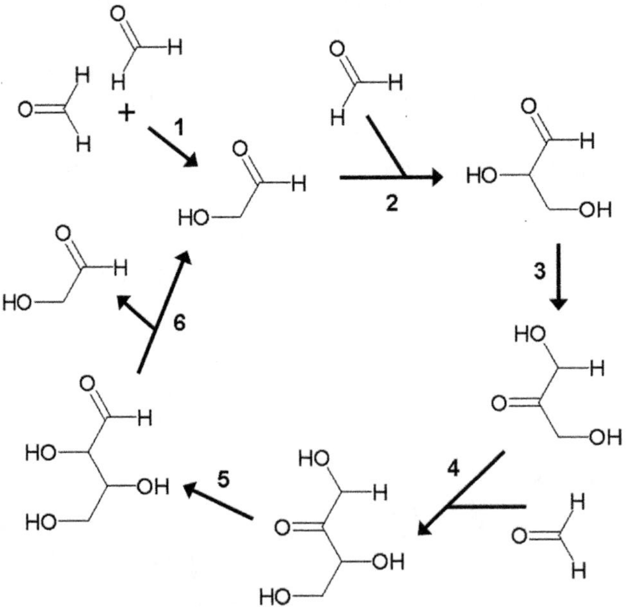

**FIGURE 2.6 The formose reaction.** The autocatalytic cycle is initiated when two formaldehyde molecules condense to glycolaldehyde (1), which forms aldotetrose through a series of aldol reactions and aldose–ketose isomerizations (2–5) that splits into two molecules of glycolaldehyde (6). Such an autocatalytic cycle acts as one composite autocatalytic entity. Note that this is just the central part of a very complex network.

steps need to be completed in order to achieve autocatalysis (Gánti, 2003; Zachár and Szathmáry, 2010). The cycle is set to motion when any of its members appears spontaneously. As any cycle member can act as an autocatalytic seed, and the appearance of any member is necessarily followed by all others, an autocatalytic cycle is actually one composite autocatalytic entity (Fig. 2.1). Let us now examine the feasibility and evolutionary capacities of such small organic cycles.

The main issue with the chemical plausibility of autocatalytic cycles is the problem of side reactions (King, 1982; Orgel, 2008; Szathmáry, 2000). As they are expected to run without enzymatic help, a certain fraction of organic material inevitably will be converted to inert by-products. The cycle can grow autocatalytically as long as the total turnover number remains positive (one molecule of autocatalyst gives rise to more than one molecules). It is easy to notice that the more reactions that make up the cycle, the less likely it is to remain functional. To be more precise, if $p$ reactions with $s_i$ specificity constitute a cycle with the turnover $t$, it grows only if

$$t\prod_{i=1}^{p} s_i > 1. \quad (3)$$

Given that the specificity cannot be too high for uncatalyzed reactions, this equation sets the upper size of cycles.

This serious limitation of nonenzymatic autocatalytic cycles might explain why they are so rare; in fact, we only have one or two clear examples apart from the formose cycle (e.g., a condensation reaction between two simple organic molecules, Patzke and von Kiedrowski, 2007), although a few hypothetical prebiotic cycles have been suggested, including an archaic reductive citric acid cycle, (Wachtershauser, 1988; Wachtershauser, 1992). As emphasized by Gánti (Gánti, 2003) and Eigen (Eigen, 1971), if *distinct, organizationally different, alternative autocatalytic* networks can coexist in the same environment, then they could compete with each other and the "fittest" would eventually prevail. The problem is, mutant copies of cycle members are generally not functional, because the chemical reactions involved in the cycle are strictly sequence specific. Even if we envision a few structurally different versions of a cycle (Wachtershauser, 1988), it only ensures flipping among the alternative variants but does not allow for heritable modifications. This feature is in sharp contrast to template replication, where, as we remember, any variant is automatically autocatalytic. Unfortunately, the lack of heritable variation excludes simple autocatalytic cycles from units of evolution, although we must stress that they might still have paramount importance for the origin of life as amplifiers of organic material and as such forming the basis of the first primitive metabolism.

If autocatalytic cycles per se cannot do the trick, the question arises whether a compartmentalized collection of cycles could be different—it seems that they might. The chemical avalanche model (Fernando and Rowe, 2007; Fernando and Rowe, 2008) envisions early evolution by autocatalytic cycles coming together to form an intermediary metabolism. The model uses a simple artificial chemistry (that nevertheless satisfies the conservation of mass and energy) enclosed in growing and dividing liposomes, with

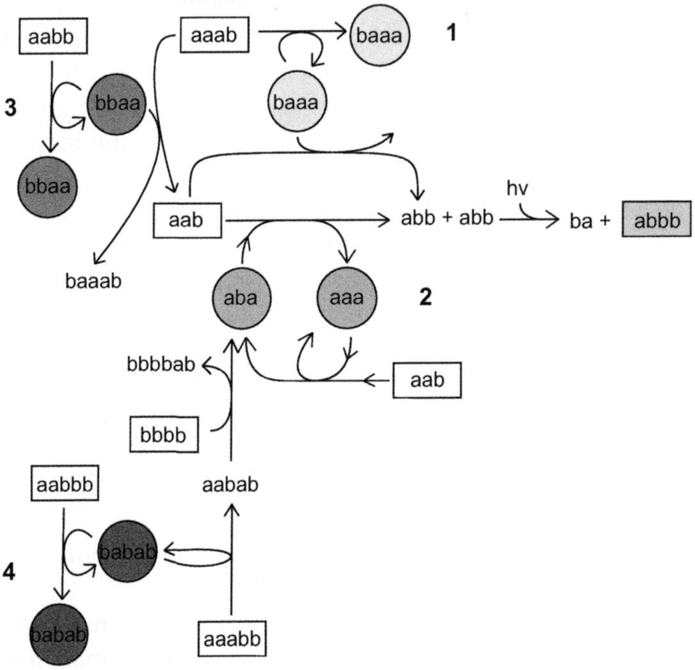


**FIGURE 2.7 The chemical avalanche model.** White rectangle: food; grey circles: autocatalytic cycles; grey rectangle: target molecule. The goal of the simple artificial chemistry of random molecular rearrangements is to produce the target molecule (abbb). Rarely occurring low-yield chemical reactions can start a chemical avalanche of novel reactions, but only autocatalytic molecules can permanently join the intermediary metabolism. Therefore, the reaction network can grow by stepwise expansions to the adjacent possible chemical space by inventing new autocatalytic cycles (the example shows four steps, 1–4).

so-called "chemical avalanches" when new molecules appear from rare, low-yield reactions and react with existing molecules. As might be expected, only those molecules that eventually catalyze their own production are able to permanently join the network (Fig. 2.7). Such molecules were typically directly autocatalytic, but there were also a few two-step and three-step autocatalytic cycles. Fitness was defined as the production of certain "growth molecules," assumed to be beneficial for the liposome, and selection was modeled by a simple hill-climbing algorithm—the mutant, produced by a chemical avalanche, replaced the parent whenever its fitness was higher.

In essence, the chemical avalanche model demonstrated that a reaction network is expected to change not by generating slightly different variants of an autocatalytic cycle, but by stepwise expansions to the adjacent possible chemical space by inventing new autocatalytic cycles. The resulting heredity is attractor based (Hogeweg, 1998). This type of heredity, which is fundamentally different from the storage implemented by DNA and RNA, arises when a random sample of molecules taken from a compartment tends to "inherit" its state. Naturally, if there is only one state, inheritance becomes meaningless:

a *chemical reaction network can only be evolvable if it has multiple attractors.* As each autocatalyst requires a seed for its production, its presence/absence is heritable, and so in the chemical avalanche model the simple combinations of which autocatalyst is present and which is missing define multiple stable states. The hill-climbing algorithm used (Fernando and Rowe, 2007; Fernando and Rowe, 2008), however, should not be accepted as a substitute for an explicit population dynamical simulation; there is so much that can go wrong as we introduce more and more reality in simulations. While explicit tests on selectability in the chemical avalanche model are still pending, results on a model exploiting the same mechanisms are promising (we will discuss this model in Section 2.4).

The weakness in proposing multiple cooperating autocatalytic cycles is, again, the lack of chemical support. Only very few nonenzymatic autocatalytic cycles are known to exist, although we hope that the recent theoretical advances will encourage experimental studies on both searching for unknown cycles and for examining the selectability of the compartmentalized formose reaction. In the formose cycle, the various pathways that can occur under different conditions are not yet fully described, and there is a chance that the full reaction network resembles the chemical avalanche model and not a simple cycle.

## 2.4 Autocatalytic Sets

If autocatalytic molecules are so rare, it is worth noting that heterocatalytic molecules can also form the basis of autocatalytic molecular networks. Understanding this principle requires becoming familiar with a lesser-known and absolutely different form of replication than the well-known template-dependent replication: ensemble replication of molecular networks.

The essential unit of organization for the so-called "reflexively autocatalytic sets" is an autocatalytic loop (Figs. 2.1 and 2.8). Autocatalytic loops are closed circular paths of any length where each molecule in the loop depends on the previous one for its production, and at least one of the steps is a catalytic dependency. Typically, the loop of reactions is coupled by catalytic interactions, but the loop maintains its autocatalytic properties—and is able to grow exponentially—as long as there is one reaction where the product is not consumed but serves as a catalyst in the next step (Eigen, 1971). In this sense, a single autocatalytic molecule or an autocatalytic cycle forms a one-member loop, and several heterocatalytic molecules can form loops of various sizes that result in indirect autocatalysis (Fig. 2.8).

The crucial difference between autocatalytic loops and cycles is that in the latter, exponential growth is the result of stoichiometric relationships. Autocatalytic loops, on the other hand, grow because their members catalyze each others' production. This difference makes them much less susceptible to the plague of side reactions; as long as a catalyst is not used up in the reaction it catalyses, its lack of specificity only means that it is wasting food. Naturally, as all molecules decay, a catalyst is still expected to

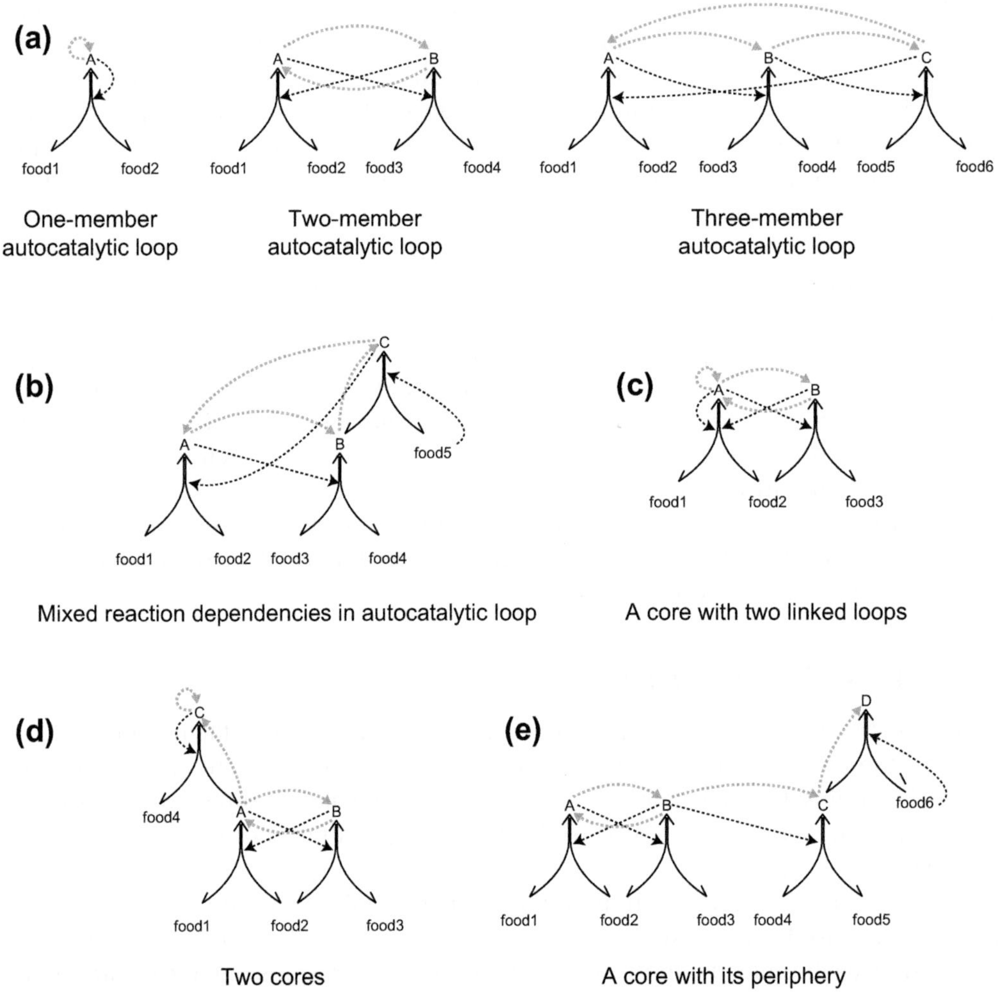


**FIGURE 2.8 Classification of network modules in reflexively autocatalytic sets.** Food1–food6: food set that is assumed to be present at all times, A–D: nonfood species generated by ligation/cleavage reactions. Solid lines: reactions; dotted lines: catalytic activities. Grey dotted lines show the superimposed autocatalytic loops. Note that molecular species can have multiple roles, i.e., can be substrates, products, or catalysts in various reactions, or even fulfill these roles in the same reaction. (a) A molecular species can be directly autocatalytic, forming a one-member autocatalytic loop, or several species can form loops of various sizes that result in indirect autocatalysis. (b) A loop is autocatalytic—and able to grow exponentially—as long as at least one of the steps is a catalytic dependency. (c) An autocatalytic core contains one or more linked loops. Note that any member of a core (molecules A or B) is sufficient to act as a seed for the core. (d) Several distinct cores can form within a catalytic reaction network. Here, the second core depends on the first for food supply. (e) An autocatalytic core is typically associated with a periphery that is dependent on the core (molecules C and D). We propose that nonfood-generated autocatalytic cores are the units of heritable adaptation in chemical networks.

act at a reasonable reaction velocity to compensate for the loss. On the other hand, a problem emerges when the product of the side reaction is actually harmful for the protocell.

Dyson (Dyson, 1985) and Kauffman (Kauffman, 1986) had developed some early ideas on how sets of mutually autocatalytic biopolymers could undergo replication, even if none of the components were individually autocatalytic. The original mathematical model (Kauffman, 1986) assumes the following: (i) there exists a large food set of abundant polymers naturally formed in the environment up to some low level of complexity and (ii) each molecule has a certain probability $P$ of catalyzing each ligation-cleavage reaction. In terms of "real" chemistry, the model envisions polypeptides, but in principle several different polymer species could potentially realize such networks. It was demonstrated that above a certain catalytic probability threshold $P_c$, a chain reaction is triggered, and a supracritical reaction network that keeps growing with accelerating speed arises. Once we have a sufficiently diverse network, due to catalytic closure, reflexively autocatalytic sets appear.

But could Kauffman's autocatalytic sets evolve into something more? We have shown recently that they cannot (Fernando and Vasas in press; Vasas et al., 2012). Although there are indeed several autocatalytic loops in the reaction network, their dynamics are coupled by catalytic interactions (the same problem described in Eigen, 1971), and this leaves no room for multiple attractors. A graph theoretical analysis of the catalytic interactions has shown that all molecular species of the sets are in the same autocatalytic core; all species catalyze the production of all other species, including themselves. This means that the provision of any one molecule of core species—including simply the food molecules—is sufficient to produce all species in the autocatalytic set; the system will always crystallize into the same attracting network which can never ever be a Darwinian unit.

However, Kauffman (1993) had speculated that the inclusion of inhibition in the network should permit the formation of autocatalytic sets having complex dynamical attractors. Interestingly, while our results substantiated this speculation, we also found that when the networks were confined into compartments undergoing a growth-splitting process, attractors could not be stably selected. Instead, the internal dynamics of the growth-splitting process completely overrode any effect of selection. This provides *a clear counterexample to the widely accepted claim that the existence of multiple attractors is sufficient to allow selectability; it is not. The attractor state of a compartment must be heritable.*

The crucial modification to the model turned out to be the simulation of rare uncatalyzed reactions among the already existing molecular species (Fernando and Vasas in press; Vasas et al., 2012). If such uncatalyzed reactions continuously produce random novel species in low copy number, a few molecules will happen to be catalysts and generate chemical avalanches of directly and indirectly catalyzed further novel molecular species. Only those species are able to permanently join the network that eventually catalyze their own production from already existing molecules and so produce a new autocatalytic core (Fig. 2.8). When we confined the reaction networks into a population of compartments, we found that networks with the novel core have an implicit selective advantage due to their higher growth rate, and so constitute the majority of the population; however, selective advantage attributed to the absence of the novel core significantly reduced its frequency.

The reader has perhaps noticed that the very same mechanism has been described previously in the case of the chemical avalanche model. Indeed, autocatalytic cycles can be regarded as one-member autocatalytic loops, and the chemical organization shown in Fig. 2.7 as the collection of such loops. Each new core that emerges from rare reactions results in a new and distinct attractor for the reaction network. Finally, every core is associated with a periphery it catalyses. Periphery molecules are not members of autocatalytic loops, and depend, as a phenotype does on its genotype, upon the core. Acquisition of cores by rare chemical events, and loss of cores at division, allows macromutation, limited heredity, and selectability.

While the theoretical advances are promising, they cannot substitute the badly needed experimental work. Autocatalytic networks of peptides have already been synthesized by Ghadiri and Ashkenazy (Ashkenasy et al., 2004), although there a direct templating effect plays a crucial role. The folding and catalytic properties of random peptides is an open experimental question, but as of today there is no known random polymer chemistry in which the probability of catalysis is ever so high as necessary for the spontaneous self-organization of autocatalytic sets.

## 2.5 Lipid World Scenario

The lipid world theory (Segré et al., 2001a) proposes that crucial steps in the origin of life might have been carried out by lipid-like molecules alone, potentially prior to the emergence of polynucleic acids and polypeptides. It envisions a type of metabolism-first scenario where lipids served as both catalysts and information carriers. The empirical support for the importance of lipids or other amphiphiles in prebiotic conditions is, indeed, impressively sound. There is a large diversity of present-day lipophilic organic compounds; it is probable that lipid-like molecules were among the first organic molecules that appeared on Earth and, above all, they have the ability to undergo spontaneous aggregation within an aqueous phase to form droplets, micelles, bilayers, and vesicles. Moreover, vesicle formation was shown to be autocatalytic (Bachmann et al., 1992; Blöchliger et al., 1998). Lipid-like amphiphiles are therefore unique among candidates of the origin of life as their self-assembly is not hypothesized, but in fact experimentally demonstrated.

In order to assess the evolutionary potential of micelle-like vesicles, Lancet and his collaborators proposed the graded autocatalysis replication domain (GARD) model (Segré et al., 2000). A basic feature in GARD is that noncovalent, micelle-like molecular assemblies capable of growing homeostatically store compositional information that can

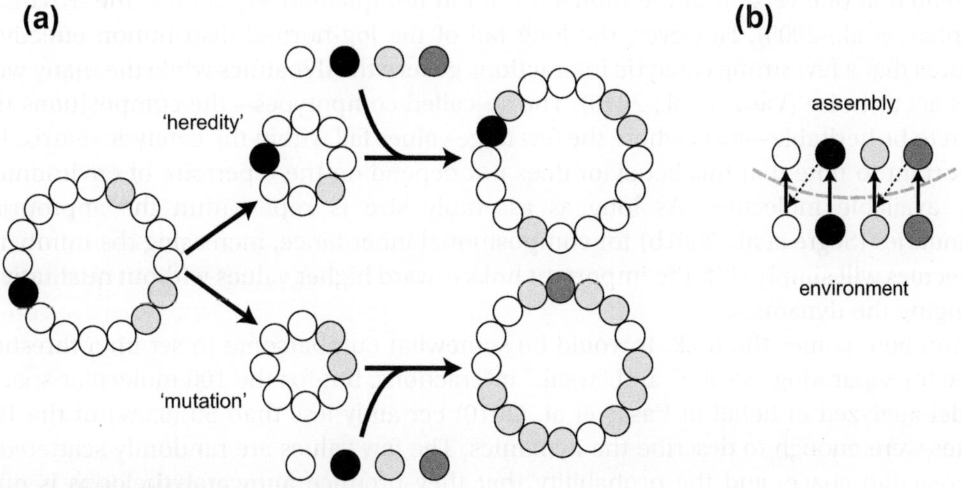


**FIGURE 2.9 The GARD model.** Micelle-like molecular assemblies of lipophils are assumed to grow by incorporating environmentally available compounds and divide after reaching a certain size. Growth is driven by catalytic interactions: molecules within the assembly catalyze the joining of others with different velocities. As these parameters are randomly drawn from a lognormal distribution, a few strong interactions govern the system. In the hypothetical example, there are two strongly catalytic molecular species (B). When the daughter compartment inherits these catalytic molecules, there is heredity; when they are lost, there is the abrupt mutation-like change to a different compotype (A).

be propagated after occasional fission (i.e., assembly splitting) (Fig. 2.9). The dynamical behavior of the system is governed by the matrix of catalytic interactions. Each element of the matrix describes the rate at which molecular species "A" enhances the inclusion of species "B" into the growing assembly (Fig. 2.9). Variation occurs due to stochastic growth and splitting of the assembly at a certain size, thus giving the potential for evolution of "composomes"—compositional genomes in which the composition of the assembly is the heritable trait. Such vesicles seem to possess the trinity needed to qualify as units of evolution (Maynard Smith, 1986): multiplication, variation, and heredity (Segré et al., 2000).

Despite the promising signs, a recent analysis has proved that the GARD model lacks selectability (Vasas et al., 2010). When selection was explicitly implemented in a population of assemblies, both analytical calculations and computer simulations have shown that it has no effect on the frequency distribution of composomes. In order to understand this result, we must have a closer look at the matrix of catalytic interactions.

These rate enhancement factors are drawn randomly from a lognormal distribution, rather than a normal distribution; in the latter case, there is no compositional inheritance (Segré et al., 2001b). Therefore, catalytic factors are always positive; in other words, all species catalyze the inclusion of themselves and all other species (inhibition was simulated in one version of the model but it did not qualitatively change the dynamics; Shenhav et al., 2005). However, the long tail of the log-normal distribution effectively ensures that a few strong catalytic interactions govern the dynamics while the many weak ones act as noise (Vasas et al., 2010). The so-called compotypes—the compositions that seem to be heritable—form where the few large values fall within the catalytic matrix. It is important to note that this behavior does not depend on the repertoire of environmentally available molecules. As long as assembly size is kept within the appropriate boundaries (Segré et al., 2001b) for compositional inheritance, increasing the number of molecules will simply shift the important links toward higher values without qualitatively changing the dynamics.

And here comes the trick. It would be somewhat cumbersome to set up a threshold value for separating "strong" and "weak" interactions, but for the 100 molecular species model analyzed in detail in Vasas et al. (2010) certainly less than 50 (0.5%) of the $100^2$ values were enough to describe the dynamics. The few values are randomly scattered in the reaction space, and the probability that they produce autocatalytic loops is quite small (see Box 2.1 and 2.2 for calculations.). What we see in the GARD model are actually not mutually catalytic lipophiles, but a few strongly catalytic molecular species that tend to give rise to different phenotypes. When the daughter compartment inherits these catalytic molecules, there is heredity; when they are lost, there is the abrupt mutation-like change to a different compotype. However, multiplication is missing—the "genetic" material is not autocatalytic. The flipping between compotypes is therefore not a mutational load that could be counterbalanced by stronger selection, but an inherent property of the model. As the original simulation only included the growth-splitting process and followed one lineage (Segré et al., 2000), it was easy to overlook this fundamental problem, and only the counterintuitive results of population dynamical simulations (Vasas et al., 2010) shed light on them—a story that, we hope, highlights the importance of more explicit approaches to selectability.

One might argue that, however unlikely, loops are still possible in the GARD model. While this is strictly true, it would require deliberate engineering, and even so such loops would be equally as unevolvable as simple autocatalytic cycles; there is no chemical story on how they could change. As molecules (along with their catalytic activities) are not synthesized by the composome but given in the environment, the only possibility left open for change is the addition or removal of lipid species. Pointing out that altering the food set of a reaction network modifies its dynamics, however, has no relevance for evolution.

In summary, despite the promising experimental plausibility, growing and splitting assemblies of lipid-like molecules cannot depart from the dynamics dictated by the available compounds in the environment. Note, however, how easily this system could be manipulated if (and only if) we assume a metabolism within the vesicle that only needs to produce a few copies of a strongly catalytic molecule to set the composition of the vesicle. Although the GARD has failed as a model for the origin of life, it could be highly relevant in the origin of membranes.

## 2.6 Conclusions

Until now, we have been discussing the origin of life while carefully avoiding what life actually is. Gánti (Gánti, 2003) proposed the chemoton as a minimal system that shows all general and principal phenomena of life, such as autonomy, reproduction, and the ability to function under a direction of a program, ensuring that it carries out work for its own benefit. He argued that living systems are a special kind of chemical automaton that consists of three stoichiometrically coupled autocatalytic subsystems—a metabolic subsystem that utilizes small organic compounds for producing building blocks for itself and others, a two-dimensional fluid membrane, and an information subsystem that grows by template polycondensation and regulates the other two.

All present-day cells follow this basic outline (Gánti, 2003). What is remarkable is that autocatalysis is present in all three subsystems: DNA replicates with enzymatic assistance; several small-molecular metabolites, including ATP, are exclusively autocatalytic (Kun et al., 2008); and lipids in a membrane not only enhance the incorporation of further lipids, but specific proteins as well, resulting in membrane inheritance (Cavalier-Smith, 1995). Which subsystem came first, and how it acquired the other two, is the question we need to understand. We have argued that evolution preceded life (Lifson, 1997), and we must understand the origin of evolvable chemical systems whose gradual increase of complexity lead to the appearance of the first living entity.

Selection happens between a given set of populated states, under fixed environmental conditions, and evolution refers to the process of further optimization procedure in a changing population and environment. It is important to note that this distinction is simply a useful abstraction (Eigen, 1971), and addressing selectability in a model is equivalent to discussing whether it can evolve or not. We found that apart from template replicators, only the generalized chemical avalanche model (applicable to small organic molecules as well as reflexively autocatalytic sets; Fernando and Rowe, 2007; Vasas et al., 2012) offers a viable proposal for chemical evolution. Simple autocatalytic cycles and lipid assemblies (Segré et al., 2000), on the other hand, "just exist" and cannot be units of evolution.

The critical properties of evolution, as outlined by Maynard Smith (Maynard Smith, 1986), translate to the world of chemical networks in the following way: (i) individuals are autonomous, that is, the units are either single molecules or compartmentalized sets of molecules; (ii) multiplication results from direct or indirect autocatalysis; (iii) variation is possible if the chemical organization allows multiple stable states; and (iv) heredity appears when the progeny inherits the parent's state.

The mechanism of template replication solves all issues in one elegant step. Template replication happens at the level of a molecule, provides autocatalytic growth, and as all possible sequences use the same copying mechanism and are therefore equally stable and transmissible, each comprises a heritable stable state. Chemical evolution by avalanches, on the other hand, is a very cumbersome process. The fact that it works by coopting a weakly or uncorrelated novel autocatalytic extension to the network means that evolution has to invent a novel autocatalytic mechanism for each macromutation. As each autocatalytic core constitutes one bit of heritable information and therefore the number of possible selectable attractors is relatively small, it can only sustain limited variation—a common feature of attractor-based inheritance systems. Even so, the potential role of these autocatalytic networks as a route to nucleotide-based template self-replicating systems should not be underestimated. As they show an intrinsic tendency to increase in complexity and can accumulate adaptations, they might have provided important functions in the RNA world or even served as a "scaffold" for its appearance (sensu Cairns-Smith, 1982). The next step toward an empirical scientific program that would address pretemplate heredity is crucial; it requires identifying and implementing a simple chemistry that could potentially realize the network structure advocated by theoretical models.

**Glossary**

**Heterocatalysis**: Molecule "A" catalyses the formation of molecule "B."

**Autocatalysis**: A molecule catalyses its own formation.

**Replicator**: An entity that passes on its structure largely intact in successive replications.

**Informational replicator**: A replicator whose variants are also replicators.

**Template-based replication:** Replication is ensured by copying a sequence of molecules from the parent to the child strand.

**Autocatalytic cycle**: A series of stoichiometric molecular reactions arranged in a cycle so that each turn of the cycle increases its mass by a factor.

**Autocatalytic loop**: Closed circular paths where each molecule in the loop depends on the previous one for its production, and at least one of the steps is a catalytic dependency.

**Storage-based heredity**: Any variant of the replicator is equally stable.

**Attractor-based heredity**: Only a few attractors of a chemical network are stable.

**Probability of Qutocatalytic Loops in the GARD Model**

Let us assume we have $N$ different molecules available in the environment. Then there are $N$ reactions that characterize the joining of each molecule, and an $N \times N$ matrix that defines the rate at which molecular species "A" enhances the inclusion of species "B." Let $P$ denote the probability of strong catalysis per molecule and per reaction. As there is only one reaction for each molecule, $P$ will indicate the probability that a given molecule strongly enhances the incorporation of another one.

**One-member loops**. The probability that a molecule catalysis its own incorporation is $P$. Therefore, the probability that any of $N$ molecules is autocatalytic in a single step is

$$\text{loop}(1) = 1 - (1 - P)^N. \quad (4)$$

**Longer loops**. For any given sequence of length $L$, the probability of a closed loop is $P^L$, and there are $\binom{N}{L}$ possible loops. Therefore, the probability that there is at least one loop of length $L$ equals

$$\text{loop}(L) = 1 - (1 - P^L)^{\binom{N}{L}}. \quad (5)$$

**Any loops**. We use the simplifying assumption that the probabilities of loops of various lengths are independent. Then, the probability that there is at least one autocatalytic loop of any length is given by

$$\text{loop} = 1 - \prod_{L=1}^{N} (1 - P^L)^{\binom{N}{L}}. \quad (6)$$

The overall probability of autocatalytic loops for $N = 100$ (the molecular repertoire usually simulated) approximates one when the probability $P$ of strong catalysis reaches 0.02. There is no clear threshold for separating "strong" and "weak" interactions, but for the 100 molecular species model analyzed in detail in Vasas et al. (2010) certainly less than 50 interactions out of $100^2$ were enough to characterize the compotypes. This corresponds to $P = 0.005$, which in turn implies a less than 30% chance of having any loops in the network.

# References


Ashkenasy, G., Jagasia, R., Yadav, M., Ghadiri, M.R., 2004. Design of a directed molecular network. Proc. Nat. Acad. Sci. USA 101, 10872–10877.

Bachmann, P.A., Luisi, P.L., Lang, J., 1992. Autocatalytic self-replicating micelles as models for prebiotic structures. Nature 357, 57–59.

Blöchliger, E., Blocher, M., Walde, P., Luisi, P.L., 1998. Matrix effect in the size distribution of fatty acid vesicles. J. Phys. Chem. B 102, 10383–10390.

Breslow, R., 1959. On the mechanism of the formose reaction. Tetrahedron Lett. 1, 22–26.

Cairns-Smith, A.G., 1982. Genetic Takeover and the Mineral Origins of Life. Cambridge University Press, Cambridge.

Cavalier-Smith, T., 1995. Membrane heredity, symbiogenesis, and the multiple origins of algae. In: Arai, R., Kato, M., Doi, Y. (Eds.), Biodiversity and Evolution. The National Science Museum Foundation, Tokyo, pp. 75–114.

Chen, X., Li, N., Ellington, A.D., 2007. Ribozyme catalysis of metabolism in the RNA world. Chem. Biodivers. 4, 633–655.

Dyson, F.J., 1985. Origins of Life. Cambridge University Press, Cambridge.

Eigen, M., 1971. Selforganization of matter and the evolution of biological macromolecules. Naturwissenschaften 58, 465–523.

Fernando, C., Rowe, J., 2007. Natural selection in chemical evolution. J. Theor. Biol. 247, 152–167.

Fernando, C., Rowe, J., 2008. The origin of autonomous agents by natural selection. Biosystems 91, 355–373.

Fernando, C., Vasas, V., 2012. Cooptive evolution of prebiotic chemical networks. In: Seckbach, J. (Ed.), Genesis - In The Beginning: Precursors of Life, Chemical Models and Early Biological Evolution. Springer, Dordrecht, pp. 35–53.

Gánti, T., 2003. The Principles of Life. Oxford University Press, Oxford.

Gilbert, W., 1986. Origin of life: the RNA world. Nature 319, 618.

Grey, D., Hutson, V., Szathmáry, E., 1995. A re-examination of the stochastic corrector model. Proc. R. Soc. B 262, 29–35.

Guerrier-Takada, C., Gardiner, K., Marsh, T., Pace, N., Altman, S., 1983. The RNA moiety of ribonuclease P is the catalytic subunit of the enzyme. Cell 35, 849–857.

Hogeweg, P., 1998. On searching generic properties of non-generic phenomena: an approach to bioinformatic theory formation. In: ALIFE: Proceedings of the Sixth International Conference on Artificial Life. MIT Press, Madison, pp. 285–294.

Joyce, G.F., 2002. The antiquity of RNA-based evolution. Nature 418, 214–221.

Kauffman, S.A., 1986. Autocatalytic sets of proteins. J. Theor. Biol. 119, 1–24.

Kauffman, S.A., 1993. The Origins of Order. Oxford University Press, New York.

King, G.A.M., 1982. Recycling, reproduction, and life's origins. Biosystems 15, 89–97.

Kruger, K., Grabowski, P.J., Zaug, A.J., Sands, J., Gottschling, D.E., Cech, T.R., 1982. Self-splicing RNA: autoexcision and autocyclization of the ribosomal RNA intervening sequence of tetrahymena. Cell 31, 147–157.

Kun, A., Papp, B., Szathmáry, E., 2008. Computational identification of obligatorily autocatalytic replicators embedded in metabolic networks. Genome Biol. 9, R51.

Lifson, S., 1997. On the crucial stages in the origin of animate matter. J. Mol. Evol. 44, 1–8.

Maynard Smith, J., 1986. The Problems of Biology. Oxford University Press, Oxford.

Nissen, P., Hansen, J., Ban, N., Moore, P.B., Steitz, T.A., 2000. The structural basis of ribosome activity in peptide bond synthesis. Science 289, 920–930.

Orgel, L.E., 2008. The implausibility of metabolic cycles on the prebiotic Earth. PLoS Biol. 6, e18.

Patzke, V., von Kiedrowski, G., 2007. Self replicating systems. Arkivoc v, 293–310.

Segré, D., Ben-Eli, D., Lancet, D., 2000. Compositional genomes: prebiotic information transfer in mutually catalytic noncovalent assemblies. Proc. Natl. Acad. Sci. USA 97, 4112–4117.

Segré, D., Ben-Eli, D., Deamer, D.W., Lancet, D., 2001a. The lipid world. Orig. Life Evol. Biosph. 31, 119–145.

Segré, D., Shenhav, B., Kafri, R., Lancet, D., 2001b. The molecular roots of compositional inheritance. J. Theor. Biol. 213, 481–491.

Shenhav, B., Bar-Even, A., Kafri, R., Lancet, D., Polymer, G.A.R.D., 2005. Computer simulation of covalent bond formation in reproducing molecular assemblies. Orig. Life Evol. Biosph. 35, 111–133.

Szathmáry, E., 1991. Simple growth laws and selection consequences. Trends Ecol. Evol. 6, 366–370.

Szathmáry, E., 2000. The evolution of replicators. Phil. Trans. R. Soc. B 355, 1669–1676.

Szathmáry, E., Demeter, L., 1987. Group selection of early replicators and the origin of life. J. Theor. Biol. 128, 463–486.

Vasas, V., Szathmáry, E., Santos, M., 2010. Lack of evolvability in self-sustaining autocatalytic networks constraints metabolism-first scenarios for the origin of life. Proc. Natl. Acad. Sci. USA 107, 1470–1475.

Vasas, V., Fernando, C., Santos, M., Kauffman, S.A., Szathmáry, E., 2012. Evolution before genes. Biol. Direct 7, 1.

Wachtershauser, G., 1988. Before enzymes and templates – theory of surface metabolism. Microbiol. Rev. 52, 452–484.

Wachtershauser, G., 1992. Groundworks for an evolutionary biochemistry – the iron sulfur world. Prog. Biophys. Mol. Biol. 58, 85–201.

Zachár, I., Szathmáry, E., 2010. A new replicator: a theoretical framework for analysing replication. BMC Biol. 8, 21.

Zintharas, E., Santos, M., Szathmáry, E., 2002. "Living" under the challenge of information decay: the stochastic corrector model vs. hypercycles. J. Theor. Biol. 217, 167–181.

# 3

# Dealing with Spatial Autocorrelation in Gene Flow Modeling

Daniela Stojanova*, Marko Debeljak*, Michelangelo Ceci†, Annalisa Appice†, Donato Malerba†, Sašo Džeroski*,‡

*DEPARTMENT OF KNOWLEDGE TECHNOLOGIES, JOŽEF STEFAN INSTITUTE, JAMOVA CESTA, LJUBLJANA, SLOVENIA; JOŽEF STEFAN INTERNATIONAL POSTGRADUATE SCHOOL, JAMOVA CESTA, LJUBLJANA, SLOVENIA, †DIPARTIMENTO DI INFORMATICA, UNIVERSITÀ DEGLI STUDI DI BARI "ALDO MORO," VIA ORABONA 4, BARI, ITALY, ‡CENTRE OF EXCELLENCE FOR INTEGRATED APPROACHES IN CHEMISTRY AND BIOLOGY OF PROTEINS

## 3.1 Introduction

Spatial data can identify the geographic locations of different kinds of ecological and environmental data. The main characteristic of spatial data is that they are geo-referenced, that is, they are usually presented with coordinates and a topology which can be visualized on a map. Spatial data are common in ecological studies. For instance, studies on population dynamics examine changes of the size and the structure of populations over space and time. Moreover, several studies on habitat modeling look for the suitability of different ecosystem environments (e.g., aquatic, arable, and forest ecosystems) for various organisms (flora and fauna) (Debeljak and Džeroski, 2011).

The main problem with spatial data is that measured at locations relatively close to each other tend to have more similar values than data measured at locations further apart (Tobler's first law of geography (Tobler, 1970)). For example, species richness at a given site is likely to be similar to that of a site nearby, but very different from that of sites far away. This is mainly due to the fact that the environment is more uniform within a shorter radius. This phenomenon is referred to as spatial autocorrelation. More specifically, the *positive* spatial autocorrelation refers to a map pattern according to which similar values of a geographic feature tend to cluster closely on the map, whereas *negative* spatial autocorrelation indicates a map pattern according to which similar values scatter throughout the map. When no statistically significant spatial autocorrelation exists, the pattern of spatial distribution is considered to be random. The inappropriate treatment of spatial autocorrelation could obfuscate important insights and the observed patterns may even be inverted when spatial autocorrelation is ignored (Kühn, 2007).

The motivation to take spatial autocorrelation into account in ecological and environmental data is manifold, but the following four factors are of particular importance (Legendre, 1993; Legendre et al., 2002):

Models of the Ecological Hierarchy. DOI: http://dx.doi.org/10.1016/B978-0-444-59396-2.00003-1

ISSN 0167-8892, 

1. Biological processes of speciation, extinction, dispersal, or species interactions are typically distance-related.
2. Nonlinear relationships may exist between species and environments, but these relationships may be incorrectly modeled as linear.
3. Classical statistical modeling may fail in the identification of the relationships between different kinds of data without taking into account their spatial arrangement (Besag, 1974).
4. The spatial resolution of data should be taken into account: Coarser grains lead to spatial smoothing of data.

Unfortunately, classical statistical models are based on the assumption that the values of observations in each sample are independent of each other. This is in contrast with spatial autocorrelation, which clearly indicates a violation of this assumption. As observed by LeSage and Pace (LeSage and Pace, 2001), "anyone seriously interested in mining sample data which exhibits spatial dependence should consider a spatial model," since this can take into account different forms of spatial autocorrelation. They showed how the inclusion of autocorrelation of the dependent variable in a predictive task provides an improvement in fitting, as well as a dramatic difference in the significance and impact of explanatory variables included in the predictive model. Even if we are not interested in analyzing the spatial structure within a data sample (i.e., the relationship between the response variable and spatial variables such as longitude and latitude), we should still care about spatial autocorrelation, as it can result in severe problems, that is, can stretch the statistics beyond the basic assumptions and, worse, still overlook important factors affecting the functioning of the ecosystems (Legendre, 1993; Legendre et al., 2002; Betts et al., 2009).

Applications of autocovariate regression, spatial eigenvector mapping, generalized least squares, conditional/simultaneous autoregressive models, and generalized estimating equations to spatial regression tasks are presented by Dormann et al. (Dormann, 2007). All these methods account for spatial autocorrelation in the analysis of species distribution data by considering both species presence/absence (binary response) and species abundance data (Poisson or normally distributed response). The suggested way to detect spatial autocorrelation effects in regression is by measuring spatial autocorrelation of the residuals of the linear model.

Other methods include spatial filtering, kriging, and geographically weighted regression (GWR). **Spatial filtering** actually transforms a variable exhibiting spatial autocorrelation into a variable that is free of the spatial dependence. The transformation is obtained by building two new synthetic variables from the original georeferenced variable. The former is a spatial filter variable capturing the latent spatial dependency that otherwise would remain in the response residuals. The latter is a nonspatial variable that is free of spatial dependence (Griffith, 2002). **Kriging** (Cressie, 1991) is a spatial statistics technique which exploits spatial autocorrelation and determines a local model of the spatial phenomenon. It applies an optimal linear interpolation method to estimate the response variable at each location across the landscape. The response variable is decomposed into a structural component, which represents a mean or a constant trend, a random (spatially correlated) component, and a random noise component, which expresses measurement errors or variations inherent in the response variable. Another standard way to take into account spatial autocorrelation in spatial statistics is the **geographically weighted regression (GWR)** (Fotheringham et al., 2002). GWR extends the traditional multiple linear regression framework so that all parameters are estimated within a local context. In this way, GWR takes advantage of positivc autocorrelation between ncighboring sites in space and provides valuable information on the nature of the processes being investigated.

In this chapter, we focus on the issues raised by spatial autocorrelation when we are interested in learning a predictive spatial regression model from a sample of geo-referenced data. In the spatial regression setting, the sample data are distributed on some domain $\boldsymbol{X}$, are geo-referenced in the space $U \times V$ (e.g., latitude × longitude), and labeled according to an unknown function $g$ with range $Y$. The domain $\boldsymbol{X}$ is spanned by $m$ independent (or explanatory) random variables $X_i$ (both continuous and categorical), while $Y$ is a subset of $\Re$, that is, the dependent (or response) variable $y$ is continuous. A learning algorithm receives a training sample $E = \{(u, v, \boldsymbol{x}, y) \in |U \times V \times \boldsymbol{X} \times Y| \; y = g(u, v, \boldsymbol{x})\}$ and returns a function $f$ close to $g$ on the domain $\boldsymbol{X}$, where closeness is often measured by means of the expected square error over a testing sample or a part of $E$ withheld for learning.

This chapter focuses on handling spatial autocorrelation when predicting gene flow from genetically modified (GM) to non-GM maize fields under real multifield crop management practices at a local scale. The gene flow measured at neighboring locations tends to have more similar values than the ones measured at locations further apart. Because this is a distance-related problem and nonlinear relationships may exist between the factors affecting gene flow environments, classical statistical modeling may fail in the identification of the true relationships between the different kinds of data if it does not take into account their spatial arrangement (Besag, 1974). Therefore, in this chapter, we present a different approach to spatial regression which accounts for spatial autocorrelation in tasks such as the one of predicting gene flow.

The proposed method, spatial predictive clustering system (SCLUS) (Stojanova et al., 2011), is an extension of the system CLUS (Blockeel et al., 1998) that builds Predictive Clustering Trees (PCTs). The SCLUS method can be used to build spatially aware regression trees by considering both local and global, spatial autocorrelation and can deal with the "ecological fallacy" problem (Robinson, 1950) according to which individual subregions do not have the same data distribution of the entire region. We also prove the usefulness of the proposed method in predicting the gene flow (outcrossing rate) from GM to non-GM maize fields under real multifield crop management practices at a local scale.

## 3.2 Modeling Method: Spatially Aware Predicting Clustering Trees

Decision tree learning is among the most popular machine learning techniques used for ecological modeling. Decision trees can be used to predict the value of one or several target (dependent) variables. They are hierarchical structures, where each internal node contains a test on an attribute, each branch corresponds to an outcome of the test, and each leaf node gives a prediction for the value of the class variable(s). Depending on whether we are dealing with a classification (discrete target) or a regression problem (continuous target), the decision tree is called a classification or a regression tree, respectively.

Predictive Clustering Trees (PCTs) (Blockeel et al., 1998) are decision trees that combine elements from both prediction and clustering. As in clustering, clusters of observations that are similar to each other are identified, but a predictive model is associated to each cluster. The predictive model can be viewed as a hierarchy of clusters complemented by a symbolic description of the clusters—the top-node corresponds to one cluster containing all data, which is recursively partitioned into smaller clusters while moving down the tree. The construction of PCTs is not very different from that of standard decision tree learning—at each internal node, a test has to be selected according to a given evaluation function. The main difference between standard decision trees and PCTs is that PCTs select the best test by maximizing the (intercluster) variance reduction with respect to an arbitrary variance function Var (.), defined as $\Delta(E,P) = Var(E) - \sum_{E_k \in P} \frac{|E_k|}{|E|} Var(E_k)$, where $E$ represents the observations in the node and $P$ defines the partition $\{E_1, E_2\}$ of $E$. When dealing with single outcrossing target $Y$, Var (.) denotes the standard variance function known from statistics $Var\ (E) = Var\ (Y)$. When dealing with several outcrossing targets, $Y_1, \dots Y_t$, $Var(E) = \sum_{i=1}^{T}(Y_i)$. The PCT framework allows different definitions of appropriate variance functions for different types of data and can thus handle complex structured data as targets. PCTs can work with complex structured data by introducing adequate variance measures, PCTs are a good candidate to appropriately deal with spatial data where the complexity comes from the need of taking autocorrelation into account. Following this idea, SCLUS extends CLUS in order to build spatially aware PCTs. In fact, the tree structure of the obtained models permits us to naturally deal with the ecological fallacy: the entire region is hierarchically partitioned into subregions such that the data distribution in each subregion is as uniform as possible across the landscape. Additionally, the tree structure permits us to model the presence of autocorrelation in the sample data at different levels of the tree: global modeling is possible at the top-levels of the tree, while local modeling is possible at the bottom levels of the tree. This ability of supporting a hierarchical modeling of spatial autocorrelation is the main difference of SCLUS with respect to the classical spatial statistical approaches (GWR and Kriging), which can only deal with autocorrelation locally and result in many different local models that make separate predictions at each site.

To consider autocorrelation, SCLUS groups examples that show high-positive autocorrelation. This means that examples in each cluster not only have similar response values, but also they fall in the same spatial neighborhood (very close to each other in space). The SCLUS method uses this rationale while building the spatially aware predictive models.

At present, SCLUS does not learn models which perform spatial splits (using the coordinates as attributes). The motivation behind this choice is that of avoiding to lose generality in the induced models. In this way, we can learn more general models that can be easily applied in the same domain, but in different spatial contexts (Ester et al., 1997). Therefore, the spatial arrangement is only used in the split evaluation. In particular, SCLUS uses the level of spatial autocorrelation among examples in the dataset. Spatial autocorrelation is measured by using standard spatial statistics, such as Global Moran's $I$ and Global Geary's $C$. Equation (1) defines the Global Moran's $I$ as

$$I = \frac{N \sum_i \sum_j w_{ij} (Y_i - \overline{Y})(Y_j - \overline{Y})}{\sum_i \sum_j w_{ij} \times \sum_j (Y_j - \overline{Y})^2} \quad (1)$$

where $N$ is the number of geo-referenced examples indexed by $i$ and $j$; $Y_i$ and $Y_j$ are the values of the variable $Y$ for the examples $o_i$ and $o_j$, respectively; $Y$ is the variable of interest (response variable in SCLUS); $\overline{Y}$ is the overall mean of $Y$; and $W = [w_{ij}]_{i=1,\ldots,N\ j=1,\ldots,N} \sum_{ij} w_{ij}$ is the matrix of spatial weights. Values of $I$ generally range from $-1$ (negative autocorrelation) to $+1$ (positive autocorrelation) and 0 indicates a random distribution of the data.

Equation (2) defines the Global Geary's $C$ as

$$C = \frac{(N-1) \sum_i \sum_j w_{ij} (Y_i - Y_j)^2}{2 \sum_i \sum_j w_{ij} \times \sum_j (Y_j - \overline{Y})^2} \quad (2)$$

Values of $C$ typically range from 0 (positive autocorrelation) to 2 (negative autocorrelation) and 1 indicates a random distribution of the data.

Both indexes (Moran's $I$ and Geary's $C$) use a spatial weight matrix $W = [w_{ij}]_{i=1,\ldots,N\ j=1,\ldots,N}$ that reflects the intensity of the spatial relationship between observations in a neighborhood. Each weight $w_{ij}$ is inversely proportional to the distance between the examples $o_i$ and $o_j$. More specifically, if the examples are far away from each other, that is, the distance between $o_i$ and $o_j$ is larger than a predefined threshold $b$ *(bandwidth)*, $w_{ij}$ tends to zero. In addition, the elements are normalized so that for each $i$, $\sum_i w_{ij} = 1$. In this work, the bandwidth $b$ is represented as a percentage of the maximum spatial distance between two examples in the dataset.

While both statistics reflect the spatial dependence of values, they do not provide identical information. $C$ emphasizes the differences in values between pairs of observations, while $I$ emphasizes the covariance between the pairs. This means that Moran's $I$ values are smoother, whereas Geary's $C$ is more sensitive to differences in small neighborhoods.

In order to evaluate the possible splits and find the best split for each internal node of the tree structure, one of these spatial statistics is linearly combined with the variance reduction. A user-defined parameter $\alpha$ regulates the relative influence of both parts of the split evaluation heuristic as reported in Eqn. (3)

$$h(E, P) = \alpha \times \Delta(E, P) + (1 - \alpha) \times S(E, P) \quad (3)$$

where $\alpha \in [0,1]$; $\Delta(E, P)$ is the variance reduction associated to the split $P$, while $S(E, P) = \sum_{E_k \in P} \frac{|E_k|}{|E|} S(E_k)$ is the autocorrelation measurement (with $S(E_k)$ computed as either $I$ or $C$) associated to the split $P$. Both $\Delta(E, P)$ and $S(E, P)$ are scaled to the same interval [0, 2] in order to be combined.

Using the split evaluation heuristic described above, the top-down induction of spatially aware PCTs proceeds recursively by taking as input a set of training examples $E$ and partitioning the descriptive space until the stopping criterion is satisfied. In the construction of the tree, at each internal node, the split that maximizes $h(E, P)$ is chosen. As stated before, splits are derived only from the nonspatial variables. Possible tests are of the form $X \leq \beta$ for continuous variables, and

$X \in \{x_{i1}, x_{i2}, \ldots, x_{ie}\}$ (where $\{x_{i1}, x_{i2}, \ldots, x_{ie}\}$ is a subset of the domain of $X$) for discrete variables. Finally, as in standard decision trees, the tree construction stops when the number of examples in a leaf is less than $\sqrt{N}$, where $N$ is the number of examples in the whole datasets. This value is considered to be a good locality threshold that does not permit to lose too much in accuracy (Gora and Wojna, 2002). As in the original CLUS method, when a leaf is constructed, the prediction associated to it is the mean of the response values of training examples clustered in the leaf.

The method, as described above, is implemented in a system that offers the opportunity to the user to define the parameters and explore their influence on the final results. The user can choose which of the spatial statistics (Global Moran's $I$ or Global Geary's $C$) to include in the model and the relative influence $\alpha$. The user can also define the neighborhood (bandwidth $b$) in which autocorrelation is considered and set the level of spatial autocorrelation to be considered in split selection when building the model.

## 3.3 Results and Discussion: Modeling Gene Flow from GM to Non-GM Fields

In this section, we describe the use of SCLUS to model gene flow from GM to non-GM fields under real muftifiled crop management practices at a local scale. We first provide a description of the data and then give the SCLUS parameter settings and the evaluation measures used for the evaluation of the effectiveness of the predictions provided by SCLUS. Then, we compare the performance of the SCLUS method to competitive regression and spatial modeling methods. Finally, we present the map of the predicted outcrossing rates constructed by using the obtained models, as a final visual product of this analysis.

### 3.3.1 Data Description

The area under study encompasses 400 ha of the Foixa region in Spain, a region with intensive production of GM and non-GM maize. The data are provided for a period of three successive years, from 2004 to 2006. A detailed description of the field setup, data used, and methods of sampling is given by Messeguer et al. (2006). The definition of the variables used in the modeling process is based on expert knowledge and previous studies (Messeguer et al., 2006; Debeljak et al., 2011) about the gene flow modeling problem.

The explanatory variables are defined by taking into account the multifield effects from the neighboring fields at a local scale. This includes the consideration of a *secureDistance*[1], that is, a distance threshold above which there is no (or negligibly small) influence from the surrounding fields and a *floweringDelay*, which expresses the difference in number of days between the flowering of the GM and the non-GM field. We use prior knowledge of previous studies to set these thresholds. The threshold for the flowering delay between the GM and non-GM fields is set to 10 days. This means that if there are a non-GM and a GM field in the same neighborhood and one of them flowers ten or more days later than the other field, the non-GM field cannot be contaminated by the GM field. The secure distance around each sampling point is set to 150 m. According to domain expertise, GM fields or parts of fields that are further away than 150 m from a conventional field cannot contaminate it significantly, even if they have an overlap in the flowering periods.

Both variables were identified by a domain expert as crucial for the analysis and further on serve as the basis for the definition of the other independent variables:

- Average (minimum) distance and average (minimum) "weighted edge" from a sampling point to surrounding GM fields
- Number of GM fields surrounding a sampling point within the *secureDistance* of 150 m considering only fields that have *floweringDelay* less or equal to 10 days
- Total GM area surrounding a sampling point within the *secureDistance* of 150 m considering only fields that have *floweringDelay* less or equal to 10 days
- Relative non-GM area defined as the proportion of the non-GM area in which the sampling point is located and the total GM area surrounding the sampling point considering both the *secureDistance* of 150 m and the *floweringDelay* less or equal to 10 days.

An image of the area under study is given in Fig. 3.1 whereas an image illustrating the definition of the independent variables used in this study is given in Fig. 3.2.

**FIGURE 3.1** An image of the area under study, the Foixa region in Spain. The legend shows the different crops grown in 2004. For color version of this figure, the reader is referred to the online version of this book.

The dataset contains observations of the outcrossing rate, the response variable, measured 420 samples points during a period of three successive years, from 2004 to 2006. The sampling points are located in 15 (total) non-GM maize fields. The outcrossing comes from the surrounding 100 (total) GM maize fields, within a secure distance of 150 m radius around a sampling point, considering only fields that have an overlap in the flowering periods, that is, flowering delay of less or equal to 10 days. The explanatory variables include the number of GM fields, the size of the surrounding GM fields, the ratio of the size of the surrounding GM fields and the size of conventional field, the average distance between conventional and GM fields, as well as the coordinates of the sampling points. Later are the two $(x, y)$ spatial coordinates that are not directly included into the models.

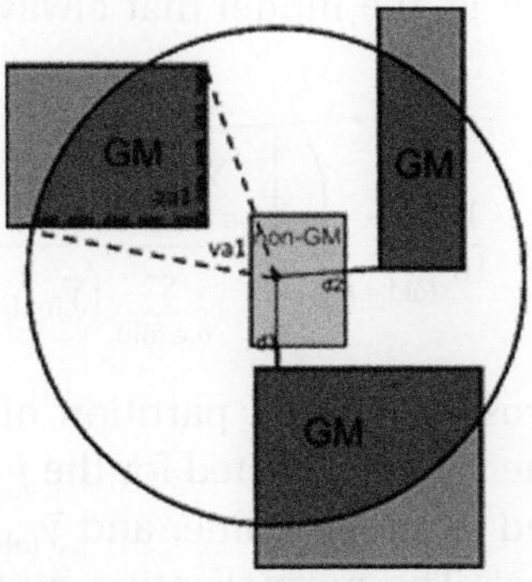


**FIGURE 3.2** The influence area (circle) defined by the influence distance from a sample point. The parts of the GM fields that are outside the circle are treated as having no or insignificantly small influence on the conventional field. The (minimal) distances from sampling point to the neighboring GM fields within the influence area are marked as $d_2$, and $d_3$, visual angle from sampling point to GM field is marked as va1 (dashed lines), and active edges are marked as aa1 (double dashed lines). For color version of this figure, the reader is referred to the online version of this book.

The response (target) variable is the gene flow (outcrossing rate) that comes from the surrounding GM fields, measured at a sampling point that is located within a non-GM field. It is a nonnegative continuous variable; it has higher values at sampling points located very close to the GM fields and lower values at sampling points located further away from the GM fields.

## 3.3.2 Data Analysis Setup: Algorithms, Parameters, Evaluation Measures

We apply SCLUS to the data described above. The performance of the SCLUS methods depends on the choice of bandwidth *b*, the measure of spatial autocorrelation, and the user-defined parameter $\alpha$. The bandwidth is represented as a percentage of the maximum distance (secure distance) between the sampling points in the dataset. Here, we use 1, 5, 10, and 20% of the maximum distance between the sampling points in the dataset. The spatial statistics are the Global Moran's *I* and Global Geary's *C*. The distance between the sampling points is calculated as the Euclidean distance between the points, using their spatial coordinates. Equal importance is given to variance reduction and spatial statistic by setting the user-defined parameter $\alpha$ to 0.5.

We compare the predictive performance of SCLUS with that of the original CLUS method. In addition, SCLUS is compared to other competitive regression methods, which include Linear Regression, M5′ Regression Trees (RT), M5′ Regression Rules (all three implemented in the WEKA system), and GWR. Only GWR and SCLUS take into account spatial autocorrelation in the data.

The predictive performance of the regression models on unseen cases is estimated using the standard 10-fold cross-validation method and evaluated according to the measure of average relative root mean squared error (AvgRRMSE). AvgRRMSE is defined by Eqn. (4) as the average of RMSE of the model predictions normalized with the

RMSE of the default model, that is, the model that always predicts (for regression) the average value of the target:

$$\text{AvgRRMSE} = \frac{1}{10}\sum_{\text{fold}_i \in E}\left(\sqrt{\frac{\sum_{o_j \in \text{fold}_i}\left(f_{E-\text{fold}_i}(x_j) - y_j\right)^2}{\sum_{o_j \in \text{fold}_i}\left(\overline{y}_{\text{fold}_i} - y_j\right)^2}}\right) \quad (4)$$

where $\{\text{fold}_1, \ldots, \text{fold}_{10}\}$ is a cross-validation partition of the sample data $E$, 10 is the number of folds, $f_{E-\text{fold}_i}(x_j)$ is the value predicted for the $j$-th test case by the model built from$E - \text{fold}_i$, $y_j$ is the observed response value, and $\overline{y}_{\text{fold}_i}$ is the average of the actual response variable on the test set. The normalization by the denominator removes the influence of the range of values of the response. AvgRRMSE typically varies in the interval [0, 1].

### 3.3.3. Results: Predictive Performance

Table 3.1 presents the predictive performance results obtained by using the described method SCLUS and the competitive methods in terms of the RRMSE measure. SCLUS results are given for both spatial statistics and different bandwidth values $b$.

From the results reported in Table 3.1 we can observe that the selection of the bandwidth may influence the accuracy of the learned regression models. In our case, the best results are obtained when the bandwidth is greater than 5%. This indicates that the autocorrelation is more evident in larger neighborhoods than in very narrow ones.

The comparison of the two statistics reveals that there is no statistically significant difference between the results. However, the obtained results confirm the distinguishing characteristics of both statistics described in Section 3.3. The errors of the SCLUS models using Geary's $C$ are more sensitive to the bandwidth, whereas the errors of the SCLUS

**Table 3.1** AvgRRMSE of the Regression Models Obtained by Using Different Methods (SCLUS, CLUS, GWR, Linear Regression, M5′ Trees, and M5′ Rules). Only GWR and SCLUS Consider Spatial Autocorrelation. The Performance of the SCLUS Models Depends on the Choice of Bandwidth, $\alpha$, and the Measure of Spatial Autocorrelation (Moran's $I$ and Geary's $C$) Used to Learn the Models

| | 1% | 5% | 10% | 20% | CLUS[1] | | | | |
|---|---|---|---|---|---|---|---|---|---|
| **SCLUS Bandwidth** | | $\alpha = 0.5$ | | | $\alpha = 1$ | **GWR** | **Linear Regression** | **M5′ Trees** | **M5′ Rules** |
| Moran $I$ | 0.902 | 0.872 | 0.872 | 0.872 | 0.904 | 1.509 | 0.944 | 0.976 | 0.948 |
| Geary $C$ | 0.891 | 0.873 | 0.878 | 0.872 | | | | | |

[1]Note that the secure distance depends on the design of the study area and it is set by a domain expert as a theoretical limit or an empirically estimated value obtained from previous studies. The bandwidth $b$, on the other hand, regulates the level of autocorrelation included into the models and depends on the data characteristics. The maximum value of the bandwidth $b$ corresponds to the secure distance.

models using Moran's $I$ are less sensitive to the bandwidth (therefore, the influence of the bandwidth is smoother).

In Table 3.1, we also report the accuracy obtained by CLUS, Linear Regression, M5' Regression Trees (RT), M5′ Regression Rules, and GWR. Besides SCLUS, only GWR (from the above listed methods) considers spatial autocorrelation. The other methods ignore this phenomenon.

The results show that SCLUS outperforms all methods. This is explained by the fact that all the competitors, except GWR, completely ignore autocorrelation when learning the regression models. GWR builds local models that consider only the local effects of spatial autocorrelation, that is, effects that are limited to only the closest neighborhood.

Beside the differences in accuracy, the obtained models may have different structure. For illustration, in Fig. 3.3a and b, we show the regression trees obtained by using CLUS and SCLUS (with Global Moran $I$ and $b = 20\%$), respectively. The trees have different splits at the root and different size (number of internal nodes, number of leaves). It is noteworthy that the spatially aware PCT (Fig. 3.3b) is smaller and more compact than the PCT learned by CLUS. Moreover, according to domain expertise, the spatially aware PCT is more interpretable and understandable than the corresponding ordinary PCT. This is also due to the fact that the spatially aware PCT is balanced.

A more detailed analysis of the training examples falling in the leaves of both the ordinary PCT and the spatially aware PCT reveals that the leaves of both trees cluster examples with similar response values (this is due to the variance reduction). But training examples falling in the leaves of the spatially aware PCT are also close in space. This

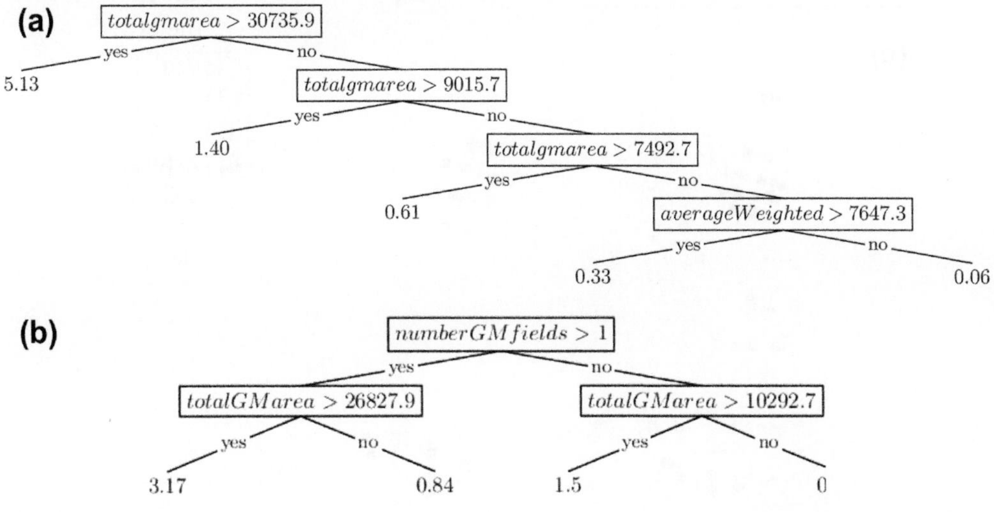


**FIGURE 3.3** Two PCTs learned on one of the training fold of Foixa data: (a) The ordinary PCT learned by CLUS and (b) the spatially aware PCT learned by SCLUS.

guarantees spatially smoothed predictions. This means that the predictions that are close to each other in space tend to have similar values. When plotted on a geographical map they form a nice, smooth continuous surface and there are not sharp edges and discontinuities.

### 3.3.4. Results: Maps of the Predicted Outcrossing Rates

The predictions for testing examples obtained by models learned with GWR, SCLUS, and CLUS (the results are obtained by using onefold as a testing set and the remaining folds as a training set) are visualized in two geographical maps of the outcrossing rates (see Fig. 3.4a–c, respectively). The maps represent the FOIXA study area, as shown in Fig. 3.1.

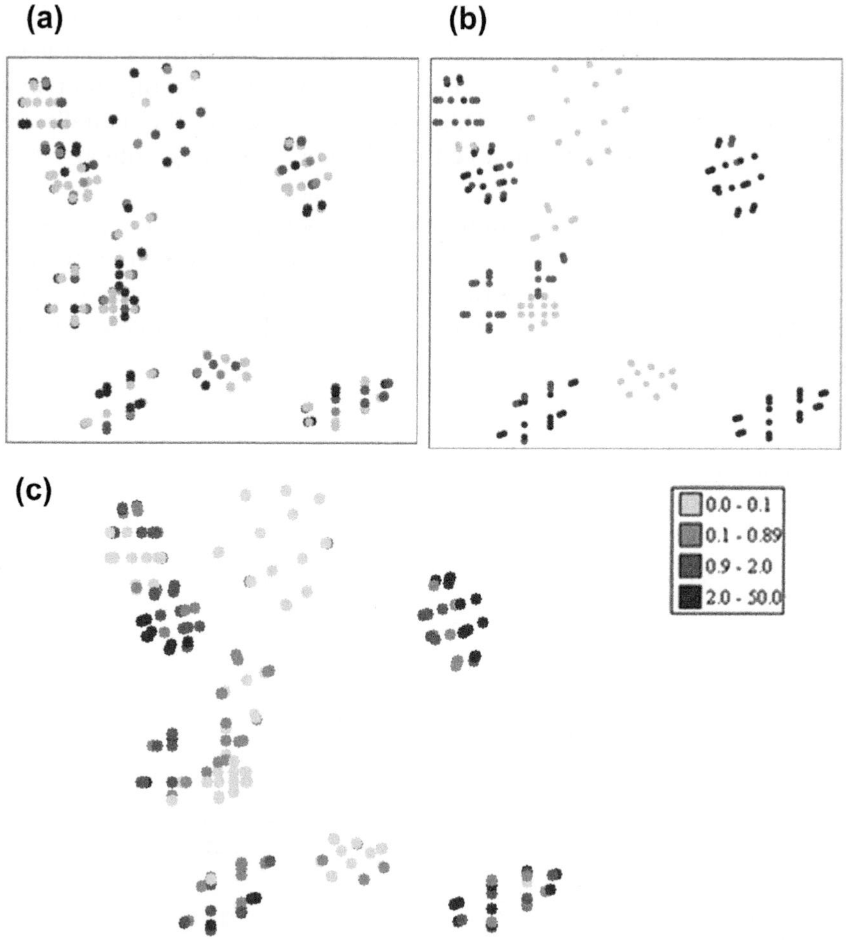


**FIGURE 3.4** Map of the predicted outcrossing rate for sampling selected of points in the FOIXA region, generated by using (a) GWR, (b) SCLUS ($b = 20\%$, Global Moran's $I$, $\alpha = 0.5$), and (c) CLUS. For color version of this figure, the reader is referred to the online version of this book.

However, here we present only the sampling points and disregard the GM and non-GM fields, in order to get a more detailed view of the predicted outcrossing rates at the sampling points.

The map obtained by using the GWR model gives "salt and pepper" results (see Fig. 3.4a). This means that the predictions that are close to each other in space tend to have very different values (very high outcrossing rates come predicted very close to very low ones). When plotted on a geographical map, they do not form a nice smooth continuous surface, but there is sharpness and discontinuity so that the map looks like a mixture of "salt and pepper." This is due to the fact that GWR builds local models at each point, which are independent of the models built in the neighboring points. On the other hand, the GWR models exploit the spatial dimension by using the positive autocorrelation between neighboring points in space and in this way accommodate stationary autocorrelation.

In contrast, the map of predictions obtained by using the model learned by SCLUS (see Fig. 3.4b) shows that the predictions are smoother. This means that the predictions that are close to each other in space tend to have similar values (very high/low outcrossing rates are very close to each other) and when plotted on a geographical map they form a nice smooth continuous surface without sharp edges and discontinuities. Hence, SCLUS models are more accurate in terms of the obtained errors (Table 3.1) and sensible than competitors' models because they are more interpretable (Fig. 3.3b) than all other models. In addition, SCLUS predictions are also spatially smoothed. In practice, these characteristics make SCLUS models more useful than (both spatial and a-spatial) competitors because the geographical mass obtained by using these models is more realistic and easier to interpret.

The map obtained by using the model learned by CLUS (see Fig. 3.4c) is smother that the map obtained by using the GWR model, but we observe that it is a bit sharper and with some discontinues that the one obtained by using the model learned by SCLUS. This is supported by the fact that the accuracy of the predictions obtained by using CLUS is better than the one of the GWR model and worse that accuracy of the model learned by SCLUS.

## 3.4 Conclusion

In this chapter, we described spatial autocorrelation phenomenon and focused on the task of predicting the outcrossing rate from GM to non-GM maize fields. We demonstrated the use of spatially aware PCTs learned by SCLUS, which explicitly take into account the spatial arrangement of the data and the positive autocorrelation coming from this spatial arrangement. In particular, spatially aware PCTs allow us to learn a predictive model and hierarchically cluster examples at the same time.

Taking spatial autocorrelation into account improves the predictive capability of the models. SCLUS clearly outperforms standard modeling techniques (CLUS, M5′) that do not consider spatial autocorrelation and GWR that takes into account autocorrelation, but can only capture local (and not global) regularities because it builds local modes at each point and these models are independent from each other. The tree structure solves the ecological fallacy problem and considers, both globally and locally, the effect of autocorrelation.

The learned spatially aware PCTs adapt to local properties of the data, providing at the same time spatially smoothed predictions. In order to take autocorrelation into account, we use well known measures, such as Global Moran's *I* and Global Geary's *C*. The split evaluation heuristic that we use in the construction of PCTs is a weighted combination of variance reduction (related to predictive accuracy) and spatial autocorrelation of the response variable. We can also consider different sizes of neighborhoods (bandwidth) when calculating spatial autocorrelation. This allows SCLUS to offer a set of design options which can be adopted to the specific application at hand.

Further work would include the use of the presented approach to other regions and in other ecological modeling problems, since the phenomenon of spatial autocorrelation is ubiquitous. We would also like to extend the use of the presented approach to classification tasks as the current approach deals only with the regression task. In addition, an automated procedure for selecting the size of the neighborhood (bandwidth) is needed to replace the manual tuning of this parameter.

In summary, spatial data measured at locations relatively close to each other tends to have more similar values than data measured at locations further apart (Tobler's first law of geography (Tobler, 1970)). Inappropriate treatment of data with spatial dependencies could obfuscate important insights (e.g., when spatial autocorrelation is ignored). Taking spatial autocorrelation into account is thus crucial, but is only performed by few regression approaches. In our work, we have extended the very popular machine learning approach of decision trees in order to take into account spatial autocorrelation. This has led to the construction of more accurate and comprehensive models for predicting gene flow, which also produces spatially smoothed predictions.

## Acknowledgments

This work is in partial fulfillment of the research objectives of the project "EMP3: Efficiency Monitoring of Photovoltaic Power Plants" funded by Fondazione Cassa di Risparmio di Puglia. Džeroski is supported by the Slovenian Research Agency (grants P2-0103, J2-0734, and J2-2285), the European Commission (grant HEALTH-F4-2008-223451), the Centre of Excellence for Integrated Approaches in Chemistry and Biology of Proteins (operation no. OP13.1.1.2.02.0005), and the Jožef Stefan International Postgraduate School. Stojanova is supported by the grant J2-2285. The authors acknowledge the support of the Slovenian–French bilateral scientific collaboration program PROTEUS 2009–2010 and the Consorci Laboratori CSIC-IRTA de Genetica Molecular Vegetal, Departament de Genetica Vegetal, Barcelona, Spain, for providing the data.

# References

Besag, J., 1974. Spatial interaction and the statistical analysis of lattice systems. J. Roy. Stat. Soc. B 36, 192–236.

Betts, M.G., Ganio, L.M., Huso, M.M.P., Som, N.A., Huettmann, F., Bowman, J., et al., 2009. Comment on "Methods to account for spatial autocorrelation in the analysis of species distributional data: a review". Ecography 32, 374–378.

Blockeel, H., De Raedt, L., Ramon, J., 1998. Top-down induction of clustering trees. In: Proceedings of the Fifteenth International Conference on Machine Learning. Morgan Kaufmann, San Francisco, pp. 55–63.

Cressie, N., 1991. Statistics for Spatial Data. Wiley, New York.

Debeljak, M., Džeroski, S., 2011. Decision trees in ecological modeling. In: Modeling Complex Ecological Dynamics. Springer, pp. 197–209.

Debeljak, M., Trajanov, A., Stojanova, D., Leprince, F., Džeroski, S., 2011. Using relational decision trees to model flexible co-existence measures in a multi-field setting. In: Proc. 7th European Conf. on Ecological Modelling, p. 53.

Dormann, C.F., 2007. Methods to account for spatial autocorrelation in the analysis of species distributional data: a review. Ecography 30, 609–628.

Ester, M., Kriegel, H., Sander, J., 1997. Spatial data mining: a database approach. In: Proc. 5th Intl. Symp. on Spatial Databases. Springer, pp. 47–66.

Fotheringham, A.S., Brunsdon, C., Charlton, M., 2002. Geographically Weighted Regression: The Analysis of Spatially Varying Relationships. Wiley.

Gora, G., Wojna, A., 2002. RIONA: a classifier combining rule induction and k- NN method with automated selection of optimal neighbourhood. In: Proc. 13th European Conf. on Machine Learning. Springer, pp. 111–123.

Griffith, D., 2002. A spatial filtering specification for the auto-Poisson model. Stat. Probab. Lett. 58, 245–251.

Kühn, I., 2007. Incorporating spatial autocorrelation may invert observed patterns. Divers. Distrib. 13 (1), 66–69.

Legendre, P., 1993. Spatial autocorrelation—trouble or new paradigm. Ecology 74, 1659–1673.

Legendre, P., Dale, M.R.T., Fortin, M.-J., Gurevitch, J., Hohn, M., Myers, D., 2002. The consequences of spatial structure for the design and analysis of ecological field surveys. Ecography 25, 601–615.

LeSage, J.H., Pace, K., 2001. Spatial dependence in data mining. In: Data Mining for Scientific and Engineering Applications. Kluwer, pp. 439–460.

Messeguer, J., Peñas, G., Ballester, J., 2006. Pollen-mediated gene flow in maize in real situations of coexistence. Plant Biotechnol. J. 4 (6), 633–645.

Robinson, W.S., 1950. Ecological correlations and the behavior of individuals. Am. Sociol. Rev. 15, 351–357.

Stojanova, D., Ceci, M., Appice, A., Malerba, D., Džeroski, S., 2011. Global and local spatial autocorrelation in predictive clustering trees. In: Proc. 15th Intl. Conf. on Discovery Science. Springer, pp. 307–322.

Tobler, W., 1970. A computer movie simulating urban growth in the Detroit region. Econ. Geogr. 46 (2), 234–240.

# 4

# Modeling *In Vitro* Cell-Based Assays Experiments: Cell Population Dynamics

José-Manuel Zaldívar Comenges*, John Wambaugh†, Richard Judson†

*EUROPEAN COMMISSION, JOINT RESEARCH CENTRE, INSTITUTE FOR HEALTH AND CONSUMER PROTECTION, SYSTEMS TOXICOLOGY UNIT, VIA E. FERMI 2749, ISPRA (VA), ITALY,
†NATIONAL CENTER FOR COMPUTATIONAL TOXICOLOGY, US EPA, RESEARCH TRIANGLE PARK, NC, USA

## 4.1 Introduction

The need to reduce and eventually replace the use of animals in toxicology testing is driving the development of Integrated Testing Strategies (ITS), which aim to predict human *in vivo* toxic doses from concentrations that cause effects *in vitro*, with a minimum of intermediate animal testing. This implies the need to consider both toxicokinetics (TK) and toxicodynamics (TD) as important, if not essential, parts in the risk assessment strategy (Adler et al., 2011). The characterization of the concentration that produces an effect (whether this is a perturbation to a molecular pathway or an apical toxic endpoint) is necessary at two levels; first, for *in vitro* experiments since "nominal" concentrations do not represent the real concentration experienced by the cell (Gülden et al., 2001); and, second, in the extrapolation of the dose for human toxicity assessment since to assess the hazard of a chemical compound, we need to know the true concentration experienced by cells within the target organ.

One possible way to solve both the problems and to compare the same concentration values from *in vitro* and *in vivo* experiments is to use TK and TD models at both levels (Escher et al., 2010). TK is essentially the study of the processes by which a substance reaches its target site. Four processes are involved in TK: Absorption (A) is the process of a substance entering the organism; Distribution (D) is the dispersion of substances throughout the fluids and tissues of the organism; Metabolism (M) is the irreversible transformation of the substances by the organism; Excretion (E) is the elimination of the substances from the organism. They are normally called ADME. The processes and interactions of an exogenous compound within an organism, including the compound's

Models of the Ecological Hierarchy. DOI: http://dx.doi.org/10.1016/B978-0-444-59396-2.00004-3

ISSN 0167-8892, 

effects on the processes at the organ, cellular, and molecular levels are called toxicodynamics (TD).

For *in vitro* experiments, we can construct a model comprising the fate of a compound in the cell-based assay, that is, its partitioning between the plastic wall, serum proteins, and lipids, and potentially the compounds dynamics within the cell; combined with a cell growth model and a toxic effects model. These together allow us to model the true concentrations causing perturbations in cells given the nominal concentrations applied in a microtiter plate well. An analogous approach *in vivo* is provided by Physiologically Based Toxicokinetics or Pharmacokinetics models—PBTK or PBPK—(Clewell et al., 2008). A PBTK model consists of a series of mathematical equations, which, based on the specific physiology of an organism and on the biophysical properties of a substance, are able to describe the absorption, distribution, metabolism, and elimination (ADME) of the compound within this organism (Andersen, 1981). The solution of these equations provides the concentration of the chemical compound and its metabolites over time in the modeled organs and allows for a sound mechanistic description of the kinetic processes including the kinetics of accumulation.

The first attempts to assess the applicability of this strategy for the safety evaluation of chemicals were based on the following elements (DeJongh et al., 1999; Gubbels-van Hal et al., 2005):

- *In vitro*/QSAR data on ADME (absorption, distribution, metabolism, excretion) as input data
- PBPK modeling (rat, human, etc.) for calculating target tissue concentration *in vivo* for the pediction of dose-response curves, NOEL (not observed effect level), LOEL (lowe*st obse*rved effect level), etc.
- *In vitro* and *in vivo* studies to validate the approach.

The application of this approach to a small set of ten substances following REACH requirements at production levels >10 tons showed that it was possible to reduce by 38% the number of animals used, but further improvement was foreseen with the refinement of the procedure. Concerning *in vitro* tests, the suggested refinement (Gubbels-van Hal et al., 2005) included the need to estimate the partitioning and bioavailability of the chemical in assay to improve the methodology used to relate *in vitro* toxic concentrations to *in vivo* target tissue concentrations.

In a more recent approach, Rotroff et al. (2010) estimated the human oral equivalent doses necessary to produce cellular concentrations equivalent to those causing bioactivity *in vitro*. They combined a simplified human pharmacokinetic (PK) model, experimental human metabolic clearance and plasma protein binding data, and *in vitro* $AC_{50}$ (half-maximal activity concentration either inhibitory or activating in a concentration response curve) values from the ToxCast database (http://www.epa.gov/ncct/toxcast/chemicals.html). They found that two compounds over the 35 chemicals analyzed showed an overlap between oral equivalent doses, estimated from *in vitro* experiments, and estimated human exposures, which indicated a potential need to focus regulatory attention on those chemicals whose overall human exposure was in the range of observed effects at *in vitro* level.

In a previous work (Zaldívar et al., 2010), we developed and implemented a compound fate model using the partitioning approach, based on HTS (high throughput screening) laboratory data. The developed fate model was able to predict the role of serum in toxicity assays as well as to provide estimates of the partitioning of a certain compound between the headspace, plastic wall, and the medium; that is, attached to serum, free dissolved, or attached to the cells. However, the partitioning approach assumes that the equilibrium is rapid in comparison with the duration of the experiments, which for some chemicals may not always be the case for the partitioning into the cells (Mundy et al., 2004). In addition, an aspect as important as the partitioning of the chemical in the cell assay is the dynamics of the cells in response to the chemical during the experiment. The fundamental process is the growth of the cell population, which will change the partitioning during the experiment and therefore the toxic effects experienced by the cells. In principle, growth may be seen as a dilution process, but it also affects the dynamics of the compound exchange between the cells and the medium. Furthermore, some cells are able to release proteins and lipids into the medium (e.g., albumin by HEPG2 cells), and this may also change the partitioning.

For these reasons, in this work, we have developed an integrated model that couples the fate model with a cell growth model and a toxic effect model to analyze the dynamics of cell populations when subject to a toxicant. The cell model is a stage-based model, whereas the toxic effect model is based on DEB/DEBTOX theory (Kooijman, 2000; Kooijman and Bedaux, 1996), which uses internal concentrations to calculate toxic effects. The results of the model have been used to analyze A549 cytotoxicity assays. This approach opens a new way of analyzing this type of data set and offers the possibility of extrapolating the values obtained to calculate *in vivo* human toxicology thresholds using reverse dosimetry in a PBTK modeling approach.

## 4.2 Methods and Approach

Normally, concentration–response curves in *in vitro* experiments are represented using the total amount of substance added to a microtiterplate well and not the dissolved (free) concentration that is the bioavailable fraction able to produce an effect. However, these curves (often characterized by their potency parameters, e.g., $IC_{50}$ values) do not properly reflect the concentrations at the site of action of the chemical in the cell, which might be a particular receptor or enzyme. The chemical partitions into a number of compartments—one of which typically includes the target molecule (e.g., the cytoplasm)—as well as other cellular compartments (e.g., the cell membranes) and extra-cellular compartments (e.g., free serum, serum proteins, plastic walls, and the headspace) [see Gülden et al. (2001); Heringa et al. (2004); Kramer (2010) amongst others]. This is illustrated in Fig. 4.1. Another aspect that should be considered when volatile compounds are tested is the possibility of evaporation.

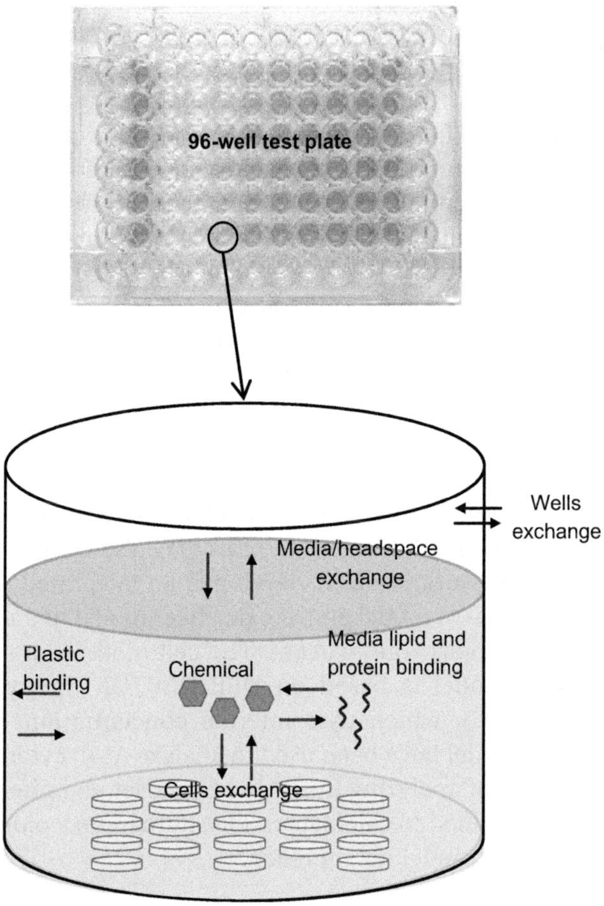


**FIGURE 4.1** Overview of the modeled systems and the processes included in the fate and transport model. For color version of this figure, the reader is referred to the online version of this book.

In addition, during the experiments, cells grow, divide, and consume nutrients. Therefore, the partitioning characteristics of the medium change with time as do the internal concentrations of the test chemical in the cells, complicating even further the comparison between different *in vitro* experiments and systems. This drives the need for an integrated modeling approach that is able to quantify all aspects of partitioning and dynamics, and therefore to "correct" the nominal concentrations as a function of the cell assay and the physicochemical properties of the tested compounds. Therefore, this integrated modeling approach must consist of the following:

- A fate and transport model
- A cell growth and division model
- A toxicity and effects model

The solution of the ordinary differential equations of the model should allow the calculation over time of the dissolved concentration of a chemical as well as the internal concentration in the cells. In this chapter, we briefly describe an implementation of this approach, including the different models and their relationships. We will illustrate and assess the performance of the growth and TD models coupled with the fate and transport model previously developed in Zaldívar et al. (2010).

### 4.2.1 Cell Assays Growth and Division Model

A typical population model described in terms of continuous ordinary differential equations ignores population structure by treating all individuals as identical. The existence of demographically important differences among individuals, however, is obvious. Matrix population models (Caswell, 1989) integrate population dynamics and population structure and are very useful when life cycle is described in terms of size classes or age classes. There are fundamentally two types of approaches: the age classified model and the stage classified model. The first one assumes that age-specific survival and fertility are sufficient to determine population dynamics. On the other hand, if the vital rates depend on body size, and growth is sufficiently plastic that individuals of the same age may differ appreciably in size, then age will provide little information about the fate of an individual (e.g., fish models, see Zaldívar and Campolongo, 2000). For the case of the A549 cell model, we have considered that the appropriate description corresponds to an approach based on four stages, each stage corresponding to one of the four cell cycle phases: G1, S, G2, M (Hartwell and Weinert, 1989).

In the stage-based type of modeling, the matrix $A$, called Leslie matrix, which describes the transformation of a population from time $t$ to time $t+1$,

$$N_{t+1} = AN_t \tag{1}$$

has the following structure:

$$A = \begin{bmatrix} P_1 & 0 & \dots & \dots & F \\ G_1 & P_2 & 0 & \dots & 0 \\ 0 & G_2 & P_3 & 0\dots & 0 \\ \dots & \dots & \dots & \dots & \dots \\ 0 & 0 & 0\dots & G_{q-1} & P_q \end{bmatrix} \tag{2}$$

where $N_t$ is a vector describing the population at each stage at time $t$, $P_i$ is the probability of surviving and staying in stage $i$; $G_i$ is the probability of surviving and growing into the next stage, and $F$ is the fecundity rate per unit time (h), $i = 1,2,\dots,q$.

Both $P_i$ and $G_i$ are functions of the survival probability $p_i$ and the growth probability $\gamma_i$ (Caswell, 1989):

$$P_i = p_i(1 - \gamma_i) \tag{3}$$

$$G_i = p_i \cdot \gamma_i \tag{4}$$

where

$$p_i = \exp(-z_i) \tag{5}$$

and

$$\gamma_i = \frac{\left(1 - p_i\right) p_i^{d_i - 1}}{1 - p_i^{d_i}} \tag{6}$$

where $z_i$ is the hourly instantaneous mortality rate and $d_i$ is the duration (h) within the $i$-th stage.

Incorporation of interaction between species at different stages can be easily done (Cushing, 1998; Zaldívar and Campolongo, 2000).

### 4.2.2 Bioconcentration and Bioaccumulation in Cells

The partitioning approach assumes that the equilibrium is rapid in comparison with the duration of the experiments. However, as it has been observed by Mundy et al. (2004) and Meacham et al. (2005) for some chemicals, several hours are needed to attain equilibrium between medium and internal cellular concentrations. Therefore, a mass balance to simulate the dynamics of internal cellular concentration has been developed. Simple mass balance models of a contaminant in an organism (Thomann, 1989) consider the organism as a single compartment. A more complete model was developed, based on the Dynamic Energy Budget (DEB) approach, by Kooijman and van Haren (1990) and van Haren et al. (1994) that takes into account changes in lipid content and size of the organism. The introduction of DEB models into matrix population models has been developed recently (Lopes et al., 2005; Klanjscek et al., 2006; Billoir et al., 2007). Following this approach, we have considered that once the chemical is taken up by the cell, it partitions instantaneously over three compartments: one aqueous fraction and two nonaqueous fractions: structural component (proteins) and the energy reserves (lipids). This approach is represented in Fig. 4.2 following Kooijman and van Haren (1990).

In this case, the total number of moles of a compound in the cell can be divided as the sum of them in the different compartments:

$$n_{\text{tot}} = n_{\text{aq}} + n_{\text{P}} + n_{\text{L}} = \left(V_{\text{aq}} \cdot C_{\text{aq}} + V_{\text{P}} \cdot C_{\text{P}} + V_{\text{L}} \cdot C_{\text{L}}\right) \tag{7}$$

where the $V_i$s refer to the compartment volumes (l) and the $C_i$s refer to the compartments concentrations (mol l$^{-1}$). Also, the total number of moles of a chemical can be expressed as:

$$n_{\text{tot}} = W \cdot C_b / \text{MW} \tag{8}$$

where $W$ is the cells weight (g), MW is the molecular weight of the chemical (g mol$^{-1}$), and $C_b$ is the contaminant concentration in the cell (g gww$^{-1}$). The chemical is assumed to be in equilibrium between the different compartments with fixed-value partition coefficients: $K_{\text{P}} = C_p/C_{\text{aq}}$ and $K_{\text{L}} = C_l/C_{\text{aq}}$. The time evolution of the concentration of this substance in the cell can be calculated by a simple mass balance, assuming that uptake and depuration rates, $r_{\text{da}}$ and $r_{\text{ad}}$ (l cm$^{-2}$ s$^{-1}$), are proportional to the surface area of the

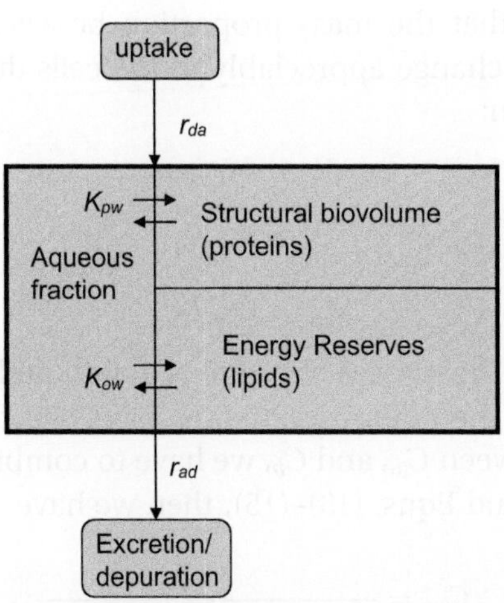


**FIGURE 4.2** Schematic representation of the chemical partitioning in the cellular compartments. For color version of this figure, the reader is referred to the online version of this book.

cell (passive diffusion), and the transfer occurs through the aqueous compartment only (Kooijman, 2000), as:

$$\frac{dn_{tot}}{dt} = V^{2/3}(r_{da} \cdot C_{diss} - r_{ad} \cdot C_{aq}) \tag{9}$$

where $C_{diss}$ and $C_{aq}$ refer to the dissolved fraction concentration in the medium (mol $l^{-1}$) and in the aqueous compartment of the cell (mol $l^{-1}$), respectively. However, it is more convenient to express the mass balance as a function of the cells concentration $C_b$ (g $gww^{-1}$). Therefore, applying the chain rule of derivation to Eqn. (8), we have

$$\frac{dn_{tot}}{dt} = \frac{1}{MW}\left(W\frac{dC_b}{dt} + C_b\frac{dW}{dt}\right) \tag{10}$$

and rearranging terms we obtain

$$\frac{dC_b}{dt} = \frac{MW \cdot V^{2/3}}{W}\left(r_{da} \cdot C_{diss} - r_{ad} \cdot C_{aq}\right) - \frac{C_b}{W}\frac{dW}{dt} \tag{11}$$

In this case, the last term represents the dilution due to the growth of the cell. This is a more realistic assumption than the linear-constant function assumed in the first bioaccumulation models (Thomann, 1989).

Since the concentration in the aqueous fraction $C_{aq}$ is not a value that is measured, then we have to convert it in terms of $C_b$ using the partitioning approach. The wet weight $W$ can also be expressed as a function of the volumes of the different compartments multiplied by the average density ($\rho$ in g $l^{-1}$):

$$W = \rho \cdot V = \rho(V_{aq} + V_P + V_L) \tag{12}$$

In addition, assuming that the mass proportion between proteins, lipids, and the aqueous fraction does not change appreciably in the cells during growth, the following relationships can be written:

$$V_P = W_P/\rho_P \tag{13}$$

$$V_L = W_L/\rho_L \tag{14}$$

$$V_{aq} = W_{aq}/\rho_{aq} \tag{15}$$

Where $W_P$, $W_L$ and $W_{aq}$ are the masses of proteins, lipids, and aqueous compartments in the cells and $\rho_P$, $\rho_L$, and $\rho_{aq}$ their densities, that is, 1350, 900, and 1000 g l$^{-1}$, respectively. To find the correlation between $C_{aq}$ and $C_b$, we have to combine $n_{tot}$ in Eqns. (7) and (8), the partition coefficients, and Eqns. (13)–(15), then we have

$$C_{aq} = \frac{C_b}{MW\left(\frac{f_{aq}}{\rho_{aq}} + \frac{f_L}{\rho_L}K_L + \frac{f_P}{\rho_P}K_P\right)} \tag{16}$$

where $f_i$ refers to the mass fractions of each compartment (aqueous, lipid, proteins) in the cell. For A549, we have: $f_{aq} = 0.72$, $f_L = 0.012$, $f_P = 0.268$ (Poulin and Krishnan, 1995). Therefore, for A549, $\rho = 1073$ g l$^{-1}$. Replacing this equation into Eqn. (11) and rearranging, we obtain a similar equation to the one proposed by Thomann (1989):

$$\frac{dC_b}{dt} = \left(\frac{MW \cdot V^{2/3}}{W} r_{da}\right) C_{diss} - \left(\frac{V^{2/3}}{W\left(\frac{f_{aq}}{\rho_{aq}} + \frac{f_L}{\rho_L}K_L + \frac{f_P}{\rho_P}K_P\right)} r_{ad}\right) C_b - \left(\frac{1}{W}\frac{dW}{dt}\right) C_b \tag{17}$$

However, in this case, the uptake and depuration rates are not constant, but depend on the status of the cell and take into account the differences in growth.

The variation of the wet weight $W$ as a function of time can be obtained, assuming constant composition and hence density, as

$$\frac{dW}{dt} = \rho \frac{dV}{dt} \tag{18}$$

If we consider spherical shape, $V = \frac{4}{3}\pi r^3$, and von Bertalanffys growth curve

$$r(t) = r_\infty - (r_\infty - r_0)e^{-\alpha_G \cdot t} \tag{19}$$

where $r_0$ and $r_\infty$ refer to the initial and final cell radius, respectively, and $\alpha_G$ is von Bertalanffy's growth rate. Then, we have

$$\frac{dV}{dt} = 4 \cdot \pi [r_\infty - (r_\infty - r_0)\exp(-\alpha_G \cdot t)]^2 (r_\infty - r_0) \cdot \alpha_G \cdot \exp(-\alpha_G \cdot t) \tag{20}$$

For A549, $r_\infty = 9.273 \times 10^{-6}$ m, $r_0 = 7.36 \times 10^{-6}$ m, and $\alpha_G = 4.17 \times 10^{-5}$ s$^{-1}$ (Jiang et al., 2010).

However, the introduction of this term considers only a single cell that develops during the simulation. To consider the whole population of the cells, instead of a single cell, we will take average values for the weight, its derivative, and the surface depending on the four stages: G1, S, G2, M.

The model in Eqn. (17) has several parameters that need to be evaluated. The uptake and depuration rates $r_{da}$ and $r_{ad}$ and the partition coefficients $K_L$ and $K_P$ depend on the compound, whereas the remaining parameters depend on the type of the cell we are modeling.

## 4.2.3 Toxicity and Effects Model

The direct effects of a chemical concentration $C$ on survival may be expressed using Eqn. (5) by the addition of a term that can be written as

$$z_i = \begin{cases} z_i + k_t(C_b - \text{NEC}) & \text{if } C > \text{NEC} \text{ and } \tau > \tau_0 \\ z_i \end{cases} \tag{21}$$

where $C_b$ is the internal concentration of the toxicant in the cell, $k_t$ is the killing rate, and NEC is the no-effect concentration term (Lopes et al., 2005; Billoir et al., 2007). This equation will modify the terms $P_i$ and $G_i$, that is, Eqns. (3) and (4), in the stage-based Leslie matrix Eqn. (2). In principle, it is possible to introduce a different expression for each cell stage. However, for simplicity reasons and due to the fact that the data to validate the model does not allow one to distinguish this aspect, we have considered global $k_t$ and NEC values. Furthermore, even though in this work we have only considered mortality effects, other effects could also be treated using a similar approach by introducing different expressions in the toxicity effects model. This was already proposed and developed by Kooijman and Bedaux (1996) who considered effects in reproduction for *Daphnia magna* and other organisms.

## 4.2.4 Experimental Set-up Characteristics and Assay Description

The performance of the integrated model has been assessed against data obtained with a recently developed technique called Real-Time Cell Electronic Sensing (RT-CES™), which incorporates an electrode that measures in real time the impedance through the layer of cells attached to the bottom of a microtiter plate well. Changes in electrode impedance are a function of the number of cells, the strength of their attachment to the well, and their morphology (Xing et al., 2005; Xing et al., 2006). This approach yields a quantity called the cell index (CI), which, importantly for the current application, is linearly related to the number of cells (Xing et al., 2005). This value is then normalized to start with a value of one when the chemical is added.

We use data for A549 cells, which are human immortalized cells with the characteristics from human alveolar type II cells (Foster et al., 1998). Cells were grown in F12k media supplemented with 10% FBS (Fetal Bovine Serum, $[S]_0 = 4.68 \times 10^{-2}$ mol protein m$^{-3}$). A nominal 3000 cells/well were seeded in the ACEA (Biosciences, San Diego, CA) 96 well

E-Plates and allowed to grow for 20–24 h to reach CI = 1 before treating with the compounds. Then, 10 μl of compound solutions (309 chemicals) was added producing a final concentrations of 100, 33.33, 11.11, 3.70, 1.23, 0.41, and 0.047 μM. In each well, there was 150 μl ($V$) of media. In addition to the selected compounds, negative control experiments using similar DMSO concentrations were performed (Judson et al., in press).

The wells in the E-plates have a total volume $V_T = 243$ μl and bottom internal diameter = 5 mm (i.e., the surface of the cell-based assay medium $A_S = 1.964 \times 10^{-5}$ m$^2$). These values give a headspace volume $V_H = 93$ μl and a surface of the well in contact with the medium $S_M = 1.194 \times 10^{-4}$ m$^2$. In summary, the data obtained consists of time-dependent curves of the CI for each chemical/concentration pair, from time $t = 0$ to 48 h. In wells with no chemical added (DMSO control), the curves are well fit with a single exponential that indicates cell doubling time of approximately 24 h. With the addition of the chemical, we see a variety of curve shapes. These shapes reflect the chemically induced balance between cell growth and cell death as a function of time.

### 4.2.5 Parameter Estimation and Optimization Procedure for Fitting Cell Population Dynamics

To run the model, it is necessary to estimate several parameters related to the fate of the chemicals in the wells as well as their effects on cell growth kinetics. Among the chemical parameters, there are those related to the chemical partitioning between the different compartments in the well, that is, plastic, proteins, lipids, cells, medium, and headspace, and those related to its dynamics: the decomposition (external to the cell)/metabolism (internal to the cell) rates and the exchange rate constants of uptake and depuration between the medium and the internal concentrations in A549 ($r_{da}$, $r_{ad}$), which are both correlated. In addition, it is also necessary to calculate those parameters related to the growth-rate effects: $k_t$, the rate constant for the mortality increase (difference between growth and death rate) as a function of the chemical and the NEC, no-effect concentration for survival, which is the threshold concentration above which the chemical effects are measurable.

In a continuous approach, the rate of change in the number of cells can be given by ${}^{dn}/_{dt} = k_g \cdot n - k_d \cdot n$, where $k_g$ is the growth rate and $k_d$ the death rate. Both of these can be the functions of the chemical concentration. From the experimental data measured, that is, the number of cells, we cannot deconvolute these two effects—it is only the assumption of different functional dependencies of $k_g$ and $k_d$ on compound concentration that allows any opportunity for independent estimation. However, $k_g$ and $k_d$ have to be greater than zero, so if we see negative growth some time, $k_d$ has to be greater than zero and cell death has to be occurring. Since there was no compartment-specific concentration data available, that is, the measurements of time-resolved concentrations in the wells, on the fate and dynamics of the chemicals, we must estimate the parameters related to the chemical and obtain them using constrained optimization relative to the observed effects on cell growth kinetics.

The estimation of the partitioning coefficients was discussed in Zaldívar et al. (2010). A general approach to describe the distribution of the organic compound is by means of the partition coefficient $K_i^c$, defined as the relationship between the concentration in a particular component—for the cell proteins and lipids—and in the aqueous component. Several of these correlations have been found in literature (DeBruyn and Gobas, 2007; Jonker and van der Heijden, 2007; Kramer, 2010) and all of them assume that the compound has a linear sorption isotherm, which is normally a good approximation at low concentrations, that is, there is no saturation. Similar correlations have been used here to calculate the partitioning inside the cell between lipids, proteins, and the aqueous fraction. In this case, the fractions of lipids and proteins have been used to calculate the cell concentrations: $[P]_c = 4.33$ mol protein m$^{-3}$ and $[L]_c = 12.88$ kg m$^{-3}$.

As a first approach, to account for degradation during the experiment, the estimated half lives ($t_{1/2}$) using the Level III fugacity multimedia model provided by EPI suite v4.0 (Mackay, 2001) were chosen and degradation constants were calculated for the medium (water) and the headspace (air) as

$$k_{\mathrm{deg}} = \frac{\ln 2}{t_{1/2}} \tag{22}$$

For the more biodegradable compounds in water, a 5% concentration decrease in 24 h was obtained, whereas for the persistent compounds less than 0.5% concentration decrease will occur. In air, degradation is normally fast, but the amount depends on the volatility of the chemical. These values are only a crude approximation since they consider degradation in a "natural" environment due to microbes, sunlight, etc., and do not take into account cell metabolism, but at least they give an indication of the chemical stability of the compound. This parameter was also considered a free parameter during the fitting of the variables and the estimated value was entered as initial guess. Further experimental research is necessary to characterize chemical stability and the amount of cellular metabolism that occurs inside the wells.

Assuming passive diffusion as the only transport mechanism, it is possible to write (Del Vento and Dachs, 2002) that the uptake constant, $k_{\mathrm{upt}}$ (l g$^{-1}$ day$^{-1}$), is given by

$$k_{\mathrm{upt}} = p \cdot S_{\mathrm{p}} \tag{23}$$

where $p$ (m d$^{-1}$) is the cell permeability and $S_{\mathrm{p}}$ is the specific surface of cells (m$^2$ kg$^{-1}$), $S_{\mathrm{p}} = V^{2/3}/W$.

There are several correlations to predict $p$ as a function of the physicochemical properties of the molecule. In this work, we have used the experimental data obtained using caco-2 cells by Yazdanian et al. (1998) to fit an expression proposed in USEPA (1992) as a function of the octanol–water partition coefficient ($K_{\mathrm{ow}}$) and the molecular weight ($MW$):

$$\log p = -1.1711 + 0.98 \log K_{\mathrm{ow}} - 0.0011\, MW \tag{24}$$

where $p$ is given in cm h$^{-1}$; to convert it to m d$^{-1}$ then

$$p = 0.24 \cdot 10^{(-1.1711 + 0.98 \log K_{\mathrm{ow}} - 0.0011\, MW)} \tag{25}$$

On the other hand, comparing Eqn. (23) with the first term in Eqn. (17) it is possible to obtain the relationship between cell permeability and uptake rate:

$$k_{\text{upt}} = \frac{V^{2/3}}{W} r_{\text{da}} \tag{26}$$

Combining Eqn. (23) with Eqn. (26), $r_{\text{da}}$ can be estimated as

$$r_{\text{da}} = 10p \tag{27}$$

where 10 is a factor to convert $\text{m d}^{-1}$ to $\text{l cm}^{-2}\,\text{d}^{-1}$.

In a similar way, the depuration constant, $k_{\text{dep}}$ ($\text{day}^{-1}$), can be obtained as (Del Vento and Dachs, 2002):

$$k_{\text{dep}} = \frac{p \cdot S_{\text{P}}}{\text{BCF}} \tag{28}$$

where BCF is the bioconcentration factor ($\text{l g}^{-1}$) defined as the ratio of concentrations of the chemical in the cell and in water—freely dissolved—at equilibrium. In addition, it is possible to observe that the relationship between the uptake and depuration constants depends only on the bioconcentration factor. By comparing Eqn. (28) with the second r.h.s. term in Eqn. (17), we can observe that the uptake and depuration rates are equivalent, $r_{\text{da}} = r_{\text{ad}}$. Furthermore, the bioconcentration factor can be defined as

$$\text{BCF} = \left( \frac{f_{\text{aq}}}{\rho_{\text{aq}}} + \frac{f_{\text{L}}}{\rho_{\text{L}}} K_{\text{L}} + \frac{f_{\text{P}}}{\rho_{\text{P}}} K_{\text{P}} \right) \tag{29}$$

## 4.3 Results and Discussion

### 4.3.1 Stage-based A459 Cell Growth Model

Since we are interested in coupling this model with a systems biology model, the model was divided in four stages (phases)—G1, S, G2, M corresponding to the cell cycle. To fit the parameters of the cell growth model, that is, $d_i$, $F$, $z_i$, and initial fraction at each stage—$N_0(i)/N_0$—we used the 166 blank (DMSO) experiments. The results with the fitted parameters are displayed in Fig. 4.3, where the average value of all experiments (in green) is plotted against the simulated sum of all cells in the four stages (in red). Also, the single experiments are plotted to show their variability. As can be observed, A549 shows an exponential growth during the duration of the experiment.

The relative abundances of each stage as a function of time during the first 24 h of the experiment are plotted in Fig. 4.4. As it can be observed, the cells have reached the relative equilibrium distribution among the four states before the treatment with the chemical.

To estimate the types of errors in the parameters that would give such experimental variability, we performed Monte Carlo analysis allowing the parameters $d_i$, $z_i$, and $F$ to change by 10% around their values. In addition, we also performed simulations assuming that the initial stages of the cell population may change by 30% with the extreme values

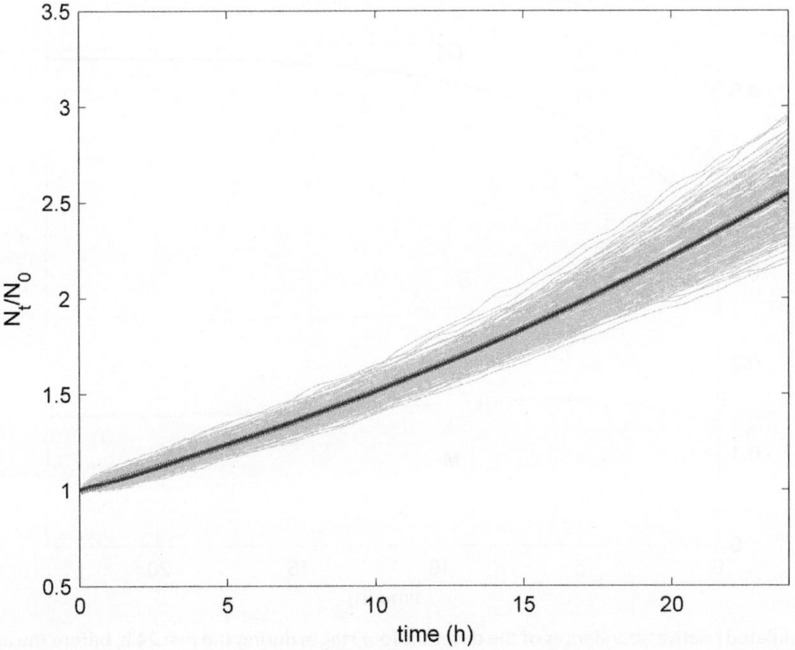


**FIGURE 4.3** Simulated relative population and experimental CI data (average value, individual experiments in gray) on A549 cell population growth. Estimated A549 parameters: $d = (12.37, 10.08, 8.83, 4.68)$ h; $F = 1.03\ \mathrm{h}^{-1}$; $z = (1.56 \times 10^{-2}, 3.75 \times 10^{-2}, 8.47 \times 10^{-4}, 1.00 \times 10^{-10})\ \mathrm{h}^{-1}$. For color version of this figure, the reader is referred to the online version of this book.

when all cells are at the G1 stage or at the M stage at the beginning of the simulation. The obtained results are represented in Fig. 4.5. The discrete time step for running the model was 1 min.

These parameters were kept constant during the subsequent simulations.

## 4.3.2 Population Dynamics in Toxicity Assays

The simulated results concerning population dynamics at several toxicant concentrations were compared with CI obtained using real-time impedance as a measure of viable cell numbers. The values of $k_t$, NEC, $r_{da}$, and $k_{deg}$ were fitted using an optimization error measured as

$$\text{error} = \sum_{i=1}^{n^o \exp} \left( CI_{\exp} - (N_t/N_0)_{\text{sim}} \right)^2 \qquad (30)$$

We used the constrained optimization procedure with lower bounds equal to zero as described in Zaldívar and Baraibar (2011). To improve the minimization procedure, we started with the estimated values for $r_{da}$ and $k_{deg}$ and then we allowed them to change if the final error was decreasing.

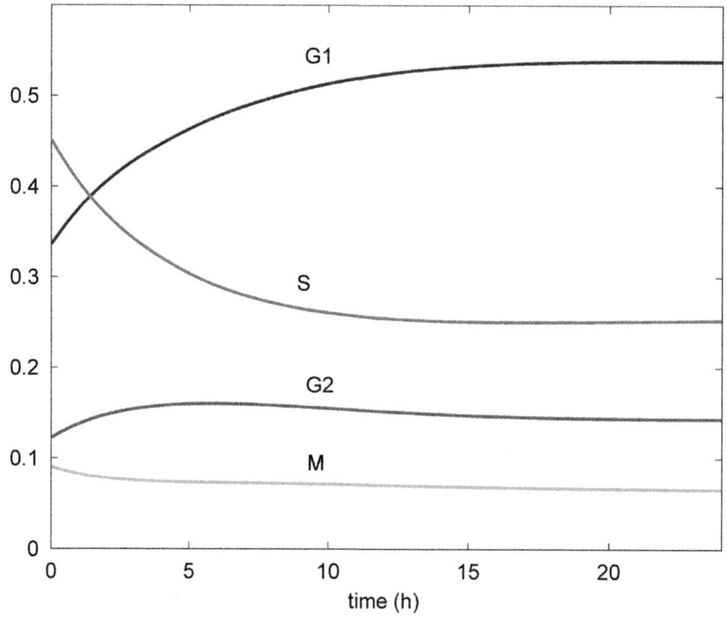


**FIGURE 4.4** Simulated relative abundances of the different four stages during the first 24 h, before the addition of the chemical. For color version of this figure, the reader is referred to the online version of this book.

As an example, different kinetic profiles have been selected from all the tested compounds. The obtained parameters are presented in Table 4.1 where the simulated and experimental results are shown in Fig. 4.6. As it can be observed, the modeling approach is able to simulate those distinct dynamics profiles as a function of time and concentration for these chemicals. Simulated results that do not show mortality converge to the same line, whereas the experimental measurements move around following the experimental variability already observed in the blank experiments (see Fig. 4.3).

Due to the different physicochemical properties of the studied compounds, the dynamics of dissolved and internal concentrations change appreciably from experiment to experiment. Like in Zaldívar et al. (2010), it was also possible to observe the differences in scale and dynamics between the nominal concentrations and the dissolved concentration time series. For hydrophilic compounds, there was practically no difference; whereas for hydrophobic compounds the differences increase, reaching up to several orders of magnitude for the more hydrophobic compounds. Concerning internal concentrations, results not shown, different profiles can also be identified depending on the uptake and depuration rates as well as the chemical stability of the compounds. In addition, there are also the effects of cell mortality on the medium composition. However, all these values should be checked against experimental concentration measurements to confirm our estimated kinetics in these reactions. In addition, once the model parameters have been obtained, it is possible to simulate viability experiments, that is, concentration response curves for each chemical. The results of this simulation are shown in Fig. 4.7.

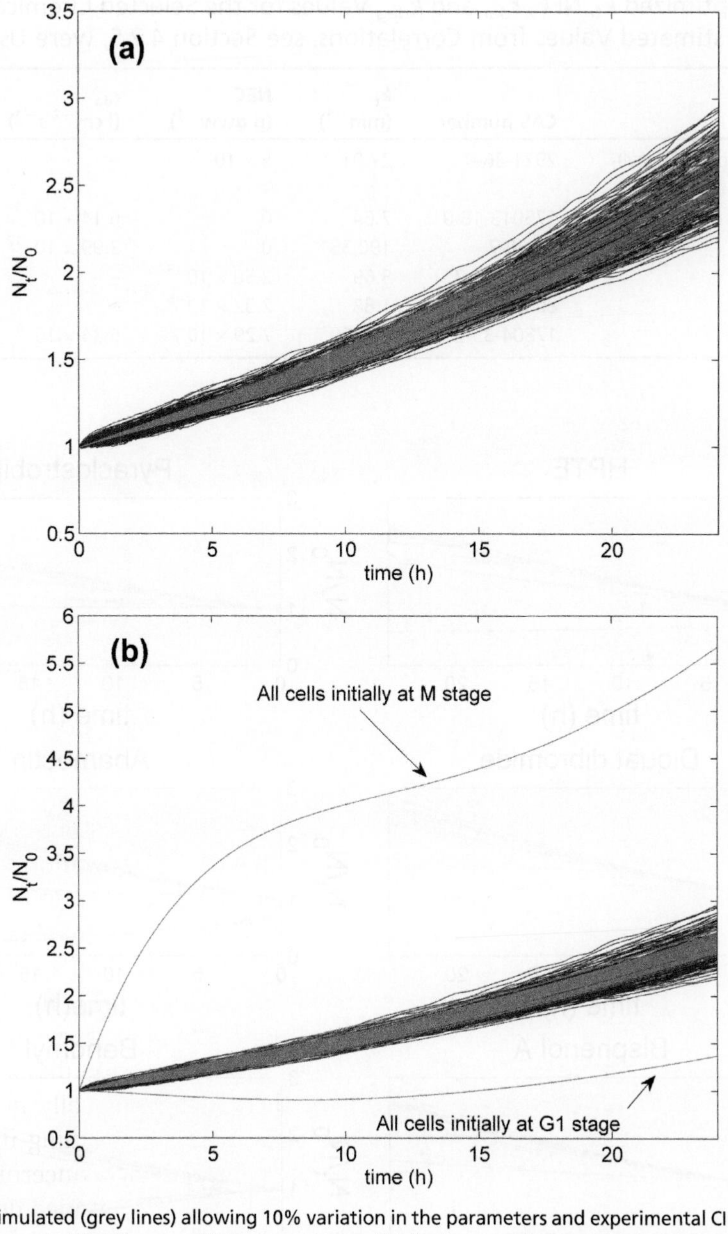


**FIGURE 4.5** (a) Simulated (grey lines) allowing 10% variation in the parameters and experimental CI data on A549 cell population growth; (b) influence of initial population distribution on A549 growth. For color version of this figure, the reader is referred to the online version of this book.

**Table 4.1** Optimized $k_t$, NEC, $r_{da}$, and $k_{deg}$ Values for the Selected Chemicals (No Value Means that Estimated Values from Correlations, see Section 4.2.5, were Used)

| Compound | CAS number | $k_t$ ($min^{-1}$) | NEC (g $gww^{-1}$) | $r_{da}$ (l $cm^{-2}$ $s^{-1}$) | $k_{deg}$ ($s^{-1}$) |
|---|---|---|---|---|---|
| HPTE 2,2-Bis(4-hydroxyphenyl)-1,1,1-trichloroethane | 2971-36-0 | 27.91 | $5 \times 10^{-7}$ | – | $3.45 \times 10^{-4}$ |
| Pyraclostrobin | 175013-18-0 | 7.64 | 0 | $6.14 \times 10^{-7}$ | $7.76 \times 10^{-9}$ |
| Diquat dibromide | 85-00-7 | 190.35 | 0 | $2.99 \times 10^{-12}$ | – |
| Abamectin | 71751-41-2 | 8.69 | $2.50 \times 10^{-5}$ | – | – |
| Bisphenol A | 80-05-7 | 1.88 | $2.32 \times 10^{-5}$ | – | – |
| Benomyl | 17804-35-2 | 264.00 | $7.29 \times 10^{-7}$ | $5.34 \times 10^{-11}$ | $1.52 \times 10^{-5}$ |

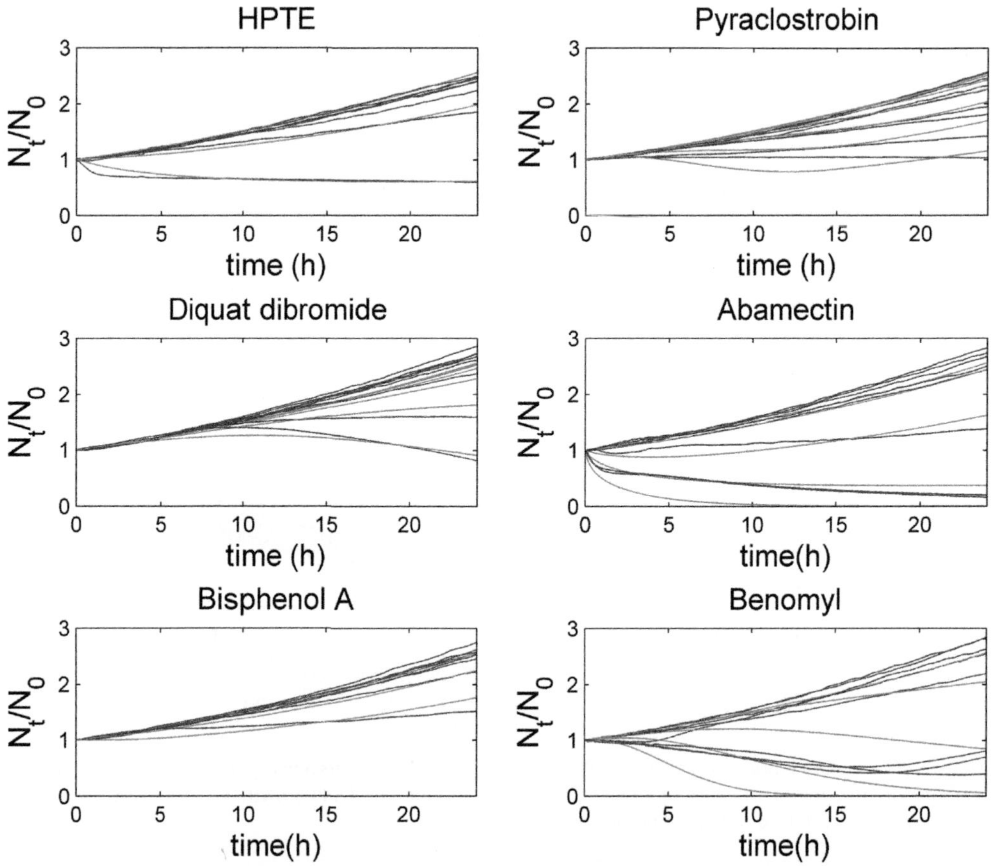


**FIGURE 4.6** Experimental CI data and simulated (grey lines) relative population at the eight experimental concentrations (0.047, 0.41, 1.23, 3.70, 11.11, 33.33, and 100 μM), for (a) HPTE, (b) Pyraclostrobin, (c) Diquat dibromide, (d) Abamectin, (e) Bisphenol A, and (f) Benomyl using the parameters from Table 4.1. For color version of this figure, the reader is referred to the online version of this book.

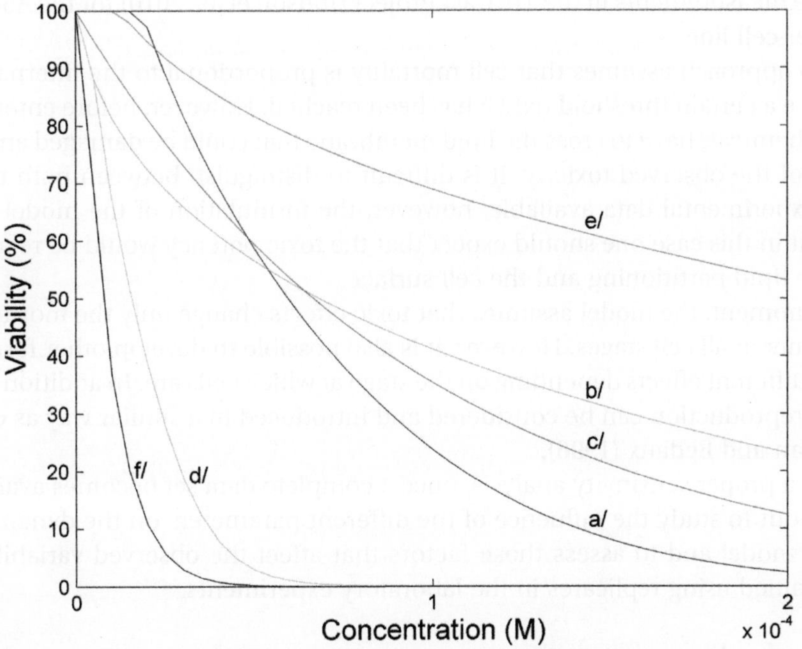


**FIGURE 4.7** Simulated 24 h viability profiles for the six chemicals: (a) HPTE; (b) Pyraclostrobin, (c) Diquat dibromide, (d) Abamectin, (e) Bisphenol A, and (f) Benomyl for nominal concentrations between 0 and $2 \times 10^{-4}$ M with a constant step of $5 \times 10^{-6}$ M. For color version of this figure, the reader is referred to the online version of this book.

## 4.4 Conclusions

A cell growth and toxicity model based on DEB theory has been developed and implemented to simulate HTS laboratory toxicity experiments using A549 cell line. In addition to predicting cell population dynamics, the model allows the estimation of dissolved concentration in the medium as well as internal concentration inside the cells. However, the validation of the predicted dissolved and internal concentrations for these experiments has not been possible since no data are available. However, parallel experimental data sets, with different cell lines, have allowed us to assess the validity of our predicted concentrations (Zaldívar et al., 2011). Further research will go in this direction since correct values concerning dissolved and internal concentrations are essential for performing *in vitro–in vivo* extrapolation. This will allow the comparison of similar values in both the cases, avoiding the comparison between $AC_{50}$ values (concentrations) and $LD_{50}$ values (doses), which do not provide a useful correlation. In addition, we are in parallel developing the PBTK models to be able to extract from $LD_{50}$ values the corresponding dissolved and internal concentrations in the target organism.

Concerning cell population dynamics under increasing concentrations of chemicals, the model is able to fit the results corresponding to the experiments performed using impedance measurements in the ToxCast Project (Judson et al., 2010) for the A549 human lung cancer cell line.

The TD approach assumes that cell mortality is proportional to the internal concentration once a certain threshold (NEC) has been reached. However, before entering in the cells, the chemicals have to cross the lipid membrane that could be damaged and become the cause of the observed toxicity. It is difficult to distinguish between both the effects with the experimental data available; however, the formulation of the model would be similar, but in this case one should expect that the toxic potency would be related to the membrane lipid partitioning and the cell surface.

At the moment, the model assumes that toxic effects change only the mortality in the same manner at all cell stages. However, it is also possible to develop other formulations assuming different effects depending on the stage at which cells are. In addition effects on growth or reproduction can be considered and introduced in a similar way as developed by Kooijman and Bedaux (1996).

Finally, a proper sensitivity analysis, once a complete data set becomes available, will be carried out to study the influence of the different parameters on the dynamics of the integrated model and to assess those factors that affect the observed variability in the values obtained using replicates in the laboratory experiments.

## Acknowledgments

JMZ would like to acknowledge the support from the EC FP7 Health 2010 COSMOS Project grant no. 407 266835.

## Disclosure

The opinions and views expressed in this chapter are purely those of the authors and may not under any circumstances be regarded as stating an official position, or reflecting the views or policies of their affiliated institutions. Mention of trade names or commercial products does not constitute endorsement or recommendation for use.

## Notation

| | |
|---|---|
| $\boldsymbol{A}$ | Leslie matrix (describes the transformation of a population from time $t$, in minutes, to time $t+1$) |
| $\boldsymbol{C}$ | concentrations (mol $l^{-1}$) |
| $\boldsymbol{C_b}$ | contaminant concentration in the cell (g gw $w^{-1}$) |
| $\boldsymbol{d_i}$ | duration within $i$-th stage (h) |
| $\boldsymbol{f_x}$ | mass fraction of compartment $x$ (dimensionless) |
| $\boldsymbol{F}$ | fecundity rate per unit time (h) |
| $\boldsymbol{G_i}$ | probability of surviving and growing into the next stage $i+1$ |
| $\boldsymbol{k_t}$ | killing rate |
| $\boldsymbol{K_L}$ | cells–lipids partitioning coefficient ($m^3$ $kg^{-1}$ lipid) |
| $\boldsymbol{K_{ow}}$ | octanol–water partition coefficient |

$K_P$ plastic-medium partitioning coefficient (m)
$K_S$ partitioning coefficient between serum proteins and medium ($m^3\ mol^{-1}$)
$k$ reaction rate constant ($s^{-1}$)
**MW** molecular weight ($g\ mol^{-1}$)
$n$ number of moles (mol)
$N_t$ vector containing number of cells at each stage (G1,S,G2, and M) at time $t$
**NEC** no-effect concentration
$P_i$ probability of surviving and staying in stage $i$
$p$ cell permeability (m)
$p_i$ survival probability
$r$ cell radius (m)
$r_{da}$ uptake rate ($l\ cm^{-2}\ s^{-1}$)
$r_{ad}$ depuration rate ($l\ cm^{-2}\ s^{-1}$)
$[S]$ concentration of proteins in medium ($mol\ m^{-3}$)
$S$ surface ($m^2$)
$t$ time (s)
$V$ volume (l)
$W$ wet weight (g)
$z_i$ instantaneous mortality rate

# Greek symbols

$\alpha_G$ von Bertalanffy's growth rate
$\gamma_i$ growing probability
$\rho$ density ($g\ l^{-1}$)

# Subscripts and superscripts

**aq** aqueous
**AW** air–water
**deg** degradation
**diss** dissolved
***L*** lipids
***M*** medium
***P*** proteins
***T*** total
***w*** well

# References

Adler, S., Basketter, D., Creton, S., Pelkonen, O., van Benthem, J., Zuang, V., Ejner Andersen, K., Angers-Loustau, A., Aptula, A., Bal-Price, A., Benfenati, E., Bernauer, U., Bessems, J., Bois, F.Y., Boobis, A., Brandon, E., Bremer, S., Broschard, T., Casati, S., Coecke, S., Corvi, R., Cronin, M., Daston, G., Dekant, W., Felter, S., Grignard, E., Gundert-Remy, U., Heinonen, T., Kimber, I., Kleinjans,.J, Komulainen, H., Kreiling, R., Kreysa, J., Batista Leite, S., Loizou, G., Maxwell, G., Mazzatorta, P., Munn, S., Pfuhler, S., Phrakonkham, P., Piersma, A., Poth, A., Prieto, P., Repetto, G., Rogiers, V., Schoeters, G., Schwarz, M., Serafimova, R., Tähti, H., Testai, E., van Delft, J., van Loveren, H., Vinken, M., Worth, A., Zaldívar, J.M., 2011. Alternative (non-animal) methods for cosmetics testing: current status and future prospects—2010. Arch Toxicol 85, 367–485.

Andersen, M.E., 1981. A physiologically based toxicokinetic description of the metabolism of inhaled gases and vapors: analysis at steady-state. Toxicol. Appl. Pharmacol. 60, 509–526.

Billoir, E., Pery, A.R.R., Charles, S., 2007. Integrating the lethal and sublethal effects of toxic compounds into the population dynamics of *Daphnia magna*: a combination of the DEBtox and matrix population models. Ecol. Model. 203, 204–214.

Caswell, H., 1989. Matrix Populations Models: Construction, Analysis and Interpretation. Sinauer, Sunderland, Massachusetts.

Clewell, H.J., Tan, Y.M., Campbell, J.L., Andersen, M.E., 2008. Quantitative interpretation of human biomonitoring data. Toxicol. Appl. Pharm. 231, 122–133.

Cushing, J.M., 1998. An Introduction to Structured Population Dynamics. SIAM, Philadelphia.

DeBruyn, A.M.H., Gobas, F.A.P.C., 2007. The sorptive capacity of animal protein. Environ. Toxicol. Chem. 26, 1803–1808.

DeJongh, J., Forsby, A., Houston, J.B., Beckman, M., Combes, R., Blaauboer, B.J., 1999. An integrated approach to the prediction of systemic toxicity using computer-based biokinetic models and biological *in vitro* test methods: overview of a prevalidation study based on the ECITTS Project. Toxicol. In Vitro 13, 549–554.

Del Vento, S., Dachs, J., 2002. Prediction of uptake dynamics of persistent organic pollutants by bacteria and phytoplankton. Environ. Toxicol. Chem. 21, 2099–2107.

Escher, B.I., Ashauer, R., Dyer, S., Hermens, J.L.M., Lee, J.-H., Leslie, H.A., 2010. Crucial role of mechanisms and modes of toxic action for understanding tissue residue toxicity and internal effect concentrations of organic chemicals. Integr. Environ. Assess. Manag. 7, 28–49.

Foster, K.A., Oster, C.G., Mayer, M.M., Avery, M.L., Audus, K.L., 1998. Characterization of the A549 cell line as a type II pulmonary epithelial cell model for drug metabolism. Exp. Cell. Res. 243, 359–366.

Gubbels-van Hal, W.M.L.G., Blaauboer, B.J., Barentsen, H.M., Hoitink, M.A., Meerts, I.A.T.M., van der Hoeven, J.C.M., 2005. An alternative approach for the safety evaluation of new and existing chemicals, an exercise in integrated testing. Regul. Toxicol. Pharma. 42, 284–295.

Gülden, M., Mörchel, S., Seibert, H., 2001. Factors influencing nominal effective concentrations of chemical compounds *in vitro*: cell concentration. Toxicol. In Vitro 15, 233–243.

Hartwell, L.H., Weinert, T.A., 1989. Checkpoints: controls that ensure the order of cell cycle events. Science 246, 629–634.

Heringa, M.B., Schreurs, R.H.M.M., Busser, F., van der Saag, P.T., van der Burg, B., Hermens, J.L.M., 2004. Towards more useful in vitro toxicity data with measured free concentrations. Environ. Sci. Technol. 38, 6263–6270.

Jiang, R.-D., Shen, H., Piao, Y.-J., 2010. The morphometrical analysis on the ultrastucture of A549 cells. Rom. J. Morphol. Embryo. 51, 663–667.

Jonker, M.T.O., van der Heijden, S.A., 2007. Bioconcentration factor hydrophobicity cutoff: an artificial phenomenon reconstructed. Environ. Sci. Technol. 41, 7363–7369.

Judson, R.S., Houck, K.A., Kavlock, R.J., Knudsen, T.B., Martin, M.T., Mortensen, H.M., Reif, D.M., Rotroff, D., Shah, I., Richard, A.M., Dix, D.J., 2010. In Vitro Screening of Environmental Chemicals for Targeted Testing Prioritization: The ToxCast Project. Environ Health Persp 118, 485–492.

Judson, R.S., Rotroff, D., Elloumi, F. et al. Extracting cellular phenotype kinetic parameters driven by chemical exposure using a real-time cell electronic sensing array, submitted.

Klanjscek, T., Caswell, H., Neubert, M.G., Nisbet, R.M., 2006. Integrating dynamic energy budgets into matrix population models. Ecol. Model. 196, 407–420.

Kooijman, S.A.L.M., van Haren, R.J.F., 1990. Animal energy budgets affect the kinetics of xenobiotics. Chemosphere 21, 681–693.

Kooijman, S.A.L.M., Bedaux, J.J.M., 1996. The Analysis of Aquatic Toxicity Data. VU University Press, Amsterdam.

Kooijman, S.A.L.M., 2000. Dynamic Energy and Mass Budgets in Biological Systems, second ed. Cambridge University Press, Cambridge, UK.

Kramer, N.I., 2010. Measuring, Modelling and Increasing the Free Concentration of Test Chemicals in Cell Assays. Utrecht University, The Netherlands. PhD Thesis.

Lopes, C., Péry, A.R.R., Chaumo, A., Charles, S., 2005. Ecotoxicology and population dynamics: using DEBtox models in a Leslie modelling approach. Ecol. Model. 188, 30–40.

Mackay, D., 2001. Multimedia Environmental Models: the Fugacity Approach, second ed. Lewis Publishers,, Boca Raton, Florida.

Meacham, C.A., Freudenrich, T.M., Anderson, W.L., Sui, L., Lyons-Darden, T., Barone Jr., S., et al., 2005. Accumulation of methylmercury or polychlorinated byphenyls in in vitro models of rat neuronal tissue. Toxicol. Appl. Pharmacol. 205, 177–187.

Mundy, W.R., Freudenrich, T.M., Crofton, K.M., DeVito, M.J., 2004. Accumulation of PBDE-47 in primary cultures of rat neocortical cells. Toxicol. Sci. 82, 164–169.

Rotroff, D.M., Wetmore, B.A., Dix, D.J., et al., 2010. Incorporating human dosimetry and exposure into high-throughput in vitro toxicity screening. Toxicol. Sci. 117, 348–358.

Poulin, P., Krishnan, K., 1995. A biologically-based algorithm for predicting human tissue: blood partition coefficients of inorganic chemicals. Hum. Exp. Toxicol. 14, 273–280.

Thomann, R.V., 1989. Bioaccumulation model of organic chemical distribution in aquatic food chains. Environ. Sci. Technol. 23, 699–707.

United States Environmental Protection Agency (USEPA), 1992. Dermal Exposure Assessment: Principles and Applications EPA/600/8–91/011B. Washington DC, USA.

van Haren, R.J.F., Schepers, H.E., Kooijman, S.A.L.M., 1994. Dynamic energy budgets affect kinetics of xenobiotics in the marine mussel *Mytilus edulis*. Chemosphere 29, 163–189.

Xing, J.Z., Zhu, L., Jackson, J.A., Gabos, S., Son, X.-J., Wang, X., 2005. Dynamic monitoring of cytotoxicity on microelectronic sensors. Chem. Tes. Toxicol. 18, 154–161.

Xing, J.Z., Zhu, L., Gabos, S., Xie, L., 2006. Microelectronic cell sensor assay for detection of cytotoxicity and prediction of acute toxicity. Toxicol. In Vitro 20, 995–1004.

Yazdanian, M., Glynn, S.L., Wright, J.L., Hawi, A., 1998. Correlating partitioning and Caco-2 cell permeability of structurally diverse small molecular weight compounds. Pharm. Res. 15, 1490–1494.

Zaldívar, J.M., Mennecozzi, M., Marcelino Rodrigues, R., Bouhifd, M., 2010. A Biology-Based Dynamic Approach for the Modelling of Toxicity in Cell-Based Assays Part I: Fate Modelling. EUR 24374 EN.

Zaldívar, J.M., Campolongo, F., 2000. An application of sensitivity analysis to fish population dynamics. In: Saltelli, Chan, Scott (Eds.), Mathematical and Statistical Methods for Sensitivity Analysis. Wiley, New York.

Zaldívar, J.M., Baraibar, J., 2011. A biology-based dynamic approach for the reconciliation of acute and chronic toxicity tests: application to Daphnia magna. Chemosphere 82, 1547–1555.

Zaldívar, J.M., Gajewska, M., Mennecozzi, M., Rodrigues, R., Sala, J.V., 2011. A Process-Based Model for In Vitro Experiments. Deliverable 4.1 COSMOS Project (Integrated In Silico Models for the Prediction of Human Repeated Dose Toxicity of Cosmetics to Optimise Safety). http://www.cosmostox.eu/home/welcome/.

# 5

# "Keystone Species" of Molecular Interaction Networks

Federico Vaggi, Attila Csikász-Nagy
*THE MICROSOFT RESEARCH-UNIVERSITY OF TRENTO CENTRE FOR COMPUTATIONAL AND SYSTEMS BIOLOGY, PIAZZA MANIFATTURA 1, 38068, ROVERETO, ITALY*

## 5.1. Introduction

One of the key paradigms in ecology is the importance of interactions among the components of ecosystems (Margalef, 1968). Focusing on the inter-relatedness of various components and ecological processes from a system's perspective is well supported by some classical methodological tools of ecological modeling. These include network analysis (Ulanowicz, 1980, 1986, 1989; Higashi and Burns, 1991), providing several metrics quantifying the fine structure of interaction networks. Some of these techniques have been recently applied also in molecular and cell biology, ranging from metabolic networks (Ma and Zeng, 2003) to protein–protein interaction networks (Nguyen and Jordán, 2010; Nguyen et al., 2011). The interest in the general (Wagner and Fell, 2001; Barabási and Oltvai 2004; Albert 2005) properties of certain cellular networks is increasing and provides some ecological view on novel biological problems (Merlo et al., 2006; Papayannopoulou and Scadden, 2008). Direct comparison of ecological and molecular biological knowledge is typically naïve but massive inspiration and methodological cross-fertilization seem to be very promising and useful.

Furthermore, both ecological networks and biological networks share important features. For example, robustness (the ability of a network to survive after the loss of a species) has been observed both in ecological (Dunne et al., 2002) networks as well as biological networks (Barabási and Oltvai, 2004). Other examples of properties common to both types of networks include modularity, scale-freeness, and redundancy (Han et al., 2004; Olesen et al., 2007). The presence of all these similarities justifies, in our view, the transplantation of tools already used in the study of ecological networks to the study of biological networks. Examples of tools that have already been "stolen" from ecology include: the concepts of centrality (see Ma and Zeng, 2003; Estrada, 2007) or deletion effects (see Pimm, 1980; Jeong et al., 2001). In this chapter, we apply a "keystone" perspective to a biological

**Models of the Ecological Hierarchy. DOI: http://dx.doi.org/10.1016/B978-0-444-59396-2.00005-5**
ISSN 0167-8892, 

problem, inspired by the notion of keystone species in ecology (Paine, 1966; Jordán and Scheuring, 2002).

Molecular interaction networks have been investigated by tools of network analysis (Yamada and Bork, 2009). A lot of effort was expended to find correlation between various network measures and lethality of removal of a given gene (Yu et al., 2004; Hahn and Kern, 2005; Batada et al., 2006; He and Zhang, 2006). These works suggest that the degree of a protein is the best predictor of its essentiality, although this result has been considered somewhat controversial recently (Yu et al., 2007). In this study, we focus on smaller networks, where we collect genes that are associated with a given biological function and investigate the interaction network of the products of these genes.

When examining networks, several different measures have been proposed to rank a node's importance (Freeman, 1977; Wolfe, 1997). The more common measure is degree, a measure of how many edges are connected to a given node. Degree measures the local importance of a node in a network, but ignores important global features of the network. A better predictor of global importance is betweenness centrality, a measure of how often a particular node/protein is found in the shortest path between two other nodes. A gene with high betweenness is thus at the "center" of the information flow of the network. In this chapter, we develop a method to identify which proteins are most important for a given biological function, focusing on proteins that reside at an important topological location, but aren't well characterized. Thus, we try to identify the nonhub, important network components, which we call "keystone species" of biological functions.

To identify genes belonging to different functional groups, we rely on the functional annotations present in Gene Ontology (GO). The Gene Ontology Consortium is an international consortium of biologists with the aim of "providing consistent descriptions for genes products in different databases" (Ashburner et al., 2000; Harris et al., 2004). This consortium provides a bioinformatics database (available at www.geneontology.org) where scientists can search for genes, categories, or relationships between genes. In order to make our analysis as free of bias as possible, we thus relied on studying the genes associated with regulation of cell cycle taken directly from GO. The methodology discussed in this chapter has also been used to study other biological processes such as polarity and cytokinesis (Vaggi and Csikasz-Nagy, 2012).

We have picked cell cycle regulation (Morgan, 2006) as the example in our study; since this might be one of the most investigated biological processes, there is cell cycle-dependent high-throughput data available in various organisms (Spellman et al., 1998; Marguerat et al., 2006; Gauthier et al., 2008), and it is a favorite system of network dynamics modeling as well (Csikász-Nagy, 2009; Ferrell et al., 2011). Budding yeast is among the most thoroughly investigated eukaryote and its distant relative fission yeast is another favorite test organism of cell cycle researchers. Both of these organisms were subject of various high-throughput studies (Goffeau et al., 1996; Wood et al., 2002), thus there is a huge amount of data that can be used in our analysis. In this work, we will focus on the GO association "GO:0051726: regulation of cell cycle" and present examples based on data on the two yeast species, but any other GO term and organism can be investigated in a similar way.

## 5.2. Materials and Methods

### 5.2.1. Generation of Networks

The pipeline that we used to generate the networks analyzed in this chapter is described in Fig. 5.1. In brief, our workflow is as shown below:

1. We connected to the MySQL GO (hosted on the EBI servers) database using custom python scripts, and downloaded all genes associated with a given biological process.
2. We take all genes downloaded and use them to query STRING, downloading all the information about protein–protein interactions in PSI-MI-TAB format. It is important to note that STRING and GO sometimes identify the same gene by a different name, so special care must be taken to use consistent nomenclature.
3. We parse the PSI-MI-TAB file and transform it into a NetworkX graph, which we can then study using both algorithms built into NetworkX as well as custom scripts.

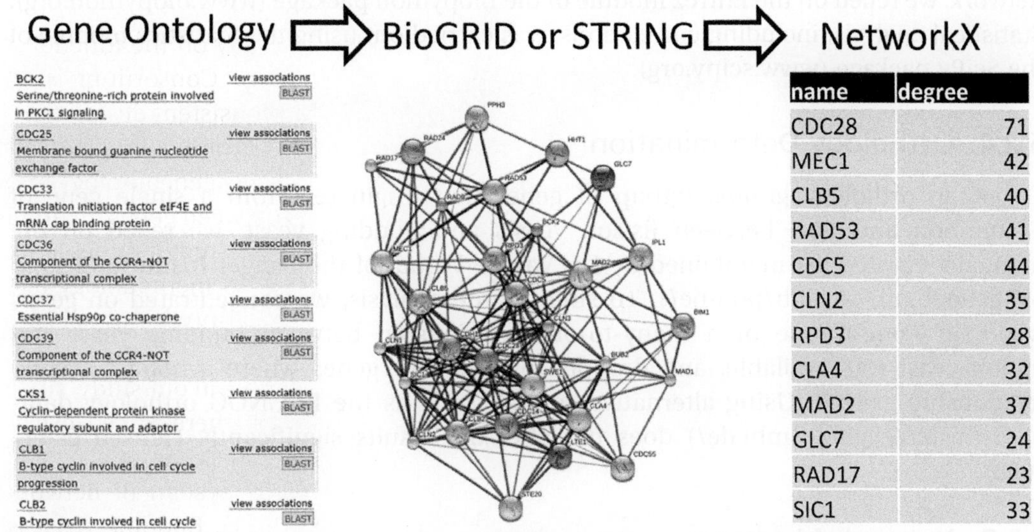


**FIGURE 5.1** Workflow of our analysis. We collect genes with a given GO annotation from the GO database (http://www.geneontology.org/), and look for interactions between the products of these genes in the STRING (http://string-db.org/) and BioGRID (http://thebiogrid.org/) databases. The networks are analyzed by NetworkX (http://networkx.lanl.gov/), which allows us to rank the genes according to their network topology measures. For color version of this figure, the reader is referred to the online version of this book.

We repeated the analysis described in the chapter using networks obtained from BioGRID, in which case, instead of using STRING in step 2, we parsed the full network of a given organism from a PSI-MI-TAB file available on BioGRID, and then extracted the subgraph containing the nodes obtained in step 1.

### 5.2.2. Cutoff Analysis

Protein interaction data in STRING (i.e., edges in the network) are characterized by a confidence score, determined by the strength of the evidence supporting a possible interaction between the two nodes (or, genes in the network). We studied the effect of varying the confidence score cutoff on network size (defined as the fraction of all genes that are connected with at least one other gene), main component size (defined as the fraction of all genes that are connected to the largest component in the network) and edge fraction (the fraction of all edges that are found compared to the theoretical maximum $E_{\max} = \frac{N * (N-1)}{2}$).

As the cutoff becomes more stringent, the network size shrinks radically—this effect is especially pronounced in fission yeast, where high-quality experimental data is less common than it is in budding yeast.

### 5.2.3. PubMed Data and Statistics

To download the number of PubMed abstracts mentioning one of the genes in the network, we relied on the Entrez module of the Biopython package (www.biopython.org). Statistical analysis, including correlations, was carried out using the Statistics module of the SciPy package (www.scipy.org).

### 5.2.4. Orthology Determination

To assign orthology groups (group of genes that originated from a single gene of a common ancestor) between fission yeasts and budding yeast, we relied on the manually curated list maintained by Dr. Valerie Wood, of the Sanger Institute (http://old.genedb.org/genedb/pombe/). To simplify the analysis, we concentrated on genes where a —one-to-one or a many-to-one relationship between budding yeast and fission yeast was available, and dropped out the few genes where a many-to-many relationship existed. Using alternative sources such as the EGGNOG orthology database (http://eggnog.embl.de/) does not alter the results significantly (Jensen et al., 2008).

### 5.2.5. Network Measures

We calculated all network measures (except Ti index) using preexisting algorithms implemented in NetworkX. The topological importance (Ti) index quantifies indirect neighborhoods (up to *n* steps) in a network, considering also the neighbors of neighbors.

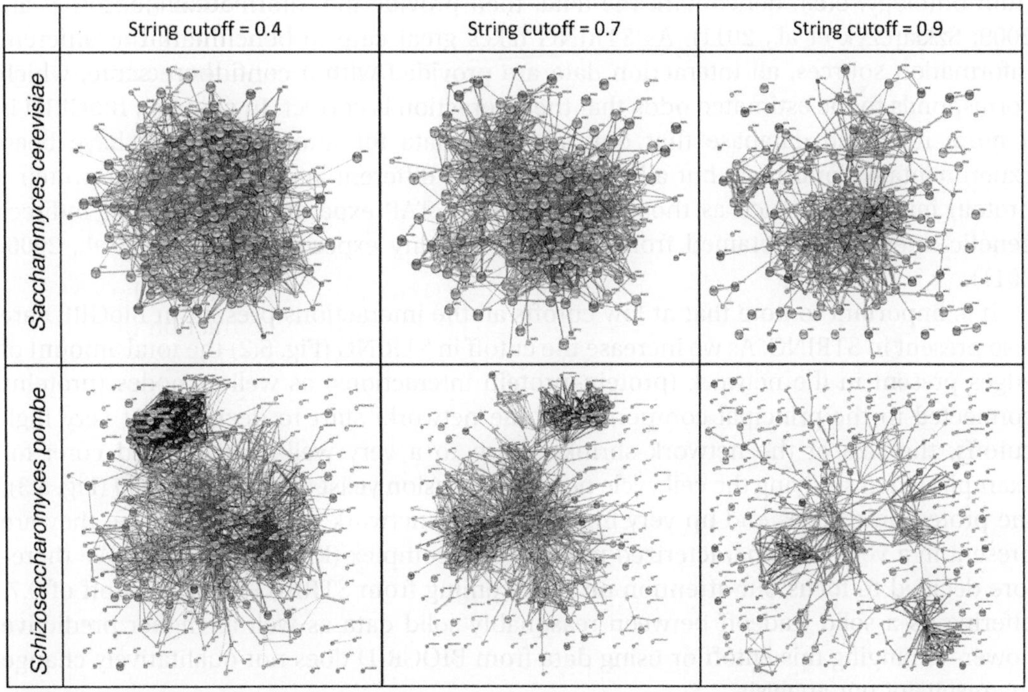


**FIGURE 5.2** Protein interaction networks at various confidence cutoffs from the STRING database. Interaction networks of proteins with a GO association "GO:0051726: regulation of cell cycle" in budding and fission yeast cells. For color version of this figure, the reader is referred to the online version of this book.

We implemented a recursive function to calculate the Ti index as described in Jordán et al. (2003). An alternative implementation using adjacency matrixes is available in CoSbiLab Graph (Valentini and Jordán, 2010).

## 5.3. Results

We collect the genes that are associated with a given biological process from the GO database (http://www.geneontology.org/) (Ashburner et al., 2000; Harris et al., 2004), feed these into the STRING database of molecular interactions (http://string-db.org/) (Jensen et al., 2009; Szklarczyk et al., 2011) and to the more restricted BioGRID protein–protein interaction database (http://thebiogrid.org/) (Stark et al., 2006, 2011) to find the interaction network of the investigated biological process. The networks are analyzed by network topology measures from NetworkX (http://networkx.lanl.gov/) (Hagberg et al., 2008). The complete workflow is explained in Fig. 5.1.

GO contains information about gene products, but not their interactions. Thankfully, several databases of protein–protein interactions are available online; STRING is an ambitious project that aims to integrate all protein–protein interaction information available from multiple sources (text mining, direct experimental evidence, inference from ontology, etc.), benchmark the data, then provide the information (Jensen et al., 2009; Szklarczyk et al., 2011). As STRING takes great care to benchmark the different information sources, all interaction data are provided with a confidence score, which corresponds to the estimated odds that the interaction is correct. In contrast, BioGRID is a more restricted database that only provides data for interactions that have been experimentally confirmed, but allows the users to differentiate between direct protein–protein interactions such as those obtained from TAP experiments and more indirect genetic interactions obtained from synthetic lethality experiments (Stark et al., 2006, 2011).

It is important to note that at low cutoffs all the interactions present in BioGRID are also present in STRING. As we increase the cutoff in STRING (Fig. 5.2) the total amount of edges present in the network (protein–protein interactions) as well as nodes (proteins connected to the principal component of the network) start to decrease. At very high cutoffs, the size of the network shrinks down to a very well-characterized core; for example, when studying the cell cycle network in fission yeast at a cutoff of 0.99 (Fig. 5.3), the proteasome genes end up very high on several network measures because they are present in a very well-characterized multiprotein complex (King et al., 1996). We therefore decided to focus our attention on data coming from STRING using a cutoff of 0.7, offering us a solid tradeoff between reasonably solid data as well as higher predictive power—changing this cutoff or using data from BIOGRID does not qualitatively change the results of our analysis.

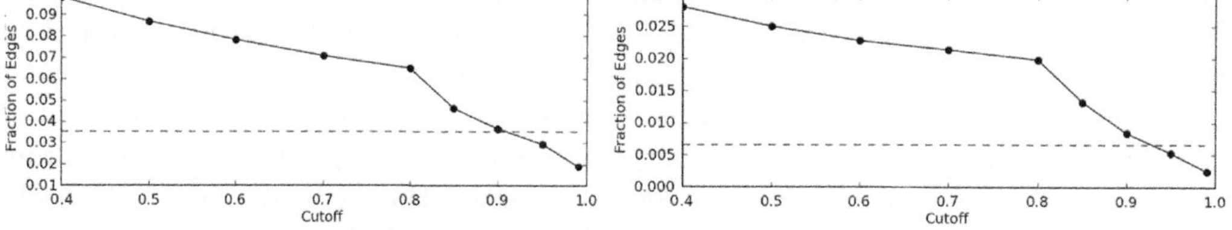

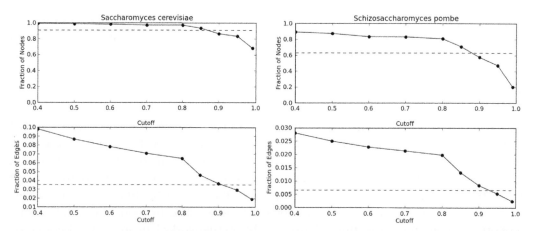


**FIGURE 5.3** Dependence of the size of the main component and number of edges on cutoff level. Ratio of proteins (compared to all proteins with the "regulation of cell cycle" GO term) that are members of the giant component of the network at a given STRING cutoff level (upper) and ratio of number of edges (compared to total possible) that are observed between nodes of the network (lower), in the two yeast species. Horizontal dashed lines show these values for the protein–protein interaction data from the BioGRID database. Note that the interactions in BioGRID are not the same as the ones we find at the corresponding cutoff in the STRING database as they use different scoring systems.

Note the difference between the two organisms we examined; in budding yeast, the network is get rid of more denser both in terms of fraction of maximum edges and in the fraction of genes connected to the network. In fission yeast, however, the amount of genes that are associated with cell cycle regulation is much greater—this might be due to the presence of a highly active fission yeast community that contributes to GO as well as the extensive use of fission yeast as a model organism to study cell cycle. Therefore, the available bioinformatic data between the two yeasts is quite different: budding yeast has a smaller amount of genes whose interactions have been highly characterized, while fission yeast has a larger amount of genes that, however, do not seem to interact with the main component of the network. Although budding yeast and fission yeast are quite genetically distant, excellent work by members of the pombe community has produced a highly accurate genome-wide orthology between the two organisms (Wood et al., 2002; Kim et al., 2010). In this chapter, we take advantage of these well-known orthology relations to try to infer new possible keystone species in fission yeast.

Degree and betweenness are somewhat correlated (Fig. 5.4) but both fission yeast and budding yeast have a number of genes with high degree/low betweenness and vice versa (Yu et al., 2007). An example of a series of genes with high degree but low betweenness are the genes that take part in the proteasome machinery—since the proteasome is a large multiprotein complex, all proteins in it interact with each other, giving them all a high degree. The Ti index is a network measure that was first used in ecology to characterize ecological networks (Jordán et al., 2003). The Ti sits somewhere in between betweenness centrality and degree—it is a mesoscale measure that is influenced by the local topology of the network, but takes into account higher-order interactions as well.

Since both fission yeast and budding yeast have been extensively studied, we can compare the number of studies on a particular gene (as measured by PubMed (http://www.ncbi.nlm.nih.gov/sites/entrez) abstracts where a gene is cited) with various network measures. If a gene more often appears in the abstracts of papers it is better characterized and probably also more important. We look for correlations between PubMed hits and the three network measures (Fig. 5.5). While there is no strong correlation between the number of abstracts in Pubmed and any of the network measures, if we limit ourselves to the genes at the very top of the list, we observe that highly ranked genes tend to have high network measures and a high number of publications. There are exceptions: the last column of Fig. 5.6 shows proteins have high Ti, but low PubMed. On the top of the lists we see the usual suspects (Fig. 5.6): the most well investigated cell cycle core regulators (CDC28, CLN2, CLB2, CDC5, SIC1 and Cdc25, Cdc2, Wee1, Plo1, Cdc13, Clp1, Cig2) and signal transductors of cell cycle checkpoints (MEC1, MAD2, BUB2, various RADs and Chk1, Cds1, Mad2, Rad3, Sty1, Rad1). It is important to note that 9 of the first 20 genes are orthologs of each other in the both budding and fission yeast (Kim et al., 2010). Since these are quite distant relatives there is a high possibility that the orthologs of the same genes are also highly

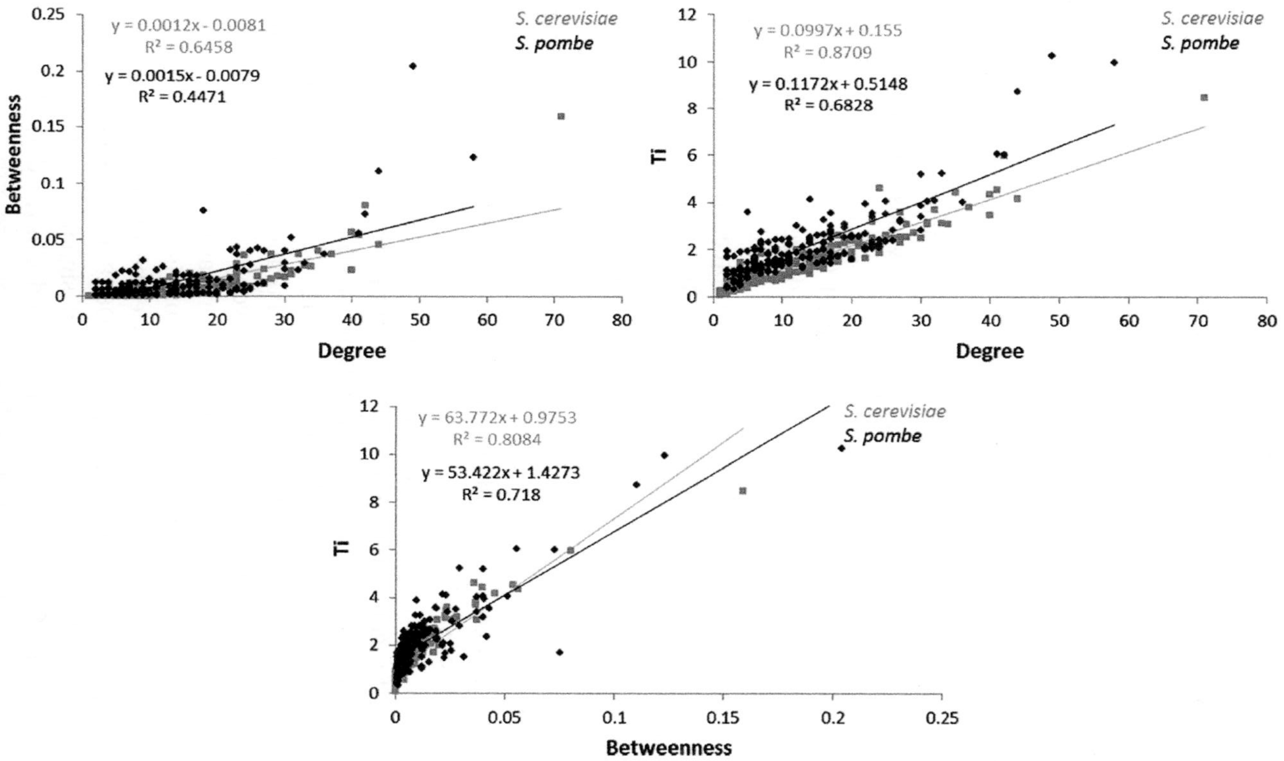


**FIGURE 5.4** Correlation between various network measures in the cell cycle interaction networks of budding and fission yeast cells. Interaction networks at the cutoff level of 0.7 from the STRING database.

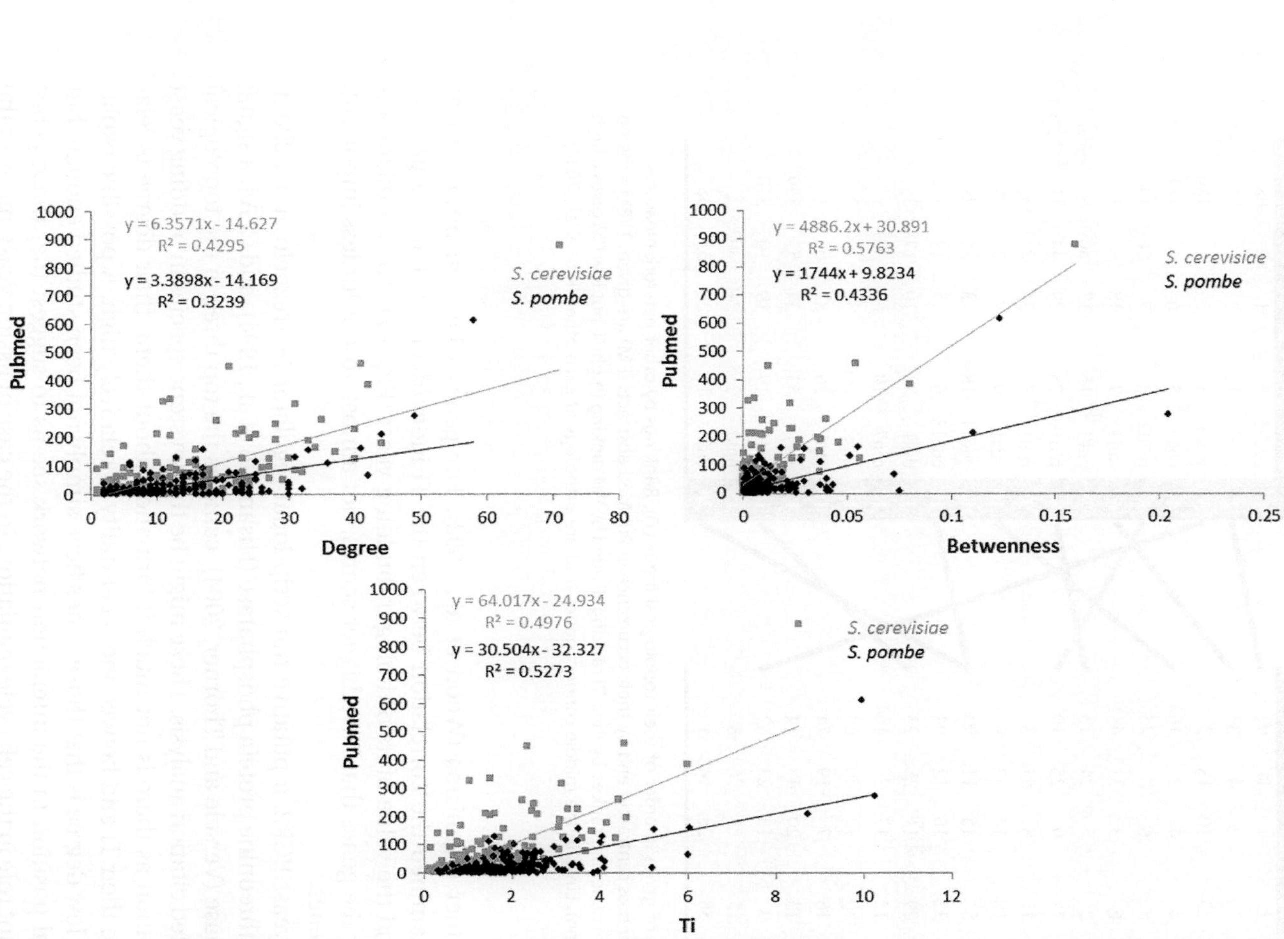


**FIGURE 5.5** Correlation between network measures of a protein and number of PubMed abstracts mentioning the name of the protein. The STRING networks at 0.7 cutoff were analyzed for this plot.

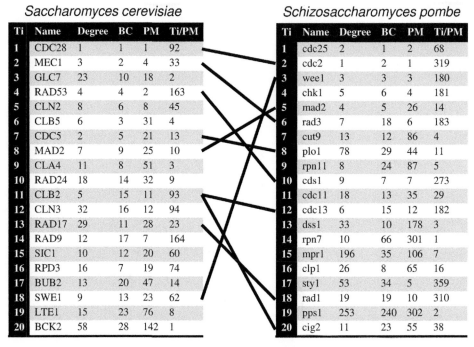

*Saccharomyces cerevisiae*

| Ti | Name | Degree | BC | PM | Ti/PM |
|---|---|---|---|---|---|
| 1 | CDC28 | 1 | 1 | 1 | 92 |
| 2 | MEC1 | 3 | 2 | 4 | 33 |
| 3 | GLC7 | 23 | 10 | 18 | 2 |
| 4 | RAD53 | 4 | 4 | 2 | 163 |
| 5 | CLN2 | 8 | 6 | 8 | 45 |
| 6 | CLB5 | 6 | 3 | 31 | 4 |
| 7 | CDC5 | 2 | 5 | 21 | 13 |
| 8 | MAD2 | 7 | 9 | 25 | 10 |
| 9 | CLA4 | 11 | 8 | 51 | 3 |
| 10 | RAD24 | 18 | 14 | 32 | 9 |
| 11 | CLB2 | 5 | 15 | 11 | 93 |
| 12 | CLN3 | 32 | 16 | 12 | 94 |
| 13 | RAD17 | 29 | 11 | 28 | 23 |
| 14 | RAD9 | 12 | 17 | 7 | 164 |
| 15 | SIC1 | 10 | 12 | 20 | 60 |
| 16 | RPD3 | 16 | 7 | 19 | 74 |
| 17 | BUB2 | 13 | 20 | 47 | 14 |
| 18 | SWE1 | 9 | 13 | 23 | 62 |
| 19 | LTE1 | 15 | 23 | 76 | 8 |
| 20 | BCK2 | 58 | 28 | 142 | 1 |

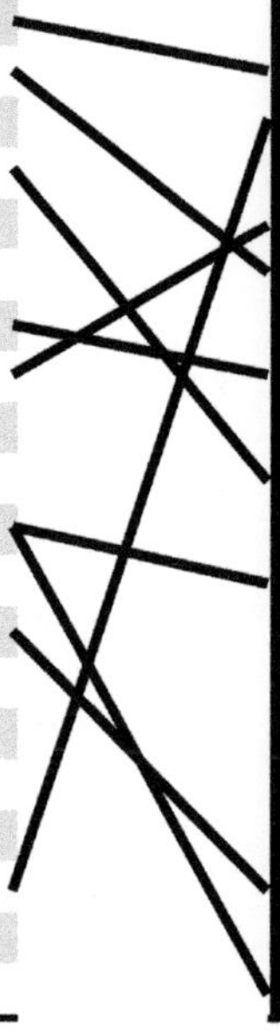

*Schizosaccharomyces pombe*

| Ti | Name | Degree | BC | PM | Ti/PM |
|---|---|---|---|---|---|
| 1 | cdc25 | 2 | 1 | 2 | 68 |
| 2 | cdc2 | 1 | 2 | 1 | 319 |
| 3 | wee1 | 3 | 3 | 3 | 180 |
| 4 | chk1 | 5 | 6 | 4 | 181 |
| 5 | mad2 | 4 | 5 | 26 | 14 |
| 6 | rad3 | 7 | 18 | 6 | 183 |
| 7 | cut9 | 13 | 12 | 86 | 4 |
| 8 | plo1 | 78 | 29 | 44 | 11 |
| 9 | rpn11 | 8 | 24 | 87 | 5 |
| 10 | cds1 | 9 | 7 | 7 | 273 |
| 11 | cdc11 | 18 | 13 | 35 | 29 |
| 12 | cdc13 | 6 | 15 | 12 | 182 |
| 13 | dss1 | 33 | 10 | 178 | 3 |
| 14 | rpn7 | 10 | 66 | 301 | 1 |
| 15 | mpr1 | 196 | 35 | 106 | 7 |
| 16 | clp1 | 26 | 8 | 65 | 16 |
| 17 | sty1 | 53 | 34 | 5 | 359 |
| 18 | rad1 | 19 | 19 | 10 | 310 |
| 19 | pps1 | 253 | 240 | 302 | 2 |
| 20 | cig2 | 11 | 23 | 55 | 38 |

**FIGURE 5.6** Ranking of genes in order of their topological index (Ti). Rankings by other network measures (Degree, BC: betweenness centrality) and by their occurrence in PubMed abstracts (PM) are given. Ti/PM ranking gives the order where genes are ranked by their Ti ranking divided by PM ranking (high Ti, but low PM gives a high ranking in this column). Lines in the middle connect genes that are orthologs of each other (Kim et al., 2010).

important in other organisms (Wood et al., 2002). For genes where an ortholog was available we examined the correlation between the Ti index/degree value of a gene in fission yeast, and the value of its ortholog in budding yeast (Fig. 5.7). The correlation is quite strong for the genes that have higher scores, but almost absent for less important, but common genes.

In budding yeast BCK2, a putative transcriptional regulator (Ferrezuelo et al., 2009), GLC7, a serine/threonine protein phosphatase (Hisamoto et al., 1994), and CLA4, a signal transducing kinase (Versele and Thorner, 2004), came as the top three in the topological index per PubMed citation analysis. These might be the keystone species in budding yeast cell cycle regulation as there is not much information about them, their degree is relatively low, while their Ti and betweenness centrality is relatively high. A possible explanation for their low degree is that those genes have simply not been studied enough, but their topological position in the interaction network seems to suggest that those genes play an important role in the cell cycle regulation. In the case of fission yeast, many of the proteasome components (rpn genes) are found with high Ti/PM values (Fig. 5.6), reflecting their role as members of the APC, a large macromolecular complex involved in

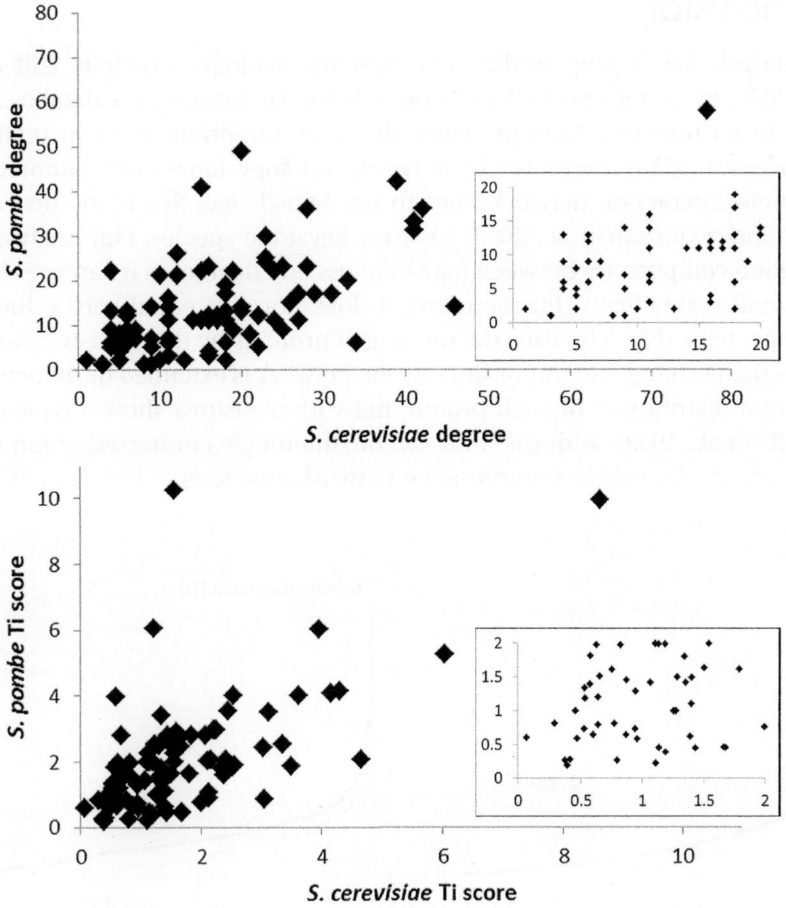


**FIGURE 5.7** Orthologs of high importance genes in budding yeast have high importance in fission yeast as well. Degrees and topological importance (Ti) indexes of 92 gene pairs, where we found one-to-one orthology between budding and fission yeast genes (Kim et al., 2010), are plotted. At large network measure values we see a correlation between network measures of the orthologs in the two organisms but at low values the correlation is absent (see inset for a zoom into this region).

regulating protein degradation throughout the cell-cycle (King et al., 1996). As they are all connected to each other in the network, they all have a high degree, but as they are not often mentioned by their specific gene name their PubMed ranking is low. Out of these genes the other interesting candidates are: the phosphatidylserine synthase Pps1 (Matsuo et al., 2007), the mRNA export protein Dss1 (Mannen et al., 2008), and the histidine-containing response regulator phosphotransferase Mpr1 (Tan et al., 2007). Importantly deletion of the mentioned genes from both fission and budding yeast leads either to total loss of viability or to serious perturbation of average cell size and growth defects, and thus we propose to investigate the role of these genes further experimentally.

## 5.4. Conclusion

Network analysis has a long tradition in systems ecology, which is still expanding (Estrada, 2007). By using network tools on existing bioinformatics databases, we can investigate, in an unbiased fashion, genes that play important roles in regulating key cellular processes. Many measures of network topology have been adopted to study protein–protein interaction networks, but to our knowledge, this is the first attempt at using the Ti index (Jordán et al., 2003) to study keystone species. Our analysis suggests that Ti is a good compromise between betweenness and degree, as it captures both large-scale and small-scale effects. Furthermore, it does not give equal zero values for low-importance proteins (Fig. 5.8); thus the ranking of proteins by this measure can be better used to investigate changes in importance as the network is extended or reduced. It is also interesting that distribution of each protein network measures show a typical keystone pattern (Mills et al., 1993), with one or a few outstandingly connected graph nodes and a vast majority of relatively low-importance network positions.

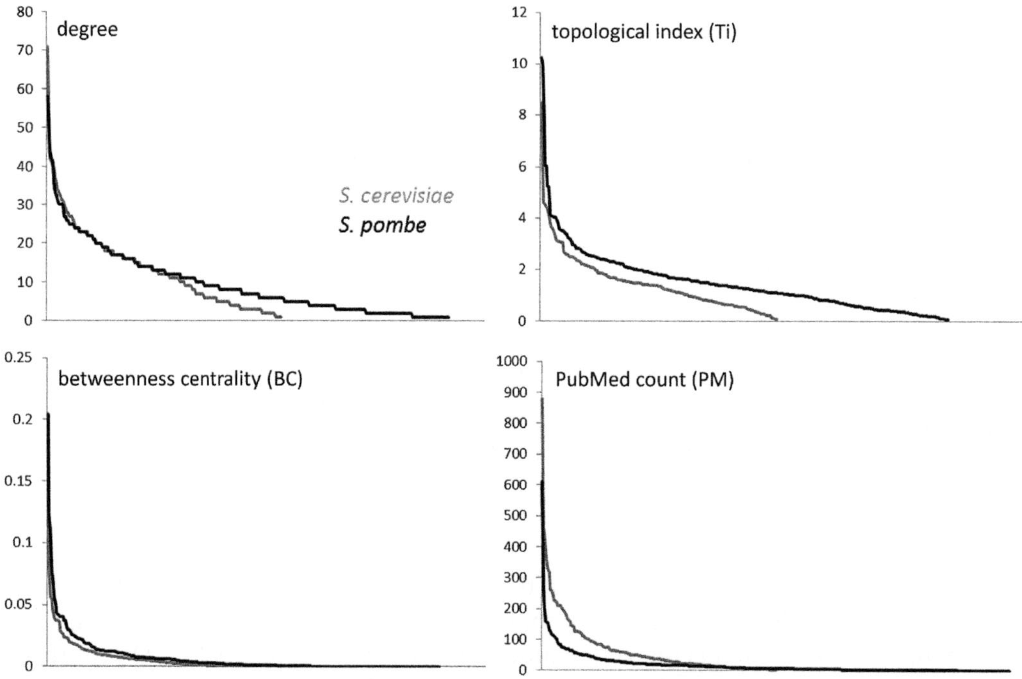


**FIGURE 5.8** Network measure distributions predict "keystone-like" network structures: Three network measures (degree, TI and BC) and the number of PubMed abstracts of cell cycle regulators in fission and budding yeast were ordered by their numerical values. All measures show that there are a few highly important molecules, while the rest of the network has minimal contribution. Betweenness centrality and PubMed literature counts have a particularly sharp drop, while degree and Ti drop less sharply to zero. Since topological importance (Ti) gave the smoothest curve with nonzero value for all genes, we will focus in our analysis onto this measure in the rest of the chapter.

By focusing on genes that have a high Ti/PubMed ratio, we discovered several genes that are located in key positions of the cell cycle network, but are not highly characterized. Investigating this property was motivated by the original and widely accepted definition of keystone species (community effect per biomass, Power et al. (1996)). As it is more likely that a species with larger biomass will be more important, it is also probable that a better studied gene with more citations will be better connected in the interaction network (as an artifact of being over-represented in the databases). We suggest that these genes are excellent candidates for further analysis. It is important to add that although the quality of the data available in bioinformatics databases keeps improving, given the huge amount of interactions and proteins under consideration, it is reasonable to expect that false positives and false negatives are present. A note of hope, however, is that as the amount of data available for all organism increases, the confidence in each of the interactions increases, as we see for budding yeast at high cutoffs. Finally, we observe that genes that have a high degree or Ti in one organism tend to have a high degree or Ti in another organism. This might be consistent with a preferential attachment model (Barabási and Albert, 1999), where genes with high scores tend to be more highly conserved, and tend to maintain more of their edges across evolution.

Similar network topology and orthology based predictions of important functional regulators and essential regulatory links could be performed on data from other databases of protein interactions (Zanzoni et al., 2002; Hermjakob et al., 2004; Salwinski et al., 2004; ). Such analysis in mammalian systems might lead to predictions of drug targets of different diseases (Hopkins, 2008; Korcsmáros et al., 2010; Farkas et al., 2011). Indeed proteins in key topological positions were found crucial and were proposed as promising novel drug targets (Hwang et al., 2008).

## Acknowledgments

The authors are thankful to L.J. Jensen, R.E. Carazo-Salas, M. Sato and F. Jordan for help and comments. The authors gratefully acknowledge support from a Human Frontier Science Program (HFSP) Young Investigator Grant (HFSP RGY0066/2009-C).

## References

Albert, R., 2005. Scale-free networks in cell biology. J. Cell Sci. 118, 4947.

Ashburner, M., Ball, C.A., Blake, J.A., Botstein, D., Butler, H., Cherry, J.M., Davis, A.P., Dolinski, K., Dwight, S.S., Eppig, J.T., 2000. Gene ontology: tool for the unification of biology. Nat. Genet. 25, 25.

Barabási, A.L., Albert, R., 1999. Emergence of scaling in random networks. Science 286, 509.

Barabási, A.L., Oltvai, Z.N., 2004. Network biology: understanding the cell's functional organization. Nat. Rev. Genet. 5, 101–113.

Batada, N.N., Hurst, L.D., Tyers, M., 2006. Evolutionary and physiological importance of hub proteins. PLoS Comput. Biol. 2, e88.

Csikász-Nagy, A., 2009. Computational systems biology of the cell cycle. Brief. Bioinform. 10, 424.

Dunne, J.A., Williams, R.J., Martinez, N.D., 2002. Network structure and biodiversity loss in food webs: robustness increases with connectance. Ecol. Lett. 5, 558–567.

Estrada, E., 2007. Characterization of topological keystone species: local, global and "meso-scale" centralities in food webs. Ecol. Complex. 4, 48–57.

Farkas, I.J., Korcsmaros, T., Kovacs, I.A., Mihalik, A., Palotai, R., Simko, G.I., Szalay, K.Z., Szalay-Beko, M., Vellai, T., Wang, S., Csermely, P., 2011. Network-based tools for the identification of novel drug targets. Sci. Signal. 4 pt3-.

Ferrell, J.E., Tsai, T.Y., Yang, Q., 2011. Modeling the cell cycle: why do certain circuits oscillate? Cell 144, 874–885.

Ferrezuelo, F., Aldea, M., Futcher, B., 2009. Bck2 is a phase-independent activator of cell cycle-regulated genes in yeast. Cell Cycle 8, 239–252.

Freeman, L.C., 1977. A set of measures of centrality based on betweenness. Sociometry 40, 35–41.

Gauthier, N.P., Larsen, M.E., Wernersson, R., De Lichtenberg, U., Jensen, L.J., Brunak, S., Jensen, T.S., 2008. Cyclebase. org—a comprehensive multi-organism online database of cell-cycle experiments. Nucleic Acids Res. 36, D854.

Goffeau, A., Barrell, B., Bussey, H., Davis, R., Dujon, B., Feldmann, H., Galibert, F., Hoheisel, J., Jacq, C., Johnston, M., 1996. Life with 6000 genes. Science 274, 546.

Hagberg, A., Swart, P., Chult, D., 2008. Exploring Network Structure, Dynamics, and Function using NetworkX. Los Alamos National Laboratory (LANL), Los Alamos, New Mexico.

Hahn, M.W., Kern, A.D., 2005. Comparative genomics of centrality and essentiality in three eukaryotic protein–interaction networks. Mol. Biol. Evol. 22, 803.

Han, J.D.J., Bertin, N., Hao, T., Goldberg, D.S., Berriz, G.F., Zhang, L.V., Dupuy, D., Walhout, A.J.M., Cusick, M.E., Roth, F.P., 2004. Evidence for dynamically organized modularity in the yeast protein–protein interaction network. Nature 430, 88–93.

Harris, M., Clark, J., Ireland, A., Lomax, J., Ashburner, M., Foulger, R., Eilbeck, K., Lewis, S., Marshall, B., Mungall, C., 2004. The Gene Ontology (GO) database and informatics resource. Nucleic Acids Res. 32, D258.

He, X., Zhang, J., 2006. Why do hubs tend to be essential in protein networks? PLoS Genet. 2, e88.

Hermjakob, H., Montecchi-Palazzi, L., Lewington, C., Mudali, S., Kerrien, S., Orchard, S., Vingron, M., Roechert, B., Roepstorff, P., Valencia, A., Margalit, H., Armstrong, J., Bairoch, A., Cesareni, G., Sherman, D., Apweiler, R., 2004. IntAct: an open source molecular interaction database. Nucleic Acids Res. 32, D452–D455.

Higashi, M., Burns, T.P., 1991. Theoretical Studies of Ecosystems: The Network Perspective. Cambridge University Press, Cambridge, MA.

Hisamoto, N., Sugimoto, K., Matsumoto, K., 1994. The Glc7 type 1 protein phosphatase of *Saccharomyces cerevisiae* is required for cell cycle progression in G2/M. Mol. Cell. Biol. 14, 3158.

Hopkins, A.L., 2008. Network pharmacology: the next paradigm in drug discovery. Nat. Chem. Biol. 4, 682–690.

Hwang, W., Zhang, A., Ramanathan, M., 2008. Identification of information flow-modulating drug targets: a novel bridging paradigm for drug discovery. Clin. Pharmacol. Ther. 84, 563–572.

Jensen, L.J., Julien, P., Kuhn, M., von Mering, C., Muller, J., Doerks, T., Bork, P., 2008. eggNOG: automated construction and annotation of orthologous groups of genes. Nucleic Acids Res. 36, D250–D254.

Jensen, L.J., Kuhn, M., Stark, M., Chaffron, S., Creevey, C., Muller, J., Doerks, T., Julien, P., Roth, A., Simonovic, M., 2009. STRING 8—a global view on proteins and their functional interactions in 630 organisms. Nucleic Acids Res. 37, D412.

Jeong, H., Mason, S.P., Barabási, A.L., Oltvai, Z.N., 2001. Lethality and centrality in protein networks. Nature 411, 41–42.

Jordán, F., Scheuring, I., 2002. Searching for keystones in ecological networks. Oikos 99, 607–612.

Jordán, F., Liu, W.C., van Veen, J.F., 2003. Quantifying the importance of species and their interactions in a host-parasitoid community. Comm. Ecol. 4, 79–88.

Kim, D.U., Hayles, J., Kim, D., Wood, V., Park, H.O., Won, M., Yoo, H.S., Duhig, T., Nam, M., Palmer, G., 2010. Analysis of a genome-wide set of gene deletions in the fission yeast *Schizosaccharomyces pombe*. Nat. Biotechnol. 28, 617–623.

King, R.W., Deshaies, R.J., Peters, J.M., Kirschner, M.W., 1996. How proteolysis drives the cell cycle. Science 274, 1652.

Korcsmáros, T., Farkas, I.J., Szalay, M.S., Rovó, P., Fazekas, D., Spiró, Z., Böde, C., Lenti, K., Vellai, T., Csermely, P., 2010. Uniformly curated signaling pathways reveal tissue-specific cross-talks and support drug target discovery. Bioinformatics 26, 2042–2050.

Korcsmáros, T., Szalay, M.S., Rovó, P., Palotai, R., Fazekas, D., Lenti, K., Farkas, I.J., Csermely, P., Vellai, T., 2011. Signalogs: orthology-based identification of novel signaling pathway components in three metazoans. PLoS ONE 6, e19240.

Ma, H.W., Zeng, A.P., 2003. The connectivity structure, giant strong component and centrality of metabolic networks. Bioinformatics 19, 1423.

Mannen, T., Andoh, T., Tani, T., 2008. Dss1 associating with the proteasome functions in selective nuclear mRNA export in yeast. Biochem. Biophys. Res. Commun. 365, 664–671.

Margalef, R., 1968. Perspectives in Ecological Theory. University of Chicago Press, Chicago.

Marguerat, S., Jensen, T.S., de Lichtenberg, U., Wilhelm, B.T., Jensen, L.J., Bähler, J., 2006. The more the merrier: comparative analysis of microarray studies on cell cycle regulated genes in fission yeast. Yeast 23, 261–277.

Matsuo, Y., Fisher, E., Patton-Vogt, J., Marcus, S., 2007. Functional characterization of the fission yeast phosphatidylserine synthase gene, pps1, reveals novel cellular functions for phosphatidylserine. Eukaryot. Cell 6, 2092.

Merlo, L., Pepper, J.W., Reid, B.J., Maley, C.C., 2006. Cancer as an evolutionary and ecological process. Nat. Rev. Cancer 6, 924–935.

Mills, L.S., Soulé, M.E., Doak, D.F., 1993. The keystone-species concept in ecology and conservation. BioScience 43, 219–224.

Morgan, D.O., 2006. The Cell Cycle: Principles of Control. New Science Press, London.

Nguyen, T.P., Jordán, F., 2010. A quantitative approach to study indirect effects among disease proteins in the human protein interaction network. BMC Sys. Biol. 4, 103.

Nguyen, T.-P., Liu, W.-C., Jordan, F., 2011. Inferring pleiotropy by network analysis: linked diseases in the human PPI network. BMC Sys. Biol. 5, 179.

Olesen, J.M., Bascompte, J., Dupont, Y.L., Jordano, P., 2007. The modularity of pollination networks. Proc. Natl. Acad. Sci. 104, 19891.

Paine, R.T., 1966. Food web complexity and species diversity. Am. Nat. 100, 65–75.

Papayannopoulou, T., Scadden, D.T., 2008. Stem-cell ecology and stem cells in motion. Blood 111, 3923.

Pimm, S.L., 1980. Food web design and the effect of species deletion. Oikos 35, 139–149.

Power, M.E., Tilman, D., Estes, J.A., Menge, B.A., Bond, W.J., Mills, L.S., Daily, G., Castilla, J.C., Lubchenco, J., Paine, R.T., 1996. Challenges in the quest for keystones. BioScience 46, 609–620.

Salwinski, L., Miller, C.S., Smith, A.J., Pettit, F.K., Bowie, J.U., Eisenberg, D., 2004. The database of interacting proteins: 2004 update. Nucleic Acids Res. 32, D449–D451.

Spellman, P.T., Sherlock, G., Zhang, M.Q., Iyer, V.R., Anders, K., Eisen, M.B., Brown, P.O., Botstein, D., Futcher, B., 1998. Comprehensive identification of cell cycle-regulated genes of the yeast *Saccharomyces cerevisiae* by microarray hybridization. Mol. Biol. Cell 9, 3273.

Stark, C., Breitkreutz, B.J., Reguly, T., Boucher, L., Breitkreutz, A., Tyers, M., 2006. BioGRID: a general repository for interaction datasets. Nucleic Acids Res. 34, D535.

Stark, C., Breitkreutz, B.J., Chatr-aryamontri, A., Boucher, L., Oughtred, R., Livstone, M.S., Nixon, J., Van Auken, K., Wang, X., Shi, X., 2011. The BioGRID interaction database: 2011 update. Nucleic Acids Res. 39, D698.

Szklarczyk, D., Franceschini, A., Kuhn, M., Simonovic, M., Roth, A., Minguez, P., Doerks, T., Stark, M., Muller, J., Bork, P., 2011. The STRING database in 2011: functional interaction networks of proteins, globally integrated and scored. Nucleic Acids Res. 39, D561.

Tan, H., Janiak-Spens, F., West, A.H., 2007. Functional characterization of the phosphorelay protein Mpr1p from Schizosaccharomyces pombe. FEMS Yeast Res. 7, 912–921.

Ulanowicz, R.E., 1980. An hypothesis on the development of natural communities. J. Theor. Biol. 85, 223–245.

Ulanowicz, R.E., 1986. Growth and Development: Ecosystems Phenomenology. Springer-Verlag, New York.

Ulanowicz, R.E., 1989. A phenomenology of evolving networks. Syst. Res. 6, 209–217.

Valentini, R., Jordán, F., 2010. CoSBiLab graph: the network analysis module of CoSBiLab. Environ. Model. Software 25, 886–888.

Vaggi, F., Dodgson, J., Bajpai, A., Chessel, A., Jordan, F., Sato, M., Carazo-Salas, R.E., Csikász-Nagy, A., 2012. Linkers of cell polarity and cell cycle regulation in the fission yeast protein interaction network. PLoS Comput. Biol.

Versele, M., Thorner, J., 2004. Septin collar formation in budding yeast requires GTP binding and direct phosphorylation by the PAK, Cla4. J. Cell Biol. 164, 701.

Wagner, A., Fell, D.A., 2001. The small world inside large metabolic networks. Proc. R. Soc. London. Ser. B. Biol. Sci. 268, 1803–1810.

Wolfe, A.W., 1997. Social network analysis: methods and applications. Am. Ethnol. 24, 219–220.

Wood, V., Gwilliam, R., Rajandream, M.A., Lyne, M., Lyne, R., Stewart, A., Sgouros, J., Peat, N., Hayles, J., Baker, S., 2002. The genome sequence of Schizosaccharomyces pombe. Nature 415, 871–880.

Yamada, T., Bork, P., 2009. Evolution of biomolecular networks: lessons from metabolic and protein interactions. Nat. Rev. Mol. Cell Biol. 10, 791–803.

Yu, H., Greenbaum, D., Xin Lu, H., Zhu, X., Gerstein, M., 2004. Genomic analysis of essentiality within protein networks. Trends Genet. 20, 227–231.

Yu, H., Kim, P.M., Sprecher, E., Trifonov, V., Gerstein, M., 2007. The importance of bottlenecks in protein networks: correlation with gene essentiality and expression dynamics. PLoS Comput. Biol. 3, e59.

Zanzoni, A., Montecchi-Palazzi, L., Quondam, M., Ausiello, G., Helmer-Citterich, M., Cesareni, G., 2002. MINT: a Molecular INTeraction database. FEBS Lett. 513, 135–140.

PART 2

# Models of Populations

# 6

# Evolutionary Transition to Complex Population Dynamic Patterns in an Age-structured Population

Efim Frisman*, Oksana Zhdanova†

*INSTITUTE FOR COMPLEX ANALYSIS OF REGIONAL PROBLEMS, FAR EASTERN BRANCH OF THE RUSSIAN ACADEMY OF SCIENCE, BIROBIDZHAN, RUSSIA, †INSTITUTE OF AUTOMATION AND CONTROL PROCESSES, FAR EASTERN BRANCH OF THE RUSSIAN ACADEMY OF SCIENCE, VLADIVOSTOK, RUSSIA

## 6.1 Introduction

The life cycles of many biological species of multicellular organisms have a distinctly pronounced time periodicity that, as a rule, coincides with the seasonal changes in the earth's climate. An overwhelming majority of such species have a clearly outlined (often very short) reproductive period, during which each local population represents a totality of discrete single-age groups. In many cases, the number of groups is determined solely by the size of the age groups in the preceding reproductive period.

Therefore, the number of sex groups and age groups in a given population may be characterized by a set of discrete magnitudes, which acquire certain values at certain fixed moments of time, such as the beginning (or end) of the usual reproductive periods. The relationship between the values of such numbers in given and subsequent moments of time may then be described by a system of recurrent equations.

## 6.2 Discrete Models of Changes in a Local Population Size

The proposed model begins with the consideration of the simplest situations, which nonetheless describe the real population processes. For many insect species, the entire preceding generation produced by adult specimens has already died during the development of one generation (from egg to imago). Hence, each individual population represents one age group (or developmental stage), and adjacent generations of this population do not overlap. The dynamics of population size for such species with non-overlapping generations may be described very simply. If the environmental conditions

Models of the Ecological Hierarchy. DOI: http://dx.doi.org/10.1016/B978-0-444-59396-2.00006-7

from generation to generation change little and they can be presumed constant, then the size of a given population would be determined solely by the size of the preceding generation. After selecting certain moments in time to record the size (e.g., the beginning of any developmental stage) and after designating the number of the $n$-th generation via $N_n$, the following deterministic equation can be used to describe the dynamics of the size of this single-age population with non-overlapping generations.

$$N_{n+1} = F(N_n) \tag{1}$$

Therefore, compared with a differential equation, Eqn (1) should provide a more realistic description of the population dynamics of biological species with non-overlapping generations.

Several scholars have attempted to describe the dynamic processes in populations using different forms of Eqn (1). The following equation was examined most frequently (Cook, 1965; Moran, 1950; Ricker, 1954; and others):

$$N_{n+1} = aN_n e^{-bN_n} \tag{2}$$

Frisman and Shapiro (1982) examined the behavior of population dynamics using Eqn (2). Some of the results of their investigation are the following.

Without restricting the existing community, one may consider $b = 1$ and assume that the population is counted not in absolute units but in relative units, which exceed the real number by $b$ times. With various values of $a$, Eqn (2) seems to lead to substantially differing types of dynamic behavior: from stable to cyclic and chaotic regimes.

A mathematical study of dimensionless discrete dynamic systems described by the general Eqn (1) shows that the essentially irregular or chaotic nature of dynamic behavior is due to the presence of cyclic trajectories of period three (or when interpreted biologically, with the presence of a number of variations with a period of three generations) among the solutions of Eqn (1). The works of the aforementioned authors prove the following important theorems. If Eqn (1) has a solution that is the cycle of a period three, then it also has solutions for any final period $k$. Moreover, an innumerable set of initial $N_0$ values exist, with which the solution of Eqn (1) does not tend to any of these cycles. The cycle of period three occurs in Eqn (2), where $a \approx 22.252$ (May, 1975; Shapiro, 1972). Therefore, a bifurcation curve may be constructed by characterizing the stable limit trajectories dependent on the value of coefficient $a$. Thus, the presence of density-dependent factors for populations with non-overlapping generations may not always be expected to lead to stable stationary size values. Moreover, precisely these density-dependent factors prove to be responsible for the oscillation modes in the dynamic behavior of the population and can even lead to chaotic (pseudorandom) dynamic behavior (Getz, 2003; Lebreton, 2006; Reluga, 2004). These results are shown in the simplest deterministic models.

The questions of how similar the regularities obtained from analyzing model (2) for the dynamics of the population of living organisms, whose adjacent generations do not overlap, and to what extent the intensity of density-dependent factors affect these regularities arise. Shapiro (1980) obtained a partial answer to this question, and it consists of the following. Let the dynamics of population size be described by the equation

$$N_{n+1} = aF(N_n) \tag{3}$$

where $F(N)$ is a unimodal differential function ($F'(0) \geq 0$) and $a$ is a certain multiplicative parameter. Let the non-zero solutions of Eqn (3) with all $a < a_1$ meet at a stable stationary point, and let $F(N)$ with sufficiently high $N$ values satisfy the inequality $F(N) < KN^{-2-\varepsilon}$. There exists a value of parameter $a = a_2$ ($a_2 > a_1$) at which the cycle of period three will be present among the solutions for Eqn (3). In addition, for the many concrete forms of $F(N)$ at the interval ($a_2$, $a_1$), the first bifurcation series are realized. Therefore, an isolated single-species population of living organisms may experience regular or chaotic oscillations in size caused by purely internal factors, namely, density-dependent limitation of population growth, which requires a sufficiently high value of parameter $a$, characterizing the reproduction intensity, and a sufficiently strong negative effect of density on population growth. With large population size, this effect should lead to a decrease in the growth rate. To establish the self-oscillating modes of population dynamics, the decline of the population growth rate should be no less than that described by the hyperbolic function $X_{n+1} = AX_n^{-2}$. The Ricker's model (2) is clear when this condition has been fulfilled.

Hence, theoretical population ecology and genetics are confronted with the quite natural task of explaining the sequence of the arrival and the role of various modes of population dynamics in the course of evolution. To resolve this problem, the quantitative relationship between the factors affecting population dynamics and those affecting changes in its genetic structure must first be revealed and described. In other words, the joint dynamics of both the size and genetic structure of populations affected by both selective and density-dependent forces must be analyzed. The next sections in the present study are devoted precisely to this analysis.

## 6.3 Density-Independent Selection in Limited Populations

A situation where selection does not depend on the population number (density) is examined. At the same time, the effect of density-dependent factors limiting population growth is considered. This approach is equivalent to the usual concept of selection as a differential survival of genotypes, assuming that this differential survival is the same for any population size. This concept appears to correspond fully to the real effect of selection on features with which natural genotypic variability is in no way correlated with the changes in population density.

Several experimental and theoretical works on density-dependent selection have, in a way, eclipsed no less than the real case where the effect of selection does not change with variations in density levels. Many authors attempted to distinguish precisely the density-dependent effects of natural selection, rarely publishing the description of situations where this form of selection was, for some reason, not evident. Numerous physiological mutations affecting the overall viability should be selected this way. In this case, the availability of density-dependent selection should be anticipated as an exception, rather than a rule.

Gaines et al. (1978) showed that no correlation exists between variations in the number and dynamics of allele frequencies for three (of five studied) polymorphous protein loci in four strongly fluctuating populations of the steppe vole, *Microtus ochrogaster*. Gottlieb (1974) demonstrated the stability of the genetic structure for five polymorphous protein loci in a geographically isolated population of the annual plant *Stephanomerica eziqua*. Despite population size variations, the allele frequencies of these loci remained virtually unchanged in four successive generations, wherein the deviation in the number of different years was close to 50:1. Such examples could be continued. However, the present study is more concerned with explaining the mechanisms of uncorrelated density changes or even the equilibrium in the genetic structure of a population, despite strong fluctuations in number.

A possible explanation for the behavior of the genetic structure lies in the assumption that the effect of natural selection (on the features discussed) is independent of population size (or density). Thus, the relative fitness values of the genetic groups of organisms do not depend on the population size and may be regarded as constants. However, the same is not true for absolute fitness. If a population is under the effect of density-dependent limiting factors, the absolute fitness values of each genotypic class must depend functionally on population size. Both requirements, namely, (a) independence of the relative fitness and (b) obligatory dependence of absolute fitness on the population size, may be satisfied if absolute fitness ($W_{ij}$) is assumed to depend on number, as in the following:

$$W_{ij}(N_n) = a_{ij} f(N_n) \tag{4}$$

where $f(N)$ is the function characterizing the density control of population dynamics similar for each genotype and $a_{ij}$ are differential coefficients equivalent to the relative genotypic fitness. Let this selective effect be the F-selection to stress the constancy of relative fitness and the presence of density-dependent factors characterized by the function $f(N)$.

Taking into account Eqn (4), the following system of equations describing the dynamics of allele frequencies and the size of an isolated Mendelian one-locus population are obtained.

$$P_{i,(n+1)} = P_{i,n} \sum_j a_{ij} P_{j,n} / \overline{a}_n \tag{5}$$

$$N_{n+1} = \overline{a}_n N_n f(N_n) \tag{6}$$

where $P_{i,n}$ is the frequency of allele $i$ in the $n$-th generation and $\overline{a}_n = \sum\sum a_{ij} P_{i,n} P_{j,n}$ is the average fitness of the population in the $n$-th generation.

The solutions of subsystem (5), which characterize the variations in the genetic structure of the population, do not depend on population dynamics and may be studied irrespective of Eqn (6). Furthermore, this subsystem coincides exactly with the designations of equations for the dynamics of the genetic structure of a "free" unlimited population (Ratner, 1973). Hence, Fisher's theorem is correct for this subsystem (Kingman, 1961). In this case, it takes the following form:

$$\overline{a}_{n+1} > \overline{a}_n \tag{7}$$

In other words, the mean value of coefficient $\overline{a}_{ij}$ in Eqn (4) could only increase in the course of evolution, irrespective of the initial state of the population and concrete values of $a_{ij}$. Naturally, this assertion will prove to be incorrect in the general case of multilocus selection, presence of mutations and recombinations, and so on. In addition, the dynamics of the genetic structure may generally decrease $a$. However, all adaptive factors, including the selection of newly occurring mutations, may only increase this value.

## 6.4 Population Dynamics with F-Selection

Equation (6) describes the variations in population size. Population dynamics is essentially related to changes in $a$. A slow and highly nonuniform time increase of parameter $\overline{a}_n$ caused by the selection of new adaptive mutations should be observed in natural populations. Moreover, the initial value of this parameter in populations of primitive organisms should be very small. The population number in certain time intervals should attain the stable stationary $N_{\overline{a}}$ because of the extremely slow growth rate of $\overline{a}_n$. However, the population dynamics may change with the growth of $\overline{a}_n$. At first, only quantitative characters may be involved in the growth of $N_{\overline{a}}$. Subsequently, with the transition of $\overline{a}_n$ via a certain bifurcative value, the dynamic pattern changes qualitatively to produce cyclic dynamic modes. If $\overline{a}_n$ continues to grow, the cycles become increasingly complex. Figure 6.1 shows the examples of population dynamics and $\overline{a}_n$ diagrams obtained by numerically solving the systems of Eqns (5) and (6) for the case of the two-allele locus and the linear form of the function $f(N)$ ($f(x) = 1-x$, where $x$ is the relative, not absolute, value of the population number), that is, in the case of

$$\left\{\begin{array}{rcl} P_{1,(n+1)} &=& P_{1,n}\left(a_{11}P_{1,n} + a_{12}P_{2,n}\right)/\overline{a}_n \\ P_{2,(n+1)} &=& P_{2,n}\left(a_{12}P_{1,n} + a_{22}P_{2,n}\right)/\overline{a}_n \\ x_{n+1} &=& \overline{a}_n x_n(1 - x_n) \\ \overline{a}_n &=& a_{11}P_{1,n}^2 + 2a_{12}P_{1,n}P_{2,n} + a_{22}P_{2,n}^2 \end{array}\right\} \tag{8}$$

with fixed values of $a_{11}$, $a_{12}$, and $a_{22}$ and $P_{1,0} = 0.05$.

As shown in the diagrams, a monotonous change in $P_{i,n}$ and $\overline{a}_n$ is accompanied by a highly complex dynamic behavior of the population size and passes through stages of converging fluctuations, monotonous behavior, diverging fluctuations with different periods, and chaotic behavior.

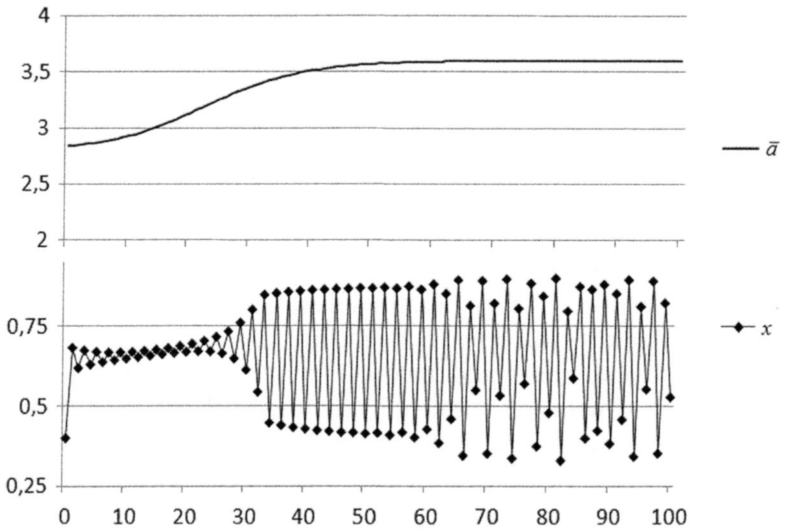


**FIGURE 6.1** Diagrams of the dynamics of $\overline{a}_n$ and $x_n$ from models (5) and (6), with $a_{11}=3.6$, $a_{12}=3.2$, $a_{22}=2.8$, $P_{1,0}=0.05$, and $x_0=0.4$. Population dynamics becomes chaotic after a series of diverging oscillations with two, four, and eight generations.

The basic difference between real population dynamics and model dynamic patterns appears to lie in the substantially slower and considerably less uniform growth of $\overline{a}_n$ and, correspondingly, in the considerably slower and less regular variations in the types of dynamic behavior of natural populations.

Thus, for limited populations of a single generation, a progressive increase in average fitness may prove to be dissonant with the stability of population growth. This result of discrete dynamic models evidently contradicts intuitive concepts on the increase of population stability with the growth of its average fitness.

## 6.5 Modeling of Natural Selection in Population with Age Structure

Previous population dynamic models consider changes of whole population number only, assuming that different generations of the population do not overlap. However, such a concept is incorrect when the lifetime of each generation is essentially longer than the time between breeding seasons. In this case, each local population consists of individuals from different age groups during the breeding season. Therefore, the number of each separated age group is naturally considered as the model's variable. The treatment for clustering the population is defined by the characteristics of biological species (Hastings, 1992; Kooi and Kooijman, 1999; Lebreton, 1996).

A model with age structure is considered. This model may be presented by a set of two age classes: junior and elder. The junior age class consists of immature individuals, and the reproductive part (elder) consists of individuals participating in the breeding process.

Let $X_n$ be the number of individuals in the junior age class in the $n$-th breeding season and $Y_n$ be the number of individuals in the reproductive part of the population. The breeding season ends by the appearance of newborn individuals in the next generation. During the time between two reproductive periods, the individuals from the junior age class are assumed to reach the age of the reproductive part, whereas newborns (or larvae) are assumed to reach the junior age class. Moreover, fitness and the reproductive potential of reproductive individuals are assumed to be independent of their ages. These assumptions are correct for organisms with not a great lifetime consisting of two or three breeding periods, such as insects, fishes, small mammals, biennials or triennials, and others.

The limitation of the junior age class number is assumed to be realized by a linear rule $f(X) = 1 - X/K$ as in the Verhulst model. In this equation, $K$ is the so-called carrying capacity or the maximum number of sustainable population. The absolute age class numbers are then changed to relative (or dimensionless) ones: $x = \frac{X}{K}, y = \frac{Y}{K}$. Hence, the dynamic equations that describe the relative numbers of the discussed age classes in neighboring generations are received as:

$$\left\{ \begin{aligned} x_{n+1} &= a y_n \\ y_{n+1} &= x_n f(x_n) + c y_n \end{aligned} \right\} \tag{9}$$

where $a$ is the reproductive potential of elders and $c$ is their survival rate.

The mechanism for the regular evolutionary complication of the dynamics of a structured population is illustrated by considering the effect of natural selection in one of the simplest model situations, when the adaptive character $a$ is encoded by one diallelic locus. The following dynamic equations relate the sizes of the age classes to their genetic structure in neighboring generations (Frisman and Zhdanova, 2009):

$$\left\{ \begin{aligned} x_{n+1} &= \bar{a}_n y_n \\ y_{n+1} &= x_n(1 - x_n) + c y_n \\ q_{n+1} &= \frac{p_n(a_{11} p_n + a_{12}(1 - p_n))}{\bar{a}_n} \\ p_{n+1} &= \frac{x_n(1 - x_n) q_n + c y_n p_n}{x_n(1 - x_n) + c y_n} \end{aligned} \right\} \tag{10}$$

where $p_n$ is the first allele frequency in the reproductive part of the population, $q_n$ is the first allele frequency in the immature part of the population, and $\bar{a}_n = a_{11} p_n^2 + 2a_{12} p_n(1 - p_n) + a_{22}(1 - p_n)^2$ is the reproductive potential (or the average fitness of embryos).

Apart from the trivial stationary point ($x = 0$, $y = 0$), model (10) has two monomorphic stationary points:

1. $\bar{q} = 0, \bar{p} = 0, \bar{x} = \frac{a_{22} + c - 1}{a_{22}}, \bar{y} = \frac{a_{22} + c - 1}{a_{22}^2}$ (which exists at $\{a_{22} > 1 - c\}$ and is stable at $\{1 - c < a_{22} < 3 - 2c$ and $a_{22} > a_{12}\}$,

**2.** $\overline{q} = 1,\ \overline{p} = 1, \overline{x} = \dfrac{a_{11} + c - 1}{a_{11}}, \overline{y} = \dfrac{a_{11} + c - 1}{a_{11}^2}$ (which exists at $\{a_{11} > 1 - c\}$ and is stable at $\{1 - c < a_{11} < 3 - 2c$ and $a_{11} > a_{12}\}$,

and a polymorphic one,

$$\overline{q} = \frac{a_{12} - a_{22}}{2a_{12} - a_{22} - a_{11}},\ \overline{p} = \frac{a_{12} - a_{22}}{2a_{12} - a_{22} - a_{11}},\ \overline{x} = \frac{\overline{a} + c - 1}{\overline{a}},\ \overline{y} = \frac{\overline{a} + c - 1}{\overline{a}^2},\ \overline{a} = \frac{a_{12}^2 - a_{11}a_{22}}{2a_{12} - a_{22} - a_{11}}$$

which exists when the following two conditions are met: $\{a_{12} > \max(a_{11}, a_{22})$ or $a_{12} < \min(a_{11}, a_{22})\}$ and $\{c > 1 - \overline{a}\}$.

The conditions for the existence and stability of the stationary points of model (10) allow the description, in general terms, of its dynamic behavior dependent on the values of population parameters $c$ and $a_{ij}$ (Table 6.1). Thus, the present study demonstrates that the genetic composition of a population, namely, whether it is polymorphic or monomorphic, is largely determined by the relative fitnesses of heterozygotes and homozygotes. In turn, the fitnesses of the genotypes present in the population determine the fitness of the embryos ($\overline{a}$, or reproductive potential). This value, together with the death rate of the reproductive age class ($1-c$), determines the population dynamic pattern of the age groups in the population. An increase in the reproductive potential ($\overline{a}$) destabilizes first the sizes, followed by the genetic structure of the age groups in the population, if the population is polymorphic (Fig. 6.2).

An analogous model for a one-locus selection in a single-age population does not show steady fluctuations of its genetic structure at any value of the population parameters. Therefore, introduction of an age structure into the model allows the unstable dynamics of the population's genetic composition to be observed.

An example of how the appearance of alleles with a higher fitness (e.g., as a result of mutations) may lead to an increase in the mean fitness of embryos ($\overline{a}$) is considered. After which, the size dynamics of the age classes become more complicated, and even chaotic.

Let the survival rate of the elder age class be fixed at $c = 0.326$, and the fitness of the "11" genotype be fixed at $a_{11} = 0.5$. Let the fitness of the other homozygote "22" be $a_{22} = 1$, and the fitness of the heterozygote be determined by the arithmetic mean of $a_{11}$

**Table 6.1** Types of Natural Selection and Model (10) Dynamics

| | |
|---|---|
| Monomorphism:<br>$a_{12} < \min(a_{11},a_{22})$ or<br>$\max(a_{11},a_{22}) > a_{12} > \min(a_{11},a_{22})$ | Stable dynamics – $1-c < a_{11}$ (or $a_{22}$) $< 3-2c$<br>Fluctuations of age group numbers – $a_{11}$ (or $a_{22}$) $> 3-2c$ |
| Polymorphism: $a_{12} > \max(a_{11},a_{22})$ | Stable dynamics – $\Phi(a_{ij})$ < stability boundary 1<br>Fluctuations of age group numbers –<br>stability boundary 1 < $\Phi(a_{ij})$ < stability boundary 2<br>Fluctuations of age group numbers and genetic composition –<br>$\Phi(a_{ij})$ > stability boundary 2 |

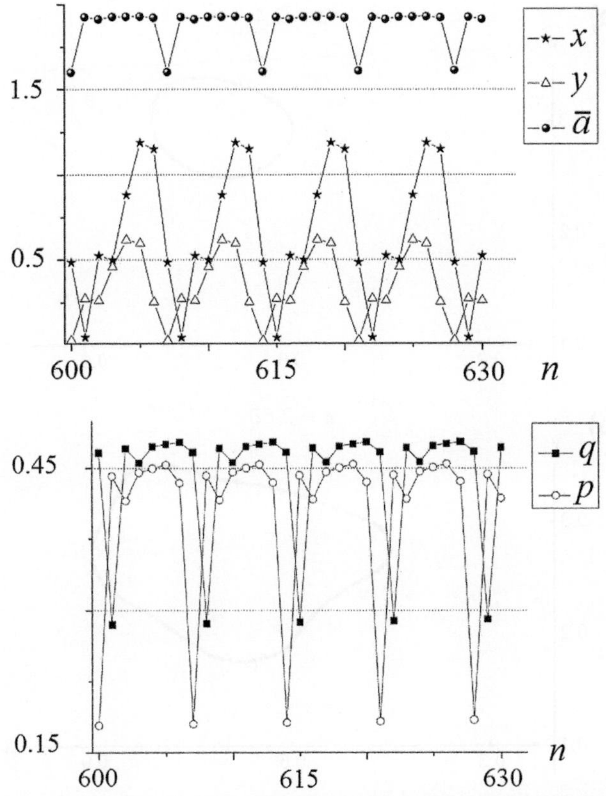


**FIGURE 6.2** Fluctuations in the sizes and genetic compositions of the age groups. (Top) Population dynamics of age groups ($x$ and $y$) and mean embryo fitness ($\bar{a}$) in a series of generations. (Bottom) Changes in the 1st allele frequency in a series of generations for the younger ($q$) and elder ($p$) age classes. Population parameters: $c = 0.8$, $a_{11} = 1.1$, $a_{12} = 2.8$, $a_{22} = 1.05$, $x_0 = 0.2$, $y_0 = 0.1$, $q_0 = 0.002$, and $p_0 = 0.001$.

and $a_{12}$. At these values, the first allele will be extinct, and the population will become monomorphic ("22"). The stationary sizes of the elder and junior age classes are $y = 0.326$ and $x = 0.326$, respectively. The population at equilibrium is monomorphic. Hence, the fitness of the embryos coincides with the fitness of the "22" homozygote: $\bar{a} = a_{22}$. Assuming that the mutations have led to the appearance of a more adapted allele 1 ($a_{11} = 1.1$), then the second allele will be displaced and the relative sizes of the two age classes will eventually become stationary at $y = 0.341$ and $x = 0.358$. A further increase in parameter $a_{11}$ would increase the mean fitness of the embryos (in the given case, at equilibrium ($\bar{a} = a_{11}$) and, initially, an increase only in the sizes of the age groups). When $a_{11}$ becomes higher than 2.348, the genetically monomorphic equilibrium ($p = 1$, $q = 1$) loses stability. In this case, the less adapted allele 2 is also displaced. However, the sizes of the age groups never reach stationary values and begin fluctuating.

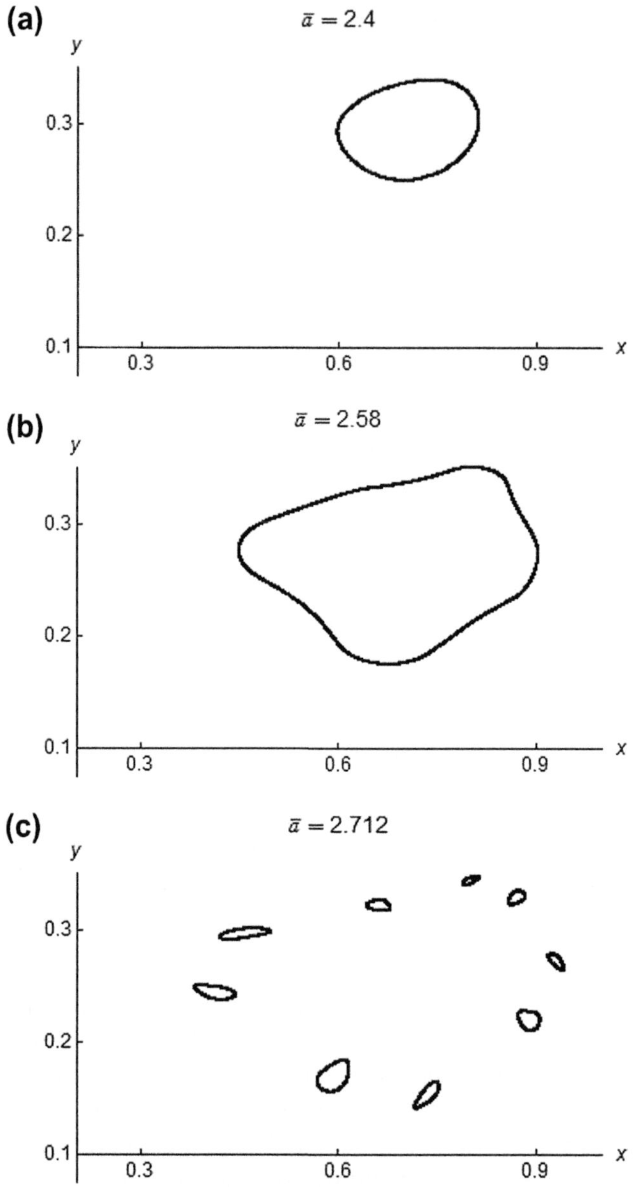


**FIGURE 6.3** Changes in the phase portrait of system (10) with increasing mean embryo fitness ($\bar{a}$) at fixed survival rate of reproductive age class $c = 0.326$. a) invariant curve; b) changed form of invariant curve; c) nine circles; d) 54 quasy-ellipses; e–g) nine pyramid-like structures; h) fusion of pyramids.

At first, the fluctuation amplitude is small but increases with increasing $a_{11}$. At $a_{11} = 2.4$, an invariant curve resembling a circle is already distinctly observed in phase space $xy$ (Fig. 6.3a). A further increase in $a_{11}$ distorts the circle (Fig. 6.3b, $a_{11} = 2.58$, followed by $a_{11} = 2.64$). A cycle with a length of nine appears, followed by a circle at every

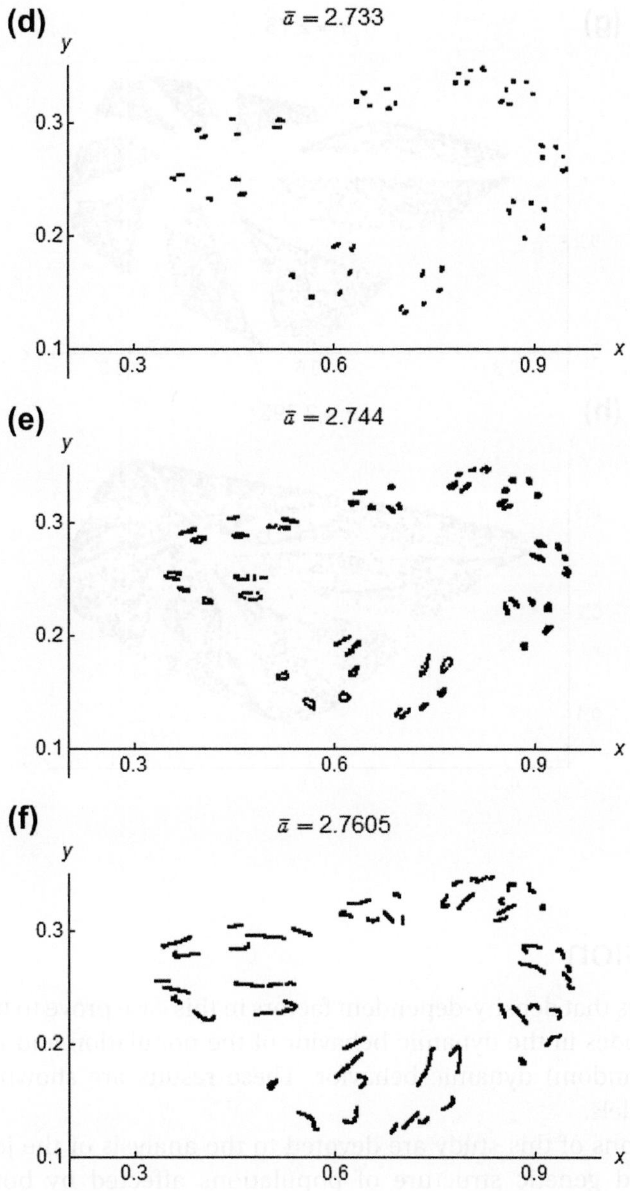


**FIGURE 6.3** (*continued*).

point of the cycle (Fig. 6.3c, $a_{11} = 2.712$; nine circles). Subsequently, each circle breaks up into nine parts (Fig. 6.3d). A further increase in this parameter leads to the formation of nine pyramid-like structures (Fig. 6.3e–g). These structures then lose shape and form an "expanding mapping," more or less densely filling the phase space (Fig. 6.3h, $a_{11} = 2.795$).

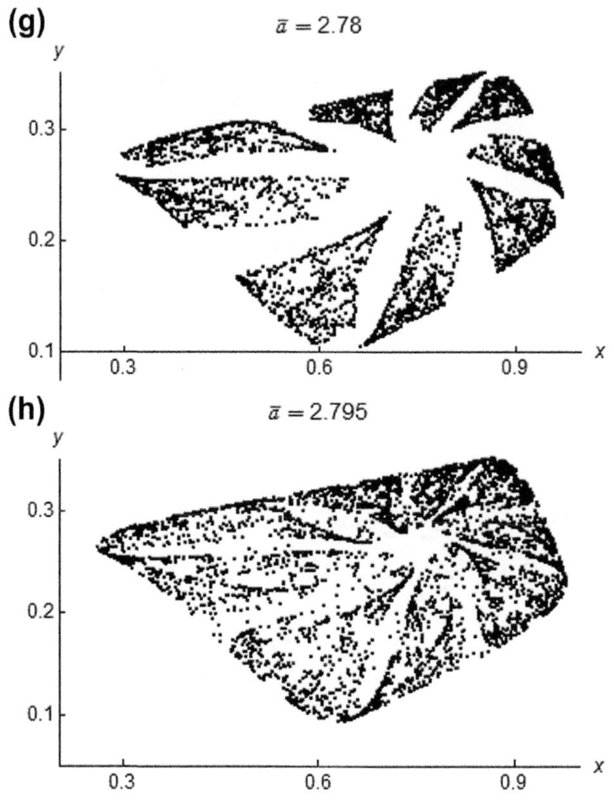


**FIGURE 6.3** (*continued*).

# 6.6 Conclusion

This chapter shows that density-dependent factors in this case prove to be responsible for the oscillation modes in the dynamic behavior of the population and may even lead to chaotic (pseudorandom) dynamic behavior. These results are shown in the simplest deterministic models.

The next sections of this study are devoted to the analysis of the joint dynamics of both the size and genetic structure of populations affected by both selective and density-dependent forces. Thus, for limited populations of a single generation, a progressive increase in average fitness may prove to be dissonant with the stability of population growth. This result of the discrete dynamic models clearly contradicts intuitive concepts on the increase of population stability with the growth of its average fitness. The mechanism of regular evolutionary complication of the dynamics of a structured population is also illustrated by considering the effect of natural selection.

## Acknowledgments

The present work was supported by the Russian Foundation for Basic Research (project 09-04-00146) and the Far Eastern Branch of the Russian Academy of Sciences in the framework of Program 25 of Presidium of RAS (project FEB RAS 11-01-98512, 11-III-B-01M-001).

## References

Cook, L.M., 1965. Oscillation in the simple logistic growth model. Nature 207, 316–317.

Frisman, E.Ya., Shapiro, A.P., 1982. Fluctuations in the size of isolated single-species populations and natural selection. Biom. J. 24 (6), 531–542.

Frisman, E.Ya., Zhdanova, O.L., 2009. Evolutionary transition to complex population dynamic patterns in a two-age population. Rus. J. Genet. 45 (9), 1124–1133.

Gaines, M.S., McClenaghay Jr., L.R., Rose, K.R., 1978. Temporal patterns in allozymic variation in fluctuating populations in *Microtus ochrogaster*. Evolution 32 (4), 723–739.

Getz, W.M., 2003. Correlative coherence analysis: variation from intrinsic and extrinsic sources in competing populations. Theor. Popul. Biol. 64 (1), 89–99.

Gottlieb, L.D., 1974. Genetic stability in a peripheral isolate of *Stephanomeriya exigus spc. coronaria* that fluctuates in population size. Genetics 76 (3), 551–556.

Hastings, A., 1992. Age dependent dispersal is not a simple process: density dependence, stability, and chaos. Theor. Popul. Biol. 41 (3), 388–400.

Kingman, J.F.C., 1961. A mathematical problem in population genetics. Proc. Camb. Philos. Soc. 57, 574–582.

Kooi, B.W., Kooijman, S.A.L.M., 1999. Discrete event versus continuous approach to reproduction in structured population dynamics. Theor. Popul. Biol. 56 (1), 91–105.

Lebreton, J.D., 1996. Demographic models for subdivided populations: the renewal equation approach. Theor. Popul. Biol. 49 (3), 291–313.

Lebreton, J.D., 2006. Dynamical and statistical models of vertebrate population dynamics. C. R. Biol. 329 (10), 804–812.

May, R.M., 1975. Biological populations obeying difference equations: stable points, stable cycles, and chaos. J. Theor. Biol. 51 (2), 511–524.

Moran, P.A.P., 1950. Some remarks on animal population dynamics. Biometrics 6, 250–258.

Ratner, V.A., 1973. A mathematical theory of evolution of Mendelian populations. In: Problems of Evolution, vol. 3. Nauka Pub, Novosibirsk, pp. 151–213 (in Russian).

Reluga, T.C., 2004. Analysis of periodic growth-disturbance models. Theor. Popul. Biol. 66 (2), 151–161.

Ricker, W.E., 1954. Stock and recruitment. J. Fisheries Res. Board Can. 11, 559–623.

Shapiro, A.P., 1972. To the question of cycles in recurring sequences. In: Management and Information. Far East Science Centre, USSR Academy of Sciences, Vladivostok, pp. 90–118 (in Russian).

Shapiro, A.P., 1980. On the existence of the first series of bifurcations in a uniparametric family of functions. In: Differential Equations and Functional Analysis. Far East Science Centre, USSR Academy of Sciences, Vladivostok (in Russian).

# 7 The Maximum Economic Yield Management of an Age-Structured Salmon Population

Anders Skonhoft
*ECONOMIC DEPARTMENT, NTNU, TRONDHEIM, NORWAY*

## 7.1 Introduction

For many years, the North Atlantic salmon (*Salmo salar*) has been one of the most important fish species in Norway because of its social, cultural, and economic importance. It was traditionally harvested for food, but is today most important to recreational anglers (NOU, 1999). The abundance has been declining during the last few decades, especially since the 1990s. There are a combination of various factors behind this development, such as sea temperature, diseases, and human activity, both in the spawning streams and through the strong growth of salmon sea farming (NASCO, 2004). As the wild stock began to decrease during the 1980s, the Norwegian government imposed gear restrictions to limit the marine harvest. Drift net fishing was banned in 1989, and the fishing season of bend net fishing, taking place in the fjords and close to the spawning rivers, has been restricted several times. At the same time, the sport fishing season in the spawning rivers has been subject to various restrictions (NOU, 1999). However, despite all these measures taken to secure and rebuild the stock, the abundance of wild salmon seems to be at only half the level experienced in the 1960s and 1970s. Today, farmed salmon is regarded as the main threat to the viability of the wild salmon population because of the spread of diseases through sea lice infection, escapees, and environmental pollution (Hindar et al., 2006; Liu et al., in press).

Wild salmon fishing has been analyzed in many papers from an economic perspective (see, e.g., Routledge, 2001; Laukkanen, 2001; Olaussen and Skonhoft, 2008). These are all studies based on a biomass approach where "a fish is a fish," while Kulmala et al. (2008) studied numerically an age-structured dynamic salmon model. The age model formulated in this chapter is much simpler than that of Kulmala et al. as we aim to say something *analytically* about the basic driving forces behind a harvest composition that maximizes the economic yield (*MEY*). For this reason, only biological equilibrium is considered. The analysis has similarities with Reed (1980) and Getz and Haight (1988),


**Models of the Ecological Hierarchy. DOI: http://dx.doi.org/10.1016/B978-0-444-59396-2.00007-9**
ISSN 0167-8892, 

but we study a different biological system in which all the spawning fish (i.e., salmon) die after spawning. This contrasts with Reed's model, where a fixed fraction of the spawning fish (e.g., cod) survives, and enters an older year class after spawning. While our analysis is directly related to Atlantic salmon, we will find that it fits various Pacific salmon species, such as pink and chum salmon, which also die after spawning (see, e.g., Groot and Margolis, 1991).

This chapter is organized as follows. In Section 7.2 the population model is formulated, where we consider two spawning, and hence two harvestable, age classes. In Section 7.3, we find the maximum sustainable economic yield fishing policy. The economic benefit of our selective harvesting scheme is compared to a uniform fishing pattern in Section 7.4.. The theoretical reasoning is numerically illustrated in Section 7.5while Section 7.6 finally summarizes and concludes this chapter.

## 7.2 Population Model

Atlantic salmon is an anadromous species that has a complex life cycle with several distinct phases. Freshwater habitat is essential in the early development stages, as this is where it spends the first 1 to 4 years from spawning to juvenile rearing before undergoing smoltification and seaward migration. Then, it stays for 1 to 3 years in the ocean for feeding and growing and, when mature, returns to its natal or "parent" river to spawn. After spawning, most salmon die, as less than 10% of the female salmon spawn twice (Mills, 1989). The Atlantic salmon is subject to fishing when it migrates back to its parent river. In Norway, most sea fishing takes place in fjords and inlets with wedge-shaped seine and bend nets. This fishing is commercial, or semicommercial. In the rivers, salmon are caught by recreational anglers with rods and hand lines. Recreational fishery is by far the most important activity from an economic point of view (NOU, 1999).

In what follows, a specific salmon population (with its native river) is considered in terms of a number of individuals at time $t$ structured into recruits $N_{0,t}$ (yr $< 1$), three young age classes, $N_{1,t}$ ($1 \leq$ yr $< 2$), $N_{2,t}$ ($2 \leq$ yr $< 3$), and $N_{3,t}$ ($3 \leq$ yr $< 4$), two adult spawning classes $N_{4,t}$ ($4 \leq$ yr $< 5$) (one sea winter, 1SW) and $N_{5,t}$ ($5 \leq$ yr $< 6$) (two sea winter, 2SW). Recruitment is endogenous and density dependent, and the 2SW has higher fertility than the 1SW. Natural mortality is fixed and density independent and, as an approximation, it is assumed that the whole spawning population dies after spawning. It is further assumed that the proportion between the two adult age classes is fixed. This proportion may be influenced by a number of factors, such as type of river ("small" salmon river vs. "large" salmon river) and environmental factors (NOU, 1999). As fishing takes place when the fish returns back to its native river, only the adult spawning classes $N_{4,t}$ and $N_{5,t}$ are subject to fishing.

With $B_t$ as the size of the spawning population, adjusted for different fertility among the two spawning classes, the stock–recruitment relationship is first given by:

$$N_{0,t} = R(B_t) \tag{1}$$

$R(B_t)$ may be a one-peaked value function (i.e., of the Ricker type) or it may be increasing and concave (i.e., of the Beverton–Holt type). In both cases, zero stock means zero recruitment, $R(0)=0$. The number of young is next defined as:

$$N_{a+1,t+1} = s_a N_{a,t}, \tag{2}$$

where $a=0$, 1, 2, and with $s_a$ as the age-specific natural survival rate, assumed to be density independent and fixed over time.

As indicated, only the spawning classes are subject to fishing mortality (marine as well as river fishing). With $0<\sigma<1$ as the fixed proportion of the adult stock that returns to spawn in the first year, the number of spawning fish of this part of the adult population (1SW) is:

$$N_{4,t+1} = s_3 N_{3,t}\sigma(1-f_{4,t}), \tag{3}$$

where $f_{4,t}$ yields the fishing mortality. Accordingly, $H_{4,t}=s_3 N_{3,t}\ \sigma\ f_{4,t}$ is the number of harvested 1SW fish in year $t$. As indicated, the parameter $\sigma$ may depend on various factors, but is considered as fixed and exogenous. The rest of this cohort $\tilde{N}_{4,t+1} = s_3 N_{3,t}(1-\sigma)$ stays one year more in the ocean. When subject to natural mortality, as well as subsequent fishing mortality, on migration back to spawning in the home river, the size of the next year's spawning population (2SW) becomes:

$$N_{5,t+1} = s_4 \tilde{N}_{4,t}(1-f_{5,t}). \tag{4}$$

Hence, $H_{5,t} = s_4 \tilde{N}_{4,t} f_{5,t}$ is the number of harvested 2SW salmon in year $t$. With $\gamma_4$ and $\gamma_5$ as the fecundity parameters of the 1SW and 2SW stocks, respectively, and where 2SW is more productive, $\gamma_5>\gamma_4$, the spawning population in year $t$ may be written as:

$$B_t = \gamma_4 N_{4,t} + \gamma_5 N_{5,t}, \text{ or}$$

$$B_t = \gamma_4 s_3 N_{3,t-1}\sigma(1-f_{4,t-1}) + \gamma_5 s_3 N_{3,t-2}(1-\sigma)s_4(1-f_{5,t-1}). \tag{5}$$

Equation (2) implies $N_{3,t+3}=s_0 s_1 s_2 N_{0,t}$, or

$$N_{3,t+3} = sR(B_t) \tag{6}$$

where $s=s_0 s_1 s_2$ comprises the previous years' survival rates. For given fishing mortalities, eqns (6) and (5) yield a system of two difference equations of degree five for the two variables $N_{3,t}$ and $B_t$.

As already indicated, we are concerned only with equilibrium fishing, or sustainable harvesting, in this chapter. The population equilibrium for fixed fishing mortalities is defined for $N_{3,t}=N_3$ and $B_t=B$ for all $t$ such that:

$$B = [\gamma_4 s_3\sigma(1-f_4) + \gamma_5 s_3(1-\sigma)s_4(1-f_5)]N_3, \tag{5'}$$

and

$$N_3 = sR(B) \tag{6'}$$

In what follows, Eqn (5′) is referred to as the *spawning constraint*, whereas Eqn (6′) represents the *recruitment constraint*. An internal equilibrium ($N_3>0$ and $B>0$) holds only if either $f_4$ or $f_5$, or both, are below one; that is, to exclude depletion, both mature classes

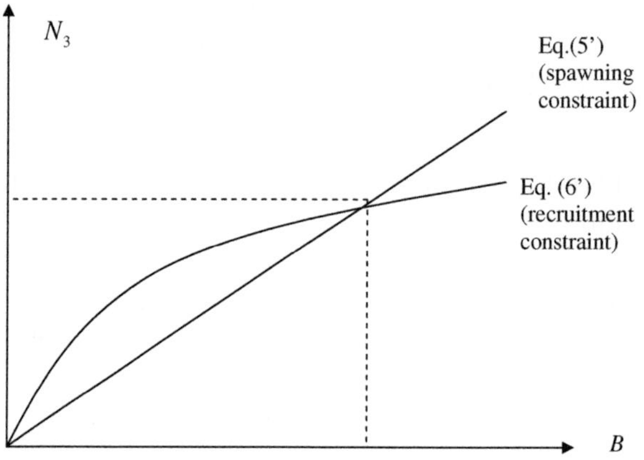


**FIGURE 7.1 Internal equilibrium for fixed fishing mortalities $0 \le f_4 \le 1$ and $0 \le f_5 \le 1$ (but not $f_4 = f_5 = 1$).** Beverton–Holt type recruitment function.

cannot be totally fished down. Notice that this is a necessary but not sufficient condition. Figure 7.1 illustrates the internal, unique equilibrium where the recruitment function is of the Beverton–Holt type, that is, $R(0) = 0$, $\partial R/\partial B_t = R' > 0$, and $R'' < 0$ (see also Section 7.5).

## 7.3 The Maximum Sustainable Economic Yield Harvesting Program

We start to analyze the optimal sustainable harvesting program under the assumption of perfect selectivity. With $w_5 > w_4$ as the fixed weights (kg/fish) of the young and old mature population, respectively, and $p_5 \geq p_4$ as the fishing values (NOK/kg), and where the 2SW is at least as valuable as the 1SW (Olaussen and Liu, 2011), $\pi = p_4 w_4 H_4 + p_5 w_5 H_5 = [p_4 w_4 s_3 \sigma f_4 + p_5 w_5 s_3 (1-\sigma) s_4 f_5] N_3$ then describes the yearly revenue in our salmon fishery. Therefore, the maximum sustainable economic yield problem is defined by finding fishing mortalities that maximize $\pi$ subject to the spawning constraint Eqn (5′) and the recruitment constraint Eqn (6′).

The Lagrangian of this problem may be written as $L = [p_4 w_4 s_3 \sigma f_4 + p_5 w_5 s_3 (1-\sigma) s_4 f_5] N_3 - \lambda[N_3 - sR(B)] - \mu\{B - [\gamma_4 s_3 \sigma (1-f_4) + \gamma_5 s_3 (1-\sigma) s_4 (1-f_5)] N_3\}$, where $\lambda > 0$ and $\mu > 0$ (both in NOK/fish) are the shadow prices of the recruitment and spawning constraints, respectively. Following the Kuhn–Tucker theorem (see, e.g., Sydsaeter et al., 2005), the first-order necessary conditions (assuming $N_3 > 0$ and $B > 0$) are:

$$\partial L/\partial f_4 = N_3(p_4 w_4 - \mu\gamma_4) \gtreqless 0; \; 0 \le f_4 \le 1, \tag{7}$$

$$\partial L/\partial f_5 = N_3(p_5 w_5 - \mu\gamma_5) \gtreqless 0; \; 0 \le f_5 \le 1, \tag{8}$$

$$\partial L/\partial N_3 = p_4 w_4 s_3 \sigma f_4 + p_5 w_5 s_3(1-\sigma)s_4 f_5 - \lambda + \mu[\gamma_4 s_3 \sigma(1-f_4) + \gamma_5 s_3(1-\sigma)s_4(1-f_5)] = 0, \quad (9)$$

and

$$\partial L/\partial B = \lambda s R'(B) - \mu = 0. \quad (10)$$

Control condition (7) indicates that the fishing mortality of the 1SW population should take place at the point where the marginal biomass value gain is equal, below, or above its marginal biomass harvest loss, determined by the fecundity parameter and evaluated by the spawning constraint shadow price. Condition (8) is analogous for the 2SW. The stock condition (9) says that the harvestable population should be managed so that the recruitment constraint shadow price $\lambda$ is equal to the total marginal harvest value gain plus the total marginal spawning biomass value gain, evaluated at its shadow price. Finally, stock condition (10) indicates that the recruitment growth, evaluated at its shadow price, should be equal to the spawning constraint shadow price $\mu$.

From the control conditions (7) and (8), it is observed that only the weight value–fecundity ratio $p_i w_i/\gamma_i$ $(i=4{,}5)$ determines the fishing mortality and the fishing composition and, hence, no other factors play a *direct* role. This outcome differs from the seminal Reed (1980) paper, who found that weight together with natural mortality ("biological discounted" value) directly determined the fishing composition. As indicated, the reason for this discrepancy is the different biological characteristics of the fish stocks. While the mature fish die after spawning in our salmon model, the spawning fish survive and enter older age classes in the Reed model.

Weight and fertility are related and larger and older fish in most instances, if not always, indicate higher fertility (e.g., Getz and Haight, 1989). According to McGinnity et al. (2003), this relationship for wild salmon is described such that fertility is an increasing, strictly concave function of weight (and age) and hence the weight–fertility ratio *increases* with weight (see also numerical section below). With $w_5/\gamma_5 > w_4/\gamma_4$ together with $p_5 \geq p_4$ and hence a higher marginal value gain–loss ratio for fishing the 2SW spawning fish, the maximum sustainable yield harvesting policy, given by conditions (7) and (8), indicates a higher fishing mortality for the 2SW than the 1SW. This is stated as the following proposition:

Proposition 1. With a higher weight–fecundity ratio for the old adult subpopulation, the maximum yield harvesting policy is governed by a higher fishing mortality of the old subpopulation.

There are three possible cases, all corner solutions, that may represent this optimal policy: (i) $f_5^* = 1$ and $0 < f_4^* < 1$, (ii) $f_5^* = 1$ and $f_4^* = 0$, and (iii) $0 < f_5^* < 1$ and $f_4^* = 0$ (superscript "*" indicates optimal values). The spawning constraint (5′), written as $N_3 = B/[\gamma_4 s_3 \sigma(1-f_4) + \gamma_5 s_3(1-\sigma)s_4(1-f_5)]$, will then be steeper in case (i) than in case (ii), which again will be steeper than that in case (iii). See Fig. 7.1.

Therefore, when taking the recruitment constraint (6′) into account (again, see Fig. 7.1), we find that the size of the spawning population $B^*$ as well as the harvestable stock $N_3^*$ will be highest with harvest option (iii) and lowest if case (i) represents the optimal policy. If $p_5w_5/\gamma_5$ is substantially higher than $p_4w_4/\gamma_4$, we may intuitively suspect that it is beneficial for the manager to invest in the salmon population by leaving the young mature population unexploited and harvesting the whole old adult population. Hence, case (ii), where $f_5^* = 1$ and $f_4^* = 0$, should maximize the sustainable yield. On the other hand, with a "small" weight value–fertility ratio difference, either case (i), with harvesting of both mature populations, or case (iii), with harvesting of the old mature population only, should possibly represent the optimal solution.

In case (i) with $f_5^* = 1$ and $0 < f_4^* < 1$, the spawning constraint shadow price is determined through condition (7) as $\mu^* = p_4w_4/\gamma_4$. A combination with Eqn (9) yields $\lambda^* = s_3[p_4w_4\sigma + p_5w_5(1-\sigma)s_4]$. When inserted into Eqn (10), the optimal spawning biomass is governed by $R'(B^*) = \mu^*/s\lambda^* = R'(B^*) = \mu^*/s\lambda^* = 1/ss_3[\gamma_4\sigma + (p_5w_5\gamma_4/p_4w_4)\ (1-\sigma)s_4]$. $N^*$ then follows from Eqn (6′), and we find the fishing mortality $f_4^*$ next through Eqn (5′), $B^* = s_3\gamma_4\sigma(1-f_4)N_3^*$. If case (ii) with $f_5^* = 1$ and $f_4^* = 0$ represents the optimal policy, the spawning constraint (5′) reads $B = \gamma_4 s_3\sigma N_3$. Therefore, in this case, Eqn (5) together with the recruitment constraint (6′) $N_3 = sR(B)$ alone determines $N_3^*$ and $B^*$. In case (iii), where $0 < f_5^* < 1$ and $f_4^* = 0$, the spawning constraint shadow price is determined through condition (8) as $\mu^* = p_5w_5/\gamma_5$, whereas Eqn (9) determines the recruitment constraint shadow price as $\lambda^* = s_3p_5w_5[\gamma_4\sigma/\gamma_5 + (1-\sigma)s_4]$. Therefore, just as in case (i), the size of the spawning biomass is found through Eqn (10) now as: $R'(B^*) = 1/ss_3[\gamma_4\sigma + \gamma_5(1-\sigma)s_4]$ while $N^*$ next follows from Eqn (6′). The optimal fishing mortality in case (iii) is again determined by the spawning constraint (5′) as $B^* = s_3[\gamma_4\sigma + \gamma_5(1-\sigma)s_4(1-f_5)]N_3^*$.

As indicated, changes in fishing prices may shift the optimal harvest policy from targeting only 2SW to targeting both stocks, and vice versa. While price shifts have no effects within case (ii) and case (iii), we find $\partial R'(B^*)/\partial p_4 > 0$ within case (i) as $R'(B^*) = 1/ss_3[\gamma_4\sigma + (p_5w_5\gamma_4/p_4w_4)(1-\sigma)s_4]$ then describes the optimal spawning stock. Therefore, $\partial B^*/\partial p_4 < 0$ and also $\partial N_3^*/\partial p_4 < 0$ hold. The effects of $p_5$ are of the opposite. Because the spawning constraint (5′) writes $B = s_3\gamma_4\sigma(1-f_4)N_3$ in case (i), a higher $p_4$ is satisfied with a higher $f_4^*$. Not surprisingly, with $p_4 = p_5$ we find zero stock and harvesting effects of price changes. There are then simply no price tradeoffs present. The price effects found here are generally different from the standard biomass (lumped parameter) fishery model (e.g., Clark, 1990) where changing harvest value, in the absence of stock-dependent harvesting costs (and other possible stock values), have no effects when determining the maximum sustainable yield policy.

## 7.4 Nonselective Fishing Pattern

It may also be of interest to find the maximum economic yield when our optimal selective fishing pattern is replaced by an optimal fishing pattern with similar, or uniform, fishing mortalities. This scheme may hence indicate a nonselective fishing situation where "a fish is a fish", as considered in biomass models used in the traditional bioeconomic analysis (e.g., Clark, 1990). Our optimizing problem is then described by the goal of maximizing

$\pi = [p_4w_4s_3\sigma f_4 + p_5w_5s_3(1-\sigma)s_4f_5]N_3$ subject to the spawning and recruitment constraints, Eqns (5′) and (6′), respectively, and $f_4 = f_5 = f$.

The Lagrangian of this problem may be written as $L = [p_4w_4s_3\sigma + p_5w_5s_3(1-\sigma)s_4]fN_3 - \lambda[N_3 - sR(B)] - \mu\{B - [\gamma_4s_3\sigma + \gamma_5s_3(1-\sigma)s_4](1-f)N_3\}$ when inserting for the uniform fishing pattern. The first-order necessary conditions (again with $N_3 > 0$ and $B > 0$) read:

$$\partial L/\partial f = N_3\{[p_4w_4\sigma + p_5w_5(1-\sigma)s_4)] - \mu[(\gamma_4\sigma + \gamma_5(1-\sigma)s_4]\} = 0;\ 0 < f < 1,;\ 0 < f < 1 \quad (11)$$

and

$$\partial L/\partial N_3 = [p_4w_4s_3\sigma + p_5w_5s_3(1-\sigma)s_4]f - \lambda + \mu[\gamma_4s_3\sigma + \gamma_5s_3(1-\sigma)s_4](1-f) = 0 \quad (12)$$

together with Eqn (10). The control condition (11) must hold as equation as stock depletion can never be beneficial under this economic yield scenario with zero discount rent. With $N_3 > 0$, this equation may also be written as $[p_4w_4\sigma/s_4 + p_5w_5(1-\sigma)] = \mu^*[\gamma_4\sigma/s_4 + \gamma_5(1-\sigma)]$ after some small rearrangements. Therefore, the optimal uniform fishing pattern may be characterized as a situation where the "biological discounted" marginal harvesting value (marginal gain) equalizes the "biological discounted" fertility (marginal loss), evaluated by the spawning constraint shadow value. This equation also determines the optimal spawning constraint shadow price, $\mu^*$.

The uniform harvesting pattern can never be more economically beneficial than the selective harvesting scheme as one more constraint is included in the nonselective maximization problem. When combining Eqns (11) and (12) we find $\lambda = \mu s_3[\gamma_4\sigma + \gamma_5(1-\sigma)s_4]$ after some small rearrangements. Inserted into condition (10), the size of the spawning biomass is next described as $R'(B^*) = 1/ss_3[\gamma_4\sigma + \gamma_5(1-\sigma)s_4]$. Therefore, we find exactly the same optimal spawning population as in the above selective harvesting case (iii) and the marginal harvesting value (marginal gain) has no influence on the optimal uniform fishing pattern. This is stated as:

Proposition 2. The marginal harvesting value (marginal gain) has no influence on the optimal uniform fishing pattern.

Different fish prices and differences in fish weight among 1SW and 2SW have therefore no influence on the optimal fishing pattern. Moreover, if prices, fish weights, and fertility are such that case (iii) represents the optimal harvesting policy under the assumption of selective harvesting, the size of the spawning biomass and the degree of exploitation will be similar under uniform harvesting.

## 7.5 Numerical Illustration

The above theoretical reasoning will now be illustrated numerically. Hansen et al. (1996) estimated a salmon recruitment function for a small river in Norway (the Imsa River) based on the Shepherd recruitment function, which includes three parameters. In our model, we choose a simpler approach and use the Beverton–Holt function (cf. Fig. 7.1). This function may be specified as $R(B) = rB/(1 + (B/K))$, with $r > 0$ as the intrinsic growth rate, or maximum number of recruits per (fertility adjusted) spawning salmon, and $K > 0$ as the stock level for which density-dependent mortality equals density-independent mortality. The size of $rK$ yields the maximum number of recruits and scales the system ("size of the river"), which is assumed to be 40,000 (number of recruits). The value of $r$ indicates the "quality" of the river, and we choose $r = 400$ (number of recruits per spawning salmon). Then, we find $K = 100$. The fertility–weight relationship is based on McGinnity et al. (2003) and given as $\gamma = 4.83w^{0.87}$ (weight $w$ is here measured in gramme). In the river Imsa in Norway, the same functional form is estimated as $\gamma = 5.10w^{0.86}$ (personal communication, senior researcher Ola Diserud, Norwegian Institute of Natural Research, Trondheim). When normalizing the fertility parameter of 1SW to one, $\gamma_4 = 1$, and using the McGinnity et al. functional form, we find $\gamma_5 = 2.4$ under the assumption of fishing weights $w_4 = 2.0$ and $w_5 = 5.5$ (kg/salmon). These weights fit a "typical" medium-sized Norwegian salmon river (NOU, 1999). The survival parameters are based on NOU (1999), whereas the fishing prices are related to recreational fishery, which, as indicated, is far more important economically than the marine fishery. The assumption here is that the fishing permit price in a reasonably good river is about 200 NOK/day (see also Olaussen and Liu, 2011). Based on average catch success, this permit price may translate into fishing prices in the range of 100–400 (NOK/kg), or even higher. We assume the same price for old and young and use $p_4 = p_5 =$ 150 (NOK/kg). We then have $p_5w_5/\gamma_5 = 343.8 > p_4w_4/\gamma_4 = 300$ (NOK/fish). Table 7.1 summarizes the baseline parameter values.

Because the weight–fertility value ratio is highest for the 2SW population, the economic yield maximizing fishing mortality will be highest for this old adult population (Proposition 1). Table 7.2 (first row) demonstrates where case (i) with $f_5^* = 1$ and $f_4^* = 0.31$ yields the optimal fishing mortality. Reducing the gain–loss ratio of the young mature subpopulation by lowering $p_4$ while keeping all other parameters at their baseline values, and hence increasing the discrepancy between $p_5w_5/\gamma_5$ and $p_4w_4/\gamma_4$, leads to an

**Table 7.1** Biological and Economic Baseline Parameter Values

| Parameter | Description | Value |
|---|---|---|
| $S$ | Natural survival rate young | 0.05 |
| $s_3$ | Natural survival rate young adult | 0.5 |
| $s_4$ | Natural survival rate old adult | 0.5 |
| $R$ | Intrinsic growth rate recruitment function | 400 (No. of recruits/fertility adjusted spawner) |
| $K$ | Scaling parameter recruitment function | 100 (No. of spawners) |
| $\sigma$ | Migration parameter | 0.5 |
| $w_4$ | Weight young adult | 2.0 (kg/fish) |
| $w_5$ | Weight old adult | 5.5 (kg/fish) |
| $\gamma_4$ | Fecundity parameter young adult | 1.0 |
| $\gamma_5$ | Fecundity parameter old adult | 2.4 |
| $p_4$ | Fish price young adult | 150 (NOK/kg) |
| $p_5$ | Fish price old adult | 150 (NOK/kg) |

**Table 7.2** Maximum Economic Yield

| | $f_4^*$ | $f_5^*$ | $N_3$(#) | *B* (#) | $H_4^*$ (#) | $H_5^*$ (#) | $\pi^*$ (1000 NOK) |
|---|---|---|---|---|---|---|---|
| Baseline values | 0.31 | 1.00 | 1420 | 245 | 110 | 178 | 179 |
| 200% reduction price young adult ($p_4 = 50$) | 0.00 | 1.00 | 1600 | 400 | 0 | 200 | 165 |
| 40% reduction natural survival rate young ($s = 0.03$) | 0.11 | 1.00 | 751 | 167 | 21 | 94 | 84 |
| Uniform fishing Baseline values | 0.70 | 0.70 | 1397 | 232 | 244 | 122 | 173 |

optimal fishing policy described by case (ii), with no harvesting of young fish (row two). Row three indicates what happens when the natural survival rate of the young *s* is reduced while all other parameters are kept at their baseline values. Such a reduction may be the result of infection through transmission of lice from farmed salmon. Indeed, as indicated in the introductory section, this is considered to be one of the most important threats to the wild Atlantic salmon (see, e.g., Verspoor et al. 2003). A 40% reduction yields quite dramatic effects. The spawning biomass declines significantly and the profit is reduced by more than 50%. Again, case (i) with harvesting of the entire old adult population represents the optimal fishing policy.

The last row in Table 7.2 finally illustrates the optimal nonselective and uniform fishing pattern under the baseline parameter values scenario and where the marginal fishing value (marginal gain) plays no role (Proposition 2).We find the fishing mortality to be 0.70 and the fish abundance reduces somewhat to the baseline selective scheme (first row). As expected, the profit is lower, but the reduction is quite insignificant. The changes in the size of the harvestable population $N_3^*$ and spawning population $B^*$ are also quite modest. However, when case (ii) with $f_5^* = 1$ and $f_4^* = 0$ represents the optimal scheme due to increased weight–fecundity discrepancy (row two), the difference in the spawning stock becomes more profound as the uniform fishing mortality is still 0.70 in the uniform case. Changes in fishing prices play no role here.

## 7.6 Concluding Remarks

In this chapter, we have from a theoretical point of view, studied the maximum sustainable yield management of an age-structured wild Atlantic salmon (*S. salar*) population with two spawning and harvestable classes. Under the assumption of perfect fishing selectivity, the basic finding is that the weight–fecundity ratio discrepancy between the harvestable classes determines the optimal fishing mortality and the fishing composition, and no other factors play a *direct* role. This is stated as Proposition 1. Our analysis and findings are based on the Atlantic salmon, but the results will also apply to the various Pacific salmon stocks, which also die after spawning.

The model also studied fishing under imperfect selectivity and similar fishing mortalities in the harvest. We find here that the marginal fishing value (gain) has no influence on the optimal fishing pattern. This is stated as Proposition 2. Only the marginal loss (fertility) counts, together with survival and composition of the 1SW and 2SW stock counts. The uniform fishing pattern yields lower profit than that under the perfect selectivity pattern. In the numerical illustration, this loss is quite small. However, as also demonstrated, the differences in the size of the spawning stock and harvestable stock size may be quite significant.

## References

Clark, C., 1990. Mathematical Bioeconomics. John Wiley Intersicience, New York.

Getz, W., Haight, R., 1988. Population Harvesting. Princeton University Press, Princeton.

Groot, G., Margolis, L. (Eds.), 1991. Pacific Salmon Life Histories. UBC Press, Vancouver.

Hansen, L.P., Jonsson, B., Jonsson, N., 1996. Overvåking av laks fra Imsa og Drammenselva (in Norwegian). NINA, Trondheim. Oppdragsmelding 401.

Hindar, K., Flemming, I., McGinnity, P., Diserud, O., 2006. Genetic and ecological effects of farmed salmon on native salmon: modeling from experimental results. ICES J. Mar. Sci. 63, 1234–1247.

Kulmala, S., Laukkanen, M., Michielsens, C., 2008. Reconciling economic and biological modelling of a migratory fish stock: optimal management of the Atlantic salmon fishery in the Baltic Sea. Ecol. Econ. 64, 716–728.

Laukkanen, M., 2001. A bioeconomic analysis of the Northern Baltic salmon fishery. Environ. Resour. Econ. 188, 293–315.

Liu, Y., Diserud, O., Hindar, K., Skonhoft, A. An ecological model on the effects of interactions between escaped and wild salmon (*Salmo salar*). Fish and Fisheries, in press.

McGinnity, P., et al., 2003. Fitness reduction and potential extinction of wild populations of Atlantic salmon, *Salma salar*, as a result of interactions with escaped farm salmon. Proc. Royal Soc. B 270, 2443–2450.

Mills, D., 1989. Ecology and Management of Atlantic Salmon. Chapman and Hall, New York.

NASCO 2004. Report on the activities of the North Atlantic salmon conservation organization. 2002–2003. (www.nasco.int).

NOU, 1999 (in Norwegian). Til laks åt alle kan ingen gjera? vol. 9. Norges Offentlige utredninger, Oslo.

Olaussen, J.O., Liu, Y., 2011. On the willingness-to-pay for recreational fishing. Escaped farmed versus wild Atlantic salmon. Aquacult. Econ. Manage. 15, 245–261.

Olaussen, J.O., Skonhoft, A., 2008. A bioeconomic analysis of a wild Atlantic salmon recreational fishery. Mar. Resour. Econ. 23, 119–139.

Reed, W.J., 1980. Optimum age-specific harvesting in a nonlinear population model. Biometrics 36, 579–593.

Routledge, R., 2001. Mixed-stock vs. terminal fisheries: a bioeconomic model. Nat. Resour. Model. 14, 523–539.

Sydsaeter, K., Hammond, P., Seierstad, A., Strom, A., 2005. Further Mathematics for Economic Analysis. Prentice Hall, Harlow England.

Verspoor, E., Stradmeyer, L., Nielsen, J. (Eds.), 2003. The Atlantic Salmon. Genetics, Conservation and Management. Blackwell Publishing, New York.

# 8

# Use of Tracking System Data for Individual-based Modeling of Sweetfish (*Plecoglossus altivelis*) Behavior

Masatoshi Denda*, Wataru Koketsu†, Junji Miwa*

*WATER ENVIRONMENT GROUP RIVER RESTORATION TEAM, PUBLIC WORKS RESEARCH INSTITUTE, 1-6 MINAMIHARA, TSUKUBA CITY, IBARAKI 305-8516, JAPAN,

†OYO CORPORATION ECOLOGY AND CIVIL ENGINEERING RESEARCH INSTITUTE, 275 AZA ISHIBATAKE OAZA NISHIKATA, MIHARU-MACHI, TAMURA-GUN, FUKUSHIMA 963-7722, JAPAN

## 8.1 Introduction

The freshwater fish sweetfish, *Plecoglossus altivelis altivelis*, is a symbolic fish that inhabits Japanese rivers. Sweetfish plays an important role in stream ecology and is an important fisheries resource. In particular, sweetfish is an important species in the Japanese rivers from the viewpoint of fisheries resource conservation and fish community conservation. Sweetfish is an important resource for inland water fisheries, and leisure fishing for sweetfish is popular in Japanese rivers. Also, sweetfish is an indicator species of habitat continuity due to its life cycle. In autumn, sweetfish spawn at the middle and lower reaches of the river. New generations drift to the sea after hatching and grow until spring. New generations of sweetfish migrate to the upper and middle reaches of the river from spring to summer. In summer, the new generation feeds on algae on rocks and grows. Sweetfish prepare for the migration to the middle and lower reaches until autumn. Habitat continuity at the basin scale is necessary for the completion of the life history. Therefore, sweetfish is a good indicator of habitat continuity.

Hence, spawning areas, breeding areas, and continuous migration routes of sweetfish have been considered for conservation.

Many studies have been conducted on sweetfish, indicating a great deal of interest in the life history and population conservation of this species. The ecology and fisheries management of sweetfish have been well researched (Kawanabe, 1957; Kuroki et al., 2006; Sekiya et al., 2005). The majority of this previous research has focused on the behavioral characteristics of sweetfish. However, conservation and habitat restoration are necessary

Models of the Ecological Hierarchy, DOI: http://dx.doi.org/10.1016/B978-0-444-59396-2.00008-0 
ISSN 0167-8892, 

to sustain populations dependable on this fish. Before we begin restoring habitats and conserving sweetfish populations, we need to determine the relationships between sweetfish behavioral characteristics and habitat suitability.

Several previous studies have focused on the relationship between behavioral characteristics and habitat suitability in other fish species. Goodwin et al. (2006) tracked trout using a sonic tag system and clarified the relationships between the environmental conditions and the trout behavior. They then modeled the behavior of trout and selected the most appropriate method for the conservation of the species. Habitat restoration for sweetfish in Japan is conducted on the basis of the results of previous studies on other species. We have faced a number of difficulties in conducting research on sweetfish. For example, we cannot use good tracking methods such as the sonic tag system because, in the rivers of Japan, bubbles prevent signals from being transmitted; thus, no effective method is available for tracking sweetfish.

Previously, sweetfish behavior has been studied using manual methods such as direct catching, ID tagging (mark recapture), visual observations by divers, and counting the fish run. These methods are not suitable in times of flood, at night, or for long-term observations because of the several limitations of tracking fish behavior.

Considering these research difficulties, the Public Works Research Institute improved the advanced telemetry system (ATS) that targets overland mammals or large freshwater fish to allow tracking of sweetfish behavior. Since 2008, we have been carrying out collaborative research as part of the program "Research for upgrading the utility of ATS" and have been improving the ATS to make it a more adaptable system for the behavioral study of sweetfish.

The purpose of this collaborative research is to track sweetfish behavior with spatially and temporally accurate data, i.e., to establish a system that can track fish every few minutes, with an error of less than 15 m. By establishing this system, detailed sweetfish behavioral data can be collected, which was not possible by previous methods. Specifically, it will be possible to collect continuous data at night, which was difficult with manual surveillance. These continuous data will help us understand the detailed behavioral characteristics of sweetfish, and when coupled with further data on sweetfish ecology, they will contribute toward river improvement.

In this study, we report the tracking results of sweetfish from a collaborative research program conducted in Chikuma River. We analyzed the behavioral characteristics of sweetfish and compared our results to those of previous studies.

## 8.2 Method

### 8.2.1 Summary of Study Area

The study was conducted at Nezumi (Fig. 8.1), a site located in the Chikuma River, Sakaki, Nagano Prefecture, Japan. The field survey was conducted from August 22 to September 15, 2009.

This study focused on the behavioral characteristics of sweetfish before the migration to the middle and lower reaches. Considering the life cycle of sweetfish, habitat condition

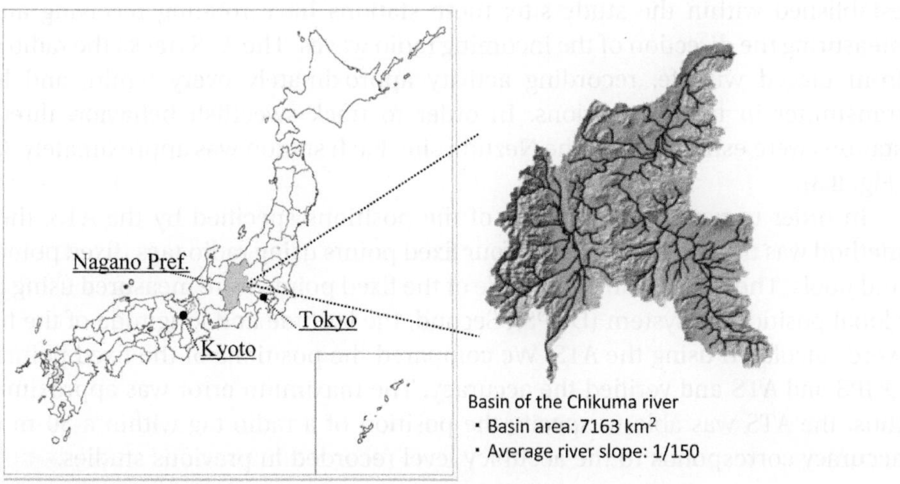


**FIGURE 8.1** General description of the study site. For color version of this figure, the reader is referred to the online version of this book.

just before the migration to the middle and lower reaches for reproduction is very important for the conservation of the sweetfish population.

Nezumi is located approximately 98 km from the prefecture boundary and flows down to the Sakaki Town agriculture area. The average river width is approximately 100 m, and the longitudinal gradient is approximately 1/200. The riverbed of the study site is mainly composed of gravel, and the landscape is mainly of the sandbar type.

## 8.2.2 Summary and Accuracy of the ATS

ATS (Fig. 8.2) is an automatic animal behavior tracking system that conducts measurements on the basis of a triangular survey. For ATS tracking, multiple receiving stations are

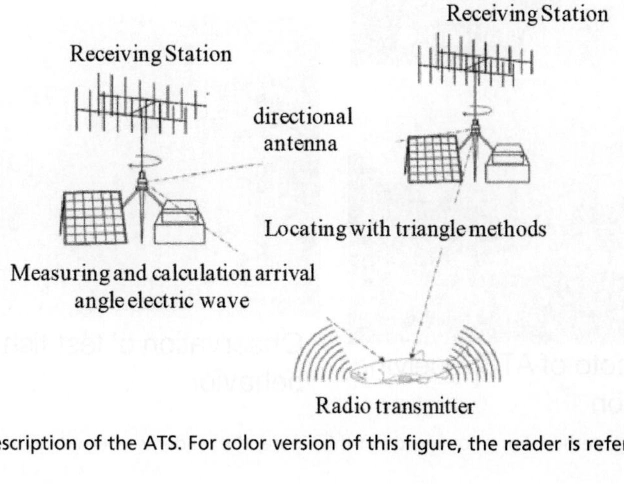


**FIGURE 8.2** General description of the ATS. For color version of this figure, the reader is referred to the online version of this book.

established within the study site; these stations have rotating receiving antennas for measuring the direction of the incoming radio waves. The ATS tracks the radio waves sent from tagged wildlife, recording activity approximately every 5 min, and locates the transmitter in two dimensions. In order to track sweetfish behavior, three receiving stations were established at the Nezumi site. Each station was approximately 350 m apart (Fig. 8.3).

In order to verify the accuracy of the positions specified by the ATS, the following method was used. First, we set up four fixed points using radio tags (fixed points) at riffles and pools. The latitude and longitude of the fixed points were measured using differential global positioning system (DGPS). Second, the latitude and longitude of the fixed points were calculated using the ATS. We compared the positions of the fixed points using the DGPS and ATS and verified the accuracy. The maximum error was approximately 30 m; thus, the ATS was able to specify the position of a radio tag within a 30-m error. This accuracy corresponds to the accuracy level recorded in previous studies.

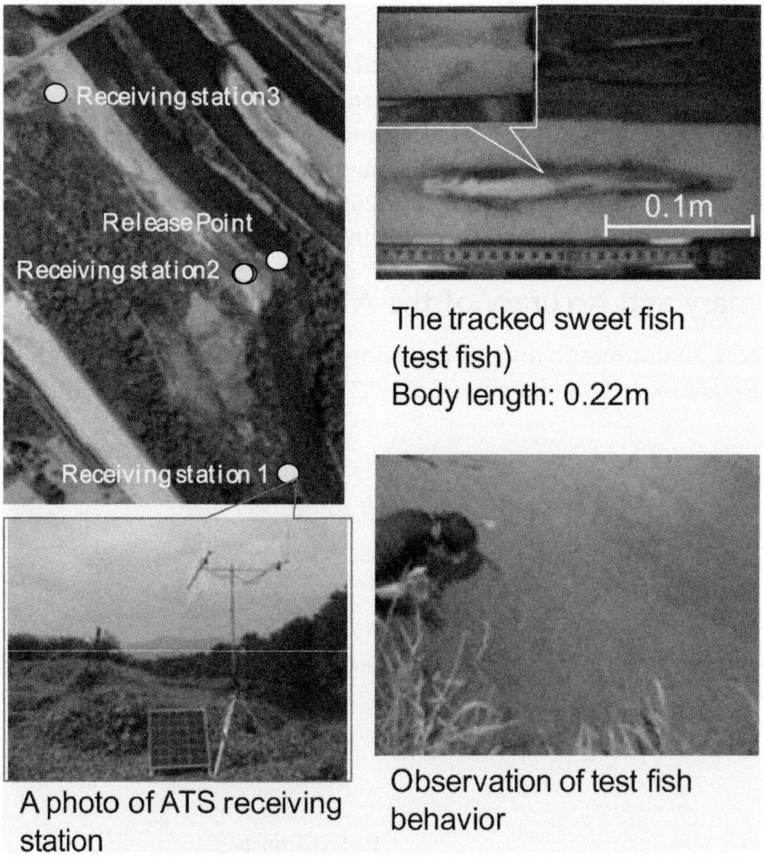


**FIGURE 8.3** Field survey method. For color version of this figure, the reader is referred to the online version of this book.

## 8.2.3 Field Survey

### *8.2.3.1 Tracking Fish Behavior using the ATS*

Each radio tag had a diameter of 8.2 mm, length of 19 mm, and weight in water of 1 g (LOTEK, MCFT-3 GM). Sweetfish (test fish) were purchased at a fish trap and harvest store located near the study site, and the test fish were immediately transported to the study site. Surgical implantation of the radio tag into each test fish was conducted at the study site. A 15-mm incision was made in the abdominal area of the sweetfish using a surgical knife, the radio tag was inserted, and the wound was sutured. Following surgery, the test fish were nursed for 20 h in a cage set up within the Chikuma River. Two test fish were released at each of the study sites on August 22, 2009.

The release point was approximately the center of each study site, and the tracking measurement began when the test fish were released. The weight of the tag was less than 2% of the test fish body weight, and the test fish behavior after release seemed normal, as per the visual observations of divers. It is estimated that the effect of radio-tagging on sweetfish behavior was very low.

## 8.2.4 Data Analysis

### *8.2.4.1 Screening of Test Fish Behavior Data*

The ATS measures the incoming direction of faint radio waves that pass through the water; therefore, it is largely affected by reflection waves from land-based buildings and structures. At the Nezumi site, there were large bridge girders, and these structural objects could have affected the accuracy of the recorded direction and angle of the incoming radio waves. Data recorded by the ATS included less-reliable data that may have been affected by structural objects. To investigate whether the data were less reliable, data screenings were undertaken.

### *8.2.4.2 Reconstructing Water-Current Distribution using Hydro Simulation*

In order to reconstruct the water-current distribution, we conducted a horizontal hydro simulation. Discharge stability should be considered for simulation; therefore, we examined the flow regime at the study area and determined that the flow regime was stable for this study. We multiplied the river basin area rate by the discharge of the nearest observation station to estimate the average discharge for the study period. Then, we conducted hydro simulations as upstream boundary discharge.

### *8.2.4.3 Analysis of the Relationship between Sweetfish Behavior and Flow Condition*

We analyzed the basic spatial preferences of test fish (current velocity and water depth). We imported the behavioral data of the test fish into a Geographical Information System and overlaid the behavioral data on flow conditions calculated using the hydro simulation. We analyzed the behavioral characteristics of the test fish in terms of their home range and preferences for current velocity and water depth. Home range was analyzed using the minimum convex polygon method. For current velocity and water depth data, we merged all the behavioral data with the current velocity and determined the preferences of the sweetfish.

A previous study reported that sweetfish change their behavior during sunrise and sunset. For the purposes of the current study, we defined four time periods: time period 1, from 0500 to 1700; time period 2, from 1700 to 2000; time period 3, from 2000 to 0000; and time period 4, from 0000 to 0500 the next day. Characteristics of sweetfish behavior were analyzed for each time period.

## 8.2.5 Data Analyses

### *8.2.5.1 Characteristics of this Study Model*

The main aim of the current study was to model the relationship between sweetfish behavior and its physical environment. Although individual-based models (IBMs) have been built to model fish behavior, they do not incorporate physical environments in detail. The trout IBMs developed by Railsback and Harvey (2002) were designed to predict how populations respond to river management. This study is a good example of how to overcome modeling problems. On the basis of the study by Railsback and Harvey (2002), we applied hydro simulation to IBMs. We modeled the relationship in two steps: the first step was to model the physical environment using the hydro simulation, and the second step was to model sweetfish behavior in response to the physical environment by using IBMs.

In the hydro simulation models, we modeled topography and calculated flow conditions (current velocity and water depth) using hydro simulation. In the IBMs, we analyzed sweetfish behavior and defined behavioral rules on the basis of these relationships. Finally, we modeled the behavior of virtual sweetfish (VSF) under varying flow conditions.

### *8.2.5.2 Application of Object Orientation to VSF and its Purpose*

Various body characteristics influence the behavioral ability of wild animals. These body characteristics differ between individuals, resulting in populations composed of individuals with different behavioral abilities. When we model a population, we can program and manage each individual; however, this is expensive and involves a great deal of effort. Therefore, we used "object-oriented" and "class-concept" technologies to efficiently model individual behavior (Fig. 8.4).

The application of object-oriented technology to ecological modeling has been dealt within detail in Bian's (2003) research, and we used both object-oriented and class-concept technologies to describe the behavior of individual sweetfish in our model. We used class as a template of the biota. When modeling sweetfish behavior, we changed the attributes of the class, creating a VSF. We then made the VSF swim in rivers modeled using hydro simulation.

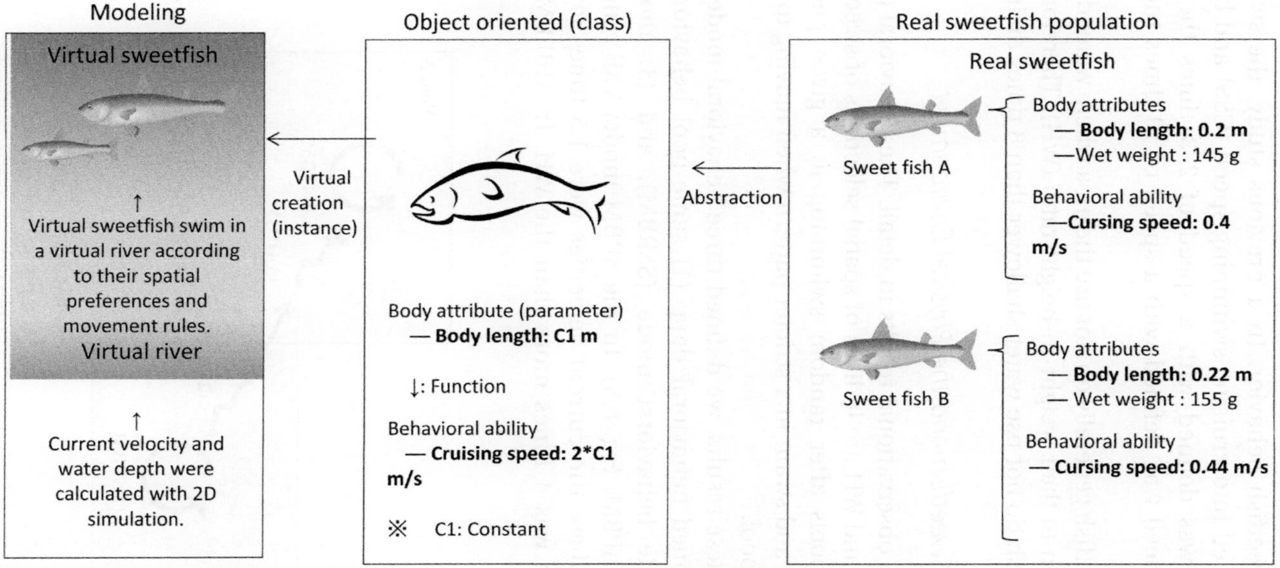


**FIGURE 8.4** Sweetfish classes. For color version of this figure, the reader is referred to the online version of this book.

### 8.2.5.3 Swimming Ability in the Model

In this study, we focused on the swimming ability and available swimming area as characteristics of sweetfish behavior. In a previous study, the swimming ability of sweetfish was classified into cruising swimming speed (CSS) and burst swing speed. The cruising speed was defined with a speed of 2–3 times the body length per second. The burst speed was defined with a speed of 10 times the body length per second.

Furthermore, sweetfish generally do not use the area where water depth is remarkably shallow in comparison to their weight-to-height ratio (WH). Therefore, we set a behavioral rule that sweetfish do not use water shallower than 8 times their WH.

### 8.2.5.4 Behavior of Sweetfish and the Physical Environment

Here, we describe the observational results in detail. The behavioral data indicated that (1) swimming ability and WH are indices of spatial selections of sweetfish, (2) sweetfish move to upstream pools after random swimming in a given period of time, and (3) sweetfish selected and swam in a stricter pattern when moving to another upstream pool or downstream pool.

On the basis of these results, we defined three behavioral modes that correspond to the above-mentioned behavioral data: (1) same-pool behavioral mode (SPBM), (2) seek-shallow-riffle behavioral mode (SSRBM), and (3) move-between-pools behavioral mode (MBPBM, Fig. 8.5). In the SPBM mode, VSF randomly selected to swim inside pools where the current velocities were 1.5 times lower than the CSS and the water depth was 4 times more than the WH. In MBPBM, VSF selected to

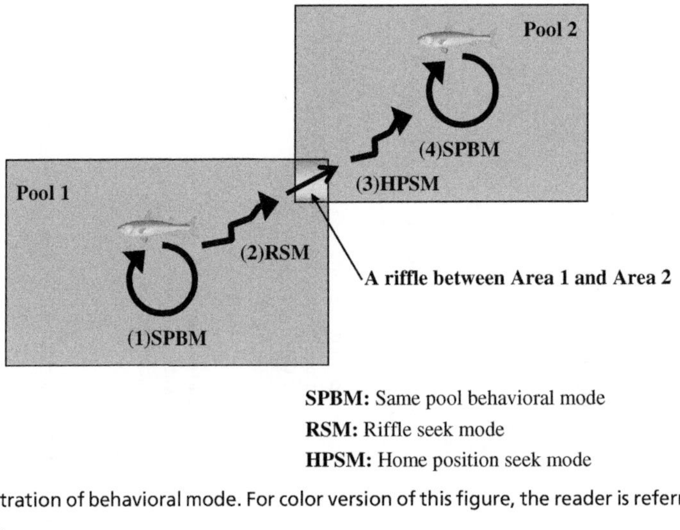


**FIGURE 8.5** Demonstration of behavioral mode. For color version of this figure, the reader is referred to the online version of this book.

swim in water 4 times shallower than the WH and with current velocities 2 times higher than CSS.

#### *8.2.5.5 Verification of the Model*

We verified the reliability of the model by comparing the ATS data and the simulation results.

## 8.3 Results

### 8.3.1 Behavioral Characteristics of Sweetfish

Figure 8.6a indicates the behavioral range of the test fish. Behavior of the test fish changed before and after September 2, 2009. In period 1 (before September 2), test fish inhabited downstream pools where the test fish were released. The area of behavioral range was approximately 6589 $m^2$ during period 1, and the test fish mainly inhabited the areas with deep water and low current velocity. The test fish moved to the upper pools through the riffle at noon on September 2, 2009. In period 2 (after September 2), the area of behavioral range was approximately 6214 $m^2$, and the test fish inhabited an area similar to that in

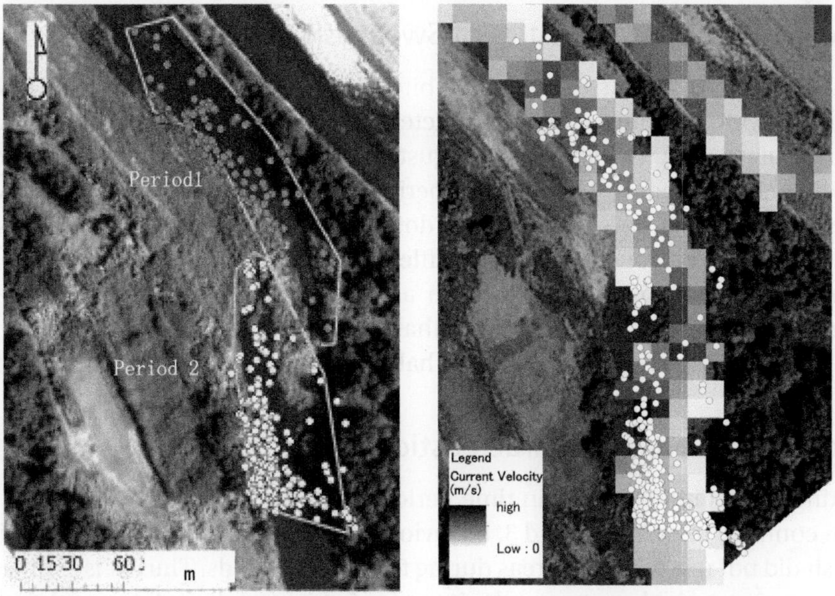


(a) behavioral range of the test fish (b) the current velocity and water depth of the areas inhabited by the test fish

**FIGURE 8.6** Demonstration of behavioral data. For color version of this figure, the reader is referred to the online version of this book.

period 1. During both periods 1 and 2, the test fish mainly inhabited areas with low current velocity and deeper than the riffle.

### 8.3.2 Spatial Characteristics of Sweetfish

Figure 8.6b indicates the current velocity and water depth of the areas inhabited by the test fish. In each time period, the test fish typically used areas where the current velocity was less than 1.5 m $s^{-1}$ and water depth was less than 0.8 m. The spatial characteristics in time period 1 differed from those in time periods 2 and 3. Thus, in time period 1, the sweetfish used areas where the water depths and current velocities were diverse. In contrast, in time periods 2 and 3, the sweetfish tended to use areas with lower current velocity and more water depth, in comparison to time period 1.

### 8.3.3 Modeling Behavior and Verification of its Accuracy

The model currently represents the home range of the test fish, particularly, the home range between upstream pools and downstream pools.

## 8.4 Discussion

### 8.4.1 Behavioral Characteristics of Sweetfish

As per the results, the test fish mainly inhabited pools. The research periods influenced the results. Previous studies generally indicted that sweetfish start feeding on algae in riffles from July to August, and after August, they inhabit low-current-velocity areas around riffles, such as pools. The research period of the current study coincided with the time when the sweetfish were preparing for downstream migration. Once sweetfish reach a certain size, they no longer need the riffles for feeding. In this period, the mature sweetfish seek refuge from predators such as birds and wait for the time to migrate downstream. Results of this study indicate that we need to conserve and restore not only riffles but also pools and their networks in habitat restoration projects for sweetfish.

### 8.4.2 Spatial Preference Characteristics of Sweetfish

The tracking results indicate that in time period 1, the test fish widely inhabited various spaces. In contrast, in periods 2 and 3, behaviors were limited to feeding and resting, and the test fish did not use high-risk areas during these two periods. Thus, it is estimated that the existence cost in high-current-velocity areas is high, and predation risk in shallow water areas is high. On the basis of these results, we conclude that sweetfish inhabit various physical environments (with different current velocities and water depths) according to their behavior. Our results indicate that a diversity of physical environments is needed to track sweetfish behavior.

### 8.4.3 Advantage of Modeling Sweetfish Behavior with ATS and the Possibility of Developing Sweetfish Behavior Models

The model represents two behavioral characteristics of the weasel (Fig. 8.7). The model represents home range and paths from downstream pool to upstream pool (red points in simulation results). Home range area (11,005 $m^2$) of VSF is almost the same as the home range area (12,803 $m^2$) of test fish. The model represents details of the paths. VSF swam the same riffle used by test fish. The results indicate efficiently advantages of the model by using ATS data. If we do not understand the spatial preference of the sweetfish before modeling, the model will not be able to represent these behaviors.

Modeling sweetfish behavior using ATS has two advantages: (a) it provides quantitative data to analyze special preferences and behavioral patterns and (b) it contributes to rapid and accurate improvement of the model. In order to develop an effective fish behavior model, quantitative behavioral and spatial data (for the physical environment, etc.) are required along with logic definition through analysis of these data. The study by Goodwin et al. (2006) is an example of such an approach. The authors tracked fish

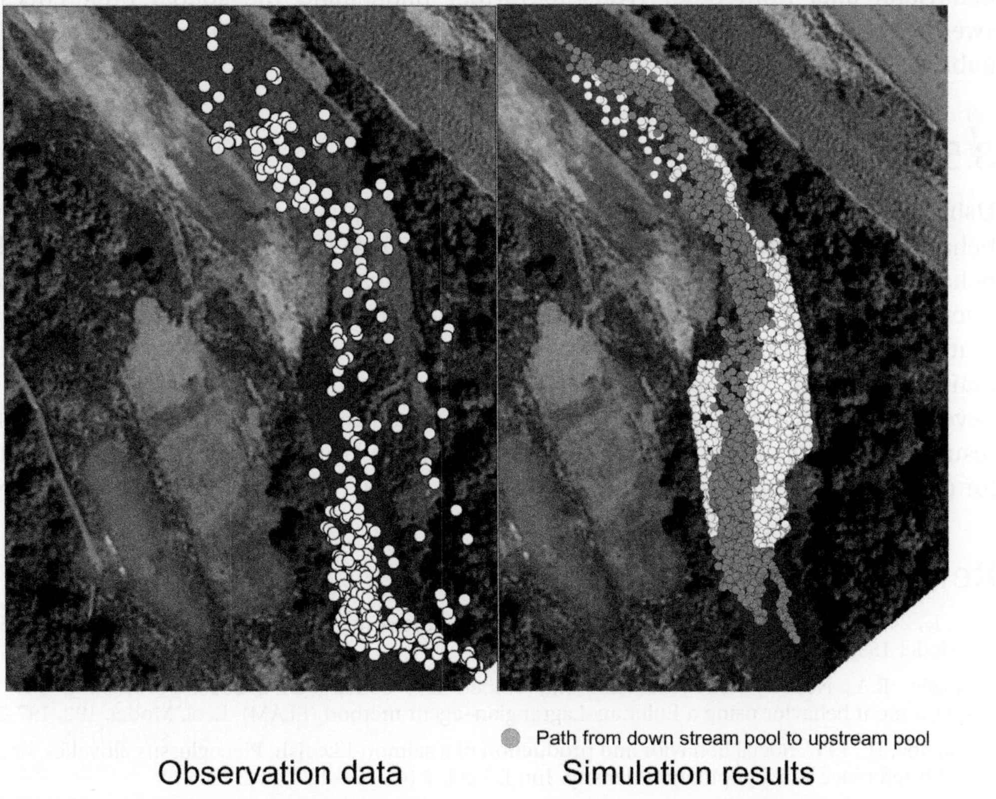


**FIGURE 8.7** Comparison of the observation data and simulation results. For color version of this figure, the reader is referred to the online version of this book.

behavior using a sonic tag telemetry system and developed simulation models. However, Japanese rivers are shallow and rapid, and bubbles in the river prevent sonic tag use.

ATS overcomes the problems experienced in previous research, and by using transmitters, it is possible to locate and track fish behavior in the middle reaches of Japanese rivers. ATS enables quantitative tracking of fish behavior and helps to accurately develop behavioral models.

By conducting similar studies in many Japanese rivers, we can develop general models that represent and predict sweetfish behavior. On the basis of the results of our study, we conclude that such a model, involving the use of basic technology and quantitative behavioral characteristics, can help in the effective habitat restoration of sweetfish.

### 8.4.4 Modeling Behavior and Verification of its Accuracy

Correct representation of the home range using the model developed in this study indicates that we succeeded in quantitatively modeling sweetfish behavior by using spatial preferences and behavioral characteristics of sweetfish during the downstream migration period. In our future work, we plan to modify the model by updating the behavioral data from other rivers to further understand the spatial preferences of sweetfish. Through this approach, we aim to obtain important information regarding habitat restoration for the conservation of sweetfish.

## 8.5 Conclusion

Using a wild-animal ATS, it is possible to accurately and continuously track animal behavior with a distance error of approximately 30 m. Using this method, we tracked behaviors of sweetfish for 2 weeks during downstream migration periods. We subsequently modeled the sweetfish behavior and verified the accuracy of the model. We found that the average home range of sweetfish was approximately 6000 $m^2$, the sweetfish mainly inhabited areas with greater water depth and low current velocity, and the developed model accurately represented the behavioral characteristics of sweetfish. The results indicate that the use of behavioral data obtained from wild-animal ATS is effective for developing accurate behavioral models such as the model developed in this study.

## References

Bian, L., 2003. The representation of the environment in the contest of individual based modeling. Ecol. Model 159, 279–296.

Goodwin, R.A., Nestler, J.M., Anderson, J.J., Weber, L.J., Loucks, D.P., 2006. Forecasting 3-D fish movement behavior using a Eulerian–Lagrangian–agent method (ELAM). Ecol. Model. 192, 197–223.

Kawanabe, H., 1957. Social behavior and production of a salmon-like fish, Plecoglossus altivelies, or ayu, with reference to its population density. Jpn J. Ecol. 7 (4), 131–137.

Kuroki, M., Ma, T., Ishida, R., Tsukamoto, K., 2006. Migratory history of wild and released ayu (Plecoglossus altivelis) in the Kurobe River, Japan. Coastal Mar. Sci. 30 (2), 425–431.

Railsback, S.F., Harvey, B.C., 2002. Analysis of habitat-selection rules using an individual based model. Ecology 83 (7), 1817–1830.

Sekiya, A., Fukui, Y., Shimomura, M., Uchida, T., 2005. Methods for preventing ayu from straying – analysis of the behavior of fishes by random walk method. J. Jpn Soc. Civil Eng. 782, 81–91.

PART 3

# Models of Interacting Species, Communities

# 9

# Formation of the Mosaic Structure of Vegetative Communities due to Spatial Competition for Life Resources

Alexey Kolobov, Efim Frisman
*INSTITUTE FOR COMPLEX ANALYSIS OF REGIONAL PROBLEMS, FAR EASTERN BRANCH OF THE RUSSIAN ACADEMY OF SCIENCE, SHOLOM-ALEYKHEM ST., 4, BIROBIDZHAN 679016, RUSSIA*

The expressed spatial structure and strong heterogeneity of plants is a complexity in biodiversity description and studies. It is expressed most vividly in phytocoenosis because most plants grow without uniformity in their habitats, forming congestions and lacunas. However, the heterogeneity of the environment conditions in corresponding areas is not the only cause of this behavior. An impressive example of this is the clear-cut spottiness of tundra phytocoenosis, although taiga plant communities appear to be very spotty and heterogeneous. The mechanisms of this heterogeneity have not been described and analyzed in terms of classical biological research.

This work offers a comparison of the research results of two mathematical models on the spatial–temporal dynamics of a plant community explaining the spotty spatial distribution through nonlinearity of the character of dynamics in a community, dynamic chaos phenomenon, and the processes of chaotic self-organization.

The first model is the analytical class, which incorporates the integral–differential equation system. Interaction of the plants located close to each other affecting both the biomass increase (new growth) and its growth restriction caused by struggle for life resources (light) was taken into account when constructing this model (Frisman et al., 2001, 2006).

The second approach employs an imitation computer model of the community dynamics of forest arboreal plants. Modeling the forest-stand dynamics develops the growth modeling of each tree using differential equations and functions, spatial arrangement, and other factors.

**Models of the Ecological Hierarchy. DOI: http://dx.doi.org/10.1016/B978-0-444-59396-2.00009-2**
ISSN 0167-8892, 

# 9.1 Plant Community Dynamics Model based on Integral–Differential Equation System

## 9.1.1 Model Description

The basic equation of the first model follows the equation developed in previous studies (Tuzinkevich, 1987, 1988, 1990; Tuzinkevich and Frisman, 1988, 1990):

$$\dot{u}_i(x,t) = \int\limits_M \alpha_i(x,y)u_i(y,t)\mathrm{d}y - u_i^{\gamma_i}(x,t)\sum_j \int\limits_M \beta_{ij}(x,y)u_j^{\rho_j}(y,t)\mathrm{d}y, \tag{1}$$

where $u_i\ (x,\ t)$ is the biomass density of the $i$-th species at point $x$ and time $t$ and $M$ is the two-dimensional physical space or the community habitat range; accordingly, points $x$ and $y$ have two dimensions. Parameter $\gamma_i$ characterizes the sensitivity of suppressed biomass from competitive impact; parameter $\rho_j$ reflects the nonlinear degree of dependence of the competitive limitation on the density of overwhelming biomass. Kernels $\alpha_i\ (x,y)$ characterize the biomass growth of the $i$-th species from point $y$ to point $x$. Kernels $\beta_{ij}\ (x,y)$ characterize the competitive impact of the $j$-th species biomass at point $y$ and the $j$-th species biomass at point $x$. Kernels $\alpha_i\ (x,y)$ and $\beta_{ij}\ (x,y)$ depend on the distance between points $x$ and $y$ and may be selected in the Gaussian curves. The kernels can also depend on time $t$ when considering the seasonal dynamics of the vegetation in the model (Tuzinkevich and Frisman, 1990; Tuzinkevich, 1989). However, in this work time dependence was not considered.

If a unispecies community is considered, the equation of the dynamics is as follows:

$$\dot{u}(x,t) = \int\limits_M \alpha(x,y)u(y,t)\mathrm{d}y - u^{\gamma}(x,t)\int\limits_M \beta(x,y)u^{\rho}(y,t)\mathrm{d}y \tag{2}$$

In the case of two species in the system, Eqn (1) can be written as:

$$\begin{aligned}
\dot{u}_1(x,t) &= \int\limits_M \alpha_1(x,y)u_1(y,t)\mathrm{d}y - u_1^{\gamma_1}(x,t)\int\limits_M \left(\beta_{11}(x,y)u_1^{\rho_1}(y,t) + \beta_{12}(x,y)u_2^{\rho_2}(y,t)\right)\mathrm{d}y \\
\dot{u}_2(x,t) &= \int\limits_M \alpha_2(x,y)u_2(y,t)\mathrm{d}y - u_2^{\gamma_2}(x,t)\int\limits_M \left(\beta_{21}(x,y)u_1^{\rho_1}(y,t) + \beta_{22}(x,y)u_2^{\rho_2}(y,t)\right)\mathrm{d}y
\end{aligned} \tag{3}$$

Consider the dynamic behavior of a single-species model (2). From the homogeneity of space, independence of the values $\alpha^* = \int\limits_M \alpha(x,y)\mathrm{d}y$ and $\beta^* = \int\limits_M \beta(x,y)\mathrm{d}y$ of $x$ arises. Therefore, the distributed model corresponds to the local model in which the biomass density in all points of space is considered identical.

$$\dot{u}(t) = \alpha^* u(t) - \beta^* u^{\gamma}(t)u^{\rho}(t) \tag{4}$$

Equation (4) shows that $\gamma + \rho > 1$ has a unique non-zero asymptotically stable stationary solution $u(t) = \overline{u}$. This behavior means that from any initial state $u(0) > 0$, variable $u(t)$ tends to be $\overline{u}$.

Accordingly, the distributed model (2) has a stationary time and spatially homogeneous solution $u(t,x) = \overline{u}$. However, this solution may not be sustainable. The loss of stability of the homogeneous solution in the model (2) may be due to the presence of an integral competitive suppression. With sufficiently high values of the parameter $\gamma$ (and fixed values of other parameters), the solution $u(t,x) = \overline{u}$ of the model (2), which is stationary in time and homogeneous in space, would be asymptotically stable. From any non-zero initial state ($u(0,x) > 0$ at least for some $x$), variable $u(t,x)$ tends to $\overline{u}$ for all $x$ over time. Thus, a plant population would fill its entire habitat space with high $\gamma$ values, covering it with a uniform "carpet," and preserve its stable spatial structure with time.

Decrease in parameter $\gamma$ (which may be interpreted as a greater intensity of competitive suppression at low densities of the suppressed biomass) would lead to bifurcation in the dynamic behavior of the model caused by the loss of stability of the spatially homogeneous solution and suppression of heterogeneous threshold distributions. With sufficiently low values of parameter $\gamma$ (almost from any heterogeneous initial state), the population would no longer distribute homogeneously such that some space points have greater vegetation density than other points.

Considering more details in the special case for $\rho = 1$, the results of the equation study of the form:

$$\dot{u}(x,t) = \int_M \alpha(x,y)u(y,t)\mathrm{d}y - u^{\gamma}(x,t)\int_M \beta(x,y)u(y,t)\mathrm{d}y \tag{5}$$

are presented with the following normalization conditions: $\int_M \alpha(x,y)\mathrm{d}y \equiv \int_M \beta(x,y)\mathrm{d}y \equiv 1$. The homogeneous stationary solution of the equation $u(x) \equiv 1$ is stable when the eigenvalues of the operator obtained by linearizing (Eqn (5)) in the neighborhood of stationary solutions are negative. Eigenvalues are formed as:

$$\lambda_i = a_i - b_i - \gamma \tag{6}$$

where $a_i$ and $b_i$ are the eigenvalues of integral operators generated by the kernels $\alpha(x,y)$ and $\beta(x,y)$, respectively.

The loss of stability of spatially homogeneous solutions bifurcates to form a heterogeneous solution.

The entry conditions of Eqn (5) can be set as spatially homogeneous $u(x,0) = const$ or spatially nonuniform $u(x,0) = f(x)$. A spatially homogeneous solution $u(x) \equiv 1$ can be attained at any entry condition if it is in a steady form. However, a distribution of the vegetation biomass density over an area depends on the initial distribution if the spatially homogeneous solution is unstable. Boundary conditions play an important role in both cases. They create approximations that essentially influence the established spatially nonuniform decision.

Based on further numerical studies, a limiting nonuniform spatial distribution of solutions Equation (2) is formed when the kernels $\alpha(x,y)$ and $\beta(x,y)$ have the form of Gaussian curves.

$$\alpha(x,y) = \begin{cases} S_\alpha \exp\left(-\dfrac{2\|x-y\|}{r_\alpha}\right), & \|x-y\| < r_\alpha \\ 0, & \|x-y\| \geq r_\alpha \end{cases}$$
$$\beta(x,y) = \begin{cases} S_\beta \exp\left(-\dfrac{2\|x-y\|}{r_\beta}\right), & \|x-y\| < r_\beta \\ 0, & \|x-y\| \geq r_\beta \end{cases} \tag{7}$$

The solutions of the equations prove to be highly heterogeneous in space under the homogeneous spatial distribution loss of stability. The solutions represent a series of discrete peaks of zero density, separated from each other by free space. Such nonuniform distribution may be either stable or unstable.

An example of a sustainable nonuniform distribution for a single-species community is described by Eqn (2) on a square habitat of side $a$ ($a = 30$). The initial distribution of biomass in space is uniform: $u(x,0) = 0.3$ for all $x$. To obtain a nonuniform spatial distribution, parameter $\gamma$ was considered sufficiently small ($\gamma = 0.1$) and parameter $\rho$ is more than 1 ($\rho = 3.9$). The radius of the kernel affects the size of germination, $r_\beta < r_\alpha$: $r_\alpha = 1$, $r_\beta = 5.9$. After some time, a nonuniform distribution was achieved (Fig. 9.1a). The density of the biomass is distributed as follows: At the edges of the field $M$, the value of $u(x,t)$ is the smallest; then the perimeter is a series of peaks coordinated with

(a)

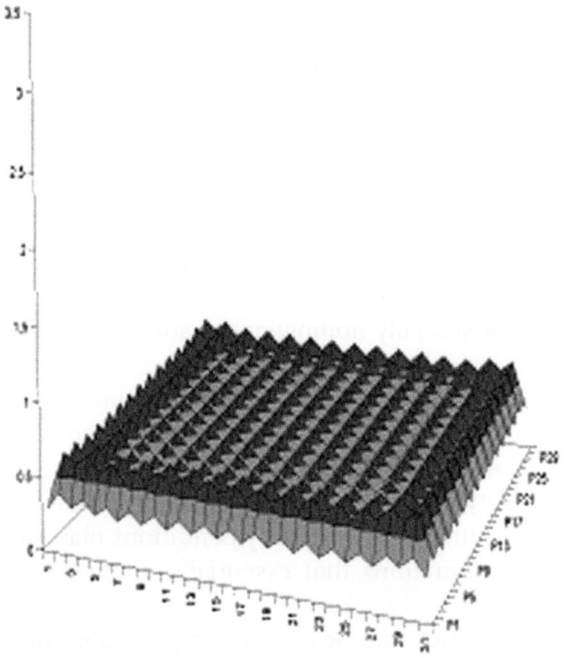

**FIGURE 9.1** Examples of stable spatially inhomogeneous solutions with a) $r_\alpha = 1$, $r_\beta = 5.9$ b) $r_\alpha = 3.9$, $r_\beta = 5.9$ c) $r_\alpha = 7.9$, $r_\beta = 5.9$ (left – side view, right – view from above).

(b)

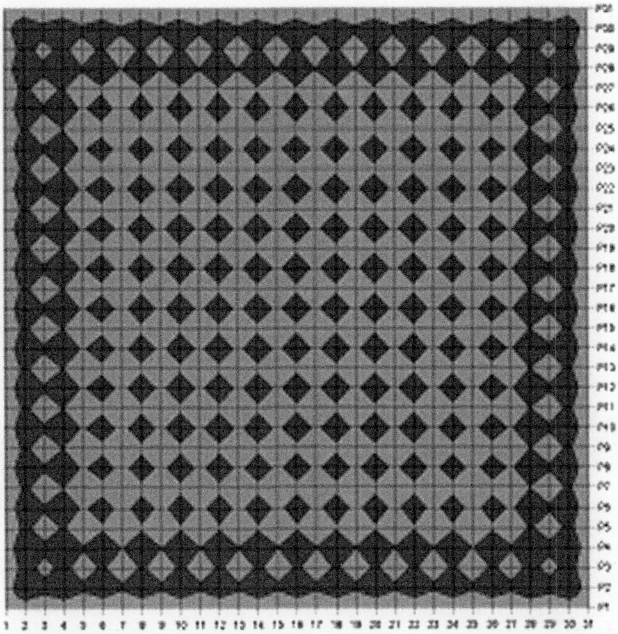

**FIGURE 9.1** (*continued*).

(c)

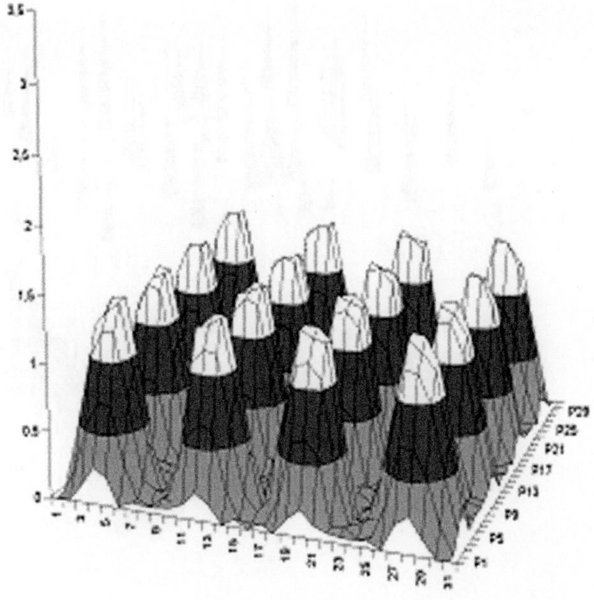


**FIGURE 9.1** (*continued*).

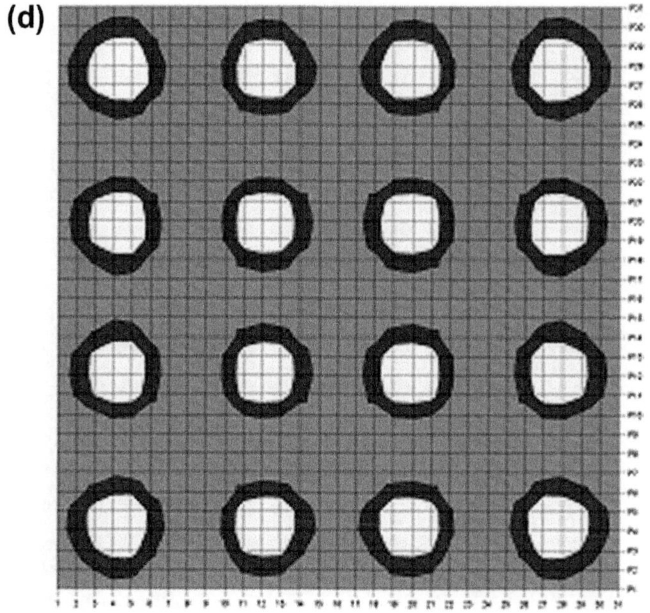


FIGURE 9.1 (*continued*).

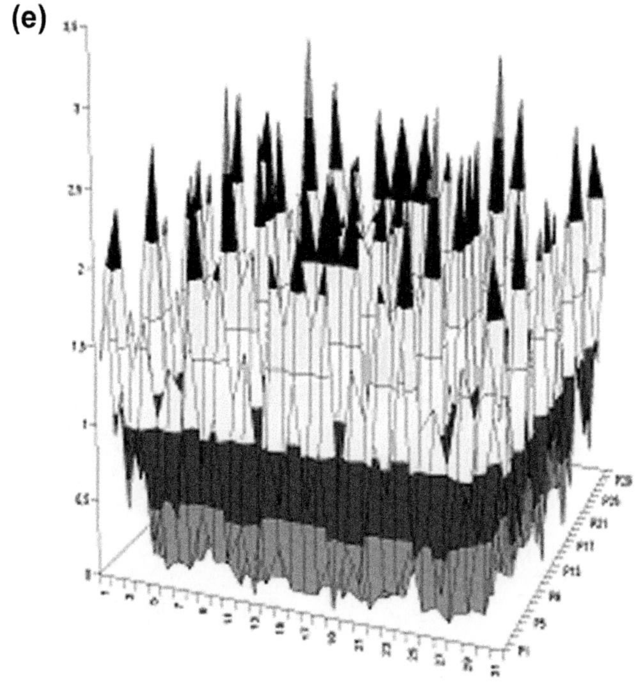


FIGURE 9.1 (*continued*).

(f)

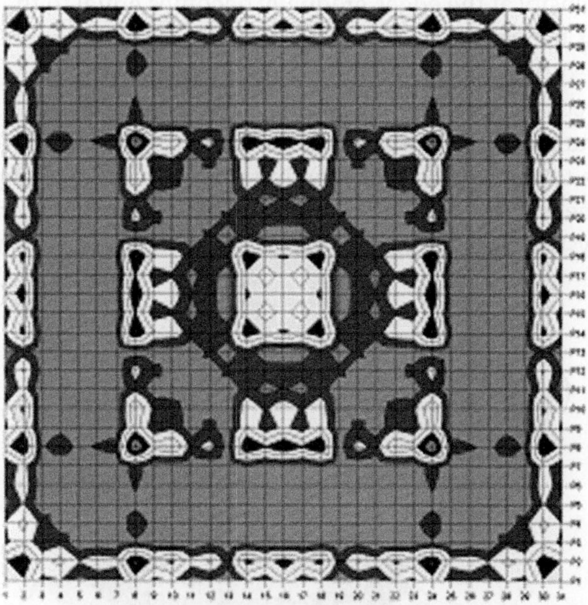

**FIGURE 9.1** *(continued)*.

the greatest value of $u(x,t)$, and the band "cavities" coordinated with the band peaks, and so on (Fig. 9.1a). The height of the peaks in the center is smaller than that at the edges. There are 15 peaks in each coordinate. The heterogeneity of distribution will increase if the value $r_\alpha$ is increased. As the number of peaks decreases, between points at which biomass is zero is observed, the amplitude of the peaks increases (biomass at these points will be almost 3 times higher than the initial), and the number will be 4 in each coordinate. When $r_\alpha = 3.9$ the number of peaks will be equal to 4 in each coordinate (Fig. 9.1b). If $r_\alpha > r_\beta$ was considered, distribution becomes highly heterogeneous (Fig. 9.1c, $r_\alpha = 7.9$), with a great number of peaks and a large amplitude (max $u(x,t)$, is 10-fold greater than $u(x,0)$). Between the peaks, zero density points remain, but will be significantly fewer compared with previous results.

These figures show that a regular density of biomass lattice appears in space. These lattice analogs can be found often in natural vegetative communities.

The unstable nonuniform distribution has the following parameters: The initial distribution of biomass is set nonuniform, $u(X,0) = 0.5$ for $X = (x_1, x_2) : x_1^2 + x_2^2 \leq (a/10)^2$ and $u(X,0) = 0$ for $X \in M$ at parameters $S_\alpha = 1.2$, $r_\alpha = 5$, $S_\beta = 1.2$, $r_\beta = 1$, $\gamma = 0.3$, and $\rho = 3$. At such parameters, the biomass does not come to any stationary distribution in space. At first, a deep hollow appears at a location where the initial density concentrates. A simultaneous, intensive growth of biomass (from $u_0 = 0.5$ to $u_t = 2.1 \div 2.4$) along the periphery (circle) of the initial distribution (Fig. 9.2a) and subsequently along the edges of the square is observed.

The biomass density at the edges gradually becomes equal to the general distribution; however, a strong heterogeneity is observed at the center (Fig. 9.2b). New peaks and hollows appear next (Fig. 9.2c), which reduce amplitude with time (Fig. 9.2d). The figure resembles the mycelium development under natural conditions.

Increasing parameter $\gamma$ or decreasing parameter $\rho$ setting for a small intraspecific competition leads to the concept that the initial distribution of population over time is distributed over the entire space with the same density at the center of $M$. If intraspecific competition is low; the biomass density in the center of the space is more than that at the

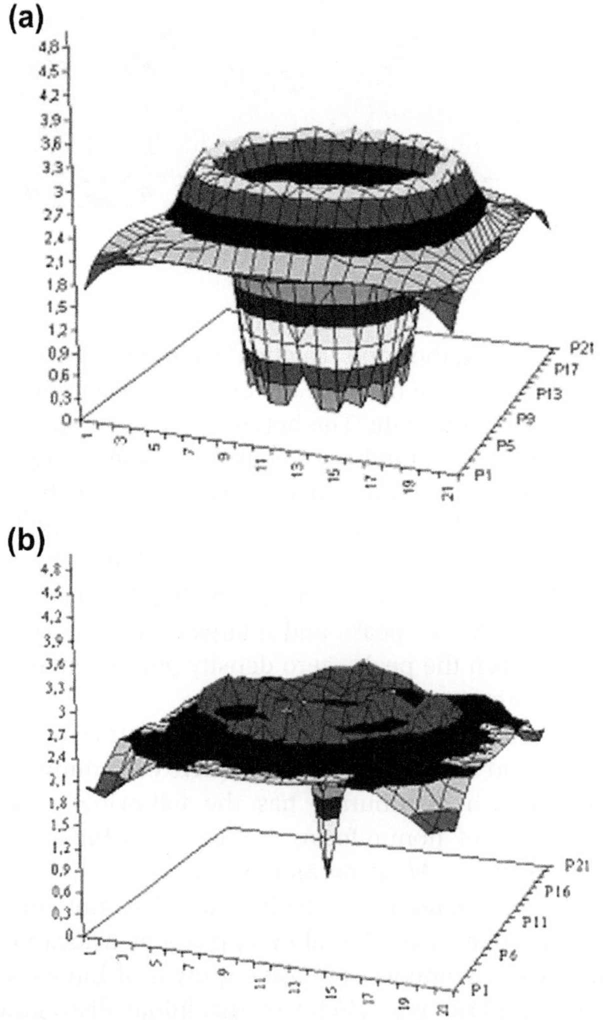


**FIGURE 9.2** Inhomogeneous density distribution at different time points.

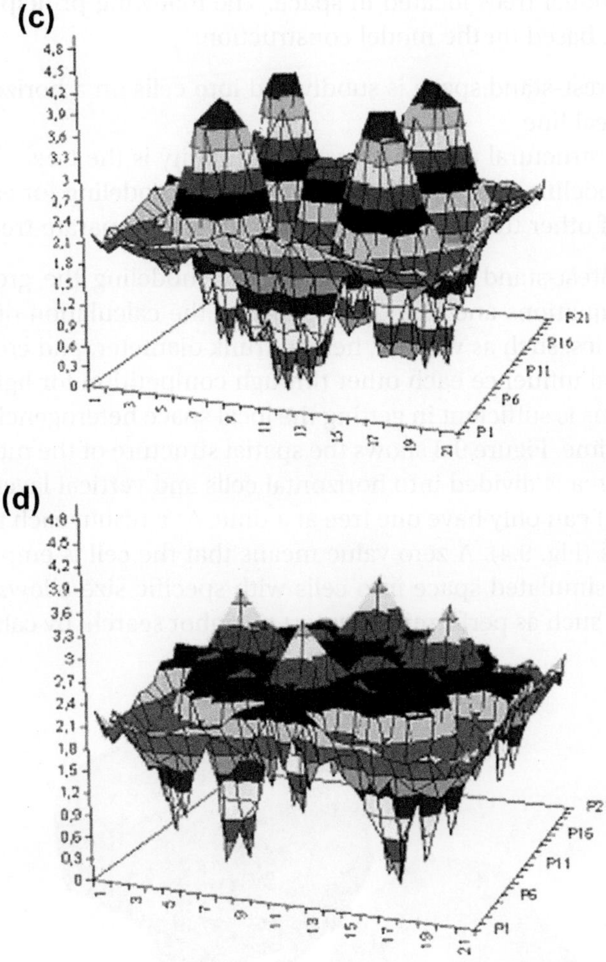


**FIGURE 9.2** (*continued*).

edges. However, high density was observed at the edges of the field in the case of strong competition.

# 9.2 Simulation Computer Model of Forest Tree Community Dynamics

## 9.2.1 Model description

The second model of the tree community dynamics using individual-based approach has emerged as the most widely applied among modern models of wood stand in the last decades (Chumachenko, 1993; Chave, 1999; Komarov et al., 2003; Huth and Ditzer, 2000, 2001). This approach considers the dynamics of the wood stand due to growth and interaction of individual trees located in space. The following principle notes of forest-stand modeling are based on the model construction:

1. The modeled forest-stand space is subdivided into cells on a horizontal plane and levels on a vertical line.
2. The elementary structural unit of a forest community is the tree.
3. Forest-stand modeling develops separate dynamics modeling for each tree.
4. The influence of other trees is considered in modeling separate tree dynamics.

Modeling the forest-stand dynamics consists of modeling the growth of each tree using differential equations and functions, allowing the calculation of basic forest estimation characteristics, such as volume, height, trunk diameter, and crown. The trees are located in space and influence each other through competition for light.

This set of notions is sufficient in getting the local space heterogeneity mosaic varying continuously with time. Figure 9.3 shows the spatial structure of the modeled forest area.

The simulated area is divided into horizontal cells and vertical layers. Each cell (with a size of $40 \times 40$ cm) can only have one tree at a time. As a result, each tree in the cell has its own coordinates (Fig. 9.4). A zero value means that the cell is empty of tree species. Partitioning of the simulated space into cells with specific size allows the algorithm to simplify the model, such as performing nearest neighbor search, by calculating the index

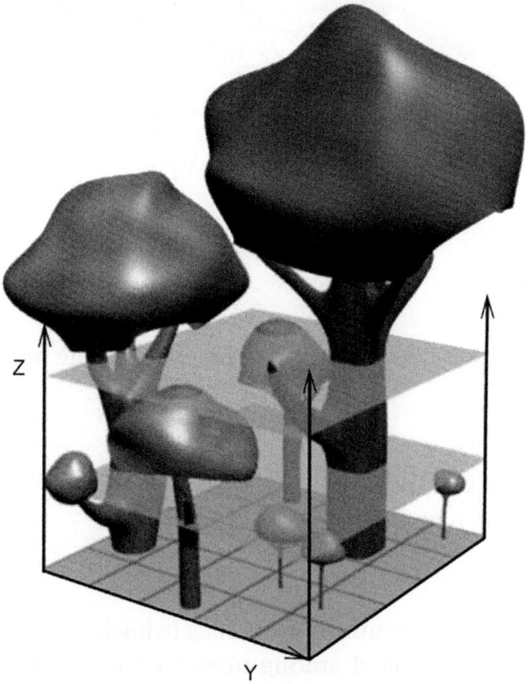


**FIGURE 9.3** Spatial structure of a modeled forest. The simulated area is divided into horizontal cells and vertical layers.

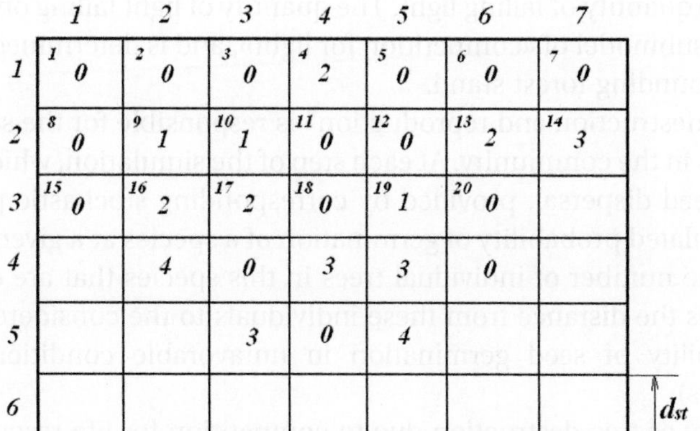


FIGURE 9.4 The horizontal plane of the simulated area is represented as a grid.

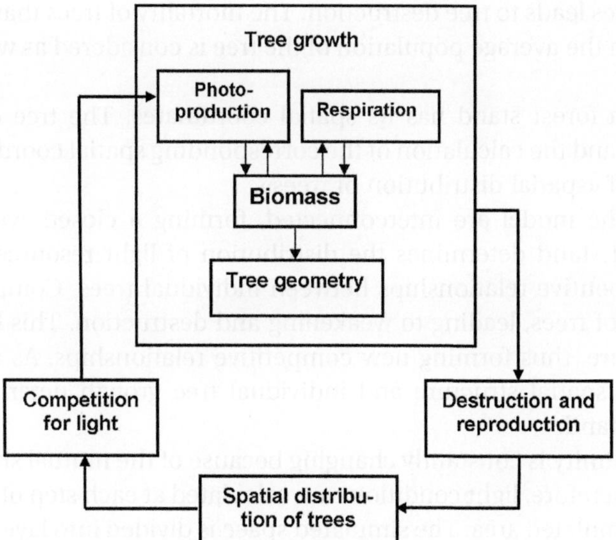


FIGURE 9.5 Model scheme for wood stand dynamics.

of competition, and reducing run time. The division of space in the vertical levels considers the relative positions of trees and their influence on each other in the calculation of light.

Stand dynamics model includes several submodels that reflect the life processes and interactions of trees. Figure 9.5 shows the submodels and their interactions.

The biomass increment and some geometrical parameters, such as volume, height, trunk diameter, and crone diameter, in the first block of «tree growth», were calculated. The biomass increment was determined by the photosynthesis intensity that is dependent on the quantity of falling light. The quantity of light falling on a separate tree is calculated in the submodel of «competition for light», and is determined by the degree of shade in the surrounding forest stand.

The block of "destruction and reproduction" is responsible for the seed reproduction and death of trees in the community. At each step of the simulation, which lasts for a year, a procedure of seed dispersal, provided by corresponding stochastic processes, is performed. The calculated probability of germination of a species at a given point in space is determined by the number of individual trees in this species that are capable of fructification, as well as the distance from these individuals to the considered point. In addition, the probability of seed germination in unfavorable conditions is considered (rodents, tall grass).

The probability of tree destruction due to competition for life resources and natural reasons are also calculated. A tree for each species is provided an average life expectancy. The greater the deviation from this average value, the higher the probability of tree destruction. Competition of trees due to lack of light results in slow growth, and falling branches and leaves leads to tree destruction. The mortality of trees that are far behind in development from the average population of the tree is considered as well (Buzikin et al., 1985, in Russian).

Every tree in a forest stand has its spatial coordinates. The tree disposal in some modeled territory and the calculation of the corresponding spatial coordinates are carried out in the block of «spatial distribution of trees».

All blocks of the model are interconnected, forming a closed system. The spatial structure of wood stand determines the distribution of light resources that can cause tension and competitive relationships between individual trees. Competitive processes affect the growth of trees, leading to weakening and destruction. This behavior changes the spatial structure, thus forming new competitive relationships. As a result, the relationship between spatial structure and individual tree growth determines the overall dynamics of the stand.

The tree community is constantly changing because of the mutual shading of incident solar radiation. Therefore, light conditions are calculated at each step of the simulation at all points of the simulated area. The simulated space is divided into layers to calculate the light regime. The tree crown projection is depicted in a corresponding layer based on tree height (Fig. 9.6).

The ability of the tree crown to absorb light depends on age and species of a tree. In calculating the light allowance at an arbitrary point, overlying layer superposition takes place to get the total luminous transmittance. The edge effect of the closure of the border area was removed using the topology of the torus.

To describe the tree growth as a basic model, the tree-free growth model was utilized, as proposed by Poletaev (1966, in Russian). The model is based on the principle of energy balance, written in the form of a law of conservation of energy. According to this model, the tree gets its energy only through photosynthesis. The free energy is spent

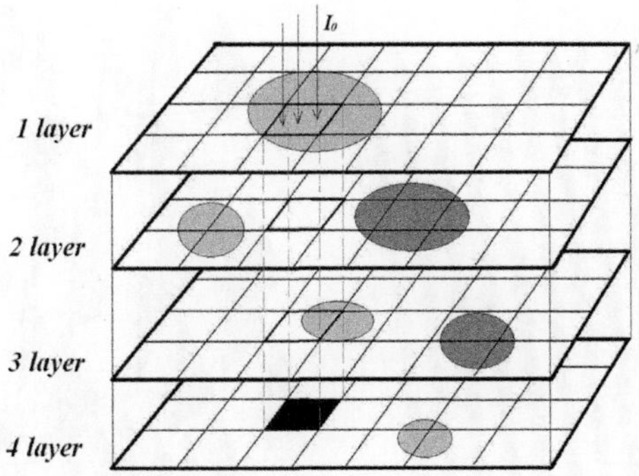


**FIGURE 9.6** Vertical structure of the modeled forest area.

on the needs of photosynthesis to build living tissues and transport the solution from the soil.

Finally, the resulting system of the equations of tree growth, with the influence of competition from around the stand, is written as follows:

$$\frac{dV}{dt} = \frac{P_m \cdot b}{p} \cdot \ln\left(\frac{P_m + a \cdot Q}{P_m + a \cdot Q \cdot \exp(-p \cdot V^d)}\right) - cVH$$

$$\frac{dH}{dt} = (k + m \cdot H) \cdot \left(R(Q) - \frac{H}{H_{\max}}\right)$$

$$D = \sqrt{\frac{4V}{\pi H f}}$$

where $V$ is the tree volume, $H$ is the tree height, $D$ is the diameter of the tree, $f$ is the form factor showing a deviation from the ideal cylinder, and $Q$ is the solar radiation proportion due to shading by the surrounding trees.

The normalized value of $Q$ varies in the range of $0 \leq Q \leq 1$, as expressed using the Monsi and Saeki model (Monsi and Saeki, 1953). In this model, the light transmission coefficient depends on the density and thickness of the vegetation. The dependence of the attenuation of radiation by vegetation can be written as: $Q_z = \exp(-kL_z)$, where $Q_z$ is the solar radiation proportion within the vegetation at height z, $L_z$ is the total leaf area of the stand above the level z, and $k$ is the extinction coefficient.

### 9.2.2 Modeling Results

A corresponding software has been utilized to put the model into practice. The software can be used to conduct computational experiments with different sets of input data, which can be different combinations of species and age structure of the wood stand.

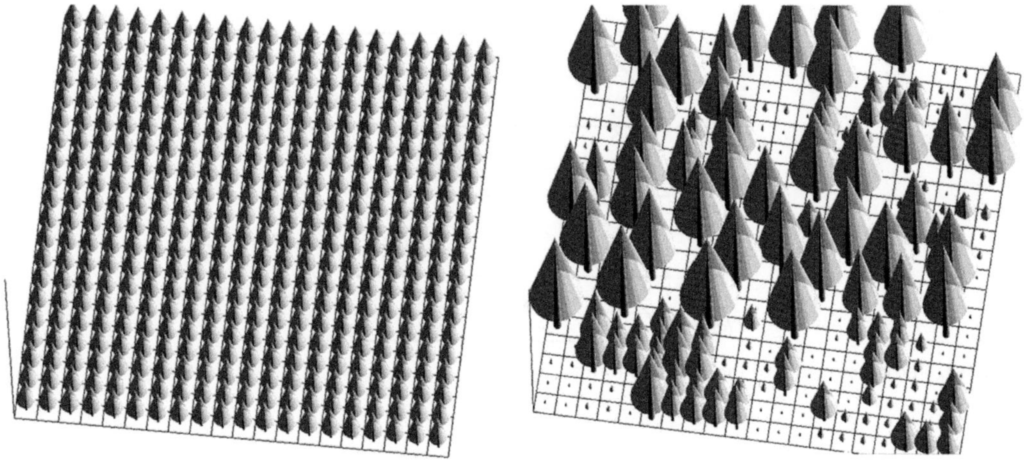

**FIGURE 9.7** Example of the formation of a spatially heterogeneous distribution of trees due to internal interactions. (left - originally, the modeled area was filled uniformly with single-species and even-aged trees, right - over time, the picture has lost its uniformity, showing uneven age distribution and density).

Various forecast scenarios of the forest can be achieved based on the simulation results. Three-dimensional visualization of trees on the coordinate plane allows the observation of the spatial–temporal dynamics of the forest.

The wood community spatial–temporal dynamics is simulated with different values of the system parameters. The influence of intraspecific competition on the formation of the stable state of the forest stand is studied. Originally, the modeled area was filled uniformly with single-species and even-aged trees (Fig. 9.7, left). Over time the picture has lost its uniformity, showing uneven age distribution and density (Fig. 9.7, right). In some locations, clusters or cavities can be observed wherein tree growth replaces these cavities, corresponding to the gap-mosaic concept of forest cenosis (Korotkov, 1991; Smirnova et al., 1990, in Russian).

The total stock volume dynamics acquires a quasi-standard level along with the complex spatial dynamics over time (Fig. 9.8a). The total stock volume does not acquire the quasi-standard level in the system with periodic oscillations if the coefficient of the competition is decreased (Fig. 9.8b). Spatial structure is coeval and uniform at any time. Death occurs as a result of aging at a small coefficient value of intraspecific competition, and a complex heterogeneous structure is not formed. Thus, competition is a major factor in ensuring the stability of tree communities.

Figure 9.9 shows the spatial distribution of trees in the real plot of the fir-spruced forest and the spatial structure of the simulated wood stand.

The modeling results showed that processes of chaotic self-organization occurs in the process of forest-stand structure formation even under homogeneous external conditions, leading to the formation of aggregately structured heterogeneous (spotty) spatial

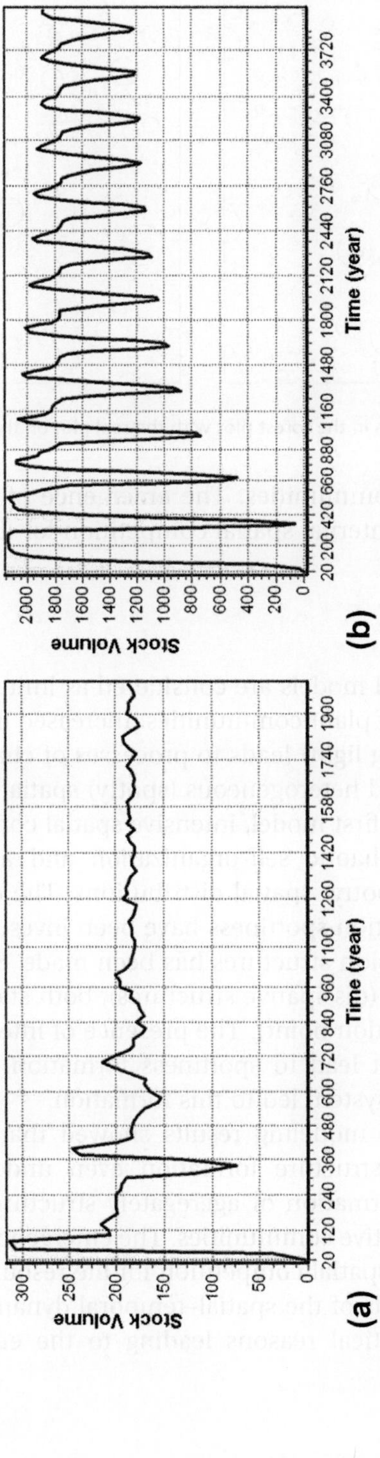


**FIGURE 9.8** The total stock volume dynamics in the forest stand at the modeled plot: (a) strong competition; (b) weak competition.

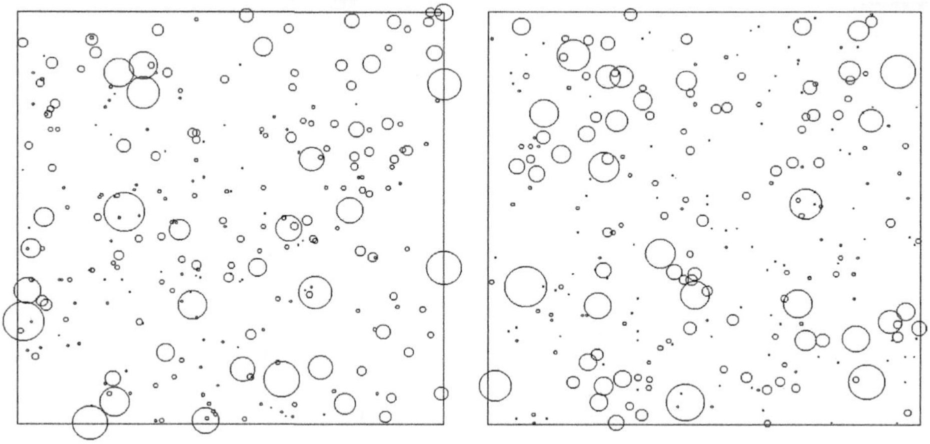

**FIGURE 9.9** Spatial distribution of trees in the forest plot with the real plot on the left and a model on the right.

distributions of vegetative communities. The emergence of this heterogeneity can be explained by the reasons of internal spatial competition for life resources.

## 9.3 Conclusion

In this chapter, mathematical models are considered as imitation and analysis tools for spatial–temporal dynamics of plant communities. Increased intensity of competition for resources (primarily involving light) leads to processes of chaotic self-organization and the rise of complex-structured heterogeneous (spotty) spatial distributions.

Based on the limits of the first model, intensive spatial competition for the resources initiates the processes of chaotic self-organization and appearance of aggregately structured heterogeneous (spotty) spatial distributions. The conditions for the appearance of spatial plant distribution spottiness have been investigated, and the analysis of spatially heterogeneous decision structures has been made. Spatial interactions lead to the appearance of spottiness (dissipative structures), both stationary and nonstationary (periodic, around the bifurcation point). The presence of interspecies, nonlocal competition in the model does not lead to spottiness formation. Instead, the intraspecies, nonlocal competition in the system led to this formation.

The analysis of imitation modeling results showed that chaotic self-organization occurs in the forest-stand structure formation even under homogeneous external conditions, leading to the formation of aggregately structured heterogeneous (spotty) spatial distributions of vegetative communities. The emergence of this heterogeneity can be attributed to the internal spatial competition for life resources.

The two modeling methods of the spatial–temporal dynamics of vegetative communities have established identical reasons leading to the emergence of spatial plant distribution spottiness.

# Acknowledgments

This work was supported in part by the Russian Foundation for Basic Research (no. 11-01-98512-r_vostok_a) and the Far Eastern Branch of Russian Academy of Sciences (no. 12-I-0-06-042).


# References


Buzikin, A.I., Gavrikov, V.L., Sekretenko, O.P., Xlebopros, R.G., 1985. Analiz strukturi drevesnix cenozov. Nauka, sib. otd-nie, Novosibirsk, 94 pp.

Chave, J., 1999. Study of structural, successional and spatial patterns in tropical rain forests using TROLL, a spatially explicit forest model. Ecol. Model. 124, 233–254.

Chumachenko, S.I., 1993. Bazovaya model dinamiki mnogovidovogo raznovozrastnogo lesnogo cenoza. V nauchnix trudaxMLTI: Voprosi ekologii i modelirovaniya lesnix ekosistem. M.: MLTI, pp. 147–180.

Huth, A., Ditzer, T., 2000. Simulation of the growth of a lowland Dipterocarp rain forest with FORMIX3. Ecol. Model. 134, 1–25.

Huth, A., Ditzer, T., 2001. Long-term impact of logging in a tropical rain forest – a simulation study. Forest Ecol. Manage. 142, 33–51.

Frisman, E.Ya., Tuzinkevich, A.V., Chernishova, N.B., 2001. Mozaichnaya struktura biologicheskogo raznoobraziya kak sledstvie dinamicheskoy neustoychivosti (na primere dvuvidovogo soobchestva rasteniy). Yubileyniy sbornik: k tridcatiletiu IAPU DVO RAN, Vladivostok, pp. 350–365.

Frisman, E.Ya., Tuzinkevich, A.V., Chernishova, N.B., Fayman, P.A., 2006. Biologicheskoe raznoobrazie i dinamika ekosistem: informacionnie texnologii i modelirovanie. Izd-vo SO RAN, Novosibirsk, pp. 393–401.

Komarov, A.S., Chertov, O.G., Zudin, S.L., et al., 2003. EFIMOD 2 – a model of growth and elements cycling of boreal forest ecosystems. Ecol. Model. 170, 373–392.

Korotkov, V.N., 1991. Novaya paradigma v lesnoy ekologii. Biol. Nauki. 8, 7–20.

Monsi, M., Saeki, T., 1953. Uber den Lichtfactor in den Pflanzengesellschaften und seine Bedeutung fur die Stoffproduction. Jpn. J. Bot. 14 (1), 22–52.

Poletaev, I.A., 1966. Problemi kibernetiki 16, 171–190.

Smirnova, O.V., Chistykova, A.A., Popaduk, R.V., et al., 1990. Populycionnay organizaciya rastitelnogo pokrova lesnix territoriy (na primere shirokolistvennix lesov evropeyskoy chasti SSSR). NCBI, Puchino, 92 pp.

Tuzinkevich, A.V., 1987. Issledovanie netradicionnoy modeli dinamiki rastitelnogo soobchestva. IAPU DVNC AN SSSR, Vladivostok, 45 pp.

Tuzinkevich, A.V., 1988. Ob odnoy tipichnoy bifurkacii v integralnix modelyx ekosistem. Prikladnie voprosi statisticheskogo analiza. DVO AN SSSR, Vladivostok, pp. 48–68.

Tuzinkevich, A.V., 1989. Integralnie modeli prostranstvenno-vremennoy dinamiki ekosistem. IAPU DVO RAN, Vladivostok, 184 pp.

Tuzinkevich, A.V., 1990. Issledovanie povedeniya integralnix modeley dinamiki biomassi. Kibernetika 4, 108–113.

Tuzinkevich, A.V., Frisman, E.Ya., 1988. Dissipativnie strukturi i pytnistost prostranstvennogo raspredeleniya organizmov. Biofizika 33 (2), 333–337.

Tuzinkevich, A.V., Frisman, E.Ya., 1990. Dissipative structures and patchiness in spatial distribution of plants. Ecol. Model. 52, 207–223.

# 10

# Complex Dynamic Modes in a Two-Sex Age-Structured Population Model

Oksana Revutskaya, Galina Neverova, Efim Frisman

*INSTITUTE FOR COMPLEX ANALYSIS OF REGIONAL PROBLEMS, FAR EASTERN BRANCH RUSSIAN ACADEMY OF SCIENCES, SHOLOM-ALEIHEM ST., 4, BIROBIDZHAN 679016, RUSSIA*

## 10.1 Introduction

Substantiation and development of the classical matrix population models (Lefkovitch, 1965; Leslie, 1945; Logofet and Belova, 2007) allow for the description and analysis of the importance of the age structure and developmental stages for the maintenance and evolution of population cycles (Hastings, 1992; Kooi and Kooijman, 1999; Lebreton, 1996). Most studies on life cycle evolution consider the formation of the age structure and its role in the development of ecologically limited populations (Charlesworth, 1993; Frisman and Zhdanova, 2009; Greenman et al., 2005). The formation of sex structure is often regarded, in this context, as an accompanying process strictly determined by differences in the survival rates of both sexes of the same age.

In the case of polygamous species, both the formation of the sex structure and the entire pattern of population dynamics are strongly related to the parameters determining the type of "mating relations" and the role of males in reproduction (Bessa-Gomes et al., 2010; Frisman et al., 1980, 1982; Jenouvrier et al., 2010).

We consider a simple mathematical model that enables the simultaneous observation of the formation of both age and sex structures and the explicit consideration of the asymmetry of the effects of females and males on population processes. Analysis of the problem allows for the description of qualitative changes in population dynamics, which depends on the difference between the characteristics of sexes that determine survival and reproduction.

## 10.2 Mathematical Model

We develop our model for situations wherein a population may be regarded as a set of three groups, namely, juveniles comprising immature individuals and two adult groups comprising mature females and males involved in the reproduction process. Increase in

Models of the Ecological Hierarchy. DOI: http://dx.doi.org/10.1016/B978-0-444-59396-2.00010-9

ISSN 0167-8892, 

the population size is regulated by density-dependent limitation of the juvenile survival rate.

We introduce the following denotations: $n$ denotes the ordinal number of the reproduction season; $P$ signifies the number of immature individuals; and $F$ and $M$ stand for the numbers of mature females and males, respectively. The model may be written as a system of three different equations:

$$\begin{aligned} P_{n+1} &= rF_n \\ F_{n+1} &= \delta w_1 P_n + sF_n \\ M_{n+1} &= (1-\delta) w_2 P_n + vM_n \end{aligned} \tag{1}$$

where $r$ is the birth rate; $\delta$ is the proportion of females among newborns; $w_1$ and $w_2$ are the survival rates of immature females and males, respectively; and $s$ and $v$ are the survival rates of mature females and males, respectively.

The birth rate $r$ is assumed to depend on the ratio of the number of males to that of females in the population. Birth rate can be described by the following function (Frisman et al., 1980, 1982): $r = \dfrac{aM}{hF+M}$, where $a$ is the reproductive potential of the population (the maximum possible mean number of offspring per fertilized female) and $h$ is the ratio of the number of males to that of females, where exactly half the females in the population are fertilized.

Indeed, if $M/F = h$, then $r = ah/(2h) = a/2$, that is, the population birth rate is two times lower than the physiological maximum. Obviously, the value of $h$ is determined by species specificity and characterizes male sexual activity. A lower parameter $h$ denotes higher activity; hence, fewer males are required to maintain a high birth rate.

The survival rates of immature females and males are assumed as the population parameters most sensitive to the population density. These rates linearly depend on the juvenile number: $w_1 = 1 - \beta_1 P$ and $w_2 = 1 - \beta_2 P$, where $\beta_1$ and $\beta_2$ are parameters describing the intensity of intrapopulation competition. This dependence has been successfully used to describe the survival of northern fur seals (Frisman et al., 1980, 1982). However, this linear dependence has a substantial drawback. If the number of offspring is extremely large, the equation yields a negative survival rate; hence, one may obtain a negative number of individuals. Given this drawback, we impose some limitations on the parameters to describe and illustrate only the dynamic modes that have a positive population size.

Let us assume that an equal number of females and males are born ($\delta = 0.5$). Substitution of variables allows us to exclude parameter $\beta_2$ (with the above assumptions taken into consideration) and to write model (1) in terms of new variables as relative numbers:

$$\begin{aligned} p_{n+1} &= af_n \frac{m_n}{hf_n + m_n} \\ f_{n+1} &= 0.5(1 - bp_n)p_n + sf_n \\ m_{n+1} &= 0.5(1 - p_n)p_n + vm_n \end{aligned} \tag{2}$$

where $b = \beta_1/\beta_2$.

The only non-zero positive equilibrium point $(\overline{p}, \overline{f}, \overline{m})$ of system (2) is

$$\overline{p} = \frac{1+C-bA}{2}\left(1-\sqrt{1-\frac{4(C-A)}{(bA-C-1)^2}}\right), \quad \overline{f} = \frac{\overline{p}(1-b\overline{p})}{2(1-s)}, \quad \overline{m} = \frac{\overline{p}(1-\overline{p})}{2(1-v)} \tag{3}$$

where $C = (2s + a - 2)/(ab)$ and $A = 2h(1 - v)/(ab)$.

The non-zero positive equilibrium (3) exists exactly if $a > 2((1 - s) + h(1 - v))$, $s \neq 1$, $v \neq 1$, and

$$a < 8(b(1-s) + h(1-v)), \quad 0 < \overline{p} < 1, \quad 0 < b \leq 1 \tag{4}$$

or

$$ab < 8(b(1-s) + h(1-v)), \quad 0 < \overline{p} < 1/b, \quad b > 1 \tag{5}$$

The stability of this solution is defined by the eigenvalues, which, in turn, are solutions to the characteristic equation

$$\lambda^3 + g\lambda^2 + u\lambda + q = 0 \tag{6}$$

where $g = -(s+v)$, $u = a(h\overline{f}^2(2\overline{p}-1) + \overline{m}^2(2b\overline{p}-1))/(2(h\overline{f}+\overline{m})^2) + sv$, and $q = -a(hs\overline{f}^2(2\overline{p}-1) + v\overline{m}^2(2b\overline{p}-1))/(2(h\overline{f}+\overline{m})^2)$.

Equilibrium instability may occur only if the solutions for characteristic Eqn (6) are conjugate at $|\lambda|$ transition through 1.

The dynamic modes of system (2) can be studied sequentially by analyzing their special versions.

## 10.3 Equality of Female and Male Survival Rates

Let us first consider the case where $s = v$ and $b = 1$, that is, females and males do not differ from one another in the survival processes, regardless of whether they are immature or not. In this situation, the expression $f_{n+1} - m_{n+1} = s(f_n - m_n)$ follows Eqn (2), that is, at $s < 0$, the number of males and females are equal.

The sexual structure dynamics is relatively constant and has approximately the same number of males and females belonging to peculiar stenobiotic animals of different taxonomic affiliations (e.g., muskrat, the European mole, beaver). These animals are usually prolific (Bolshakov and Kubantsev, 1984).

If the number of mature females and that of males are equal, system (2) can be written as one of the following two-component models:

$$\begin{aligned} p_{n+1} &= kf_n \\ f_{n+1} &= 0.5(1-p_n)p_n + sf_n \end{aligned} \tag{7}$$

or

$$\begin{aligned} p_{n+1} &= rq_n \\ q_{n+1} &= (1-p_n)p_n + sq_n \end{aligned} \tag{8}$$

where $q_n = f_n + m_n$ is the size of a population's reproductive part and $k = a/(1+h)$ and $r = k/2$ are parameters characterizing the reproductive potential of the population (they are equal to the average number of newborn individuals per one female and one adult, respectively).

The non-zero positive equilibrium point of system (8) is $\overline{p} = (s+r-1)/r$, $\overline{q} = (s+r-1)/r^2$. It exists if $r+s>1$ and

$$r + 4s < 4 \tag{9}$$

Note that inequality (4) reduces to inequality (9) if $k = a/(1+h)$, $r = k/2$, $s = v$, and $b = 1$.System (8) was investigated in detail by Frisman (1994) and Frisman and Skaletskaya (1994), where for a large range of admissible (biologically meaningful) values of population parameters, the occurrence of limit sets of phase trajectories can be strange attractors. We investigated systems (2) and (7) and obtained similar results, thus showing that when parameters (attractors) cross the stability boundary,

$$a = 2(1+h)(3-2s)$$

a limit invariant curve appears in the phase space, that is, as the parameters change (shifting deeper into the unstable zone), the curve breaks and complex limit structures are formed.

Figure 10.1(a and b) shows examples of strange attractors in the phase space $(p, m, f)$ for system (2) and the phase plane $(p, f)$ for system (7).

Notably, the strange attractor of system (7) (Fig. 10.1b) coincides with the projection of the phase portrait of system (7) (Fig. 10.1a) in the phase plane $(p, f)$. This coincidence results from the equality of coordinates $f$ and $m$ for the attractor (Fig. 10.1c). This equality results in a "flat" limit set. Moreover, the values of the parameters for visualizing the limit sets of the models were not chosen by chance, that is, the strange attractor, shown in Fig. 10.1b, was obtained by investigating system (8) at values of $r = 3.1$ and $s = 0.15$ (Frisman, 1994; Frisman and Skaletskaya, 1994).

These values were selected based on the analysis of the population dynamics of gray-sided voles (Frisman et al., 2010).

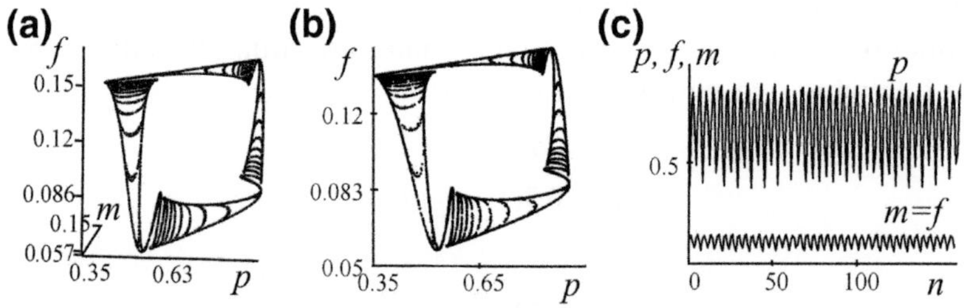


**FIGURE 10.1** Strange attractors of system (2) at $b = 1$, $v = s = 0.15$, $h = 0.01$, $a = 6.262$ (a); and system (7) at $s = 0.15$, $k = 6.2$ (b); trajectories of system (2) at $b = 1$, $v = s = 0.15$, $h = 0.01$, $a = 6.262$ (c).

## 10.4 Population Dynamic Modes at Different Survival Rates of Immature Males and Females

Let us now consider the behavior of system (2) when the survival rates of mature females and males are equal ($s = v$). Differences between sexes in the equilibrium number of individuals may result from only a difference in the survival rate between immature females and males.

The stability of equilibrium point (3) at $s = v$ is determined by eigenvalues satisfying the characteristic equation:

$$(\lambda - s)(\lambda^2 - s\lambda + G) = 0$$

where $G = a(h\overline{f}^2(2\overline{p} - 1) + \overline{m}^2(2b\overline{p} - 1))/(2(h\overline{f} + \overline{m})^2)$.

The equilibrium solution may lose stability only if a pair of complex conjugate roots of this characteristic equation intersects a unit circumference.

The critical value of birth rate $a_c$, at which the equilibrium loses its stability, can be obtained from the expression

$$a_c = 2(h\overline{f} + \overline{m})^2/(h\overline{f}^2(2\overline{p} - 1) + \overline{m}^2(2b\overline{p} - 1))$$

Let us consider the situation in more detail, where $b = 0$ ($\beta_1 = 0$), that is, the population growth is regulated only by a decrease in the survival rate of immature males accompanying an increase in juvenile number.

In this case, the non-zero equilibrium solution (3) exists at $a > 2(1 + h)(1 - s)$ and is stable if

$$a < h + 2(1 + h)(1 - s) + (h((1 + h)(3 - 2s)^2 - 1))^{1/2} \tag{10}$$

If the system's parameters are changed in such a way that inequality (10) is violated, quasi-periodic oscillations of the population number appear in the system. These oscillations become chaotic if the system's parameters are further changed.

For example, an increase in the reproductive potential $a$ results in the instability of the system's equilibrium solution and the appearance of quasi-periodic oscillations (Fig. 10.2a).

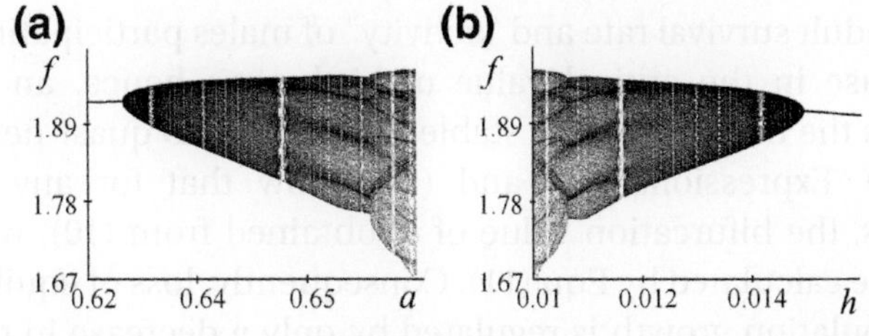


**FIGURE 10.2** Scenarios of changes in the limit distributions of the number of females, *f*, in the attractors of system (2) that are dependent on parameters *a* at *h* = 0.01 (a) and *h* at *a* = 0.66 (b) at fixed values of *b* = 0, *s* = *v* = 0.75. The behavior of the limit distributions of the number of mature males, *m*, and immature individuals, *p*, are similar.

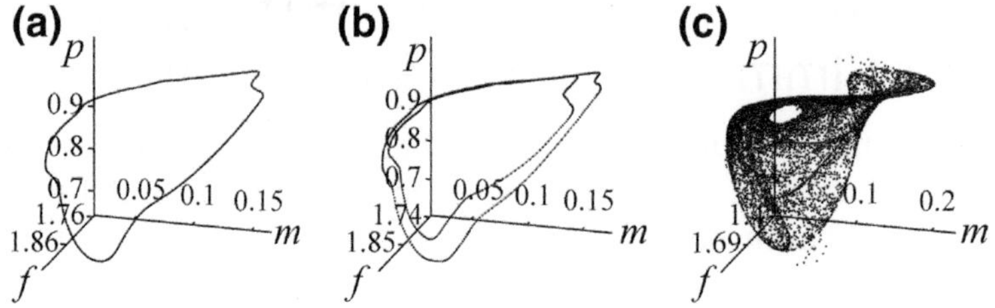


**FIGURE 10.3** Limit trajectories (attractors) of system (2) at $b = 0$, $h = 0.01$, $s = 0.75$, and (a) $a = 0.655$, (b) $a = 0.656$, (c) $a = 0.668$.

Notably, the bifurcation value of $a$ decreases as the adult survival rate $s$ increases and may even become less than 1 at large values of $s$ satisfying the following condition: $0.5(1 + (3h(1 + h))^{1/2}/(1 + h)) < s < 1$, where $h < 0.5$.

A decrease in parameter $h$, characterizing the dependence of birth rate on the sex ratio (in fact, a decrease in the role of males or an increase in their sexual activity), also results in the loss of equilibrium stability and a transition to cyclic or chaotic modes (Fig. 10.2b).

Bifurcation transitions may be followed in more detail using the portraits of the attractors in the phase space of system (2), as shown in Fig. 10.3. For example, at fixed values of parameters $h = 0.01, s = 0.75, b = 0$, and $a_c \approx 0.628$, oscillations begin in system (2), and a limit invariant curve appears in the phase space (Fig. 10.3a). As the reproductive potential is further increased, the curve breaks, forming a complex structured attractor (Fig. 10.3b and c). Similar scenarios of the transition to chaotic dynamics are observed in other parameters of the system, which determine the formation of various limit structures.

Let us now consider the case when $b_1 = \beta_2/\beta_1 = 0$ ($\beta_2 = 0$), that is, population growth is regulated only by a decrease in the survival of immature females accompanying an increase in the number of juveniles.

Similar to the previous case, a non-zero equilibrium exists when $a > 2(1 + h)(1 - s)$

Loss of equilibrium (3) stability occurs when the parameters of system (2) change while crossing over the stability boundary

$$a = 1 + 2(1 + h)(1 - s) + (1 + 4(1 + h)(1 - s)(2 - s))^{1/2} \quad (11)$$

An increase in the adult survival rate and "activity" of males participating in reproduction results in a decrease in the critical value of birth rate; hence, an increase in these parameters hastens the transition from stable equilibrium to quasi-periodic fluctuations in population size. Expressions (10) and (11) show that for any relation between parameters $h$ and $s$, the bifurcation value of $a$, obtained from (10), will be less than its corresponding value calculated by Eqn (11). Consequently, loss of equilibrium stability in the case where population growth is regulated by only a decrease in the survival rate of immature males occurs at a much less reproductive potential than in the case of a reduction in survival rate of immature females (Fig. 10.4).

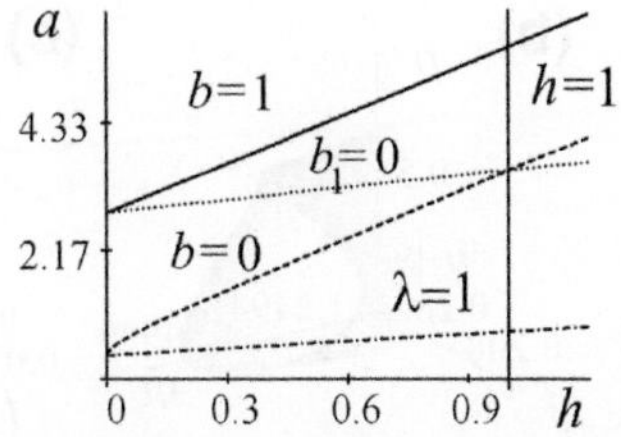


**FIGURE 10.4** Graphs corresponding to the stability boundaries of $|\lambda| = 1$ for cases when $b = 0$, $b_1 = 0$, and $b = b_1 = 1$ at $s = 0.8$. The stability boundaries of $\lambda = 1$ are equal for all cases.

Thus, the density-dependent regulation of the growth of immature females ($b_1 = \beta_2/\beta_1 = 0$, $\beta_2 = 0$) provides a significantly greater parametric range of population stability and, in this sense, is more effective than the limiting of immature males ($b = 0$), which could result in unstable modes at small values of the reproductive potential.

Naturally, a simultaneous decrease in the survival rate of immature males and females with the population size growth ($b = 1$) yields an even greater range of stability compared with restrictions for only females or males (Fig. 10.4). This condition appears to be the most effective mechanism to withstand population destabilization.

## 10.5 Development of the Two-Sex Population at the Maximum Equilibrium Size of Mature Females or Males

Expression (3) shows that the equilibrium numbers of mature males, $\overline{m}$, and females, $\overline{f}$, depend nonlinearly on the equilibrium number of juveniles, $\overline{p}$. In connection with this concept, special versions are naturally considered, namely, (a) $\overline{p} = 1/(2b)$ and (b) $\overline{p} = 1/2$, which provide the maximum equilibrium numbers of males and females, respectively.

Equality $\overline{p} = 1/(2b)$ is equivalent to a dependence of the parameter $b$ on the other parameters of system (2):

$$b = (a - 4(1 - s))/(2(a - 2(1 - s)(h + 2)))$$

For the sex ratio to have $\overline{m}/\overline{f} = (2b - 1)/b$, parameter $b$ must be greater than 1/2 ($b > 1/2$). The equilibrium population structure is stable at $\overline{p} = 1/(2b)$ if the inequality is correct

$$a < h + 2(1 + h)(1 - s) + (h(h(3 - 2s)^2 + 8(1 - s)))^{1/2} \qquad (12)$$

Similar to the particular case where $b = 0$, the critical value of $a_c$ can be less than 1 at large values of s, satisfying the inequalities: $(7h + 6 + (h(9h + 8))^{1/2})/(8(h + 1)) < s < 1$, where $h < 0.5$.

To observe the maximum number of males, that is, $\overline{p} = 1/2$, the following equality must be satisfied: $b = 2(a - 2(1 - s)(1 + 2h))/(a - 4h(1 - s))$. To ensure that $\overline{m}/\overline{f} = 1/(2 - b)$, $b$ must be greater than 2 ($b < 2$).

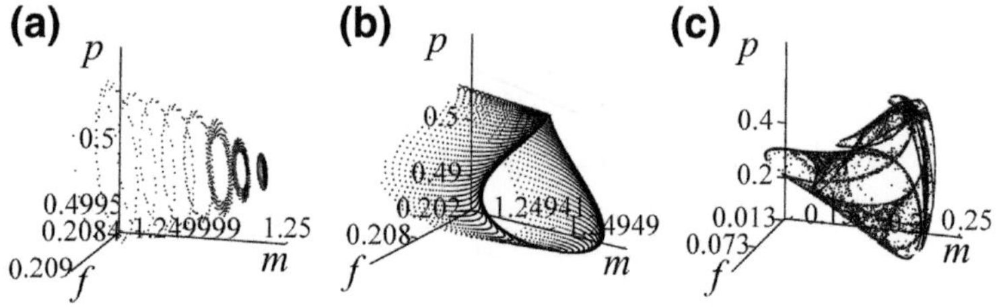


**FIGURE 10.5** Trajectories and attractors of system (2) at $\overline{p} = 1/2, h = 0.01$ and (a) $a = 2.3999, s = 0.9$; (b) $a = 2.409, s = 0.9$; and (c) $a = 5.15, s = 0.5$.

The equilibrium population size is stable at $\overline{p} = 1/2$ if parameter $a$ satisfies:

$$a < 1 + 2(1 + 2h)(1 - s) + ((3 - 2s)^2 + 8h(1 - s))^{1/2} \tag{13}$$

Stability is lost when inequality (13) is incorrect. The result is a chaotic quasi-periodic oscillation when the system's parameters are changed (Fig. 10.5).

Figure 10.5(a and b) shows that some variations of the attractors in phase space, which result when the equilibrium point becomes unstable, and an invariant curve appear. Thus, the equilibrium solution at the critical value of parameter $a$ is a set of points close to one another. The shape of this point set resembles a "funnel." By further increasing parameter $a$, the "funnel" is destroyed and the attractor points are attracted to the limit cycle.

# 10.6 Dynamic Modes at Different Survival Rates of Mature Females and Males

Males of many predators are usually (and sometimes significantly) numerically dominant among adult breeding animals, given that females in this age group die earlier than males. This type of sexual population dynamics is peculiar to animals, which usually do not form large clusters. Feeding and taking care of their offspring are associated with high energy costs and difficulties, especially for females (Bolshakov and Kubantsev, 1984).

On the contrary, numerous other species have higher male mortality, and the females have numerical dominance. This type of sex structure is typical for nomad mammals, which are polygamous with relatively slow reproduction, comparatively high endurance to unfavorable conditions, and sufficiently long life expectancy (e.g., wild donkeys, deer, antelope, wild goat and sheep, elephants, seals, and many others). In addition, this type of population dynamics is typical for some species of prolific animals, which are also polygamous mammals but with short life expectancy. Such prolific animals are incapable of enduring unfavorable conditions and have a high density of population (some lemmings). For these animals, the average life expectancy for males is less than that for females, and in the event of unfavorable conditions, males tend to die in large numbers. The survival rate of females is lower than that of males only in some cases (Bolshakov and Kubantsev, 1984). Moreover, the predominance of females in mammals is the result of the social behavior of animals, which exclude subordinate males from the group (harem).

In this regard, the behavior of system (2) must be considered when the survival rates of mature females and males differ ($s \neq v$) and the survival rates of immature individuals are equal ($b = 1$).

Stability of equilibrium solution (3) at $s \neq v$ and $b = 1$ depends on the eigenvalues determined by the characteristic equation

$$\lambda^3 - (s+v)\lambda^2 + \left(\frac{a(2\overline{p}-1)(h\overline{f}^2 + \overline{m}^2)}{2(h\overline{f}+\overline{m})^2} + sv\right)\lambda - \frac{a(2\overline{p}-1)(hs\overline{f}^2 + v\overline{m}^2)}{2(h\overline{f}+\overline{m})^2} = 0 \quad (14)$$

The critical value of birth rate, $a_c$, at which point the equilibrium becomes unstable, may be obtained from the following equation:

$$\frac{(2\overline{p}-1)^2(hs+v\overline{\sigma}^2)^2}{4(h+\overline{\sigma})^4}{a_c}^2 + \frac{(2\overline{p}-1)((h+\overline{\sigma}^2)-(s+v)(hs+v\overline{\sigma}^2))}{2(h+\overline{\sigma})^2}a_c + sv = 1, \quad \overline{\sigma} = \overline{m}/\overline{f} \quad (15)$$

As parameter $a$ crosses the stability boundary (15), an invariant curve appears, which breaks up as $a$ increases, and forms a rather complex limit structure.

Let us compare two situations where the survival rates of mature females, $s$, and males, $v$, are equal (Fig. 10.6a and b) and different (Fig. 10.6c and d). A comparison of the images of the limit structures and the values of the attractor dimension exemplifies that the difference in the male and female survival rates gives "volume" to resultant limit trajectories, thereby increasing the dimension of the attractors. In addition, the patterns of changes in the first Lyapunov exponent and the attractor's dimension indicate that the sequences of bifurcations in the two cases are similar. However, an increase in the

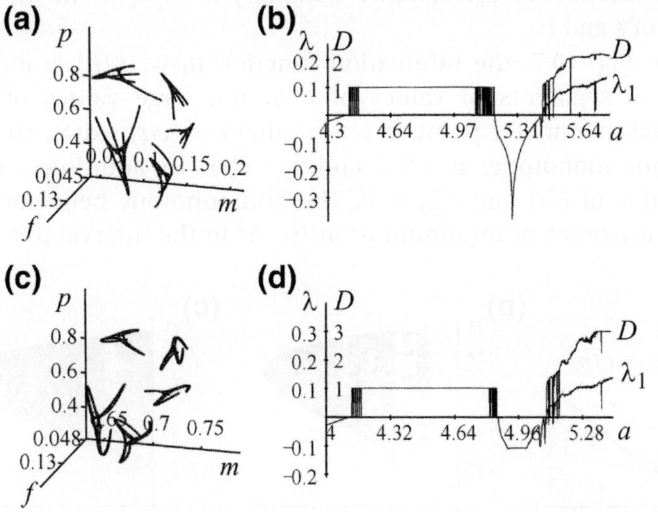


**FIGURE 10.6** Limit trajectories (attractors) for system (2) at $h = 0.1$, $b = 1$, and $a = 5.503$, $s = v = 0.5$ (a) and $a = 5.12$, $s = 0.5$, and $v = 0.9$ (c); changes in the first Lyapunov exponent ($\lambda_1$) and the attractor's dimension ($D$), dependent on parameter $a$ at $s = v = 0.5$ (b) and $s = 0.5$ and $v = 0.9$ (d). Benettin's algorithm was used to calculate the Lyapunov exponent, and the attractor's dimension was calculated using Kaplan–Yorke's formula (Kaplan and Yorke, 1979).

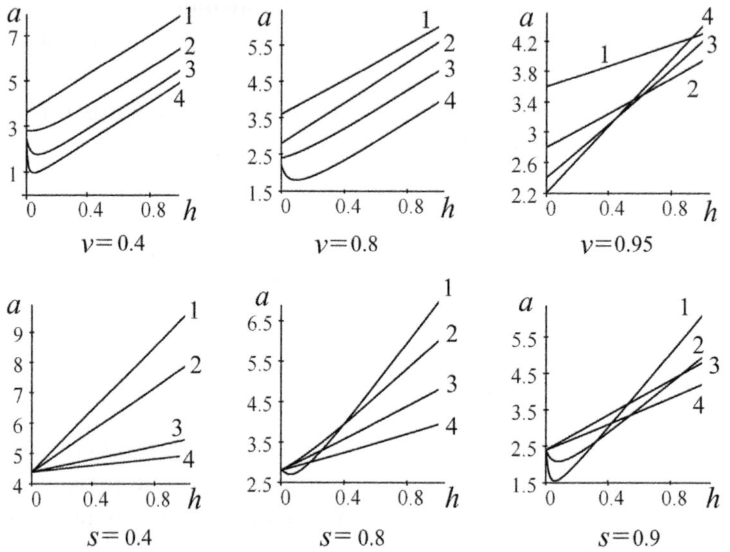


**FIGURE 10.7** Graphs of the function $a_1 = a_c(h)$ limiting the stability domain of the non-zero equilibrium of system (2) on the segment $0 \leq h \leq 1$, with $b = 1$ and fixed values of the parameters $s$ and $v$. Top row: line 1 corresponds to the graph $a_1 = a_c(h)$ at $s = 0.6$; line 2 – $s = 0.8$; line 3 – $s = 0.9$; line 4 – $s = 0.95$. Bottom row: line 1 – $v = 0.2$; line 2 – $v = 0.6$; line 3 – $v = 0.9$; line 4 – $v = 0.95$. Domain of stability is under the bifurcation curves.

survival rate of adult males (at average female survival rates) results in a considerable decrease in the corresponding bifurcation values of the reproductive potential $a$.

Function (15) is fairly complicated for two fixed parameters. Figure 10.7 shows the variation in the location of the stability boundary $a_1 = a_c(h)$ in the plane $(h, a)$ for different values of $s$ and $v$.

As seen from Fig. 10.7, the bifurcation function $a_1 = a_c(h)$ is monotone for any value of $v$ on all segments of values of $h$ at not large values of $s$ $(s < \tilde{s} \approx 0.74)$. However, for each parameter point of $s > \tilde{s}$, value $v = \tilde{v}(s)$ exists, such that function $a_1 = a_c(h)$ is not monotone at $s > \tilde{s}$ and $v > \tilde{v}(s)$. In addition, increasing $s$ also increases the value of $\tilde{v}(s)$ and $\tilde{v}(\tilde{s}) = 0$. The nonmonotone behavior of the function $a_1 = a_c(h)$ (the existence of minimum $a^*$ at $h = h^*$ in the interval $0 < h < 1$; Fig. 10.8a

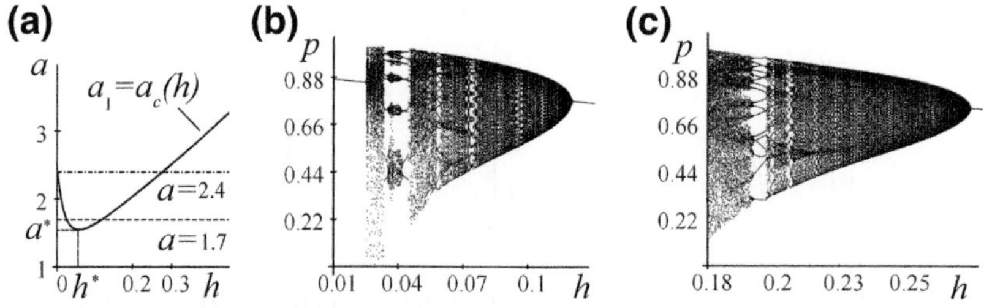


**FIGURE 10.8** Graph of the function $a_1 = a_c(h)$ limiting the stability domain for $v = 0.2$ and $s = 0.9$ (a). Bifurcation diagrams of the dynamic variable $p$ of system (2) dependent on parameter $h$ at $b = 1$, $s = 0.9$, and $v = 0.2$ with $a = 1.7$ (b) and $a = 2.4$ (c).

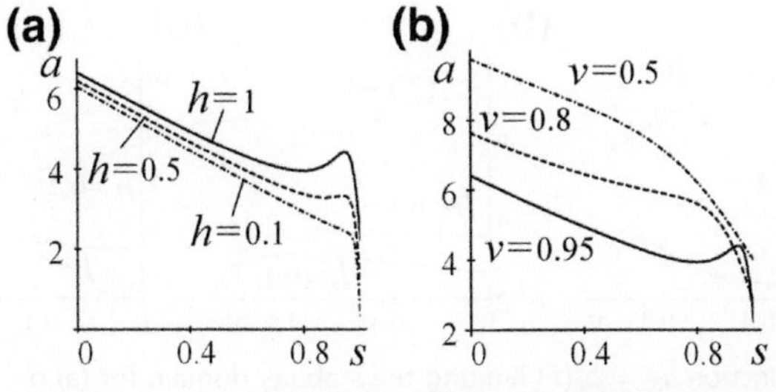


**FIGURE 10.9** Graph of the function $a_2 = a_c(s)$ limiting the stability domain for (a) $v = 0.95$ and (b) $h = 1$.

facilitates very complex scenarios of population dynamics by varying parameters $a$ and $h$.

If the reproductive potential is sufficiently small ($1 < a < a^*$), the equilibrium solution is stable for all values of $h$. If $a^* < a < a_0(a_0 = a_c(0))$, the equilibrium is stable for small values of $h$. However, as $h$ increases, the parameter point passes through a critical value of $h_1$ ($a_c(h_1) = a$), and nonstationary dynamic modes appear (Fig. 10.8b).

As parameter $h$ further increases, a complex pattern of interchange between irregular dynamic behavior and "windows" of cycles with finite length was observed. In addition, when parameter $h$ once again crosses the critical value of $h_2$ ($a_c(h_2) = a$), a limit cycle contraction to point occurs, and the equilibrium becomes stable. If the reproductive potential value is sufficiently large ($a > a_0$), the equilibrium solution is unstable for small $h$ (Fig. 10.8b).

Thus, a decrease in $h$, characterizing the dependence of birth rate on the sex ratio (in fact, a decrease in the "role" of males or an increase in their sexual activity), also results in the instability of the equilibrium and in a transition to cyclic or chaotic modes. However, this decrease in $h$ may eventually promote recovery of stability equilibrium at a large survival rate of females, $s$; relatively small values of reproductive potential, $a$; and a small survival rate of males, $v$ (Fig. 10.8b).

Figure 10.7 shows that at small values of $v$, stability boundaries $a_1 = a_c(h)$ with the larger $s$ are below the curves with less $s$ and do not intersect on segment $h \in [0, 1]$. Therefore, in this case, the stability boundary $a_2 = a_c(s)$ for fixed values of $v$ and $h$ is

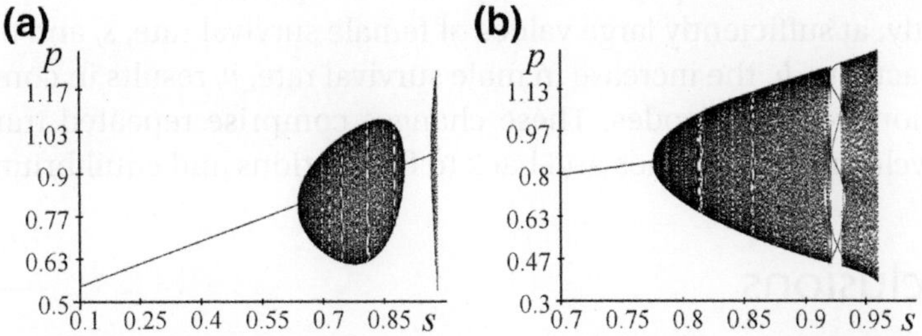


**FIGURE 10.10** Bifurcation diagrams of the dynamic variable $p$ of system (2) dependent on parameter $s$ at $b = 1$ and $v = 0.95$ at (a) $h = 1$, $a = 4.2$, and (b) $h = 0.1$, $a = 3$.

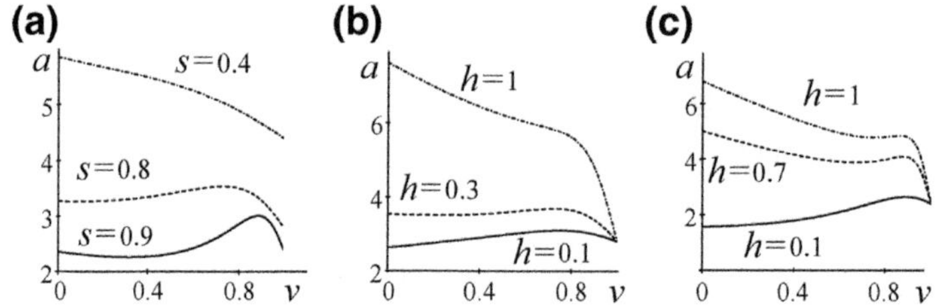


**FIGURE 10.11** Graph of the function $a_3 = a_c(v)$ limiting the stability domain for (a) $h = 0.25$, (b) $s = 0.8$, and (c) $s = 0.9$.

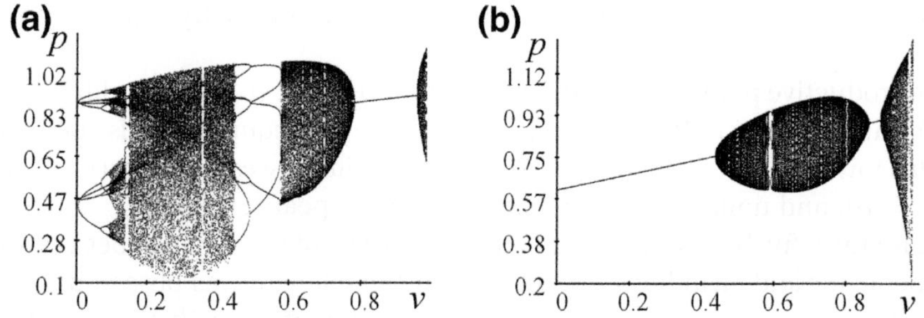


**FIGURE 10.12** Bifurcation diagrams of dynamic variable $p$ of system (2), dependent on parameter $v$ at $b = 1$ and (a) $s = 0.9$, $h = 0.25$, and $a = 2.8$, and (b) $s = 0.9$, $h = 0.7$, and $a = 4.1$.

monotone in parameter plane $(s,a)$. However, for sufficiently large $h$ and $v$, the stability boundaries $a_2 = a_c(s)$ do not become monotone (Fig. 10.9a and b). As parameter $s$ varies, exotic scenarios of population dynamics occur, characterized by the appearance of the so-called "bubbles of instability," which is limited by stationary modes in the bifurcation diagrams (Fig. 10.10a). A decrease in parameter values of $h$ (i.e., an increase in male sexual activity) reduces the stability domain and eliminates stationary modes for large values of $s$ (Fig. 10.10b).

Similar to the case when $a_2 = a_c(s)$, the stability boundary $a_3 = a_c(v)$ for small values of $s$ is monotone in parameter plane $(a,v)$, but as $s$ increases, this monotony is lost (Fig. 10.11a). This condition occurs in a wide range of values of $h$ (Fig. 10.11b and c) and results in exotic scenarios for population dynamics as parameter $v$ varies (Fig. 10.12).

Consequently, at sufficiently large values of female survival rate, $s$, and different levels of male sexual activity, $h$, the increase in male survival rate, $v$, results in complex changes in the population dynamic modes. These changes comprise repeated transitions from stationary or cyclic modes to chaos and back to fluctuations and equilibrium (Fig. 10.12).

## 10.7 Conclusions

We have demonstrated that an increase in birth and survival rates in the course of natural evolution in ecologically limited populations may result in instability and the appearance of chaotic attractors, the structures and dimensions of which change when model parameters are changed (Inchausti and Ginzburg, 1998). This transition to chaotic population dynamics is unsurprising given the large reproductive potentials of many animals, including insects, fish, birds, and small mammals (Frisman, 1994).

The possibility of the appearance of chaotic population dynamic rhythms with an increase in the sexual potency of males (e.g., upon transition to polygamous reproduction) and a decrease in the proportion of males necessary for successful reproduction is discussed.

An interesting result is the irregular population dynamics in species with low fertility when the population growth is regulated by a reduction in the survival rate of immature males with an increase in incremental population size. This regulatory mechanism is characteristic of polygamous species, where irregular long-period oscillations of population sizes are often observed (Frisman et al., 1980, 1982).

## Acknowledgments

This work was supported in part by the Russian Foundation for the Fundamental Research (project no. 11-01-98512-r_vostok_a).

## References

Bessa-Gomes, C., Legendre, S., Clobert, J., 2010. Discrete two-sex models of population dynamics: on modeling the mating function. Acta Oecol. 36, 439–445.

Bolshakov, V.N., Kubantsev, B.S., 1984. Polovaya struktura populyatsiy mlekopitayushchikh i ee dinamika. M.: Nauka, 233 pp. (in Russian).

Charlesworth, B., 1993. Natural selection on multivariate traits in age-structured populations. Proc. R. Soc. Lond. B 251, 47–52.

Frisman, E.Ya, 1994. Strannye attraktory v prosteyshikh modelyakh dinamiki chislennosti populyatsiy s vozrastnoy strukturoy. Dokl. Akad. Nauk 338, 282–286 (in Russian).

Frisman, E.Ya., Last, E.V., Lazutkin, A.N., 2010. The mechanism and peculiar characters of seasonal and long-term dynamics of voles *Clethrionomys rufocanus* and *Cl. rutilus*: a quantitative study and mathematical modeling. Bull. of the N.-E. Sc. Center, RAS FEB, 2010 2 (22), 43–47 (in Russian).

Frisman, E.Ya., Skaletskaya, E.I., 1994. Strange attractors in elementary models for dynamics of the quantity of biological populations. Surv. Appl. Ind. Math. 1008–1988 (in Russian).

Frisman, E.Ya., Skaletskaya, E.I., Kuzin, A.E., 1980. Mathematic modeling of the abundance dynamics of the northern fur seal, *Callorhynus ursinus*. The simplest model of a local population. J. Gen. Biol. 41 (2), 270–278 (in Russian).

Frisman, E.Ya., Skaletskaya, E.I., Kuzyn, A.E., 1982. A mathematical model of the population dynamics of a local northern fur seal with seal herd. Ecol. Model. 16, 151–172.

Frisman, E.Y., Zhdanova, O.L., 2009. Evolutionary transition to complex modes of population number dynamics. Genetics 45 (9), 1277–1286.

Greenman, J.V., Benton, T.G., Boots, M., White, A.R., 2005. The evolution of oscillatory behavior in age-structured species. Am. Nat. 166 (1), 68–78.

Hastings, A., 1992. Age dependent dispersal is not a simple process: density-dependence, stability, and chaos. Theory Popul. Biol. 41 (3), 388–400.

Inchausti, P., Ginzburg, L.R., 1998. Small mammals cycles in northern Europe: patterns and evidence for the maternal effect hypothesis. J. Anim. Ecol. 67, 180–194.

Jenouvrier, S., Caswell, H., Barbraud, C., Weimerskirch, H., 2010. Mating behavior, population growth, and the operational sex ratio: a periodic two-sex model approach. Am. Nat. 175 (6), 739–752.

Kaplan, J.L., Yorke, J.A., 1979. Chaotic behavior of multi-dimensional difference equation. Lect. Notes Math. 730, 204–227.

Kooi, B.W., Kooijman, S.A.L.M., 1999. Discrete event versus continuous approach to reproduction in structured population dynamics. Theory Popul. Biol. 56 (1), 91–105.

Lebreton, J.D., 1996. Demographic models for subdivided populations: the renewal equation approach. Theory Popul. Biol. 49 (3), 291–313.

Lefkovitch, L.P., 1965. The study of population growth in organisms grouped by stages. Biometrics 21, 1–18.

Leslie, P.H., 1945. On the use of matrices in certain population mathematics. Biometrika 33 (3), 183–212.

Logofet, D.O., Belova, I.N., 2007. Nonnegative matrices as a tool to model population dynamics: classical models and contemporary expansions. Fund. Prikl. Mat. 13 (4), 145–164 (in Russian).

PART 4

# Models of Ecological Networks, Food Chains and Food Webs

# 11 Influence of Intra-Seasonal Variability of Metabolic Rates on the Output of a Steady-State Food Web Model

Brigolin Daniele[1], Pastres Roberto
[1]DEPARTMENT OF ENVIRONMENTAL SCIENCES, INFORMATICS AND STATISTICS, UNIVERSITÀ CA' FOSCARI VENEZIA, DORSODURO 2137, 30123 VENEZIA, ITALY

## 11.1 Introduction

The estimation of energy and mass fluxes which flow through food webs is an important topic in system ecology, since they allow one to calculate ecological indexes, which, besides providing insights into the ecosystem functioning, can also be used for assessing the status of marine ecosystems (CEC, 2000). Steady-state linear food web models are regarded as useful tools for providing the above estimates on the basis of partial information embodied in field data. Therefore, in most instances, they represent the only way to derive ecological indexes from incomplete data sets. Furthermore, these models play a crucial role in nonlinear dynamic simulations also, since they can be used for supplying consistent initial conditions (Belgrano et al., 2005). Unfortunately, the application of steady-state food web models requires comprehensive data sets. In principle, these data should be representative of the "average" condition of the system within the selected time frame. However, in the large majority of cases, data are being collected at a given point in time which: (i) may not be the same for all compartments; and (ii) may not provide data which are representative of the whole time frame (e.g., summer or spring–summer in Degré et al. (2006) and Savenkoff et al. (2007)). Furthermore, care should be taken in using literature data for setting important input parameters, since these estimates may be obtained in environmental conditions rather different from those of the studied ecosystem. In particular, specific production rates (*P/B*) are among the most important input parameters of a food web model. These production rates are strongly affected by water temperature, since metabolic rates are strongly dependent on the thermal requirements of organisms (Kooijman, 2000) and also depend on the size of individuals.

Models of the Ecological Hierarchy. DOI: http://dx.doi.org/10.1016/B978-0-444-59396-2.00011-0

ISSN 0167-8892, 

Furthermore, in both cases the relationship is nonlinear: therefore estimating the the production rate from an average value of temperature and size, to be regarded as representative of the time frame of application of the steady-state model, may lead to incorrect estimates of energy and matter fluxes. Thus far, the potential role of the above factors in determining the result of food web models has not been investigated. The present study aims at filling this gap, by proposing the use of dynamic individual bioenergetic models for estimating the specific production of single-species compartments. The approach proposed here was tested on a simplified food web, with the following objectives:

- Assessing the effect of seasonal fluctuations of water temperature on the estimation of energy and matter flows across a food web.
- Testing the sensitivity of food web model results to interannual variability of water temperature and size structure of the population.

The work was carried out by combining two models:

- the individual model presented in Solidoro et al. (2000), which is based on a Scope for Growth formulation;
- a linear food web model, in which the constrained inverse problem was solved by means of two open-access R packages (van Oevelen et al., 2009).

The main features of these models and the methodology for their coupling are presented in the following section. Details concerning their application are given in Section 11.2.2, followed by the presentation and discussion of our main findings.

## 11.2 Methods

The approach proposed here is based on the use of dynamic bioenergetic models for estimating some of the most important input parameters of a food web model, namely the specific productions *P/B*, which is the ratio between the rate of increase of the biomass and the biomass standing stock of a given compartment. In fact, the specific production of most marine organisms depends strongly on water temperature (Mann and Lazier, 2006) and, in general, decreases with the size/age of individuals (Kooijman, 2000). Therefore, the average *P/B* over the time window to which the steady state model is applied should take into account the actual evolution of water temperature and that of the size structure of the population. Bioenergetic individual models provide a straightforward way to compute the specific production, if one assumes that this is not density dependent and that the size of an individual is representative of the average size of the population. In this case, the ratio *P/B* can be approximated by the specific growth rate, which, in turn, is the result of an energy balance:

$$P/B = \frac{1}{w}\frac{\mathrm{d}w}{\mathrm{d}t} = \frac{E_{in}(w,T,F) - E_{out}(w,T)}{w \cdot \varepsilon} \tag{1}$$

where $E_{\text{in}}$ and $E_{\text{out}}$ are the energy input and output per unit of time, $w$ is the wet weight of the individual, $\varepsilon$ is a coefficient which converts the energy gain/loss into weight gain/loss, $F$ represents the food availability, and $T$ the water temperature. According to Eqn (1), the rate of weight variation depends also on food availability. However, in this preliminary investigation, we assumed that this was not a limiting factor. Eqn (1) can then be used for estimating the average value of *P/B* over a given time window:

$$P/B = \frac{1}{t^*}\int_0^{t^*} \frac{\mathrm{d}w}{w\,\mathrm{d}t}\,\mathrm{d}t \tag{2}$$

The value of *P/B* thus obtained, besides being consistent with the steady hypothesis, is site specific and can take into account year-to-year variation of water temperature and size structure of the population. Therefore, this method can be used for assessing the effect of the interannual variability of these factors on the energy and matter flows.

We tested the sensitivity of the model results to a variation of these factors by using a simplified food web model and comparing the model output obtained by using the following estimates of the *P/B* of the top predator:

1. The literature value.
2. A value of *P/B* estimated from Eqn (1), using the average temperature and the average size.
3. Equation (2).

The approach was applied to six nodes of linear food, represented in Fig. 11.1. As one can see, the model represents a simplified benthic-pelagic food web, in which the top predator is the filter feeder *R. philippinarum*. The main features of the models are presented below.

### 11.2.1 Model Features

The steady-state food web model was designed on the basis of the findings presented in Sorokin and Giovanardi (1995), which provide a good representation of the food items digested and assimilated by *R. philippinarum*, namely, green algae, cyanobacteria, diatoms, bacterioplankton, microzooplankton, and dead, dissolved, and/or particulate organic matter. The resulting linear system of nine mass-balance equations, reported in Table 11.1, is under-determined, since it includes 24 unknown energy fluxes. The system was solved using the inverse methodology initially adapted from the physical sciences by Vézina and Platt (1988), and improved by van Oevelen et al. 2009. The R packages LIM (Soetaert and Van Oevelen, 2008) and LIMSolve (Soetaert et al., 2008) were used to carry out the analysis. Based on the notation by Van den Meersche et al. (2009), the inverse problem was defined according to the general formulation:

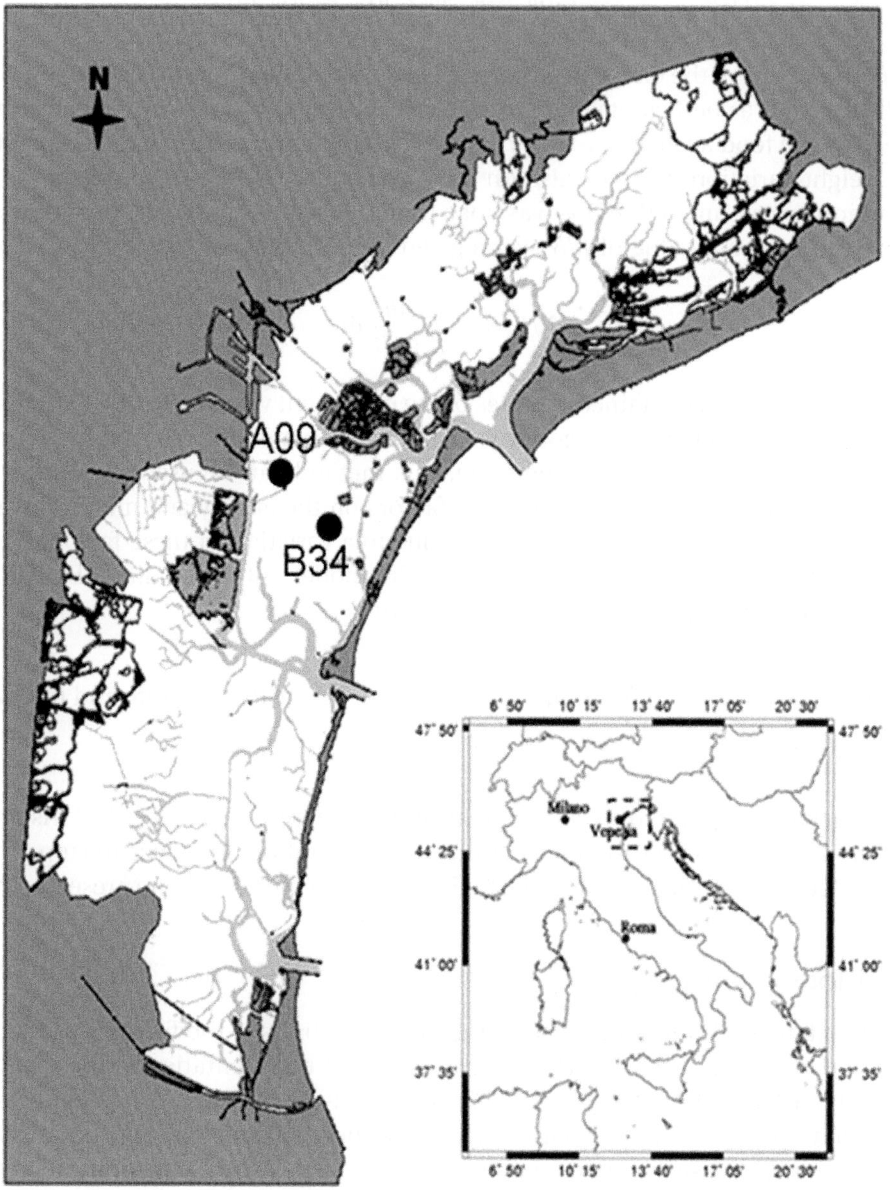


**FIGURE 11.1** Map of the study site. (a) Station A09—water temperature data, and (b) station 34, sampled by Sorokin et al. (1996)—plankton community data.

**Table 11.1** Food Web Model Details

| Compartments | Constraint Equalities [cal m$^{-2}$ day$^{-1}$] | Constraint Inequalities |
|---|---|---|
| PHP = phytoplankton | PHPProduction = 460 | MIZRespiration > 0.2 * MIZConsumption |
| BPL = bacterioplankton | BPLProduction = 900 | MEZRespiration > 0.2 * MEZConsumption |
| MIZ = microzooplankton | BPLRespiration = 2,800 | TAPRespiration > 0.2 * TAPConsumption |
| MEZ = mesozooplankton | MIZProduction = MIZBiomass*0.8 | TAPProduction < 0.3 * TAPConsumption |
| TAP = *R. philippinarum* | MEZProduction = MEZBiomass*0.04 | TAPProduction > 0.09 * TAPConsumption |
| DET = Organic detritus | TAPProduction = TAPBiomass*$P/B_{Rp}$ | PHP → MIZ > 0.1 * MIZConsumption |
| | MIZAssimilation = MIZConsumption*0.5 | PHP → MEZ > 0.1 * MEZConsumption |
| **Biomasses [cal m$^{-2}$]** | MEZAssimilation = MEZConsumption*0.5 | PHP → TAP > 0.1 * TAPConsumption |
| PHPBiomass = 160 | TAPAssimilation = TAPConsumption*0.5 | MIZ → MEZ > 0.1 * MEZConsumption |
| BPLBiomass = 1,600 | | MIZ → TAP > 0.1 * TAPConsumption |
| MIZBiomass = 170 | | BPL → MIZ > 0.1 * MIZConsumption |
| MEZBiomass = 3,000 | | BPL → MEZ > 0.1 * MEZConsumption |
| TAPBiomass = 1,463 | | BPL → TAP > 0.1 * TAPConsumption |
| | | DET → TAP > 0.1 * TAPConsumption |

**Equations****

PHP

PHPProduction = $CO_2$ → PHP

PHPLoss = PHP → MIZ + PHP → MEZ + PHP → DET + PHP → TAP

BPL

BPLAssimilation = BPLConsumption

BPLRespiration = BPL → $CO_2$

BPLProduction = BPLAssimilation − BPLRespiration

BPLLoss = BPL → MEZ + BPL → MIZ + BPL → TAP

MIZ

MIZConsumption = PHP → MIZ + BPL → MIZ + MIZ → MIZ

MIZFaeces = MIZ → DET

MIZAssimilation = MIZConsumption − MIZFaeces

MIZRespiration = MIZ → $CO_2$

MIZProduction = MIZAssimilation − MIZRespiration

MIZLoss = MIZ → MIZ + MIZ → MEZ + MIZ → TAP

MEZ

MEZConsumption = PHP → MEZ + BPL → MEZ + MIZ → MEZ + MEZ → MEZ

MEZFaeces = MEZ → DET

MEZAssimilation = MEZConsumption − MEZFaeces

MEZRespiration = MEZ → $CO_2$

MEZProduction = MEZAssimilation − MEZRespiration

MEZLoss = MEZ → MEZ

DET

DETConsumption = PHP → DET + MIZ → DET + MEZ → DET + input → DET

DETloss = DET → BPL + DET → TAP + DET → Export

*(continued)*

**Table 11.1—*cont'd***

| Compartments | Constraint Equalities [cal $m^{-2}$ $day^{-1}$] | Constraint Inequalities |
|---|---|---|
| TAP | | |
| TAPConsumption = PHP → TAP + BPL → TAP + MIZ → TAP + DET → TAP | | |
| TAPFaeces = TAP → DET | | |
| TAPAssimilation = TAPConsumption – TAPFaeces | | |
| TAPRespiration = TAP → $CO_2$ | | |
| TAPProduction = TAPAssimilation – TAPRespiration | | |
| TAPLoss = TAP → Harvesting | | |

Formulations and parameters describing the relationships between anabolic and catabolic processes and water temperature (based on the work by Solidoro et al. (2000)):

$$f_{gT}(T) = \left(\frac{T_{mg} - T}{T_{mg} - T_{og}}\right)^{\beta_g(T_{mg}-T_{og})} e_g^{\beta(T-T_{og})}$$

$T_{mg} = 32\ °C$

$T_{og} = 22.7\ °C$

$\beta_g = 0.2$

$$f_{rT}(T) = \left(\frac{T_{mr} - T}{T_{mr} - T_{or}}\right)^{\beta_r(T_{mr}-T_{or})} e_r^{\beta(T-T_{or})}$$

$T_{mr} = 35\ °C$

$T_{or} = 20.5\ °C$

$\beta_r = 0.17$

Notation follows general indications specified in the documentation of the R package LIM (Soetaert and van Oevelen, 2008).
**For all groups production = loss.

$$\begin{cases} Ax = b + \varepsilon \\ Ex = f \\ Gx \geq h \end{cases} \tag{3}$$

in which the first row expresses the continuity equation, Table 11.1, with $\varepsilon$ representing an error vector, while the second and the third rows respectively set equality and inequality constraints of the problem.

The individual growth model for *R. philippinarum* adopted here is based on a Scope for Growth formulation, in which gonadic tissues are not modeled explicitly; the structure of state equations and functional expressions relating metabolic rates to environmental forcings is described in detail in Solidoro et al. (2000). The model was calibrated and validated on field data collected at three Northern Adriatic lagoons, including the lagoon of Venice. For this model, Eqn (1) reads as:

$$\frac{1}{w}\frac{dw}{dt} = \frac{F}{F^*} G_{\max} f_{gT}(T)\, w^{-1/3} - r_{\max} f_{rT}(T) \tag{4}$$

where $w$ is the wet weight of the individual, $G_{\max}$ and $R_{\max}$ are the species-specific anabolic and catabolic rates, $F$ represents the food availability, and $F^*$ the maximum ingestible food. The relationships between the anabolic and catabolic processes and water temperature are modeled using the bell-shaped functions $f_{gT}(T)$ and $f_{rT}(T)$

(see Table 11.1), which reach a maximum at the optimal temperature and present one inflection point at temperatures below the optimum.

## 11.2.2 Model Setup

The approach outlined in the previous section was implemented using data collected in the lagoon of Venice, a shallow water body covering an area of nearly 500 $km^2$ and connected to the Northern Adriatic Sea through three inlets. The models were applied during the period May–October 1993, based on two main reasons:

- At the beginning of the 1990s, *R. philippinarum*, which was introduced in the Venice lagoon in 1983 (GRAL website), started to play a primary role as a macrobenthic filter feeder, out-competing the native species *Tapes decussatus.* Due also to the simultaneous decline of macroalgal population, the trophic interactions among the plankton community, the organic detritus, and *R. philippinarum* represented, in those years, the dominant pattern of energy exchange in the central areas of the Venice lagoon (Pranovi et al., 2003).
- Accurate literature data concerning the state of the planktonic community in 1993 (Sorokin et al., 1996) and site-specific estimations of the *P/B* ratio for *R. philippinarum*, based on a set of short-term experiments (Sorokin and Giovanardi, 1995), were available.

Field data used to construct the food web model were collected by Sorokin et al. (1996) from September 24 to October 11, 1993, at 36 sampling stations located in the lagoon of Venice. Sampling and analytical methodologies are described in detail in the original paper. The energy balance presented here was carried out for station 34, located in the Venice-Lido area (see Fig. 11.2). Based on the literature quoted above, it was possible to quantify the biomass densities of all plankton compartments (cal $m^{-2}$), while *R. philippinarum* biomass was set according to Libralato et al. (2002). Phytoplankton net primary production, bacterioplankton production, and respiration were imposed as equality constraints (line 2 in the system of Eqn (3)), based on the rates measured by Sorokin et al. (1996) by means of $^{14}C$ methods. Micro- and meso-zooplankton production and assimilation were constrained by using *P/B* coefficients and assimilation efficiencies reported by Sorokin et al. (1996), Table 6 in the original paper), while assimilation efficiencies for *R. philippinarum* was set to 0.5, based on laboratory measurements of assimilation reported by Sorokin and Giovanardi (1995). Additional constraints, set as inequality bounds (line 3 in the system of Eqn (3)), were imposed on metabolic expenses due to respiration and on the growth efficiency of Manila clam (see Savenkoff et al., 2007, and references cited therein). The lack of specific data (e.g., stable isotopes signatures) prevented us from setting up equality constraints for diet preferences; nonetheless, the laboratory experiments performed by Sorokin and Giovanardi (1995) provided a qualitative list of the food items assimilated by this species. On the basis of this information, a lower limit was set for each trophic interaction, corresponding to 10% of the total consumption, in order to guarantee that Manila clam was effectively using the whole potential diet spectra.

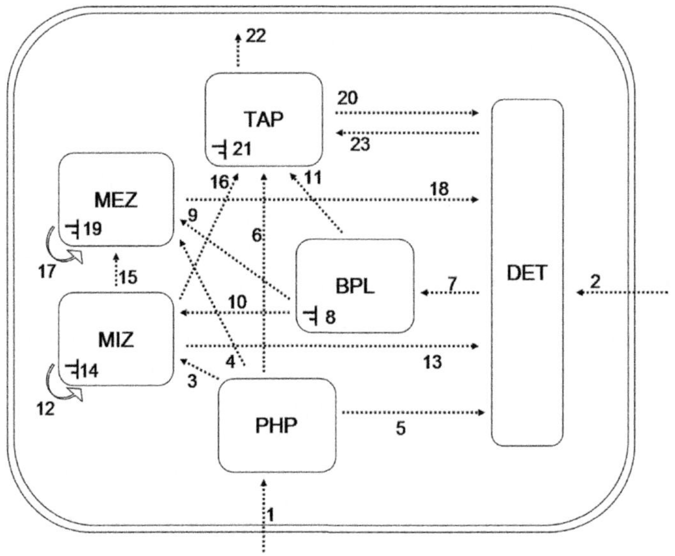


**FIGURE 11.2** Structure of the simplified "phytoplankton-clam" food web. PHP, phytoplankton; MIZ, microzooplankton; MEZ, mesozooplankton; BPL, bacterioplankton; TAP, *R. philippinarum*; DET, organic detritus. Numbered flows are listed in Table 11.2.

The dynamic individual model of *R. philippinarum* requires as an input the daily water temperature data. These were not available for the year 1993 and were estimated on the basis of hourly air temperature data (http://www.ismar.cnr.it/), using the site-specific correlation model presented in Pastres et al. (2004). The water temperature input data used for exploring the sensitivity of the model output to interannual variations were estimated from site-specific monthly data collected during the two time windows 1986–1990 and 1995–1998 at station A09 (see Fig. 11.2), in the framework of the surveys carried out by the Venice Water Authority. Monthly values of temperature representative for the whole water column were linearly interpolated to daily values.

The specific productivity for *R. philippinarum* compartment, $P/B_{\mathrm{Rp}}$, was estimated in two ways: (i) by means of Eqn (4), using the average temperature and the average size; and (ii) by means of Eqn (4), using as input the time series of water temperature, estimated as described above, and two different initial conditions for the shell length on May 1: 10 mm and 30 mm. The first value was representative of the individuals that settled at the end of the previous summer, 0+ year class, while the second value was corresponding to the average length reported for a 1+ year class (Pellizzato et al., 2011).

## 11.3 Results

Average daily values of air temperature for the period May 1–October 31, 1993, and water temperatures estimated from the site-specific correlation model are shown in Fig. 11.3a. In accordance with the typical evolution in the Lagoon of Venice (Pastres et al., 2004), the water temperatures ranged between 13.6 and 29.1 °C. The highest values were reached at

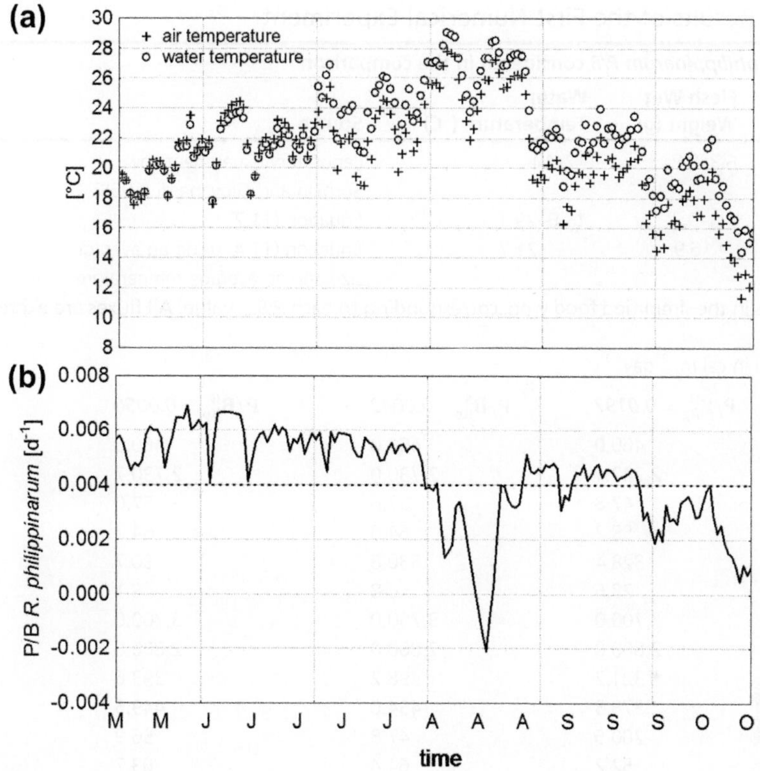


**FIGURE 11.3** (a) Daily values of air temperatures and water temperatures estimated from the site-specific correlation model, for the period May 1 October 31, 1993, and (b) $P/B_{Rp}$ values estimated from the *R. philippinarum* bioenergetic model.

the beginning of August. Afterward, the temperature steadily decreased, falling below 18 °C at the end of October. Equations (4) and (2) were solved using as input this time series, thus obtaining the daily $P/B_{Rp}$ values shown in Fig. 11.3b, and the average $P/B_{Rp}$ (see Table 11.2a). Mean weight and water temperature values for the same period were used as input in Eqn (4), to obtain an alternative estimate of $P/B_{Rp}$, based on the assumption that average thermal conditions and size were representative of the whole semester. As one can see in Table 11.2a, the value thus obtained is approximately 20% higher than the Eqn (2) estimate. Both values are four times lower than the average of the values estimated by Sorokin and Giovanardi (1995) through a set of short-term experiments performed on individuals of the same weight range, but acclimatized at temperatures between 12 and 14 °C.

The effect of changes in metabolic rates induced by water temperature fluctuations on the food web model was tested using as input parameters of the three values of $P/B_{Rp}$ reported in Table 11.2a. The inverse problem was solved three times by minimizing the norm of the vector of unknown energy fluxes in correspondence to each $P/B_{Rp}$ value. As a direct consequence of model formulation, variations between fluxes in the network

**Table 11.2** Results of the First Numerical Experiment

**a) Values of *R. philippinarum P/B* considered in the comparison**

| $P/B_{Rp}$ ($day^{-1}$) | Flesh Wet Weight (g) | Water Temperature (°C) | Source |
|---|---|---|---|
| 0.0197 | 6.3–8.8 | 12–14 | Laboratory experiments by Sorokin and Giovanardi (1995) |
| 0.0042 | 4.3–9.3 | 13.6–29.1 | Equation (11.2) |
| 0.0050 | 6.9 | 21.7 | Equation (11.4) using an average size and an average temperature |

**b) Energy fluxes in the simplified food web, corresponding to each $P/B_{Rp}$ value. All fluxes are expressed in cal $m^{-2}$ $day^{-1}$.**

Values expressed in cal $m^{-2}$ $day^{-1}$

| | $P/B^O_{Rp} = 0.0197$ | $P/B^A_{Rp} = 0.0042$ | $P/B^B_{Rp} = 0.0050$ | # |
|---|---|---|---|---|
| $CO_2 \rightarrow$ PHP | 460.0 | 460.0 | 460.0 | 1 |
| input → DET | 2,742.1 | 2,730.0 | 2,730.7 | 2 |
| PHP → MIZ | 47.8 | 57.6 | 57.0 | 3 |
| PHP → MEZ | 55.2 | 64.8 | 64.2 | 4 |
| PHP → DET | 328.4 | 330.8 | 330.7 | 5 |
| PHP → TAP | 28.6 | 6.8 | 8.1 | 6 |
| DET → BPL | 3,700.0 | 3,700.0 | 3,700.0 | 7 |
| BPL → $CO_2$ | 2,800.0 | 2,800.0 | 2,800.0 | 8 |
| BPL → MEZ | 321.2 | 398.2 | 393.6 | 9 |
| BPL → MIZ | 378.3 | 454.0 | 449.5 | 10 |
| BPL → TAP | 200.5 | 47.8 | 56.9 | 11 |
| MIZ → MIZ | 52.2 | 64.4 | 63.7 | 12 |
| MIZ → DET | 239.2 | 288.0 | 285.1 | 13 |
| MIZ → $CO_2$ | 103.2 | 152.0 | 149.1 | 14 |
| MIZ → MEZ | 55.2 | 64.8 | 64.2 | 15 |
| MIZ → TAP | 28.6 | 6.8 | 8.1 | 16 |
| MEZ → MEZ | 120.0 | 120.0 | 120.0 | 17 |
| MEZ → DET | 275.8 | 323.9 | 321.0 | 18 |
| MEZ → $CO_2$ | 155.8 | 203.9 | 201.0 | 19 |
| TAP → DET | 143.2 | 34.1 | 40.6 | 20 |
| TAP → $CO_2$ | 114.4 | 28.0 | 33.3 | 21 |
| TAP → Higher TLs | 28.8 | 6.1 | 7.3 | 22 |
| DET → TAP | 28.6 | 6.8 | 8.1 | 23 |

are linearly related to the ones between the $P/B_{Rp}$ input parameters. Energy fluxes connected with the TAP node experienced the highest relative variations, of the same size as that between the $P/B_{Rp}$ parameters. As an example, in the case of energy flow from TAP to higher trophic levels (#22), the solution obtained by imposing the $P/B_{Rp}{}^A$ is nearly five times lower than the one obtained with $P/B_{Rp}{}^O$. It is interesting to note that changes in fluxes were significant also for those ones which are not directly interconnected with the TAP node, such as the MIZ → $CO_2$ flux, which represents the respiration term of the micro-zooplankton compartment and presents an increase of approximately 50%. The relative variations among results obtained with $P/B^A_{\mathrm{Rp}}$ and $P/B^B_{\mathrm{Rp}}$ values are lower, but still significant for the fluxes directly connected with the TAP node (around 16%). Fluxes no. 1, 7, 8, and 17 remained fixed, since their values were constrained a priori (second row of Eqn (3)).

The effect of interannual variability of water temperature and the size structure of the population on the energy and matter flows were assessed by forcing Eqn (2) with the nine time series of water temperature available at station A09. The results, summarized in the first three columns of Table 11.3a, show the $P/B_{\mathrm{Rp}}$ values corresponding to cohort 0+ and 1+, starting respectively from initial shell lengths of 10 and 30 mm. $P/B_{\mathrm{Rp}}$ estimates change with the years, presenting minimum values in 1989, and maximum in 1986. As expected, the specific production of younger individuals was higher (approximately four times in average) than that of older ones. Three scenarios were compared, by considering alternative size structures of the population, and in each case values of $P/B_{\mathrm{Rp}}$ obtained for the two cohorts in the same year were averaged by weighting them by their relative frequency (see Table 11.3b). Results of the food web model for each scenario are reported in Table 11.3b. As in the previous simulation, relative variations among fluxes are linearly proportional to changes in $P/B_{\mathrm{Rp}}$ and, considering the temperature factor only, 30% of the fluxes are over 10%. These variations are slightly higher in the case of scenario 3, in which the 1+ class dominates the population. When two factors are mixed, relative variations exceed 25% for one-third of the fluxes, as is the case for variations between fluxes corresponding to the minimum production in scenario 1 and those relative to the maximum production in scenario 3.

## 11.4 Discussion

In the first numerical experiment carried out in this work, we compared the results obtained, using as input two $P/B$ values computed by means of the growth model and a reference value taken from the literature. Such comparison shows that the use of $P/B$ deduced from dynamic bioenergetic models leads to markedly different estimates of energy fluxes in the network. The literature value considered here was the result of a detailed set of short-term feeding and respiration measurements, carried out by Sorokin and Giovanardi (1995) with the aim of studying the trophic ecology of the Manila clam in the lagoon of Venice. The site and species specificity, and the range of diets considered in the experiments, were all good reasons for considering this as a robust estimate for the parameter, which was used in different food web models (Brigolin et al., 2011; Libralato et al., 2002; Pranovi et al., 2003). However, metabolic rates reported by Sorokin and Giovanardi (1995) were measured on individuals acclimatized at temperatures of 12–14 °C, while water temperatures in the shallow waters of the lagoon of Venice varied between 13.6 and 29.3 °C during the spring–summer semester considered here. This interval of temperatures includes thermal conditions which are less favorable for the Manila clam metabolism. According to the parameters of the individual growth model by Solidoro et al. (2000), metabolic activity of the Manila clam progressively decreases above

**Table 11.3** $P/B_{Rp}$ Values and Results of the Food Web Model for Each Scenario

a) $P/B_{Rp}$ values corresponding to cohort 0+ and 1+, and scenarios compared by considering alternative size structures of the population

| | Cohort 0+ $P/B_{Rp}$ | Cohort 1+ $P/B_{Rp}$ | Scenario 1 50% 0+ 50% 1+ | Scenario 2 75% 0+ 25% 1+ | Scenario 3 25% 0+ 75% 1+ |
|---|---|---|---|---|---|
| Values expressed in $day^{-1}$ | | | | | |
| 1986 | 0.0177 | 0.0047 | 0.0112 (max) | 0.0145 (max) | 0.0079 (max) |
| 1987 | 0.0169 | 0.0042 | 0.0106 | 0.0138 | 0.0074 |
| 1988 | 0.0176 | 0.0045 | 0.0111 | 0.0143 | 0.0078 |
| 1989 | 0.0162 | 0.0040 | 0.0101 (min) | 0.0131 (min) | 0.0070 (min) |
| 1990 | 0.0171 | 0.0044 | 0.0108 | 0.0140 | 0.0076 |
| 1995 | 0.0170 | 0.0044 | 0.0107 | 0.0139 | 0.0075 |
| 1996 | 0.0176 | 0.0045 | 0.0111 | 0.0144 | 0.0078 |
| 1997 | 0.0177 | 0.0046 | 0.0111 | 0.0144 | 0.0078 |
| 1998 | 0.0168 | 0.0044 | 0.0106 | 0.0137 | 0.0075 |

b) Results of the food web model for each scenario. All fluxes are expressed in cal $m^{-2}$ $day^{-1}$

| | Scenario 1 | | Scenario 2 | | Scenario 3 | |
|---|---|---|---|---|---|---|
| | min($P/B_{Rp}$) | max($P/B_{Rp}$) | min($P/B_{Rp}$) | max($P/B_{Rp}$) | min($P/B_{Rp}$) | max($P/B_{Rp}$) |
| Values expressed in cal $m^{-2}$ $day^{-1}$ | | | | | | |
| $CO_2$ → PHP | 460.0 | 460.0 | 460.0 | 460.0 | 460.0 | 460.0 |
| input → DET | 2,735.3 | 2,736.3 | 2,738.1 | 2,739.3 | 2,732.5 | 2,733.4 |
| PHP → MIZ | 53.3 | 52.5 | 51.1 | 50.1 | 55.6 | 54.9 |
| PHP → MEZ | 60.5 | 59.8 | 58.4 | 57.4 | 62.8 | 62.1 |
| PHP → DET | 329.7 | 329.5 | 329.2 | 328.9 | 330.3 | 330.1 |
| PHP → TAP | 16.4 | 18.2 | 21.3 | 23.6 | 11.4 | 12.8 |
| DET → BPL | 3,700.0 | 3,700.0 | 3,700.0 | 3,700.0 | 3,700.0 | 3,700.0 |
| BPL → $CO_2$ | 2,800.0 | 2,800.0 | 2,800.0 | 2,800.0 | 2,800.0 | 2,800.0 |
| BPL → MEZ | 364.4 | 358.1 | 347.2 | 339.1 | 382.1 | 377.0 |
| BPL → MIZ | 420.7 | 414.5 | 403.8 | 395.9 | 438.2 | 433.1 |
| BPL → TAP | 114.9 | 127.4 | 149.1 | 165.0 | 79.7 | 89.9 |
| MIZ → MIZ | 59.0 | 58.0 | 56.3 | 55.0 | 61.9 | 61.0 |
| MIZ → DET | 266.5 | 262.5 | 255.6 | 250.5 | 277.8 | 274.5 |
| MIZ → $CO_2$ | 130.5 | 126.5 | 119.6 | 114.5 | 141.8 | 138.5 |
| MIZ → MEZ | 60.5 | 59.8 | 58.4 | 57.4 | 62.8 | 62.1 |
| MIZ → TAP | 16.4 | 18.2 | 21.3 | 23.6 | 11.4 | 12.8 |
| MEZ → MEZ | 120.0 | 120.0 | 120.0 | 120.0 | 120.0 | 120.0 |
| MEZ → DET | 302.7 | 298.8 | 292.0 | 287.0 | 313.8 | 310.6 |
| MEZ → $CO_2$ | 182.7 | 178.8 | 172.0 | 167.0 | 193.8 | 190.6 |
| TAP → DET | 82.1 | 91.0 | 106.5 | 117.9 | 56.9 | 64.2 |
| TAP → $CO_2$ | 67.3 | 74.6 | 87.3 | 96.6 | 46.7 | 52.7 |
| TAP → Higher TLs | 14.8 | 16.4 | 19.2 | 21.2 | 10.2 | 11.6 |
| DET → TAP | 16.4 | 18.2 | 21.3 | 23.6 | 11.4 | 12.8 |

the optimal temperature of 22.7 °C, it stops feeding at 32 °C, and the upper temperature of 35 °C is lethal. In such thermal conditions, the use of a dynamic model which takes into account explicitly the dependence of metabolic rates on water temperature seems to be an appropriate choice to set a *P/B* value representative for the entire season. Moreover, an advantage of using *P/B* values provided by the model is that species-specific maximum growth rate $G_{max}$ in Eqn (4) was calibrated on a time series of length with age data collected in the field and, for our purpose, can be regarded as a more representative estimate of the variability of the metabolism of this species than short-term filtration and respiration measurements performed under controlled laboratory conditions.

Differences between the results obtained by applying Eqn (2) and Eqn (1) with the average values of temperature and size of the clam can be interpreted in relation to the nonlinearity of the individual growth model. This is clearly visible from the trajectory of $P/B_{Rp}$ (Fig. 11.3b), which presents a dramatic decrease in August in correspondence with the unfavorable thermal condition due to high water temperatures. These results indicate that metabolic rate estimates obtained from an average water temperature and body size may lead to energy flux estimates which are not representative for the whole time frame of application of a steady-state model. Although solving this with a dynamic model equation could have partly overcame this bias, an important assumption in our first screening experiment was that the specific production is not density dependent and that the size of an individual at the beginning of the simulation is representative of the average size of the population. In the second set of simulations, we removed this hypothesis. The interesting aspect of this approach is that it allows one to consider effects on the food web which are induced by external fluctuations of physical environmental factors, mixed with internal modifications in the structure of the population, which could be a consequence of abrupt changes in the social environment, such as those caused by the introduction of a new competitor. The age/size distribution, which was roughly represented here by comparing three different scenarios, can be described more realistically by simulating explicitly the dynamics of the population, coupling the individual-based growth model with a population dynamic model (Gurney and Nisbet, 1998). Considering the two factors together, approximately one-third of the fluxes in the network presented variations of more than 25%. Comparable values were obtained by Vézina and Savenkoff (1999), who tested the sensitivity of the solutions of an inverse model by imposing arbitrary random perturbations, ranging between ±10% of the nominal value on measured fluxes, and between ±50% of the nominal value on micro- and meso-zooplankton biomasses. The difference in the approach presented here is given by the range of variability imposed for the parameters, which is not the result of an arbitrary choice, nor of literature analysis or educated guess, but is the result of a deterministic dynamic model forced by realistic water temperature values. In this way, the growth model can be used to test the sensitivity of the food web model with respect to interannual fluctuations of water temperature and population age/size structure.

## 11.5 Conclusions

The comparison carried out shows that fluctuations of water temperature can induce remarkable effects on food web model estimations, and that metabolic rate estimates obtained from an average water temperature and body size may lead to estimate energy fluxes which are not representative for the whole time frame of application of a steady-state model. Offline coupling of dynamic growth models and food web models provides a way to test the sensitivity of the result of the food web model to interannual fluctuations of water temperature and population age/size structure. The methodology tested in this work should not be seen as an isolated example, in which a validated bioenergetic model of the studied species is available. Research efforts since the 1990s have remarkably broadened the range of species for which bioenergetic models or dynamic energy budget models are available (see, e.g., http://www.bio.vu.nl/thb/), potentially allowing to extend this methodology to many single-species compartments present in coastal and marine food webs.

## References


Belgrano, A., Scharler, U.M., Dunne, J., Ulanowicz, R.E., 2005. Aquatic Food Webs: An Ecosystem Approach. Oxford University Press, Oxford.

Brigolin, D., Savenkoff, C., Zucchetta, M., Pranovi, F., Franzoi, P., Torricelli, P., Pastres, R., 2011. Inverse analysis of the structure of the Venice lagoon ecosystem. Ecol. Model. 222, 2404–2413.

CEC, 2000. Council directive of 23 October 2000, establishing a framework for community action in the field of water policy (2000/60/EC). L327 of 22.12.2000. Off. J. Eur. Commun. 1–72.

Degré, D., Leguerrier, D., du Chatelet, E.A., Rzeznik, J., Auguet, J.C., Dupuy, C., Marquis, E., Fichet, D., Struski, C., Joyeux, E., Sauriau, P.G., Niquil, N., 2006. Comparative analysis of the food webs of two intertidal mudflats during two seasons using inverse modelling: Aiguillon Cove and Brouage Mudflat, France. Estuar. Coast. Shelf Sci. 69, 107–124.

Gurney, W.S.C., Nisbet, R.M., 1998. Ecological Dynamics. Oxford University Press, New York.

GRAL website. <http://www.gral.venezia.it> (in Italian), (accessed 17.11.11).

Kooijman, S.A.L.M., 2000. Dynamic Energy and Mass Budgets in Biological Systems, 419. Cambridge University Press, Cambridge.

Libralato, S., Pastres, R., Pranovi, F., Raicevich, S., Granzotto, A., Giovanardi, O., 2002. Comparison between the energy flow networks of two habitats in the Venice Lagoon. Mar. Ecol. 23, 228–236.

Mann, K.H., Lazier, J.R.N., 2006. Dynamics of Marine Ecosystems. Biological–Physical Interactions in the Oceans, third ed. Blackwell publishing.

Pastres, R., Brigolin, D., Petrizzo, A., Zucchetta, M., 2004. Testing the robustness of primary production models in shallow coastal areas: a case study. Ecol. Model. 179, 221–233.

Pellizzato, M., Galvan, T., Lazzarini, R., Penzo, P., 2011. Recruitment of *Tapes philippinarum* in the Venice Lagoon (Italy) during 2002–2007. Aquacult. Int. 19, 541–554.

Pranovi, F., Libralato, S., Raicevich, S., Granzotto, A., Pastres, R., Giovanardi, O., 2003. Mechanical clam dredging in Venice Lagoon: ecosystem effects evaluated with a trophic mass-balance model. Mar. Biol. 143, 393–403.

Savenkoff, C., Swain, D.P., Hanson, J.M., Castonguay, M., Hammill, M.O., Bourdages, H., et al., 2007. Effects of fishing and predation in a heavily exploited ecosystem: comparing periods before and after the collapse of groundfish in the southern Gulf of St. Lawrence (Canada). Ecol. Model. 204, 115–128.

Soetaert, K., Van Oevelen, D., 2008. LIM: Linear Inverse Model Examples and Solution Methods. R package version 1.4. http://lib.stat.cmu.edu/R/CRAN/web/packages/LIM/index.html. (accessed Oct 2011).

Soetaert, K., Van den Meersche, K., Van Oevelen, D., 2008. LIM-Solve: Solving Linear Inverse Models. R Package Version 1.5.1. http://lib.stat.cmu.edu/R/CRAN/web/packages/limSolve/index.html. (accessed Oct 2011).

Solidoro, C., Pastres, R., Melaku Canu, C., Pellizzato, M., Rossi, R., 2000. Modelling the growth of *Tapes philippinarum* in Northern Adriatic lagoons. Mar. Ecol. Prog. Ser. 199, 137–148.

Sorokin, Y.I., Giovanardi, O., 1995. Trophic characteristics of the Manila clam (*Tapes philippinarum* Adams and Reeve). ICES J. Mar. Sci. 52, 853–862.

Sorokin, Y.I., Sorokin, P.Y., Giovanardi, O., Dalla Venezia, L., 1996. Study of the ecosystem of the lagoon of Venice with emphasis on anthropogenic impact. Mar. Ecol. Prog. Ser. 141, 247–261.

Van den Meersche, K., Soetaert, K., Van Oevelen, D., 2009. xsample(): an R function for sampling linear inverse problems. J. Stat. Softw. 30, 1–15.

Van Oevelen, D., Van den Meersche, K., Meysman, F.J.R., Soetaert, K., Middelburg, J.J., Vézina, A.F., 2009. Quantifying food web flows using linear inverse models. Ecosystems 13, 32–45.

Vézina, A.F., Platt, T., 1988. Food web dynamics in the ocean. I. Best-estimates of flow networks using inverse methods. Mar. Ecol. Prog. Ser. 42, 269–287.

Vézina, A.F., Savenkoff, C., 1999. Inverse modeling of carbon and nitrogen flows in the pelagic food web of the northeast subarctic Pacific. Deep Sea Res. II 46, 2909–2939.

# 12

# Trophic Network Analysis: Comparison of System-Wide Properties

Oksana Y. Buzhdygan*,[1], Bernard C. Patten[†], Svitlana S. Rudenko*

*DEPARTMENT OF ECOLOGY AND BIOMONITORING, CHERNIVTSI NATIONAL UNIVERSITY, CHERNIVTSI 58012, UKRAINE, [†]ODUM SCHOOL OF ECOLOGY AND FACULTY OF ENGINEERING, UNIVERSITY OF GEORGIA, ATHENS, GA 30602, USA

[1]CORRESPONDING AUTHOR: E-MAIL: OKSANA.BUZH@GMAIL.COM

## 12.1 Introduction

The focus on complex structure and behavior of ecosystems has derived many attempts in their assessment and provided a myriad of system performance measures (Jørgensen, 1986; Lindeman, 1942; Margalef, 1963; Odum, 1969; Patten, 1991; Ulanowicz, 1980,1986). The understanding of relationships between ecosystem properties has been in the forefront of system research in ecology for years (Elton, 1958; Gardner and Ashby, 1970; Odum, 1969; MacArthur, 1955; May, 1973).

Empirical as well as theoretical approaches in ecosystem study led to the considering of network organization as the reflection of ecosystem structure and functions. Numerous works consider trophic (food-web) networks specifically as the essential in ecosystem organization (Arii et al., 2007; Dell et al., 2005; Garlaschelli et al., 2003; Martinez, 1994; Paine, 1988; Pimm, 1982; Warren, 1994; Williams and Martinez, 2000). Ecological networks capture the organisms within the system as well as multiple interactions between them. Such interactions serve as the exchange of conservative substances, such as energy and matter, and create complex structures and behaviors within ecological systems.

Ecological network analysis (ENA), environmental application of Leontief's economic input–output analysis (1936, 1966), has been adopted as a way to bring multicompartment modeling (Matis et al., 1979) into ecological system analysis and theory. Several advanced approaches of ENA theory are available: input–output analysis (Hannon, 1973, 1985); network environ analysis (NEA) (Patten, 1978, 1981, 1982, Holoecology, in prep.; Fath and Patten, 1999); energy intensity analysis (Herendeen, 1981); ascendency study (Ulanowicz, 1980, 1986, 1997); embodied energy analysis (Brown andHerendeen, 1996; Herendeen, 1989).

Models of the Ecological Hierarchy. DOI: http://dx.doi.org/10.1016/B978-0-444-59396-2.00012-2

ISSN 0167-8892, 

Many system-level organizational properties have been provided by ENA that drive several attempts to connect ENA output variables and to determine their interrelatedness (Jørgensen, 2002). Despite the plurality of these approaches, the simplification associated with data availability produces limitations associated with assessing the network properties (Cohen et al., 1993; Paine, 1988). Because of differences in network construction and aggregation (Ulanowicz, 1986) and also because currency and time-scales differ among networks, direct comparisons of their properties are difficult and limited.

To overcome the above-stated limitations, we compared seven empirical trophic networks of geographically close pastoral ecosystems. In order for comparative network analysis to be valid, the study food webs were constructed similarly with no variation in aggregation, currency, or time scale.

Another current limitation in network analysis is associated with a focus on few particular indices, even though a multiplicity of such quantitative studies remains necessary. Our work covers a wider range of network analysis output variables that reflect system-wide organizational properties, such as degree of system connectance, link density, total system throughflow, cycling degree, ascendency, developmental capacity, indirect effects dominance, system aggradation, system synergism, and mutualism. Current investigation brings these system-level measures together for the assessment of their interrelations.

## 12.2 Materials and Methods

### 12.2.1 Study Area

The research objects are food webs of pastoral ecosystems located in the Chernivtsi Region (west of the Ukraine). The Chernivtsi Region is divided into the three main physic-geographical zones: **Plain, Foothills**, and **Karpaty Mountains**, and each zone has specific climate conditions, landscape and relief characteristics, and age and type of rocks. Current comparative analysis is based on the food webs of the seven pastoral ecosystems located throughout the Karpaty mountain zone, which is determined by high elevation (Fig. 12.1). The soils of the Karpaty Mountains have strong acidity that is caused by high concentration of $Al^{3+}$ in the soil solution. The close geographic locations of the study plots gave us the opportunity to avoid significant influences of climate conditions on comparative analysis results. Table 12.1 depicts geographic characteristics and basic soil properties of the seven study plots we focused on. All of the study grasslands, unmanaged since 1992, are communal grazing lands.

### 12.2.2 Sampling

The sampling and analysis methods were performed similarly for each of the compared ecosystems. Biological samples for food-web analysis were gathered during peak growing

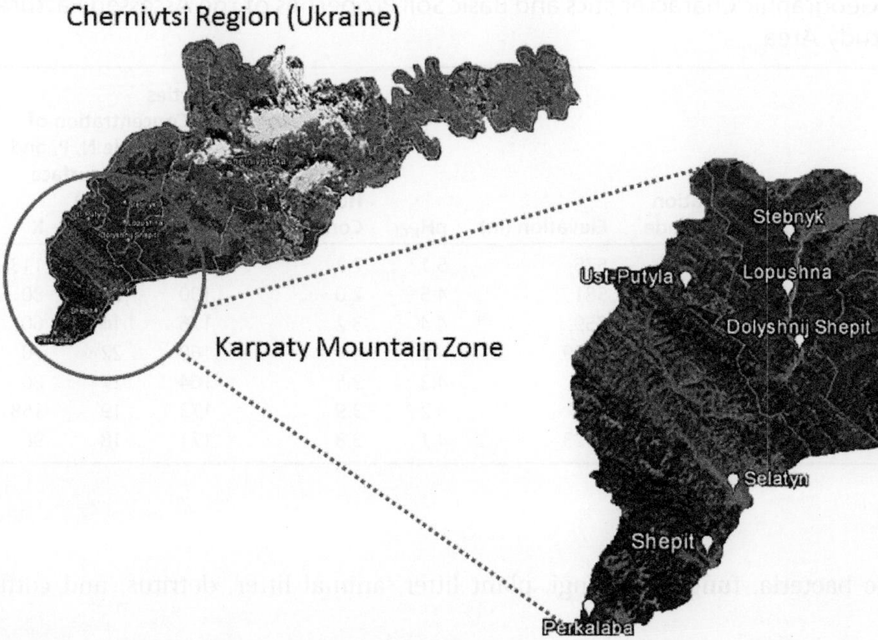


**FIGURE 12.1** Study plots (pastures) located throughout the Karpaty Mountain Zone of Chernivtsi Region (west of the Ukraine). For color version of this figure, the reader is referred to the online version of this book.

seasons (June–July) in 2005, 2006, and 2007. Study plots for each of the compared pastures were 10 m × 10 m.

Plant and insect specimens were identified as much as possible to the species level. Earthworms (subclass Oligochaeta) were separated from the 1 $m^3$ plots by quantitative hand-sorting. Microbiological soil analysis was based on cell counts of three microbial groups: Heterotrophic Bacteria, Fungi (Micromycetes), and Ray Fungi (Actinomycetes). Cells were cultured on specific substrates under controlled temperature ($T$) conditions—Heterotrophic Bacteria: meat-peptone agar, $28 < T < 30$ °C; Fungi (Micromycetes): modified Czapek-Dox substrate with streptomycin, $20 < T < 25$ °C; Ray Fungi (Actinomycetes): starch-ammonium agar, $28 < T < 30$ °C. Cattle density was counted according to number of animals/100 $m^2$.

Several faunal groups were omitted because of available resource limitations. This affects our description of food webs, but still allows their valid comparative study as our sampling and analysis methods were standardized.

## 12.2.3 Food-Web Construction

We defined trophic compartments based on distinct feeding roles in the studied pastures. Our basic categories for compartments were plant species, their pollen and nectar, cattle, ontogenetic stages and sexes of insects reflecting distinct trophic roles, earthworms,

**Table 12.1** Geographic Characteristics and Basic Soil Properties of the Assessed Pastures within the Study Area

| Study Plots (Pastures) | Geographic Location | | Elevation (m) | $pH_{KCl}$ | Soil Properties | | | |
|---|---|---|---|---|---|---|---|---|
| | | | | | Humus Content (%) | Concentration of available N, P, and K on the surface soil (mg/kg) | | |
| | Latitude | Longitude | | | | N | P | K |
| Stebnyk | 48° 09′ N | 25° 17′ E | 575 | 5.1 | 2.2 | 122 | 32 | 113 |
| Lopushna | 48° 05′ N | 25° 17′ E | 581 | 4.5 | 2.0 | 100 | 21 | 80 |
| DolyshnijShepit | 48° 02′ N | 25° 18′ E | 759 | 4.4 | 3.2 | 176 | 11 | 60 |
| Ust-Putyla | 48° 05′ N | 25° 06′ E | 1039 | 4.2 | 4.1 | 160 | 22 | 70 |
| Selatyn | 47° 52′N | 25° 11′ E | 831 | 4.3 | 3.5 | 164 | 27 | 80 |
| Shepit | 47° 48′ N | 25° 08′ E | 917 | 4.2 | 3.9 | 173 | 19 | 158 |
| Percalaba | 47° 44′N | 24° 56′ E | 1423 | 4.1 | 3.8 | 171 | 18 | 96 |

heterotrophic bacteria, fungi, ray fungi, plant litter, animal litter, detritus, and cattle excrement.

To construct and portray our food webs, we used the software Pajek (Batagelj and Mrvar, 2010) for large network analysis, and *Ucinet 6* (Borgatti, et al., 2002) for social network analysis.

We formed a square adjacency matrix, $A_{n\times n} = (a_{ij})$, where $i, j = 1, \ldots, n$ compartments, oriented from rows ($i$) to columns ($j$). A matrix entry $a_{ij} = 1$ signifies a biomass [M (mass)–L (length)–T (time) dimensions = *M*] feeding flow, $f_{ij}$ [$ML^{-2}T^{-1}$ (mass/unit area · time)], directed from row compartment $i$ to column compartment $j$; $a_{ij} = 0$ indicates no $i$ to $j$ food transfer ($f_{ij} = 0$). We used network environ theory (Patten, 1981, 1982, Holoecology, in prep.) to construct the food-web networks. Thus, each compartment $i$ has a boundary input $z_i$ [$ML^{-2}T^{-1}$], and output $y_i$ [$ML^{-2}T^{-1}$].

To quantify adjacency-based relations from qualitative digraphs, we transformed the adjacency matrix, $A_{n\times n}$ in a flow matrix $F_{n\times n} = (f_{ij})$, where $i, j = 1, \ldots, n$ compartments, oriented from rows ($i$) to columns ($j$), and used the equiprobability concept from probability theory. According to Laplace's principle of indifference, a matrix entry $f_{ij}$ (a biomass feeding flow [$ML^{-2}T^{-1}$]) as well as the boundary output $y_i$ are assigned the probability $1/N_i$, where $N_i$ signifies a number of mutually exclusive feeding flows directed from row compartment $i$ to column compartments ($1, \ldots, n$) including a boundary output $y_i$. Boundary inputs $z_i$ and standing stocks $x_i$ are equal to 1.

## 12.2.4 Food-Web Simulation

For simulation of study networks, we used a dynamic web-based simulation and network analysis software, EcoNet 2.1 *Beta* (available at http://eco.engr.uga.edu/).

Network analysis was performed based on the final state of the solution when systems reached a static steady-state ($dx_i/dt = 0$ as the system inputs and outputs are equal at steady state).

The simulation flow type was based on donor-controlled mass-action kinetics. Thus, the rate of the flow $f_{ij}$ is computed by EcoNet as the product of the flow coefficient $c_{ij}$ and the stock value $x_i$ of the originating compartment $i$. The rate of the flow from $i$ to $j = c_{ij} \times x_i$.

A differential mass–energy balance equation for donor-controlled flow type is as follows:

$$\frac{dx_i}{dt} = z_i + \Sigma j(\neq i)c_{ji} \times x_j - \Sigma i(\neq j)c_{ij} \times x_i - y_i \times x_i$$

where $z_i$ and $y_i$ are boundary inputs and outputs, respectively.

For a more comprehensive introduction to simulation and network analysis in EcoNet refer to http://eco.engr.uga.edu/DOC/econet1.html, Kazanci (2007), and Schramski et al. (2010).

## 12.2.5 System-Wide Properties

As discussed earlier EcoNet drives the system from the given initial conditions to steady state and outputs the system-wide organizational properties based on the final state of the solution.

Ten system-level indices assessed are fully documented in literature, but brief descriptions are provided in the following. While our focus is on these whole system variables, we incorporated the **system size** (**N**, number of nodes), **number of links** (**L**, $\sum_{i \text{ or } j} a_{ij}$), and **number of trophic classes** (**Cl**) as additional network properties appropriate to comparative interpretations. The definition of trophic classes is described in Section 12.2.6.

**Link density (LD)** is assessed as the ratio of the number of links ($L$) to the network size (number of nodes $N$) (Bersier et al., 2002; Dell et al., 2005): **LD** $= L/N$.

**Connectance** (**C**) is the ratio of actual to possible links (Gardner and Ashby, 1970; Martinez, 1992; Parrott, 2010): **C** $= L/N^2$.

**Total system throughflow** (**TST**) is the sum of compartment throughflows (total amount of flows within a network); it is dependent on ecosystem structure (Hannon, 1973; Finn, 1976; Han, 1997):

$$\textbf{TST} = \sum T_i$$

where $T_i$ is the total amount of flow through compartment $i = 1, \ldots, n$.

**Finn cycling index (FCI)** is the fraction of TST that cycles (Finn, 1976):

$$\textbf{FCI} = \frac{\text{TST}_\text{c}}{\text{TST}}$$

where $TST_c$, the cycled portion, is the weighted sum of cycling efficiencies of all compartments (Kazanci et al., 2009):

$$TST_c = C_1T_1 + C_2T_2 + \cdots + C_nT_n$$

Cycling efficiency is $C_i = n_{ii} - 1/n_{ii}$, where $n_{ii}$ is the number of times a flow quantity will return to $i$ before being lost from the system (Finn, 1976; Fath and Borrett, 2006).

**Indirect effects index (IEI)** is the amount of flow that occurs over indirect versus direct connections (Higashi and Patten, 1989):

$$\mathbf{IEI} = \frac{\sum(N - I - G)z}{\sum Gz}$$

$N$ is the dimensionless integral (boundary + direct + indirect) flow matrix:

$$N = I + G^1 + G^2 + \cdots + G^m + \cdots = (I - G)^{-1}$$

$G$ is the matrix of dimensionless direct flow intensities from $i$ to $j$:

$$G = (g_{ij}) = \left(\frac{f_{ij}}{T_i}\right)$$

where $T_i$ is the total amount of flow through compartment $i$,

$I = G^0$ is the boundary input flow intensity;

as stated earlier $G^1$ is the direct flow intensity matrix, $G^2 \ldots G^m$ are the indirect flow intensity matrices (fractions of boundary flow that travels from node $i$ to $j$ over all pathways of length $m$, where $m$ shows the orders given by the divergent power series $m = 2, \ldots, \infty$).

The integral matrix $N$ multiplied by boundary input vector $z$ returns the throughflow vector $T$:$T = Nz$.

**Synergism index (SI)** is the benefit–cost ratio ($b/c$) (Patten, 1991, 1992) of total positive utility $\sum(+U)$ to total negative utility $\sum(-U)$ in the system specifying pairwise compartment relations (Fath and Borrett, 2006; Fath and Patten, 1998; Patten, 1991, 1992):

$$\mathbf{SI} = \left|\frac{b}{c}\right| = \left|\frac{\sum(U^+)}{\sum(U^-)}\right|$$

where $U^{\pm}$ are positive and negative partition matrices of the dimensionless integral (boundary + direct + indirect) utility matrix $U$ (Patten, 1991, 1992):

$$U = I + D^1 + \cdots + D^m + \cdots = (I - D)^{-1}$$

$D$ is a direct utility matrix (net-flow intensity matrix) where:

$$D = (d_{ij}) = \left(\frac{f_{ij} - f_{ji}}{T_i}\right)$$

where $d_{ij}$ can be positive or negative ($-1 \leq d_{ij} < 1$) as it represents the direct utility between compartments $j$ and $i$ (net-flow between $j$ and $i$ is expressed relative to the total amount of flow through compartment $i$ ($T_i$);

$I = D^0$ is the initial intensive utility input matrix;

$D^2 \ldots D^m$ are the indirect utilities that correspond to the flows of the same power $m = 2, \ldots, \infty$.

**Mutualism index (MI)** is ratio of number of positive (+) to negative (−) signs in network utility analysis matrices that specify kinds of pairwise interactions (Fath, 2007; Fath and Borrett, 2006; Fath and Patten, 1998, 1999; Patten, 1991, 1992):

$$\mathbf{MI} = \frac{\sum \mathrm{sign}(U^+)}{\sum \mathrm{sign}(U^-)}$$

where $U$, $U^+$ and $U^-$ are as described above (Patten, 1991, 1992).

**Ascendency (AS)** is the degree of network development that includes average mutual information (AMI) and TST (Patrício et al., 2004; Morris et al., 2005; Ulanowicz, 1986, 1997; Ulanowicz et al., 2006):

$$\mathbf{AS} = \mathrm{TST} \times \mathrm{AMI},$$

where AMI (bits) is the degree of organization with which the exchanges between compartments are processed:

$$AMI = \sum p(T_{ij}) \times \log_2[\{p(T_{ij})/p(T_j)\}/p(T_i)]$$

where $T_{ij}$ is the flow from $j$ to $i$;

$p(T_{ij})$ is the joint probability given by:

$$p(T_{ij}) = \frac{\mathrm{T}_{ij}}{\mathrm{TST}}, p(T_i) = \sum jp(T_{ij}), \text{ and } p(T_j) = \sum ip(T_{ij}).$$

**Full development capacity (DC)** is the network flow organization. It is the upper bound of ascendancy. It is calculated as the product of TST by diversity of flow structure ($H_f$) that is estimated using the Shannon (1948) information formula (Christian et al., 2005; Ulanowicz, 1986):

$$\mathbf{DC} = TST \times H_f$$

where Shannon flow diversity $H_f$ is based on the individual joint probabilities of flows from each species $j$ to each species $i$:

$$H_f = \sum (-p(T_{ij}) \times \log_2(p(T_{ij}))$$

where $\mathrm{T}_{ij}$ is the flow from $j$ to $i$; and $p(T_{ij})$ is as stated above.

**Network aggradation index (AI)** can also be found in literature as a multiplier effect (Samuelson, 1948), or an average path length (Finn, 1976), or a flow multiplying ability (Han, 1997). All of these reflect the average number of times a unit of input flow passes through a system before exiting (Patten and Fath, 1998; Fath and Patten, 2001; Ulanowicz et al., 2006):

$$\mathbf{AI} = \frac{\mathrm{TST}}{\sum Z_i}$$

where $z_i$ is a boundary input of compartment $i = 1, \ldots, n$;

TST is total system throughflow.

### 12.2.6 Definition of Trophic Classes

The presence or absence of feeding relations between food-web compartments was determined by feeding history of species under study. In order to investigate the number of trophic classes (**Cl**) in our food-webs, we used the regular equivalence method drawn from social network theory (Borgatti and Everett, 1993) and applied as a tool in *Ucinet 6* (Borgatti, et al., 2002). Regular equivalence algorithm assesses similarity of the trophic roles of compartments through the use of binary presence–absence feeding relations between them (adjacency matrix $A_{n\times n}$). Johnson's hierarchical clustering (Johnson, 1967) of equivalence similarity values allows us to define the separate trophic classes. For more details of regular equivalence algorithm and current aggregation methods see Borgatti and Everett (1993) and Luczkovich et al. (2003).

Based on the trophic role that species play in pastoral ecosystems, the following 15 trophic classes were derived by employing the above-stated method: (1) Gazing Plants; (2) Grazing-Tolerant Plants; (3) Pollen and Nectar; (4) Cattle; (5) Herbivorous Insects; (6) Pollinators; (7) Bloodsucking Parasites of Cattle; (8) Carnivorous Insects; (9) Omnivores Insects; (10) Nonspecific Insects (consumed in other ecosystems); (11) Decomposers; (12) Plant Litter; (13) Animal Litter; (14) Cattle Excrement; and (15) Detritus.

The network property **Cl** (trophic class number) demonstrates how many trophic classes from the 15 above-stated possibilities are presented in each food web. The variation in **Cl** results from the presence/absence of species of such classes as Nonspecific Insects; Carnivorous Insects; Bloodsucking Parasites of Cattle; and Omnivorous Insects. All other classes have been found in each food web from the seven studied ones.

### 12.2.7 Abbreviations

The following abbreviations are used throughout this chapter:

| | | | |
|---|---|---|---|
| ENA | ecological network analysis | SI | synergism index |
| N | system size (number of nodes) | MI | mutualism index |
| L | number of links | AS | ascendency measure |
| Cl | number of trophic classes | DC | full development capacity |
| LD | link density measure | AI | network aggradation index |
| C | network connectance | CV | coefficient of variation |
| TST | total system throughflow | $r$ | Pearson correlation coefficient |
| FCI | Finn cycling index | $r_s$ | Spearman correlation coefficient |
| IEI | indirect effects index | | |

### 12.2.8 Statistical Analysis

We performed the Spearman rank order correlation test and the Pearson product-moment correlation analysis in order to determine the strength of statistically significant

($p < 0.05$) relationships between system-wide properties pairwise. Then we used cluster analysis to examine the correlation-based similarity between network properties.

## 12.3 Results

### 12.3.1 Network Properties Range and Variability

The seven empirical pastoral food webs we focused on in this study exhibited a range of **sizes** ($49 \leq N \leq 113$), **link numbers** ($98 \leq L \leq 856$), and **trophic class numbers** ($12 \leq Cl \leq 15$).

Table 12.2 summarizes the ranges and variability of network properties across the study area. Figure 12.2 depicts the values of network properties for each of the studied ecosystems.

### 12.3.2 Network Properties Interrelations

An analysis of relationships between the network properties demonstrates few different behavioral patterns. Most of network indices (**LD, C, TST, FCI, IEI, AS, DC, AI**, and **MI**) are positively related to each other, and negatively related to **SI**. Thus, network synergism measure runs counter to other trends in our research. Further statistical analysis of the observed relationships will be discussed in the following.

Figure 12.3 demonstrates statistically significant Spearman correlations between the whole-system properties assessed. It reveals that **LD** and **IEI** have the largest number of statistically significant relationships (eight significant of nine possible interrelations) in comparison with the other variables. They are followed by **TST, FCI, AS, DC**, and **AI**. Each of these has seven of nine possible statistically significant interrelations. **C** is significantly related with two of the nine indices, while **MI** shows only one significant correlation.

Cluster dendrogram (Fig. 12.4) depicts the similarity between network properties based on their pairwise correlations. It shows that **TST, DC**, and **AS** were the most similar to each other and were joined to form the first cluster. The Spearman correlation test

**Table 12.2** Ranges and Variability of Network Properties Assessed across the Pastoral Food Webs ($n = 7$) within the Study Area

| Network Properties | Minimum | Maximum | Mean | CD | CV (%) |
|---|---|---|---|---|---|
| LD | 2.00 | 7.58 | 4.47 | 2.08 | 46 |
| C | 0.03 | 0.07 | 0.05 | 0.01 | 28 |
| TST | 109.4 | 342.6 | 245.6 | 80.3 | 33 |
| FCI | 0.03 | 0.06 | 0.04 | 0.01 | 28 |
| IEI | 1.05 | 1.62 | 1.35 | 0.19 | 14 |
| AS | 178.6 | 582.9 | 412.2 | 137 | 33 |
| DC | 476.8 | 1604.6 | 1127.6 | 382.6 | 34 |
| AI | 2.23 | 3.07 | 2.69 | 0.30 | 11 |
| SI | 1.69 | 2.10 | 1.80 | 0.15 | 8 |
| MI | 0.37 | 0.96 | 0.56 | 0.20 | 38 |

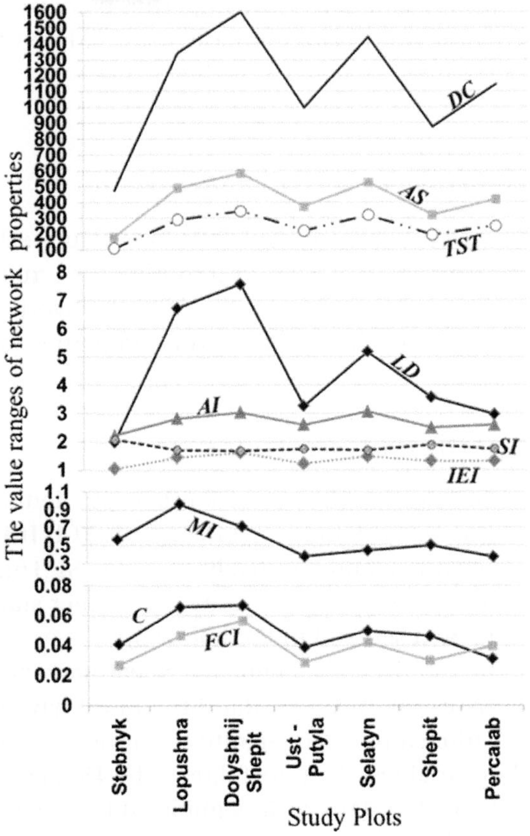


**FIGURE 12.2** The value ranges of system-wide properties assessed across the seven pastoral food webs within the study area.

shows the statistically significant perfect positive relations ($r_s = 1$, $p = 0.000$) between each of them pairwise. This reveals that each of the current three variables being compared here is the perfect monotone function of the others. Moreover, the statistically significant positive Pearson correlation fits the data with nearly perfect linear dependence ($r = 0.999$, $p = 0.000$ for each pair being compared).

**TST, AS,** and **DC** also show perfect monotonically increasing association with $\boldsymbol{N}$ ($r_s = 1$, $p = 0.000$ for each pair being compared). The product-moment correlations of $\boldsymbol{N}$ with **TST** ($r = .98$), **AS** ($r = 0.99$), and **DC** ($r = 0.99$) are also strongly significant ($p = 0.000$ for each pair being compared). Simple linear regression fits the data with high significance as follows:

$$Y_{TST} = 3.7 \times X_N - 81.8; r^2 = 0.99, F(1,5) = 157.7, p = 0.0001$$

$$Y_{AS} = 6.3 \times X_N - 147.7; r^2 = 0.97, F(1,5) = 157.7, p = 0.0004$$

$$Y_{DC} = 17.5 \times X_N - 436.89, r^2 = 0.97, F(1,5) = 7656.4, p = 0.000.$$

The number of feeding links demonstrates strong increasing relations ($r_s = 0.93$, $p = 0.003$) with each of **TST, AS,** and **DC** variables.

We also found that **TST, AS, DC,** and **N** have perfect negative Spearman correlation with **SI** ($r_s = -1$, $p = 0.000$). These results imply that **SI** has monotonically decreasing association with each of these four indices that are compared. **SI** also shows the strong

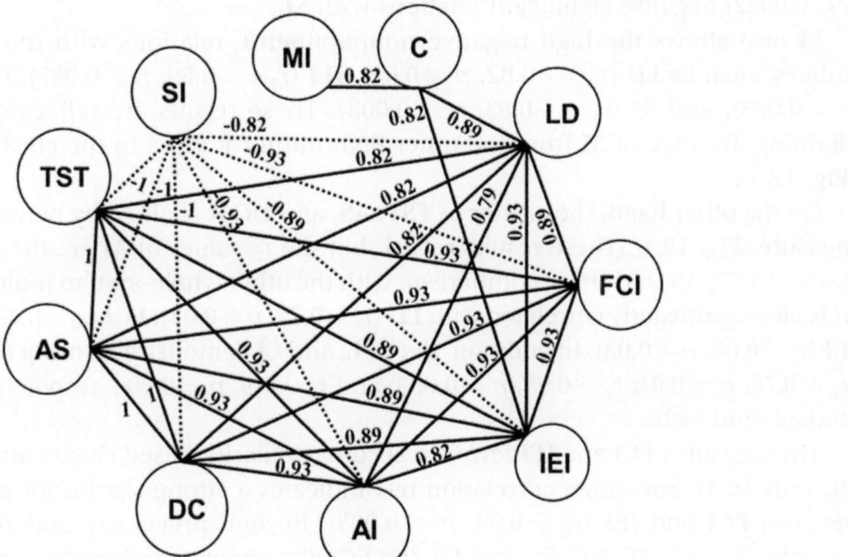


**FIGURE 12.3** Statistically significant relationships ($p < 0.05$) between whole-system properties assessed within pastoral food webs ($n = 7$) of the study area. Dotted and solid lines illustrate the decreasing and increasing directions of Spearman correlations ($r_s$), respectively.

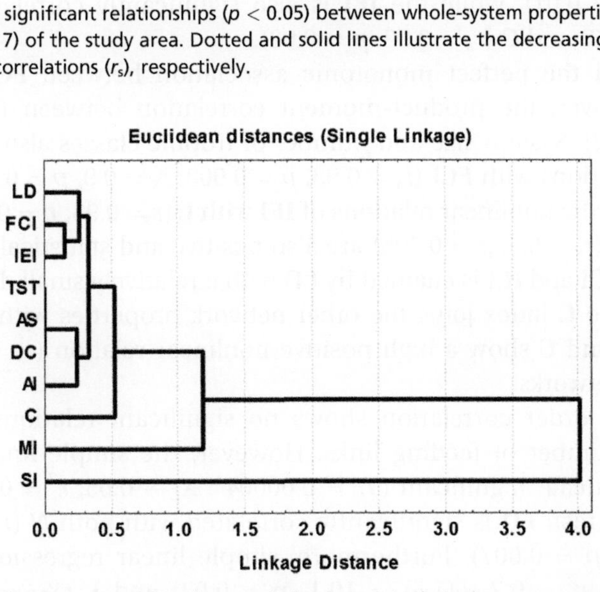


**FIGURE 12.4** Cluster dendrogram containing the correlation-based similarity between assessed network properties.

negative linear relations with **TST** ($r = -0.92$, $p = 0.003$), **AS** ($r = -0.93$, $p = 0.003$), **DC** ($r = -0.93$, $p = 0.003$), and ***N*** ($r = -0.96$, $p = 0.003$).

Pearson product-moment correlation analysis shows no significant linear association between **SI** and **L**. Despite this result, Spearman correlation test demonstrates high degree of negative nonparametric association between them ($r_s = -0.93$, $p = 0.003$). In contrast, **Cl** measure shows both linear ($r = -0.86$, $p = 0.002$) and nonlinear ($r = -0.97$, $p = 0.0002$) negative significant relations with **SI**.

**SI** also shows the high negative nonparametric relations with the other network indices, such as **LD** ($r_s = -0.82$, $p = 0.02$), **IEI** ($r_s = -0.89$, $p = 0.007$), **FCI** ($r_s = -0.93$, $p = 0.003$), and **AI** ($r_s = -0.93$, $p = 0.003$). These results are reflected as the largest chaining distance of **SI** from the other ENA output indices in the cluster dendrogram (Fig. 12.4).

On the other hand, the cluster of **TST, AS**, and **DC** is chained by network aggradation measure (Fig. 12.4). These results reveal that the $r_S$ values of **AI** are the most similar to those of **TST, AS**, and **DC** in comparison with the other whole-system indices. In addition, **AI** is also significantly correlated with **LD** ($r_s = 0.79$, $p = 0.04$), **FCI** ($r_s = 0.79$, $p = 0.04$), and **IEI** ($r_s = 0.82$, $p = 0.02$). In addition, the **N**, **L**, and **Cl** demonstrate strong relations with **AI** ($r_s = 0.79$, $p = 0.04$; $r_s = 0.93$, $p = 0.003$; and $r_s = 0.9$, $p = 0.006$, respectively) within the studied food webs.

The measures **FCI** and **IEI** form the second similarity-based cluster after **TST, AS**, and **DC** (Fig. 12.4). Spearmen correlation test indicates a strong significant positive relation between **FCI** and **IEI** ($r_s = 0.93$, $p = 0.003$). Beyond previously underlined measures (specifically **TST, AS, DC, SI**, and **AI**), the **FCI** also shows a high positive correlation with **LD** ($r_s = 0.89$, $p = 0.01$), while the **IEI** is also significantly correlated with both **LD** ($r_s = 0.93$, $p = 0.003$) and **C** ($r_s = 0.82$, $p = 0.02$).

Our data reveal the perfect monotonic association between **FCI** and **L** ($r_s = 1$, $p = 0.000$). Moreover, the product-moment correlation between them is also high ($r = 0.95$, $p = 0.002$). System size and number of trophic classes also show the positive nonparametric relations with **FCI** ($r_s = 0.93$, $p = 0.003$; $r_s = 0.9$, $p = 0.006$, respectively). On the other hand, the nonlinear relations of **IEI** with **L** ($r_s = 0.93$, $p = 0.003$), **N** ($r_s = 0.89$, $p = 0.007$), and **Cl** ($r_s = 0.9$, $p = 0.006$) are also positive and statistically significant.

The cluster of **FCI** and **IEI** is chained by **LD** with a relatively small distance (Fig. 12.4). Opposite to **LD**, the **C** index joins the other network properties with 0.1 more linkage distance. The **LD** and **C** show a high positive nonlinear relation ($r_s = 0.89$, $p = 0.007$) within the study networks.

Spearman rank order correlation shows no significant relations between **C** and network size or number of feeding links. However, the simple linear regression for **C** versus **L** is statistically significant ($Y_C = 0.00004 * X_L + 0.03$, $r^2 = 0.77$, $F(1,5) = 17.1$, $p = 0.009$). In contrast, **LD** is significantly correlated with both ***N*** ($r_s = 0.82$, $p = 0.02$) and *L* ($r_s = 0.89$, $p = 0.007$). Furthermore, simple linear regressions for **LD** versus **N** ($Y_{\mathbf{LD}} = 0.08{*}X_{\mathbf{N}}$, $r^2 = 0.7$, $F(1,5) = 10.1$, $p = 0.02$) and **L** ($Y_{\mathrm{LD}} = 0.008{*}X_L + 1.19$, $r^2 = 0.99$, $F(1,5) = 347.6$, $p = 0.00001$) are also statistically significant. Our data show that neither **LD** nor **C** has statistically significant associations with the number of trophic classes.

Despite a strong correlation between them, **LD** shows 70% more relations with the other network indices than **C** does. On the other hand, of the ENA indices we focus on, only **C** shows a statistically significant correlation with **MI** ($r_s = 0.82$, $p = 0.02$; and $r = 0.85$, $p = 0.02$). These results are depicted by cluster dendrogram as a relatively large distance of **MI** to all the other clusters (Fig. 12.4). Furthermore, both the nonparametric and the linear correlation tests show no significant relations of **MI** with **L**, **N**, and **Cl**.

## 12.4 Discussion and Conclusions

### 12.4.1 Network Properties: Robustness and Sensitivity

Borrett and Osidele (2007) suggest that more robust network indicator properties are those that are less variable. The variability (CV) of network properties assessed across the pastoral food webs ($n$=7) within the study area ranged from 8% to 46% (see Table 12.2). Network synergism index had most robustness, followed by aggradation measure and indirect effects dominance. Less robust were network connectance and **FCI**, followed by **TST**, ascendency index, and development capacity measure. Greatest value variability was found in the network mutualism index and **LD** measure, illustrating them to be the least robust of the assessed indices.

The assessed network properties also differed in degree of sensitivity to each other. Most sensitive were **IEI** and **LD** measures. Close behind these were **TST, FCI, AS, DC, AI,** and **SI**. The lowest sensitivities were found for **C** and **MI** respectively.

### 12.4.2 System-Wide Properties: Interrelations

A number of independent theoretical and empirical investigations provide the basis for comparing interrelations between network properties. Our results provide evidence for many existing hypotheses as well as offer additional explanations and suggestions.

Some theoretical studies report a hyperbolic decline in connectance with an increasing number of compartments (reviewed by Fonseca and John, 1996). In contrast, Martinez (1992) found that connectance of different-sized food webs was almost constant. Several comparative analyses of empirical food webs showed no clear association between **C** and **N** (Christian and Luczkovich, 1999; Garlaschelli, 2004; Martinez,1991).

Our results confirm this "constant connectance hypothesis," showing no clear interrelations between **C** and **N**. Connectance appears to be a scale-invariant property. However, our observed significant strong associations of **C** with **IEI** and with **MI** demonstrate the complex nature of this topological food web property.

Several studies have generalized the idea that **LD** tends to remain constant across networks of varying size (Cohen and Briand, 1984; Sugihara et al., 1989; Yodzis, 1980). But other research does not confirm this finding (Havens, 1992; Martinez, 1992; Deb, 1995; Winemiller, 1990). Our analyses do not support scale invariance of **LD** across the studied food webs. Despite the fact that **C** and **LD** are both measures of system complexity, we found **LD** to be much more sensitive to variations in other network properties. The definitions of link density (**LD** $= L/N$) and connectance (**C** $= L/N^2$) clearly show that the resolution of **N** causes a smaller range of connectance values across the study food webs, which may explain the small sensitivity of **C** measure.

Several investigations have shown the effect of network size on the behavior of other network properties (Fath, 2004) assessed here. Finn (1976) suggested that **TST** is sensitive to the number of compartments. Our results provide strong evidence for the increasing monotonic association between **TST** and **N**. Also, **ASC** and **DC** measures demonstrate the same tendency in relation to network size. The definitions of these two variables show that they are both driven by total system throughflow and limited by the number of compartments, as **TST** increases with **N**.

Higashi and Patten (1989) also gave an algebraic proof that indirect effects also increase with network size. Using synthesized large-scale cyber-networks, Fath (2004) supported these results and also demonstrated a strong direct association between cycling index and network size. This author also indicated that the indirect effects ratio is more strongly dependent on **N** than is the cycling index. Our data confirm these findings in the case of linear relations between the assessed indices. Curvilinear correlations show, however, that **FCI** increases more than **IEI** as network size increases.

Network synergism occurs in all models regardless of the system size (Fath and Patten, 1998). However, Fath (2004) illustrated that synergism decreases with increased **N**. These results agree with our investigations as well. Moreover, we observed decreasing **SI** in relation to the other system-wide properties assessed.

There is the evidence that network properties are strongly dependent on degree of cycling (Fath, 2004). Comparative study of aquatic ecosystems by Christensen (1995) showed strong increasing correlation of cycling index with network aggradation but not with **TST**. However, Borrett and Osidele (2007) observed strong correlations between **FCI, AI**, and **TST**. Fath et al. (2001) also observed directly increasing relations in network cycling and **TST**. As stated previously, **FCI** is the fraction of all flows in a system that cycles. Besides **FCI**, network aggradation is an alternative way to assess cycling (Finn, 1976) and correlates positively with **FCI** (Thomas and Christian, 2001). Our comparative study of terrestrial food webs confirms strong correlations between **FCI, AI**, and **TST**. We also found that **AI** tends to increase with increasing the network size.

On the other hand, the positive association of indirect effect ratio with several other system-wide properties has been reported in available literature. Patten and Higashi (Higashi and Patten, 1986; Patten, 1991) showed that network connectance, **FCI**, and **TST** increase the dominance of indirect effects. Fath (2004) provides strong evidence for the increasing relations between indirect effects index and network cycling measure. Furthermore, a strong association between **IEI** and **AI** was observed by Borrett et al. (2006) in the study of estuarine nitrogen networks. Our results support all these findings and additionally show strong increasing relations of **IEI** with **LD, AS, DC**, and decreasing association with **SI**.

Network mutualism measure (Fath, 2007; Patten, 1991) has received little attention and comparisons of its relationships with the other whole-system properties are quite rare. Surprisingly, our results reveal no significant relations of **MI** with any other network measures assessed, except degree of network connectance.

Clearly, the increase of species (nodes) in trophic network leads to an increase of relations (links) between them. In this study, we found positive relations of **LD, TST, FCI, AS, DC**, and **AI** and negative relations of **SI** with **L**. Interestingly, **FCI** appears to be the perfect monotonic function of link number. The number of trophic classes shows direct significant associations with **TST, FCI, AS, DC**, and **AI**, and inverse relation with **SI**.

### 12.4.3 General Conclusions

The broader results of our comparative study of a set of system-wide properties of empirical pastoral food webs can be summarized through the following general conclusions:

Objective measures of system-wide properties express a general intuition about the nature of network organization in different ways. Most of the assessed network indices such as **LD, C, TST, FCI, IEI, AS, DC, AI**, and **MI** demonstrate positive associations with one other, but negative relations with **SI**. Incongruously, network synergism appears to be negatively associated with all the other indices, which will bear further examination in subsequent studies.

The sensitivity of the assessed system-wide properties in interrelationships between each other is in the following descending order: **IEI = LD > TST = FCI = AS = DC = AI = SI > C > MI.**

Degree of robustness of the assessed system-wide properties is in the following descending order:

$$\mathbf{SI > AI > IEI > FCI = C > TST = AS > DC > MI > LD}$$

In subsequent studies we plan to use a larger set of empirical food webs for comparative analysis and also investigate the interrelations of system-wide variables with different empirical parameters, such as taxonomic diversity, plant biomass, cattle density, community urbanization, and pastoral degradation.

## Acknowledgments

Qianqian Ma wrote script to convert adjacency matrices to flow matrices in MATLAB and EcoNet. Discussions with Caner Kazanci, Stuart J. Whipple, Lindsey K. Tuominen, and Ernest W. Tollner helped in the clarifying the interpretation of results. We would like to thank the editor for a constructive review.

# References


Arii, K., Derome, R., Parrott, L., 2007. Examining the potential effects of species aggregation on the network structure of food webs. Bull. Math. Biol. 69, 119–133.

Batagelj, V., Mrvar, A., 2010. Pajek. http://vlado.fmf.uni-lj.si/pub/networks/pajek/.

Bersier, L.-F., Banašek-Richter, C., Cattin, M.-F., 2002. Quantitative descriptors of food web matrices. Ecology 83, 2394–2407.

Borgatti, S.P., Everett, M.G., 1993. Two algorithms for computing regular equivalence. Soc. Netw. 15, 361–376.

Borgatti, S.P., Everett, M.G., Freeman, L.C., 2002. UCINET VI. Software for Social Network Analysis. Analytic Technologies, Natick.

Borrett, S.R., Osidele, O.O., 2007. Environ indicator sensitivity to flux uncertainty in a phosphorus model of Lake Sidney Lanier, USA. Ecol. Model 200, 371–383.

Borrett, S.R., Whipple, S.J., Patten, B.C., Christian, R.R., 2006. Indirect effects and distributed control in ecosystems. Temporal variability of indirect effects in a seven-compartment model of nitrogen flow in the Neuse River Estuary (USA)—Time series analysis. Ecol. Model 194, 178–188.

Brown, M.T., Herendeen, R., 1996. Embodied energy analysis and emergy analysis: a comparative view. Ecol. Econ. 19, 219–236.

Christensen, V., 1995. Ecosystem maturity—towards quantification. Ecol. Model 77, 3–32.

Christian, R.R., Luczkovich, J.J., 1999. Organizing and understanding a winter's seagrass food web network through effective trophic levels. Ecol. Model 117, 99–124.

Christian, R.R., Scharler, U.S., Dunne, J., Ulanowic, R.E., 2005. Role of network analysis in comparative ecosystem ecology of estuaries. In: Belgrano, A., Scharler, U.M., Dunne, J., Ulanowicz, R.E. (Eds.), Aquatic Food Webs. An Ecosystem Approach. Oxford University Press, New York, pp. 25–40.

Cohen, J.E., Briand, F., 1984. Trophic links of community food web. Proc. Nat. Acad. Sci. USA 81, 4105–4109.

Cohen, J.E., et al., 1993. Improving food webs. Ecology 74, 252–258.

Deb, D., 1995. Scale-dependence of food web structures, tropical ponds as paradigm. Oikos 72, 245–262.

Dell, A.I., Kokkoris, G.D., Banašek-Richter, C., Bersier, L.-F., Dunne, J., Kondoh, M., et al., 2005. How do complex food webs persist in nature? In: Ruiter, P.C., Wolters, V., Moore, J.C. (Eds.), Dynamic Food Webs: Multispecies Assemblages, Ecosystem Development and Environmental Change. Theoretical Ecology Series, 3. Academic Press, London, UK, pp. 425–436.

Elton, C.S., 1958. Ecology of Invasions by Animals and Plants. Chapman & Hall, London.

Fath, B.D., 2004. Network analysis applied to large-scale cyber-ecosystems. Ecol. Model 171, 329–337.

Fath, B.D., 2007. Network mutualism, positive community-level relations in ecosystems. Ecol. Model 208, 56–67.

Fath, B.D., Borrett, S.R., 2006. AMATLABfunction for network environ analysis. Environ. Model. Softw. 21, 375–405.

Fath, B.D., Patten, B.C., 1998. Network synergism, emergence of positive relations in ecological systems. Ecol. Model 107, 127–143.

Fath, B.D., Patten, B.C., 1999. Review of the foundations of Network Environ Analysis. Ecosystems 2, 167–179.

Fath, B.D., Patten, B.C., 2001. A progressive definition of network aggradation. In: Ulgiati, S., Brown, M.T., Giampietro, M., Herendeen, R.A., Mayumi, K. (Eds.), Proceedings of the Second International

Workshop on Advances in Energy Studies, Exploring Supplies, Constraints and Strategies. Porto Venere, Italy, pp. 551–562. May 23–27, 2000.

Fath, B.D., Patten, B.C., Choi, J.S., 2001. Complementarity of ecological goal functions. J. Theor. Biol. 208, 493–506.

Finn, J.T., 1976. Measures of ecosystem structure and function derived from analysis of flows. J. Theor. Biol. 56 (2), 363–380.

Fonseca, C.R., John, J.L., 1996. Connectance, a role for community allometry. Oikos 77, 353–358.

Gardner, M.R., Ashby, W.R., 1970. Connectance of large (cybernetic) systems, critical values for stability. Nature 228, 784.

Garlaschelli, D., 2004. Universality in food webs. Eur. Phys. J. B 38, 2.277–2.285.

Garlaschelli, D., Caldarelli, G., Pietronero, L., 2003. Universal scaling relations in food webs. Nature 423, 165–168.

Han, B.P., 1997. On several measures concerning flow variables in ecosystems. Ecol. Model 104, 289–302.

Hannon, B., 1973. The structure of ecosystems. J. Theor. Biol. 41, 535–546.

Hannon, B., 1985. Linear dynamic ecosystems. J. Theor. Biol. 116, 89–110.

Havens, K.E., 1992. Scale and structure in natural food webs. Science 257, 1107–1109.

Herendeen, R.A., 1981. Energy intensities in ecological and economic systems. J. Theor. Biol. 91, 607–620.

Herendeen, R.A., 1989. Energy intensity, residence time, exergy, and ascendency in dynamic ecosystems. Ecol. Model 48 (1–2), 19–44.

Higashi, M., Patten, B.C., 1989. Dominance of indirect causality in ecosystems. Am. Nat. 133, 288–302.

Higashi, M., Patten, B.C., 1986. Further aspects of the analysis of indirect effects in ecosystems. Ecol. Model 31, 69–77.

Johnson, S.C., 1967. Hierarchical clustering schemes. Psychometrika 32, 241–253.

Jørgensen, S.E., 1986. Structural dynamics model. Ecol. Model 31, 1–9.

Jørgensen, S.E., 2002. Integration of Ecosystem Theories: A Pattern, third ed. vol. 432. Kluwer Academic, Dordrecht, The Netherlands.

Kazanci, C., 2007. EcoNet: a new software for ecologicalmodeling, simulation and network analysis. Ecol. Model 208, 3–8.

Kazanci, C., Matamba, L., Tollner, E.W., 2009. Cycling in ecosystems: an individual based approach. Ecol. Model 220, 2908–2914.

Leontief, W.W., 1936. Quantitative input–output relations in the economic system of the United States. Rev. Econ. Stat 18, 105–125.

Leontief, W.W., 1966. Input–Output Economics. Oxford University Press, London/New York.

Lindeman, R., 1942. The trophic-dynamic aspect of ecology. Ecology 23, 399–418.

Luczkovich, J.J., et al., 2003. Defining and measuring trophic role similarity in food webs using regular equivalence. J. Theor. Biol. 220 (3), 303–321.

MacArthur, R.H., 1955. Fluctuations of animal populations and a measure of community stability. Ecology 36, 533–536.

Margalef, R., 1963. On certain unifying principles in ecology. Am. Nat. 97, 357–374.

Martinez, N.D., 1991. Artifacts or attributes? Effects of resolution on the Little Rock Lake food web. Ecol. Monogr 61, 367–392.

Martinez, N.D., 1992. Constant connectance in community food webs. Am. Nat. 139, 1208–1218.

Martinez, N.D., 1994. Scale-dependent constraints on food-web structure. Am. Nat. 144, 935–953.

Matis, J.H., Patten, B.C., White, G.C. (Eds.), 1979. Compartmental Analysis of Ecosystems Models. International Co-operative Publishing House, Fairland, MD.

May, R.M., 1973. Stability and complexity in model ecosystems. Princeton University Press, Princeton, NJ.

Morris, J.T., Christian, R.R., Ulanowicz, R.E., 2005. Analysis of size and complexity of randomly constructed food webs by information theoretic metrics. In: Belgrano, A., Scharler, U.M., Dunne, J., Ulanowicz, R.E. (Eds.), Aquatic Food Webs. An Ecosystem Approach. Oxford University Press, New York, pp. 73–85.

Odum, E.P., 1969. The strategy of ecosystem development. Science 164, 262–270.

Paine, R.T., 1988. Food webs: road maps of interactions or grist for theoretical development. Ecology 69, 1648–1654.

Parrott, L., 2010. Measuring ecological complexity. Ecol. Indicators 10, 1069–1076.

Patrício, J., Ulanowicz, R., Pardal, M.A., Marque, J.C., 2004. Ascendency as an ecological indicator: a case study of estuarine pulse eutrophication. Estuar. Coast. Shelf Sci. 60, 23–25.

Patten, B.C., 1978. Systems approach to the concept of environment. Ohio J. Sci. 78, 206–222.

Patten, B.C., 1981. Environs: the super niches of ecosystems. Am. Zool. 21, 845–852.

Patten, B.C., 1982. Environs: relativistic elementary particles for ecology. Am. Nat. 119, 179–219.

Patten, B.C., 1991. Network ecology, indirect determination of the life-environment relationship in ecosystems. In: Higashi, M., Burns, T. (Eds.), Theoretical Studies of Ecosystems, The Network Perspective. Cambridge University Press, New York, pp. 288–351.

Patten, B.C., 1992. Energy, emergy and environs. Ecol. Model 62, 29–69.

Patten, B.C., Fath, B.D., 1998. Environ theory and analysis, relations between aggradation, dissipation, and cycling in energy-matter flow networks at steady-state. In: Ulgiati, S., Brown, M.T., Giampietro, M., Herendeen, R.A., Mayumi, K. (Eds.), Advances in Energy Studies, Energy Flows in Ecology and Economy. MUSIS Publisher, Rome, Italy, pp. 483–497.

Patten, B.C., in prep., Holoecology: The Unification of Nature by Network Indirect Effects.

Pimm, S.L., 1982. Food webs. Chapman & Hall, London.

Samuelson, P.A., 1948. Economics: An Introductory Analysis. McGraw–Hill Book, New York.

Schramski, J.R., Kazanci, C., Tollner, E.W., 2010. Network environ theory, simulation and EcoNet 2.0. Environ. Model. Softw. 26, 419–428.

Shannon, C.E., 1948. A mathematical theory of communication. Bell Syst. Tech. J. 27, 379–423.

Sugihara, G., Schoenly, K., Trombla, A., 1989. Scale invariance in food web properties. Science 245, 48–52.

Thomas, C.R., Christian, R.R., 2001. Comparison of nitrogen cycling in salt marsh zones related to sea-level rise. Mar. Ecol. Progr. Ser. 221, 1–16.

Ulanowicz, R.E., 1980. An hypothesis on the development of natural communities. J. Theor. Biol. 85 (2), 223–245.

Ulanowicz, R.E., 1986. Growth and Development, Ecosystems Phenomenology. Springer-Verlag, New York.

Ulanowicz, R.E., 1997. Ecology, the Ascendant Perspective. Columbia University Press, New York.

Ulanowicz, R.E., Jorgensen, S.E., Fath, B.D., 2006. Exergy, information and aggradation: an ecosystems reconciliation. Ecol. Model 198, 520–524.

Warren, P.H., 1994. Making connections in food webs. Trends Ecol. Evol. 9, 136–141.

Williams, R.J., Martinez, N.D., 2000. Simple rules yield complex food webs. Nature 6774, 180–183.

Winemiller, K.O., 1990. Spatial and temporal variation in tropical fish trophic networks. Ecol. Monog. 60, 331–367.

Yodzis, P., 1980. The connectance of real ecosystems. Nature 284, 544–545.

# 13

# An Individual-Based Approach for Studying System-Wide Properties of Ecological Networks

Qianqian Ma*,[1], Caner Kazanci[†]

*DEPARTMENT OF BIOLOGICAL AND AGRICULTURAL ENGINEERING, UNIVERSITY OF GEORGIA, ATHENS, GA 30602, USA, [†]FACULTY OF ENGINEERING AND DEPARTMENT OF MATHEMATICS, UNIVERSITY OF GEORGIA, ATHENS, GA 30602, USA, EMAIL: MAQIAN12@UGA.EDU

## 13.1 Introduction

Ecological networks are widely used to represent the biotic interactions in an ecosystem, such as the movement of biomass in consumer-resource systems, transfer of energy in food webs, and carbon cycling in the biosphere. Ecological network models are built to describe the flow of biomass or energy within ecosystems. Network Environ Analysis (NEA) (Patten, 1978; Fath and Patten, 1999) formulates system-wide properties to describe various relations in an ecosystem. For example, cycling index quantifies how much of the energy or biomass is recycled and throughflow analysis measures how the environmental inputs contribute to the throughflow of each compartment. Computation of most of these properties relies on formulations based on linear algebra. These algebraic formulations are only applicable to ecosystems at steady state, which are rarely seen in the environment. Dynamic ecosystems are more common as ecosystems are often driven by seasonal changes and as unexpected regime shifts or environmental impacts can occur. The purpose of this work is to extend the applicability of these useful but limited properties to dynamic ecosystem models.

Network Particle Tracking (NPT) is an individual-based method that simulates system dynamics. It works like an *in silico* version of tracer experiments by simulating pathways of traced elements. As an alternative to algebraic formulations of NEA properties, we construct simulation-based formulations that compute the same NEA properties. While these new parallel definitions for NEA properties agree with conventional methodologies on steady-state systems, they are also applicable to dynamic models.

In this chapter, we first provide a brief introduction for NEA, followed by a description of how NPT simulates ecosystem models. We then focus on formulating NPT-based

Models of the Ecological Hierarchy. DOI: http://dx.doi.org/10.1016/B978-0-444-59396-2.00013-4
ISSN 0167-8892, 

definitions for various ecosystem properties such as cycling index, throughflow analysis, and storage analysis. Finally, we discuss how NPT-based definitions can be used to compute dynamic network properties, and demonstrate this methodology on storage analysis.

## 13.2 Network Environ Analysis

Leontief (1951, 1966) developed the input–output analysis to analyze the interdependence of different branches of national economy. This analysis was introduced to ecology by Hannon (1973). NEA (Fath and Patten, 1999; Patten, 1978) uses the same idea of economic input–output analysis to study environmental systems. Ecosystem models are represented as individual compartments connected by pairwise links. NEA methodology formulates various measures that describe the relationship between these compartments and the environment. NEA consists of two main analyses: (i) structural analysis and (ii) functional analysis. The former investigates the direct and indirect connections from one compartment to another, and is also the basis for the functional analysis. The functional analysis quantifies the relationship between the compartments and the environment and includes throughflow, storage, and utility analyses.

NEA works with compartmental models representing ecosystems. Figure 13.1 shows a hypothetical system with three compartments. Each circle represents one compartment, which corresponds to a part of the system. The arrows represent environmental inputs/outputs or flows between compartments. Depending on what ecological process is modeled, the flow currency could be biomass, energy, or a specific nutrient or element.

NEA stores both the qualitative and quantitative information of the network in matrices and vectors. For the simple model in Fig. 13.1, the environmental inputs ($z$), outputs ($y$), and storage values ($x$) are defined as follows:

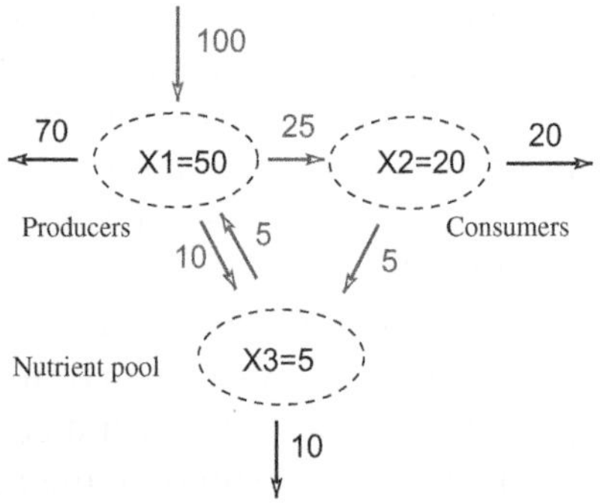


**FIGURE 13.1** A hypothetical three-compartment ecosystem model consisting of *Producers*, *Consumers*, and *Nutrient pool* with stocks X1 = 50, X2 = 20 and X3 = 5 units, respectively. These three compartments are connected by four inter-compartmental flows. Only the first compartment has an environmental input, whereas all compartments have environmental outputs. For color version of this figure, the reader is referred to the online version of this book.

$$z = \begin{bmatrix} 100 \\ 0 \\ 0 \end{bmatrix} \quad y = \begin{bmatrix} 70 \\ 20 \\ 10 \end{bmatrix} \quad x = \begin{bmatrix} 50 \\ 20 \\ 5 \end{bmatrix}$$

$z_i$ : Rate of enviornmental input to compartment $i$

$y_i$ : Rate of environmental output leaving compartment $i$

$x_i$ : Storage value of compartment $i$

**Adjacency matrix** $A$ consists of zeros and ones. $A_{ij}$ indicates if there exists a direct flow from compartment $j$ to compartment $i$. $A$ matrix represents all connections within the system. In addition to the connectivity information, **flow matrix** $F$ describes how strong the connections are. $F_{ij}$ denotes the rate of direct flow from $j$ to $i$. $A$ and $F$ matrices for the simple ecosystem model in Fig. 13.1 are:

$$A = \begin{bmatrix} 0 & 0 & 1 \\ 1 & 0 & 0 \\ 1 & 1 & 0 \end{bmatrix} \quad F = \begin{bmatrix} 0 & 0 & 5 \\ 25 & 0 & 0 \\ 10 & 5 & 0 \end{bmatrix}$$

**Throughflow** $T_i$ is the rate of material (or energy) moving through compartment $i$. It is defined as the sum of flows to compartment $i$ from other compartments and the environment ($T_i^{\text{in}}$). For a system at steady state, it equals the sum of flows from compartment $i$ to other compartments and the environment ($T_i^{\text{out}}$):

$$T_i^{\text{in}} = \sum_{j=1}^{n} F_{ij} + z_i \quad T_i^{\text{out}} = \sum_{j=1}^{n} F_{ji} + y_i$$

For this simple model,

$$\mathbf{T}^{\text{in}} = \mathbf{T}^{\text{out}} = \begin{bmatrix} 105 \\ 25 \\ 15 \end{bmatrix}$$

Finally, **total system throughflow** (TST) is defined as the sum of all compartmental throughflows in the system:

$$TST = \sum_{i=1}^{n} T_i = 105 + 25 + 15 = 145$$

Current NEA methodology and various properties involved are developed for steady-state systems. All the matrices and vectors involved in the computation of these properties are required to be constants, not functions of time. Little work exists on NEA applied to evolving systems (Hippe, 1983; Shevtsov et al., 2009). Hippe (1983) states "it would be desirable to develop a nonlinear environ analysis for either general or specific classes of nonlinear systems. Although such an analysis would not be isomorphic to the analysis presented herein, certainly there would exist analogous formulations of concepts between these two types of analysis." Shevtsov et al. (2009) describes a methodology applicable to dynamic systems; however, there are limitations in its accuracy and convergence. The methodology presented in our work is general, and is valid for models with linear or nonlinear flows, and time-invariant or time-dependent coefficients.

## 13.3 Network Particle Tracking: An Individual-Based Methodology

NPT is an individual-based simulation method. Energy (or mass) entering the system is divided into small packets, and each packet is labeled and tracked in time as it flows through the network. Figure 13.2 uses a simple three-compartment model to illustrate the working of NPT. It starts with breaking initial stocks or input flows into discrete packets, which we call "particles". For example, for a carbon flow model, a particle could represent a single C atom, or 1 gram of Carbon. This choice is left to the modeler. The smaller the particle, the more the computational resources required for the simulation. In principal, it is preferred to use the smallest possible unit allowed by computational resources. Based on flow rates, NPT determines which flow is likely to occur and when. A particle is then chosen randomly from the donor compartment and introduced to the recipient compartment. As ecosystems are open systems, new particles will enter the system. So if the chosen flow is an environmental input, a new particle is labeled and introduced to the recipient compartment. NPT simulation output includes the history of the movements of all particles within the system.

Figure 13.3 shows a partial NPT simulation output for the three-compartment system (from Fig. 13.2), which includes the pathway, flow time, and residence time data. The

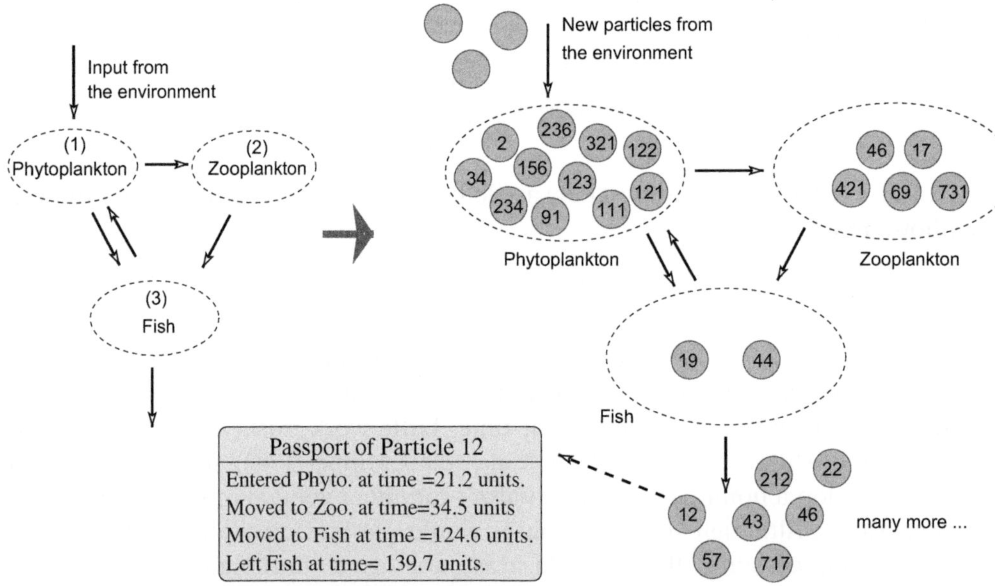


**FIGURE 13.2** Three-compartment model depicting particles and their flow information in the system. For color version of this figure, the reader is referred to the online version of this book.

pathway of a particle is defined as an ordered list of compartments visited by that particle. Flow time data shows when each particle flows from one compartment (or the environment) to another compartment (or the environment). Residence time data records how long a particle stays in a compartment it visits. For example, Fig. 13.3 represents the fact that "Particle 12" entered *Phytoplankton* (compartment 1) at time = 21.2, left at time = 34.5, and spent 13.3 time units in *Phytoplankton*. The time unit is determined by the modeler.

NPT is a discrete and stochastic method with several advantages over ordinary differential equation (ODE) models. However, ecological networks are often modeled using ODEs, which are deterministic and continuous. The same initial conditions for an ODE model will always generate exactly the same outcome, which is usually not the case in real life. On the other hand, stochastic models represent the inherent fluctuations in real-life systems by adding noise derived from a probability distribution, and therefore outcomes are not unique.

Another issue with the ODE models is that all flows are treated as continuous processes, which is not always appropriate. Take the simple food chain as an example.

$$\text{Tree} \rightarrow \text{Deer} \rightarrow \text{Wolf}$$

Assuming the deer could obtain leaves almost anytime, the flow from *Tree* to *Deer* can be approximated by a continuous process. However, it is not appropriate to regard the flow from *Deer* to *Wolf* as continuous since wolves attack deers occasionally. Once they succeed, a large amount of energy is moved from *Deer* to *Wolf* in a short period of time. It is preferable to model such rare and instantaneous processes with discrete, rather than continuous models.

NPT is based on the Gillespie algorithm (Gillespie, 1977), which is an exact stochastic simulation method. As described earlier, it discretizes material flows as movements of particles among compartments. For the simple food chain above, it is possible to represent the flow from *Tree* to *Deer* and *Deer* to *Wolf* with small and large-sized particles, respectively. In other words, the flow from *Tree* to *Deer* could be modeled using small-size particles (e.g., 1 gram of Carbon), while the flow unit from *Deer* to *Wolf* could be modeled

| | | |
|---|---|---|
| Particle 12 | Pathway | * → 1 → 2 → 3 → * |
| | Flow time | 21.2 34.5 124.6 139.7 |
| | Residence time | 13.3 90.1 15.1 |
| Particle 43 | Pathway | * → 1 → 3 → 1 → 2 → 3 → * |
| | Flow time | 30.4 50.3 65.8 88.4 115.3 140.2 |
| | Residence time | 19.9 15.5 22.6 26.9 24.9 |
| Particle 212 | Pathway | * → 1 → 2 → 3 → 1 → 3 → * |
| | Flow time | 31.2 50.3 79.8 98.0 123.5 131.6 |
| | Residence time | 19.1 29.5 18.2 25.5 8.1 |

**FIGURE 13.3** Partial NPT output for three-compartment ecosystem in Fig. 13.2, including pathway, flow time, and residence time data. The numbers of 1, 2, and 3 correspond to compartments *Phytoplankton*, *Zooplankton*, and *Fish*. For color version of this figure, the reader is referred to the online version of this book.

as numerous small particles bound together, to represent the entire deer. This modified model represents the food chain better than just moving all particles independently.

NPT is a stochastic method compatible with the master equation (Gillespie, 1992, 2000). In other words, the mean of many NPT results agrees with differential equation solution. Therefore, studies based on NPT simulations, which are stochastic and discrete, are compatible with the conventional methodologies, which are often continuous and deterministic.

## 13.4 Cycling Index

Cycling index is an essential ecological indicator used for ecosystem analysis. It measures how much of the environmental input is cycled before exiting the system. Finn's cycling index (Finn, 1978) is widely accepted as an accurate measure of cycling within an ecosystem. The definition of Finn's cycling index relies on throughflow analysis. Therefore, here, we briefly introduce the throughflow generating matrix $N$, but postpone a more detailed description to the next section. $N_{ij}$ represents the throughflow generated at $i$ by per-unit input at $j$.

By definition, diagonal values of throughflow generating matrix $N_{ii}$ represent the amount of throughflow generated at compartment $i$ by one unit input into compartment $i$. This one unit input into compartment $i$ contributes to its throughflow ($T_i$) at least once. Therefore, if $N_{ii}$ is larger than 1, then $N_{ii} - 1$ represents the amount of throughflow generated at compartment $i$ through cycling. Finn (1977, 1978) defines his cycling index based on this idea as the fraction of the total system throughflow (TST) that is due to cycling ($\mathrm{TST_c}$). For an ecosystem model at a steady state, **Finn's cycling index** (FCI) is defined as follows:

$$\mathrm{FCI} = \frac{\mathrm{TST_c}}{\mathrm{TST}} = \sum_{i=1}^{n}\left(\frac{T_i}{\mathrm{TST}}\frac{N_{ii}-1}{N_{ii}}\right) \tag{1}$$

Here, $n$ is the total number of compartments in the ecosystem model. By definition, the range of $\mathrm{TST_c}$ values are between zero and TST, therefore FCI varies between zero and one. This definition of FCI is applicable only to systems at steady state, where $T$, $N$, and TST are constants. We define an alternative, simulation-based definition for FCI, which is applicable to dynamic ecosystems as well. This definition is more intuitive, and does not require throughflow analysis or linear algebra.

NPT simulations generate the pathway data of individual particles within the system. Each compartment visit of a particle contributes to TST. Therefore, using the pathway data, TST represents the total number of visits to compartments by particles. And $\mathrm{TST_c}$ is computed as the number of compartments visited by a particle more than once. Therefore, we can quantify cycling in an intuitive manner using particle pathways. Table 13.1 demonstrates the computation of FCI based on the partial pathway data provided in Fig. 13.3.

In Table 13.1, the second column lists the compartments visited by each particle, and we underline compartments visited by a particle more than once. TST values are computed by counting the number of compartments visited by each particle. $TST_c$ values are computed by counting the number of underlined compartments, which represent throughflows due to cycling. Therefore, the simulation-based cycling index is computed as: $FCI = \frac{TST_c}{TST} \approx \frac{4}{13} = 0.31$.

The result of this computation is based on the specific pathway data used, and therefore different pathway data sets will produce different results. In general, the larger the pathway data set, the higher the accuracy of the computed cycling index. Since NPT is based on a stochastic algorithm, the pathway data generated for each simulation will be different. However, as the pathway data size increases, the simulation-based FCI converges to its conventional definition in Eqn (1).

Kazanci et al. (2009) demonstrates the accuracy and convergence of the NPT-based FCI using the Hubbard Brook temperate forest ecosystem model (Finn, 1980). The flow currency in this model is Calcium, which explains the relatively high cycling index of 0.797. Figure 13.4 shows NPT-based FCI values computed for three different simulations.

**Table 13.1** Computing of an Approximate Cycling Index for the Three-Compartment Model, Using Partial Pathway Data in Fig. 13.3

| Particle | Pathway | TST | $TST_c$ |
|---|---|---|---|
| 12 | 1 2 3 | 3 | 0 |
| 43 | 1 3 $\underline{1}$ 2 $\underline{3}$ | 5 | 2 |
| 212 | 1 2 3 $\underline{1}$ $\underline{3}$ | 5 | 2 |
| Total | | 13 | 4 |

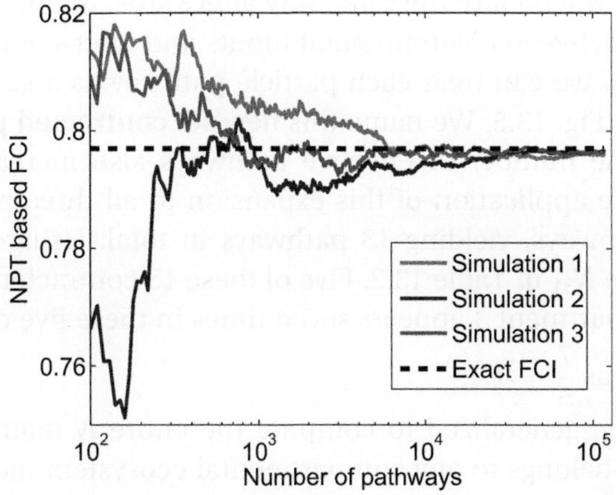


**FIGURE 13.4** Finn cycling index is computed using the pathway data generated by three different NPT simulations for Hubbard Brook temperate forest ecosystem. For color version of this figure, the reader is referred to the online version of this book.

We observe that as pathway data size increases, the NPT-based FCI converges to the value computed with the conventional methodology, shown with the horizontal dotted line. The agreement of the two definitions reveals the meaning of FCI on an individual-based context.

## 13.5 Throughflow Analysis

**Throughflow analysis** investigates the relationship between environmental inputs and compartmental throughflows. Throughflow analysis consists of the **throughflowgenerating matrix** $N$, which is a linear mapping from the input vector $z$ to the throughflow vector $T$, as shown in $Nz = T$. Throughflow analysis is an essential part of NEA, and is the basis for other system-wide properties such as cycling index (Finn, 1977, 1978, 1982) and indirect effects ratio (Higashi and Patten, 1986; Patten, 1986, 1985).

$N_{ij}$ represents the throughflow generated at compartment $i$ by one unit environmental input at compartment $j$. Using NPT, we could regard this unit input as one particle, and simulate its movement within the system. The definition then becomes the number of times a particle goes through compartment $i$ given that it enters the system at $j$. Since NPT simulations provide pathways of particles as shown in Fig. 13.3, such computation becomes feasible. To compute $N_{ij}$, we use all pathways starting at $j$, which correspond to those particles entering the system at $j$. Then, we count the occurrences of compartment $i$ in these pathways. The average number of occurrences of $i$ for all particles defines the NPT-based $N_{ij}$.

However, for many ecosystem models, not every compartment receives environmental input. This is a problem because $N_{ij}$ cannot be computed unless compartment $j$ receives environmental input. To compute the full $N$ matrix, we use the fact that whether a particle enters a compartment from the environment or from another compartment, it does not matter and it will behave the same way afterwards. In other words, we need not make any distinction between environmental inputs and inter-compartmental inputs to compute $N$. Therefore, we can treat each particle pathway as a set of multiple particle pathways as shown in Fig. 13.5. We name this new set **contracted pathways**.

This increase in the number of effective pathways also increases the accuracy of computation of $N$. The application of this expansion to all three pathways in Fig. 13.3 results in 10 new pathways, yielding 13 pathways in total. Using these 13 contracted pathways, we compute $N_{31}$ in Table 13.2. Five of these 13 contracted pathways start with compartment 1. Compartment 3 appears seven times in these five contracted pathways, then $N_{31}$ is computed as $\frac{7}{5} = 1.4$.

This method can be generalized to compute the entire $N$ matrix based on a larger pathway data set that belongs to any compartmental ecosystem model, as follows:

$$N_{ij} = \frac{1}{|P_j|} \sum_{p \in P_j} \text{number of occurances of compartment } i \text{ in } p \qquad (2)$$

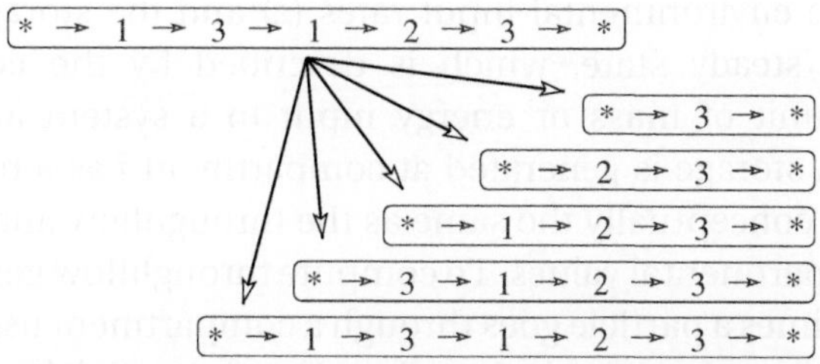


**FIGURE 13.5** Pathway of particle 43 from Figure 13.3 is expanded into five separate contracted pathways. Such expansion enables us to compute $N_{ij}$ and $S_{ij}$ (in next section) even if there is no environmental input into some compartments. For color version of this figure, the reader is referred to the online version of this book.

**Table 13.2** Computation of $N_{31}$ Using the Partial NPT Simulation Output from Fig. 13.3

| Particle | Pathways | Contracted Pathways Starting with 1 | Number of Occurrences of 3 |
|---|---|---|---|
| 12 | 1 2 3 | 1 2 3 | 1 |
| 43 | 1 3 1 2 3 | 1 3 1 2 3 | 2 |
| | | 1 2 3 | 1 |
| 212 | 1 2 3 1 3 | 1 2 3 1 3 | 2 |
| | | 1 3 | 1 |
| Total | | 5 | 7 |

where $P_j$ is a list of contracted pathways that starts with compartment $j$, and $|P_j|$ represents the number of contracted pathways on this list. The larger the value $|P_j|$, the more accurate the computation of NPT-based throughflow generating matrix $N$.

Similar to NPT-based FCI, the NPT-based N matrix will converge to its conventional definition as more pathway data is used in its computation. Matamba et al. (2009) includes similar convergence and accuracy information for NPT-based $N$ as we show for cycling index in Fig. 13.4. However, the method presented in this chapter is based on contracted pathways, and is an improved version of the methodology presented in Matamba et al. (2009). This new method converges much faster using less pathway data, increasing accuracy. Similar to FCI, conventional throughflow analysis is only applicable to steady-state ecosystem models. The NPT-based methodology can compute $N$ matrix for pathway data that belong to dynamic systems, eliminating this limitation. Furthermore, this alternative definition of the $N$ matrix is more intuitive, and confirms the conventional algebraic formulation of $N$.

## 13.6 Storage Analysis

**Storage analysis** (Matis and Patten, 1981) investigates the relation between input flows and compartment storage values. The **storage generating matrix** $S$ represents a linear mapping between the environmental input rates ($z$) and the storage values (x) of each compartment at the steady state, which is described by the equation $Sz = x$. For instance, given one unit of mass or energy input to a system at compartment $j$, $S_{ij}$ represents how much storage is generated at compartment $i$ as a result.

Storage analysis is conceptually the same as the throughflow analysis. Both measures map the input to compartmental values. To compute throughflow generating matrix $N$, we count the number of times a particle goes through a compartment using the pathway data. In storage analysis, the number of visits is replaced by the particle's total residence time in that compartment. The longer a particle resides in a compartment, the larger its contribution to the storage of that compartment. NPT simulations provide not only the list of compartments a particle visits, but also the duration of each visit. This additional information enables the computation of the $S$ matrix in a similar way to that of the $N$ matrix.

To compute $S_{ij}$, we use all pathways starting at $j$, which correspond to those particles entering the system at $j$. Then, we sum up their residence times at $i$. We compute $S_{ij}$ as the average residence time at $i$ for all particles.

Similar to throughflow analysis, the process of computing the $S$ matrix is slightly more involved because an ecosystem model will rarely have environmental inputs into all its compartments. For example, we cannot compute $S_{32}$ if no particle enters compartment 2 from the environment. To solve this problem, we carry out the computation of $S$ on contracted pathways like what we did for computing $N$ matrix. Since all compartments receive an environmental input and/or at least one inter-compartmental input, it is possible to compute the full $S$ matrix for any ecosystem model.

To demonstrate the NPT-based definition of $S$, we compute $S_{31}$ using the pathway data provided in Fig. 13.3. First, we determine how many contracted pathways start with compartment 1. Table 13.2 shows that five contracted pathways start with compartment 1. Using the residence time information in Fig. 13.3, we compute the total residence time at compartment 3 for these five contracted pathways. For example, the first contracted pathway ($\rightarrow 1 \rightarrow 2 \rightarrow 3 \rightarrow *$) goes through compartment 3 once, and its residence time at compartment 3 is 15.1 time units. The second contracted pathway ($\rightarrow 1 \rightarrow 3 \rightarrow 1 \rightarrow 2 \rightarrow 3 \rightarrow *$) visits compartment 3 twice, staying for 15.5 and 24.9 time units, respectively. So, its residence time is the sum of these two values. Then, $S_{31}$ is computed as follows:

$$S_{31} = \frac{(15.1) + (15.5 + 24.9) + (24.9) + (18.2 + 8.1) + (8.1)}{5} = 22.96$$

In general, we compute $S$ as follows:

$$S_{ij} = \frac{1}{|P_j|} \sum_{p \in P_j} \text{sum of residence times at compartment } i \text{ in } p \tag{3}$$

where $P_j$ is a list of contracted pathways that start with compartment $j$. $|P_j|$ represents the number of contracted pathways on this list. The larger the value of $|P_j|$, the more accurate the computation of NPT-based $S$. Convergence of the NPT-based $S$ matrix to the conventional $S$ matrix is discussed in Kazanci and Ma (2012), and the results are similar to that of the throughflow analysis and FCI.

## 13.7 Dynamic Network Environ Analysis

In previous sections, we demonstrated the NPT-based methodology to study several network properties. NPT-based definitions of network properties depend solely on the output of an NPT simulation, which consists of particle pathways and residence times. This simulation may belong to an ecosystem at a steady state, in transition to a steady state, oscillating between states, or even a chaotic system. Therefore, NPT-based definitions can be computed for dynamic ecosystems, unlike the conventional definitions of these measures, which are only limited to steady-state systems.

The NPT-based definitions introduced previously only allow us to compute constant values that represent the "average" behavior of a dynamic system over a time interval. In this section, we describe how to extend NPT-based definitions to compute time-varying network properties. We demonstrate this new technique on storage analysis. However, the process described here to extend the applicability of storage analysis to dynamic systems can be adapted to any NPT-based ecosystem property, including throughflow analysis and cycling index.

Our aim is to compute a true dynamic storage analysis matrix function $S(t)$, where $S(a)$ represents the instantaneous storage generating matrix at time $t = a$. To construct an NPT-based formulation for $S(t)$, a rigorous definition for $S_{ij}(t)$ is needed. Let $S_{ij}([a, b])$ represent the amount of storage created at compartment $i$ over time $([a, \infty])$, by per-unit input into compartment $j$ over the time interval $[a, b]$. Then, we define the dynamic storage-generating matrix function $S(t)$ as follows:

$$S(t) = \lim_{h \to 0} S([t - h,\ t + h]) \tag{4}$$

In this definition, the storage contribution can occur at anytime. However, since environmental models are open and dissipative systems, all particles leave the system sooner or later. Therefore, we can revise the definition of the above $S_{ij}([a, b])$ and use an arbitrary large number $M$ instead of $\infty$ ($[a, M]$instead of $[a, \infty]$).

Utilizing this new definition for dynamic storage analysis, it is now possible to compute $S(t)$ based on the output of NPT simulations. Similar to numerical differential equation solutions (e.g., Euler, Runge-Kutta, etc.), the NPT-based computation of $S(t)$ will employ a discrete time-step value $h$. Smaller $h$ values are preferred for higher accuracy. To compute an approximate value of $S_{ij}(t)$, we first set a time interval $[t - h,\ t + h]$. Then, we label those particles that enter compartment $j$ during this time period, and sum up their storage contribution to $i$. A detailed description of this process is as follows:

**Step 1:** Simulate the model with NPT until all the particles that move during the time window $[t - h,\ t + h]$ leave the system.

**Step 2:** To compute $S_{ij}(t)$, we find out all contracted pathways of the form

$$\to j \to \cdots \to i \to \cdots \to *$$

where the first flow "$\to j$" occurs during $[t - h,\ t + h]$.

**Step 3:** To get an estimated value for $S_{ij}(t)$, we add up all residence times at $i$ for each contracted pathway, and divide this sum by the number of contracted pathways. This value is the storage contribution (at $i$) of a unit environmental input at $j$.

To demonstrate this process, we compute $S_{31}$ over the time interval [30, 32] using the sample NPT output provided in Fig. 13.3. The result will be an approximate value for $S_{31}$ at $t = 31$. As described in Step 2, we focus on the contracted pathways of the form $\rightarrow 1 \rightarrow \cdots \rightarrow 3 \rightarrow \cdots \rightarrow *$. Based on the three pathways shown in Fig. 13.3, we totally get 13 contracted pathways (three for particle 12, five for particle 43, and five for particle 212). Five of these contracted pathways start with compartment 1 (one for particle 12, two for particle 43, and two for particle 212), as shown in Table 13.2. For these five contracted pathways, the first flow ($\rightarrow 1$) occurs at 21.2, 30.4, 65.8, 31.2, and 98.0 time units, respectively. Only the second ($\rightarrow 1 \rightarrow 3 \rightarrow 1 \rightarrow 2 \rightarrow 3 \rightarrow *$) and the fourth ($\rightarrow 1 \rightarrow 2 \rightarrow 3 \rightarrow 1 \rightarrow 3 \rightarrow *$) contracted pathways have their first flow time in the interval [30, 32]. For the second contracted pathway, Particle 43 enters compartment 1 at 30.4 time units and visits compartment 3 twice, staying for 15.5 and 24.9 time units, respectively. For the fourth contracted pathway, Particle 212 enters compartment 1 at 31.2 time units and visits compartment 3 twice also, staying for 18.2 and 8.1 time units, respectively. Therefore:

$$S_{31}(t = 31) \approx \frac{(15.5 + 24.9) + (18.2 + 8.1)}{2} = 33.35$$

In general, we compute $S(t)$ as follows:

$$S_{ij}(t) \approx S_{ij}([t-h,\ t+h]) = \frac{\sum_{p \in P_j(t-h,t+h)} \left( \begin{array}{c} \text{Sum of residence times at} \\ \text{compartment } i \text{ for pathway } p \end{array} \right)}{|P_j(t-h,\ t+h)|} \tag{5}$$

where $P_j(t-h,\ t+h)$ is a list of contracted pathways that start with compartment $j$. This list contains the particles that enter compartment $j$ during the time interval $[t-h,\ t+h]$. $|P_j(t-h,\ t+h)|$ represents the number of items on this list.

We demonstrate the dynamic storage analysis computation using a simplistic lake model with three compartments: Phytoplankton ($P$), Zooplankton ($Z$), and Fish ($F$). The model accounts for the biomass flow among these compartments, simulating the annual changes of lake biomass. The differential equation model is as follows:

$$\begin{aligned} \dot{P} &= 1000 + 600\sin(t/8.3) - 2.5\cdot10^{-5}PZ - 2.5\cdot10^{-5}PF - 0.15P \\ \dot{Z} &= 2.5\cdot10^{-5}PZ - 2.5\cdot10^{-5}ZF - 0.1Z \\ \dot{F} &= 2.5\cdot10^{-5}PF + 2.5\cdot10^{-5}ZF - 0.08F \end{aligned} \tag{6}$$

The predatory relations among the three compartments are modeled using Lotka-Volterra type (Lotka, 1925; Volterra, 1926) predator-prey equations (Berryman, 1992). The Phytoplankton compartment gets time-varying environmental input ($1000 + 600\sin(t/8.3)$), which represents the fluctuating availability of nutrients and sunlight with seasonal variation. Figure 13.6a shows the time course of compartmental storage values. Successive peaks of storage values for the three compartments are in

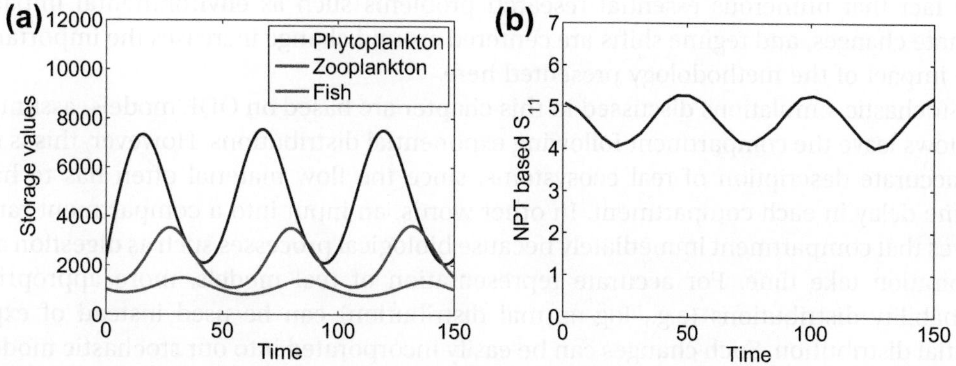


**FIGURE 13.6** (a) Time course of biomass storage, simulated using the differential equations (see Eqn. (6)) for the simplistic lake model. The oscillations are indicative of the seasonal changes. (b) The first entry of the dynamic storage analysis matrix $S_{11}(t)$ computed for this model. For color version of this figure, the reader is referred to the online version of this book.

accordance with their trophic level in the lake ecosystem. Figure 13.6b shows one element of the dynamic storage analysis matrix $S(t)$, computed using $h = 10$ time units. We observe that the period of oscillations in Fig 13.6a and b coincide with each other (about 50 time units). $S_{11}(t)$ represents how much storage is generated at compartment 1 (Phytoplankton) by a unit input into itself. Comparing Figs 13.6a and b, we observe a negative correlation between $S_{11}$ and the storage value of Phytoplankton. This is because any input received by the Phytoplankton at its lowest storage value is more likely to be retained longer as its storage value increases over time. Similarly, at its peak storage value, any input that the Phytoplankton receives is more likely to be lost as its storage value declines. This inverse relation indicates that more complicated and insightful results might be observed for larger systems with feedback cycles.

## 13.8 Conclusion

Individual-based studies of ecosystem flows are now within reach, thanks to sophisticated mathematical methods, efficient computational algorithms, and advances in computer technology. NPT simulations, although costly for extremely large models, provide new ways to analyze ecosystems that were not possible before.

In this chapter, we presented an individual-based methodology to compute various well-known system-wide ecological network properties. It turns out that this pathway-based methodology is simpler and more intuitive as opposed to the conventional algebraic methods based on matrix power series. The agreement of both methodologies verifies that the rather complicated algebraic formulations of the current NEA measures do accurately reflect their intended meanings. However, this new approach is not an alternative, but a parallel development to the existing methodology, extending the applicability of useful ecosystem properties to include dynamic, time-varying systems.

The fact that numerous essential research problems such as environmental impacts, climate changes, and regime shifts are centered around change increases the importance and impact of the methodology presented here.

Stochastic simulations discussed in this chapter are based on ODE models, assuming all flows leave the compartment following exponential distributions. However, this is not an accurate description of real ecosystems, since the flow material often has to have a time delay in each compartment. In other words, an input into a compartment rarely leaves that compartment immediately because biological processes such as digestion and respiration take time. For accurate representation of real models, more appropriate probability distributions (e.g., log-normal distribution) can be used instead of exponential distribution. Such changes can be easily incorporated into our stochastic models.

Another advantage of the NPT methodology is its ability to represent a wider range of scenarios than ODE or PDE models. Since NPT is an individual-based method, hybrid models where particles actually represent individual organisms, displaying intelligent behavior, combining both thermodynamical and animal behavior can be built. However, the compatibility between ODE and the NPT methodology will be broken as the ODE models do not have such flexibility.

Not everyone has quick access to NPT simulations. Therefore, simulation-based results are integrated into an experimental version of EcoNet (http://eco.engr.uga.edu), a free online software for modeling, simulation, and analysis (Kazanci, 2007, 2009; Schramski et al., 2010). This integration makes this rather computationally heavy method accessible to a wide range of scientists.

For steady-state models, the conventional methodology is still preferable because it is easier to use, simulations are not necessary, and the results are exact values. However, the conventional method only computes average behavior, whereas the NPT-based method incorporates noise and randomness, allowing the computation of standard deviation over multiple simulations. The focus of the work presented here is to provide a more capable alternative methodology still compatible with the currently available methods. For this purpose, we only use the mean of stochastic simulations. However, the additional information provided by the standard deviation and higher-order moments is an interesting topic to investigate for future studies.

# References

Berryman, A.A., 1992. The origins and evolution of predator-prey theory. Ecology 73 (5), 1530–1535.

Fath, B.D., Patten, B.C., 1999. Review of the foundations of network environ analysis. Ecosystems 2, 167–179.

Finn, J., 1977. Flow analysis: a method for tracing flows through ecosystem models. University of Georgia. Ph.D. thesis.

Finn, J.T., 1978. Cycling index: a general definition for cycling in compartment models. In: Adriano, D., Brisbin, I. (Eds.), Environmental Chemistry and Cycling Processes, U.S. Dep. Energy Symp., vol. 45. National Technical Information Center, Springfield, VA, pp. 148–164.

Finn, J.T., Jun 1980. Flow analysis of models of the Hubbard Brook ecosystem. Ecology 61 (3), 562–571.

Finn, J.T., 1982. Ecosystem succession, nutrient cycling and output-input ratios. J. Theor. Biol. 99 (3), 479–489.

Gillespie, D.T., 1977. Exact stochastic simulation of coupled chemical reactions. J. Phys. Chem. 81 (25), 2340–2361.

Gillespie, D.T., 1992. A rigorous derivation of the chemical master equation. Physica A 188 (3), 404–425.

Gillespie, D.T., 2000. The chemical Langevin equation. J. Chem. Phys. 113 (1), 297–306.

Hannon, B., 1973. The structure of ecosystems. J. Theor. Biol. 41 (3), 535–546.

Higashi, M., Patten, B.C., 1986. Further aspects of the analysis of indirect effects in ecosystems. Ecol. Model. 31 (1), 69–77.

Hippe, P.W., 1983. Environ analysis of linear compartmental systems: the dynamic, time-invariant case. Ecol. Model. 19 (1), 1–26.

Kazanci, C., 2007. Econet: a new software for ecological modeling, simulation and network analysis. Ecol. Model. 208 (1), 3–8.

Kazanci, C., 2009. Handbook of Ecological Modelling and Informatics. WIT Press. Chapter. Network Calculations II: A user's Manual for EcoNet. 325–350.

Kazanci, C., Ma, Q., 2012. Extending ecological network analysis measures to dynamic ecosystem models, Ecol. Model. doi:10.1016/j.ecolmodel.2012.05.021.

Kazanci, C., Matamba, L., Tollner, E.W., 2009. Cycling in ecosystems: an individual based approach. Ecol. Model. 220, 2908–2914.

Leontief, W., 1951. The Structure of American Economy, 1919–1929: An Empirical Application of Equilibrium Analysis. Harvard University Press, Cambridge, MA.

Leontief, W., 1966. Input–Output Economics. Oxford University Press, USA.

Lotka, A.J., 1925. Elements of Physical Biology. Williams and Wilkins, Baltimore.

Matamba, L., Kazanci, C., Schramski, J.R., Blessing, M., Alexander, P., Patten, B., 2009. Throughflow analysis: a stochastic approach. Ecol. Model. 220, 3174–3181.

Matis, J.H., Patten, B.C., 1981. Environ analysis of linear compartmental systems: the static, time invariant case. Bull. Int. Stat. Inst. 48, 527–565.

Patten, B.C., 1978. Systems approach to the concept of environment. Ohio J. Sci. 78 (4), 206–222.

Patten, B.C., 1985. Energy cycling, length of food chains, and direct versus indirect effects in ecosystems. Can. Bull. Fish. Aquat. Sci. 213 119–113.

Patten, B.C., 1986. Energy cycling, length of food chains, and direct versus indirect effects in ecosystems. Ecosystem Theory for Biological Oceanography. Can. Bull. Fish. Aquat. Sci. 213, 119–138.

Schramski, J., Kazanci, C., Tollner, E., 2010. Network environ theory, simulation, and econet® 2.0. Environ. Model.Soft. 26, 419–428.

Shevtsov, J., Kazanci, C., Patten, B., 2009. Dynamic environ analysis of compartmental systems: a computational approach. Ecol. Model. 220 (22), 3219–3224.

Volterra, V., 1926. Variazioni e fluttuazioni del numero d'individui in specie animali conviventi. Mem. Acad. Lincei 2, 31–113.

# 14

# Trophic Interactions in Lake Tana, a Large Turbid Highland Lake in Ethiopia

Ayalew Wondie*, Seyoum Mengistou†, Tadesse Fetahi†,[1]

*DEPARTMENT OF BIOLOGY, BAHIR DAR UNIVERSITY, ETHIOPIA, †ADDIS ABABA UNIVERSITY, COLLEGE OF NATURAL SCIENCE, DEPARTMENT OF ZOOLOGICAL SCIENCES, P.O. BOX 1176, ADDIS ABABA, EHIOPIA

[1]CORRESPONDING AUTHOR: T.FETAHI@YAHOO.COM

## 14.1 Introduction

Knowledge on the components of ecosystem is very important in order to maximize resource utilization in a sustainable manner as well as to preserve biodiversity. Furthermore, the trophic fluxes and efficiency of energy assimilation, transfer, and dissipation are also basic to understand ecosystem structure and functioning so as to maximize utilization without compromising the basic functioning of the ecosystem (Baird and Ulanowicz, 1993). Food web analysis may also play a key role in understanding the effects of climate change on lakes (Straile, 2005). For instance, the impact of global warming on the food web structure of Lake Tanganyika has been demonstrated to be accompanied with fishery reduction (O'Reilly et al., 2003; Verburg et al., 2003; Verschuren, 2003). A clear insight of ecosystem relationships can in turn help to devise a sound and correct lake management. To this end, food web analysis requires the quantification of the trophic relationships between the different functional groups in the system, which can be easily handled by computer models.

In recent years, a user-friendly software model called Ecopath with Ecosim (EwE) (Christensen et al., 2005; freely available at www.ecopath.org) has been widely used to describe the trophic relationships in aquatic ecosystems on quantitative bases. Determining the main components of the ecosystem and their feeding network in the system are the first steps in constructing Ecopath model. In Ecopath, a system is partitioned into boxes (functional groups) comprising species having a common physical habitat, similar diet, and life history characteristics or ecological similarities (same prey, same predator), rather than taxonomic classification. All trophic levels should be represented at a similar level of detail. Ecopath bases the parameterization on an assumption of mass balance in

Models of the Ecological Hierarchy, DOI: http://dx.doi.org/10.1016/B978-0-444-59396-2.00014-6

ISSN 0167-8892, 

that the production of any given prey is equal to the biomass consumed by predators, the biomass caught (e.g., in fisheries), and any exports from the system (Christensen and Walters, 2000), which is in agreement with the universal first law of thermodynamics that matter and energy are constant and irreducible.

In general, it can be said that holistic lake studies in Africa are scarce although the literature on temperate lakes is quite extensive. The few holistic lake studies done in African lakes have come up with interesting findings. For instance, Mavuti et al. (1996) applied Ecopath model to Lake Naivasha (Kenya) and Lake Ihema (Rwanda) and identified misutilization of food sources in the ecosystem. Very recently, Fetahi and Mengistou (2007) and Fetahi et al. (2011) applied Ecopath model to Lake Hawasa and Lake Haqy in Ethiopia and reported trophic interactions and energy flow of different functional groups. The authors demonstrated that the model could be used as a tool for management purposes. For example, the authors demonstrated that in Lake Hawasa, large *Labeobarbus* (*Labeobarbus intermedius intermedius*; Rüppell, 1835) was underexploited and tilapia fish (*Oreochromis niloticus*) overexploited, suggesting that utilizing the former will give sufficient time for stock recovery of the latter.

Lake Tana (Ethiopia, East Africa), the focus of the present study, is the largest lake in the country and is the source of Blue Nile. Many aspects of fishery biology of the lake were studied by various workers. The possible competition and ecological overlap existing in terms of habitat and food availability between different fish populations were reported by Nagelkerke (1997), Wudneh (1998), and de Graaf (2003). Dejen (2003) also indicated the importance of small barbs in the lake ecology. Thus, sufficient literature data is available to quantify parameters of the Ecopath model such as fish biomass, production/biomass ratio, and feeding matrix (Nagelkerke, 1997; Wudneh, 1998; Dejen, 2003; de Graaf, 2003). Furthermore, data on phytoplankton and zooplankton biomass and production rates of the major plankton groups are available (Wondie and Mengistou, 2006; Wondie et al., 2007). Despite the fact that there are enough data obtained from different trophic groups in the lake, this is the first attempt to put all these information together in order to assess the food web relationship and energy flow in the lake.

The objective of the study was to assemble the available data and employ Ecopath to characterize the food web interaction of the different functional groups and to analyze the flow of energy using mass balance **EwE** version 5 model. This study will be a reference point for future investigation on biomass and production of different trophic groups in the system, because many changes may occur in and around the lake as a result of intense anthropogenic impact as well as climatic factors.

# 14.2 Materials and Methods

## 14.2.1 The Study Site: Lake Tana

Lake Tana, the largest lake (area 3156 $km^2$) in Ethiopia is the source of the Blue Nile and is located between latitude 10°58′–12°47′N and longitude 36°45′–38°14′E (Fig. 14.1). Lake

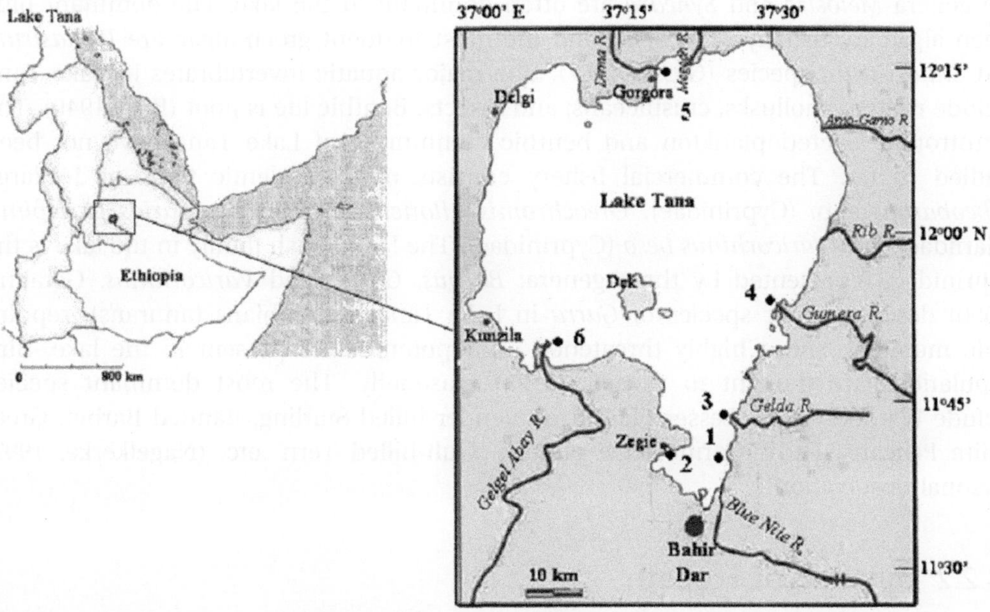


**FIGURE 14.1** Map of Lake Tana in Northern Ethiopia and the sampling locations within the lake. Numbers refer to sampling stations in the littoral and open water regions of the lake.

Tana is shallow (average depth 8 m, maximum 14 m) and meso-oligotrophy. It has a maximum length of 78 km, width of 67 km, and a shoreline of 385 km. The community utilizes the lake for various purposes such as source of water (domestic, irrigation), navigation and transportation, recreation, and fisheries. Lakenet (1999) reported that Lake Tana is one of the top 250 lake regions that are most important for biological diversity in the world.

Five major permanent rivers—Gelgel Abbay (Small Blue Nile), Gumara, Rib, Megech, and Dirma—as well as more than 30 seasonal streams feed the lake, whereas the Blue Nile River is the only outflow. A study during 1990/2000 reported that the average retention time of the lake water was 6.5 years (Wudneh, 1998).

The climate of Lake Tana is characterized by a main-rainy season with heavy rains during July–September, a dry season during December–April, and two minor rainy seasons during May–June (pre-rainy season) and October–November (post-rainy season) (Wudneh, 1998).

The eastern and southern shores of Lake Tana are covered with swamps dominated by papyrus (*Cyperus papyrus*), *Typha latifolia*, water lily (*Nymphaea* sp.), and *Ceratophyllum* sp. Close to the shore, two species of trees are common: *Ficus* sp. and *Milettia ferruginea* (Nagelkerke and Sibbing, 1996). Macrophytes are more common in the rainy season. The phytoplankton is dominated by diatoms and previous records show that the genera *Melosira* and *Synedra* are often dominant in the lake. The dominant blue green algae are usually *Microcystis* and the most frequent green algae are *Pediastrum* and *Staurastrum* species (Gasse, 1987). The major aquatic invertebrates in Lake Tana include rotifers, mollusks, crustaceans, and insects. Benthic life is poor (Bini, 1940). The hetrotrophic bacterioplankton and benthic community of Lake Tana have not been studied so far. The commercial fishery consists of an endemic flock of 15 large *Labeobarbus* spp. (Cyprinidae), *Oreochromis niloticus* (Cichlidae), *Clarias gariepinus* (Clariidae), and *Varicorhinus beso* (Cyprinidae). The largest fish family in the lake is the Cyprinidae, represented by three genera: *Barbus, Garra,* and *Varicorhinus.* Getahun (2000) described four species of *Garra* in Lake Tana. Amphibians (anurans), reptiles (Nile monitor), and a highly threatened hippopotamus are present in the lake. Bird populations are thought to exceed 20,000 seasonally. The most dominant species include Wattled Crane, Lesser Flamingo, Slender-billed Starling, Banded Barbet, Great White Pelicans, Great Cormorants, Herons, Gull-billed Tern, etc. (Nagelkerke, 1997, personal observation).

## 14.2.2 Equation of Ecopath

The basic equation, which applies to any box (functional group) in the model, is

$$B_i \times P/B_i \times EE_i = \sum B_j \times Q/B_j \times DC_{ij} + Y_i \quad (1)$$

which implies that the biomass $B_i$ of a group (or a box) $i$ multiplied by the Production/Biomass ratio ($P/B$) and by what is called "ecotrophic efficiency" (EE) is equal to the biomass consumed by every predator ($j$). This group is then multiplied by the relative consumption of food of any predator $j(Q/B_j)$ and by the percentage of this box in the diet of the predator $j$. When suitable, the fish yield $Y_i$ has to be added. In Ecopath model, one of the four parameters $B$, $P/B$, $Q/B$, or EE can be left unknown and is computed by the software. Detail description of Ecopath can be found in Christensen and Pauly (1992, 1993), Christensen et al. (2000), and Pauly et al. (2000).

## 14.2.3 Functional Groups

### 14.2.3.1 Phytoplankton

Average Chl-a concentration in Lake Tana measured over 2 years was taken as 10.82 mg Chl-a $m^{-3}$. This value was converted into depth integrated Chl-a over $z_{eu}$ (27.05 mg Chl-a $m^{-2}$) by multiplying with the euphotic depth (2.5 m). This value was then converted to wet weight using two successive conversion factors (Jones, 1979). The first was to change Chl-a to Carbon: 40:1—carbon to Chl-a ratio, then Carbon = 10% of wet weight. Hence average phytoplankton biomass (wet weight) of Lake Tana was estimated at 10.82 g fw $m^{-2}$.

Depth-integrated gross primary production in the open water column was 0.8 g C $m^{-2}$ $day^{-1}$ assuming a 10 h production period per day (Wondie et al., 2007). Therefore, the annual photosynthetic production becomes 292 g C $m^{-2}$ $year^{-1}$. When this value is changed into wet weight, using a conversion factor of 10% wet weight to C, the annual primary production is 2920 g WW $m^{-2}$ $year^{-1}$. The $P/B$ ratio of phytoplankton is estimated at 270 per year.

#### *14.2.3.2 Macrophytes Biomass*

Biomass of macrophytes was estimated based on preliminary survey made by Bahir Dar Fishery Research Center and informed guess of Dr. Vijverberg (2005, personal communication). Fresh weight estimate of macrophytes biomass was averaged as 200 t $km^{-2}$.

#### *14.2.3.3 Detritus Biomass (D)*

There were no available literature data regarding the biomass of detritus in Lake Tana. Hence, the detrital biomass was calculated as a function of primary production and euphotic depth by employing the relationship as suggested by Christensen and Pauly (1993):

$$\log D = 0.954 \log \mathrm{PP} + 0.863 \log E - 2.41 \tag{2}$$

where $D$ is the detrital biomass (g C $m^{-2}$); PP, the primary production (in g C $m^{-2}$ $year^{-1}$); and $E$, euphotic depth in meters.

Euphotic depth was estimated from the Secchi depth transparency measurement as 2.5 m, hence detrital biomass was estimated as 2.37 g C $m^{-2}$.

A carbon to wet weight ratio of 10% was used to obtain wet weight detritus biomass of 23.7 g WW $m^{-2}$. However, estimation of detritus biomass in oligotrophic ecosystems (like Lake Tana and based on primary production) may underestimate the actual value.

#### *14.2.3.4 Zooplankton*

In the present study, we estimated the biomass and production of the major zooplankton at species level and functional groups based on field measurements for two years. The biomass of herbivorous and carnivorous zooplankton was estimated as 41.93 and 6.18 mg DW $m^{-3}$, and converted into depth integrated biomass when multiplied by the mean depth of the lake (8 m), which gives 335.44 and 49.44 mg $m^{-2}$, respectively. Then, using the relationship of Hall et al. (1976) and Burgis (1974), values in dry weight were converted into the corresponding wet weights: dry weight = 20% of wet weight.

Hence wet weight values of herbivorous and carnivorous zooplankton are 1677.2 mg $m^{-2}$ (1.7 g WW $m^{-2}$) and 0.25 g $m^{-2}$(247 mg WW $m^{-2}$), respectively.

The production of herbivorous and carnivorous zooplankton was estimated as 1335.9 and 156.95 mg DW m$^{-}$3 year$^{-1}$, and converted into depth integrated production when multiplied by the mean depth of the lake (8 m), which gives 10687.2 and 1255.6 mg DW m$^{-2}$ year$^{-1}$, respectively (Wondie and Mengistou, 2006). Using the above relationship, the wet weight values of herbivorous and carnivorous zooplankton are 53.44 g m$^{-2}$ year$^{-1}$ and 6.3 g m$^{-2}$ year$^{-1}$, respectively.

The production/ biomass ratio ($P/B$) for herbivorous and carnivorous zooplankton were 31.86 and 25.34 per year, respectively. Consumption/biomass ratio ($Q/B$) for zooplankton was estimated based on assumed gross food conversion efficiency ($P/Q$) of 0.2. Hence $Q/B$ for herbivorous and carnivorous zooplankton are 157.2 and 126 per year, respectively.

### *14.2.3.5 Zoobenthos*

Zoobenthos include insect larvae (Chironomidae, Ephemeroptera, and Odonata), annelids, mollusca (bivalves and gastropods), macrocrustaceans (e.g., crabs), and benthic microcrustacean (e.g., Ostracoda). The biomass and production estimation of zoobenthos were obtained from a preliminary study of Dr. Vijverberg (2005, personal communication) as 8.5 t km$^{-2}$ and 35 t km$^{-2}$ year$^{-1}$, respectively.

$P/B$ ratios for benthic organisms were estimated using the empirical formula of Brey (1999 as cited in Christensen et al. (2000)):

$$\log P/B = 1.672 + 0.993 \times \log (1/A_{\text{max}} - 0.035 \times \log (M_{\text{max}})\ 300.447 \times 1/(T + 273) \qquad (3)$$

where $A_{\text{max}}$ is the maximum age (per year); $M_{\text{max}}$, maximum individual body mass (g DM); and $T$, bottom water temperature ($^{\circ}$C)

Maximum individual body weight and maximum age for zoobenthos was determined as 0.2 g wet weight and 1.5 year, respectively. Hence $P/B$ was calculated as 3.1.

$Q/B$ for zoobenthos was estimated based on assumed gross food conversion efficiency ($P/Q$) of 0.2 (Christensen et al., 2000). Hence benthos have an annual $Q/B$ of 20.58.

### *14.2.3.6 Fisheries*

Average fish catch for the whole of Lake Tana is very low, 4 kg ha$^{-1}$ year$^{-1}$ (0.4 t km$^{-2}$), in comparison with other African lakes (45–180 kg ha$^{-1}$ year$^{-1}$). However, potential fish yield estimates for Lake Tana using various productivity indicators is 32 kg ha$^{-1}$ year$^{-1}$ (3.2 t km$^{-2}$) (Wudneh, 1998; de Graaf, 2003). No previous detail analysis of production and $P/B$ for most of the commercial fishes was available. According to Wudneh (1997), it is assumed that each major fish taxa (Cyprinids, Cichlids, and Clarids) in Lake Tana contributes one-third of the total fish biomass. However, Dejen (2003) estimated the biomass of the most abundant large *Labeobarbus* species, *brevicephalus* (30% by WW) as 4 kg ha$^{-1}$.

Fifteen large *Labeobarbus* spp. (piscivores, non-piscivores), *Oreochromis niloticus*, *Clarias gariepinus*, and *V. beso*, which are important fish species for the commercial fishery and unexploited fish species (juveniles, small barbs, and *Garra* sp.) of Lake Tana, were considered in this modeling analysis. All of the fish species share the $P/B$ and $Q/B$ empirical relationship but their estimation were made separately.

#### 14.2.3.7 Production/Biomass Ratio (P/B)

Production /biomass ratios are difficult to estimate directly and were taken as equivalent to total mortality ($Z$) (Pauly et al., 2000) assuming steady-state ecosystem (Allen, 1971). Total mortality ($Z$) of the exploited fish species was determined by summing the value of fishing mortality ($F$) and natural mortality ($M$) as $Z = F + M$. $F$ will be determined subsequently. However, $M$ was estimated using the following empirical equation (Pauly, 1984):

$$\ln(M) = -0.0152 - 0.279\ln(L_{\text{inf}}) + 0.6543\ln(K) + 0.463\ln(T) \quad (4)$$

where $L_{\text{inf}}$ and $K$ are parameters of VBGF and $T$ is the annual mean surface water temperature of the lake, taken as 22 °C. These parameters of finding $M$ of fishes were obtained from the literature as mentioned above.

Natural mortality ($M$) for unexploited fish species, including juveniles' of exploited fish species, was estimated using Pauly's empirical formula as in Eqn (3) and considered as equivalent to total mortality ($Z$), as they do not suffer from fishing mortality ($F$). This $M$ value was directly taken as $P/B$.

Natural mortality and $Q/B$ values of exploited and unexploited fish species were estimated using parameters compiled in Table 14.1.

#### 14.2.3.8 Consumption/Biomass Ratio (Q/B)

This parameter expresses food consumption ($Q$) per unit biomass for a conventional period of one year, which means number of times a given population consumes its own weight per year. This quantity, for each consumer eco-group, was estimated using the following empirical relationship suggested by Palomares and Pauly (1998).

$$\log(Q/B) = 7.964 - 0.204 \times \log W_{\text{inf}} - 1.965 \times T + 0.083 \times A + 0.532 \times h + 0.398 \times d \quad (5)$$

where $W_{\text{inf}}$ is the asymptotic weight (g); $T$ is an expression for the mean annual temperature of the water body expressed using $T = 1000/\text{Kelvin}$ (Kelvin = °C + 273.15); $A$ is aspect ratio ($Ar = h^2/s$ of the caudal fin of fish, given height ($h$) and surface area ($s$); h is a dummy variable expressing food type (1 for herbivorous, and 0 for detritivores and carnivores); and $d$ is a dummy variable also expressing food type (1 for detritivores and 0 for herbivores and carnivores).

**Table 14.1** Growth and Other Parameters Compiled for *Q*/*B* and *M* Estimation

| | Catfish | Juv. catfish | L. labeo: Piscivores | L. labeo: Herbivores | L. labeo: Benthivores | L. labeo: Zooplanktivore | *V. beso* | Juv. L. Cyprinids | Tilapia | Juv. tilapia | Small barbs | *Garra* |
|---|---|---|---|---|---|---|---|---|---|---|---|---|
| $L_{inf\ (cm)}$ | $90^T$ | $30^T$ | $58^T$ | $42^T$ | $45^T$ | $42^T$ | $42^G$ | $22^T$ | $35.7^T$ | $18^T$ | $7.6^E$ | $13^G$ |
| $W_{inf\ (gm)}$ | $12{,}000^G$ | $540^G$ | $3000^G$ | $2000^G$ | $2500^G$ | $2000^G$ | $2000^G$ | $40^G$ | $950^G$ | $170^G$ | $15^G$ | $20^G$ |
| $T$ (°C) | 22 | 22 | 22 | 22 | 22 | 22 | 22 | 22 | 22 | 22 | 22 | 22 |
| $K$ (year$^{-1}$) | $0.2^T$ | $1.64^T$ | $0.25^T$ | $0.3^T$ | $0.33^T$ | $0.3^T$ | $1.5^G$ | $0.5^T$ | $0.5^T$ | $0.7^T$ | $1.65^E$ | $1.5^G$ |
| $A$ | 1.5 | 1.5 | 3 | 3 | 3 | 3 | 3 | 2.5 | 1.8 | 1.8 | 2.5 | 1.4 |
| $h$ | 0 | 0 | 0 | 1 | 0 | 0 | 1 | 0 | 1 | 1 | 0 | 1 |
| $d$ | 0 | 0 | 0 | 0 | 0 | 0 | 0 | 1 | 0 | 0 | 0 | 0 |

Source: Wudneh (1998), Dejen (2003).

It is assumed that for *Q*/*B*, in a population under equilibrium conditions, young individuals are more in number than old ones and consume much more food (in relation to their weight) than old fish.

#### 14.2.3.9 Commercial Fishes and Juveniles

Growth and mortality parameters were estimated earlier by Wudneh (1998). Growth parameters were based on back calculated and length frequency analysis. The resulting growth parameter values were $L_\infty = 90$ cm and $K = 0.2\ \text{year}^{-1}$ for *C. gariepinus*, $L_\infty = 35.7$ cm, and $K = 0.5\ \text{year}^{-1}$ for *O. niloticus* and $L_\infty = 45$ cm and $K = 0.33\ \text{year}^{-1}$ for *B. tsanensis*.

The instantaneous total ($Z\ \text{year}^{-1}$) and natural ($M\ \text{year}^{-1}$) mortality of Lake Tana's commercial fishes were estimated using Pauly's empirical formula. *C. gariepinus, O. niloticus*, and *B. tsanensis* were fully vulnerable to the fishery at 3.9, 2.3, and 9 years, respectively. The estimates of *Z* were 0.9 for *C. gariepinus* and 1.8 $\text{year}^{-1}$ for other species (African tilapia, large barbs, and *V. beso*).

#### 14.2.3.10 Small Barbs

Small barbs form the main biomass of fish in the lake and are the most important link between zooplankton and piscivorous species (Dejen, 2003). Growth, mortality, and productivity of the two small barbs of Lake Tana were studied by Dejen (2003). Small *Labeobarbus* is one of the unexploited fish species in Lake Tana. Production estimates were 49 kg $\text{ha}^{-1}\ \text{year}^{-1}$ (16.8 for *B.h* and 32.4 for *B.t*). The annual maximum sustainable yield of *B. tanapelagius* was 3850 tons (Dejen, 2003).

Resulting growth parameter values were $L_\infty = 9.1$ cm and $K = 1.2\ \text{year}^{-1}$ for *B. humilis* and $L_\infty = 7.6$ cm and $K = 1.65\ \text{year}^{-1}$ for *B. tanapelagius*. Total mortality coefficient (*Z*) was estimated as 3.94 and 3.88 $\text{year}^{-1}$ for *B. humilis* and *B. tanapelagius*, respectively.

#### 14.2.3.11 Garra Species

Since *Garra* is a small fish which does not reach table size in its life time in Lake Tana, Estimates of $L_\infty$, *K*, and $W_\infty$ were made following the procedures of Lake Hawasa (Fetahi, 2005). Values of these parameters are tabulated in Table 14.1.

#### 14.2.3.12 Fish and Other Mixed-Diet Eating Birds, and Hippo

There are many fish-eating birds around Lake Tana and some of them include Great White Pelicans, Great Cormorants, Herons, African Fish Eagle, African Jacana, Lesser Black-backed Gull, Gull-billed Tern, Pied Kingfisher, Malachite Kingfisher, Giant Kingfisher, Yellow Wagtail, etc. (Nagelkerke, 1997). Despite the birds being important functional groups placed in different trophic levels and compartments, we did not include them in this holistic analysis simply due to lack of any data on them. This includes the African hippo also, which is known to consume macrophytes in Lake Tana. We realize that this omission will somehow influence the final results and discussions in some way or another.

#### *14.2.3.13 Diet Composition (DC)*

The average composition of the food of each consumer group was assembled from literature data and some data was generated from the present study. The 15 large *Labeobarbus* are well diversified in their distribution and feeding ecology (de Graaf, 2003). On the basis of gut content data, Nagelkerke (1997) described different (approximately five) trophic groups. Eight of them (*B. acutirostris, B. dainelli, B. gorguari, B. longissimus, B. macrophthalmus, B. megastoma, B. platydorsus, and B truttiformis*) feed mainly on fish (small barbs, *Garra* sp., and juvenile barbs species and tilapia) and little on benthic organisms (insect larvae and detritus). Three species (*B. crassibarbus, B. nedgia, and B. tsanensis*) feed on benthic invertebrates (insect larvae and gastropod mollusks) and detritus. Two species (*B. gorgorensis* and *B. surkis)* feed on a mixture of macrophytes and molluscs. One species (*B. brevicephalus)* feeds mainly on zooplankton, and one other species (*B. intermedius)* has diversified diet. Three small barbs species (*B. pleurogramma, B. tanapelagius,* and *B. humilis*) were identified in Lake Tana (Dejen, 2003). However, *B. tanapelagius and B. humilis* are the most common small barbs in the Lake. Zooplankton was the major diet component of *B. tanapelagius, B humilis,* and juvenile large barbs.

## 14.3 Results and Discussion

Estimates of all basic input parameters ($B$, $P/B$, and $Q/B$) as well as biomass, ecotrophic efficiencies, food consumptions, and trophic levels are presented in Table 14.2. The relevant feeding matrix of all functional groups is given in Table 14.3. In general, estimated basic input parameters and result of mass balance are in agreement with each other. That is, most of the data collected on the various groups are mutually compatible as confirmed by the fact that all values of EE are lower than 1.

With 0.84, the pedigree index of the present analysis falls within the range given for the overall model quality (Christensen et al., 2005). Indeed, several functional groups are represented at species level, and most of the input data were system-specific obtained from the present study.

### 14.3.1 Trophic Structure and Flows

The balanced flow diagram for the data sets is given in Fig. 14.2. Flows from detritus were as important as flows from phytoplankton, suggesting the importance of both detritus and grazing food chain in Lake Tana (Table 14.4). The main sources for flow to detritus are primary producers including phytoplankton and macrophytes, which

**Table 14.2** Basic Estimates (Input Data) Used for Analysis of Lake Tana Ecosystem

| Species/Group | Trophic level | *B* (t km$^{-2}$) | *P/B* (year$^{-1}$) | *Q/B* (year$^{-1}$) | EE | GE |
|---|---|---|---|---|---|---|
| *Clarias gariepinus* | 2.82 | 1.06(1.8) | 0.67(1.2) | 3.94 | 0.1 | 0.170 |
| LLb-piscivores | 3.46 | 0.70(1.2) | 0.63 | 6.96 | 0.818 | 0.091 |
| LLb–benthivores | 2.91 | 0.40 | 0.93 | 25.74 | 0.582 | 0.036 |
| LLb-zooplanktivore | 2.81 | 0.50 | 0.93 | 7.56 | 0.422 | 0.123 |
| LLb–herbivore | 2.00 | 0.25 | 1.13 | 7.22 | 0.188 | 0.157 |
| Small barbs | 2.66 | 3.67 | 1.47 | 18.64 | 0.809 | 0.079 |
| *Oreochromis niloticus* | 2.00 | 2.00 | 1.30 | 23.82 | 0.151 | 0.055 |
| *Varicorhinus beso* | 2.00 | 0.30 | 1.70 | 25.74 | 0.383 | 0.066 |
| *Garra* sp. | 2.00 | 0.20 | 1.00 | 33.84 | 0.037 | 0.021 |
| Juvenile fish | 2.00 | 1.2 | 1.60 | 48.50 | 0.99 | 0.047 |
| Zoobenthos | 2.00 | 8.5 | 3.10 | 20.58 | 0.171 | 0.151 |
| Carnivore zooplankton | 2.50 | 0.25 | 25.34(40) | 126.00 | 0.247 | 0.201 |
| Herbivores zooplankton | 2.00 | 1.7 | 31.86(58) | 157.20 | 0.98 | 0.203 |
| Phytoplankton | 1.00 | 10.82 | 270.00 | – | 0.162 | – |
| Sessile algae | 1.00 | 5.00 | 5.50 | – | 0.527 | – |
| Macrophytes | 1.00 | 200 | 2.00 | – | 0.185 | – |
| Detritus | 1.00 | 23.70 | 1.00 | – | 0.043 | – |

**Table 14.3** Diet Composition (% of Weight of Stomach Contents) of Consumers in Lake Tana. Estimates Were Obtained from Literature (Nagelerke, 1997; de Graaf, 2003; Dejen, 2003)

| Prey | Predators | | | | | | | | | | | | |
|---|---|---|---|---|---|---|---|---|---|---|---|---|---|
| | 1 | 2 | 3 | 4 | 5 | 6 | 7 | 8 | 9 | 10 | 11 | 12 | 13 |
| 1 *Clarias gariepinus* | 0.048 | 0.024 | – | – | – | – | – | – | – | – | – | – | – |
| 2 LLb-piscivores | 0.052 | – | – | – | – | – | – | – | – | – | – | – | – |
| 3 LLb-benthivores | 0.031 | – | – | – | – | – | – | – | – | – | – | – | – |
| 4 LLb-zooplanktivores | 0.052 | – | – | – | – | – | – | – | – | – | – | – | – |
| 5 LLb-herbivores | 0.016 | – | – | – | – | – | – | – | – | – | – | – | – |
| 6 Small barbs | 0.105 | 0.848 | 0.134 | – | – | – | – | – | – | – | – | – | – |
| 7 *Oreochromis niloticus* | 0.105 | 0.016 | – | – | – | – | – | – | – | – | – | – | – |
| 8 *Varicorhinus beso* | 0.001 | 0.08 | – | – | – | – | – | – | – | – | – | – | – |
| 9 *Garra* sp. | 0.001 | 0.002 | – | – | – | – | – | – | – | – | – | – | – |
| 10 Juvenile fish | 0.11 | 0.011 | 0.624 | – | – | – | – | – | – | – | – | – | – |
| 11 Zoobenthos | 0.107 | – | 0.239 | 0.242 | – | 0.044 | – | – | – | – | – | – | – |
| 12 Carnivore zooplankton | – | 0.001 | – | 0.01 | – | 0.056 | – | – | – | – | 0.001 | – | – |
| 13 Herbivores zooplankton | – | 0.018 | – | 0.546 | – | 0.507 | – | – | – | – | 0.001 | 0.494 | – |
| 14 Phytoplankton | – | – | – | 0.101 | 0.1 | 0.2 | 0.9 | 0.3 | 0.55 | 0.9 | 0.249 | 0.203 | 0.9 |
| 15 Sessile algae | – | – | – | – | – | – | – | 0.5 | 0.3 | – | 0.049 | – | – |
| 16 Macrophytes | – | – | – | – | 0.75 | – | 0.05 | – | 0.05 | – | 0.4 | – | – |
| 17 Detritus | 0.293 | – | – | 0.101 | 0.15 | 0.193 | 0.05 | 0.2 | 0.1 | 0.1 | 0.3 | 0.304 | 0.10 |

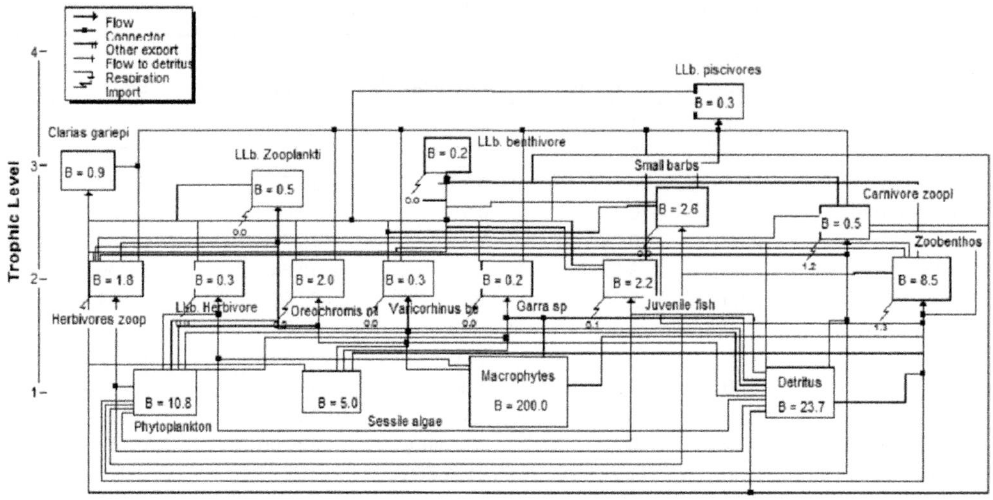


**FIGURE 14.2** Flow diagram showing trophic flows in Lake Tana, Ethiopia. Flows are expressed in t km$^{-2}$ year$^{-1}$.

**Table 14.4** Transfer Efficiency (%) from Producers and Detritus in Lake Tana

| Source/TL | II | III | IV | V | VI |
|---|---|---|---|---|---|
| Producer | 10.7 | 5.2 | 4.4 | 2.1 | 0.1 |
| Detritus | 7.1 | 5.1 | 3.9 | 1.8 | |
| All flows | 10.0 | 5.2 | 4.3 | 2.1 | 0.1 |

Proportion of total flow originating from detritus: 0.43
Transfer efficiencies (calc. as geometric mean for TL II–IV)
From primary producers: 6.2%
From detritus 5.2%
Total: 6.1%

contribute more than 75% of the total flow. Transfer efficiencies along the different trophic levels are presented in Table 14.4. The net transfer efficiency in total flow originating from producers (6.2%) is slightly higher than detritus (5.2%) (Table 14.4). The importance of detrital food chain in Lake Tana can be explained by the occurrence of zoobenthos and zoobenthivorous fish. Transfer efficiencies show similar trend, i.e., a decrease in both producers and detritus along the trophic levels. It appears that most of the biomasses and flows are concentrated on trophic levels I and II.

The values of trophic levels are a function of the diet composition of a group and the trophic level of its prey. The largest trophic value is 3.46, which is slightly higher than Lake Hawasa (3.37) indicating that almost there is no specialized top predator in both systems.

The Lake Tana ecosystem possesses more than one major trophic pathway. The following food chain paths could be taken as examples:

- Phytoplankton → herbivorous zooplankton → zoobenthos → small barbs → LLb-piscivores
- Phytoplankton → herbivorous zooplankton → small barbs → LLb-piscivores
- Phytoplankton → herbivorous zooplankton → LLb-zooplanktivore → catfish
- Phytoplankton → tilapia → catfish
- Sessile algae → *Garra*/*V. beso* → catfish

Large piscivorous labeobarbs (LLb-piscivores) may not gain the required amount of energy in the first food chain as most of the energy is lost in the long process originating from producers. However, the second link is much better for LLb-piscivores. Indeed, the major food stuff of small barbs is herbivorous zooplankton not zoobenthos, demonstrating the second link is the major means of energy acquisition for the fish. For catfish, phytoplankton → tilapia → catfish is very efficient because of the shorter food chain and the greater energy transfer from producers.

### *14.3.2 Ecotrophic Efficiencies (EE)*

Ecotrophic efficiency, the proportion of the production that is either consumed by predators or harvested/exported within the system, varied considerably in Lake Tana (Table 14.2). Top predator fishes such as LLb-benthivores and primary producers were lightly exploited, i.e., they were not much consumed in the system by the relevant functional groups. For example, phytoplankton was not efficiently utilized by zooplankton. However, zooplankton growth was largely controlled by high predation pressure of small barbs, indicating the top-down control of fish on zooplankton. On the other hand, small barbs were highly consumed by piscivorous labeobarbs, contributing almost 75% for its mortality. This suggests that any fishery activity on small barbs should consider their high mortality, a view contrary to Dejen (2003) who recommended artisanal fishing on small barbs (like clupeids of Lake Tanaganyika).

The bulk of phytoplankton and macrophytes biomass was underutilized showing that a large amount of primary production goes to detritus. In fact, the present study as well as other studies in several lakes has shown that detritus is an important food source for a number of fish species (Teferra and Fernando, 1989; Tadesse, 1999, Bowen, 1984; Nagelkerke, 1997). The observed lower utilization of macrophytes could be due to the exclusion of African hippo and the usage of macrophytes by the community for ceremonials and roof thatching in the analysis.

Carnivorous zooplankton consumes herbivorous zooplankton and both of them are utilized by the exploited fish species. Hence, zooplankton is a major link to the higher trophic levels although most live phytoplankton was not transferred through them. This could be due to the dominance of copepods in Lake Tana over cladocerans, high filter feeders, suggesting that top-down effect on phytoplankton could be minimal, similar to what was reported for Lake Hawasa (Fetahi, 2005).

In mass balance ecosystem models, all energy flows and biomasses can be shown in a single flow diagram (Fig. 14.2) (Christensen et al., 2000).

Among the food sources, phytoplankton, detritus, and macrophytes were the least consumed in the ecosystem whereas zooplankton and small sized fishes were highly consumed. Therefore, energy flows to higher trophic levels are very high from zooplankton. This is in agreement with the findings of Dejen (2003) who emphasized that zooplankton is a critical food web link in the Lake Tana ecosystem. However, the EE values of most functional groups indicate that they are fairly utilized in the ecosystem, which according to Dickie (1972) is sustainable. He stated that ecotrophic efficiency in nature is unlikely to exceed a value of about 0.5.

Phytoplankton and macrophytes are of little use as direct food in the ecosystem probably due to small biomass of herbivorous groups. This indicates that the producers could support more herbivores. *O. niloticus* and herbivorous zooplankton are too low to ensure full utilization of phytoplankton communities. Furthermore, ingestion of detritus from decomposing blue-green cells during the post-rainy season was suggested as an important food source for zooplankton in Lake Tana (Wondie and Mengistou, 2006). It was also suggested that silt (turbidity) may inhibit the filtration efficiency of herbivores or the large size structure of *Melosira*, which forms an almost inedible soup in the lake seasonally.

The inefficient consumption of zooplankton community on phytoplankton can be compensated by the alternative/complementary feeding of zooplankton on suspended organic material as well as bacteria. A similar pattern was noted in Lake Tanganyika (Moreau et al., 1993), Lake Turkana (Christensen and Pauly, 1993), and Lake Kariba.

In general, the estimates of the ecotrophic efficiency of the functional group are in agreement with what is known about their importance in the whole ecosystem, with the exception of small barbs, which were believed to be less exploited. In addition, *O. niloticus* appears to be much less exploited than would be expected.

### *14.3.3 Analysis of Mixed Trophic Impact*

The result of mixed trophic impact for Lake Tana ecosystem is shown in Fig. 14.3. Based on input and output analysis, we were able to assess the impact of an individual group on other groups in the system (Christensen and Walters, 2000). The mixed trophic impact routine provides useful information in explaining or predicting short-term variations but it cannot be an instrument for medium- and long-term predictions.

As a prey, a group causes a positive impact on its predators. As a direct predator, it has a negative impact on its prey. Some examples corroborate this view (Fig. 14.3).

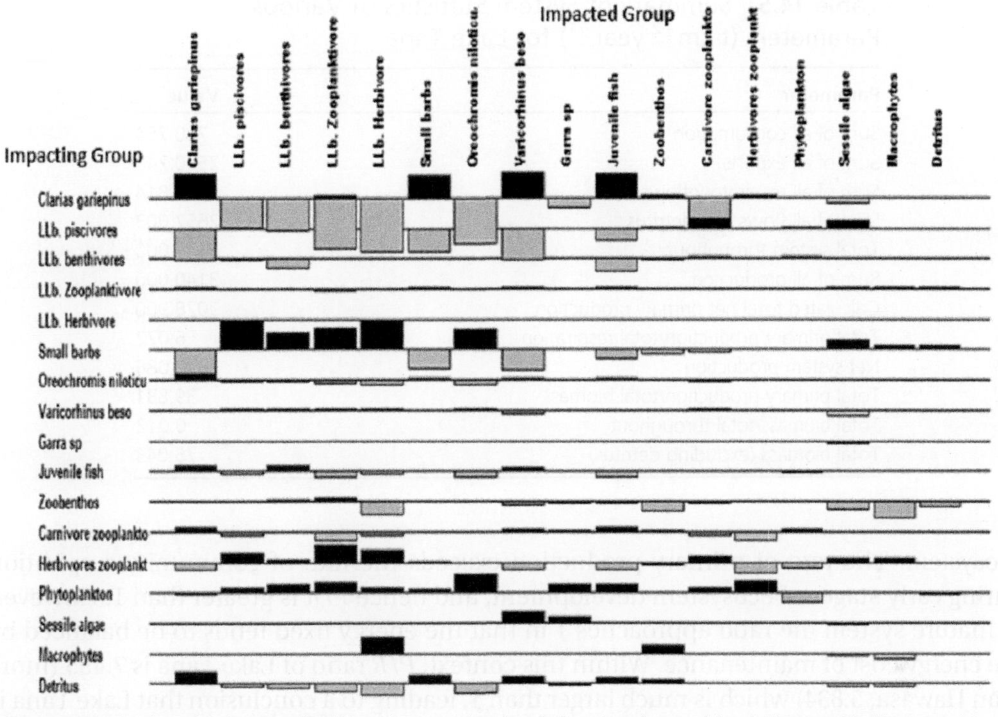


**FIGURE 14.3** Mixed trophic impacts (MTI) in Lake Tana food web showing the combined direct and indirect trophic impacts. Positive impacts are shown above the baseline (black bar) and negative below (gray bar). The impacts are relative and comparable between groups.

Phytoplankton has a positive impact on most other groups. The impact is greatest on their direct predators. For instance, the impact of phytoplankton was large on *O. niloticus* and herbivorous zooplankton. Similarly, macrophytes had positive effect on LLb-herbivores and LLb-benthivores. On the other hand, herbivorous zooplankton had negative impact on phytoplankton and juvenile fish.

LLb-piscivore (an apex predator for this particular model) and *C. gareipnus* have direct negative impact on most of their prey such as small barbs, *V. beso*, and *Garra*. However, LLb-piscivore has a cascading positive impact on herbivorous zooplankton, which is a prey of small barbs. The feeding of *C. gariepinus* on large labeobarbs has a positive effect on small barbs and *V. beso*.

# 14.4 System Statistics

The result of Ecopath system statistics is shown in Table 14.5.

## 14.4.1 System Primary Production/Respiration (*P*/*R*)

This is the ratio between total primary production and total respiration ($P/R$ ratio) in a system. Odum (1975) demonstrated that $P/R$ ratio describes the maturity of an

**Table 14.5** Summary of System Statistics of Various Parameters (t $km^{-2}$ $year^{-1}$) for Lake Tana

| Parameter | Value |
|---|---|
| Sum of all consumption | 760.754 |
| Sum of all exports | 2530.248 |
| Sum of all respiratory flows | 498.814 |
| Sum of all flows into detritus | 2657.003 |
| Total system throughout | 6447.000 |
| Sum of all production | 3140.000 |
| Calculated total net primary production | 3028.900 |
| Total primary production/total respiration | 6.072 |
| Net system production | 2530.086 |
| Total primary production/total biomass | 39.831 |
| Total biomass/total throughout | 0.012 |
| Total biomass (excluding detritus) | 76.043 |

ecosystem. The rate of primary production exceeds the rate of community respiration during early stages of ecosystem development, and hence $P/R$ is greater than 1. However, in mature system the ratio approaches 1 in that the energy fixed tends to be balanced by the energy cost of maintenance. Within this context, $P/R$ ratio of Lake Tana is 7.495 (more than Hawasa, 5.834) which is much larger than 1, leading to a conclusion that Lake Tana is relatively at developmental stage, with high autotrophy. This could be due to the small water retention time of the lake leading to insufficient time for heterotrophic development in the plankton. On the other hand, despite the low primary production in the system, there appears to be room for eutrophication trend and with high human impacts in the lake catchment, this should be given serious attention in the lake management plan.

## 14.5 Conclusion

Ecopath with Ecosim model was applied in Ethiopia for the third time for Lake Tana after Lake Hawasa (Fetahi and Mengistou, 2007) and Lake Hayq (Fetahi et al., 2011). Our present knowledge of the trophic functioning of this lake has been significantly improved by investigations based on simultaneous sampling of some of the lower trophic levels, which was the major gap to run ecosystem analysis in the previous times. Although the model depicts scenarios averaged over the whole year, some of our findings could not be incorporated into the overall results. For instance, the seasonal cycle of both phytoplankton and zooplankton abundance and production were not considered by the model (Wondie and Mengistou, 2006; Wondie et al., 2007).

In the present analysis, we have been able to identify that the producers are less exploited by the consumers in the system and energy transfer to the second trophic level is very low. This indicates that it is not only the primary production that supports the next trophic groups but also that of detritus originating either from autochthonous or allochthonous source or both. Therefore, catchment management options should be implemented that will help to optimize the energy balance for lake ecosystem. Besides, future study should consider using stable isotope to ascertain the correct sources of organic matter in the lake.

However, the lake at present is threatened both from nonpoint and point source pollution (from Bahir Dar and Gondar regions) that could result in total alteration of the biological productivity of the lake. The impact of this anthropogenic effect on the lake ecosystem has not been adequately examined. Given the high system $P/R$ ratio, the threat of eutrophication appears imminent, as human and climatic impacts are enhancing changes that facilitate this, such as clearance of macrophytes stands, warmer water temperature, catchment alterations, etc. EwE has features that can be used for management purpose such as habitat changes, policy exploration and fisheries management, but these applications were not exploited in this study. Further extension work on food web structure and simulations should be done to understand factors that regulate the productivity of the lake as well as extend management options for Lake Tana ecosystem.

## Acknowledgments

We thank the School of Graduate Studies of Addis Ababa University and the Bahir Dar Fisheries Research Center for fund and data acquisition.

## References

Allen, K.F., 1971. Relation between production and biomass. J. Fish. Res. Bd. Can. 20, 1573–1581.

Baird, D., Ulanowicz, R.E., 1993. Comparative study on the trophic structure, cycling and ecosystem properties of four tidal estuaries. Mar. Ecol. Prog. Ser. 99, 221–237.

Bini, G., 1940. l peci del Lago Tana (The fishes of Lake Tana). Missione di Studio at Lago Tana richerge limnologiche. B. Chemica e Biologia, Reale Accadamia d'italia 3, 137–179.

Burgis, M.J., 1974. Revised estimates for the biomass and production of zooplankton in Lake George, Uganda. Freshw. Biol. 4, 535–541.

Christensen, V., Pauly, D., 1992. Ecopath II a software for balancing steady state ecosystems models and calculating network characterstics. Ecol. Model. 61, 169–185.

Christensen, V., Pauly, D. (Eds.), 1993, Trophic models of aquatic ecosystems, No. 26, p. 390. ICLARM Conference Proceedings.

Christensen, V., Walters, C., 2000. Ecopath with Ecosim: methods, capabilities and limitations. In: Pauly, D., Pitcher, T.J. (Eds.), Methods for Assessing the Impact of Fisheries on Marine Ecosystems of the North Atlantic, 8 (2), 120–140. Fisheries Center Research Reports.

Christensen, V., Walters, C.J., Pauly, D., 2000. Ecopath with Ecosim: A User's Guide. Fisheries Center, Univeristy of British Columbia, Vancouver and ICLARM, Malaysia.

Christensen, V., Walters, C.J., Pauly, D., 2005. Ecopath with Ecosim: A User's Guide. Fisheries Center, University of British Columbia/ICLARM, Vancouver/Malaysia.

de Graaf, M., 2003. Lake Tana's piscivorous Barbus (Cyprinidae, Ethiopia): ecology, evolution, exploitation. Ph.D. Thesis. Wageningen University, Wageningen, the Netherlands.

Dejen, E., 2003. Ecology and potential for fishery of the small barbs (Cyprinidae, Teleostei) of Lake Tana, Ethiopia. Ph.D. Thesis. Wageningen University, Wageningen, the Netherlands.

Dickie, L.M., 1972. Food Chains and Fish Production, vol. 8. ICNAF Special Publication. 201–219.

Fetahi, T., 2005. Trophic analysis of Lake Hawasa using Mass-balance Ecopath model. M.Sc. Thesis. Addis Ababa University, Ethiopia.

Fetahi, T., Mengistou, S., 2007. Trophic analysis of Lake Hawasa (Ethiopia) using mass-balance Ecopath model. Ecol. Modell. 201, 398–408.

Fetahi, T., Schagerl, M., Mengistou, S., Libralto, S., 2011. Food web structure and trophic interactions of the tropical highland lake Hayq, Ethiopia. Ecol. Modell. 222, 804–813.

Gasse, F., 1987. Ethiopie et Djibouti. In: Burgis, M.J., Symoens, J.J. (Eds.), African Wetlands and Shallow Water Bodies, 211. Travaux et documents/Institut Francais de Recherche Scientifique pour le Development en cooperatin, Paris, ORSTOM, pp. 300–311.

Hall, D.J., Threlkeld, S.T., Burns, C.W., Crowley, P.H., 1976. The size efficiency hypothesis and the size structure of zooplankton communities. Ann. Rev. Ecol. Syst. 7, 177–208.

Jones, J.G., 1979. A guide to methods for estimating microbial numbers and biomass in fresh water, No.39. Freshwater Biological Association, Scientific Publication, pp. 112.

Mavuti, K., Moreau, J., Munyadrorero, J., Plisnier, P.D., 1996. Analysis of trophic relationships in two shallow equatorial lakes Lake Naivasha (Kenya) and Lake Ihema (Rwanda) using a multispecifies trophic model. Hydrobiologia 321, 89–100.

Moreau, J., Christensen, V., Pauly, D., 1993. A trophic ecosystem model of Lake George, Uganda. In: Christensen, V., Pauly, D. (Eds.), Trophic Models of Aquatic Ecosystems. ICLARM Conference Proceedings 26. ICLARM, Manila, Philippines, pp. 124–129.

Nagelkerke, L.A.J., 1997. The barbs of Lake Tana, Ethiopia. Morphological diversity and its implications for taxonomy, trophic resource partitioning, and fisheries. pp.296. Doctoral Thesis. Wageningen, Experimental Animal Morphology and Cell Biology, Wageningen Agricultural University.

Nagelkerke, L.A.J., 1997. The barbs of Lake Tana, Ethiopia: morphological diversity and its implications for taxonomy, trophic resource partitioning and fisheries. PhD Thesis. Wageningen University, Wageningen, The Netherlands.

Nagelkerke, L.A.J., Sibbing, F.A., 1996. Reproductive segregation among the large barbs (*Barbus* sp., Cyprinidae, Teleostei) of Lake Tana, Ethiopia. Neth. J. Zool. 45, 431–454.

Odum, E.P., 1975. Ecology: The Link Between the Natural and the Social Sciences, second ed. Holt, Rinehart and Winston, New York.

O'Reilly, C.M., Alin, S.R., Plisnier, P.-D., Cohen, A.S., McKee, B.A., 2003. Climate change decreases aquatic ecosystem productivity of Lake Tanganyika, Africa. Nature 424, 766–768.

Pauly, D., 1984. Fish population dynamics in tropical waters: manual for use with programmable calculators. ICLARM Stud. Rev. 8, 325.

Pauly, D., Christensen, V., Walters, C., 2000. Ecopath, ecosim, and ecospace as tools for evaluating ecosystem impact of fisheries. ICES J. Mar. Sci. 57, 697–706.

Straile, D., 2005. Food webs in lakes—seasonal dynamics and the impact of climate variability. In: Belgrano, A. (Ed.), Aquatic Food Webs—An Ecosystem Approach. Oxford University Press, pp. 41–50.

Teferra, G., Fernando, C.H., 1989. The food habits of an herbivorous fish (*Oreochromis niloticus* Linn.) in Lake Hawasa, Ethiopia. Hydrobiologia 174, 195–200.

Verburg, P., Hecky, R.E., Kling, H., 2003. Ecological consequences of a century of warming in Lake Tanganyika. Science 301, 505.

Verschuren, D., 2003. The heat on Lake Tanganyika. Nature 424, 731–732.

Wondie, A., Mengistou, S., 2006. Duration of development, biomass and rate of production of the dominant copepods (calanoida and cyclopoida) in Lake Tana, Ethiopia. SINET. Ethiop. J. Sci. 29, 107–122.

Wondie, A., Mengistu, S., Vijverberg, J., Dejen, E., 2007. Seasonal and spatial variations in primary production of a large tropical lake (Lake Tana, Ethiopia): effects of nutrient availability and turbidity. Aquat. Ecol. 41, 195–207.

Wudneh, T., 1998. Unpublished Ph.D. Thesis. Biology and management of fish stocks in Bahir Dar Gulf, Lake Tana, Ethiopia, 143. Wageningen Agricultural University.

PART

5

# Models of Element Cycling in Ecosystems

# 15

# Modeling the Mercury Cycle in the Marano-Grado Lagoon (Italy)

Donata M. Canu*, Alessandro Acquavita†, Christopher D. Knightes‡, Giorgio Mattassi†, Isabella Scroccaro†, Cosimo Solidoro*

*OGS, THE NATIONAL INSTITUTE OF OCEANOGRAPHY AND EXPERIMENTAL GEOPHYSICS, BORGO GROTTA GIGANTE 42C, SGONICO (TS), ITALY,

†ARPA FVG FRIULI VENEZIA-GIULIA REGIONAL ENVIRONMENTAL PROTECTION AGENCY, TECHNICAL SCIENTIFIC MANAGEMENT DEPARTMENT, VIA CAIROLI 14, PALMANOVA (UD), ITALY,

‡UNITED STATES ENVIRONMENTAL PROTECTION AGENCY, OFFICE OF RESEARCH AND DEVELOPMENT, NATIONAL EXPOSURE RESEARCH LABORATORY, ECOSYSTEMS RESEARCH DIVISION, ATHENS, GA, USA

## 15.1 Introduction

The Marano-Grado Lagoon (MGL) is one of the most important and best conserved wetland environments in the whole Mediterranean area. It is located between the Tagliamento and Isonzo River deltas (Northern Adriatic Sea, Italy), covering an area of approximately 160 km$^2$. The system is defined as a coastal microtidal lagoon with large dimension (Italian Decree 131/08) and is protected by the Ramsar Convention since 1971. Following the implementation of the Habitat Directive (92/43/EC), concerning the protection of biodiversity, the MGL was identified as a Site of Community Importance (SCIs—IT3320037). The area hosts economic, tourist, and industrial services. Nowadays, fishing, clam harvesting (mainly *Tapes philippinarum*) and fish-farming represent important economic resources for the local community. However, the MGL has been declared a polluted site of national interest (SIN; Ministerial Decree 468/01) because of its high levels of sanitary and environmental risks (Ramieri et al., 2011).

Trace elements distribution in lagoon sediments was first investigated in the early 1970s (Stefanini, 1971), but only recently there has been an evaluation of the anthropogenic influence on Hg contamination levels (Mattassi et al., 1991). These authors reported elevated Hg concentrations in sediments, ranging from 0.61 to 14.01 μg/g with a spatial distribution showing a clear eastward positive gradient. This has resulted in elevated Hg concentrations in the biota and throughout the food web (Brambati, 2001).

Covelli et al. (2008) reported a broad range of water column total Hg (HgT) concentrations, from 2 ng/L at an undisturbed natural environment to 70 ng/L close to a fish

Models of the Ecological Hierarchy. DOI: http://dx.doi.org/10.1016/B978-0-444-59396-2.00015-8

ISSN 0167-8892, 

farm. Non-negligible diurnal and seasonal variations of Hg and MeHg concentrations in the water column (Bloom et al., 1999; Muresan et al., 2007; Covelli et al., 2008; Emili et al., 2011) have been observed in several lagoon environments.

Hg occurs in the marine environment mainly in three chemical forms: elemental, Hg(0); divalent ionic, Hg(II), present in a variety of both inorganic and organic complexes; and methylated forms that include mono- and di-methylmercury, MeHg, DMHg. All these species are linked intricately through the Hg(II) pool (Fitzgerald et al., 2007).

Hg is a ubiquitous environmental pollutant of great concern because of the toxicity of its methylated form (MeHg) and bioaccumulative and biomagnifying properties (Clarkson, 1999). Methylmercury exposure in humans may suppress the immune system, cause neurodevelopmental delays in children, and impair cardiovascular health in adults (Crespo-Lopez et al., 2007; Mergler et al., 2007).

The widespread Hg contamination has been caused by both natural and anthropogenic sources of historical and more recent origins. During 500 years of Hg mining activity in Idrija (Slovenia)—the world's second largest Hg mining area—more than 12 million t of Hg ore was mined, and approximately 37,000 t was released into the environment (Dizdarevič, 2001). Releases from waste products of smelting activities deposited along the Idrijca River were washed downstream by flooding, carried through the Soča/Isonzo River, and transported to the Gulf of Trieste. Several investigations on coastal sediments of the Gulf of Trieste have shown high concentrations of total Hg, decreasing exponentially from the fluvial source (Covelli et al., 2001). The contamination extends its effects to the nearby lagoon, especially when peak river discharges and Bora—ENE wind—conditions simultaneously occur (Covelli et al., 2007). In addition, a chlor-alkali plant was active since 1949, discharging about 186 t of Hg into the Aussa–Corno River system which outflows within the lagoon system (Piani et al., 2005). Although direct Hg discharges stopped in 1984, Hg still reaches the MGL in the form of dissolved and particulate phases and accumulates in sediments (Covelli et al., 2009). As a consequence of these two high loading sources of Hg, the MGL has elevated concentrations of Hg in the sediments.

The European Union (2000), with the Water Framework Directive—WFD 2000/60/EC—gives special attention to Hg, which is considered a priority pollutant, setting as environmental quality standard for the average annual concentrations in transitional and marine water bodies, the limit of 10 ng/L for the dissolved form. For the application of the WFD, ARPA FVG implemented a sampling strategy for the whole lagoon basin, with a monthly field sampling of dissolved HgT concentration in the 17 water bodies identified in agreement with the WFD (Bettoso et al., 2010).

Concurrent application of environmental modeling tools can be used to complement the field observations. However, the development of a Hg model is not a straightforward task since some of the processes governing its fate and transport and bioaccumulation are still not well understood, especially the processes affecting Hg speciation, such as microbiological and diagenetic processes. Therefore, there are still many challenges confronting, representing, and understanding mercury exposure dynamics at the MGL, despite having several literature mercury modeling examples. Regardless, the understanding of the biogeochemical cycle of Hg needs to be supported by speciation measures, but, at the moment, within the MGL, there is a paucity of data for the different Hg species in different media.

This chapter serves as a foundation for the development of a comprehensive, modeling framework for Hg exposure in the MGL ecosystem. Therefore, a short review of the modeling tools applied to coastal-marine environments is presented and a conceptual site-specific mercury model has been drawn. Site-specific data related to Hg processes have been collected for the MGL system and boundaries. Then, several simple models, which in future might be combined in a complete modeling framework, have been identified and applied to the MGL.

The Land Ocean Interactions in the Coastal Zone (LOICZ) budgeting procedure has been extended to compute the Hg budget and the effective volume of water exchanged at the sea–lagoon interface. This information, together with available data, has been used to make tests and simulations with SERAFM (Spreadsheet-based Ecological Risk Assessment for the Fate of Mercury) a screening-level, steady-state mixed-batch reactor surface water model of Hg transformation processes (Knightes and Ambrose, 2006; Knightes, 2008). The model was adapted to simulate the specific features of a transitional system environment. First tests were made to describe the response of the zero-dimensional system to Hg loading, in terms of HgT concentration in the water column and bioconcentration on target biota. Sensitivity tests have been conducted to select the most sensitive parameters and processes. Finally, SHYFEM (Umgiesser et al., 2004; Ferrarin et al., 2010), an hydrodynamic model particularly suited for lagoon systems and shallow water environments, has been run to describe the spatial distribution of a passive tracers release in a contamination hotspot—the chlor-alkali plant located close to the Aussa-Corno River system—and model ability to address this issue, also in comparison with published data.

### 15.1.1 Hg Modeling Cycle Applications

Hg modeling tools have been developed for several aquatic environments: lakes, freshwaters, rivers, estuarine and marine waters.

A review of Hg modeling applications can be found in Wang et al. (2004) and in Massoudieh et al. (2010). The latter discusses models developed for freshwater/lake environments, giving, among others, some examples of coastal-marine environment applications. The authors identify the key processes of Hg cycling and also highlight the need for further investigations to potentially important and often less understood processes, such as microbial and digenetic. Wang et al. (2004) review nine papers of modeling applications in different environments. Among these, Abreu et al. (1998), Sirca et al. (1999), and Rajar et al. (2000) addressed Hg cycling in coastal systems. In more detail, the first paper focuses on the Hg water/sediment interaction in the Ria de Aveiro Coastal Lagoon (Portugal) by applying ECoS (the Estuarine Contaminant

Simulator model). Pato et al. (2008) also focus on this system, but used a two-dimensional vertically integrated hydrodynamic model coupled with Hg measures to estimate the seasonal and intertidal variability of dissolved and particulate Hg fluxes between the Ria de Aveiro (Portugal) and the Atlantic Ocean. They estimated that the relative contribution and the magnitude of Hg transported within each phase varies both seasonally and with tidal range, while the net Hg export is mostly due to the contribution of the particulate phase. Širca et al. (1999) and Rajar et al. (2000) applied the Hg module STATRIM (STAtionary TRIeste gulf Hg model) in order to investigate the Hg cycle in the Gulf of Trieste, simulating the fate and transport of both Hg(II) and MeHg. STATRIM is a two-dimensional steady-state model using a hydrodynamic model driven primarily by wind (PCFLOW2D-HD) and a two-dimensional sediment transport model (MIKE 21 MT). It calculates the transport of Hg(II) and MeHg, while partition constants are used to compute the variable fractions associated with the dissolved, particulate, and plankton phases. The model accounts for the average annual exchange with the boundary systems (i.e., atmosphere, rivers, and sediments) and incorporates Hg transformation processes such as methylation, demethylation, and reduction. In Rajar et al. (2000), STATRIM is updated in its hydrodynamic and sediment transport modules by using a three-dimensional spatial representation (PCFLOW3D), thus improving the description of the dispersion processes and enabling the model to reproduce the wind induced resuspension of Hg. Žagar et al. (2007) updated and applied the PCFLOW3D at a Mediterranean basin scale, introducing the Hg(0) in the mass budget. In this application, the model was linked with the RAMS-Hg atmospheric model, providing real-time meteorological data, Hg concentration in the atmosphere, and Hg deposition. Modeling results were in good agreement for Hg(0) surface water concentrations, while simulated MeHg surface water concentrations were less satisfactory, thus suggesting the need of further improvements. Rajar et al. (2007) estimated an Hg mass balance for the Mediterranean Sea with the inclusion of water/air exchange (both deposition and evasion).

The review of scientific literature evidences that standard Hg modeling applications generally represent the Hg cycle in the dissolved and in one or more different phases (e.g., silt, clay, detritus, plankton, and organic carbon), which is generally calculated using partition coefficients. The transformation processes are generally of first-order and in some cases (Ambrose et al., 1993) temperature-dependent. Bioaccumulation processes in the trophic chain are often not considered and in some cases they are modeled via bioaccumulation factors (BAFs). Few models, MCM and BASS, for example, simulate bioaccumulation using fish energetics (Barber et al., 1991). Mercury speciation processes are generally ignored and Hg compounds are lumped into to three groups: Hg(II), Hg(0), and MeHg, where Hg(II) includes all the species such as Hg-Hum, $HgCl_2$, Hg(OH), and HgClOH, Hg-Hum, HgS(HS), and $Hs(HS)_2$ (see Fig. 15.1 for details). Moreover, the models used in the aforementioned examples are not suited for implementation in very shallow water, such as a lagoon.

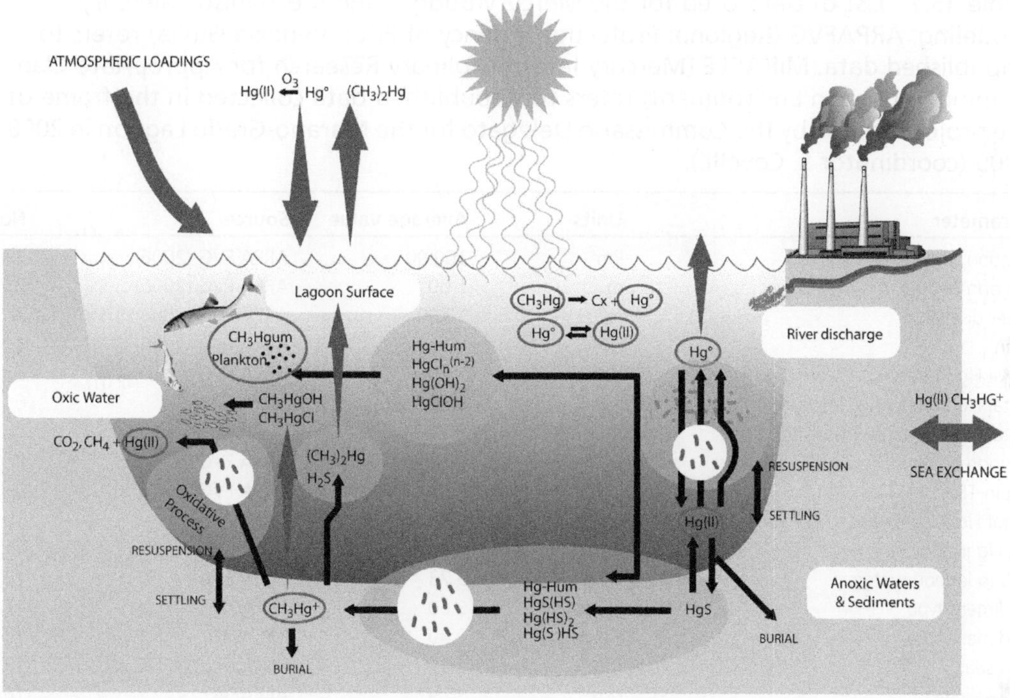


**FIGURE 15.1** The biogeochemical cycle of mercury in the lagoon environment. Blue arrows indicate the fluxes across the boundaries. Black arrows indicate the Hg fluxes in the lagoon system. SERAFM simulates the Hg variables circled in red, while the transformation processes highlighted in green have been parameterized in the model using specific parameters. *(Redrawn from Barkay et al., 2004).* For interpretation of the references to color in this figure legend, the reader is referred to the online version of this book.

# 15.2 Material and Methods

## 15.2.1 Available Data

The information needed to set up the Hg mass balance and to parameterize the Hg transformation processes was selected from previous studies conducted in the Marano-Grado Lagoon or other similar environments. When available, as in the case of methylation rate constants (parameters 26 and 27; Table 15.1), local information was used. In some other cases, for example, in the case of atmospheric loading, a value estimated for the whole Gulf of Trieste (Sirca et al., 1999) has been rescaled to compute the local loading of the MGL. In other cases, parameters have been taken from similar aquatic ecosystems (parameters 34–36; Table 15.1) (e.g., Mediterranean Sea or coastal systems) or set at the default model values.

## 15.2.2 LOICZ Budget

A first-order estimate of Hg budget has been obtained by modifying and applying the LOICZ budgeting approach. This methodology was applied in order to estimate the Hg

**Table 15.1** List of Data Used for the Mercury Budget and Mechanistic Mercury Modeling. ARPAFVG (Regional Protection Agency of Friuli Venezia Giulia) refers to unpublished data. MIRACLE (Mercury Interdisciplinary Research for Appropriate Clam farming in Lagoon Environment) refers to unpublished data collected in the frame of the project funded by the Commissario Delegato for the Marano-Grado Lagoon in 2008–2009 (coordinator S. Covellli).

| Parameter | Units | Average value | Source | No. |
|---|---|---|---|---|
| Lagoon surface | $km^2$ | 160.00 | ARPA FVG (2006) | 1 |
| Lagoon depth | m | 1.00 | ARPA FVG (2006) | 2 |
| River discharge | $m^3/s$ | 79.00 | ARPA FVG (2006) | 3 |
| Rain | cm/year | 87.00 | ARPA FVG (2006) | 4 |
| Total Hg in freshwater (dissolved) | ng/L | 1.05 | ARPA FVG | 5 |
| Total Hg in seawater (dissolved) | ng/L | 5.40 | ARPA FVG | 6 |
| Total Hg in lagoon (dissolved) | ng/L | 6.15 | ARPA FVG | 7 |
| Lagoon salinity | adimensional | 28.30 | ARPA FVG | 8 |
| Hg in TSS | μg/g | 2.34 | ARPA FVG | 9 |
| Total Hg in sediment | μg/g | 4.33 | ARPA FVG | 10 |
| MeHg in sediment | ng/g | 1.61 | ARPAFVG | 11 |
| TSS in lagoon waters | μg/g | 5.23 | ARPA FVG | 12 |
| Sediment type | %silt | 0.60 | MIRACLE | 13 |
| Sediment type | %sand | 0.40 | MIRACLE | 14 |
| TSS sea | mg/L | 4.27 | Solidoro et al. (2010) | 15 |
| Sea salinity | psu | 35.92 | ARPAFVG | 16 |
| DIC lagoon | mg/L | 26.70 | ARPA FVG | 17 |
| DOC lagoon | mg/L | 4.17 | ARPA FVG | 18 |
| DIC sea | mg/L | 130 | Solidoro et al. (2010) | 19 |
| DOC sea | mg/L | 81.67 | Solidoro et al. (2010) | 20 |
| POM sea | mg/L | 1.21 | Solidoro et al. (2010) | 21 |
| MeHg in seawater (dissolved) | ng/L | 0.005 | Horvat et al. (1999) | 22 |
| MeHg pore water (0 – 1 cm) | ng/L | 0.005 | MIRACLE | 23 |
| MeHg pore water | ng/L | 0.75 | MIRACLE | 24 |
| Hg pore water (0 – 1 cm) | ng/L | 5.10 | MIRACLE | 25 |
| Sediment methylation rate constant average [Hg(II) → MeHg] | %/d | 0.83 | MIRACLE | 26 |
| Sediment demethylation rate constant average [MeHg → Hg(II)] | %/d | 11.87 | MIRACLE | 27 |
| Water column methylation Rate constant average [Hg(II) → MeHg] | %/d | 0. | Model default | 28 |
| Water column demethylation Rate constant average [MeHg→Hg(II)] | %/d | 0. | Model default | 29 |
| Water column biotic reduction rate constant [Hg(II) → Hg(0)] | %/d | 3 | Mason et al. (1995), Monperrus et al. (2007) | 30 |
| Photo-degradation rate constant at water surface [MeHg → Hg(0)] | %/d | 10 | Monperrus et al. (2007) | 31 |
| Photo-oxidation rate constant at water surface [Hg(0)→ Hg(II)] | %/d | 10 | Lalonde et al. (2001) | 32 |

*(continued)*

**Table 15.1**—*cont'd*

| Parameter | Units | Average value | Source | No. |
|---|---|---|---|---|
| Photo-reduction rate constant at water surface [Hg(II) → Hg(0)] | %/d | 10 | Monperrus et al. (2007) | 33 |
| Water column dark oxidation rate constant [Hg(0) → Hg(II)] | %/d | 6 | Lalonde et al (2001) | 34 |
| Hg(II) Kd water/particulate average | log L/kg | 6 | MIRACLE | 35 |
| Hg(II) Kd water/DOC | log L/kg | 5.4 | Allison and Allison (2000) | 36 |
| MeHg Kd water/particulate average | log L/kg | 5.23 | Allison and Allison (2000) | 37 |
| MeHg Kd water/DOC | log L/kg | 5 | Allison and Allison (2000) | 38 |
| Ulva Hg bioconc. | mg/kg fw | 0.096 | Brambati (1996) | 39 |
| Gracilaria bioconc. | mg/kg fw | 0.080 | Brambati (1996) | 40 |
| Chaetomorpha bioconc. | mg/kg fw | 0.130 | Brambati (1996) | 41 |
| Tapes philippinarum (clam) bioconc. | mg/kg fw | 0.114 | MIRACLE | 42 |
| Sparus aurata bioconc. | mg/kg fw | 0.725 | ARPA FVG | 43 |
| Mugil cephalus bioconc. | mg/kg fw | 0.190 | ARPAFVG | 44 |
| Dicentrarchus labras bioconc. | mg/kg fw | 2.064 | ARPA FVG | 45 |
| Mustelus mustelus bioconc. | mg/kg fw | 0.753 | ARPA FVG | 46 |
| M. galloprovincialis bioconc. | mg/kg fw | 0.132 | ARPA FVG | 47 |
| BAF: Trophic level 1 | L/kg | $1.98 \times 10^5$ | This work | 48 |
| BAF: Trophic level 2 | L/kg | $5.48 \times 10^5$ | This work | 49 |
| BAF: Trophic level 3 | L/kg | $1.51 \times 10^6$ | This work | 50 |
| BAF: Trophic level 4 | L/kg | $4.18 \times 10^6$ | This work | 51 |

budget taking into account only the exchanges between the system of the MGL and the boundary systems: land, Adriatic Sea, and atmosphere. This approach has been chosen to describe and consider the exchange with the Adriatic Sea that, due to the tidal activity, cannot be estimated without proper monitoring activities or without a coupled model of the Hg fate and transport.

Further information about the effective lagoon–seawater exchange is also necessary to properly load the SERAFM Hg model. The simplest version of the LOICZ model, commonly employed to estimate the fluxes of non-conservative elements (i.e., nutrients, carbon) at the local coastal scale (Gordon et al., 1996), was adapted to the computation of the total dissolved Hg ($HgT_D$) fluxes between the lagoon and Adriatic Sea. Following this approach the lagoon is treated as a single well-mixed reactor at steady-state and the Hg budget is estimated using water and salt budget results. The water budget and the system residence time were calculated using various sources of water: runoff, $V_q$; precipitation, $V_p$; evaporation, $V_e$; and by assuming that the net volume of the lagoon basin is constant. Therefore the residual volume, $V_r$, Eqn (1), is calculated as

$$V_r = V_e - (V_p + V_q) \quad (1)$$

The salt budget was calculated and then used for the estimation of the horizontal exchange volume, $V_x$, Eqn (2), that, in a single-layer system, represents the horizontal mixing between the budgeted system and the adjacent sea.

$$V_x = \frac{\sum_i V_i \times S_i}{S_{\log} - S_{\text{sea}}} \tag{2}$$

Here, $V_i$ are the terms of the water budget, including the calculated $V_r$.

The value of Hg exchanged with the Adriatic Sea $\Delta Hg_s$, Eqn (3), has been computed using the variable values of river discharges, $V_q$, and Hg values in riverine $Hg_q$, lagoon $Hg_{lag}$, and seawaters $Hg_{sea}$:

$$\Delta Hg_s = -V_q Hg_q - V_r Y_r - V_x(Hg_{sea} - Hg_{lag}) \tag{3}$$

The LOICZ budget method gives the estimation of a residual value of the non-conservative compounds, which is an indication of the contribution of $\Delta Hg_i$, the fraction of Hg involved in the internal system dynamics which has been calculated as

$$\Delta Hg_i = -V_r\, Hg_q - V_x\, (Hg_{Sea} - Hg_{lag}) - V_q Hg_q \tag{4}$$

## 15.2.3 SERAFM Model

The aquatic cycling of Hg is simulated using the SERAFM model, which is an updated version of the IEM-2M model used by US EPA to assess cycling of Hg released from coal-fired utilities in the Hg Study Report to Congress (US EPA, 1997). Details of the development and evaluation of the SERAFM model can be found elsewhere (Knightes and Ambrose, 2006; Knightes, 2008). SERAFM is a steady-state model that predicts Hg concentrations for three principal Hg species, Hg(0), Hg(II), and MeHg, in the water column, the sediment layer, and trophic food-web components (phytoplankton, zooplankton, benthic invertebrates, and trophic levels 3 and 4 fish). Trophic level 3 fish are forage fish—predators feeding opportunistically on fish, invertebrates, and vertebrates; trophic level 4 fish are top predator fish. The body burdens due to ingestion of Hg-contaminated aquatic species are calculated for wildlife and humans. The SERAFM model includes current information on speciation of Hg(II), photo-reactions, reaction rates for methylation and demethylation in water and sediments, and partitioning of Hg(II) and MeHg to different types of suspended solids and complexation with dissolved organic carbon. Transformation processes among Hg species in the model are generally represented by first-order rate constants for methylation of Hg(II) to MeHg in water and sediments, demethylation of MeHg to Hg(II) in water and sediments, biotic reduction and photo-reduction of Hg(II) to Hg(0) in water only, photo-oxidation and dark oxidation of Hg(0) to Hg(II) in water, and photo-degradation of MeHg to Hg(0) in water (Knightes, 2008). In other applications, the SERAFM model has been used to model Hg cycling in constructed wetland mesocosms (Brown et al., 2007) and has been applied in a dynamic manner to evaluate the response time of ecosystems to changes in atmospheric deposition (Knightes et al., 2009; Selin et al., 2010).

In this application, the original SERAFM representation of the trophic chain has been changed using a site-specific parameterization which gives the exposure risk of selected target species living in the lagoon. The specific parameters used are the body weight, the trophic level, and the BAF. The BAF has been calculated considering the bioaccumulation data of aquatic biota and the MeHg concentration in pore water as a proxi of MeHg concentration exposure (Table 15.1).

For this application, an exchange process has been added to the model in order to take into account the sea–lagoon exchange, which was not included in the original version.

### 15.2.4 Hydrodynamic Model

The two-dimensional version of the finite element model SHYFEM (Umgiesser et al., 2004) has been used in order to reproduce the spatial distribution of a passive tracer released at the chlor-alkali plant site. The model consists of a hydrodynamic module suited for shallow waters with a transport-diffusion module. Model setup followed Ferrarin et al. (2010). Meteo-marine forcing for the year 2007 was used as real forcing. Data for boundary conditions were mainly provided by ARPA FVG, ISPRA, and Unità Operativa Idrografica di Udine—Servizio Idrografico Regionale. The time step of the simulation is 300 s. Further details of the model implementation for the MGL are discussed in Ferrarin et al. (2010)

## 15.3 Results and Discussion

### 15.3.1 Hg Input

The atmospheric input was calculated proportionally using the value of Širca et al. (1999) for the adjacent Gulf of Trieste. This input (1.44 kg/year) fits well with data from previous studies for the Mediterranean Sea. Wängberg et al. (2008) calculated an input of approximately 115 kmol/year for the whole Mediterranean area with 1.45 kg/year for the MGL. In order to be conservative, the Hg discharged by the river network was computed from the available data of dissolved total Hg in the rivers entering the lagoon system and highest average annual values of freshwater discharge that give an annual average input of total dissolved Hg from the river network of 2.8 kg/year. The computation of the exchanges with the Adriatic Sea is less straightforward, due to the tidal action, and it would be better addressed by using specific measurements, with a combined use of measurements and modeling, as made by Pato et al. (2007) or with a coupled sea–lagoon model. In our application this was estimated using the LOICZ budget method.

### 15.3.2 Dissolved Hg Sea–Lagoon Exchange Estimates with the LOICZ Approach

The budget analysis was done using the available data of the following:

- Total dissolved Hg ($THg_D$) measured monthly for one year—from August 2009 to August 2010—at the monitoring stations (rivers, lagoon, and coastal Adriatic sea)
- Atmospheric loading of Hg from Širca et al. (1999)
- Riverine water discharges (180 measures relative to the year 2007)
- Precipitation data (one year cumulative value)
- Bimonthly salinity data in the Adriatic sea (year 2007)
- Bimonthly salinity data inside the lagoon (year 2007)

The total river discharges relative to year 2007 measures are 53.01 $m^3/s$; however, other estimates (ARPA FVG, 2006) report a value of 77.5 $m^3/s$. The budget analysis was therefore conducted for the whole year and for the two cases of river discharge (low and high) (Table 15.2). For the two scenarios, it emerges that the system exports annually 10.6–15.51 kg of Hg in the dissolved form, while 17.01–25.55 kg are involved in the internal transformation processes and dynamics. The estimated volume of water exchanged between the two systems is 32112.82–46979.86 $10^3$ $m^3$/day and the residence time is 6.6–4.5 days.

### 15.3.3 SERAFM Application

The SERAFM model was applied for 12 different scenarios, and the results are reported in Table 15.3.

Scenario 1, "Base case" is the base simulation performed using the set of parameters shown in Table 15.1 and using, for the sea exchange, the lowest value of Table 15.2, using the results of the LOICZ calculation. This scenario represents the current condition, with an observed total HgT concentration in the MGL sediment, Hg loading coming into the MGL via the river, and Hg loading coming in from the Adriatic Sea via tidal exchange.

**Table 15.2** LOICZ Budget Results (Residence Time, Volume of Effective Mixing Water Exchanged at the Inlets, Fraction of Dissolved Hg Exchanged with the Sea $\Delta Hg_s$, and Fraction of Mercury Used in the Internal Process $\Delta Hg_s$).

| | Riverine Discharge | | Residence time | | Exchange Volume | | $\Delta Hg_s$ Sea exchange | | $\Delta Hg_i$ Non cons. | |
|---|---|---|---|---|---|---|---|---|---|---|
| | Low | High | Low | High | Low | High | Low | High | Low | High |
| | $m^3/s$ | $m^3/s$ | day | day | $10^3$ $m^3$/day | $10^3$ $m^3$/day | kg/year | | kg/year | kg/year |
| Winter | 48.74 | 71.30 | 5.94 | 4.06 | 33155.24 | 48504.89 | −9.24 | −13.52 | 14.59 | 5.50 |
| Spring | 46.41 | 67.89 | 7.5 | 5.13 | 17906.75 | 26196.91 | −13.15 | −19.23 | 19.82 | 7.42 |
| Summer | 54.82 | 80.19 | 3.7 | 2.53 | 49865.82 | 72951.85 | −9.19 | −13.45 | 13.87 | 5.24 |
| Autumn | 62.07 | 90.81 | 9.27 | 6.33 | 27523.47 | 40265.81 | −10.84 | −15.85 | 19.76 | 7.39 |
| Average | 53.01 | 77.55 | 6.6 | 4.51 | 32112.82 | 46979.86 | −10.6 | −15.51 | 17.01 | 25.55 |

**Table 15.3** SERAFM Simulation Results: Hg(0), Hg(II), MeHg, and HgT in Water Column (Filtered and Unfiltered); MeHg in Fish Tissue, and Hazard Quotients for Exposed Wildlife. Hazard quotients greater than unity are in bold.

| | | Scen 1 Base scenario | Scen 2 No river Hg load | Scen 3 No tidal Hg load | Scen 4 High sea exchange | Scen 5 High sea exchange, low particulate Hg | Scen 6 Low Hg particulate Hg | Scen 7 Hg particulate =0 | Scen 8 Low Hg sediment | Scen 9 High Hg Sediment | Scen 10 Highest demeth and meth rates | Scen 11 High methylation rate | Scen 12 High demethylation rate |
|---|---|---|---|---|---|---|---|---|---|---|---|---|---|
| Water | Hg(0) (ng/L) | 0.58 | 0.56 | 0.1 | 0.73 | 0.51 | 0.42 | 0.26 | 0.51 | 0.65 | 0.58 | 0.58 | 0.58 |
| | Hg(II) (ng/L) | 3.49 | 3.37 | 0.58 | 4.38 | 3.07 | 2.50 | 1.54 | 3.10 | 3.87 | 3.49 | 3.49 | 3.49 |
| | MeHg (ng/L) | 0.4 | 0.39 | 0.14 | 0.46 | 0.34 | 0.31 | 0.23 | 0.29 | 0.52 | 0.38 | 0.38 | 0.34 |
| | HgT (ng/L) | 4.47 | 4.32 | 0.82 | 5.57 | 3.92 | 3.23 | 2.02 | 3.91 | 5.03 | 4.45 | 4.44 | 4.41 |
| | Hg(0) (ng/L) | 0.58 | 0.56 | 0.1 | 0.73 | 0.51 | 0.42 | 0.26 | 0.51 | 0.65 | 0.58 | 0.58 | 0.58 |
| | Hg(II) (ng/L) | 4.07 | 3.93 | 0.68 | 5.12 | 3.58 | 2.92 | 1.80 | 3.62 | 4.52 | 4.07 | 4.07 | 4.07 |
| | MeHg (ng/L) | 0.49 | 0.47 | 0.17 | 0.55 | 0.41 | 0.38 | 0.27 | 0.35 | 0.62 | 0.45 | 0.45 | 0.41 |
| | HgT (ng/L) | 5.14 | 4.97 | 0.94 | 6.39 | 4.51 | 3.71 | 2.33 | 4.49 | 5.78 | 5.11 | 5.11 | 5.07 |
| Sediment, dry | Hg(II) (ug/g) | 4.05 | 4.05 | 4.05 | 4.05 | 4.05 | 4.05 | 4.05 | 0.47 | 7.57 | 4.11 | 3.92 | 4.18 |
| | MeHg (ug/g) | 0.28 | 0.28 | 0.28 | 0.28 | 0.28 | 0.28 | 0.28 | 0.03 | 0.53 | 0.22 | 0.41 | 0.15 |
| | HgT (ug/g) | 4.33 | 4.33 | 4.33 | 4.33 | 4.33 | 4.33 | 4.33 | 0.50 | 8.10 | 4.33 | 4.33 | 4.33 |
| Fish | Trophic 3 (ug/g) | 0.43 | 0.42 | 0.15 | 0.49 | 0.37 | 0.34 | 0.24 | 0.31 | 0.55 | 0.40 | 0.50 | 0.37 |
| | Trophic 4 (ug/g) | 1.19 | 1.16 | 0.41 | 1.35 | 1.02 | 0.93 | 0.67 | 0.86 | 1.52 | 1.11 | 1.37 | 1.02 |
| Hazard quotient | Wildcat | 0.2 | 0.2 | 0.1 | 0.3 | 0.2 | 0.2 | 0.1 | 0.2 | 0.3 | 0.2 | 0.3 | 0.2 |
| | Shrew mouse | **20.7** | **20.2** | **7.1** | **23.5** | **17.6** | **16.1** | **11.6** | **14.9** | **26.4** | **19.3** | **23.8** | **17.7** |
| | Common stilt | **13.9** | **13.5** | **4.8** | **15.7** | **11.8** | **10.8** | **7.8** | **10.0** | **17.7** | **12.9** | **15.9** | **11.8** |
| | Kingfisher | **16.6** | **16.2** | **5.7** | **18.9** | **14.1** | **12.9** | **9.3** | **12.0** | **21.2** | **15.5** | **19.1** | **14.2** |
| | Osprey | **6.7** | **6.5** | **2.3** | **7.5** | **5.7** | **5.2** | **3.7** | **4.8** | **8.5** | **6.2** | **7.6** | **5.7** |
| | Hooded merganser | **8.7** | **8.5** | **3.0** | **9.9** | **7.4** | **6.8** | **4.9** | **6.3** | **11.1** | **8.1** | **10.0** | **7.4** |
| | Wood duck | **1.6** | **1.6** | 0.6 | **1.9** | **1.4** | **1.3** | 0.9 | **1.2** | **2.1** | **1.5** | **1.9** | **1.4** |

The unknown input of particulate HgT coming from the sea has been extrapolated from the $HgT_D$, dissolved, assuming that the dissolved is made up about one-third of the total (Faganeli et al., 2003).

Scenario 2, "No river Hg load" is the same as the "Base case" except that the incoming river flow has a total Hg concentration of 0 ng/L.

Scenario 3, "No tidal Hg load" is the same as the "Base case" except that the tidal exchange with the Adriatic Sea has a total Hg concentration of 0 ng/L.

Scenario 4, "High sea exchange" is the same as "Base case," but using the highest value calculated in Table 15.2 for the Adriatic Sea exchange rate.

Scenarios 5–12 have been developed in order to assess the model sensitivity to specific parameters. The parameterization of the conversion factor of $HgT_D$/HgT total in seawater is addressed in Scenario 5, assuming the total is twice the dissolved, and in Scenario 6 where the total is assumed to be equal to the dissolved. In Scenarios 7 and 8 the HgT concentration in sediment has been set to the minimum and maximum values reported in ARPA FVG data, namely to 0.5 and to 8.1 mg/kg. Scenarios 9–11 address changes in methylation and demethylation rates, set one at a time and simultaneously (Scenario 9) at the highest summer values. Clearly, Scenarios 2 and 3 are extreme and unrealistic, while the others explore a local variability of the subspaces of the space of parameters which can be realistic.

The MeHg surface water concentrations is most sensitive to the methylation and demethylation rates; it is the ratio of the methylation rate constant to the demethylation rate constant that drives the fraction of HgT present as MeHg. The fish tissue Hg concentrations and wildlife exposure risks are most sensitive to the BAFs. In this system, the exchange with the sea is a large driving force for the surface water MeHg, fish tissue Hg, and wildlife exposure risk, and therefore these variables are sensitive to the Hg concentrations in the sea and the value of the exchange rate with the sea. The HgT, Hg(II), and Hg(0) surface water concentrations are sensitive to the oxidation and reduction rate constants and partition coefficients, and the MeHg, fish tissue Hg, and wildlife exposure risks are less sensitive to these parameters. The model results are not as sensitive to the surface area and depth of the lagoon, the river inflow, and river concentrations.

The model predicts, for the Base case, filtered HgT a value of 4.47 ng/L, which, considering the simplicity of the model, is an acceptable indication of the state compared to the observed value of 6.15 ng/L.

Among the 12 scenarios, the HgT dissolved value changes between 0.82, for the case of "No tidal Hg load", and 5.57, when the exchange volume of water is assumed to be the highest value calculated with the LOICZ method.

Given the current parameterization of loadings, the SERAFM model suggests that the dominant source of Hg loading into the MGL is via tidal exchange with the Adriatic Sea. Removal of Hg load from the rivers resulted in only a small reduction in exposure concentrations (decrease in unfiltered HgT of approximately 3%) and wildlife exposure risk, suggesting that management strategies directed at reducing river loading may not have a pronounced impact on Hg exposure concentrations in the MGL. Similarly, as shown in Scenario 8, a remediation of the Hg in sediment may not have an appreciable impact on exposure risk unless the historic Hg concentrations in the Adriatic Sea are reduced.

The wildlife exposure risk represented by the hazard quotients demonstrates how the shrew mouse, common stilt, and the kingfisher are the most sensitive species at the MGL. Through this screening-level analysis, we see that the tidal exchange is the dominant source of Hg, but even when this is removed, the exposure risk of these three species is such that they still have hazard quotients in excess of unity.

A rigorous sensitivity analysis goes beyond the purposes of this chapter; nevertheless, first indication can be derived from this simple analysis. Using a simple measure of the model sensitivity to changes in parameters, obtained by comparing the "Base case" scenario with the others, and expressed as sensitivity value SV, it can be observed that the sediment dry concentration is not sensitive to changes in seawater exchange or by decreasing the HgT concentration in the sea. The other model variables are sensitive to the latter: Hg concentration in water (filtered and unfiltered) has SV = 78/79%, fish bio-accumulation SV = 64%, and hazard coefficients SV = 65%. By changing seawater exchange, for Hg concentration in water, the SV is 48–49% and for biota bioaccumulation and hazard coefficients SV ~ 30%. The sensitivity of model variables to changes in contaminated sediment concentration is 18% for Hg concentration in water, 100% for Hg sediment concentration, and 30% for fish bioconcentration and hazard coefficients. Sensitivity to changes in summer sediment methylation rate constants is 4% for Hg concentration in water, 29% for sediment concentration, and 30% for fish bio-accumulation and hazard index. Sensitivity to the sediment demethylation rate constant is lower than the sensitivity to sediment methylation rate constant. This analysis confirms that, among the explored parameters, the system is highly sensitive to parameters that induce changes in mercury exchange with the Adriatic Sea.

### 15.3.4 SHYFEM Simulations

In order to explore the importance of hydrodynamic processes affecting spatial variability of Hg, and the potential of hydrodynamic model to address this issue, we performed a numerical simulation of the space–time distribution of a generic tracer in the Marano-Grado Lagoon representing the inputs of particle-bound Hg. The simulation takes into account the chlor-alkali plant located close to the Aussa-Corno River system, as a passive source discharging directly into the lagoon. The concentration of the tracer applied at the source is 1000 generic units as at present stage we are interested only in the relative spatial pattern. Instantaneous values of tracer dispersion are obtained at each time step as shown in Fig. 15.2.

The tracer moves mainly westward and accumulates in front of the Zellina River and around the Aussa-Corno River mouth, driven by the water circulation and clearly influenced by the wind action. At the eastern sector, the influence of the tracer is quite lower and is confined around the eastern part of the Porto Buso channel.

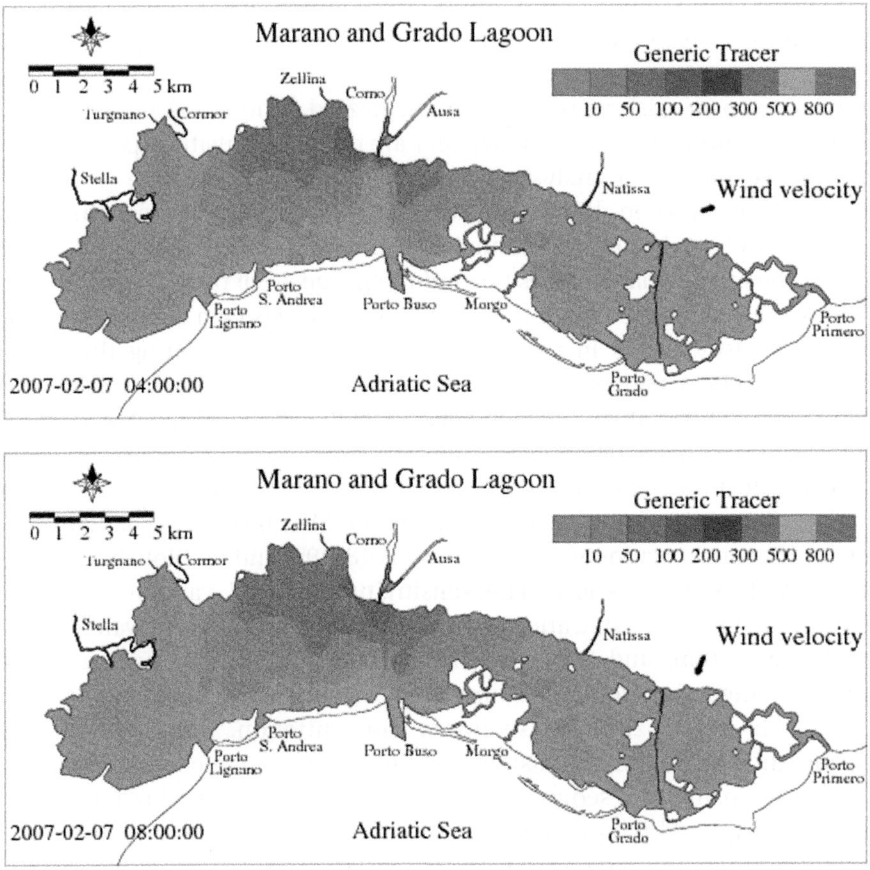


**FIGURE 15.2** Generic passive tracer released from the chlor-alkali site. The distribution is clearly influenced by the wind action. For color version of this figure, the reader is referred to the online version of this book.

Simulation results are in agreement with Hg field measurement (Piani et al., 2005) that found anomalously high concentrations of Hg, bound in non-cinnabar form, in the surface sediments near the Zellina and Aussa-Corno river mouths and on tidal flats in the eastern part of the Porto Buso channel. The presence of Hg bound in non-cinnabar form means that it originates from recent industrial inputs. This encourages further modeling investigations, also in the direction of developing a coupled hydrodynamic-Hg process model.

## 15.3.5 Gaps and Future Work

This modeling application attempts to make a synthesis of the available information concerning the Hg budget and dynamics in the MGL. The SERAFM model is a steady-state, zero-dimensional model, and therefore has the necessary simplifying assumptions inherent in these types of models. There are many advantages for using a simple modeling application over a more complex one. By modeling the entire lagoon as one uniform water body at a steady state, processes governing the fluxes into and out of the lagoon, the bulk fractions of each species present, and the exposure risks for wildlife can be investigated on a systems-level, which is something hardly done with more complex models. Increasing model complexity results in increased model sensitivity. To increase the spatial and temporal resolution requires a detailed understanding of the hydrology and solids transport within the system, necessitates detailed parameterization, and further details on the exposure concentrations throughout the system.

The SERAFM model, in its zero-dimensional application, suggests that the tidal exchange with the Adriatic Sea is the dominant loading source of the elevated Hg exposure concentrations, suggesting that management strategies that solely address the historical contamination delivered by the river network to the MGL will not be effective in sufficiently reducing exposure risk overall at the lagoon thus indicating priorities in terms of monitoring and modeling needs. This is, in our opinion, a very relevant result from an environmental conservation perspective.

While on the whole the tidal exchange is the most important loading source, there may be specific zones where Hg exposure is greater and specific and focused management strategies may be useful. For example, the river loadings may be most important where the river ends at the MGL, suggesting management strategies in these locations. Applying a temporally varying and spatially explicit model may identify where specific processes and loadings may be more important.

This modeling experiment showed knowledge gaps and needs for future research and monitoring work. Most of the environmental data, such as river freshwater discharges, temperature, and salinity in lagoon/freshwater and in seawater are currently monitored by the regional laboratories during the institutional monitoring activities. However, regular measurements of Hg concentrations in the environment are not available yet. Synoptic measures of Hg concentration as total Hg and MeHg in the main subsystems (water, particulate, sediment, pore water, freshwater, and seawater) are needed to assess the system state at one point in time or during the four seasons.

Another gap that needs to be filled is the current knowledge of the transport of Hg associated with the sediment transported through the river network and at the sea/lagoon interface. The introduction of this term in the budget calculation could modify the results, in particular under ENE wind conditions which increase both the sediment resuspension and the inflows.

Upon data availability, future modeling should therefore address the spatial variability, for example, by applying SERAFM in a multi-box system with temporal variability or applying a dynamic, spatially resolved model like WASP (Ambrose et al., 1993). Clearly, a spatial and temporal resolution would be obtained by implementing a fully coupled hydrodynamic-sediment transport-Hg fate model.

## 15.4 Conclusion

A zero-dimensional mercury model has been set up for the MGL, reproducing essential processes that govern the mercury exchange with the boundaries and regulate the fate and transformation of mercury in the lagoon through water, sediment, and biotic components. This required the collection of all the available information such as environmental parameters, system parameters, Hg loadings, transformation process parameters, and observed Hg concentrations in the water column, sediments, and biota.

We used alternative and complementary tools, such as the SHYFEM model, the LOICZ budget model, and the SERAFM model. The SHYFEM model gave a spatial representation of a passive tracer dispersion released from the internal Hg-contaminated site. The LOICZ budget computed the effective volume of water mixing at the lagoon–sea interface and calculated the amount of total dissolved Hg exchanged with the sea and the fraction of Hg used in the internal processes. The current state of knowledge has not allowed the application of a model with temporal and/or spatial resolution which would require extensively more data for calibration and parameterization. With its simple approach, the work presented in this study serves as a foundation for further research by providing a general understanding of the entire system.

The zero-dimensional SERAFM application, based on 12 scenarios and sensitivity analysis, already provided important management indication, by showing the importance of the Hg loading coming from the Adriatic Sea and the negligible effect of the total reduction of the Hg loading coming from the river network.

Our work also enabled us to indicate data gaps and identify priorities for developing a temporally varying and spatially explicit model.

## Disclaimer

This paper has been reviewed in accordance with the U.S. Environmental Protection Agency's peer and administrative review policies and approved for publication. Mention of trade names or commercial products does not constitute endorsement or recommendation for use.

## Acknowledgments

The authors thank Valentina Mosetti for the graphic composition of Fig. 15.1.

## References

Abreu, S.N., Pereira, M.E., Duarte, A.C., 1998. The use of a mathematical model to evaluate mercury accumulation in sediments and recovery time in coastal lagoon (Ria De Aveiro, Portugal). Water Sci. Technol. 37 (6–7), 33–38.

ARPA FVG, 2006. Agenzia Regionale per l'Ambiente Friuli Venezia Giulia-ARPA FVG, 2006. Rapporto sullo Stato dell'Ambiente. Aggiornamento 2005.

Ambrose, R.B., Wool, T.A., Martin, J.L., 1993. The Water Quality Analysis Simulation Program, WASP5. Model Documentation Environmental Research Laboratory, Athens, GA.

Barber, M.C., Suárez, L.A., Lassiter, R.R., 1991. Modelling bioaccumulation of organic pollutants in fish with an application to PCBs in Lake Ontario salmonids. Can. J. Fish. Aquat. Sci. 48, 318–337.

Barkay, T., Miller, S.M., Summers, A.O., 2003. Bacterial mercury resistance from atoms to ecosystems. FEMS Microbiol. Rev. 27, 355–384.

Bettoso, N., Aleffi, I.F., Faresi, L., Rossin, P., Mattassi, G., 2010. Evaluation on the ecological status of the macrozoobenthic communities in the Marano and Grado Lagoon (Northern Adriatic Sea). Annales 20, 193–206.

Bloom, N.S., Gill, G.A., Cappellino, S., Dobbs, C., Mc Shea, L., Driscoll, C., Mason, R., Rudd, J., 1999. Speciation and cycling of mercury in Lavaca bay, Texas, sediments. Environ. Sci. Technol. 33, 7–13.

Brambati, A., 2001. Coastal sediments and biota as indicators of Hg contamination in the Marano and Grado Lagoons. RMZ Mat. Geoenviron. 48, 165–171.

Brambati, A., 1996. Metalli pesanti nelle Lagune di Marano e Grado. Regione Autonoma Friuli Venezia Giulia.

Brown, S., Saito, L., Knightes, C.D., Gustin, M., 2007. Calibration and evaluation of a mercury model for a western stream and constructed wetland. Water Air Soil Pollut. 182, 275–290.

Clarkson, T.W., 1999. Human toxicology of mercury. J. Trace Elem. Exp. Med. 11, 303–317.

Covelli, S., Acquavita, A., Piani, R., Predonzani, S., De Vittor, C., 2009. Recent contamination of mercury in an estuarine environment (Marano lagoon, Northern Adriatic, Italy). Estuar. Coast. Shelf Sci. 82, 273–284.

Covelli, S., Faganeli, J., De Vittor, C., Predonzani, S., Acquavita, A., Horvat, M., 2008. Benthic fluxes of mercury species in a lagoon environment (Grado Lagoon, Northern Adriatic Sea, Italy). Appl. Geochem. 23, 529–546.

Covelli, S., Faganeli, J., Horvat, M., Brambati, A., 2001. Mercury contamination of coastal sediments as the result of long-term cinnabar activity (Gulf of Trieste, Northern Adriatic Sea). Appl. Geochem. 16, 541–558.

Covelli, S., Piani, R., Acquavita, A., Predonzani, S., Faganeli, J., 2007. Transport and dispersion of particulate Hg associated with a river plume in coastal Northern Adriatic environment. Mar. Pollut. Bull. 55, 436–450.

Crespo-Lopez, M.E., Lima de Sa, A.L., Herculano, A.M., Burbano, R.R., Martins do Nascimento, J.L., 2007. Methylmercury genotoxicity: a novel effect in human cell lines of the central nervous system. Environ. Int. 33, 141–146.

Dizdarevič, T., 2001. The influence of mercury production in Idrija mine on the environment in the Idrija region and over a broad area. RMZ Mat. Geoenviron. 48 (1), 56–64.

Emili, A., Koron, N., Covelli, S., Faganeli, J., Acquavita, A., Predonzani, S., De Vittor, C., 2011. Does anoxia affect mercury cycling at the sediment-water interface in the Gulf of Trieste (northern Adriatic Sea)? Incubation experiments using benthic flux chambers. Appl. Geochem. 26, 194–204.

European Union, 2000. Directive 2000/60/EC of the European Parliament and of the Council of 23 October 2000 establishing a framework for Community action in the field of water policy (Water Framework Directive). Off. J. Eur. Commun. C L 327 22/12/2000.

Faganeli, J., Horvat, M., Covelli, S., Fajon, V., Logar, M., Lipej, L., Cemelj, B., 2003. Mercury and methylmercury in the Gulf of Trieste (Northern Adriatic Sea). Sci. Total Environ. 304, 315–326.

Ferrarin, C., Umgiesser, G., Bajo, M., Bellafiore, D., De Pascalis, F., Ghezzo, M., Mattassi, G., Scroccaro, I., 2010. Hydraulic zonation of the lagoons of Marano and Grado, Italy. A modelling approach. Estuar. Coast. Shelf Sci. 87, 561–572. Copyright © 2010 Elsevier Ltd.

Fitzgerald, W.F., Lamborg, C.H., Hammerschmidt, C.R., 2007. Marine biogeochemical cycling of mercury. Chem. Rev. 107, 641–662.

Gordon Jr., D.C., Boudreau, P.R., Mann, K.H., Ong, J.-E., Silvert, W.L., Smith, S.V., Wattayakorn, G., Wulff, F., Yanagi, T., 1996. LOICZ Biogeochemical Modelling Guidelines. LOICZ Reports & Studies No. 5, second ed., LOICZ, Texel, the Netherlands, vi + 96 pp.

Horvat, M., Covelli, S., Faganeli, J., Logar, M., Mandic, V., Rajar, R., Sirca, A., Zagar, D., 1999. Mercury in contaminated coastal environments: a case study: the Gulf of Trieste. Sci. Total Environ. 237, 238, 43–56.

Knightes, C.D., Sunderland, E.M., Barber, M.C., Johnston, J.M., Ambrose Jr., R.B., 2009. Application of ecosystem-scale fate and bioaccumulation models to predict fish mercury response times to changes in atmospheric deposition. Environ. Toxicol. Chem. 28 (4), 881–893.

Knightes, C.D., 2008. Development and test application of a screening-level mercury fate model and tool for evaluating wildlife exposure risk for surface waters with mercury-contaminated sediments (SERAFM). Environ. Model Software 23, 495–510.

Knightes C.D., Ambrose R. B. Jr., 2006. Development of an Ecological Risk Assessment Methodology for Assessing Wildlife Exposure Risk Associated with Mercury-Contaminated Sediments in Lake and River Systems, EPA Athens, GA, USA.

Lalonde, J., Amyot, M., Kraepiel, A., Morel, F., 2001. Photooxidation of Hg(0) in artificial and natural waters. Environ. Sci. Technol. 35, 1367–1372.

Mason, R.P., Morel, F.M.M., Hemond, H.F., 1995. The role of microorganisms in elemental mercury formation in natural waters. Water Air Soil Pollut. 80, 775–787.

Massoudieh, A., Zagar, D., Green, P.G., Cabrera-Toledo, Horvat M., Ginn, T.R., Barkouki, T., Weathers, T., Bombardelli, F.A., 2010. Modeling mercury fate and transport in aquatic systems. In: Mihailovic, Dragutin T., Gualtieri, Carlo (Eds.), Advances in Environmental Fluid Mechanics. World Scientific, London, New York, Singapore.

Mattassi G., Daris, F., Nedoclan, G., Crevatin, E., Modonutti, G.B., Lach, S. 1991. La qualità delle acque della Laguna di Marano. Regione Autonoma Friuli Venezia Giulia – U.S.L. n.8 "Bassa Friulana", pp. 101.

Mergler, D., Anderson, H.A., Chan, L.H.M., Mahaffey, K.R., Murray, M., Sakamoto, M., Stern, A.H., 2007. Methylmercury exposure and health effects in humans: a worldwide concern. Ambio 36 (1), 3–11.

Monperrus, M., Tessier, E., Amouroux, D., Leynaert, A., Huonnic, P., Donard, O.F.X., 2007. Mercury methylation, demethylation and reduction rates in coastal and marine surface waters of the Mediterranean Sea. Mar. Chem. 107, 49–63.

Muresan, B., Cossa, D., Jézéquel, D., Prévot, F., Kerbellec, S., 2007. The biogeochemistry of mercury at the sediment–water interface in the Thau lagoon. 1. Partition and speciation. Estuar. Coast. Shelf Sci. 72, 472–484.

Pato, P., Lopes, C., Valega, M., Lillebø, A.I., Dias, J.M., Pereira, E., Duarte, A.C., 2008. Mercury fluxes between an impacted coastal lagoon and the Atlantic Ocean. Estuar. Coast. Shelf Sci. 76, 787–796.

Piani, R., Covelli, S., Biester, H., 2005. Mercury contamination in Marano Lagoon (Northern Adriatic sea, Italy): source identification by analyses of Hg phases. Appl. Geochem. 20, 1546–1559.

Rajar, R., Četina, M., Horvat, M., Žagar, D., 2007. Mass balance of mercury in the Mediterranean Sea. Mar. Chem. 107, 89–102.

Rajar, R., Zagar, D., Sirca, A., Horvat, M., 2000. Three-dimensional modelling of mercury cycling in the Gulf of Trieste. Sci. Total Environ. 260, 109–123.

Ramieri, E., Barbanti, A., Picone, M., Menchini, G., Bressan, E., Dal Forno, E., 2011. Integrated plan for the sustainable management of the Lagoon of Marano and Grado. Littoral. doi:10.1051/litt/201105008.

Selin, N.E., Sunderland, E.M., Knightes, C.D., Mason, R.P., 2010. Sources of mercury exposure for U.S. seafood consumers: implications for policy. Environ. Health Perspect 118 (1), 137–143.

Sirca, A., Rajar, R., Harris, R.C., Horvat, M., 1999. Mercury transport and fate in the Gulf of Trieste (Northern Adriatic)—a two-dimensional modelling approach. Environ. Model. Software 14, 645–655.

Solidoro C., Del Negro, P., Libralato, S., Melaku Canu, D., (eds.), 2010. Sostenibilità della mitilicoltura triestina, Istituto Nazionale di Oceanografia e Geofisica Sperimentale - OGS. p. 88. Rapporto di sintesi: http://echo.ogs.trieste.it/echo/siti/mitili/Sostemits.htm.

Stefanini, S., 1971. Distribuzione del Li, Na, K, Mg, Ca, Sr, Cr, Mn, Fe, Ni, C, Zn, Pb, C e N organici nei sedimenti superficiali delle Lagune di Marano e Grado (Adriatico Settentrionale). Studi Trentini di Scienze Naturali, Sez A 48, 39–79.

Umgiesser, G., Melaku Canu, D., Cucco, A., Solidoro, C., 2004. A finite element model for the Venice Lagoon. Development, set up, calibration and validation. J. Mar. Syst. 51, 123–145.

U.S. Environmental Protection Agency (US EPA). 1997. Mercury Study Report to Congress. EPA-452/R-97–005. Final Report. U.S. Government Printing Office, Washington, DC.

Wang, Q., Kim, D., Dionysiou, D.D., Ag. A., Sorial, Timberlake, D., 2004. Sources and remediation for mercury contamination in aquatic systems—a literature review. Environ. Pollut. 131, 323–336.

Wängberg, I., Munthe, J., Amouroux, D., Andersson, M.E., Fajon, V., Ferrara, R., Gårdfeldt, K., Horvat, M., Mamane, Y., Melamed, E., Monperrus, M., Ogrinc, N., Yossef, O., Pirrone, N., Sommar, J., Sprovieri, F., 2008. Atmospheric mercury at Mediterranean coastal stations. Environ. Fluid Mech. 8 (2008), 101–116.

Žagar, D., Petkovšek, G., Rajar, R., Sirnik, N., Horvat, M., Voudouri, A., Kallos, G., Četina, M., 2007. Modelling of mercury transport and transformations in the water compartment of the Mediterranean Sea. Mar. Chem. 107, 64–88.

# 16

# Impact of Global and Local Pressures on the Ecology of a Medium-Sized Pre-Alpine Lake

Elisa Carraro*, Nicolas Guyennon†, Gaetano Viviano*, Emanuela C. Manfredi*, Lucia Valsecchi*, Franco Salerno*, Gianni Tartari*, Diego Copetti*

*IRSA-CNR, VIA DEL MULINO 19, 20861 BRUGHERIO (MB), ITALY, †IRSA-CNR, VIA SALARIA KM 29,300 10 00015 MONTEROTONDO ST., ROME, ITALY

## 16.1 Lake Ecosystems Between Global and Local Pressures

Global warming has different impacts on the ecology of lacustrine environments. The progressive increase of water column stability leads to a reduction of exchanges between surface and deep layers, an expansion of the anoxic/hypoxic layer in productive environments (Verburg et al., 2003), and an increase of the nutrients release from sediments (Jeppesen et al., 2009). In general, lake warming is expected to enhance many biogeochemical processes, exacerbating the process of eutrophication (Schindler, 2001) and altering the trophic web (Schindler, 2001).

Alterations of the phytoplankton phenology have been observed earlier in spring or later in autumn, with extensions of the growing season allowing bloom (Thackeray et al., 2008). Loss of regularity in spring–autumn pattern may result in irregular pulses of biomass and changes in the phytoplankton assemblages (Winder and Cloern, 2010). These changes may favor the establishment of invasive or harmful species, such as cyanobacteria (Paerl and Huisman, 2008; Salmaso, 2010).

The ecological lake dynamics are strongly influenced by local impacts due to human modifications in the catchment. A mitigation of external loading pressure often leads to a general improvement of the lake water quality, even though high rates of internal recycling of nutrients between the sediments and the overlying water column could lessen or delay the benefits of management actions at the catchment level (Søndergaard et al., 2003; Jeppesen et al., 2005).

Due to the scarcity of long time series of observations, there are only a limited number of publications which unequivocally measure the impact of climate change on nutrient

Models of the Ecological Hierarchy. DOI: http://dx.doi.org/10.1016/B978-0-444-59396-2.00016-X
ISSN 0167-8892, 

loads or concentrations in water bodies (e.g., Parmesan and Yohe, 2003; Thackeray et al., 2008). Conversely, a vast number of physical and ecological modeling studies deal with the effects of increasing temperatures on lakes thermal stability, nutrients supply, and ecosystem responses (Peeters et al., 2002; Malmaeus et al., 2006; Trolle et al., 2011). Recently, many different mathematical models have been implemented to investigate the lake responses to environmental pressures improving the cost-effective management options, but no profound scientific understanding on the ecosystem responses to multiple anthropogenic modifications has been achieved so far (Blenckner, 2008).

Some authors (Malmaeus et al., 2006; Trolle et al., 2011) tried to link global or regional climate models to ecological lake models, simulating the impacts of meteorological alterations on lakes with different morphometry or trophic status. Nevertheless Global Climate Models (GCMs) and Regional Climate Models (RCMs) are still far from matching the real variability due to complex atmosphere-land processes at local scale, thus their outputs cannot be expected to exactly reproduce the observed local dynamics and may produce larger bias as higher the frequency of local simulations (Elliott et al., 2005; Portoghese et al., 2011). An integrated lake-catchment modeling approach is presented here to differentiate the impacts of the atmospheric temperature increase (globally mediated) from the local impact by external loads in a pre-alpine lake. The separation of the global and the local pressures is gained by analyzing four scenarios with different degrees of local and global weights. Pristine conditions were simulated coupling model outcomes with paleolimnological analysis, while a specific statistical technique (Spectral Singular Analysis) was performed to flatten the atmospheric temperature increasing trend detected on daily values between 1960 and 2010, thus isolating the human-induced global warming.

## 16.2 A Case Study for Long-Term Impacts

Lake Pusiano is a medium-sized (surface area 5 $km^2$) pre-alpine lake (Fig. 16.1a), with maximum depth around 25 m, residence time of about 1 year, and a typical warm monomictic behavior, the complete overturn occurring between January and February. The catchment area (about 95 $km^2$) is characterized by a typical pre-alpine steep slope (mean slope = 39%, maximum altitude = 1453 m a.s.l, median altitude = 683 m a.s.l.) which produces a rapid hydrological response in the Lambro River, the main tributary with a mean daily discharge of 2.6 $m^3\,s^{-1}$ (Salerno and Tartari, 2009). Other two minor inflows and outflows complete the hydrological network. Natural lands cover about the 82% of the catchment area, while only 1% is devoted to agriculture. Urbanized areas (17%) are principally distributed along the river and lake shores. Hydrometric and water quality data were taken on Lambro River during 1974–1982, 2002–2004, and 2010 campaigns. The evolution of external loads to Lake Pusiano has been described in Vuillermoz et al. (2006).

A long-term time series of physical and chemical lake measurements (temperature, transparency, pH, oxygen, phosphorus, nitrogen, silica, and principal ions) during winter overturn is available from 1984 to 2011. Between 1973 and 2010 six limnological

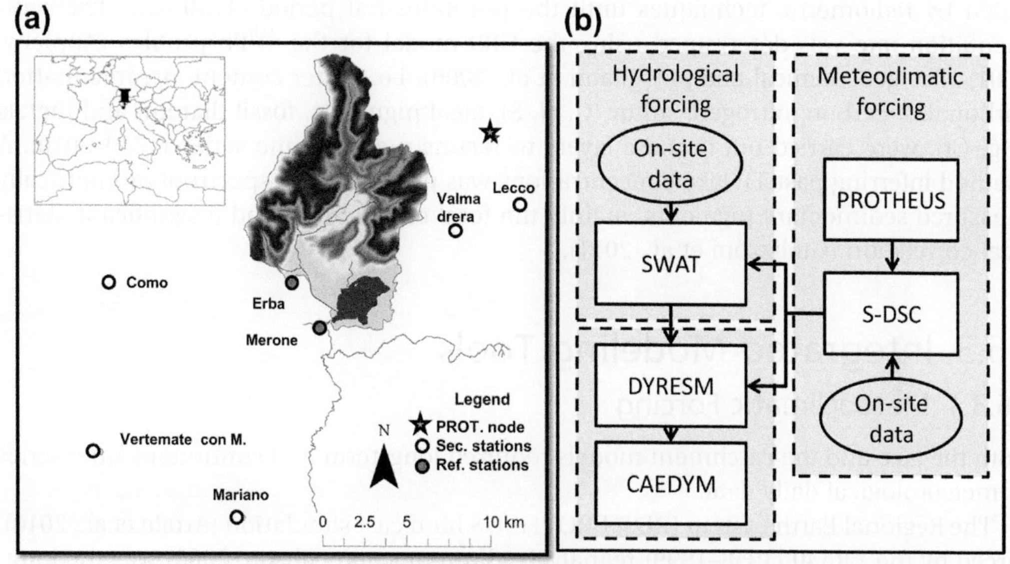


**FIGURE 16.1 Panel (a): Lake Pusiano with the catchment area, the reference and secondary meteorological stations, and the closest PROTHEUS node. Panel (b): the method outline illustrating data flux to build the hydrological and meteorological forcing for the integrated modelling purpose.**

campaigns were carried out (Gerletti and Marchetti, 1977; Chiaudani and Premazzi 1992; Legnani et al., 2005; Vuillermoz et al., 2006; 2010, unpublished data). The temperature trend at winter overturn shows a yearly increase of about 0.015 °C year$^{-1}$, in line with other studies conducted in European alpine area (Ambrosetti and Barbanti, 1999; Livingstone, 2003).

The lake is phosphorus-limited (Legnani et al., 2005; Vuillermoz et al., 2006). Total phosphorus (TP) concentrations increased until mid-1980s (0.2 mg P L$^{-1}$ at 1984 winter overturn) and progressively decreased toward a mesotrophic condition (0.04 mg P L$^{-1}$ at 2010 winter overturn) after the construction of the sewage plant system in 1985 (Vuillermoz et al., 2006).

The main phosphorus contribution to the lake comes from Lambro River and shows a great variability, ranging from 7 to 20 t P year$^{-1}$ during 2000–2003, which clearly relates to the rainfall interannual variability (Salerno and Tartari, 2009).

By contrast, total nitrogen (TN) lake concentration during winter overturn ranged between 1.90 and 2.33 mg N L$^{-1}$ during the period 1984–2010.

Despite the improvement of the trophic conditions, several cyanobacterial blooms caused a bad ecological status, particularly due to the filamentous cyanobacterium *Planktothrix rubescens* (Legnani et al., 2005), which has been strongly dominating the phytoplankton assemblage for the last decade.

Sediment cores of about 80 cm were collected on several stations, from the river entry to the lake center, during the experimental campaign in spring 2010 for paleolimnological investigations (Lami, personal communication). The cores were cut into 1 cm layers and dated by radiometric techniques until the pre-industrial period (1730 AC). The sedimentation rate was determined using the CRS model for the $^{210}Pb$ profiles (Appleby, 2001). The geochemical analyses (Lami et al., 2000), i.e., water content, organic matter, carbonates, carbon, nitrogen, sulfur (C, N, S), algal pigments, fossil diatom, and metals (Hg, Cr), were carried out on core layers to reconstruct the time series (1730–2010). A method inferring past TP lake concentrations was performed by spectrophotometrically measured sedimentary pigments, mainly the total carotenoids, and a significant statistical correlation (Guilizzoni et al., 2011).

# 16.3 Integrated Modeling Tools

## 16.3.1 Meteoclimatic Forcing

Both the lake and the catchment models required long-term and continuous time series of meteorological daily data.

The Regional Earth System (RES) PROTHEUS hind cast simulation (Artale et al., 2010), forced by the ERA40 (1958–1999) reanalysis (Simmons and Gibson, 2000) was the long-term climate reference for the available local historical series. A statistical downscaling (S-DSC) was performed to correct the RES output for local bias (mainly due to the raw approximation of land use and topography in RES), thus obtaining a realistic meteorological forcing for local impacts analysis.

The S-DSC technique is based on the estimation of the inverse of the Cumulative Distribution Function (CDF) or quantiles function. The quantiles of predictor and predictand was used to define a Quantile–Quantile (Q–Q) correction algorithm (Déqué, 2007), which was computed monthly and applied to the predictor.

Two meteorological stations (Erba and Merone), close to Lake Pusiano, were chosen as reference for cumulated daily rainfall, air temperature (minimum, maximum, and daily average values), short-wave solar radiation, vapor pressure, and wind speed (Fig. 16.1a). Data were supplied by the Lombardy Regional Agency for Environmental Protection (ARPA Lombardia http://ita.arpalombardia.it/meteo/dati/richiesta.asp) covering the period between 1991 and 2010. Cloud cover data were available at the National Oceanic and Atmospheric Administration website (http://cimss.ssec.wisc.edu/clavr/) for the entire period.

A first S-DSC was applied to secondary stations distributed on the catchment area to fulfill missing data in the primary time series (Fig. 16.1a). The resulting dataset was then used to downscale the RES hind cast simulation, allowing to reconstruct a continuous daily dataset covering the 1960–2010 period with all the meteorological variables required for the integrated modeling approach (Fig. 16.1b).

A trend analysis was performed on the atmospheric temperatures with a statistical approach based on Singular Spectrum Analysis (SSA) proposed by Broomhead and King (1986). It was computed over a 20-year window, as a compromise between the maximum theoretical window of 25 years (half of the overall period) and the minimal time lag required to encompass the interannual and the interdecadal oscillations due to natural variability (4–10 years) in the temperatures trend (Ghil and Vautard, 1991), avoiding the effects of periodic climatic fluctuations, such as El Niño Southern Oscillation (ENSO) and the North Atlantic Oscillation (NAO). The resulting SSA trend was then removed from the time series using the mean value for the 1960–1980 time window, thus obtaining detrended atmospheric temperatures.

## 16.3.2 Catchment Modeling

Hydrological and nutrients transport models can be distinguished by different approaches, i.e., lumped or distributed and physical or stochastic processes, often linked to water quality models. The choice of a suitable model depends on the availability of data and the aims to be achieved. In this study we chose the Soil and Water Assessment Tool (SWAT, Srinivasan and Arnold, 1994) a deterministic model encompassing the spatial heterogeneity due to weather, soil, topography, and vegetation variability through the support of a Geographical Information System (GIS) interface (Neitsch et al., 2005). Model input and output are organized into hydrologic response units (HRUs) or areas of homogeneous land use, management, and soil characteristics while the hydrological modeling is based on the water balance equation, including surface runoff, precipitation, evapotranspiration, infiltration, and subsurface runoff.

The model was previously calibrated and validated (Salerno and Tartari, 2009) using available discharge data for different sections of Lambro River over several years (1974–1982; 1998–2004). In this study, a further validation was performed using new data collected in 2010; thus, the model was applied to the other smaller inlets, determining the direct drainage basin and the whole daily inflow to the lake.

The estimation of the pristine conditions (or reference conditions) is crucial in any ecological research, providing the baseline to determine human-induced changes and thus drawing conclusions about the weight of the current or future human impact. Different proposed approaches have been summarized in the line-guide about reference conditions for the Water Framework Directive programs (Anonymous, 2003).

Here the natural phosphorus load to Lake Pusiano was estimated comparing SWAT model and the use of the export coefficients approach (Johnes, 1996; Bennion et al., 2005), both calibrated using field data by a selected natural sub-basin (deciduous forests), representative of the catchment pristine condition.

Currently, the main source of phosphorus pollution originates by the point sources, consisting in a network of combined sewage overflows (CSOs), which release polluted water into the rivers during rainfall events (Boguniewicz et al., 2006). Since SWAT is not able to model the point sources loading, the CSOs phosphorus load was estimated computing the exported amount by the specific unitary coefficients on the basis of the population census (Barbiero et al., 1991) and the management interventions (i.e., the construction of the sewage plant in 1986). The actual annual load (Fig. 16.2a) was finally computed by the Vollenweider equation (OECD, 1982) and the lake TP concentration at

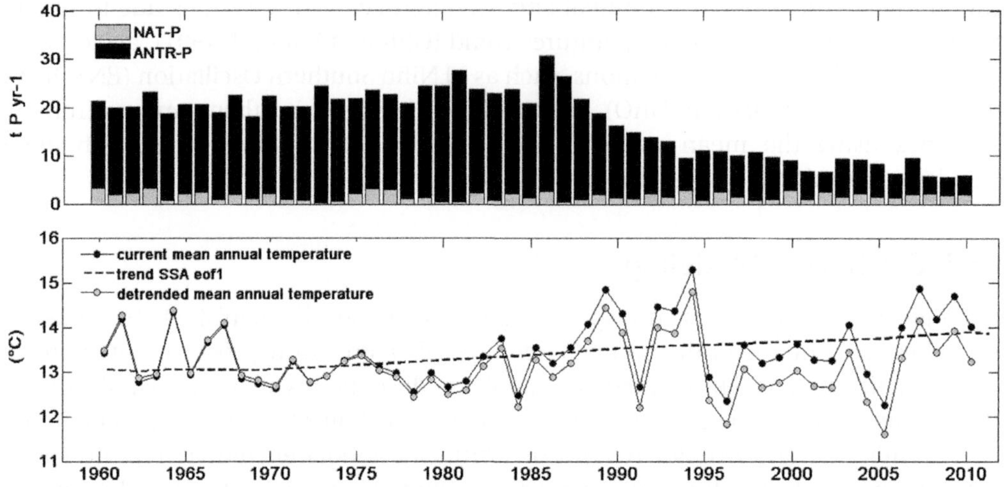


**FIGURE 16.2 Panel (a): natural (simulated by SWAT) and anthropogenic (estimated out of the model) annual P-load to Lake Pusiano. Panel (b): mean annual atmospheric real temperature associated to the trend computed with SSA and annual series after a detrending computation on the real temperature.**

overturn during 1973–2010, applying the reduction coefficients along the river axes (Tartari et al., 2002). Daily loads were estimated by the anthropic and natural annual ratio.

The nitrogen and carbon loads were computed in a simplified way. Nitrogen river concentration was marked as two discharges (0–5 $m^3 s^{-1}$ and > 5 $m^3 s^{-1}$) according to the historical data while carbon river concentration was considered to be a constant value according to the paleolimnological analysis. Nitrogen and carbon natural levels were referred to the analysis on sediment core in the pre-industrial age (ca. 1730).

### 16.3.3 Lake Modeling

The Computational Aquatic Ecosystem Dynamics Model (CAEDYM; Hamilton and Schladow, 1997; Hipsey et al., 2006) was coupled with the 1D hydrodynamic driver DYnamic REServoir Simulation Model (DYRESM, Imerito, 2007). The SWAT model was integrated with the DYRESM-CAEDYM (DYCD) model running from 1960 to 2010 at daily time step. DYCD simulations were performed to study the long-term evolution of lake temperature, biogeochemical variables, and phytoplankton. In DYRESM, the lake is represented as a series of homogeneous horizontal layers of variable thickness (Yeates and Imberger, 2004; Imerito, 2007). The main modeled processes are surface heat, mass and momentum transfers, mixed layer dynamics, hypolimnetic mixing, benthic boundary layer mixing, and inflows and outflows balance. CAEDYM dynamically simulates a range of chemical and biological variables commonly associated with water quality, such as nutrients, metals and oxygen concentrations, and phytoplankton and zooplankton biomass (Hipsey et al., 2006). Mass balance of ecological variables is performed at each time step and at each layer via DYRESM as the layers expand, contract, merge, or are affected by inflows and outflows (Bruce et al., 2006). DYRESM simulations are initialized through salinity and temperature profiles and require both meteorological and hydrological data input. Meteorological (daily average values) data include air temperature, short- and long-wave radiation, vapor pressure wind speed, and rainfall. Hydrological data encompass daily discharge, daily average water temperature, and salinity for the main inflows and daily discharge for the main outflows to the lake. To run the coupled DYCD, daily averaged data of all the biogeochemical simulated variables are also required (Hipsey et al., 2006).

In this study the meteorological and hydrological forcing were set from 1960 to 2010. The Lambro River discharges were supplied by SWAT simulations. The river temperature was computed through a regression with air temperature (Copetti et al., 2006), while salinity, oxygen (DO), and carbon (POC; DOC) and nitrogen (PON; DON; N-$NH_4$; N-$NO_3$) river concentrations (mg N $L^{-1}$) were set as constant values, according to the time series and paleolimnological data. The phosphorus river concentration (POP; DOP; P-$PO_4$) was derived by the simulated or estimated loading. The input was used to feed DYCD and to simulate both long-term nutrients and phytoplankton dynamics. The phytoplankton biomass was expressed as chlorophyll-a concentration (μg Chl-a $L^{-1}$) and the simple Michealis–Menten model was chosen to simulate the nutrient uptake. Phytoplankton biomass was divided into two groups. The first group (Cyanobacteria) was parameterized by the specific eco-physiological parameters of *P. rubescens* (Dokulil and Teubner, 2000; Omlin et al., 2001; Walsby and Schanz, 2002), while the second group was parameterized by the features of a general competitor similar to Chlorophyta, one of the largest taxonomic groups detected among the Lake Pusiano algal community in the last 50 years (Vuillermoz et al., 2006). The main parameters for algal groups are summarized in Table 16.1. The biological model was completed with a zooplankton herbivorous general group (mg C $L^{-1}$). A comprehensive description of DYCD setup and calibration for Lake Pusiano can be found in Copetti et al. (2006).

The lake was initialized in 1960 with mean values of carbon, nitrogen, phosphorus, and total Chl-a content from the paleolimnological survey. The phytoplankton and zooplankton groups were initialized according to data from the historical lake surveys. Model simulated with a 1 h time step and the outputs were assessed by available time series of measurements.

### 16.3.4 Scenarios Construction

To quantify the effects of temperature warming and phosphorus load on the ecology of Lake Pusiano, inputs for P-loads and air temperature driving the anthropic and natural pressure were combined into a set of four different scenarios, running over the same period (1960–2010). Table 16.2 summarizes the possible combinations of the drivers. The current scenario (CUR) represents the real world, used here for the model assessment, and included both the local and the global human impact. The global pressure scenario (GPS) was cleaned by the anthropogenic phosphorus load, representing only the effects of global impact. The local pressure scenario (LPS) was cleaned by the temperature

**Table 16.1** Two Phytoplankton Groups, CYANO and CHLOR Are the Identification Names Used in CAEDYM, Have Been Parameterized to Simulate the Keystone Species *P. rubescens* and a General Competitor as Representative of Algal Evolution (1960–2010) in Lake Pusiano

| **CYANO** | | **CHLOR** | |
|---|---|---|---|
| Cold stenotherm species | 10–22.5 °C (Dokulil and Teubner, 2000) | Warm euryterm species | 20–35 °C |
| Low growth rate | 0.12 $day^{-1}$ (Omlin et al., 2001) | Higher growth rate | 0.9 $day^{-1}$ |
| Low respiration rate | 0.0128 $day^{-1}$ (Omlin et al., 2001) | Higher respiration rate | 0.1 $day^{-1}$ |
| Low light threshold | 10 $\mu E\ m^{-2}\ s^{-1}$ (Walsby and Schanz, 2002) | Higher light threshold | 90 $\mu E\ m^{-2}\ s^{-1}$ |
| Low phosphorus threshold | $K_p = 0.004$ mg P-$PO_4$ $L^{-1}$ (Dokulil and Teubner, 2000) | Higher phosphorus threshold | $K_p = 0.030$ mg P-$PO_4$ $L^{-1}$ |
| Vertical position regulation | 0.1–0.8 m $day^{-1}$ (Walsby et al., 2006) | No vertical position regulation | – |
| Grazing defense | (Dokulil and Teubner, 2000) | No grazing defense | – |

**Table 16.2** The Possible Combinations Resulting by the Inputs Differentiation: CUR = Current scenario (it corresponds to reality), GPS = Global pressure scenario; LPS = Local pressure scenario; UND = Undisturbed scenario; 0 indicates no distance to reality (both the measured pressures are inside); 1 indicates a partial reality isolating both the global and the local (measured) pressures; 2 indicates maximum distance to reality (without human presence).

| Scenario code | Global pressure | Local pressure | Distance to reality |
|---|---|---|---|
| CUR | + | + | 0 |
| GPS | + | – | 1 |
| LPS | – | + | 1 |
| UND | – | – | 2 |

warming, thus representing only the effects of local impact. The undisturbed scenario (UND) was cleaned by the global and the local impacts, implementing the model only with the natural loads and detrended air temperature.

# 16.4 Results

## 16.4.1 Global and Local Pressures

Inflow discharges were calibrated and validated ($R^2 = 0.58$; $R^2 = 0.45$) for a period of about 4 years, when all the measurements were available, to indirectly compute the weekly water balance of Lake Pusiano. The lake volume change was derived by the daily measured lake levels and compared to the simulated lake volume change, resulting by the water balance of the simulated inflow discharges, the measured rainfall and outflow discharges, and the estimation of the other terms (i.e., evaporation or the groundwater contribution). The simulated pristine mean annual P-load was confirmed by the load estimation with the export coefficients methods, resulting respectively 2.4 and 2.6 t P $year^{-1}$ and converted into lake TP concentrations (during winter overturn) through the Vollenweider equation (OECD, 1982), resulting respectively in around 12 and 14 μg P $L^{-1}$. The outcome was confirmed by data from the paleololimnological survey (Guilizzoni et al., 2011) and the application of the Morphoedaphic Index (MEI) predicting the pristine lake conditions (Vighi and Chiaudani, 1985) which have pointed out similar concentrations (respectively around 13 and 10 μg P $L^{-1}$). The actual or anthropic annual P-load estimation was converted into lake TP concentrations and validated by the time series of TP concentrations taken in the lake from 1973 (Fig. 16.2a).

The SSA was chosen to represent the air temperature trend dynamics, which resulted in a mean annual increase of 0.015 °C $year^{-1}$ (Fig. 16.2b), in line with the considerations issued in Ambrosetti and Barbanti (1999) for alpine and pre-alpine lakes in North Italy. The resulting detrended air temperature series showed no increasing trend but slightly higher mean annual values in the first 10 years, as the best compromise between the method requirement of long series to stabilize the outcomes, data availability for this study, and the identification of a cold period in 60s (Ghil and Vautard, 1991). The temperature differentiation began in 1980s and strongly separated in the last 15 years by almost 1 °C in mean annual values.

## 16.4.2 Lake Model Assessment

After calibration DYCD was re-initialized and validated over 1960–2010 period for lake level, water temperature, TP, TN, and total Chl-a (Fig. 16.3). The lake level was the only daily series (except for the period 1973–1985 with no data). The simulated lake levels matched rather well with the observed data, providing confidence of an accurately balance of water fluxes, showing a low mean bias of 0.16 m in relation to the vertical model resolution (1 m), while the root-mean-square error (RMSE) was of 0.46 m on 13,674 observed data (Fig. 16.3a). Simulated temperatures as mean value by daily January values matched the observed temperature at winter overturn, though the model generally showed an underestimation (mean bias 0.98 °C and RMSE 1.24 °C on 28 observed data, Fig. 16.3b). Simulated TP lake concentration, showed as January monthly mean, followed the interannual trend of respective observed series as instantaneous recorded data at the lake overturn, with an increase from 1970s to 1990s and a decrease after the sewage plant construction and load reduction, (mean bias 0.0184 mg P $L^{-1}$ and RMSE 0.0403 mg P $L^{-1}$ on 30 observed data, Fig. 16.3c). Simulated TN lake concentration resulted within the range of the observed series, with a mean concentration of about 2 mg N $L^{-1}$, by January monthly mean values and also quite reproduced the interannual variability though

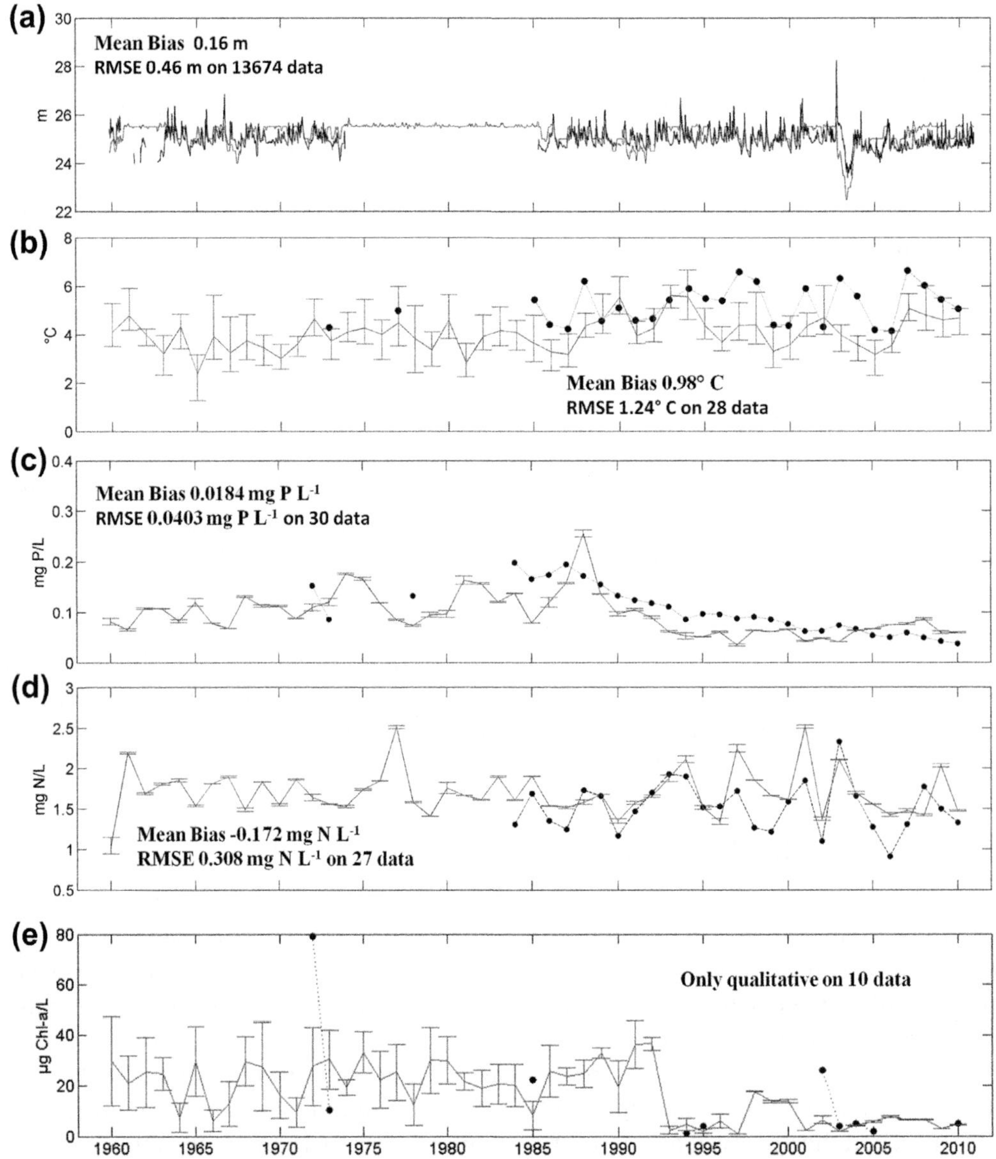


**FIGURE 16.3 (a) Measured (black line or circles) and simulated (gray solid line) daily values for the lake levels, (b) the winter values for temperature, (c), TP, (d), TN, and (e) the early spring values for total Chl-a.** The error bars on simulated monthly mean values represent the daily maximum and minimum.

a general overestimation (mean bias $-0.172\ \text{mg N L}^{-1}$ and RMSE $0.308\ \text{mg N L}^{-1}$ on 27 observed data, Fig. 16.3d). Despite very few available measurements of Chl-a lake concentration, the simulated biological response, as mean value of total Chl-a at early spring (April) in the upper layers, qualitatively caught higher productivity before the

1990s and an abrupt drop when the available phosphorus started limiting the algal growth. The model reproduced the peak due to *P. rubescens* first bloom at the end of 1990s while it was first measured in April 2002 (Fig. 16.3e).

### 16.4.3 Scenarios Comparison

Simulated scenarios have been compared for temperature, TP, TN, and Chl-a as proxies tracking the lake changes. The evolution of the mean annual temperature, as lake volume-weighted, the winter mean concentration of TP and TN, and the lake productivity as early spring Chl-a concentration are showed in Fig. 16.4.

Generally the local pressures had more marked effects than the global pressure considering the scenarios comparison. On the whole, the mean annual current water

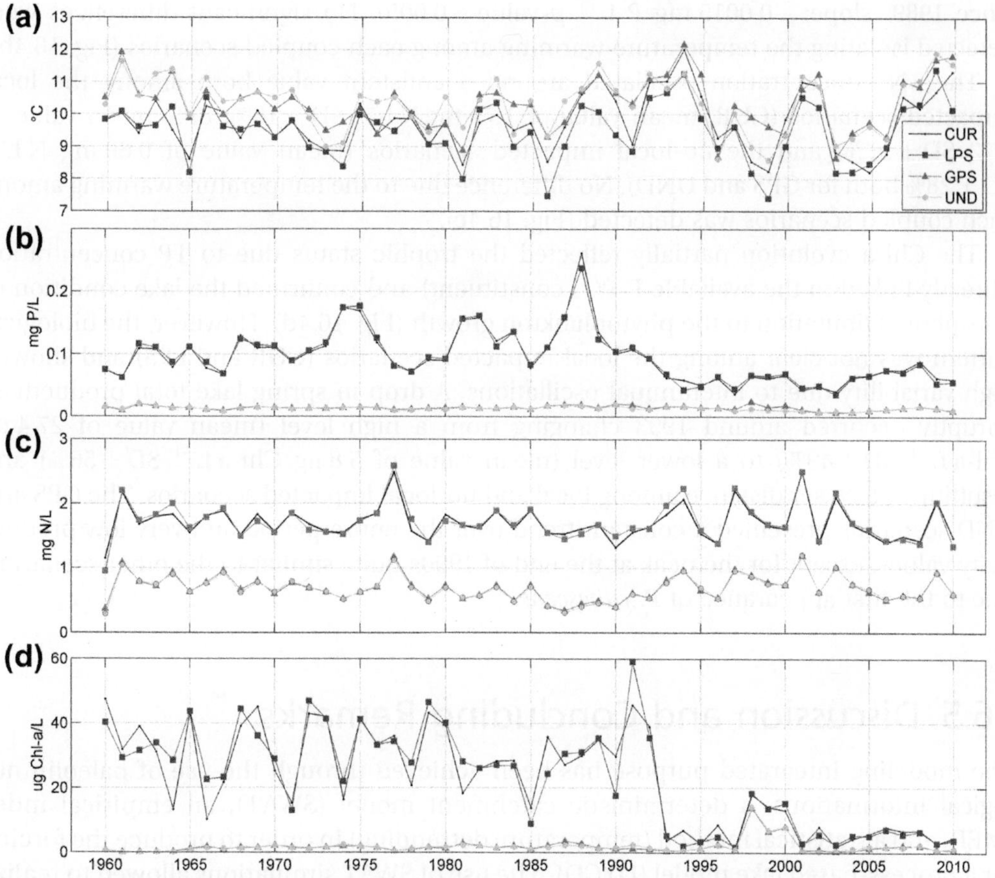


**FIGURE 16.4 Scenarios comparisons: (a) simulated mean annual temperatures, representing the lake heat content; (b) simulated total phosphorus (TP) and (c) total nitrogen (TN) concentration at winter overturn; (d) simulated Chl-a concentration at early spring.** CUR = current scenario, GPS = global pressure scenario; LPS = local pressure scenario; UND = undisturbed scenario.

temperatures resulted lower than the undisturbed ones (10.3 °C ± 13% and 9.8 °C ± 11% between UND and CUR *p*-value: <0.001, Fig. 16.4a) but only the global impacted scenarios (CUR and GPS) produced a significant increasing trend by SSA. The SSA curve smoothed the annual series toward a first-order polynomial, thus it was linearized by a linear model. The analysis of covariance (ANCOVA) was used to test the different slopes for the scenarios considering the smoothed lake temperature as the covariate: the slope for the CUR scenario (0.012 °C $year^{-1}$, *p*-value: <0.001) significantly differed from GPS scenario (0.004 °C $year^{-1}$, *p*-value: <0.001).

The TP concentration had not a significant trend among GPS and UND scenarios: the mean value on 50 years of simulation was respectively 0.0136 mg P $L^{-1}$ (SD ± 25%) and 0.0127 mg P $L^{-1}$ (SD ± 23%), The TP concentration in the CUR and LPS scenarios clearly reflected the dynamics of the anthropogenic P-load evolution: an increasing trend occurred until 1988 (slope: 0.0025 mg P $L^{-1}$ *p*-value: <0.001), while a decrease started since 1989 (slope: −0.0015 mg P $L^{-1}$ *p*-value = 0.005). No significant differences were detected isolating the temperature warming among each coupled scenarios (Fig. 16.4b).

The TN concentration oscillated around a constant value both among the local impacted scenarios (CUR: mean value of 1.71 mg N $L^{-1}$ SD ± 16%; LPS: mean value of 1.72 SD ± 17%) and the no-local impacted scenarios (mean value of 0.69 mg N $L^{-1}$ SD ± 28% both for GPS and UND). No difference due to the temperature warming among each coupled scenarios was detected (Fig. 16.4c).

The Chl-a evolution partially reflected the trophic status due to TP concentration (directly linked to the available P-$PO_4$ constituent) and confirmed the lake condition of phosphorus limitation to the phytoplankton growth (Fig. 16.4d). However, the biological pattern was not clear among the local impacted scenarios (CUR and LPS) and showed high variability due to interannual oscillations. A drop in spring lake total productivity abruptly occurred around 1993 changing from a high level (mean value of 27.4 μg Chl-a $L^{-1}$ SD ± 49%) to a lower level (mean value of 5.8 μg Chl-a $L^{-1}$ SD ± 56%) and resulting in a closer distance among local and no-local impacted scenarios. The GPS and UND scenarios presented a constant trend over the entire period and very low productivity values, except for the peak at the end of 1990s quite similar to the other scenarios, due to the first appearance of *P. rubescens*.

## 16.5 Discussion and Concluding Remarks

The modeling integrated purpose has been achieved through the use of paleolimnological informations, a deterministic catchment model (SWAT), an empirical index (MEI), and a statistical method (temperature detrending) in order to produce the forcing for a process-based lake model (DYCD). The use of SWAT simulations allowed to realize a low P-load pressure extending a supposed natural sub-basin to the whole catchment, the so-called "no-human local impact" whose mean values were confirmed by MEI and paleolimnological results for the pristine lake condition. However, the model cannot reproduce the load from the point sources, thus the urban pollution have been computed out of the model and processed to equilibrate the lake trend concentrations by the Vollenweider equation. The global impact was represented by air temperatures trend computed by SSA and cleaned by the measured time series to obtain a sort of "no-human global impact." The different drivers were properly combined into the lake model (DYCD) simulating four different scenarios characterizing a gradient in both locally and globally mediated pressures. This model is adequate to investigate both the daily or seasonal lake fluctuations for a long time and the annual aggregated mean values or a particular monthly mean values. In this preliminary study, only the last option was considered to analyze the assessment with available time series and establish an integrated methodology for the study of reference conditions. Nevertheless, further steps are going on to further investigate the model outcomes through apposite statistical analyses capturing both the long-term trend and daily or seasonal variability.

Considering the lake after the winter overturn, the first macroscale-analysis confirmed the local pressures provided more marked lake responses over long time changes, determining significant differences on the water quality and phytoplankton biomass. Nitrogen lake concentrations remained relatively constant as the catchment nitrogen loads were inferred by rivers quality data and hardly influenced by wet and dry deposition amount (Balestrini et al., 2000). The phosphorus pattern showed a clear response to the load reduction occurred after sewage plant building leading to a decline of the phytoplankton production in spring, as shown by the abrupt drop in the Chl-a concentrations after the 1990s. No effects on annual phosphorus content were detected isolating the warming temperature driver, but a variability on primary production was revealed and thus a possible response on biogeochemical seasonal behavior could emerge by daily investigation. The lake temperatures underwent a low differentiation. Warmer temperatures occurred removing the local impact (anthropogenic P-load); a possible explanation should refer to a lower lake productivity, a greater light transmission, a weaker thermal stratification during the summer, and thus an anticipated overturn or warmer hypolimnetic waters. The current scenario showed a significant increase in water warming, while the respective scenario with no-global-change (without air temperature warming) resulted generally colder with a flat trend.

This preliminary analysis provides an indication to further investigate whether the climate change could produce a gentler effect on natural (or pristine) lakes. From a methodological point of view, the integrated approach presented in this study appears very promising, as it provided a quantitative transferable methodology to separate pressures at different scales and it resulted in a suitable tool supporting other tried methods for the establishment of reference conditions. Finally it should be noted that a warmer climate may increase the nutrient loading from the catchment due to rainfall variability or temperature-depending soil processes (Jeppesen et al., 2009), and it should be considered also in the catchment modeling, in the future.

# Acknowledgments

The authors wish to thank the EPSON Meteo Centre (http://www.meteo.it/) for the computational support, the Centre for Water Research (http://www.cwr.uwa.edu.au/) for kindly supplying the in-lake hydrodynamic-ecological model DYRESM-CAEDYM, Andrea Lami (CNR-ISE) for the precious paleolimnological analysis and Sandro Calmanti (ENEA) for kindly providing meteorological RES PROTEUS data forcing.

# References

Ambrosetti, W., Barbanti, L., 1999. Deep water warming in lakes: an indicator of climatic change. J. Limnol. 58, 1–9.

Anonymous, 2003. River and Lakes–Typology, Reference Conditions and Classification Systems. Guidance No 10. CIS Working Group 2.3 REFCOND, European Communities, Luxembourg, p. 87.

Appleby, P.G., 2001. Chronostratigraphic techniques in recent sediments. In: Last, W.M., Smol, J.P. (Eds.), 2001. Tracking Environmental Change using Lake Sediments, vol. 1. Basin Analysis, Coring, and Chronological Techniques, Kluwer, Dordrecht, pp. 171–203.

Artale, V., Calmanti, S., Carillo, A., Dell'Aquila, A., Herrmann, M., Pisacane, G., Ruti, P.M., Sannino, G., Struglia, M.V., Giorgi, F., Bi, X., Pal, J.S., Rauscher, S., 2010. An atmosphere-ocean regional climate model for the Mediterranean area: assessment of a present climate simulation. Clim. Dynam. 35, 721–740.

Balestrini, R., Galli, L., Tartari, G., 2000. Wet and dry atmospheric deposition at prealpine and alpine site in Northern Italy. Atmos. Environ. 4, 1455–1470.

Barbiero, G., Carone, G., Cicioni, G., Puddu, A., E Spaziani, F.M., 1991. Valutazione dei carichi inquinanti potenziali per i principali bacini idrografici italiani: Adige, Arno, Po e Tevere, vol. 90. Quaderni Istituto di Ricerca sulle Acque, pp. 233.

Bennion, H., Johnes, P., Ferrier, R., Phillips, G., Haworth, E.Y., 2005. A comparison of diatom phosphorus transfer functions and export coefficient models as tools for reconstructing lake nutrient histories. Freshwater Biol. 50, 1651–1670.

Blenckner, T., 2008. Models as tools for understanding past, recent and future changes in large lakes. Hydrobiologia 599, 177–182.

Boguniewicz, J., Capodaglio, A., Salerno, F., Tartari, G., 2006. Approach to support water quality catchment project. WSEAS Transactions on Environment and Development 2, 1079–1084.

Broomhead, D.S., King, G.P., 1986. Extracting qualitative dynamics from experimental data. Physica D20, 217–236.

Bruce, L.C., Hamilton, D.P., Imberger, J., Gal, G., Gophen, M., Zohary, T., Hambright, K.D., 2006. A numerical simulation of the role of zooplankton in C, N and P cycling in Lake Kinneret, Israel. Ecol. Model. 193, 412–436.

Chiaudani, G., Premazzi, G., 1992. I laghi briantei. Situazione trofica e soluzioni scientifiche per il risanamento Report EUR 14548 IT (In Italian).

Copetti, D., Tartari, G., Morabito, G., Oggioni, A., Legnani, E., Imberger, J., 2006. A biogeochemical model of the Lake Pusiano (North Italy) and its use in the predictability of phytoplankton blooms: first preliminary results. J. Limnol. 65, 59–64.

Déqué, M., 2007. Frequency of precipitation and temperature extremes over France in an anthropogenic scenario: model results and statistical correction according to observed values. Global Planet. Change 54, 16–26.

Dokulil, M.T., Teubner, K., 2000. Cyanobacterial dominance in lakes. Hydrobiologia 438, 1–12.

Elliott, J.A., Thackeray, S.J., Huntingford, C., Jones, R.G., 2005. Combining a regional climate model with a phytoplankton community model to predict future changes in phytoplankton in lakes. Freshwater Biol. 50, 1404–1411.

Gerletti, M., Marchetti, R., 1977. Indagini sui laghi della Brianza. Water Research Institute, CNR Italy. Books n. 19 (In Italian).

Ghil, M., Vautard, R., 1991. Interdecadal oscillations and the warming trend in global time series. Nature 350, 324–327.

Guilizzoni, P., Marchetto, A., Lami, A., Gerli, S., Musazzi, S., 2011. Use of sedimentary pigments to infer past phosphorus concentration in lakes. J. Paleolimnol. 45, 433–445.

Hamilton, D.P., Schladow, S.G., 1997. Prediction of water quality in lakes and reservoirs. Part I. Model description. Ecol. Model 96, 91–110.

Hipsey, M.R., Romero, J.R., Antenucci, J.P., Hamilton, D., 2006. Computational Aquatic Ecosystem Dynamics Model. v. 2.3 Science Manual. Centre for Water Research, the University of Western Australia (online). http://www.cwr.uwa.edu.au/services/models/legacy/model/dyresmcaedym/dyresmcaedymdocumentation.html.

Imerito, A., 2007. Dynamic Reservoir Simulation Model v4.0 Science Manual (online). Centre for Water Research, the University of Western Australia. http://www.cwr.uwa.edu.au/services/models/legacy/model/dyresmcaedym/dyresmcaedymdocumentation.html.

Jeppesen, E., Søndergaard, M., Jensen, J.P., Havens, K., Anneville, O., Carvalho, L., Coveney, M.F., Deneke, R., Dokulil, M., Foy, B., Gerdeaux, D., Hampton, S.E., Kangur, K., Köhler, J., Körner, S., Lammens, E., Lauridsen, T.L., Manca, M., Miracle, R., Moss, B., Nõges, P., Persson, G., Phillips, G., Portielje, R., Romo, S., Schelske, C.L., Straile, D., Tatrai, I., Willén, E., Winder, M., 2005. Lake responses to reduced nutrient loading–an analysis of contemporary long-term data from 35 case studies. Freshwater Biol. 50, 1747–1771.

Jeppesen, E., Kronvang, B., Meerhoff, M., Søndergaard, M., Hansen, K.M., Andersen, H.E., Lauridsen, T.L., Liboriussen, L., Beklioglu, M., Özen, A., Olesen, J.E., 2009. Climate change effects on runoff, catchment phosphorus loading and lake ecological state, and potential adaptations. J. Environ. Qual. 38, 1930–1941.

Johnes, P.J., 1996. Evaluation and management of the impact of land use change on the nitrogen and phosphorus load delivered to surface waters: the export coefficient modelling approach. J. Hydrol. 183, 323–349.

Lami, A., Guilizzoni, P., Marchetto, A., 2000. High resolution analysis of fossil pigments, carbon, nitrogen and sulphur in the sediment of eight European Alpine lakes: the MOLAR project. J. Limnol. 59, 15–28.

Legnani, E., Copetti, D., Oggioni, A., Tartari, G., Palumbo, M.T., Morabito, G., 2005. *Planktothrix rubescens* seasonal and vertical distribution in Lake Pusiano (North Italy). J. Limnol. 64, 6–73.

Livingstone, D.M., 2003. Impact of secular climate change on the thermal structure of a large temperate central European lake. Climatic Change 57, 205–225.

Malmaeus, J.M., Blenckner, T., Markensten, H., Persson, I., 2006. Lake phosphorus dynamics and climate warming: a mechanistic model approach. Ecol. Model 190, 1–14.

Neitsch, S.L., Arnold, J.G., Kiniry, J.R., Williams, J.R., 2005. Soil and Water Assessment Tool Theoretical Documentation and User's Manual, Version 2005. GSWR Agricultural Research Service & Texas Agricultural Experiment Station, Temple Texas.

Omlin, M., Reichert, P., Foster, R., 2001. Biogeochemical model of Lake Zürich: model equations and results. Ecol. Model 141, 77–103.

Organisation for Economic Co-operation and Development (OECD), 1982. Eutrophication of waters. Monitoring, Assessment and Control. OECD, Paris, pp. 154.

Paerl, H.W., Huisman, J., 2008. Blooms like it hot. Science 320, 57–58.

Parmesan, C., Yohe, G., 2003. A globally coherent fingerprint of climate change impacts across natural systems. Nature 421, 37–42.

Peeters, F., Livingstone, D.M., Goudsmit, G.H., Kipfer, R., Forster, R., 2002. Modelling 50 years of historical temperature profiles in a large central European lake. Limnol. Oceanogr. 47, 186–197.

Portoghese, I., Bruno, E., Guyennon, N., Iacobellis, V., 2011. Stochastic bias-correction of daily rainfall scenarios for hydrological applications. Nat. Hazards Earth Syst. Sci. 11, 2497–2509.

Salerno, F., Tartari, G., 2009. A coupled approach of surface hydrological modelling and wavelet analysis for understanding the baseflow components of river discharge in karst environments. J. Hydrol. 376, 295–306.

Salmaso, N., 2010. Long-term phytoplankton community changes in a deep subalpine lake: responses to nutrient availability and climatic fluctuations. Freshwater Biol. 55, 825–846.

Schindler, D.W., 2001. The cumulative effects of climate warming and other human stresses on Canadian fresh-waters in the new millennium. Can. J. Fish Aquat. Sci. 58, 18–29.

Simmons, A.J., Gibson, J.K., 2000. The ERA-40 Project Plan, ERA-40 project report series no.1 ECMWF, pp. 62.

Søndergaard, M., Jensen, J.P., Jeppesen, E., 2003. Role of sediment and internal loading of phosphorus in shallow lakes. Hydrobiologia. 506–509. 135–145.

Srinivasan, R., Arnold, J.G., 1994. Integration of a basin-scale water quality model with GIS. Water Resour. Bull. 30, 453–462.

Tartari, G., Copetti, D., Barbiero, G., Tatti, S., Pagnotta, R., 2002. Contribute of GIS technique in the evaluation of anthropic impact change in a sub-alpine shallow lake. Proc. Int. Congr. iEMSs 2002, Integrated Assessment and Decision Support Lugano June 24–27, 2002, 610.

Thackeray, S.J., Jones, D., Maberly, C., 2008. Long-term change in the phenology of spring phytoplankton: species-specific responses to nutrient enrichment and climatic change. J. Ecol. 96, 523–535.

Trolle, D., Hamilton, D.P., Pilditch, C.A., Duggan, I.C., 2011. Predicting the effects of climate change on trophic status of three morphologically varying lakes: implications for lake restoration and management. Environ. Modell. Softw.

Verburg, P., Hecky, R.E., Kling, H., 2003. Ecological consequences of a century of warming in Lake Tanganyika. Science 301, 505–507.

Vighi, M., Chiaudani, G., 1985. A simple method to estimate lake phosphorus concentrations resulting from natural, background, loadings. Water Res. 19, 987–991.

Walsby, A.E., Schanz, F., 2002. Light-dependent growth rate determines changes in the population of *Planktothrix rubescens* over the annual cycle in Lake Zurich, Switzerland. New Phytol. 154, 671–687.

Walsby, A.E., Schanz, F., Schmid, M., 2006. The Burgundy-blood phenomenon: a model of buoyancy change explains autumnal waterblooms by *Planktothrix rubescens* in Lake Zurich. New Phytol. 169, 109–125.

Winder, M., Cloern, J.E., 2010. The annual cycles of phytoplankton biomass. Philos. Trans. R. Soc. B 365, 3215–3226.

Yeates, P.S., Imberger, J., 2004. Pseudo two-dimensional simulations of internal and boundary fluxes in stratified lakes and reservoirs. Int. J. River Basin Res. 1, 1–23.

17

# Biogeochemical 1D ERSEM Ecosystem Model Applied to Recent Carbon Dioxide and Nutrient Data in the North Sea

Khalid Elkalay*[,1], Karima Khalil*, Helmuth Thomas[†], Yann Bozec[‡], Piet Ruardij[‡], Hein de Baar[‡]

*SAEDD LABORATORY, ECOLE SUPÉRIEURE DE TECHNOLOGIE D'ESSAOUIRA KM 9, ROUTE D'AGADIR, ESSAOUIRA ALJADIDA BP. 383, ESSAOUIRA. MOROCCO AND ENSA LABORATORY (FP SAFI), MOROCCO, [†]DALHOUSIE UNIVERSITY, DEPARTMENT OF OCEANOGRAPHY, 1355 OXFORD STREET, HALIFAX, NOVA SCOTIA, B3H 4J1, CANADA, [‡]STATION BIOLOGIQUE DE ROSCOFF UMR CNRS–UPMC 7144–EQUIPE CHIMIE MARINE. PLACE GEARGES TEISSIER 29682 ROSCOFF, FRANCE, [1]FAX. +212 44 66 93 57, E-MAIL ADDRESS: ELKALAY_KHALID@YAHOO.FR (K. ELKALAY)

## 17.1 Introduction

Understanding the marine organic carbon and nutrients cycles is fundamental to identify and understand present, past and future ocean-climate linkages. The transfer of carbon from surface waters to deep waters can exert a major influence on the long-term atmospheric $CO_2$ levels and storage of anthropogenic excess atmospheric $CO_2$. During the last 30 years various large international research projects allowed to identify and quantify the major vertical fluxes in the oceans. Modeling at global scale allows to apprehend these vertical fluxes in a consistent temporal and spatial frame and to refine the understanding of these fluxes with a mechanistic approach. Also, models allow reconstructing past fluxes and predicting their future evolution, in particular with respect to climate change and ocean-climate feedbacks.

Numerous models use only one chemical element for the model currency (e.g. Andersen and Nival, 1989; Carlotti and Radach, 1996). However, several studies have demonstrated that biogeochemical cycles and elemental ratios are neither independent, nor parallel, but deeply interconnected and complementary (e.g. Droop, 1973; Elser and Hassett, 1994; Baretta et al., 1995, Ruardij et al., 1995; Andersen, 1997; Thomas et al., 2003). This relatively new theory of elemental stoichiometry is mainly concerned with the


Models of the Ecological Hierarchy. DOI: http://dx.doi.org/10.1016/B978-0-444-59396-2.00017-1
ISSN 0167-8892, 

spatial and temporal variability of elemental ratios, as affected by chemical and biological interactions. Hence the emergence of a new generation of ecosystem models such as ERSEM (European Regional Seas Ecosystem Model), where the variability and interplay of the most important elemental ratios is explicitly parameterized.

In order to build realistic dynamical and biological coupled models of the world ocean, each physical and biogeochemical regime must be understood. In this respect, the North Sea can be considered as an interesting target since several different physical and biogeochemical regimes can be found in a rather small and well studied area.

This work presents preliminary application of ERSEM, validated with recent field data from the CANOBA project whose first aim is to help understand the role of coastal seas in the global carbon cycle as assessed in a pilot study in the North Sea (Thomas, 2002; Bozec et al., 2004; Thomas et al. 2004). The hypothesis of a "continental shelf pump" for the uptake of $CO_2$ from the atmosphere by a coastal sea with subsequent transport to the open ocean was tested. Interactions of the pools of carbon (C), nitrogen (N), phosphorous (P), silicate (Si) and oxygen ($O_2$) were assessed in this study to describe the "biological $CO_2$ pump". This was achieved using a high resolution analysis of the carbon cycle in the North Sea with a coupled hydrodynamic-ecosystem model, budgeting of carbon and related nutrient fluxes in, into and out of the North Sea. This paper presents the preliminary steps: introduction of a $CO_2$ submodel in the 1-D ERSEM; testing the applicability of the biogeochemical 1D-ERSEM setup at two stations in the southern and northern North Sea; validation of the simulations with recent field data-sets (Thomas 2002; Thomas et al. 2004). The prognostic potential of the model setup is tested with a sensitivity analysis based on the feed-back between $CO_2$ increase and primary production.

## 17.2 Methods

### 17.2.1 Ecological model

ERSEM was developed to simulate the annual cycles of carbon, nitrogen, phosphorus and silicon in the pelagic and benthic components of the North Sea (Baretta et al., 1995). The model requires detailed data inputs and focuses on the phytoplankton and zooplankton groups, with detailed representation of microbial, detrital and nutrient regeneration dynamics. The model is driven by a wide range of forcing factors including irradiance and temperature data, atmospheric inputs of nitrogen, suspended matter concentration, hydrodynamical information and inorganic and organic river load data (Lenhart et al., 1997). ERSEM has many advantages, the first one, he use a variety of biological formulations based on trophic identity to capture 98 the critical performance of the different components which include most of the major functional groups and processes thought to be important in coastal marine systems and state-of-the-art biogeochemical models. The second most advantage of ERSEM that he could be easily adapted for other regions as it is essentially a generic model coupled to an appropriate physical model for a region, such as the General Ocean Turbulence Model (GOTM). The one big disadvantage of ERSEM, that he use fixed sets of parameters (Jørgensen, 1994). It is important to note that within the last two decades, there has been a proliferation of ERSEM applications in specialized journals essentially in Ecological modeling with literally more than 20 articles, of varying scope and quality (see e.g., Aveytua-Alcázar, et al., 2008; Vanhoutte-Brunier et al., 2008, and references therein). ERSEM has been applied in 18 locations in a variety of physical contexts from 1D to 3D including the North Sea (Patsch and Radach, 1997; Lenhart et al., 1997), the Adriatic Sea (Allen et al., 1998; Vichi et al., 1998a, b), the Baltic Sea (Vichi et al., 2004) and the Mediterranean Sea (Allen et al., 2001). Adapting it to other systems requires a fair amount of data.

ERSEM applied in this work describes in an appropriate way the pelagic and benthic processes as well as the sediment-water exchange in the North Sea (Fig. 17.2 il faut commencer par figure 1) (Ebenhöh et al., 1995; Ruardij et al., 1995). A detailed mathematical description of ERSEM can be found in the two ERSEM special issues (Baretta et al., 1995; Ebenhöh et al., 1997). The structure of ERSEM-North Sea has been described in detail by Blackford (1997) and, a technical description of the modules can be found in Radford and Blackford (1993). The main characteristics of the physical model are summarized by Vichi et al. (1998b). The physical model is a mixed-layer model in which vertical transport is induced by shifting of layers ("boxes"). The model runs within the simulation environment SESAME (Ruardij et al., 1995). For the coupling of the physical model to the ecological model the possibility of a two-step integration is introduced in the SESAME environment at each time step. The details of the physical model and the steps of the SESAME environment see (Ruardij et al., 1997).

ERSEM is conceived as a generic model, which applied to a basin scale, should be capable to correctly simulate the spatial pattern of ecological fluxes throughout the seasonal cycle and across eutrophic to oligotrophic gradients. The model describes C, N and P cycles and with each functional group containing these three pools. The inclusion of the Si cycle allows the distinction between diatom and non-diatom phytoplankton compartments. Inorganic nitrogen is sub-divided into ammonium and nitrate compartments enabling the distinction between new and recycled primary production.

ERSEM uses a "functional" group approach to describe the ecosystem where the biota is grouped together according to their trophic level (subdivided according to size classes or feeding methods). State variables have been chosen in order to keep the model relatively simple without omitting any components that may significantly influence the energy balance of the system. Biological functional dynamics are described by both physiological (ingestion, respiration, excretion, egestion, grazing, etc.) and population processes (growth and mortality). The biologically driven C dynamics are coupled to the chemical dynamics of N, P, Si and O2. (Ebenhöh et al., 1995; Blackford et al., 2004).

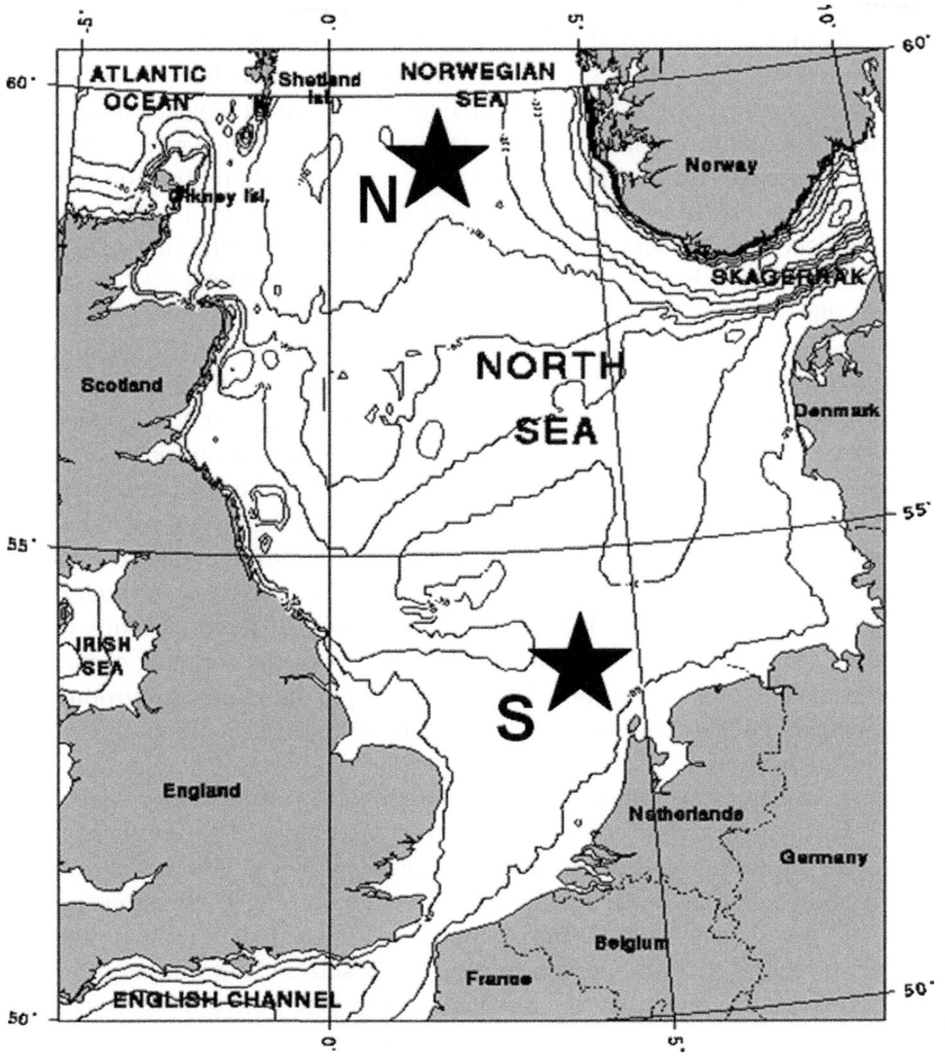


**FIGURE 17.1** Study area of the North Sea. The stars indicate the stations (S: southern station, N: northern station) for validation of the 1-D ERSEM model. For color version of this figure, the reader is referred to the online version of this book.

### 17.2.2 $CO_2$ module

The $CO_2$ module implemented into ERSEM follows the OCMIP (Ocean Carbon Cycle Modeling Intercomparaison Project) guidelines. Several $CO_2$ models exist and but only few are treated with different nutrients and other variables of the ecosystem as e.g. phytoplankton or zooplankton. The dissolved inorganic carbon ($DIC = [CO_2] + [HCO_3^-] + [CO_3^{2-}]$) integrates thermodynamic equilibrium reactions of the $CO_2$ system, the air-sea exchange of $CO_2$, and the $CO_2$ sink and source terms due to biological activity,

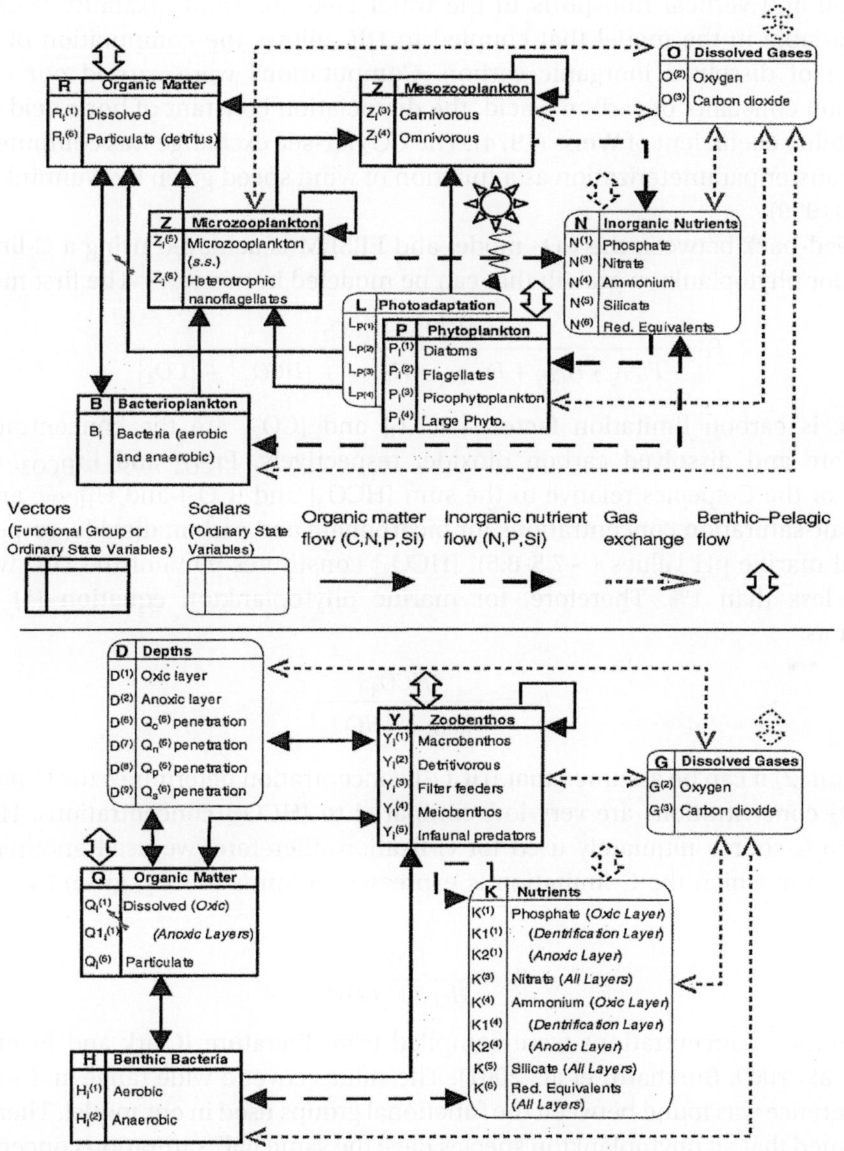


**FIGURE 17.2** ERSEM conceptual diagram.

and lateral and vertical transports in the water column. Total Alkalinity (TA) is also a state variable in the model that coupled to DIC allows the computation of the full speciation of dissolved inorganic carbon. Computations were carried out with the dissociation constants of carbonic acid, the dissociation constant of boric acid and the $CO_2$ solubility coefficient of Weiss (1974). The $CO_2$ air-sea exchange was computed using the gas transfer parameterization as a function of wind speed given by Wanninkhof and McGillis (1999).

The feed-back between the $CO_2$ model and ERSEM is achieved using a C-limitation function for phytoplankton growth that can be modeled in two ways. The first method is

$$F_c = \frac{[HCO_3^-] + [CO_2]}{Fr_{CO_2} * H_{CO_2} + Fr_{HCO_3^-} * H_{HCO_3^-} + [HCO_3^-] + [CO_2]} \quad (1)$$

where Fc is carbon limitation factor, $[HCO_3^-]$ and $[CO_2]$ are the concentrations of bicarbonate and dissolved carbon dioxide, respectively, $Fr_{CO2}$ and $Fr_{HCO3-}$ are the fractions of the C-species relative to the sum $[HCO_3^-]$ and $[CO_2]$ and $H_{HCO3-}$ and $H_{CO2}$ are the half-saturation concentrations for bicarbonate and carbon dioxide, respectively. At typical marine pH values (~7.5-8.5), $[HCO_3^-]$ constitutes 90 % of the DIC where as $[CO_2]$ is less than 1%. Therefore, for marine phytoplankton equation (1) can be rewritten as:

$$F_c = \frac{[HCO_3^-]}{H_{HCO_3^-} + [HCO_3^-]} \quad (2)$$

In equation (2) it can be assumed that $[HCO_3^-]$ concentration determines the C-limitation since $CO_2$ concentrations are very low compared to $[HCO_3^-]$ concentrations. However, $CO_2$ is the C-source ultimately used for C-fixation, therefore, we used another model formulation in which the C-limitation is expressed in terms of $CO_2$, given by a Monod function:

$$F_c = \frac{[CO_2]}{H_{CO_2} + [CO_2]} \quad (3)$$

Half-saturation concentrations were compiled from literature (Clark and Flynn, 2000; Tortell et al., 2000; Burkhardt et al., 2001). The values cover a wide range and no significant difference was found between the functional groups used in our model. Therefore, it was assumed that all phytoplankton species have the same half-saturation concentration, and the model was run with values within the range found in literature.

### 17.2.3 Overall model set-up

Each module consists of a coupled set of ordinary differential equations, which may be solved by Euler integration or by an implicit higher-order Runge-Kutta method. The simulations were made using SESAME (Software Environment for Simulation and Analysis of Marine Ecosystems; Ruardij et al., 1995). The physical model uses a semi-implicit finite difference scheme forward in time and centered in space. The net transport of the biological variables is calculated using the physical model and then transferred into SESAME.

### 17.2.4 Field data-sets for validation

The data used in this study are from a consistent data set which was collected for the North Sea during 1994 (Ruardij et al., 1997) and during 2000 and 2001 (Thomas, 2002). The latter data set was collected during four cruises: 18.08.2001-13.09.2001; 06.11.2001-29.11.2001; 11.02.2002-05.03.2002; 06.05.2002-26.05.2002. Each cruise consistent of 12 vertical profile stations for inorganic nutrients, DIC and TA and surface water partial pressure of $CO_2$ was measured continuously on the sub-surface waters (details in Thomas, 2002; Bozec et al., 2004; Thomas et al. 2004).

## 17.3 Results and discussion

### 17.3.1 Initialization

The variables used for the calibration run include the DIC, TA, temperature, chlorophyll *a*, nitrate, ammonium, phosphate and silicate from the first CANOBA cruise in August 2001.Two sites were chosen to compare the behavior of the model ecosystem, one in the south, and one in the northern part of the North Sea (Fig. 17.1). In order to simulate a standard state of the ecosystem annual evolution, a run was performed using initial conditions and physical forcing constraints corresponding to mean values for the study area.

### 17.3.2 Standard run

To show general features of the standard simulation we present the results for several variables. Fig. 17.3a shows the simulated annual evolution of the vertical temperature profile in the southern North Sea. The model reliably reproduces the development of the seasonal thermocline observed in the study area. Temperature stratification starts to develop in mid-April, forming a surface layer extending down to about 20 m, deepening slowly during remaining of spring and summer, and the water column is again fully mixed in October. In winter, temperature is below 12 °C and uniform in the water column.

Figure 17.3b–g show the temporal evolution of the simulated chlorophyll *a*, DIC, ammonium, nitrate, phosphate and silicate, respectively. As shown in Fig. 17.3b, the first phytoplankton bloom occurs in late April in the upper 25 m, and the maximal chlorophyll *a* value reaches 4.5 $\mu gl^{-1}$. The second bloom phytoplankton starts early June at the surface layer and deepens with time and the maximum chlorophyll *a* value (3 $\mu gl^{-1}$) occurs at around 25 m depth. Phytoplankton moves progressively deeper, and in summer significant biomasses can be observed close to the nitracline.

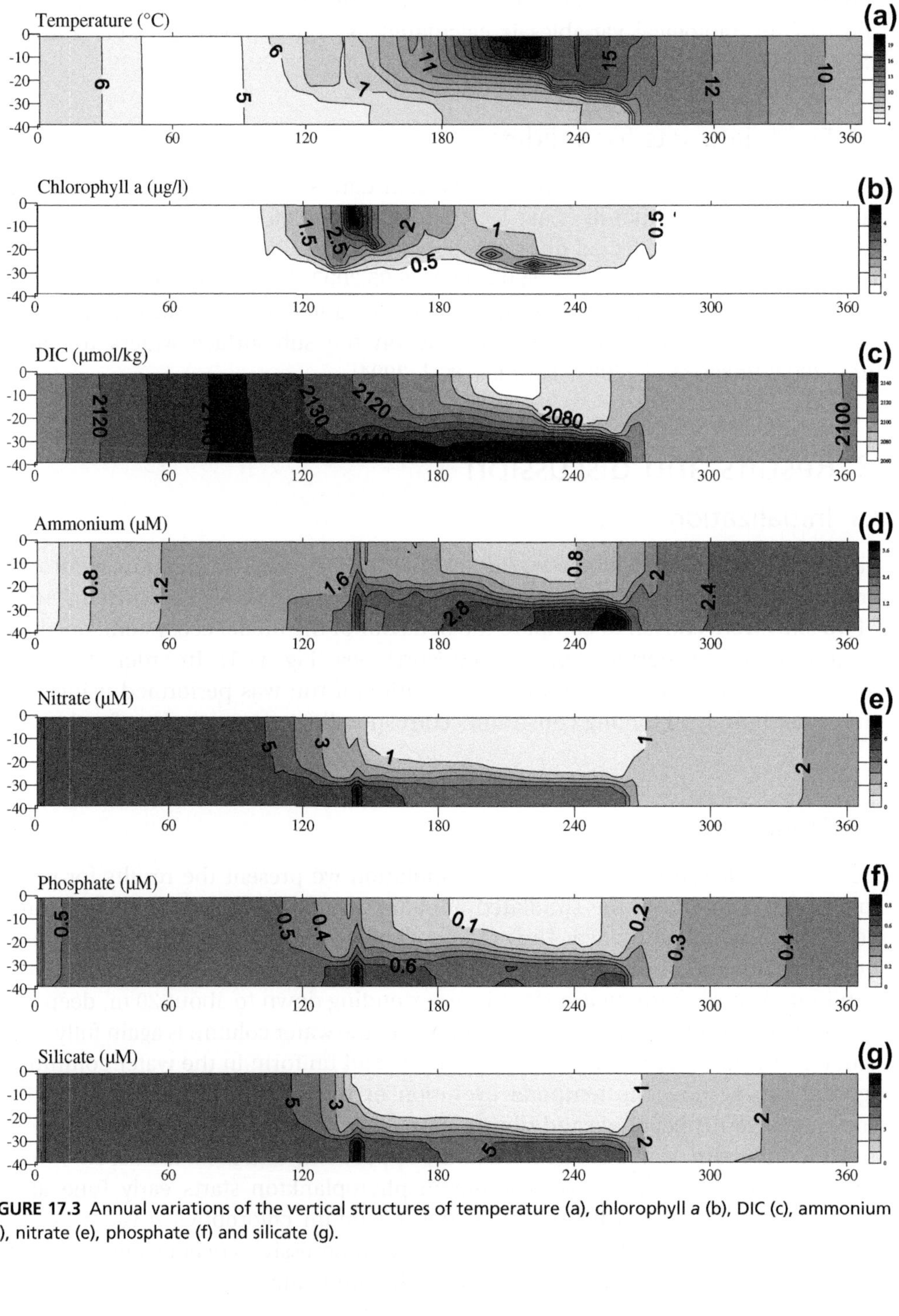

**FIGURE 17.3** Annual variations of the vertical structures of temperature (a), chlorophyll *a* (b), DIC (c), ammonium (d), nitrate (e), phosphate (f) and silicate (g).

In spring, the phytoplankton biomass and primary production rise as a result of stratification and increase in solar radiation. The early spring phytoplankton bloom is followed by a dramatic decrease of nutrients in the mixed layer.

The annual nitrate simulation is shown in Fig. 17.3e. In winter, there is replenishment of nitrate up to the surface caused by strong mixing, followed by depletion in the surface layer due to the uptake by phytoplankton. From autumn nitrate increases regularly in the water column due to destratification, diffusion from the sediments and decrease of uptake by phytoplankton. Nitrate reaches maximum values in April. The other nutrients follow the same trends as nitrate (Fig. 17.3).

In general, we note that the North Sea constitutes of two biogeochemical provinces (Thomas et al., 2004): In the shallow southern North Sea, biological uptake and release of DIC occur in a single compartment with a mixed water column throughout the year. As a result, after the initial DIC drawdown during the spring phytoplankton bloom the DIC remains at intermediate levels throughout the mixed water column.

In the seasonally stratified northern part, primary production draws down DIC in the surface mixed layer. Organic material sinks into the subsurface layer where remineralization releases DIC with no contact to the atmosphere. Low DIC levels prevail in the surface layer, while the DIC-enriched deeper waters are exported to the adjacent North Atlantic. In fall, mixing and remineralization restore uniform high winter DIC levels in both regions (Bozec et al., 2006). Weak annual net air–sea CO2 fluxes have been reported for the southern regions, while the North has been identified as a strong sink for atmospheric CO2 (Thomas et al., 2004).

Schiettecatte and al. (2007) showed that on an annual scale, the pCO2 dynamics appeared to be controlled by biological processes (primary production in springtime and respiratory processes in summer), rather than temperature (in summer). The comparison with measurements carried out in 2001 and 2002 shows that the inter-annual variability of pCO2 was close to the range of the spatial variability and mostly observed in spring, associated to biological processes (primary production) (Schiettecatte et al., 2007).

## 17.3.3 Validation

To estimate the model performance, in the following sections results are compared to the recent 2001-2002 data set (Thomas, 2002; Bozec et al. 2004, Thomas et al., 2004), older data sets from 1994 (Ruardij et al., 1997) and from July 96 to September 99 (Borges and Frankignoulle, 2002).

## 17.3.4 Seasonal evolution

Simulated and measured seasonal evolutions of nutrients in surface waters (southern station) are shown in Fig. 17.4. Nitrate, ammonium, phosphate and silicate reach respectively 5.5, 1.5, 0.55 and 5 μM during the early spring phytoplankton bloom. Simulated ammonium concentrations never exceed 2.5 μM while simulated nitrate concentration can reach 6 μM. The high nitrate concentrations occur from Januay to

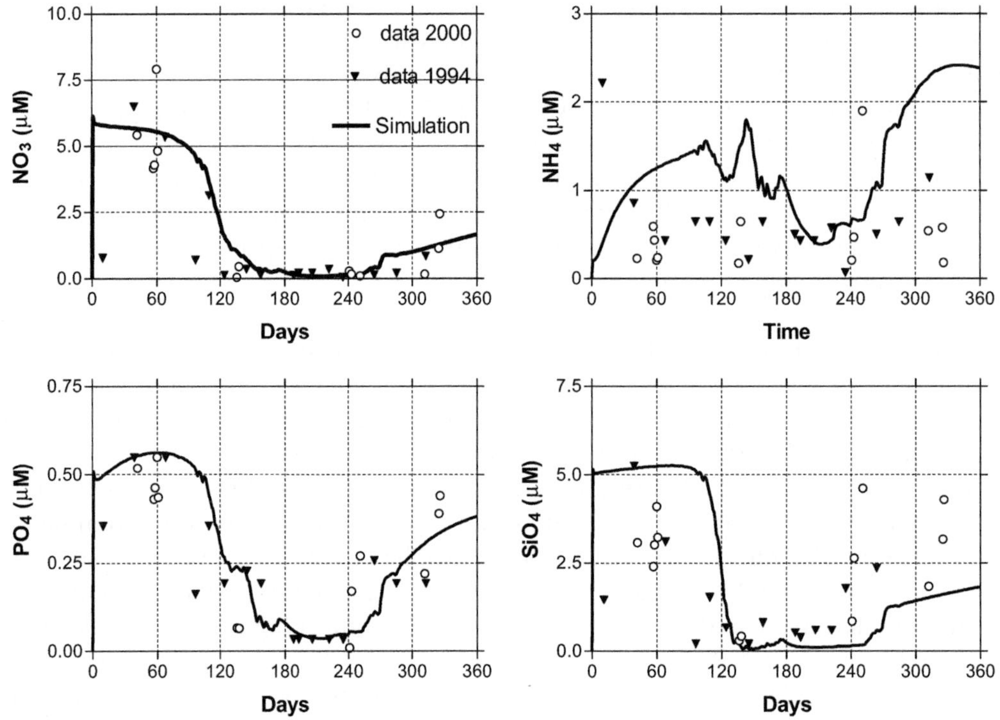


**FIGURE 17.4** Time variations of nutrients simulated by the model (line) and field data (symbols) in surface waters (southern station).

February, while from November to December ammonium is more abundant than nitrate. The observations suggest that the four nutrients concentrations are low in early-summer probably due to a continuation of the primary production phase. The model correctly simulates winter replenishment but fails in maintaining the summer decay. After the spring bloom, the nitrogen concentration increases because of local remineralisation of ammonium and the eventual nitrification that leads to an accumulation of unutilized nitrate. The modeled time courses of surface water nitrate, ammonium, phosphate and silicate are generally in reasonable agreement with concentrations but the model overestimates ammonium concentrations.

It should, however, be stressed that we are comparing observations and model data from different years; inter annual variability of river run off might play an important role in causing parts of these differences (Radach and Patsch, 2007). Generally, the quality of the model results depends strongly on the availability and quality of the outer sources of nutrient (riverine inputs, input from the Baltic and from the North Atlantic).

Year-to-year changes in meteorological forcing can have profound impacts on ocean temperature, chemistry, vertical mixing and circulation, all of which impact the ocean carbon cycle (McKinley et al., 2004). Quantifying inter-annual variability and the mechanisms that control this variability are therefore a critical component of the estimation of

$CO_2$ fluxes on continental shelves and to understand the current and future role of continental shelves in the oceanic carbon sink. This can be achieved using biogeochemical models such as the one presented in this paper.

## 17.3.5 Vertical gradients

The vertical distributions of the model output and field data of the states variables for the southern station are presented in Fig. 17.5. The physical model simulates the establishment of the thermocline but mixing processes are not well represented as shown by the offset in temperature below 30 m depth (Fig. 17.5a). This is inherent to a 1-D model, because mixing processes and high turbulent kinetic energy in shallow seas are partly caused by advective processes (in particular tidal currents) and relative shear stress. In the same southern region, Ruardij et al. (1997) shows a good validation of the 1D model for the data from 1994, which suggests that the situation encountered in the southern North Sea in summer 2001 may be an exceptional situation.

Fig. 17.5b shows the simulation of DIC profiles with an increase of values below the simulated thermocline. In general, the model is able to reasonably reproduce the *in situ*

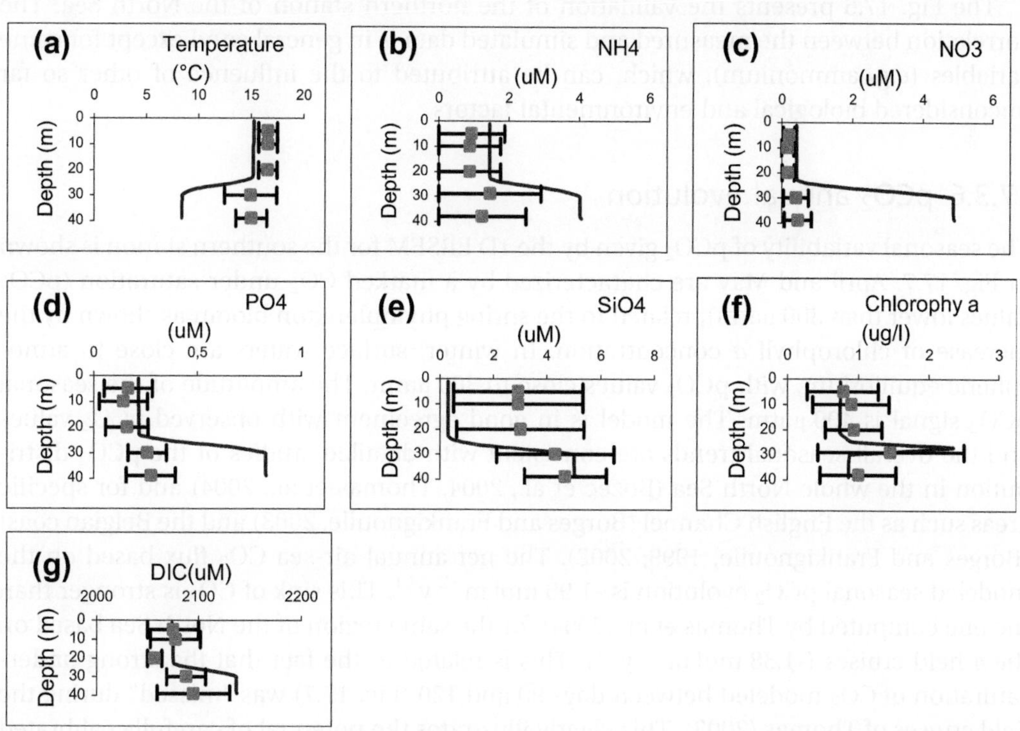


**FIGURE 17.5** Vertical distribution of temperature, DIC, chlorophyll *a*, nitrate, ammonium, phosphate and silicate for the southern station (simulation day 03.09.01). Field data (symbols) are averages from several stations. The horizontal bars represent the standard error on the means. The solid line represents the model values. For color version of this figure, the reader is referred to the online version of this book.

vertical structure of DIC. The largest difference between the simulations and field data, below 30 m depth can be related to the same cause as the offset described for temperature. The absence of stratification prevents export of organic matter from the mixed layer which means that the organic matter is remineralized in the whole water column not allowing a net biological $CO_2$ uptake after the productive period in contrast with the stations in the northern North Sea (Bozec et al., 2004). In the northern North Sea, organic carbon is efficiently exported from the mixed layer below the thermocline and after remineralization exported as DIC to the adjacent deep ocean (Bozec et al., 2004; Thomas et al., 2004).

We investigated the vertical distribution given by the model for other variables such as chlorophyll *a*, nitrate, ammonium, phosphate and silicate (Fig. 17.5). The model provides a very satisfactory simulation of the chlorophyll *a* profile (Fig. 17.5f). In general, the modeled vertical profiles of nutrients are similar to the observed vertical profiles. In the top 20 m, the model slightly over-estimates nitrate, ammonium and phosphate concentrations but under-estimates silicate concentrations. Below 20 m, the model systematically over-estimates all nutrient concentrations. However, the qualitative vertical patterns of nutrients and chlorophyll *a* concentrations are reasonably well simulated by the model.

The Fig. 17.6 presents the validation of the northern station of the North Sea. The correlation between the measured and simulated data is in general good except for some variables (e.g. ammonium), which, can be attributed to the influence of other so far unconsidered biological and environmental factors.

### 17.3.6 $pCO_2$ annual evolution

The seasonal variability of $pCO_2$ given by the 1D ERSEM for the southern station is shown in Fig. 17.7. April and May are characterized by a marked $CO_2$ under-saturation ($pCO_2$ values lower than 300 μatm), related to the spring phytoplankton bloom as shown by the increase of chlorophyll *a* concentration. In winter, surface waters are close to atmospheric equilibrium, with $pCO_2$ values close to 360 μatm. The amplitude of the seasonal $pCO_2$ signal is 400 μatm. The model is in good agreement with observed $pCO_2$ values and the overall seasonal trends are consistent with detailed studies of the $pCO_2$ distribution in the whole North Sea (Bozec et al., 2004, Thomas et al., 2004) and for specific areas such as the English Channel (Borges and Frankignoulle, 2003) and the Belgian coast (Borges and Frankignoulle, 1999; 2002). The net annual air-sea $CO_2$ flux based on the modeled seasonal $pCO_2$ evolution is -1.96 mol $m^{-2}$ $y^{-1}$. This sink of $CO_2$ is stronger than the one computed by Thomas et al. (2004) for the same region of the North Sea based on the 4 field cruises (-1.38 mol $m^{-2}$ $y^{-1}$). This is related to the fact that the strong under-saturation of $CO_2$ modeled between days 90 and 120 (Fig. 17.7) was "missed" during the field cruises of Thomas (2002). This clearly illustrates the potential of carefully calibrated models to provide high resolution flux estimates in coastal ecosystems that are characterized by intense but highly variable biogeochemical cycling that cannot always be fully apprehended by field work.

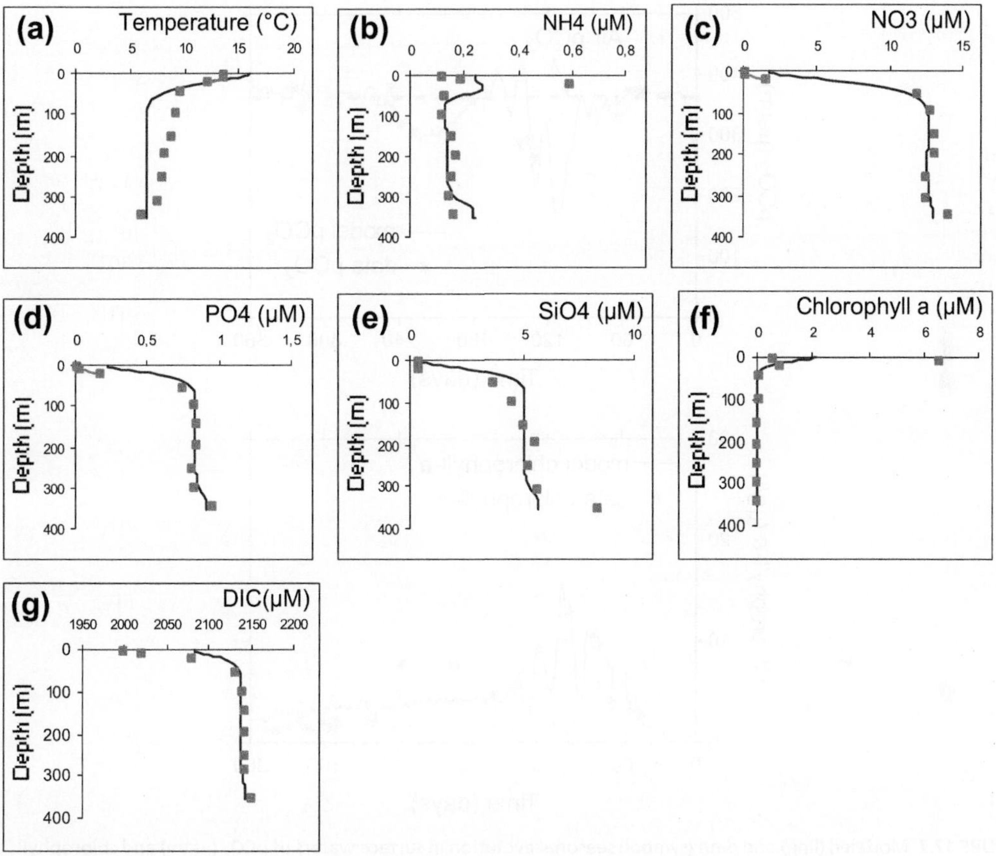


**FIGURE 17.6** Vertical distribution of temperature, DIC, chlorophyll a, nitrate, ammonium, phosphate and silicate for the northern station (simulation day 08.28.01). Field data are shown by symbols and the solid line represents the model values. For color version of this figure, the reader is referred to the online version of this book.

We note that the range of 180-240 days (Jun and July) there is the decrease of CO2 concentration, but there are no corresponding increase of chlorophyll-a, which is observed in real data and in the model simulation inside 60-180 days range. This difference between data and model results in the range of 180-240 days may be explained in our model by a phytoplankton sinking factor used during the summer period. So as an improvement of our model is to test the magnitude and the period of this sinking factor.

## 17.3.7 Sensitivity analysis to predicted future increases of atmospheric $pCO_2$

Atmospheric $pCO_2$ is predicted to continue to increase in the present century according to the scenarios published by IPCC (2007). At the southern station a sensitivity analysis

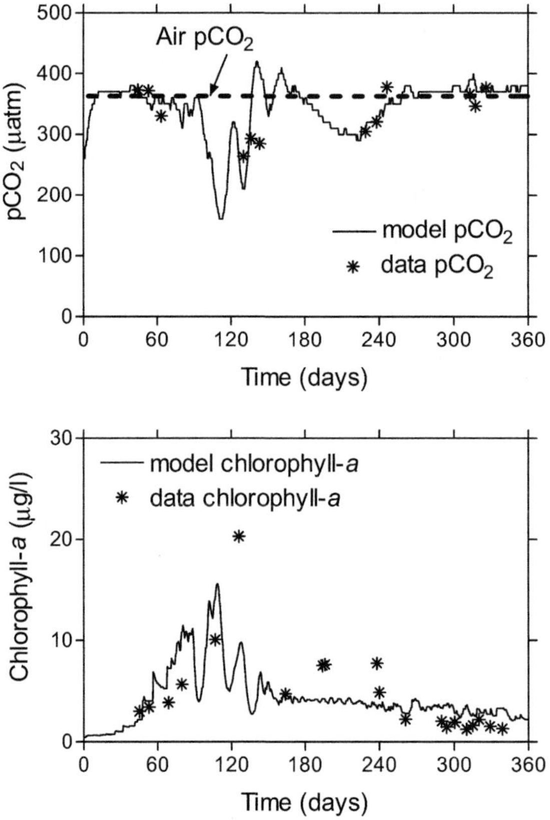


**FIGURE 17.7** Modeled (line) and data (symbol) seasonal evolution in surface waters of $pCO_2$ (µatm) and chlorophyll *a*. (µg $l^{-1}$; data from July 1996 to September 1999 from Borges and Frankignoulle, 2002)

was carried out by increasing the atmospheric $pCO_2$ values and testing the effect of this increase on the ecosystem using various scenarios values (365, 730, 1095, 1460 and 1825 µatm).

The chlorophyll *a* concentration and the biomass of different phytoplankton and zooplankton classes are presented in Fig. 17.8. In early spring, at each atmospheric $pCO_2$ levels, chlorophyll *a*, as well as the biomass of the picophytoplankton is similar for the three $pCO_2$ values (365, 1095 and 1825 µatm). So, the changes in time of blooming may be caused by increased predation on picophytoplankton by microzooplankton (Fig. 17.8c), which has higher biomass in winter and earlier spring. Because the factors that usually play a direct role in biomass changes as light and temperature do not change, and nutrient concentrations are even higher at elevated air $pCO_2$, in spring.

For the present day atmospheric $pCO_2$ level, the diatoms start growing in late April. But under elevated atmospheric $pCO_2$ levels diatoms start growing in early April. The

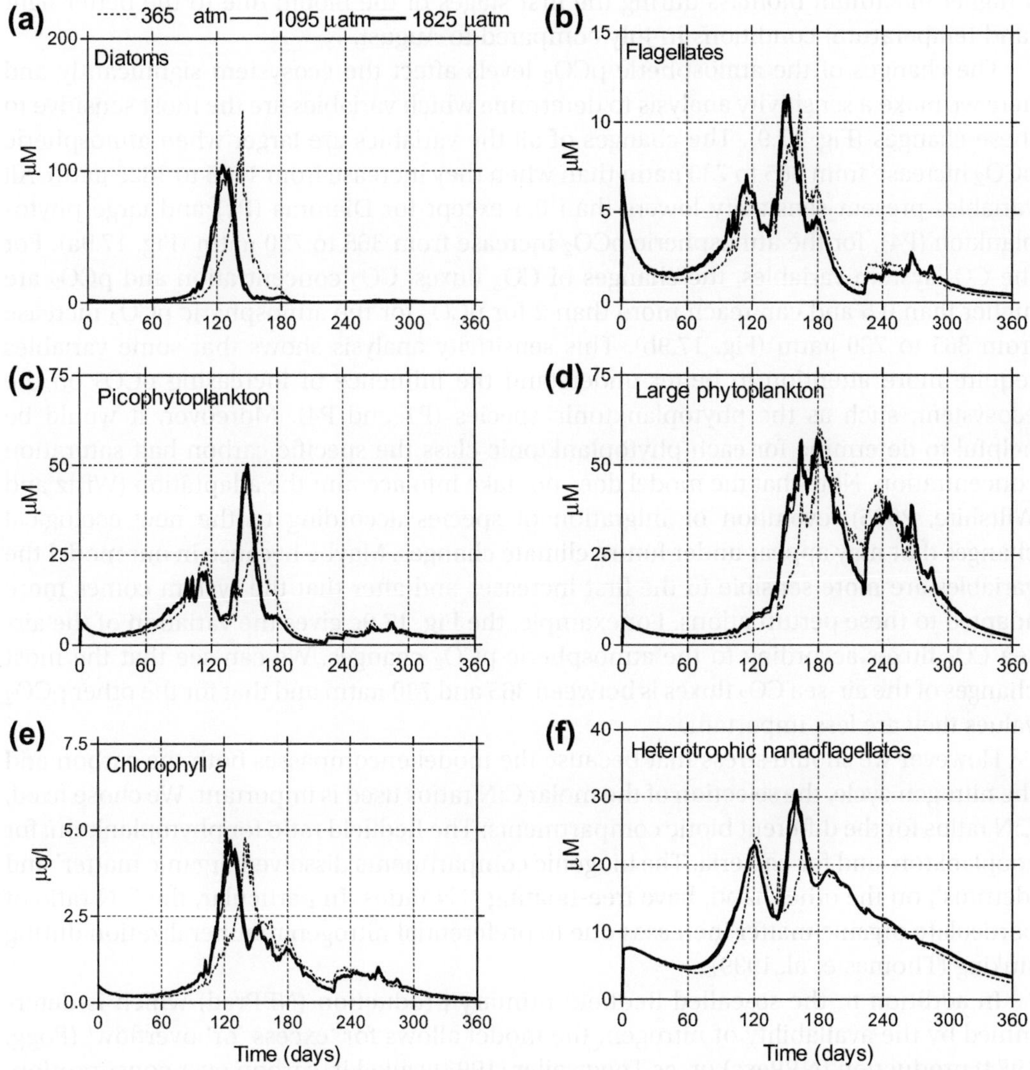


**FIGURE 17.8** Scenarios of atmospheric $pCO_2$ changes and their impact on diatoms, flagellates, picophytoplankton, large phytoplankton, chlorophyll *a* and heterotrophic nanoflagellates.

overall period of the bloom and the maximum biomass (production) reached by diatoms and flagellates are very similar between the runs with different air $pCO_2$ levels. This may be due to the zooplankton species that are feeding on diatoms and flagellates because also become abundant earlier in the year at elevated air $pCO_2$. In late summer, the large phytoplankton species start growing. Under elevated air $pCO_2$ values, the onset of the large phytoplankton bloom occurs earlier as they may take advantage of the earlier decline in diatom and nanoflagellates biomass. The large phytoplankton species reach a higher maximum biomass during the first stages of the bloom due to the better light (and temperature) conditions in July compared to August.

The changes of the atmospheric $pCO_2$ levels affect the ecosystem significantly and here we make a sensitivity analysis to determine which variables are the most sensitive to these changes (Fig. 17.9). The changes of all the variables are larger when atmospheric $pCO_2$ increase from 365 to 730 μatm than when they increase from 1425 to 1825 μatm. All variables present sensitivity lowers than 0.5 except for Diatoms (P1) and large phytoplankton (P4), for the atmospheric $pCO_2$ increase from 365 to 730 μatm (Fig. 17.9a). For the $CO_2$ system variables, the changes of $CO_2$ fluxes, $CO_2$ concentration and $pCO_2$ are higher than 0.5 and can reach more than 2 for $pCO_2$ for the atmospheric $pCO_2$ increase from 365 to 730 μatm (Fig. 17.9b). This sensitivity analysis shows that some variables require more attention to better understand the influence of increasing $pCO_2$ on the ecosystem, such as the phytoplanktonic species (P1 and P4). Moreover, it would be helpful to determine for each phytoplanktonic class the specific carbon half saturation concentration. Note that the model does not take into account the adaptation (Wirtz and Wiltshire, 2005), evolution or migration of species according to the new ecological changes that may appear under future climate changes. Maybe because in our model the variables are more sensible to the first increases and after that the system comes more adapted to these perturbations. For example, the Fig. 17.9c gives the variation of the air-sea $CO_2$ fluxes according to the atmospheric $pCO_2$ changes. We can see that the most changes of the air-sea $CO_2$ fluxes is between 365 and 730 μatm and that for the other $pCO_2$ values they are less important.

However we should stress that because the model encompasses both the carbon and the nitrogen cycle, the selection of the molar C:N ratios used is important. We chose fixed, C:N ratios for the different biotic compartments: The Redfield ratio for phytoplankton, for zooplankton, and for bacteria. The biogenic compartments dissolved organic matter' and 'detritus', on the other hand, have free-floating C:N ratios. In particular, the C:N ratio of particulate organic matter increases due to preferential nitrogen remineralization during sinking (Thomas et al.,1999).

In addition to the so-called Redfield primary production (NPPred) which is determined by the availability of nitrogen, the model allows for 'excess 'or' overflow' (Fogg, 1983) production (NPPexc) or, as Toggweiler (1993) called it, carbon over consumption, defined as carbon fixation by photosynthesis during periods when surface layers are depleted in bio available nitrogen (Thomas et al., 1999). This additionally fixed carbon is released in the form of dissolved or colloidal extracellular carbohydrates which tend to coagulate forming transparent exopolymer particles (TEP) (Schartau et al., 2007). In the model the excess carbon is immediately channeled into the pool of slowly degradable semi-labile dissolved organic carbon, which is eventually metabolized by the bacteria. Thus, the model differentiates between 'normal' exudation by phytoplankton, the result of which is labiled is solved organic matter with Redfield composition (Fogg, 1983), and an excess exudation of semi labile organic carbon with higher C:N ratios.

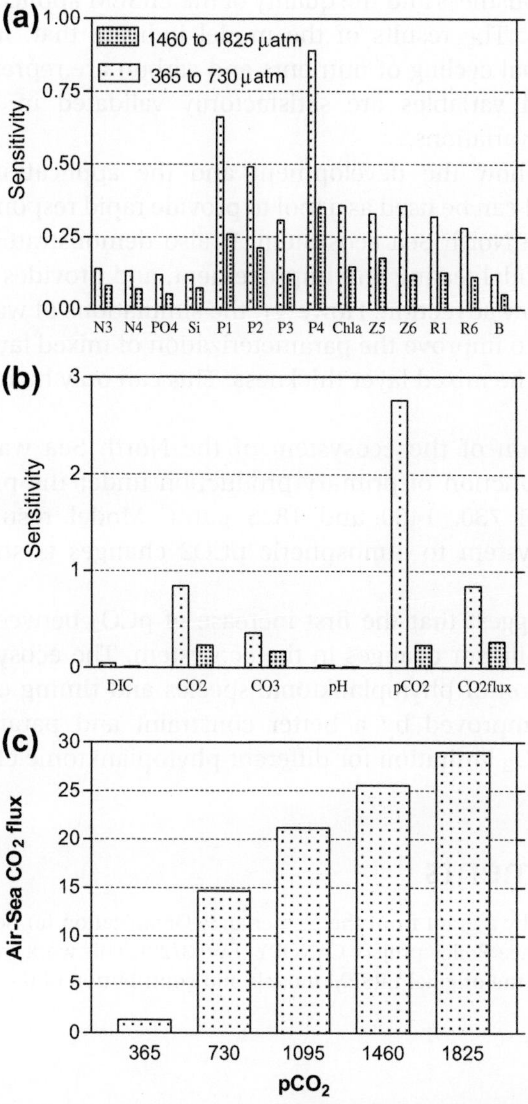


**FIGURE 17.9** Sensitivity analysis for some selected variables calculated when atmospheric $pCO_2$ is increased from 370 to 730 µatm and from 1460 to 1825 µatm. $CO_2$ air-sea flux in mmol m-2 y-1, a positif flux corresponds to a transfer of $CO_2$ from the atmosphere to the sea. N3 (nitrate), N4 (ammonium), P1 (Diatoms), P2 (Flagellates), P3 (Picophytoplankton), Z5 (Microzooplankton), Z6 (Heterotrophic Flagellates), R1 (Dissolved organic matter), B (Pelagic bacteria).

## 17.4 Conclusions

The complex biogeochemical 1-D ERSEM coupled physical model was applied to describe DIC dynamics in the southern and the northern North Sea with the incorporation of a $CO_2$ submodel that was validated using a recently obtained data set. This exercise proves the robustness and the quality of the ERSEM applicability to a large range of different situations. The results of the model indicate that, in general, the major controls on the seasonal cycling of nutrients and carbon are represented by the model, since most simulated variables are satisfactorily validated in the magnitude, and temporal and vertical variations.

This work shows how the development and the application of an ERSEM 1D biogeochemical model can be used as a tool to provide rapid response and knowledge on the functioning of the North Sea ecosystem. It also demonstrates the utility of a 1-D model as a tool for model testing and improvement, and provides acceptable results in a region of relatively low advection. However, the simulations of water column structure have shown the need to improve the parameterization of mixed layer dynamics in order to correctly hind cast the mixed layer thickness. This can only be resolved using the 3-D set-up of ERSEM.

The future evolution of the ecosystem of the North Sea was investigated using a carbon-limitation function of primary production under the projected increases of atmospheric pCO2 of 730, 1460 and 1825 μatm. Model results suggest that the response of the ecosystem to atmospheric pCO2 changes is strongly controlled by C-limitation.

The results also suggest that the first increase of $pCO_2$ between 365 and 730 μatm induces the most significant changes in the ecosystem. The ecosystem would react by changing the succession of phytoplanktonic species and timing of the blooms. Model simulations can be improved by a better constraint and parameterization of half-saturation values of $CO_2$ limitation for different phytoplanktonic classes.

# Acknowledgements

This work was supported by a grant from the Netherlands Organization for Scientific Research (NOW) and Belgium Federal Science Policy project CANOPY (EV/03/20). This work would not be possible also without the foundation of the original ERSEM models and contribution of the many scientists involved in the ERSEM program.

# References

Allen, J.I., Blackford, J.C., Radford, P.J., 1998. A 1-D vertically resolved modelling study of the ecosystem dynamics of the middle and southern Adriatic Sea. J. Mar. Sys 18, 265–286.

Allen, J.I., Blackford, J.C., Ashworth, M.I., Proctor, R., Holt, J.T., Siddorn, J.R., 2001. A highly spatially resolved ecosystem model for the north-west European continental shelf. Sarsia 86, 423–440.

Andersen, T., 1997. Pelagic nutrient cycles. Herbivores as source and sinks. In: Caldwell, R.M.M., Heldmaier, G., lange, O.L., Mooney, H.A., Shulze, E., Sommer, D.U. (Eds.), 1997. Ecological studies, analysis and synthesis, vol. 129. Springer,, Berlin, p. 280.

Andersen, V., Nival, P., 1989. Modelling of phytoplankton population dynamics in an enclosed water column. Journal of the Marin Biological Association of the united Kinggdom 69, 625–646.

Baretta, J.W., Ebenhöh, W., Ruardij, P., 1995. The European Regional Seas Ecosystem Model, a complex marine ecosystem model. Netherlands J. Sea Res. 33, 233–246.

Blackford, J.C., 1997. An analysis of benthic biological dynamics in a North Sea ecosystem model. J. Sea Res. 38, 213–230.

Blackford, J.C., Allen, J.I., Gilbert, F.J., 2004. Ecosystem dynamics at six contrasting sites: a generic modeling study. J. Mar. Sys 52, 191–215.

Borges, A.V., Frankignoulle, M., 1999. Daily and seasonal variations of the partial pressure of $CO_2$ in surface seawater along Belgian and southern Dutch coastal areas. J. Mar. Sys 19, 251–266.

Borges, A.V., Frankignoulle, M., 2002. Distribution and air-water exchange of carbon dioxide in the Sheldt plume off the Belgian coast. Biogeochemistry 59 (1-2), 41–67.

Borges, A.V., Frankignoulle, M., 2003. Distribution of surface carbon dioxide and air-sea exchange in the English Channel and adjacent areas. J. Geoph. Res. 108, 3140.

Bozec, Y., Thomas, H., Elkalay, K., de Baar, H.J.W., 2004. The continental shelf pump in the North Sea-evidence from summer observation. Mar. Chemist 93, 131–147.

Bozec, Y., Thomas, H., Schiettecatte, L.-S., Borges, A.V., Elkalay, K., de Baar, H.J.W., 2006. Assessment of the processes controlling the seasonal variations of dissolved inorganic carbon in the North Sea. Limnology and Oceanography 51 (6), 2746–2762.

Burkhardt, S., Amoroso, G., Riebesell, U., Sultemeryer, D., 2001. $CO_2$ and $HCO_3$ uptake in marine diatoms acclimated to different $CO_2$ concentrations. Limnol. Oceanogr 46 (6), 1378–1391.

Carlotti, F., Radach, G., 1996. Seasonal dynamics of phytoplankton and *Calanus finmarchicus* in the North Sea as revealed by a coupled one-dimensional model. Limnol. Oceanogr 41, 522–539.

Clark, D.R., Flynn, K.J., 2000. The relationship between the dissolved inorganic carbon concentration and growth rate in marine phytoplankton. Proc. R. Soc. Lond 267, 953–959.

Droop, M.R., 1973. Some thoughts on nutrient limitation in algae. J. Phycol 9, 264–272.

Ebenhöh, W., Kohlmeier, C., Radford, P.J., 1995. The benthic biological model. Netherlands J. Sea Res. 33, 423–452.

Ebenhöh, W., Baretta-Bekker, J.G., Baretta, J.W., 1997. The primary production module in the marine ecosystem model ERSEM II, with emphasis on the light forcing. J. Sea Res. 38, 173–193.

Elser, J.J., Hassett, R.P., 1994. A stoichiometric analysis of the zooplakton-phytoplankton interaction in marine and freshwater ecosystem. Nature 370, 211–213.

IPCC, 2007. Climate Change 2007: Synthesis Report, This underlying report, adopted section by section at IPCC Plenary XXVII (Valencia, Spain, 12-17 November 2007).

Fogg, G.E., 1983. The ecological significance of extracellular products of phytoplankton photosynthesis. Botanica Marina 26, 3–14.

Jorgensen, S.E., 1994. Fundamentals of ecological modeling. Elsevier Science, Amsterdam.

Kohlmeier, C., Ebenhöh, W., 2007. Modelling the ecosystem dynamics and nutrient cycling of the Spiekeroog back barrier system with a coupled Euler–Lagrange model on the base of ERSEM. Ecol. Model. vol. 202, 3–4 (10), 297–310.

Lenhart, H.J., Radach, G., Ruardij, P., 1997. The effects of river input on the ecosystem dynamics in the continental coastal zone of the North Sea using ERSEM. J. Sea Res. 38, 249–274.

Leslie, Aveytua-Alcázar, Victor, F., Camacho-Ibar, Alejandro, J., Souza, J.I., Allen, Ricardo Torres, 2008. Modelling Zostera marina and Ulva spp. in a coastal lagoon. Ecol. Model. 218, 3–4 (10), 354–366.

McKinley, G.A., Follows, M.J., Marshall, J., 2004. Mechanisms of air–sea CO2 flux variability in the equatorial Pacific and the North Atlantic. Global Biogeochemical Cycles 18, GB2011. http://dx.doi.org/10.1029/2003GB002179.

Pätsch, J., Radach, G., 1997. Long term simulation of the eutrophication of the North Sea. Temporal development of nutrients, chlorophyll and primary production in comparison to observations. J. Sea Res. 38, 275–310.

Radach, G., Pätsch, J., 2007. Variability of continental riverine fresh water and nutrient inputs into the North Sea for the years 1977–2000 and its consequences for the assessment of eutrophication. Estuaries and Coasts 30 (1), 66–81.

Radford, P.J., Blackford, J.C., 1993. Interdisciplinary methods for successful ecological simulation. In: McAleer, M., Jakeman, A. (Eds.), International Congress on modelling and simulation. Publ. Uni. Western Australia, pp. 139–144.

Ruardij, P., Van Haren, H., Ridderinkhof, H., 1997. The impact of thermal stratification on phytoplankton and nutrient dynamics in shelf seas, a model study. J. Sea Res. 38, 311–331.

Ruardij, P., Baretta, J.W., Baretta-Bekker, J.G., 1995. SESAME, Software Environment for Simulation and Analysis of Marine Ecosystems. Netherlands J. Sea Res. 33, 261–270.

Schartau, M., Engel, A., Schroter, J., Thoms, S., Volker, C., Wolf-Gladrow, D., 2007. Modelling carbon overconsumption and the formation of extracellular particulate organiccarbon. Biogeosciences 4, 422–454.

Schiettecatte, L.-S., Thomas, H., Bozec, Y., Borges, A.V., 2007. High temporal coverage of carbon dioxide measurements in the Southern Bight of the North Sea. Marine Chemistry 106, 161–173.

Thomas, H., Ittekot, V., Osterroht, Ch., Schneider, B., 1999. Preferential recycling of nutrients—the ocean's way to increase new production and to pass nutrient limitation? Limnology and Oceanography 44, 1999–2004.

Thomas, H., 2002. "shipboard report of the RV Pelagia cruies 64PE184, 64PE187, 64PE195". Royal Netherlands Institute for Sea Research, Den Burg, Texel, Netherlands. 2002.

Thomas, H., Pempkpwaik, J., Wulff, F., Nagel, K., 2003. Autotrophy, nitrogen accumulation and nitrogen limitation in the Baltic Sea, a paradox or a buffer for eutrophication? Geophysical Research Letters 30 (21), 2130. http://dx.doi.org/10.1029/2003GL017937.

Thomas, H., Bozec, Y., Elkalay, K., de Baar, H.J.W., 2004. Enhanced open ocean storage of $CO_2$ from shelf sea pumping. Science 304 (5673), 1005–1008.

Toggweiler, J.R., 1993. Carbon overconsumption. Nature 363, 210–211.

Tortell, P.D., Rau, G.H., Morel, F.M.M., 2000. Inorganic carbon acquisition in coastal Pacific phytoplankton communities. Limnol. Oceanogr 47 (7), 1485–1500.

Vichi, M., Zavatarelli, M., Pinardi, N., 1998a. Seasonal modulation of microbial-mediated carbon fluxes in the Northern Adriatic Sea. Fisheries Oceanography. GLOBEC Special Issue 7 (3/4), 182–190.

Vichi, M.N., Pinardi, M., Zavatarelli, G., Matteucci, M.Marcaccio, Bergamini, M.C., Frascari, F., 1998b. One-dimensional ecosystem model tests in the Po Prodelta Area (Northern Adriatic Sea). Environmental Model. Soft 13, 471–481.

Vichi, M., Ruardij, P., Baretta, J.W., 2004. Link or sink, a modeling interpretation of the Baltic Proper biogeochemistry. Biogeosciences 1, 79–100.

Weiss, R.F., 1974. Carbon dioxide in water and seawater: the solubility of a non-ideal gas. Mar. Chemi 2, 203–215.

Wanninkhof, R., McGillis, W.R., 1999. A cubic relationship between air–sea CO2 exchange and wind speed. Geophy. Rese. Letters 26 (13), 1889–1892.

Wirtz, K., Wiltshire, K., 2005. Long-term shifts in marine ecosystem functioning detected by inverse modeling of the Helgoland Roads times-series. J. Mar. Sys 56, 262–282.

PART 6

# Models of Ecosystems

# 18

# Cities as Ecosystems: Functional Similarities and the Quest for Sustainability

Antonio Bodini*,[1], Cristina Bondavalli*, Stefano Allesina[†]

*DEPARTMENT OF ENVIRONMENTAL SCIENCES, UNIVERSITY OF PARMA, VIALE USBERTI 11/A, 43100 PARMA, ITALY, [†]DEPARTMENT OF ECOLOGY & EVOLUTION, COMPUTATION INSTITUTE, THE UNIVERSITY OF CHICAGO, CHICAGO IL, USA, [1]BODINI A. DEPARTMENT OF ENVIRONMENTAL SCIENCES, UNIVERSITY OF PARMA, VIALE USBERTI 11/A, 43100 PARMA, ITALY. E-MAIL: ANTONIO.BODINI@UNIPR.IT

## 18.1 Cities and Ecosystems

The ecosystem concept, that was initially a fascinating metaphor has turned, in the last few years, into a reference model for the urban environment (Alberti et al., 2003; Gragson and Grove, 2006; Grimm et al., 2000; Pickett et al., 1997, 2001; Pickett et al., 2008). Two major projects, the Baltimore Ecosystem Study (BES) and the Central Arizona Phoenix (CAP), which are both part of the Long Term Ecological Research (LTER, http://www.lternet.edu) program, translated the ecosystem model into usable tools to investigate metropolitan areas as integrated systems (Grimm and Readman, 2004; Grove and Burch, 1997). The parallelism between urban systems and ecosystems has led scientists to compare city development and ecological succession, but what has come out (Decker et al., 2000) seems more a collection of sparse evidences rather than a systematic classification of differences and similarities. A major consequence is that a coherent synthesis that clarifies to what extent urban system shares the features of the ecosystem has not been derived yet.

As a contribution in this direction we propose a shift in focus—from the complexity of the multiple interactions and their complicated feedbacks, which characterizes the ecosystem model (Grimm et al., 2008; Hope et al., 2003; Kinzig et al., 2005; Pickett and Grove, 2009; Roach et al., 2008), to the complex metabolism from which growth and development emerge as holistic attributes. This is possible through the use of network analysis that has been extensively used to explore ecosystems. By analyzing the structure and topology of ecological networks, in particular, Ulanowicz (1986, 1997) derived a suite of indices to quantify ecosystem growth and development; by these metrics he also contributed to substantiating the process of ecological

Models of the Ecological Hierarchy. DOI: http://dx.doi.org/10.1016/B978-0-444-59396-2.00018-3

ISSN 0167-8892, 

succession (Ulanowicz, 1980; Weber et al., 1989). The use of ecological flow networks guarantees the congruity of the methods in respect to the objective—we need to apply the same tools used for the ecosystem study so that the outcomes of the analysis of urban systems could be properly interpreted from an ecological point of view. In this way similarities between urban systems and ecosystems could be revealed.

Three urban ecosystems are investigated as water flow networks. If one is limited to data on a single medium, one would choose the medium that is limiting to most processes. Usually this medium is energy or carbon. In economics it is often currency. But projections for the near future show that scarcity of water will have more of a limiting effect on human activities than either energy or capital. This will be especially true in urban systems where most human activities concentrate.

Studying how cities grow and develop offers further opportunities to address sustainability because these two features encompass what this trait is all about. We address this question by contrasting the flow networks of the three selected urban systems with alternative flow configurations conceived on the basis of actions targeted to improve sustainability.

With this chapter we hope to contribute to an understanding of (a) to what extent urban expansion is similar to ecosystem development; and (b) what the implications of framing sustainability in a whole ecosystem perspective are while current practices for implementation are piecemeal.

## 18.2 Cities as Water Flow Networks

Three urban settlements that are located in the Emilia Romagna Region (Northern Italy) are the focus of this case study. They have been selected to represent different levels of urbanization, which can also be thought of as stages of a hypothetical urban developmental trajectory. The city called Albareto (E1, from now on) is a little mountain city with less than 2000 inhabitants. It has maintained a primarily rural character—its residential portion is fragmented into small settlements dispersed in the natural landscape, which clearly predominates over the urbanized environment. No significant expansion, both residential and productive, has occurred in the last few decades. Industry is not present at all and the main economic activities are agriculture, commerce, and tourism.

Sarmato (E2), too, is rather small. Although it is comparable with E1 as regards population size (2589 inhabitants), this city, which was formerly a typical rural settlement of the Po river plain, has undergone a fast urban growth, with rapid industrial expansion mainly due to its closeness to the capital of the province. Its industrial area, more than its residential portion, and the level of infrastructures (i.e., road network) define it as a city in transition from a rural configuration, well integrated in the surrounding habitat, to a new expansion that is becoming progressively separated from the natural context and which predominates over it.

Finally, the city of Ravenna (E3) stands at the other extreme of the urbanization gradient. It is a provincial capital and an old, extended city. It has continuously developed over the course of the years and it underwent an urban explosion, like many other towns in Italy, after the World War II until the 1980s. Although it has not stopped growing, its residential and industrial core areas are now well consolidated and the city can be considered as having a quasi definitive shape. The different degrees of urban expansion of the three settlements can be appreciated from the aerial maps that are reported in the images Figs 18.1 and 18.2

Considering that cities develop while they grow larger (Alberti et al., 2003; Batty, 2008; Blank and Solomon, 2003; Gabaix, 1999), E1, E2, and E3 can be viewed as three sequential steps of an urban developmental trajectory. E1, with its built up environment of limited extension and its rather simple urban landscape, can represent an initial phase in this developmental process. Rapid urbanization with new edges being developed in the periphery accompanied by a progressively increasing density of the built up environment characterizes E2. As a more composite configuration than E1 it can be viewed as a successive step in the urban developmental trajectory. Finally, E3 shows the most complex characterization as regards size and shape, transportation network, and internal processes (i.e., complexity of the productive system, transformation of the natural landscape). If the three systems under investigation represent different stages of urban evolution, certainly E3 would be a later, more mature configuration.

The main sectors of human activities, namely, industry, agriculture, tertiary sector (including families and commercial activity), and public services (such as schools, hospitals, sport facilities), were designated as the components of these systems; water consumption data were gathered from public reports or information provided by private enterprises. These data served to design water flows. For every system, we analyzed two networks—one basic setting and an alternative scheme in which flows were modified according to actions conceived to foster sustainability. Considering sustainability as "a global process of development that minimizes the use of natural resources and reduces the impacts on the environment" (WECD, 1987), one basic prescription is that resource consumption must be reduced. Accordingly, alternative networks for the three systems were conceived on the basis of very simple straightforward criteria, such as eliminating leakage from pipelines and elongating pathways of water use so that the same resource is shared by more compartments before it is discharged. The networks are annual snapshots of the water flows in the three cities. They are represented in Figs 18.3 to 18.5.

## 18.3 Network Analysis and System Level Indices

Network analysis comprises several techniques for the systematic analysis of ecological flow networks. The four primary methods are assembled into a single software package, WAND (Allesina and Bondavalli, 2004), that includes input–output analysis, trophic analysis, cycling analysis, and the calculation of whole system indices. In what follows, an

**FIGURE 18.1** Images representing the aerial maps of systems E1 (Albareto) and E2 (Sarmato). For color version of this figure, the reader is referred to the online version of this book.

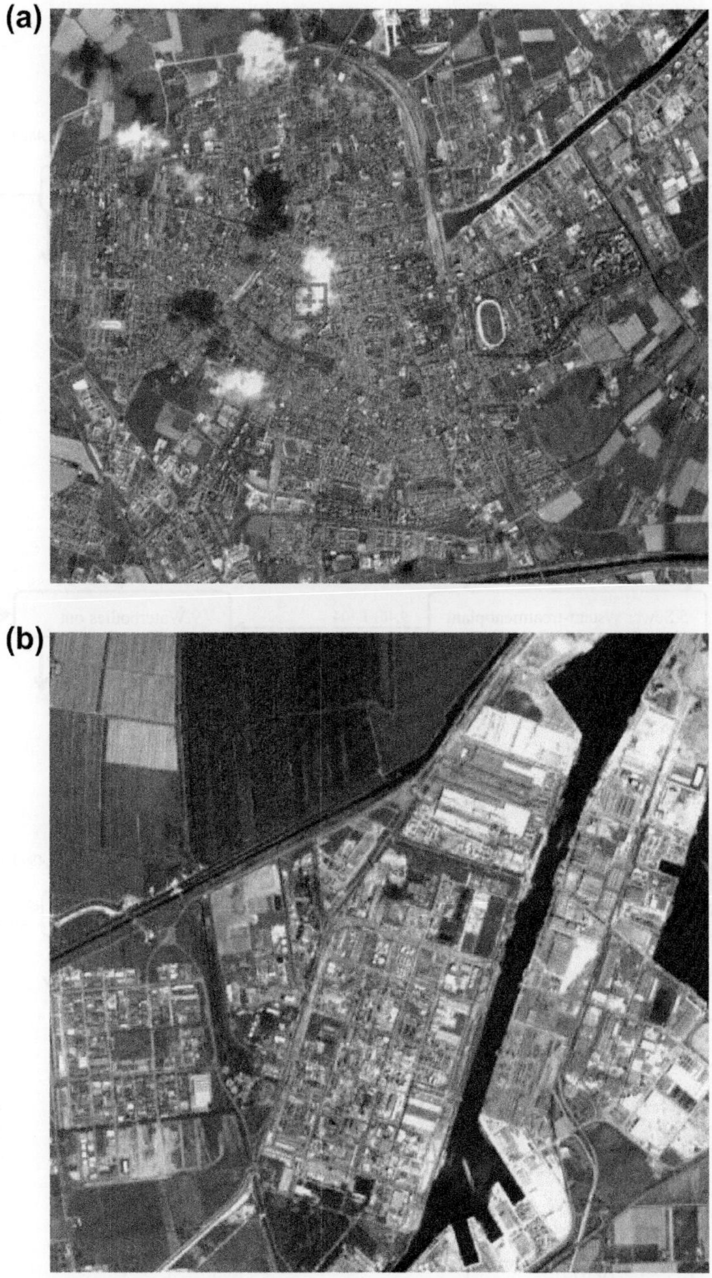


**FIGURE 18.2** Images representing the aerial map of system E3 (Ravenna): (a) The residential core of the city, (b) The industrial part of the city. For color version of this figure, the reader is referred to the online version of this book.

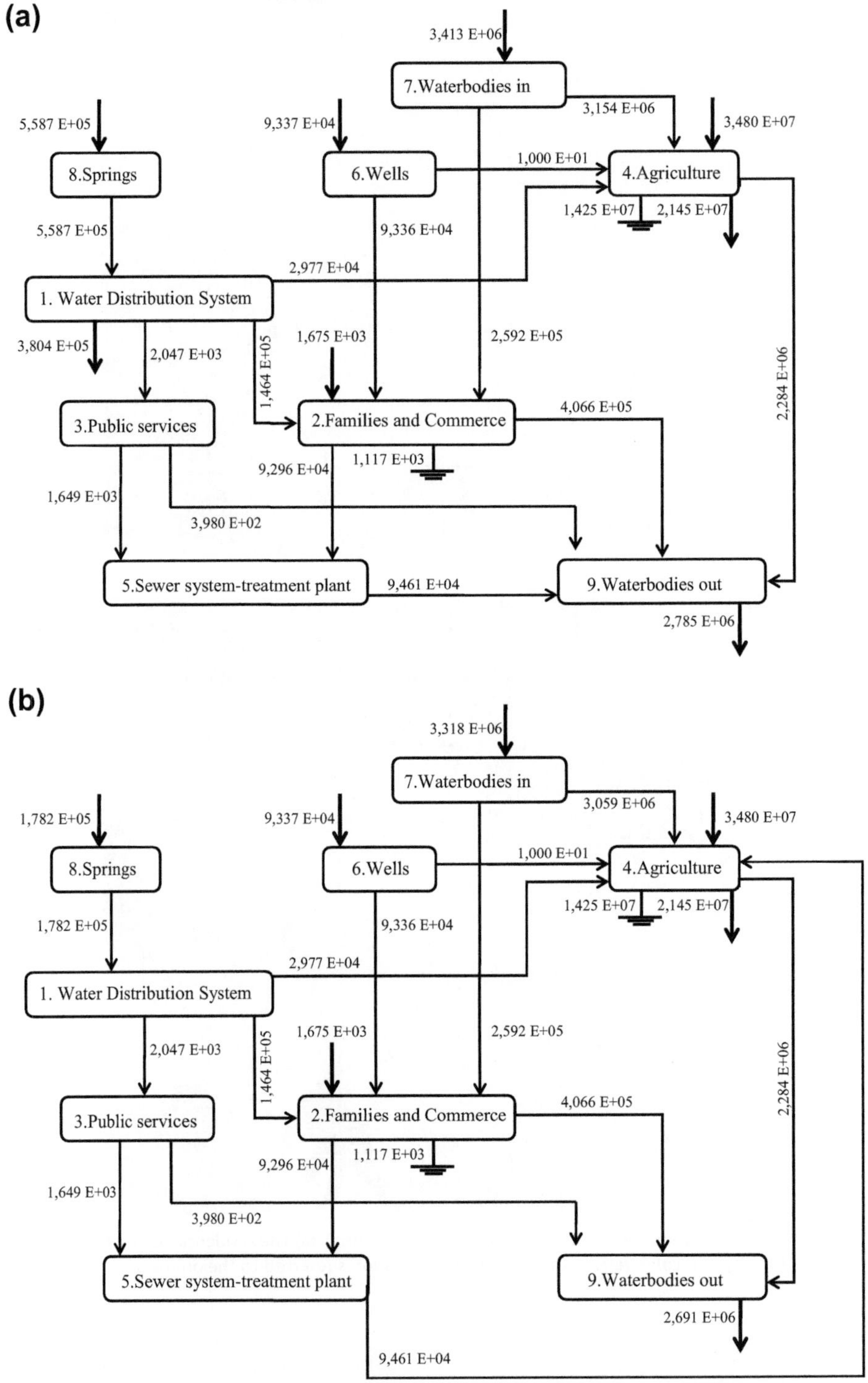


**FIGURE 18.3** The urban system E1 represented as water flow networks. (a) Present configuration E1 and (b) modified network. Details about the types of flows are in Ulanowicz (1986, 1997, 2004). Flow values are in $m^3$ $year^{-1}$.

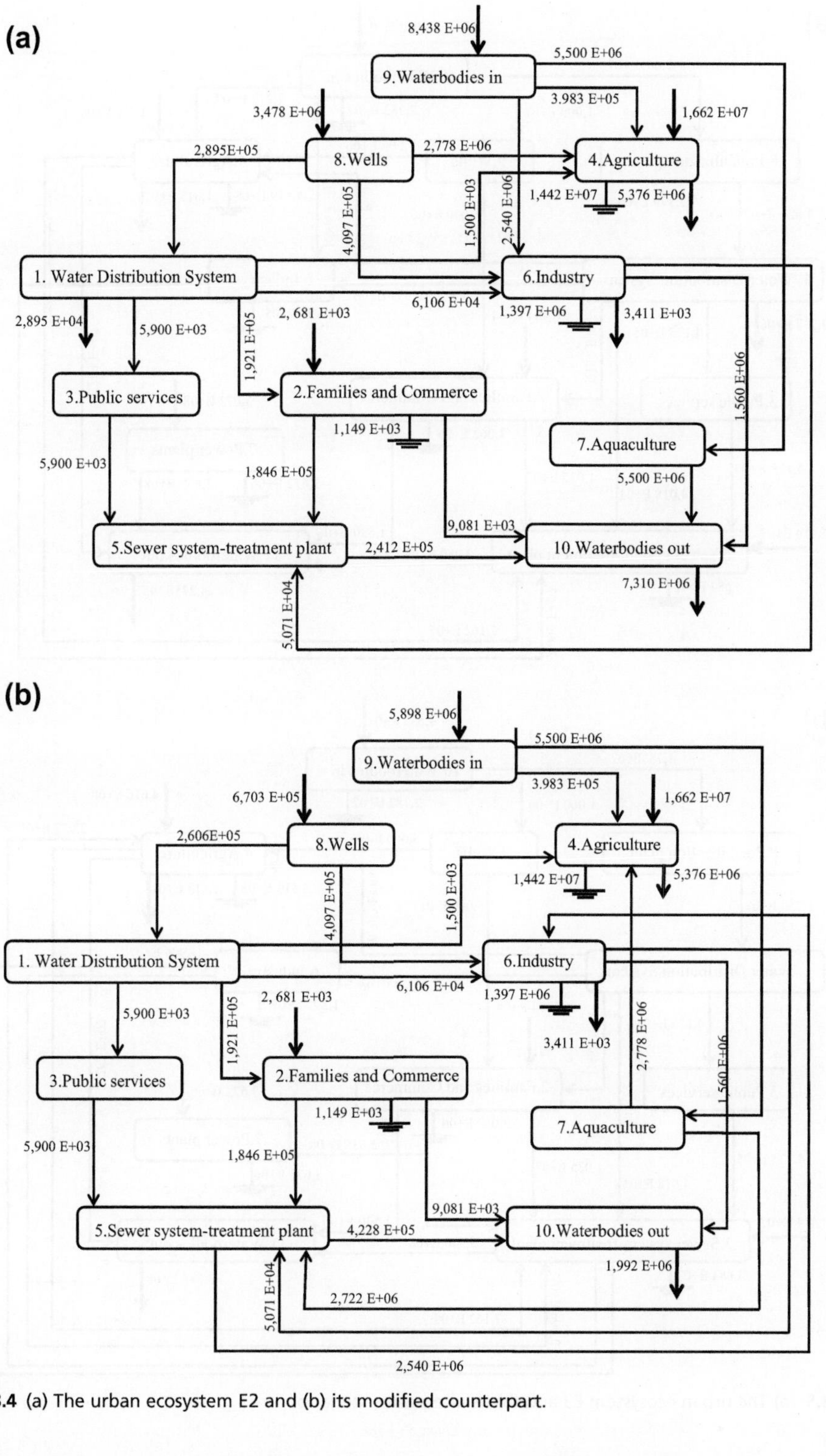


**FIGURE 18.4** (a) The urban ecosystem E2 and (b) its modified counterpart.

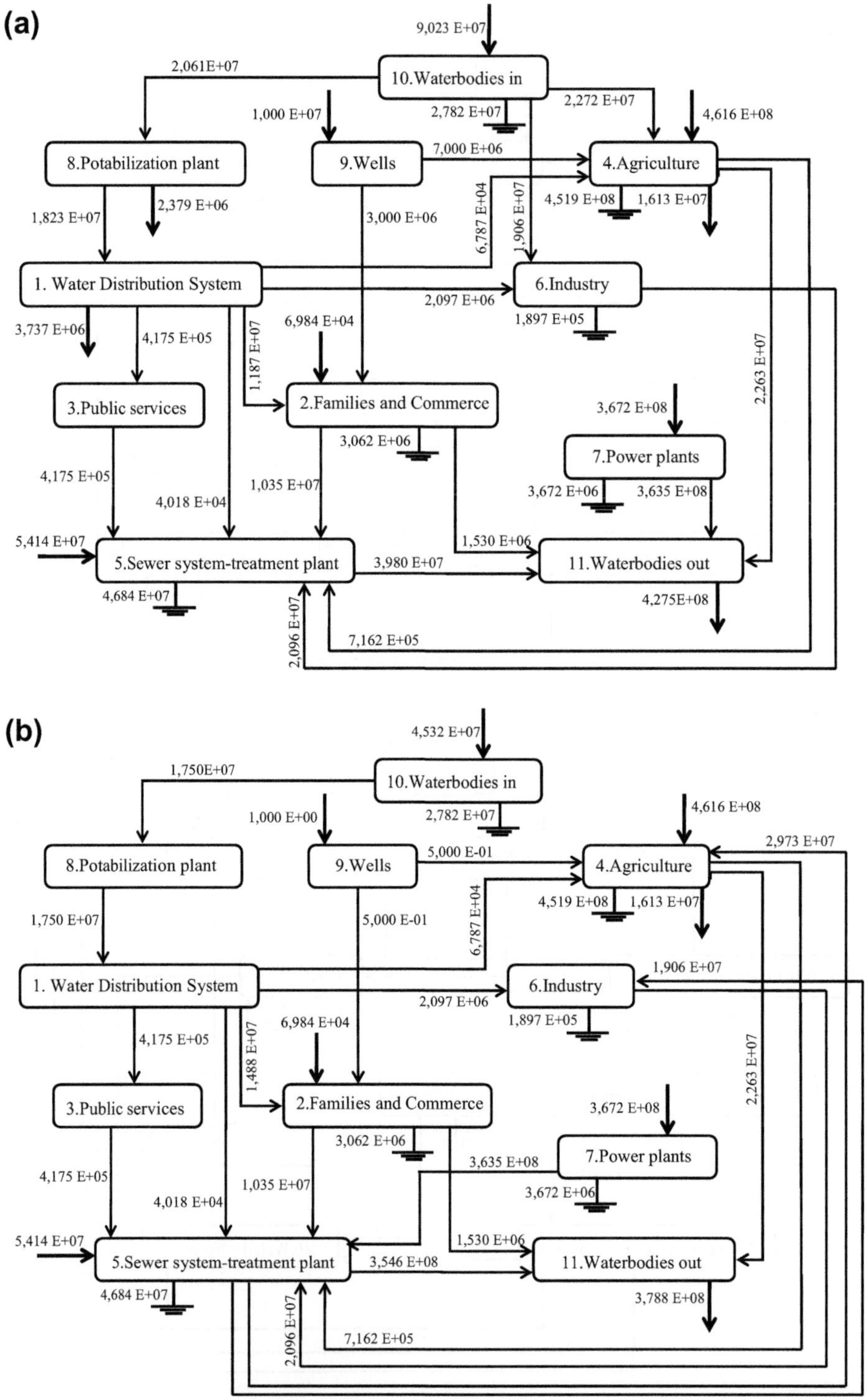


**FIGURE 18.5** (a) The urban ecosystem E3 and (b) its alternative scenario.

essential sketch of the algorithms for system level indices is provided while a more detailed description can be found in various pieces of specialized literature (Ulanowicz, 1986, 2004).

The Total System Throughput (TST) is the sum of all flows in an ecosystem. It quantifies how much medium (in this case water) is processed by a system and, as such, it is a measure of its size and it quantifies the level of activity of an ecosystem, exactly the same way as the sum of monetary flows (GNP) estimates the overall economic activity of a nation. (Ulanowicz, 1986).

$$\mathrm{TST} = \sum_{i=0}^{N+2}\sum_{j=0}^{N+2} T_{ij} \tag{1}$$

The term $T_{ij}$ identifies a flow from compartment *i* to compartment *j*. Summations are extended to include the external environment as: (a) source of input (compartment 0) for the system; (b) receiver of usable medium (compartment N + 1) from the system; and (c) sink of medium (compartment N + 2) dissipated by the system. Labels 1,2,......N identify the system's compartments. Both the number of compartments and the magnitude of flows contribute to TST which is a quantitative measure for growth. TST is extensive in nature. It pertains to the "size" of a system but it does not provide details about how the currency is distributed within the system.

The average mutual information (AMI) quantifies the amount of flow diversity that is encumbered by structural constraints and estimates how orderly and coherently flows are connected. AMI depends on flows in the following way:

$$\mathrm{AMI} = k\sum_{i=1}^{N+2}\sum_{j=1}^{N+2} \frac{T_{ij}}{T_{..}} \log_2\left(\frac{T_{ij}T_{..}}{T_{i.}T_{.j}}\right) \tag{2}$$

High values for AMI identify structures that are tightly constrained in respect to the movement of the currency. The information hidden in this index pertains to the fate of the unit of medium that circulates in a network. In complete constrained structures, knowing that a particle is leaving one compartment, automatically informs us that it is entering another given compartment. So the information is complete and no uncertainty exists. These structures are also said to be highly organized. Highly redundant networks instead are said to be less organized. They possess lower AMI values. Since development goes in the direction of pruning away the less efficient connections (Ulanowicz, 1997), AMI quantifies ecosystem development.

Ascendency (ASC) is defined as the product of the AMI and the TST (Tribus and McIrvine, 1971; Ulanowicz, 1986).

$$\mathrm{ASC} = \mathrm{AMI}\cdot\mathrm{TST} = T_{..}\sum_{i=1}^{N+2}\sum_{j=1}^{N+2} \frac{T_{ij}}{T_{..}} \log_2\left(\frac{T_{ij}T_{..}}{T_{i.}T_{.j}}\right) = \sum_{i=1}^{N+2}\sum_{j=1}^{N+2} T_{ij} \log_2\left(\frac{T_{ij}T_{..}}{T_{i.}T_{.j}}\right) \tag{3}$$

It measures the fraction of medium that an ecosystem distributes in an efficient way (i.e., along efficient routes); also, by combining ecosystem activity and organization, it provides a single measure of ecosystem growth and development.

The overall complexity of the flow structure is completed by another coefficient that quantifies the disorder and freedom that is preserved:

$$\mathrm{H} = -k\sum_{i=1}^{N+2}\sum_{j=1}^{N+2}\frac{T_{ij}}{T_{..}}\log_2\left(\frac{T_{ij}^2}{T_{i.}T_{.j}}\right) \tag{4}$$

Scaling H by the TST, we obtain the development capacity (DC) of the system:

$$\mathrm{DC} = T_{..}\sum_{i=1}^{N+2}\sum_{j=1}^{N+2}\frac{T_{ij}}{T_{..}}\log_2\left(\frac{T_{ij}}{T_{..}}\right) = \sum_{i=1}^{N+2}\sum_{j=1}^{N+2}T_{ij}\log_2\left(\frac{T_{ij}}{T_{..}}\right) \tag{5}$$

DC quantifies the maximum potential that a system has at its disposal to achieve further development, intended at improving organization of flows so that a higher quantity of currency is handled efficiently. DC serves as an upper boundary for ecosystem organization (ascendency).

The index (H) of residual diversity of flows can be scaled by TST, yielding the so-called Total System Overhead (TSO), which pertains to redundant flows. We define the overhead as:

$$\begin{aligned}\mathrm{TSO} = \mathrm{DC} - \mathrm{ASC} &= \sum_{i=1}^{N+2}\sum_{j=1}^{N+2}T_{ij}\log_2\left(\frac{T_{ij}}{T_{..}}\right) - \sum_{i=1}^{N+2}\sum_{j=1}^{N+2}T_{ij}\log_2\left(\frac{T_{ij}T_{..}}{T_{i.}T_{.j}}\right)\\ &= -\sum_{i=1}^{N+2}\sum_{j=1}^{N+2}T_{ij}\log_2\left(\frac{T_{ij}^2}{T_{i.}T_{.j}}\right)\end{aligned} \tag{6}$$

It is also made of different contributions: overhead on imports ($\mathrm{OI} = -\sum_{j=1}^{N} T_{0j}\log\left(\frac{T_{0j}^2}{T_0 T_j}\right)$), on exports ($\mathrm{OE} = -\sum_{j=1}^{N} T_{i,n+1}\log\left(\frac{T_{i,N+1}^2}{T_{i.}T_{.,\ N+1}}\right)$), dissipative overhead ($\mathrm{DO} = -\sum_{j=1}^{N} T_{i,n+2}\log\left(\frac{T_{i,N+2}^2}{T_{i.}T_{.,\ N+2}}\right)$), and redundancy ($\mathrm{RED} = -\sum_{i=1}^{N}\sum_{j=1}^{N} T_{ij}\log\left(\frac{T_{ij}^2}{T_{i.}T_{.j}}\right)$) of internal exchanges. While dissipative overhead is naturally related to what any ecosystem dissipates through internal processes, the other terms are strongly tied to the effective multiplicity of parallel flows by which energy is imported, exported as usable medium, or exchanged between the compartments.

Summarizing, the total system throughput quantifies the entire amount of currency that the system handles. Ascendency measures the fraction of this currency that is exchanged efficiently through optimized connections. What remains is the encumbered complexity or overhead, that is, currency dissipated or channeled through redundant connections, either internally or as import and export flows. The summation of ascendency and overhead defines the development capacity, that is, the potential of a system to become a completely organized whole. The overhead and its component parts have conflicting interpretations. At a first glance it represents the system's inefficiencies at processing material and energy; on the other hand it is "strength-in-reserve" of degrees of freedom which the system can call upon to adapt to a new threat (flexibility) (Ulanowicz, 1997). Ulanowicz (1997) stated that the interplay between efficiency (specialization) and protection (redundancy) is crucial for maintaining ecosystem health and the two conflicting functions are both necessary to make the system thrive.

Because any network provides a single value for each index, network comparison on the basis of system level indices is not possible due to the lack of a sufficient sample size needed to make any statistical analysis.

To overcome this problem we performed a sensitivity analysis on flow matrices to obtain distributions for some of the indices that quantify ecosystem status. We imposed random variations to flow values according to an estimated range of variability for yearly water consumption obtained from ample water consumption data sets (from year 2000). Results from this meta-analysis revealed an average variation between years of about 20%. An interval of ±20% for each and every flow was thus considered and as many as 5000 networks were built for the six urban systems by randomly changing the flow values within this interval. The general algorithm used can be summarized by the formula

$$M[i,j] = M[i,j] + (\text{uniform}[-C, C])M[i,j] \tag{7}$$

in which $M[i,j]$ is the original coefficient (flow value) and $[-C,C]$ is its interval of variability (as a fraction of the original value), and uniform is a random number generator that extracts values from a uniform distribution. The simulations yielded distributions for TST, DC, AMI, and TSO (18 distributions overall, six for each index). The results were not significantly different to a Gaussian by the Shapiro–Wilks and the Kolmogorov–Smirnov tests.

## 18.4 Patterns of Growth and Development in City Networks

The values obtained for system level indices are organized in groups and presented in Table 18.1. Their values are given as absolute coefficients (E1, E2, E3, columns on the left); percentage of the Development Capacity (central columns); and intensive components, the part of any index that is inherent to the diversity of flows only (columns on the right). For all and every index, the values obtained from present and modified networks are given.

Differences between indices (between sites and networks) were tested using the T statistics. Table 18.2 synthesizes the result of this analysis. All the pair-wise comparisons produced significant differences that substantiate what follows.

In the present urban networks and their modified counterparts all indices progressively increase from E1 to E3 except TST, where E2 < E1 < E3. E1 is located in a mountain area characterized by a high precipitation level, a diffused presence of springs, and being

**Table 18.1** System level indices for the three systems in their present and modified network configuration. Values are given as absolute coefficients (left), percentage of the development capacity (center) and intensive components, the fractions pertaining to flow topology only (right), which are obtained by dividing the indices by the TST, their scaling factor.

| | | | | | | | | | | | |
|---|---|---|---|---|---|---|---|---|---|---|---|
| TST | *E1* | *E2* | *E3* | | | | | | | | |
| Present | 8.4847E + 07 | 7.6806E + 07 | 2.5319E + 09 | | | | | | | | |
| Modified | 8.3422E + 07 | 6.3486E + 07 | 2.7330E + 09 | | | | | | | | |
| DC | *E1* | *E2* | *E3* | | | | | DC/TST | *E1* | *E2* | *E3* |
| Present | 2.0225E + 08 | 2.5945E + 08 | 8.3253E + 09 | | | | | Present | 2.3837E + 00 | 3.3780E + 00 | 3.2882E + 00 |
| Modified | 1.9136E + 08 | 2.0455E + 08 | 8.9690E + 09 | | | | | modified | 2.2939E + 00 | 3.2220E + 00 | 3.2817E + 00 |
| ASC | *E1* | *E2* | *E3* | ASC% | *E1* | *E2* | *E3* | AMI | *E1* | *E2* | *E3* |
| Present | 9.6699E + 07 | 1.4442E + 08 | 4.8424E + 09 | present | 48 | 56 | 58 | present | 1.1397E + 00 | 1.8803E + 00 | 1.9126E + 00 |
| Modified | 9.2358E + 07 | 1.1850E + 08 | 5.4774E + 09 | modified | 48 | 58 | 61 | modified | 1.1071E + 00 | 1.8666E + 00 | 2.0042E + 00 |
| TSO | *E1* | *E2* | *E3* | TSO% | *E1* | *E2* | *E3* | TSO/TST | *E1* | *E2* | *E3* |
| Present | 1.0555E + 08 | 1.1503E + 08 | 3.4830E + 09 | present | 52 | 44 | 42 | present | 1.2440E + 00 | 1.4977E + 00 | 1.3756E + 00 |
| Modified | 9.9006E + 07 | 8.6050E + 07 | 3.4916E + 09 | modified | 52 | 42 | 39 | modified | 1.1868E + 00 | 1.3554E + 00 | 1.2776E + 00 |
| OI | *E1* | *E2* | *E3* | OI% | *E1* | *E2* | *E3* | OI/TST | *E1* | *E2* | *E3* |
| Present | 2.6189E + 07 | 4.2606E + 07 | 1.7090E + 09 | present | 13 | 16 | 21 | present | 3.0866E−01 | 5.5472E−01 | 6.7499E−01 |
| Modified | 2.3276E + 07 | 2.7314E + 07 | 1.5848E + 09 | modified | 12 | 13 | 18 | modified | 2.7902E−01 | 4.3024E−01 | 5.7988E−01 |
| OE | *E1* | *E2* | *E3* | OE% | *E1* | *E2* | *E3* | OE/TST | *E1* | *E2* | *E3* |
| Present | 3.3195E + 07 | 2.3054E + 07 | 2.4806E + 08 | present | 16 | 9 | 3 | present | 3.9123E − 01 | 3.0016E − 01 | 9.7974E − 02 |
| Modified | 2.9854E + 07 | 1.6390E + 07 | 1.7674E + 08 | modified | 16 | 8 | 2 | modified | 3.5787E−01 | 2.5817E−01 | 6.4669E−02 |
| DO | *E1* | *E2* | *E3* | DO% | *E1* | *E2* | *E3* | DO/TST | *E1* | *E2* | *E3* |
| Present | 2.0180E + 07 | 1.4981E + 07 | 6.1861E + 08 | present | 10 | 6 | 7 | present | 2.3784E−01 | 1.9505E−01 | 2.4433E−01 |
| Modified | 2.0180E + 07 | 1.4981E + 07 | 7.0233E + 08 | modified | 11 | 7 | 8 | modified | 2.4190E−01 | 2.3597E−01 | 2.5698E−01 |
| RED | *E1* | *E2* | *E3* | RED% | *E1* | *E2* | *E3* | RED/TST | *E1* | *E2* | *E3* |
| Present | 2.5983E + 07 | 3.4392E + 07 | 9.0731E + 08 | present | 13 | 13 | 11 | present | 3.0623E−01 | 4.4778E−01 | 3.5835E−01 |
| Modified | 2.5696E + 07 | 2.7365E + 07 | 1.0277E + 09 | modified | 13 | 13 | 11 | modified | 3.0802E−01 | 4.3104E−01 | 3.7603E−01 |

**Table 18.2** Summary of the T test applied to all pair wise comparisons of index distributions.

| TST | | E2 | E3 | E1M | E2M | E3M | TSO | | E2 | E3 | E1M | E2M | E3M |
|---|---|---|---|---|---|---|---|---|---|---|---|---|---|
| E1 | t= | 90.71 | -1371.099 | 13.41 | 254.24 | -1324.48 | E1 | t= | 60.57 | -1536.82 | 26.98 | 162.91 | -1290.59 |
| | df= | 9161 | 5015 | 9998 | 8339 | 5012 | | df= | 1315 | 5122 | 1410 | 12020 | 5073 |
| | p= | 0 | 1 | 6.06E-41 | 0 | 1 | | p | 0 | 1 | 6.4E-130 | 0 | 1 |
| E2 | t= | - | -1376.108 | -75.19 | 194.66 | -1328.90 | E2 | t= | - | -1552.33 | -60.63 | 212.34 | -1300.37 |
| | df= | - | 5008 | 9143 | 9730 | 5006 | | df= | - | 5018 | 9846 | 9579 | 5010 |
| | p | - | 1 | 1 | 0 | 1 | | p | - | 1 | 1 | 0 | 1 |
| E3 | t= | - | - | 1371.85 | 1383.74 | -74.58 | E3 | t= | - | - | 1547.93 | 1564.71 | -78.42 |
| | df= | - | - | 5015 | 5005 | 9873 | | df= | - | - | 5023 | 5011 | 9404 |
| | p | - | - | 0 | 0 | 1 | | p | - | - | 0 | 0 | 1 |
| E1M | t= | - | - | - | 237.42 | -1325.16 | E1M | t= | - | - | - | 261.99 | -1297.08 |
| | df= | - | - | - | 8320 | 5012 | | df= | - | - | - | 9043 | 5014 |
| | p | - | - | - | 0 | 1 | | p | - | - | - | 0 | 1 |
| E2M | t= | - | - | - | - | -1335.69 | E2M | t= | - | - | - | - | -1309.81 |
| | df= | - | - | - | - | 5004 | | df= | - | - | - | - | 5006 |
| | p | - | - | - | - | 1 | | p | - | - | - | - | 1 |
| DC | | E2 | E3 | E1M | E2M | E3M | AM1 | | E2 | E3 | E1M | E2M | E3M |
| E1 | t= | -263.84 | -1528.989 | 57.13 | -12.23 | -1399.10 | E1 | t= | -1717.71 | -2814.9 | 170.20 | -1554.12 | -2771.48 |
| | df= | 9474 | 5005 | 9991 | 9994 | 5004 | | df= | 6317 | 8525 | 9673 | 6100 | 7690 |
| | p | 1 | 1 | 0 | 1 | 1 | | p= | 1 | 1 | 0 | 1 | 1 |
| E2 | t= | - | -1517.919 | 316.24 | 255.20 | -1389.774 | E2 | t= | - | -68.29 | 1827.64 | 21.21 | -253.29 |
| | df= | - | 5009 | 9367 | 9391 | 5006 | | df= | - | 7939 | 2918 | 9913 | 8792 |
| | p | - | 1 | 0 | 0 | 1 | | p= | - | 1 | 0 | 5.54E-98 | 1 |
| E3 | t= | - | - | 1531.03 | 1528.58 | -78.51 | E3 | t= | - | - | 3077.00 | 89.04 | -256.68 |
| | df= | - | - | 5005 | 5005 | 9737 | | df= | - | - | 7634 | 7516 | 9714 |
| | p | - | - | 0 | 0 | 1 | | p= | - | - | 0 | 0 | 1 |
| E1M | t= | - | - | - | -70.1151 | -1400.83 | E1M | t= | - | - | - | -1650.03 | -2983.04 |
| | df= | - | - | - | 9998 | 5003 | | df= | - | - | - | 5764 | 6937 |
| | p | - | - | - | 1 | 1 | | p= | - | - | - | 1 | 1 |
| E2M | t= | - | - | - | - | -1398.75 | E2M | t= | - | - | - | - | -261.66 |
| | df= | - | - | - | - | 5003 | | df= | - | - | - | - | 8329 |
| | p | - | - | - | - | 1 | | p= | - | - | - | - | 1 |

rich in water bodies. This explains why this system possesses higher TST than E2. This excess throughput becomes less important when the systems are very different in the level of activity (size). That is why E3, whose urban part is much more extended than in the other two systems, handles more medium than E1 and has, consequently, a higher TST. The modifications proposed for the three networks do not change this trend although the TST is predicted to diminish in the new architectures for E1 and E2 while for E3 this index is expected to increase. Although in this latter system the volume of imported water diminishes, as shown by the overhead on import (OI), pathway elongation increases the number of flows so that TST in the alternative configuration is higher than in the present configuration.

The DC mirrors TST. Whenever there is growth, that is, an augmented TST, a greater potential for development inevitably follows (Ulanowicz, 1986, 1997). The intensive component of this index (DC/TST) quantifies at what level the diversity of a finite total flow can increase either through greater evenness among the individual flows or by finer partitioning into an ever larger number of constituent flows. The contribution of the flow diversity is sufficient to compensate for the unbalance due to the different values for TST between E1 and E2 (as this latter index is higher in E2 than E1) and DC shows a trend that parallels the gradient of urban expansion. To clarify the meaning of DC in respect to its intensive component it can be said that DC/TST quantifies how much potential a system has to improve its topology whereas the absolute coefficient pertains to the overall amount of currency that the system would potentially manage efficiently due to its topological improvement.

The level of development that a system reaches depends on how much of its potential becomes realized development and which fraction remains as overhead. ASC and TSO, respectively, quantify these two aspects. The former increases with urban expansion (passing from E1 to E3) in both the present and modified networks, although for E1 and E2 the new configurations show lower ascendency, whereas this index increases for E3 after network reshaping. TSO mirrors ASC both in respect to urban expansion and alternative network configurations. Ascendency and overhead are opposite in nature because the former appraises organized complexity by gauging how well the system is performing at processing medium (i.e., it quantifies the fraction of TST that is distributed by the ordered part of the network) whereas the latter is the disorganized residual, and quantifies the fraction of currency that moves along the less efficient portion of the network (i.e., redundant connections between compartments or multiple imports and exports). Accordingly, one would expect ascendency to increase at the expense of overhead, and the two indices to be negatively associated. This expected trend remains hidden because of the disproportionate effect that TST has upon the values of the two indices, so that they increase along the urban gradient both in the present and alternative networks. However, when expressed as a fraction of DC they track with the urban expansion in an opposite fashion and their complementary nature clearly emerges. Network modifications do not change these trends.

Once the TST is taken apart and only the intensive fraction of the indices, inherent to the diversity of flows, is considered, realized development and overhead are quantified by AMI and TSO/TST, respectively. The former follows urban expansion (i.e., increases from E1 to E3) whereas TSO/TST goes in the opposite direction. In E3 the new architecture is more organized (higher AMI) than the present network, whereas the opposite occurs in E1 and E2. As for TSO/TST the new networks for the three systems maintain the same trend shown by their present configurations. The level of the residual disorganized flow diversity in its intensive fraction (TSO/TST) mirrors the intensive form of the development capacity (DC/TST). These results show that a greater potential for development does not automatically translate into realized development. E2 has the greatest topological development capacity but much of it remains in the overhead. Transforming the networks according to the proposed changes reflects on a lower AMI, with the exception of E3, and lower TSO/TST. However, the trends shown by these metrics do not change in the new configurations. TSO as a fraction of DC decreases with the level of urbanization but its component parts do not contribute homogeneously to its overall trend. The overhead on import, in fact, increases with urban expansion, while the other fractions of overhead diminish from E1 to E3. The modified networks maintain this trend along the gradient of urbanization. In respect to their present counterparts they all dissipate a higher fraction of DC and maintain same level of redundancy. When these indices are expressed as intensive fractions the overhead on import (OI/TST) increases with urban expansion, and that on exports (OE/TST) progressively diminishes from E1 to E3; redundancy and dissipative overhead do not follow urban expansion and show complementary trends—E2 shows the highest value for RED/TST and the lowest for DO/TST. From the present networks to the modified networks, trends in relation to urbanization gradient remain the same but OI/TST and OE/TST become lower whereas DO/TST increases and so does RED/TST.

As absolute coefficients, two out of the four components of overhead (OI and RED) increase along with urbanization. OE and DO, on the contrary, do not track with the urban gradient. From the present networks to the modified networks, changes are such that OI and OE diminish in all the three systems. Redundancy gets lower for E1 and E2 but increases in E3. This latter system also increases its DO, which for E1 and E2 does not change passing from the present to the alternative configurations.

## 18.5 Cities as Ecosystems?

A concomitant increase of system activity (TST) and the mutual information of the exchange configuration (AMI) would mark an ecosystem as it grows and develops (Ulanowicz, 1997, 2004). Results of this analysis show a continuing increment in size from E1 to E3 (TST) and in the mutual information of the exchange configuration (AMI). This in turn reflects on a progressive boost of ascendency. TST, ASC, and AMI indicate that while cities develop in their urban structure and complexity they also do from an ecosystem point of view.

To further substantiate this statement the outcomes of the network analysis can be interpreted in relation to classical attributes that would signal the course of ecosystem development. Odum (1969, 1985) provided the most complete summary of these features and Ulanowicz (1986, 1997) reinterpreted them in the framework of ecological networks so that Odum's predictions about the trends of many of these 24 characteristics appear implicit in the propensity to increase in ascendency. However, a one-to-one correspondence between Odum's attributes and system level indices would shed further light on the similarities and differences between urban systems and ecosystems. Such correspondence, however, is not easy to find because of two reasons: (a) classical attributes of ecosystem evolution cannot be easily associated with network properties quantified by system level indices; and (b) the ecosystem's features proposed by Odum pertain to ecological traits that are difficult to translate in the urban context because adequate proxies to represent them are still to be found. It is the case for gross production/community respiration (P/R ratio) and other community energetic attributes for which a proper definition and quantification for urban systems should be identified. For example, in this exercise, while respiration can intuitively be associated to water dissipation (DO), no production function can be identified. In ecosystems, production is carried out by primary producers, and this function is associated to carbon as currency. Production as a function exists also in urban systems but only in association with currencies that take part in the transformation processes through which material goods are obtained.

Nonetheless, there are interesting issues emerging form this study—they should be taken as stimulating observations rather than definitive statements. One of Odum's expectations is that the amount of total organic matter (attribute #6 in his list, 1969) would increase during ecosystem development. In its broadest sense this prediction would apply to every medium that is essential to systems. So the observed increase in TST from E1 to E3 would confirm this expectation in the urban context. Ecosystems would also show an increasing tendency to conserve nutrients (attribute #21). Translated in the framework of this study this would mean conserving water inside the system. The overhead on export (OE) can be seen as a proxy for this attribute because it indicates the propensity of a system to convey medium to the outside environment. If urban systems would follow Odum's prediction OE would decrease along with city size. This does not correspond to what this index reveals when taken as an absolute coefficient. However, if it is normalized in respect to system size (OE/TST), or expressed as a percentage of DC, the observed trend is exactly what one would expect according to Odum.

This author included information (#24), entropy (#23), and diversity (#8, 9, 10) in the list of attributes to describe ecosystem development. Information can be unambiguously quantified by the AMI, which show that the values obtained for the city networks increase with the level of urban expansion. This would be in agreement with Odum's postulate that later successional stages would possess higher information content.

As for entropy and diversity some clarification is needed to find a correspondence between what Odum meant and the indices of the network analysis. Odum's diversity pertains to the variety (and equitability) of species and of biochemical compounds. In the network context this variety can be seen in relation to the number of compartments (both living and nonliving) and their standing stocks. System level indices are calculated using flows and a possibility that Odum's diversities fit in the framework of network analysis would require standing stocks to be included in the formulation of system level indices (Ulanowicz and Abarca Arenas, 1997). However, in the water flow networks discussed here there are no standing stocks for the compartments because the water is used right away and the resource does not accumulate in the system. Accordingly, no assessment of Odum's diversity in a network fashion is possible at this stage. Also, the lack of standing stocks precludes the exploration of certain known attributes of developing systems, such as the possibility that during development, compartments with slower than average turnover time and larger (body) size would take over and predominate in the system dynamics (Ulanowicz and Abarca Arenas, 1997). Exploring this issue in urban systems would have a strong impact because compartments in these networks represent sectors of human activities, and dominance of larger entities with slower turnover range, if confirmed, would imply an evolution of the economic system into a few large entities (i.e., companies).

Diversity of flows, instead, can be more properly associated to Odum's entropy and stability. Information theory applied to flow networks (Latham II and Skully, 2002; Rutledge et al., 1976; Ulanowicz, 2001) signals that DC and TSO correspond to entropy terms—the former identifies the joint entropy of flows while the latter is the sum of conditional entropies of the same flows (see Allesina and Bodini, 2008 for details). Which of these two terms corresponds to Odum's idea of entropy has to be established. This author (Odum, 1953) pointed out that when multiple connections between components exist, it is likely that a disturbance along one path would be compensated for by the variables relying on alternative pathways to perform the same function (i.e., transfer of currency). This idea links immediately stability with system overhead; so if TSO becomes a proxy of Odum's stability then DC (joint entropy) would express Odum's entropy.

TSO is defined as the residual disorganization due to the redundancy of all types of flow: imports, exports, dissipative flows, and internal exchanges. This redundancy can be seen as a quota of inefficiency in processing medium but that gives flexibility to overcome perturbations acting on the exchange flows. In the ecosystem framework TSO becomes a proxy for resilience (the capability to overcome external stress), which is one of the factors for sustainability (Ulanowicz et al., 2009). Mageau et al. (1995) offers convincing arguments about the possibility to extend these concepts and their meaning to all complex systems, from cells to ecosystems to economic systems. Thus an extension to urban systems is straightforward, although no experimental test has been conducted so far to establish whether these concepts have a practical impact on the design and management of urban systems.

Odum hypothesized that stability would augment during ecosystem development. The systems discussed here confirm this prediction if one considers overhead expressed as absolute coefficient (TSO). But if this index is normalized by system size (TSO/TST) a clear trend accompanying urban gradient cannot be observed. If this index is taken as a fraction of the development capacity (TSO%) it seems that cities would reduce their stability while they grow larger. As for entropy, the expectation is that it would be lower in the mature than in the early stages of ecosystem succession. This study shows that DC, the proxy for entropy, would be increasing with urban size, contrary to the expectation.

The largest number of Odum's attributes can be interpreted to imply that mature systems exhibit more cycling and greater internalization of medium (i.e., P/R ratio, #1 in the Odum's list; P/B ratio #2; mineral cycles # 15; nutrient exchange rate # 16; role of detritus in nutrient regeneration # 17; internal symbiosis #20). In the urban networks described in this study cycling remains a marginal component of water exchanges because the amount of currency involved in this function, measured by the Finn's index (Allesina and Ulanowicz, 2004), remains close to zero (results that are not presented in detail here).

Another trait that appears in the Odum's list is niche specialization. Whenever there is greater specialization in the exchange flow structure, the compartments tend to be more specialized in selecting their sources of medium. That is, niche specialization would coincide with greater articulation of the exchange flows and AMI can be used as a proxy for this attribute. Niche specialization is expected to increase during succession (Odum, 1969). AMI values for the three urban systems show that something like niche specialization would occur during urban development.

According to Decker et al. (2000), efficiency would diminish during urban development, whereas the opposite is expected during ecological succession (Ulanowicz, 1980; Weber et al., 1989). The amount of currency that is dissipated by networks gets closer to a measure for efficiency. Evidences emerging from this study are contrasting. The DO does not seem to track with the urban expansion but E3, the most developed configuration, shows the highest level of dissipation both as absolute coefficient (DO) and intensive component (DO/TST). This does not apply to DO as a percentage of the development capacity, which makes it hard to find a convergence of these outcomes so that a clear interpretation is possible.

Network analysis also looks at efficiency through AMI. It measures how efficiently the currency is distributed in the ecosystem (Ulanowicz, 1986, 1997). In the systems discussed here AMI increases with the level of urban expansion, thus indicating that as city development proceeds, a progressively greater efficiency in exchanging medium arises. AMI thus speaks in favor of Decker's idea (Decker et al., 2000) that cities would increase efficiency as they grow larger, if this attribute pertains to distributing medium. When, instead, efficiency is in connection with the extent to which cities dissipate currencies, then the three systems do not form a clear pattern, although E3 seems to be the least efficient system overall.

Alternative network configurations for these three urban systems were conceived with the objective of reducing water consumption. Although this result was easily achieved, as demonstrated by the values of OE, reshaping the networks yielded different outcomes, observed in the whole system performance as well. A dichotomous pattern clearly emerges—for most of the indices, in fact, the alternative networks for E1 and E2 possess lower values than their present structures. The opposite holds for E3.

Considering the indices of network analysis as related to Odum's descriptor of ecosystem development, it seems that new networks conceived to reduce water consumption would induce for E1 and E2 a sort of retrogression in their evolutionary pathway, when viewed as ecosystems. In fact total amount of matter (TST), information, niche specialization and efficiency (AMI), stability (TSO), and entropy (DC) are all decreasing in the new configurations, thus bringing these two systems toward earlier or less mature stages of ecosystem development. On the contrary, with all these descriptors expected to augment, network reshaping would push E3 toward a more developed condition, that is a more mature stage from an ecological point of view.

According to conceptual models explaining ecosystem succession, (Golley, 1974; Holling, 1986; Norton and Ulanowicz, 1992; Ulanowicz, 1986, 1997), a cyclic trajectory with phases called renewal, exploitation, conservation, and destruction would be expected also for urban systems, if they are to mimic ecosystems. Retrogression of E1 and E2 toward earlier stages of the urban evolution would imply moving backward along this trajectory whereas E3 would move forward. With what consequences it is impossible to predict because it is not known at what point of this hypothetical pathway E1, E2, and E3 presently are, and that even oldest cities might be in the earliest stages of their succession, a process that may take centuries to complete (Decker et al., 2000).

Retrogression for E1–E2 and progression for E3 predicted according to Odum's descriptors seem to be confounded when the cyclic pathway of ecosystem evolution is described in terms of what the theory expects for system level indices (Ulanowicz, 1986, 1997). At the beginning, resource availability would guarantee an intense activity—TST would be increasing and so would the system's development capacity. As long as succession proceeds, DC would plateau and ascendency, that lags behind it, would get closer to it—as DC augments an ever greater part of it is hijacked to organization rather than encumbered complexity. According to this and considering the values of indices as percentage of the development capacity (Table 18.1, central columns), it comes out that E2 and E3 seem to progress along the ecosystem developmental trajectory while no change is expected for E1 because ASC and TSO share the same percentage of DC in both configurations.

In this exercise the comparison between urban systems and ecosystems has been conducted in the framework of Odum's characterization of ecosystem development. Other important features should be investigated to characterize urban systems in their similarities with living systems, such as self-organization, autocatalytic processes, and dissipative structures (Kauffman, 1993). What is presented here simply responds to a criterion of congruity of the methods in relation to the objectives. Urban systems, like ecosystems, can be represented as networks, and this offers the opportunity to apply network analysis. However, because the indices of the network analysis have a clear meaning only in the ecosystem framework, the outcomes of this methodology have been used here to exploit the ecosystem nature of cities.

## 18.6 The Quest for Sustainability

Reducing water consumption is commonly intended as fostering sustainability because the resource is preserved. Accordingly, all the three systems would improve their sustainability as regards water demand through network reshaping. Ulanowicz (2009) however showed that sustainability could be expressed as a whole system attribute that depends on the balance between two opposing features, which are efficiency and flexibility. In practice, sustainability requires that systems must not only have sufficient ascendency to maintain their integrity over time but also they must possess a reserve of flexibility to overcome the effects of novel disturbances. Indices, taken as absolute coefficients and intensive components, indicate that new networks for E1 and E2 would be both less efficient and less flexible—their propensity to sustainability would be lower. E3 on the contrary seems to become more efficient and flexible after network modifications, although TSO/TST would speak of a less flexible system from a pure topological point of view. The interpretation of these outcomes is further complicated by the fact that the development capacity changes as well. So a more parsimonious interpretation of the evidences provided by system level indices advice to consider the indices as fraction of the development potential (DC). In so doing, what emerges is that both E2 and E3 would become more developed but show a tendency to become "brittle" in the sense of Holling (1986) or "senescent" in the sense of Salthe (1993), conditions that are typical of highly managed systems, which are characterized by high organization (AMI, ASC) but less flexibility (TSO).

These results strongly advise that single-issue criteria may have contrasting effects on the whole system scale and that the final outcomes very much depend on the network structure of the system and how it varies under the pressure of management actions.

Further evidences must be accumulated to substantiate these findings and to draw reliable general conclusions. Should these conclusions survive further scrutiny, establishing quantitative limits would be an additional requirement. Systems that are sustainable require that an intermediate balance between flow organization and flexibility is reached but at what point optimal conditions occur is not easy to say. Ulanowicz (2009) provided some evidence that natural ecosystems would be more sustainable within the boundaries of a "window of vitality," defined by values of parameters obtained by further elaborating system level indices. To incorporate this analysis in the study of urban systems we need first to consolidate the meaning of these indices when used in an urban context to make them more intelligible, reliable, and of practical support.

## References

Alberti, M., Marzluff, J.M., Shulenberger, E., Bradley, G., Ryan, C., Zumbrunnen, C., 2003. Integrating humans into ecology: opportunities and challenges for studying urban ecosystems. BioScience 53, 1169–1179.

Allesina, S., Bodini, A., 2008. Ascendency. In: Jørgensen, . E., S., Fath, B.D. (Eds.), Ecological Indicators. Volume 1 of Encyclopedia of Ecology, vol. 5. Elsevier, Oxford, pp. 254–263.

Allesina, S., Bondavalli, C., 2004. WAND: an ecological network aqnalysis user-friendly tool. Environ. Model. Soft. 19, 337–340.

Batty, M., 2008. The size, scale, and shape of cities. Science 319, 769–771.

Blank, A., Solomon, S., 2003. The Emergence of Cities: Complexity and Urban Dynamics. http://www.casa.ucl.ac.uk/working papers/paper64.pdf UCl, Working paper series. Published online URL.

Decker, E.H., Elliot, S., Smith, F.A., Blake, D.R., Rowland, F.S., 2000. Energy and material flow throuhgh the urban ecosystem. Ann. Rev. Energy Environ 25, 685–740.

Gabaix, X., 1999. Zip's law for cities. An explanation. Q. J. Econ. 114, 739–767.

Golley, F.B., 1974. Structural and functional properties as they influence ecosystem stability. In: Cave, A.J. (Ed.), Proceedings of the First International Congress of Ecology. Centre for Agricultural Publishing and Documentation, Wageningen, The Netherlands, pp. 97–102.

Gragson, T.L., Grove, J.M., 2006. Social science in the context of the long term ecological research program. Soc. Nat. Resour. 19, 93–100.

Grimm, N.B., Redman, C.L., 2004. Approaches to the study of urban ecosystems: the case of central Arizona-Phoenix. Urban Ecosyst. 7, 199–213.

Grimm, N.B., Grove, J.M., Pickett, S.T.A., Redman, C.L., 2000. Integrated approaches to long-term studies of urban ecological systems. Bioscience 50, 571–584.

Grimm, N.B., Faeth, S.H., Golubiewski, N.E., Redman, C.R., Wu, J., Bai, X., et al., 2008. Global change and the ecology of cities. Science 319, 756–760.

Grove, J.M., Burch Jr., W.R., 1997. A social ecology approach and applications of urban ecosystem and landscape analyses: a case study of Baltimore, Maryland. Urban Ecosyst. 1, 259–275.

Holling, C.S., 1986. The resilience of terrestrial ecosystems: local surprise and global change. In: Clark, W.C.,, Munn, R.E. (Eds.), Sustainable Development of the Biosphere. Cambridge University Press, Cambridge, pp. 292–317.

Hope, D., Gries, C., Zhu, W., Fagan, W.F., Redman, C.L., Grimm, N.B., et al., 2003. Socioeconomics drive urban plant diversity. PNAS 100, 8788–8792.

Kauffman, S.A., 1993. The Origins of Order: Self Organization and Selection in Evolution. Oxford University Press, Oxford. 709.

Kinzig, A., Warren, P.S., Martin, C., Hope, D., Katti, M., 2005. The effects of human socio-economic status and cultural characteristics on urban patterns of biodiversity. Ecol. Soc. 10, 23–36.

Latham II, L.G., Scully E.P, 2002. Quantifying constraint to assess development in ecological networks. Ecol. Modell. 154, 25–44.

Mageau, M.T., Costanza, R., Ulanowicz, R.E., 1995. The development, testing and application of a quantitative assessment of ecosystem health. Ecosystem Health 1, 201–213.

Norton, B.G., Ulanowicz, R.E., 1992. Scale and biodiversity policy: a hierarchical approach. Ambio 21, 244–249.

Odum, E.P., 1953. Fundamental of Ecology. Saunders, Philadelphia, 384 pp.

Odum, E.P., 1968. Ecology and our Endangered Life-Support Systems. Sinauer, Sunderland MA. 283.

Odum, E.P., 1969. The strategy of ecosystem development. Science 164, 262–269.

Odum, E.P., 1985. Trends expected in stressed ecosystems. BioScience 35, 419–422.

Pickett, S.T.A., Grove, P.M., 2009. Urban ecosystems: what would Tansley do? Urban Ecosyst. 12, 1–8.

Pickett, S.T.A., Burch Jr., W.R., Dalton, S.E., Foresman, T.W., Grove, J.M., Rowntree, R., 1997. A conceptual framework for the study of human ecosystems in urban areas. Urban Ecosyst. 1, 185–199.

Pickett, S.T.A., Cadenasso, M.L., Grove, J.M., Nilon, C.H., Pouyat, R.V., Zipperer, W.C., et al., 2001. Urban ecological systems: linking terrestrial ecological, physical, and socioeconomic components of metropolitan areas. Ann. Rev. Ecol. Syst. 32, 127–157.

Pickett, S.T.A., Cadenasso, M.L., Grove, J.M., Groffman, P.M., Band, L.E., Boone, C.G., 2008. Beyond urban legends: an emerging framework of urban ecology, as illustrated by the Baltimore Ecosystem Study. BioScience 58 (2), 139–150.

Roach, W.J., Heffernan, J.B., Grimm, N.B., Arrowsmith, J.R., Eisinger, C., Rychener, T., 2008. Unintended consequences of urbanization for aquatic ecosystems: a case study from the Arizona Desert. BioScience 58, 715–727.

Rutledge, R.W., Basorre, B.L., Mulholland, R.J., 1976. Ecological stability: an information theory viewpoint. J. Theor. Biol. 57, 355–371.

Tribus, M., McIrvine, E.C., 1971. Energy and information. Sci. Am. 225, 179–188.

Ulanowicz, R.E., 1980. An hypothesis on the development of natural communities. J. Theor. Biol. 85, 223–245.

Ulanowicz, R.E., 1986. Growth and Development: Ecosystem Phenomenology. Springer-Verlag, New York.

Ulanowicz, R.E., 1997. Ecology, the Ascendent Perspective. Complexity in Ecological Systems. Columbia University Press, New York.

Ulanowicz, R.E., 2001. Information theory in ecology. Comput. Chem. 25, 393–399.

Ulanowicz, R.E., 2004. Quantitative methods for ecological network analysis. Comput. Biol. Chem. 28, 321–339.

Ulanowicz, R.E., 2009. The dual nature of ecosystem dynamics. Ecol. Modell. 220, 1886–1892.

Ulanowicz, R.E., Goerner, S., Lietaer, B., Gomez, R., 2009. Quantifying sustainability: resilience, efficiency and the return of information theory. Ecol. Complex. 6, 27–36.

Ulanowicz, R.E., Abarca-Arenas, L.G., 1997. An informational synthesis of ecosystem structure and function. Ecol. Modell 95, 1–10.

Weber, B.H., Depew, D.J., Dyke, C., Salthe, S.N., Schneider, E.D., Ulanowicz, R.E., 1989. Evolution in thermodynamic perspective. Biol. Philos. 4, 373–405.

WCED, 1987. Our Common Future. Oxford University Press,, Oxford. (for the UN World Commission on Economy and Environment).

# 19

# Three-Dimensional Modeling of Pollutant Dispersion in Lake Garda (North Italy)

Tomas Lovato*,†,1, Giovanni Pecenik†

*CENTRO EURO-MEDITERRANEO SUI CAMBIAMENTI CLIMATICI—CMCC, VIALE A. MORO 44, I-40127 BOLOGNA, †DEPARTMENT OF ENVIRONMENTAL SCIENCES, INFORMATICS AND STATISTICS, UNIVERSITÀ CA' FOSCARI VENEZIA, DORSODURO 2137, I-30123 VENEZIA, 1CORRESPONDING AUTHOR: TOMAS LOVATO, CENTRO EURO-MEDITERRANEO SUI CAMBIAMENTI CLIMATICI—CMCC, VIALE A. MORO 44, I-40127 BOLOGNA. E-MAIL: TOMAS.LOVATO@CMCC.IT

## 19.1 Introduction

Lake Garda is the largest fresh water supply in Italy: its water is used for domestic consumption and agricultural and industrial purposes as well for supporting a small, but productive fishery activity (Magni et al., 2008). In addition, due to its mild climate conditions, attractive landscapes, and historic heritage, this is one of the most popular touristic destinations of northern Italy, with nearly 20 million visitors per year (Pasotti, personal communication).

Despite the recreational and socioeconomic value of the basin (Pilotti et al., 2011), limnological research does not appear to have been carried out with the thoroughness that this water body deserves, especially when compared to other large subalpine lakes of similar importance (Salmaso and Mosello, 2010).

Since 1995, monitoring activities have been routinely performed by local environmental authorities (Regional Environmental Protection Agencies of Trento, Lombadia, and Veneto Regions), while only fragmented research campaigns have been carried out by national institutions (Chiaudani and Premazzi, 1990; Mosello et al., 1997; Zilioli, 2004). These surveys were mostly aimed at detecting early warnings of eutrophication, and provided basic limnological information on major structural aspects of the ecosystem (Büsing, 1998; Gilebbi et al., 2000; Defrancesco et al., 2004; Salmaso and Padisák, 2007). Attempts were also made to relate the trophic status to relevant limnological features such as the thermal structure, while accounting for morphometric and hydraulic parameters and annual variation of global heat budget (Ambrosetti and Barbanti, 2002a,b; Salmaso, 2005). Recent investigations were mostly oriented toward the detection

Models of the Ecological Hierarchy. DOI: http://dx.doi.org/10.1016/B978-0-444-59396-2.00019-5
ISSN 0167-8892, 

of long-term changes in the planktonic community due to climatic fluctuations (Salmaso, 2010), as well as the application of new bio-optical methodologies to monitor the water quality by means of satellite observations (Giardino et al., 2007; Bresciani et al., 2009).

Conversely, functional characteristics of in-lake physical processes, which may help in interpreting the observed limnological conditions, are still unexplored (Salmaso and Mosello, 2010). In fact, with the exception of a flux-gradient estimation of vertical eddy diffusivities (Sacco, 1996), the scientific works dealing with the lake do not account for *in situ* determinations of relevant hydrodynamic features, like current velocities and mixed-layer dynamics.

The long-term environmental surveys indicate that the lake is in an oligotrophic condition, a state which is naturally favored by the low catchment-to-surface ratio and the limited industrial emissions (Parati et al., 2011). Moreover, to cope with the ever-increasing resident and tourist population, an inter-municipality subsurface sewage collector was built in the 1970s (see Fig. 19.1) for the safeguard of the ecosystem. Nevertheless, due to the

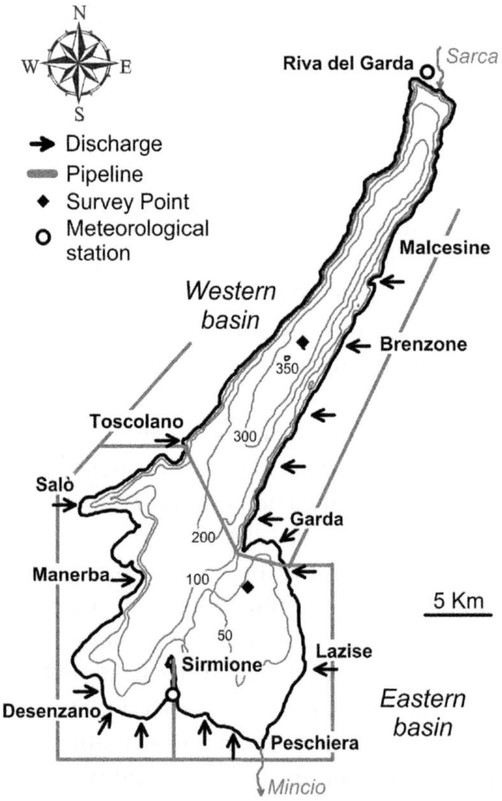


**FIGURE 19.1** Bathymetric map of the Lake Garda. The sewage collector pipeline is schematically represented, together with the existing discharge sites (arrows). Monitoring stations of the water quality (diamonds) and meteorological parameters (circle) maintained by the local environmental authorities are indicated. For color version of this figure, the reader is referred to the online version of this book.

ageing of the structure and to the increased anthropic pressure, after heavy rainfall, storm water and sewage are occasionally discharged into the lake (Egiatti, 2011). Concerns about this practice have been raised, particularly during the most critical summer periods, when access to localized coastal resort facilities is temporarily forbidden as a consequence of the sudden deterioration of the water quality (ARPAV, 2010).

With regard to this issue, a model-based approach is urged, as, besides providing a comprehensive integration of all relevant information so far separately obtained by specific limnological disciplines, it would permit to supersede the descriptive phase and directly support management decision policies (Jørgensen, 2010; Mooij et al., 2010). However, modeling efforts to couple the hydrodynamic characteristics of Lake Garda with physicochemical and ecological processes are still in the very early stages. So far, only a three-dimensional lake model has been used to describe the larval transport taking into account the underlying lake hydrodynamics (Mari et al., 2009). In addition, a one-dimensional turbulent-eutrophication model was developed for the reproduction of mixed layer and phytoplankton dynamics (Pecenik and Lovato, 2011).

Therefore, to meet with both scientific needs and environmental problems, the three-dimensional hydrodynamic model TRIM-3D (Casulli and Cattani, 1994) was applied to Lake Garda.

The objective of the present work is to investigate the short-term dispersion of a degradable organic pollutant, under realistic meteoclimatic forcing in presence of stable stratification conditions. The hydrodynamic model was proved to be robust and computational efficient through its successful applications to a variety of marine and lacustrine situations exhibiting rather different topographic features, such as the Lagoon of Venice (Casulli and Cheng, 1992), San Francisco Bay (Cheng et al., 1993), and Lake Mangueira (Fragoso et al., 2011).

Section 19.2 presents a brief description of the study site. Section 19.3 gives the details of the three-dimensional hydrodynamic model and the numerical experiment setup. Section 19.4 is divided in two parts: we first present and discuss the space-time variations of velocity fields, emphasizing on the prevailing hydrodynamic structures under different forcing conditions. The second part focuses on the analysis of pollutant distribution patterns, as induced by the periodicity of local winds and topographic characteristics. In Section 19.5, conclusions on the consistency of the model and promising improvements are drawn.

## 19.2 Study Site

Lake Garda is located in the Italian subalpine area, at an altitude of 65 m above mean sea level (a.m.s.l.) and covers a surface of 368 km$^2$, with a coastline of 185 km. The maximum depth is 350 m, the major axis measures 51.6 km while the minor one is 17.2 km, and the total volume is close to 49 km$^3$. The basin is geographically sited among the following extreme points: North: 45° 53′ 52″ Lat N, 10° 50′ 32″ Lon E (Riva del Garda); South: 45° 26′ 23″ Lat N, 10° 41′ 35″ Lon E (Peschiera).

The water body is subdivided into two naturally defined basins: a northern narrow, an elongated part with a mean depth of 200 m delimited by a submerged ridge directed SW–NE, and a southern basin with an average depth of 60 m. The ridge rises up to 5 m below the surface (facing Sirmione) and delimits an eastern, shallower circular basin (Gerletti, 1974).

The hydrological balance of the lake is governed by the contributions of the Sarca River at the extreme North (yearly discharge rate of 30 $m^3/s$) and of 18 minor tributaries with an annual inflow <1 $m^3/s$ each (see Fig. 19.1). The only effluent is the Mincio River (at the south side of the lake), whose outflow (annual average ~60 $m^3/s$) is regulated by a dam system to provide irrigation water (Egiatti, 2009). The lake level is constrained to oscillate within −0.05 m and +1.75 m from a fixed datum of 64 m a.m.s.l. The lake retention time was estimated as 26.8 years (Ambrosetti et al., 1992).

Meteoclimatic information from two monitoring stations located in Sirmione and nearby Riva del Garda for the period 2005–2010 was provided by the Environmental Protection Agencies of the Lombardia Region and the Trento Province, respectively. Average annual precipitation varies from 950 mm in the southern area up to 1090 mm in the northern one, while yearly mean temperature ranges from 14.1 °C to 13.2 °C. The lake surface receives an average energy input ranging from 853 $W\,m^2\,day^{-1}$ (January) to 5040 $W\,m^2\,day^{-1}$ (July). Winds are characterized by regular daily periodicity at all seasons, and exhibit a well-defined fetch: the "Peler" wind blows from NE to SW during the night, while the "Ora" wind blows in the opposite directions in the early afternoon at an average speed of 10 $ms^{-1}$ and 5 $ms^{-1}$, respectively.

*In situ* data of water temperature were collected in the period 2005–2010 at two survey points (see Fig. 19.1) by the Environmental Protection Agency of Veneto Region (Parati et al., 2011). Surface water temperature normally ranges between 7 and 25 °C, up to values close to 29 °C, occasionally observed in the shallower southern area.

The water column exhibits a nearly homothermous condition from the late autumn till March. Thermocline deepening begins in early April from a depth of 5 m with a gradient value equal to 0.1 $°C\,m^{-1}$ and reaches a maximum of 0.5 $°C\,m^{-1}$ at a depth of 25 m in August–September. Thermal variations affect the upper 60 m, below which a constant temperature of ~8 °C is constantly observed. Stable stratified conditions are observed from May to late September, clearly indicating that the thermal hysteresis should be sufficiently high to dissipate wind-induced turbulence and reestablish normal temperature profiles. The lake is oligomictic and the more recent overturns have been recorded in the years 1999–2000 and 2004–2006 (Salmaso, 2010).

## 19.3 TRIM Hydrodynamic Model

The TRIM model is based on the three-dimensional Reynold's averaged Navier-Stokes equations, under the assumption of hydrostatic pressure, and accounts also for baroclinic effects (Cheng et al., 1993). The equations involved are:

$$\frac{\partial u}{\partial t}+u\frac{\partial u}{\partial x}+v\frac{\partial u}{\partial y}+w\frac{\partial u}{\partial z}=-g\frac{\partial \eta}{\partial x}-\frac{g}{\rho_0}\int_z^{\eta}\frac{\partial \rho}{\partial x}d\xi+\mu\left(\frac{\partial^2 u}{\partial x^2}+\frac{\partial^2 u}{\partial y^2}\right)+\frac{\partial}{\partial z}\left(\gamma\frac{\partial u}{\partial z}\right)+fv \qquad (1)$$

$$\frac{\partial u}{\partial t}+u\frac{\partial v}{\partial x}+v\frac{\partial v}{\partial y}+w\frac{\partial v}{\partial z} = -g\frac{\partial \eta}{\partial y}-\frac{g}{\rho_0}\int_z^{\eta}\frac{\partial \rho}{\partial y}d\xi+\mu\left(\frac{\partial^2 v}{\partial x^2}+\frac{\partial^2 v}{\partial y^2}\right)+\frac{\partial}{\partial z}\left(\gamma\frac{\partial v}{\partial z}\right)-fu \tag{2}$$

$$\frac{\partial \eta}{\partial t}+\frac{\partial}{\partial x}\left(\int_{-h}^{\eta} udz\right)+\frac{\partial}{\partial y}\left(\int_{-h}^{\eta} vdz\right) = 0 \tag{3}$$

$$\frac{\partial w}{\partial z} = -\left(\frac{\partial u}{\partial x}+\frac{\partial v}{\partial y}\right) \tag{4}$$

$$\frac{\partial \phi}{\partial t}+u\frac{\partial \phi}{\partial x}+v\frac{\partial \phi}{\partial y}+w\frac{\partial \phi}{\partial z} = \mu\left(\frac{\partial^2 \phi}{\partial x^2}+\frac{\partial^2 \phi}{\partial y^2}\right)+\frac{\partial}{\partial z}\left(\gamma\frac{\partial \phi}{\partial z}\right)+R+I \tag{5}$$

In the equations, $u$, $v$, and $w$ are the velocity components in the horizontal $x$, $y$, and the vertical $z$, directions. Free surface elevation is denoted by $\eta$, $g$ represents the constant of gravitational acceleration, $f$ is the Coriolis parameter assumed constant, $\mu$ and $\gamma$ are the horizontal and vertical eddy viscosity coefficients, $\phi$ is a scalar property of the water column, $R$ denotes the associated rate of change due to either physicochemical or biological processes, and $I$ the source term.

Boundary conditions at the free surface are specified through the prescription of wind stress, $\tau^{\mathrm{w}}$,

$$\tau_{\mathrm{x}}^{\mathrm{w}} = v\frac{\partial u}{\partial z} = C_{\mathrm{W}}\frac{\rho_{\mathrm{a}}}{\rho}(u_{\mathrm{W}}-u)^2,\ \tau_{\mathrm{y}}^{\mathrm{w}} = v\frac{\partial v}{\partial z} = C_{\mathrm{W}}\frac{\rho_{\mathrm{a}}}{\rho}(v_{\mathrm{W}}-v)^2 \tag{6}$$

and bottom friction is described by means of a quadratic formulation

$$v\frac{\partial u}{\partial z} = \frac{g\sqrt{(u^2+v^2)}}{C_{\mathrm{z}}^2}u,\ v\frac{\partial v}{\partial z} = \frac{g\sqrt{(u^2+v^2)}}{C_{\mathrm{z}}^2}v \tag{7}$$

where, $C_{\mathrm{W}}$ and $C_{\mathrm{z}}$ are, respectively, the wind drag and Chezy coefficients, $\rho_{\mathrm{a}}$ is the density of the air and $\rho$ the one of the water, and $u_{\mathrm{W}}$ and $v_{\mathrm{W}}$ the absolute wind speed.

Model equations are solved over vertical layers to yield three-dimensional velocity fields and free surface elevation, together with the temperature and pollutant distributions. The numerical solution of Eqns (1–4) involves an implicit treatment of those terms implying too-restrictive stability conditions on the time step, whereas all other terms are treated explicitly (Casulli, 1990). Thus, the barotropic pressure gradient and the baroclinic term in the momentum equations, the velocity divergence in the continuity equations, and the vertical viscosity are taken implicitly, while convective, horizontal viscosity, and Coriolis terms are taken explicitly. This semi-implicit realization allows for an extremely efficient numerical scheme that was proved to be unconditionally stable, while permitting to adopt a relatively fine mesh and a large-time step (Casulli and Cattani, 1994). The convective terms for momentum and scalar variables are treated using an Eulerian-Lagrangian scheme (Casulli and Cheng, 1992).

The lake bathymetry was digitized from a nautical map (scale 1:25000) on a horizontal grid composed by 330*515 nodes, with a uniform spacing of $\Delta x = \Delta y = 100$ m. Along the vertical axis, 12 layers were imposed: the consecutive separation between layers were set at 1, 5, 10, 15, 20, 25, and 50 m, while a regular spacing of 50 m was set down to the bottom. On each layer, a temperature value corresponding to the average August stratification profile was assigned: namely 23, 19, 15, 14 ,11.5, 11, and 10 °C. A constant value of 8 °C was set in the deeper layers below 100 m.

The Chezy coefficient was described through the Manning formula (Cheng et al., 1993) with a bottom roughness coefficient varying exponentially from 0.024 s $m^{1/3}$ near the surface to 0.015 s $m^{1/3}$ at a depth of 15 m, below which it was held constant.

Winds were simulated by alternating the direction along the NE–SW axis every 12 h, with a timing of 6 h forcing and 6 h calm. Maximum speeds for the Peler and Ora winds were set to 12 m/s and 10 m/s, respectively, and the wind stress was computed by means of a constant drag coefficient equal to $1.4 \times \cdot 10^{-3}\ m^2 s^{-1}$.

A continuous discharge of 1 $m^3$/s was used to simulate a day-long disposal of the pollutant close to the shore of the following urban settlements: Garda, Sirmione, Desenzano, and Riva del Garda (see Fig. 19.1). An additional source, located in the middle of the lake, was used to reproduce the tracer dispersion outflow from the submerged part of the sewage collector pipeline (at a depth of 50 m). The reaction term for the organic pollutant was assumed to undergo a first order decay kinetic $R = -k\phi$ (see Eqn. (5)), with $k = 1.08^{T-15}$.

A time step of 12 s was adopted in the numerical experiment, which was initiated with the hydrodynamic fields produced after a spin-up period of 24 h.

## 19.4 Results and Discussion

Complying with the observed thermal conditions of the lake from the month of August, numerical experiments have been performed to visualize the major flow patterns under varying wind conditions. The surface circulation induced by south-westerly wind ($\sim$10 m/s) over the northern and southern basins are shown in Fig. 19.2 after 48 h of simulation.

A prominent feature of circulation in the northern basin is represented by a persistent gyre rotating counterclockwise in front of Riva del Garda. Under this condition, stronger downwind velocities are generated in the central part of the basin, while return flows are created in the deeper layers. As regards the southern basin, the simulation discloses the importance that the topography has over the currents: two distinct subbasins emerge, a shallower eastern basin characterized by uniform and lower velocities (max. velocity 0.12 m/s), and a deeper western basin, extending from South to North, with higher velocities (max. velocity 0.36 m/s) and a highly irregular distribution. Even though the extreme end of the peninsula of Sirmione appears to be the most exposed to the wind action, the effect of wind stress significantly enhances the circulation even in the two contiguous bays. Velocity fields are characterized by circulation structures that develop, stabilize, and gradually disappear following the periodicity of the wind. This characteristic is particularly felt in all small bays such as Salò, Desenzano, Peschiera, and Garda.

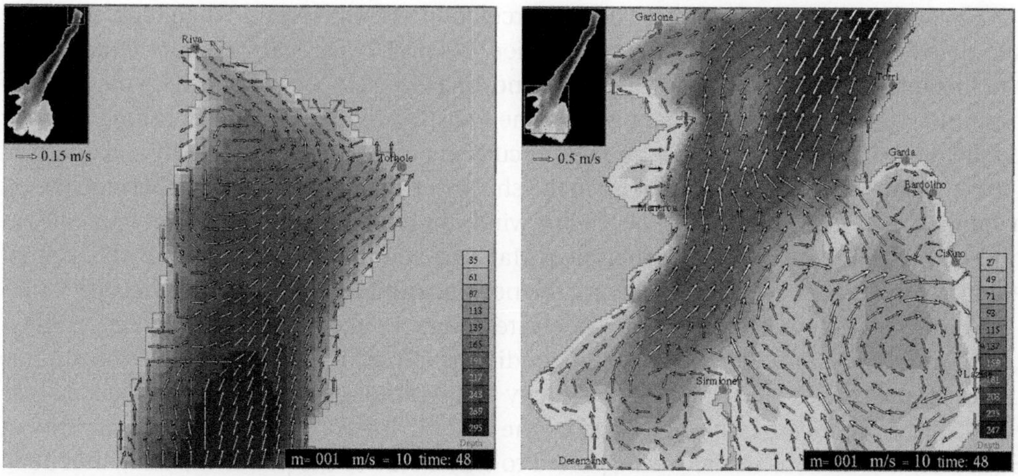


**FIGURE 19.2** Surface circulation after at 48 h of simulation in the northern (left) and southern (right) areas of the lake, under SE–NW wind forcing (~ 10 m/s). For color version of this figure, the reader is referred to the online version of this book.

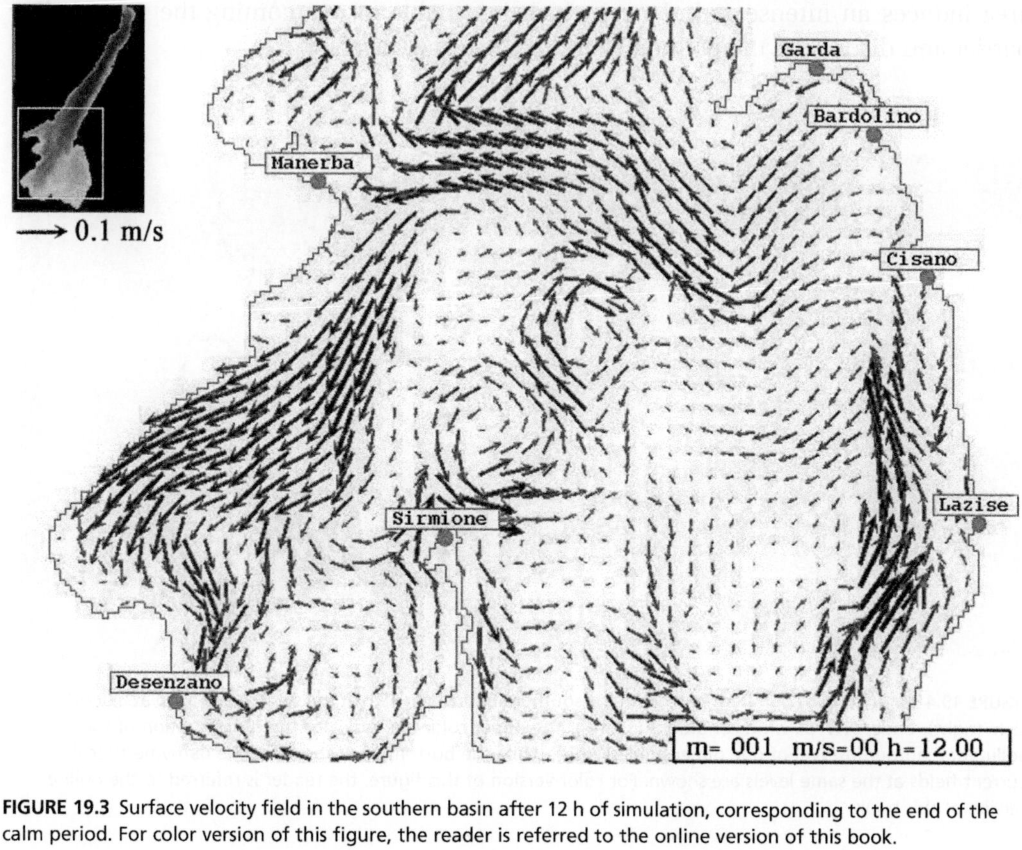


**FIGURE 19.3** Surface velocity field in the southern basin after 12 h of simulation, corresponding to the end of the calm period. For color version of this figure, the reader is referred to the online version of this book.

The high spatial heterogeneity of the circulation is emphasized while looking at the simulated velocity fields during the calm periods (Fig. 19.3): vertical structures emerge over the shallow area of the submerged ridge connecting Garda to Sirmione, where the effect of bottom stress becomes dominant. Along the coast of the eastern basin strong velocity gradients are also observed, due to countercurrents converging from north and south of Lazise. In fact, countercurrents set up in the shallower basin just below the thermocline (at a depth of $\sim$25 m), in relation to the leading wind conditions. A surface current originates in the central basin, which diverges in front of Manerba, thus generating a water displacement along the western coast directed toward the northern and the southern subbasins.

The simultaneous discharge of wastewater from the four urban settlements is illustrated in Fig. 19.4, under NE–SW wind conditions (12 m/s). As one can see, the surface pollutant patches move according to velocity fields, although in the shallow coastal areas the dispersion process is strengthened by the bottom stress. In the northern extremity of Sirmione, the tracer is initially transported to either side of the peninsula and, after 18 h, the pollution affects the entire coastline from Desenzano to Peschiera. From the comparison between the pollutant distribution at the surface and 25 m depth, it can be seen that the discharge originating from the submerged portion of the pipeline (in the middle of the basin) emerges at the surface. The presence of a large wind fetch over the area induces an intense turbulence, which is capable of overcoming the thermocline barrier and dispersing the pollutant toward the upper layers.

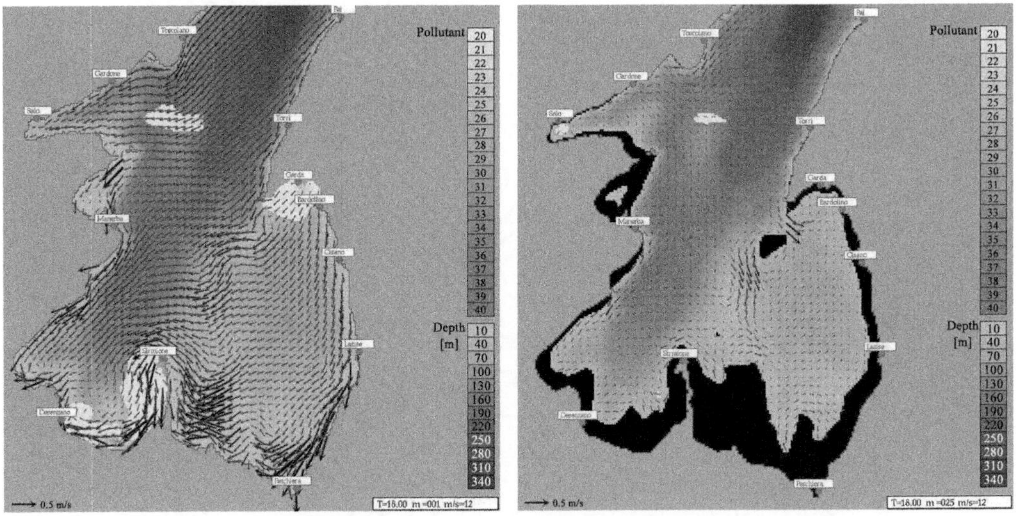


**FIGURE 19.4** Simulation of pollutant patches in the southern basin after 18 h at the surface (left) and at a depth of 25 m (right), under NW–SE wind conditions (12 m/s). The upper color bar indicates the concentration of the pollutant (contour interval of one dimensionless unit), while the bottom one represents the bathymetry scale. Current fields at the same levels are shown. For color version of this figure, the reader is referred to the online version of this book.

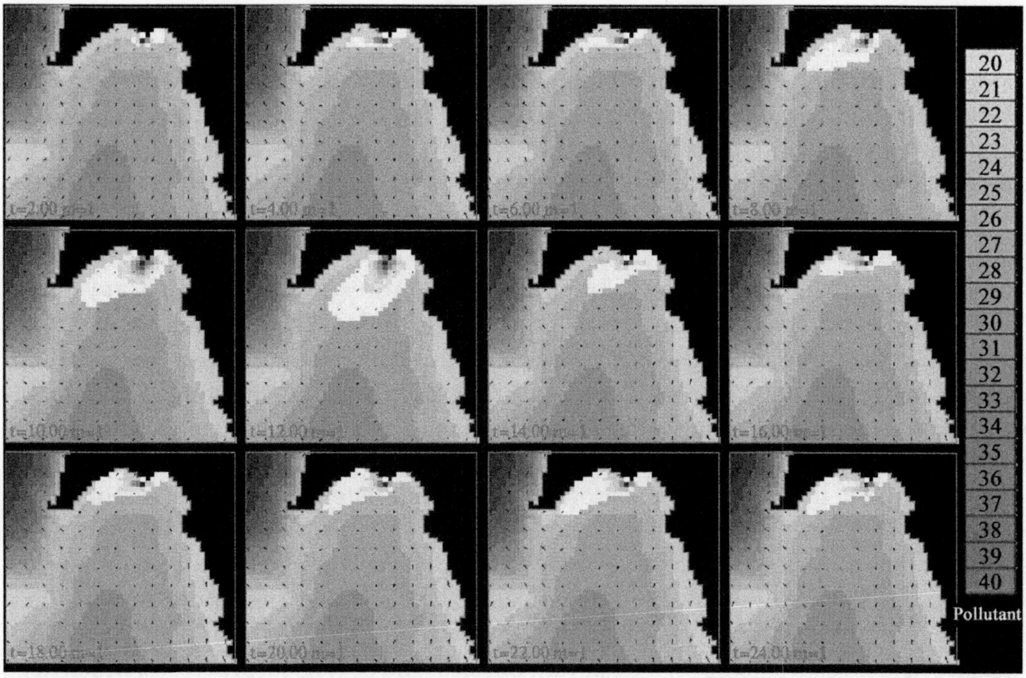


**FIGURE 19.5** Surface daily evolution of the pollutant patch on the surface in the Garda bay (eastern basin), with a 2-h lag representation. The velocity fields and bathymetry at the same level are also reported, while the contour interval of the tracer is a one dimensionless unit. For color version of this figure, the reader is referred to the online version of this book.

In the south-eastern basin, the pollutant is strongly diluted and its dispersion is localized over a limited portion of the coastal region as a consequence of the reduced circulation. In particular, the eastern coasts between Garda and Lazise, where the water motion essentially depends on the weaker residual currents, are the most impacted by the pollutant discharges. Figure 19.5 shows the surface daily evolution, with a 2-h lag representation, of the pollutant patch in the small Garda bay. Despite the variability of wind conditions, the local circulation pattern prevents the pollutant patch from dispersing toward the central area of the lake, and from being transported by the main northward current. Moreover, the occurrence between 8 and 10 h of north-easterly winds (mean speed of 8 m/s) leads the highly concentrated near-shore patch to spread over the entire bay. Conversely, when wind direction reverses (after 20 h), the tracer is transported back to the coast, where a small clockwise gyre develops (see also Fig. 19.2).

## 19.5 Conclusion

The herein described application of the TRIM-3D hydrodynamic model to Lake Garda allowed to realistically simulate the prevailing features of the wind-driven circulation in the system, under stratified conditions. In fact, model outcomes exhibited a consistent and coherent behavior in response to all imposed forcing, while permitting to adequately reproduce the high spatial heterogeneity of the velocity fields at the basin scale. The temporal evolution of circulation structures in small bays was principally driven by the alternation of wind directions, as in the case of Riva del Garda. Moreover, intense residual flows were observed along the eastern and western coasts of the southern basin during calm periods.

The description of the resulting local dispersion patterns of the pollutant was shown to be strongly confined near the coast, like in the case of Garda and between Desenzano and Sirmione. In particular, it was possible to identify those coastal areas that are mostly affected by temporary pollution events, and where the occurrence of trapping zones represents a potential threat to lake water quality. An interesting feature was also represented by the presence of the pollutant outflowing from the submerged part of the sewage collector on the surface layer: the simulated internal dynamics of the lake leads to an upward transport, thus breaking the thermocline barrier.

At present, a thorough comparison between simulated and measured data was precluded, due the lack of experimental measurements on relevant hydrodynamic features, and only a qualitative estimation was possible. Nevertheless, evidences addressed the potentialities of the model in achieving the objective of the application.

Further advances in the development of the model are expected from the realization of extensive field measurements on the relevant physical and ecological processes, which need to be performed on similar spatial and temporal scales. In this context, the role of a combined transport-water quality model is emphasized as it could provide a unifying framework for a closer interdisciplinarity of limnological research in Lake Garda and it might constitute an invaluable instrument to plan more targeted and coordinated fieldworks.

## Acknowledgments

Authors are grateful to Andrea Ragnolini, Tiziano Bullio, and Antonio Pasotti for their contribution, support, and suggestions.

## References

Ambrosetti, W., Barbanti, L., 2002a. Physical limnology of Italian lakes. 1. Relationship between morphometry and heat content. J. Limnol. 61 (2), 147–157.

Ambrosetti, W., Barbanti, L., 2002b. Physical limnology of Italian lakes. 2. Relationships between morphometric parameters, stability and Birgean work. J. Limnol. 61 (2), 159–167.

Ambrosetti, W., Barbanti, L., Mosello, R., Pugnetti, A., 1992. Limnological studies on the deep southern Alpine lakes Maggiore, Lugano, Como, Iseo and Garda. Mem. Ist. ital. Idrobiol. 50, 117–146.

Agenzia Regionale per la Protezione dell'Ambiente del Veneto (ARPAV), 2010. Zone vietate temporaneamente alla balneazione (anni 1997-2010) [cited 2011 Sep 30]; [about 2 p.]. Available from. http://www.arpa.veneto.it/acqua/htm/balneazione.asp.

Bresciani, M., Stroppiana, D., Fila, G., Montagna, M., Giardino, C., 2009. Monitoring reed vegetation in environmentally sensitive areas in Italy. Riv. Ital. Telerilevam. 41, 125–137.

Büsing, N., 1998. Seasonality of phytoplankton as an indicator of trophic status of the large perialpine Lago di Garda. Hydrobiologia 370, 153–162.

Casulli, V., 1990. Semi-implicit finite difference methods for the two-dimensional shallow water equations. J. Comp. Phys. 86 (1), 56–74.

Casulli, V., Cattani, E., 1994. Stability, accuracy and efficiency of a semi-implicit method for three-dimensional shallow water flow. Comp. Maths with Appl. 27 (4), 99–112.

Casulli, V., Cheng, R.T., 1992. Semi-implicit finite difference methods for three-dimensional shallow water flow. Int. J. Numer. Meth. Fl. 15, 629–648.

Cheng, R.T., Casulli, V., Gartner, J.W., 1993. Tidal, Residual, Inter-tidal Mud-flat (TRIM) model with applications to San Francisco Bay. Estuar.Coast. Shelf Sci. 36, 235–280.

Chiaudani, G., Premazzi, G., 1990. Il Lago di Garda. Evoluzione trofica e condizioni ambientali attuali, 196. Commissione delle Comunità, Luxemburg, Europee. EUR 12925.

Defrancesco, C., Bombardelli, D., Cimonetti, E., Costaraoss, S., Fedrizzi, F., Fravezzi, L., 2004. Quaderni del Garda 2001/2004, raccolta dati monitoraggio. Agenzia Provinciale per la Protezione dell'Ambiente, Trento, p. 106 (in Italian).

Egiatti, M., 2009. Considerazioni sul bilancio idrico del lago di Garda. Agenzia Provinciale per la Protezione dell'Ambiente Veneto, Padova. Report No. 05/09 (in Italian).

Egiatti, M., 2011. I livelli del lago di Garda negli anni 2009–2010. Agenzia Provinciale per la Protezione dell'Ambiente Veneto, Padova. Report No. 03/11 (in Italian).

Fragoso, C.R., Motta Marques, D.M.L., Finkler Ferreira, T., Janse, J.H, van Nes, E.H.,, 2011. Potential effects of climate change and eutrophication on a large subtropical shallow lake. Environ. Model. Soft. 26, 1337–1348.

Gerletti, M., 1974. Indagini sul lago di Garda. Quaderni dell'Istituto di Ricerca Sulle Acque. Consiglio Nazionale delle Ricerche, Rome. Report No. 18.

Giardino, C., Brando, V.E., Dekker, A.G., Strömbeck, N., Candiani, G., 2007. Assessment of water quality in Lake Garda (Italy) using Hyperion. Remote Sens. Environ. 109, 183–195.

Gilebbi, S., Cavolo, F., Salmaso, N., Cordella, P., 2000. Underwater light measurements and phytoplankton content in Lake Garda. Verh. Int. Verein. Limnol. 27, 2908–2912.

Jørgensen, S.E., 2010. A review of recent developments in lake modelling. Ecol. Model. 221, 689–692.

Magni, D., Chinaglia, N., Maffiotti, A., Borasi, L., Lefebvre, P., Zanella, D., 2008. Alpine Lakes. A common approach to the characterization of lakes and their catchment area. In: INTERREG III B JTS Alpine Space (Ed.), p. 248.

Mari, L., Biotto, C., Decoene, A., Bonaventura, L., 2009. A coupled ecological–hydrodynamic model for the spatial distribution of sessile aquatic species in thermally forced basins. Ecol. Model. 220, 2310–2324.

Mosello, R., Calderoni, A., De Bernardi, R., 1997. Research on the evolution of the deep southern subalpine lakes performed by the CNR Istituto Italiano di Idrobiologia. Doc. Ist. Ital. Idrobiol. 61, 19–32 (in Italian).

Mooij, W.M., Janse, J.H., De Senerpont Domis, L.N., Hülsmann, S., Ibelings, B.W., 2010. Predicting the effect of climate change on temperate shallow lakes with the ecosystem model PCLake. Hydrobiologia 584, 443–454.

Parati, P., Brotto, E., Cason, M., Pinton, S., Ragusa, F., Tanduo, I., et al., 2011. Stato delle acque superficiali del Veneto. Corsi d'acqua e laghi 2010. Agenzia Provinciale per la Protezione dell'Ambiente Veneto, Padova, p. 323(in Italian).

Pecenik, G., Lovato, T., 2011. A combined 1-D turbulence-eutrophication model for Lake Garda. In: Proceedings of the 7th ECEM Congress. Riva del Garda, Italy 30 May–2 June 2011.

Pilotti, L., Tedeschi-Toschi, A., Apa, R., 2011. Long tail and destination management: the impact of market's diversification on competitiveness in touristic services, 17. University of Milan. The case of Garda Lake. Working Paper No. 2011–07.

Sacco, P., 1996. Modello monodimensionale chimico-fisico del Lago di Garda [dissertation]. University Ca' Foscari Venice, Venice (IT) (in Italian).

Salmaso, N., 2005. Effects of climatic fluctuations and vertical mixing on the interannual trophic variability of Lake Garda. Italy. Limnol. Oceanogr. 50 (2), 553–565.

Salmaso, N., 2010. Long-term phytoplankton community changes in a deep subalpine lake: responses to nutrient availability and climatic fluctuations. Freshw. Biol. 55, 825–846.

Salmaso, N., Mosello, R., 2010. Limnological research in the deep southern subalpine lakes: synthesis, directions and perspectives. Adv. Oceanogr. Limnol. 1 (1), 29–66.

Salmaso, N., Padisák, J., 2007. Morpho-functional groups and phytoplankton development in two deep lakes (Lake Garda, Italy and Lake Stechlin, Germany). Hydrobiologia 578, 97–112.

Zilioli, E., 2004. Experimentation of a remote sensing integrated system for lake water monitoring. Ninfa Technical Report 2, 29 November 2002–01 April 2004, ASI-CNR n. I/R/073/01.

# 20

# DLES: A Component-Based Framework for Ecological Modeling

Maria Bezrukova*, Vladimir Shanin*, Alexey Mikhailov*, Natalia Mikhailova†, Yulia Khoraskina*, Pavel Grabarnik*, Alexander Komarov*

*INSTITUTE OF PHYSICO-CHEMICAL AND BIOLOGICAL PROBLEMS IN SOIL SCIENCE OF THE RUSSIAN ACADEMY OF SCIENCES, 142290 INSTITUTSKAYA UL., 2, PUSHCHINO, MOSCOW REGION, RUSSIA, †INSTITUTE OF MATHEMATICAL PROBLEMS OF BIOLOGY OF THE RUSSIAN ACADEMY OF SCIENCES, 142290 INSTITUTSKAYA UL., 4, PUSHCHINO, MOSCOW REGION, RUSSIA

## 20.1 Introduction

Evolution of knowledge about the structure and driving forces of ecosystem change has led to great advances in ecological modeling. Progress in computing techniques, software engineering, and mathematical tools has simplified the development of methods of investigation of its specific problems. In the past, models were usually created by small research groups or even individual researchers in an *ad hoc* manner (Wilson, 2006). In contrast, models are now being developed by large groups from different institutions. These models are based on state-of-the-art findings and data from ecology, plant physiology, pedology, earth sciences, climatology, and so on. Moreover, a holistic approach to ecological modeling is necessary, i.e., the researcher should consider the ecological system as a whole instead of simulating its individual parts.

The traditional approach to software development assumes that there is a finite set of requirements for a software solution, which can change only insignificantly, in contrast to exploratory scientific programming which can be characterized by a high degree of variability in its requirements. Another distinction is that commercial software has a well-defined life cycle with periodical release of new versions. In contrast, the process of scientific software development is continuous, often with several working solutions or even rollbacks to previous versions. To avoid duplicating work and "reinventing the wheel," it is necessary to take advantage of code reuse. Code reuse, i.e. including earlier-created components in the system, is one of the basic techniques to reduce labor contribution while developing software (Wenderholm, 2005).

As ecosystem processes have a very wide range of spatial and temporal scales, it is possible to develop a large number of models to simulate these processes. The

Models of the Ecological Hierarchy. DOI: http://dx.doi.org/10.1016/B978-0-444-59396-2.00020-1
ISSN 0167-8892, 

development of a simulation model is a quest for compromise between being true to the nature of the simulated object and simplifying its description. On one hand, it is to be wished that the model will be able to represent as many characteristics of the object as possible. However, the amount of necessary input data increases with increased model refinement. Many of these data are hard-measured or not measured at all. Therefore, the researcher sets a goal to make the model representing the significant features of the simulated object as simple as possible. No one can say *a priori* what type of research model is better or final. The choice in each specific case should depend on the modeling goals, and sometimes the next step depends on the results of a previous version. Therefore, it is convenient to have a model system that allows combining of different types of models.

Development of integral programs with source code compiled into a single executable file is a traditional approach. It does not allow the creation of a flexible system of models. It was noted (He et al., 2002; Liu et al., 2002) that the solution of such a problem can be achieved by so-called "modular-based" applications. Modularity is one of the principles of system construction, where functionally related parts are separated into modules. These modules can be further united into the whole system in runtime, i.e., after program start.

Recently, more and more scientists are paying attention to module structure, reusability, and integration of environmental models (He et al., 1999; Knox et al., 1997; Robinson et al., 1998; Voinov et al., 1999). A component-based approach is the next stage of modular development after object-oriented ecosystem modeling (He et al., 1999).

There are two approaches to designing component-based systems:

- The development of model systems as a combination of already existing models (Liu et al., 2002; Peng et al., 2002).
- The development of special frameworks (de Coligny et al., 2002; Knox et al., 1997; May and Conery, 2003; Wenderholm, 2005;Moore and Tindall, 2005; Argent et al., 2006; van Ittersum et al., 2008; Vorotyntsev, 2009; and others) or common protocols (Grimm et al., 2006) for data exchange which allow models created by different research groups to be combined.

Systems of the first type are easier to develop and have fewer overheads for data exchange. However, they can be characterized by low capacity for further development due to the necessity of full architecture reconstruction at every change. In contrast, model systems based on specialized frameworks have almost unlimited possibilities for upbuilding. It is necessary to mention that ecological processes have to be described taking into account spatial aspects, such as finding the nearest neighbor object, for example, for describing competitive processes between similar objects placed randomly on a two-dimensional grid, such as a tree in a forest with local interactions—shadowing and redistributing of soil nutrition. The modular system has to simulate such complicated ensembles with specific local spatial interdependencies.

## 20.2 Advantages and Disadvantages of Modular Design

Software development based on a modular approach may have many benefits:

- The development and modification of individual components is a much easier task than development of the whole model system.
- Modular design provides the possibility of gradual software development.
- Different modules can be written by different research groups working independently. It is necessary only to keep within predesigned standards which specify the implementation of component interaction to ensure compatibility. The intrinsic logic, i.e., the operation algorithm of the submodel, may be arbitrary.
- Different modules can be written in different programming languages.
- Modules can be reused in other model systems.
- Modules can be easily replaced by others without recompilation of the whole model system. This allows comparative testing of different submodels which simulate the same ecosystem process (e.g., different submodels of tree productivity).
- It is possible to create a system that changes its own functionality according to specified requirements. Clearly, a system used in an educational process is distinguished from one used in forest management planning.
- Modular design facilitates the computing process. It can be implemented by simultaneous execution of several submodels as well as by running several copies of the same submodel. Moreover, parallel computing can be also facilitated.

However, when implementing a component-based approach, certain difficulties can arise:

- It is necessary to construct, additionally, a common framework to unify different modules.
- Different submodels may be implemented with different spatial and temporal resolution. Therefore, it is necessary to provide synchronizing.
- Component-based applications usually show lower performance because additional computational resources are wasted on data exchange (Liu et al., 2002).
- As a rule, model initialization requires some input data which should be read from files. Therefore a broad-based and flexible file format is needed.

When using a modular approach it is necessary to achieve a rational balance between requirements for unification of component interfaces and optimization of each module for a certain goal (Veryard, 2000). Additionally, one of the most important features of the model system is its decomposability/composability (Voinov et al., 2004), i.e. the balance between the possibility of separate analysis of each module (submodel) and the ability to combine submodels into a more complex system.

To be versatile and easy to use, any component-based system must meet certain requirements:

- The ability to properly link submodels that differ in dimensionality, type (discrete or continuous), spatial/temporal resolution, spatial representation (point-like, network, lattice, polygonal, etc.), programming language, platform, and so on.
- The necessity for minor changes in existing models to facilitate their linking into the system, and easy development of new submodels.
- Convenient tools for describing the structure of a system and interactions between its components.
- Full separation of simulation algorithms, user interface, and data input interface from files or other data sources in terms of both source code and executable files.
- Convenient tools for data access and management of the system of models.
- Comparable performance in software based on the integral concept.
- Complete and well-written documentation, distinct for different categories of users (component developers, researchers, decision makers, students, etc.).
- Open source code and automatic version control.

When developing our framework for component-based modeling, we tried to take most of these requirements into account.

## 20.3 DLES Description

### 20.3.1 Objectives

The main objective was to create a framework for the assemblage of individual simulation models with discrete spatial and temporal resolution. Our task was to facilitate combining and comparative testing of different approaches for simulations of forest ecosystems on a local scale. Moreover, the system should be flexible and scalable enough to allow for modeling of other natural systems with various spatial scales. It should be noted that DLES wasn't developed as an all-purpose system. Our task was to create a framework to facilitate some of the routines faced by model developers. Our new concept was designed to make this easy, like playing LEGO®. So the model should be reconstructed into standardized model components, which are then pieced together to form a new model system with the desired characteristics. The development of a model system can be greatly facilitated by using libraries of reusable model components.

According to our idea, the system has its own field between model systems based on integral concepts and those based on mathematical frameworks. In the first case, the model developer should be a programmer who writes the whole model system. Systems of the second type are completely prepackaged and, as a rule, do not allow any additions to the set of methods provided. Examples include MATLAB–Simulink (Gray, 2011), Wolfram Mathematica (Wolfram, 2003), and Maple (Lynch, 2010).

## 20.3.2 Basic Concepts

DLES is focused on the simulation of complex ecological systems with discrete spatial and temporal resolution. DLES can take into account some specific problems related to the development of individual forest stand models, such as programming of local interactions between trees (shadowing, redistribution of soil nutrition), matching of spatial and temporal scales, and so on. The simulation is carried out on a two-dimensional grid that is divided into cells. The cell and grid sizes can be set by the user. The use of a two-dimensional grid simplifies the development of models with localized objects (e.g., trees) by means of simpler and faster neighbor-search algorithms to describe the interactions between objects. Such structure also simplifies the representation of spatial discontinuity.

The system is written in Delphi programming language using Borland® Turbo Delphi® 2005 IDE for Microsoft® Windows™ platform. DLES has open modular architecture and uses COM interfaces. DLES allows the combination of various submodels into a unified system, presetting the system with sets of initial data, carrying out simulation experiments using different scenarios, and saving the results of these experiments.

Any DLES module (called "component") is a functionally complete and self-sufficient system unit. There are components of the following types:

- The system *kernel* is the basic component that provides interaction between components and facilitates their "team-work" through performance control and supporting data exchange.
- The *shell* provides a user interface; it transfers user commands to the system and receives the modeling results for display.
- *Models* (*submodels*) are the implementations of algorithms which simulate certain processes in the ecosystem; the submodels receive initial data and return the results of simulation.
- *Tools* are shell extensions (e.g., tools for statistical analysis and data visualization).

The structure of DLES is shown in Fig. 20.1.

Each component is implemented as a dynamic-link library (DLL) or executable file (EXE) which contains an operating algorithm implemented in machine-language code. The distinctive feature of the system is that components can be written in different programming languages. The only requirement is that the language should support COM interfaces.

Thus, the *model system* is the complex of components and relationships between them.

The kernel integrates all components into the unified system using information that is contained in the *scheme*, or description, of the model system. The scheme includes the list of all submodels and their execution orders and a description of relationships between components. *Scheme editor* is a special program for easy and safe editing of the model system scheme.

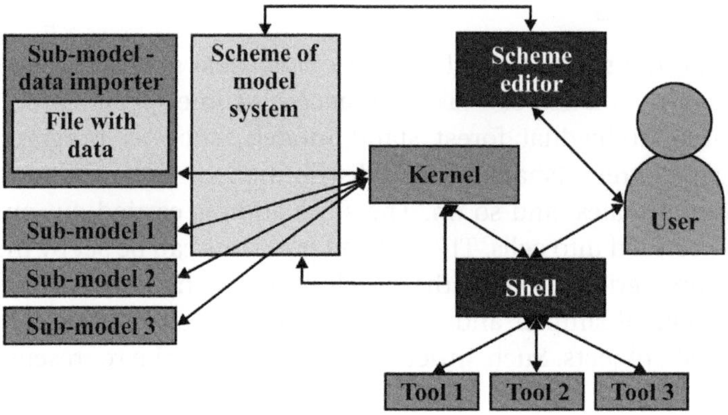


**FIGURE 20.1** Structure of DLES.

We assume three levels of DLES users:

- *Model developers* create submodels based on specifications provided.
- *Scheme administrators* construct new schemes or modify existing ones. They can, for example, add or remove components.
- *Users* can use existing schemes, changing only files with input data to customize simulation experiments.

## 20.3.3 Program Implementation: Classes and Functions

As mentioned in Section 3.2, DLES uses interfaces to combine submodels. An interface is a set of abstract functions and properties that make the components compatible with one another (Rogerson, 1997). Interface declarations as well as declarations and implementation of some common procedures are contained in a common unit with the program code. This file must be included in the source code of each DLES component. A brief overview of interfaces and methods provided to developers is presented in the following. In Fig. 20.2, these interfaces are presented in the form of a class diagram.

The kernel is an implementation of the interface *IKernel*. It provides some function prototypes. *LoadScheme* loads the specified scheme. *Init* initializes the model system, i.e., it calls the *Init* method for all submodels and builds the list of state variables. *GetStateVariables* returns the list of state variables. *GetComponents* returns the list of submodels. *DoStep* forces the model system to make a single time step. When this method is called, the kernel successively calls the *StepModel* method for each submodel and runs the procedures of data exchange synchronization, if necessary.

To work properly, each submodel must implement the interface *IComp*. This interface has some basic methods. *GetModuleInfo* provides service information on the submodel in the form of the structure *TModuleInfo* which contains the submodel name, version, and

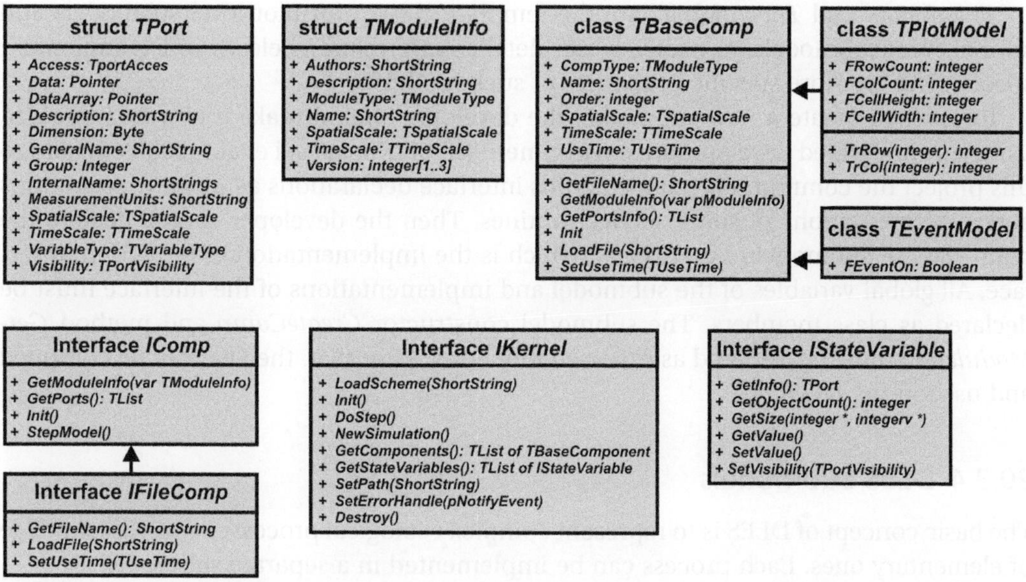


**FIGURE 20.2** Unified Modeling Language (UML) class diagram of DLES.

spatial and temporal resolution. *GetPorts* returns the list of the submodel's ports. *Init* is called by the kernel for each submodel on the system initialization step. The model developer must implement into this procedure the operations for initialization of the submodel, for example, setting the initial value for all variables. *StepModel* is called by the kernel for each submodel during the simulation. The model developer can implement into this procedure the operations which must be done at the submodel step.

As an input file or database can be represented as a submodel, the interface *IFileComp* is also declared, inheriting all methods of the previous interface in addition to its own method *LoadFile.* It loads the specified file and represents it as a submodel so other components can access data stored in the file as ports of this submodel. The constructor of the submodel must be named *CreateComp.*

The *IStateVariable* interface is a prototype of the *TStateVariable* object, which facilitates access to the values of all state variables by means of some methods. *GetSize* returns the size of the simulation grid. *GetObjectCount* returns the value of the object counter. *GetInfo* returns the service information on the state variable in the form of structure *TPort,* which contains port name, measurement units, port type (single value, linear array, or matrix), variable type (number, string, Boolean, or list), and so on. *GetValue* returns the current value of the state variable. *SetValue* sets the new value of the given state variable.

The procedure *AddPort* is also listed in the common unit. It must be called inside the *CreateComp* function when creating a submodel. This procedure adds port to the system list of state variables.

*TPlotModel* and *TEventModel* are the template classes for plot-level submodels and special event submodels, respectively (see detailed information below), which implement special fields to store specific properties of such models.

In order to create a new submodel, the developer should make a new DLL-project, using any integrated development environment (IDE) (Konsynski et al., 1985), and add to this project the common unit that contains interface declarations as well as declarations and implementations of some service routines. Then the developer should declare the main class of submodel (e.g., *TModel*) which is the implementation of the *IComp* interface. All global variables of the submodel and implementations of the interface must be declared as class members. The submodel constructor *CreateComp* and method *GetModuleInfo* must be declared as exported functions. After that, the DLL can be compiled and used in the DLES.

## 20.3.4 DLES Submodels

The basic concept of DLES is to represent complex ecological processes as an assemblage of elementary ones. Each process can be implemented in a separate submodel.

The framework imposes no restriction on the complexity of the submodel code itself. Submodels are independent and not directly referenced to other components; interaction is indirect via updates to state variables.

Each submodel should provide some information: its name, spatial scale, and temporal resolution. Also, the required input variables and available output variables should be listed with a declaration of measuring units. The variables that are external and available for other submodels are called "ports."

Three spatial levels are implemented in the system: cell level, object level, and plot level. Models with different spatial levels have different ways of running and have some specific features of memory allocation for their variables. Plot-level submodels run one time per step, and the corresponding variables are related to the whole simulation plot. Cell-level submodels run iteratively for each cell of the simulation grid at each step. The variables that correspond to this kind of submodel are two-dimensional data arrays equal in size to the simulation grid. For object-level submodels, the variables are linear arrays. Submodels of this kind are designed to simulate objects of which there may be more than one on the simulation grid but which have an irregular spatial distribution, for example, trees. The storage of such data in cell-level variables is unpractical. Accordingly, the object-level submodel runs iteratively for each object on the simulation grid during one time step of the model system.

The possibility of development of submodels with different spatial resolutions allows the most effective use of memory and computational resources. For example, suppose that we are developing a system of models that simulates the forest ecosystem and doing a computational experiment on the gridded plot of $100 \times 100$ m size, with cell size $0.5 \times 0.5$ m. The climate in such a small plot is homogeneous enough; therefore, the submodel of climate can be of plot-level spatial resolution. However, when we have to deal with submodels of soil or ground vegetation, the spatial inhomogeneity is of great importance; therefore, it is advisable to use cell-level resolution. The trees, which usually have irregular distribution on the simulated plot, should be presented by object-level submodels.

These spatial levels are hierarchical. A plot-level submodel can access the whole array of variables of object-level and cell-level submodels. Plot-level submodels can also manage their own cell- or object-level variables.

Such organization allows a range of spatially dependent routines, such as searching for the nearest neighbors or estimating the range between objects on the grid. These actions are required, for example, for computation of the degree of competitive interaction between trees. Embedded routines also include cell grouping on the grounds of neighborhood (or any other feature), which can be necessary, for example, when determining nutrition zones for each tree. All routines mentioned above will be implemented in the system kernel and will be run on demand.

Often the problem arises of developing models which must be run by the user only at certain steps instead of working at each step of the model system, for example, models of fire events which do not occur at each step. Such models must be parameterized by data entered by the user directly in runtime. To facilitate the development of such models, the *TEventModel* class template is provided. It contains the *FEventOn* field, which is the submodel execution flag. Since the shell is allowed to modify the state variables, it is possible to set the initial parameters of "event" submodels in runtime.

*ToolFile* is a specialized submodel. It allows loading of data from external sources, e.g., text files, datasheets, databases, and so on. *ToolFile* automatically imports all variables determined by the user and represents them as its ports.

Users are constantly faced with the necessity for various transformations of simulation results, for example, if a model produces data on stand productivity with a monthly time step but a researcher needs to have the annual values of this output. Another example is the requirement to know the dynamics of total carbon stock in a forest ecosystem, whereas outputs represent values of carbon stock only in separate pools. Of course, one could make the necessary changes to the source code of models or make recalculations after the simulation, using additional software. However, the DLES framework provides an easier way to implement these operations with a special component called *Outputs*. This is a virtual component which does not implement any algorithms. The values of all ports of this component are calculated from the values of ports of other submodels, using logical or arithmetic expressions which are determined by the user in the scheme of the model system. The values of expressions are calculated by the kernel in runtime.

When dealing with a simulation grid, a so-called "edge effect" is inevitable. This effect is due to the fact that cells on the edge of the grid have no neighbors on one side. One method to eliminate this effect is "mapping onto a torus" where the cells from the opposite edge of the simulation grid act as wanted neighbors. This operation is implemented in the special procedure of the system kernel. Such spatial organization frees developers of submodels from the implementation of some routine procedures. However, developers may implement other approaches to eliminate the edge effect, e.g., "tiling", "mirroring", or others (Haefner et al., 1991).

Submodels included in the scheme can work with different temporal resolution. To provide the proper "team-work" of submodels, the system will operate with a time step equal to the minimal temporal resolution among all submodels.

DLES allows simulations to be carried out not only in two-dimensional but also in three-dimensional space by using the two-dimensional plot-level array of lists of variables, which allows 3D data storage.

### 20.3.5 Interconnection of Submodels: The Scheme of the Model System

When all submodels are ready, it is necessary to combine them into the unified system. To describe such a system, an XML-format scheme is used, as mentioned above. Each scheme consists of three sections: (1) general description of the scheme; (2) description of components; and (3) description of relationships between components.

The general scheme description contains the characteristics of the simulation grid (its size and the size of cells), minimal and maximal time steps, and user comment to the scheme. The description of a component contains its name, a path to the file from which it should be loaded, its version, an execution order, and spatial and temporal resolution. The description of a relationship contains the relationship type, names of connected ports, parameters of spatial and temporal synchronization, and conversion of measurement units.

For the system to work properly, the information stored in the scheme must be correct. Therefore, before running a simulation the scheme must be checked for errors. A special procedure checks the actual presence of components listed in the scheme, the correctness of port connections, and many other options.

The XML document (http://www.w3.org/XML/) is considered as the scheme. It can be edited in any text editor, but this is laborious and can be a source of potential errors. In order to simplify the work with schemes, a special scheme editor was developed. This scheme editor allows: (1) viewing, editing, and saving of schemes; (2) validation of schemes; (3) adding/removing of components; (4) changing the execution order of components; and (5) connecting component ports by relationships and setting their parameters (execution time, spatial and temporal synchronization, conversion of measurement units).

### 20.3.6 How It Works

All components are combined into the model system under the management of the kernel. At the initial stage, the kernel loads the scheme and analyzes it. The kernel loads all submodels listed in the scheme and requests service information (spatial and temporal resolution of submodels) and information about ports of submodels.

The kernel allocates memory according to the spatial scale and type (numeric, string, Boolean, or list) of variable. All ports of all submodels on the united list of state variables are managed by the system kernel. Each submodel can access any of these state variables and get their values through pointers. A conceptual sketch of data exchange is shown in Fig. 20.3.

Some ports of a submodel can be combined into a group of coherent ports called a "*port-array*." Each port-array has a controlling variable which specifies the number of elements in the array. The kernel automatically calculates the value of the controlling variable on the basis of the port-array length during initialization. The length of the port-array cannot be changed after the initialization of the model system.

An *object* is a specific type of port-array. It is a group of coherent ports-arrays whose length can be changed during modeling. Such ports also have the controlling variable which maintains array length. For example, let's consider a model of a forest stand with individual trees. Each tree has its own characteristics: height, diameter, crown size, and others. A possible situation during the simulation is the introduction of new trees when simulating natural regeneration/planting and the removal of some trees when simulating mortality. Thus, the most convenient way to keep the characteristics of individual trees is by using ports-arrays. The variable defining a number of trees will automatically change during modeling as well as change the length of corresponding port-objects. This feature allows the integrity of the array to be maintained for all operations of adding or removing the objects.

Each object represented in the model system has a unique identifier. Usually, there are a number of cells on the simulation grid associated with each object, for example, to represent crown projection or nutrition area of an individual tree. Obviously, such areas may overlap, so some cells will be related to several trees. The ability to store such data is implemented via the *List* type, which is a specialized linear array. Each element of such an array is related to one cell of the simulation grid and stores the identifiers of objects occupying this cell. *List* also can be used, for example, to represent the spatial distribution

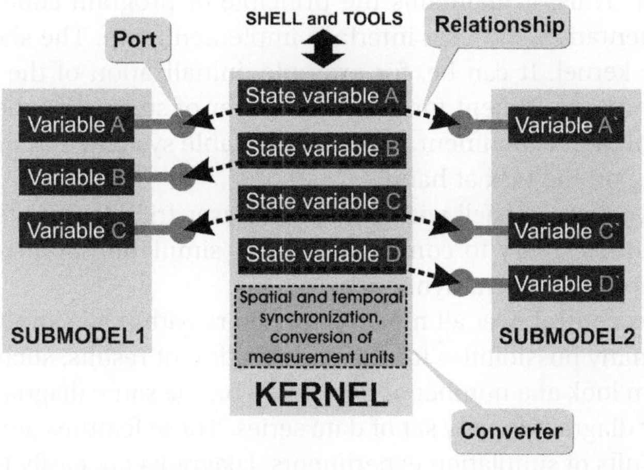


**FIGURE 20.3** Conceptual sketch of data exchange in DLES.

of different resources, such as available solar radiation under the canopy or the nutrient content of the soil.

After allocating memory for variables of all submodels, the kernel begins the analysis of relationships listed in the scheme. Then it connects ports according to descriptions of relationships. As submodels can work with different spatial and temporal resolution, and the ports of these submodels may have different units of measurement, it is necessary to provide correct synchronization during data exchange. To this end, the kernel takes account of the coefficients of conversion for each of ports. The simplest situation is when it is necessary to convert measurement units. It is possible to convert not only simple (e.g., [m] to [km]) but also complex (e.g., [t $ha^{-1}$] to [kg $m^{-2}$]) measurement units.

Moreover, the kernel takes care of spatial and temporal synchronization. For example, consider the relationship between ports of a plot-level submodel with a monthly time step and those of a plot-level submodel with an annual time step. According to the scheme, the plot-level submodel should be run first. The kernel runs the first submodel 12 times and then runs the second submodel. The synchronization method can be set by the user and saved in the scheme. For example, it may be the computation of a sum for 12 months, or an average value. The synchronization between submodels with different spatial resolution is carried out in the same manner.

At the final stage, the kernel initializes all submodels, i.e., it sets the initial values for all input variables of submodels. Now the system is ready to work. On user command, the kernel runs the simulation for the given number of steps, successively running all submodels in the same order as they are listed in the scheme.

### 20.3.7 DLES User Interface

The DLES user interface is implemented as a *Shell* component which is an executable Win32-application. Thus, it maintains the principle of program code partitioning into algorithm implementation and user interface implementation. The shell translates user commands to the kernel. It can be, for example, initialization of the model system or running a simulation experiment for a given number of steps. The shell also shows the results of the simulation experiment. DLES is a scalable system, and different shells can be used depending on the task at hand.

For investigative tasks, a shell with a graphical user-friendly interface (GUI) is most suitable. It is often necessary to compare different simulation approaches or simulate different scenarios of ecosystem dynamics.

The shell allows control over all model parameters within all simulation steps.

The shell has many possibilities for the visualization of results, such as diagrams and maps. The user can look at a number of data series on the same diagram. It is possible to create a particular diagram for any set of data series. These features are very useful when illustrating the results of simulation experiments. Diagrams are easily formatted without any external editors. Modeling results are easy to export to various textual or graphical formats. In other words, this kind of shell is the most suitable for simulation experiments on individual plots with "manual" control under experimental procedure.

Using a shell with GUI, several simulations can be carried out simultaneously with different sets of models and/or files with initial data, and the results of these simulations can be combined on one graph plot for comparative analysis. This feature is implemented by using "Project groups" which are special XML-based files combining information on schemes and initial data sources.

Another type of shell is a console shell, which is the most suitable for rapid computations due to high performance. The model system is run with command line parameters. It allows DLES to be integrated with other applications, e.g., decision support systems or GIS. Also, the batch files can be used to automate some experimental procedures.

For landscape-level modeling, it is necessary to create a database-oriented shell. This kind of shell facilitates simulations on large territory, and the results can be shown on thematic maps.

## 20.4 Exemplary Applications

DLES allows simplification of routine work, e.g., comparison of different simulation scenarios or different versions of submodels, site-specific calibration of parameters, research on the theoretical basis of submodels, and sensitivity analysis and verification.

We are developing a large simulation model of nutrient turnover in forest ecosystems based on a previous version (EFIMOD2, Komarov et al., 2003). This system will consist of a number of submodels, including a submodel of soil organic matter dynamics (ROMUL), submodels of the productivity of individual trees, a mortality and reallocation of biomass increment, a spatially explicit individual-based submodel of forest stand, and a dynamic model of forest ground vegetation. Some of these models have already been developed and tested.

### 20.4.1 Simulation Experiment with Different Litter Fractions

Our approach makes it easy to compare different versions of a submodel. For example, there are different approaches to calculating nitrogen leaching in soil. Different implementations may have structural or functional differences. Also, users can compare different initial values of variables for the same submodel, using only one scheme and several sets of initial values. These actions can be done directly in runtime by means of the DLES user interface. The example of comparison of two versions of the submodel ROMUL is discussed below.

ROMUL (Chertov et al., 2001) allows calculation of the dynamics of soil organic matter (SOM) in forest soils and the corresponding dynamics of nitrogen, including the evaluation of the amount of mineral nitrogen which is available for plants.

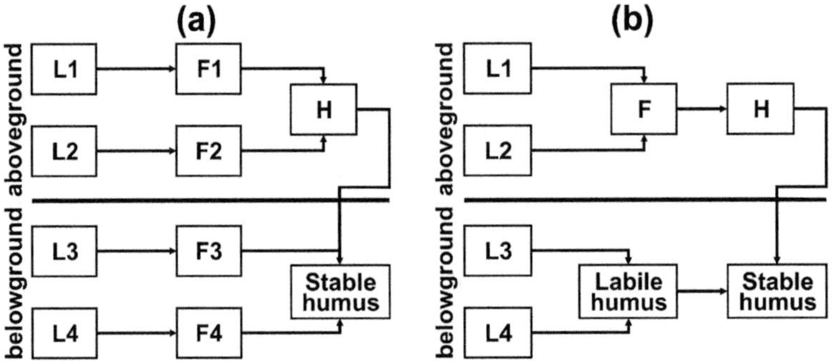


**FIGURE 20.4** Scheme of submodel ROMUL. The older version (a) has separate pools for SOM in F horizon of the forest floor, while in the newer version (b), these pools are combined into one. L1–L4, F1–F4, and H are horizons of the forest floor.

We developed two versions of the ROMUL submodel and, correspondingly, two model schemes. In an older version, the organic matter in F horizon of the forest floor was divided into several fractions related to the corresponding fractions of litter fall, the same as in L horizon. In this version, SOM was united in one pool only in H horizon of the forest floor (Fig. 20.4a). In a newer version, the organic matter in F horizon of the forest floor was united into one pool (Fig. 20.4b). We tried to prove that dividing F horizon of the forest floor into fractions is senseless due to the extreme scarcity of input data on SOM content in different fractions of the forest floor. So, it would be more convenient to use the newer version of the submodel.

The data on SOM content in different fractions of the forest floor in a spruce forest (Lukina et al., 2008) were used for the simulation experiments. The same set of data was used for both versions of the submodel. The simulation period was 50 years. Some results are shown in Fig. 20.5.

Comparison of versions showed no significant difference between the dynamics of pools of organic matter. The newer version indicated lower SOM content in the forest floor and slightly higher SOM content in the humus pool compared with the older version. The most significant difference was observed in the forest floor. By the end of the simulation period, this pool was 10.6% lower than in the older version. In contrast, the amount of humus predicted by the newer version was approximately 2% higher. This was probably due to a higher decomposition rate of the forest floor. Thus we can compare versions and make a decision about the model's structure.

Another example is a comparison of soil organic matter dynamics on long-term monitoring sites in Alice Holt Forest, United Kingdom (Vanguelova et al., 2011). Two climatic scenarios were used: (1) a semistationary climatic scenario compiled from the measurements for the years 1951–2000 and (2) the most "dramatic" climate change scenario based on the HadCM3 model and an A1Fi emission scenario. In this case, we used one model scheme and one set of initial data for the ROMUL model of SOM

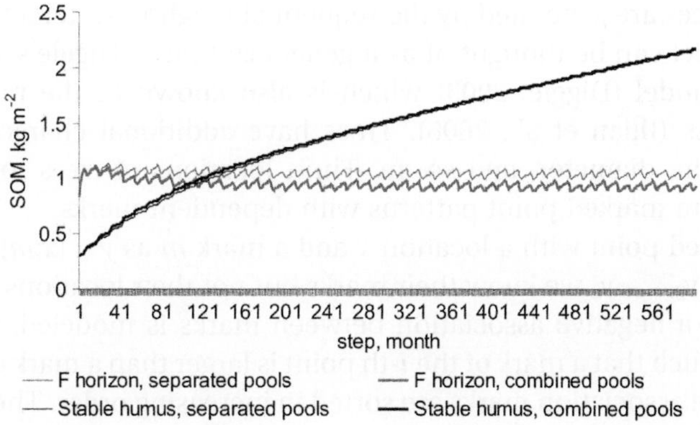


**FIGURE 20.5** Simulated dynamics of soil organic matter produced from different versions of the submodel. The older version (a) has separate pools for SOM in F horizon of the forest floor, while in the newer version (a), these pools are combined into one.

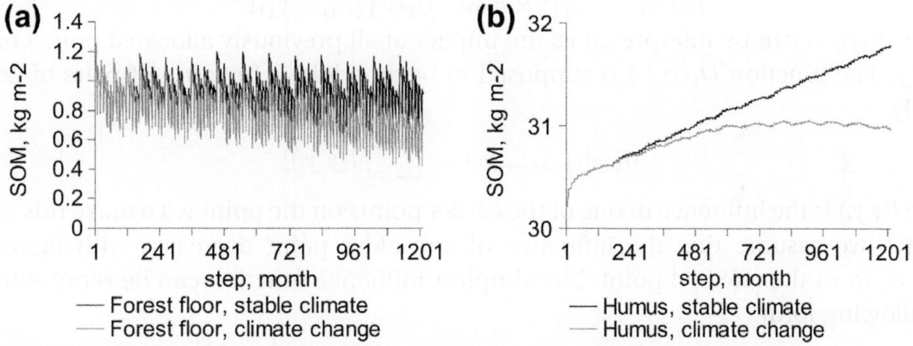


**FIGURE 20.6** Simulated dynamics of SOM in pools of forest floor (a) and humus (b) under different climate scenarios.

dynamics. As in the previous example, DLES allowed us to run both simulations simultaneously and compare the dynamics of carbon stocks in soil under different climatic scenarios (Fig. 20.6). This feature is useful when comparing different sets of initial data or calibrating submodels.

## 20.4.2 An Algorithm for Generating the Spatial Structure of a Forest Stand by a Marked Point Process with Hierarchical Interactions

Generation of tree locations is part of the initialization block of individual-based models of forest dynamics. Tree patterns are generated by a model describing a system of interacting objects, locations of which are specified in space. Spatial interrelations of modeled trees determine interactions between the objects through partitioning of common resources, and the relations are distance- and size-dependent.

Locations of trees are generated by the sequential neighbor-interaction marked point process. This model can be thought of as a generalization of Diggle's simple sequential inhibition (SSI) model (Diggle, 2003) which is also known as the random sequential adsorption process (Illian et al., 2008). Trees have additional characteristics, such as species, age, height, diameter, and so on. Thus, locations of trees together with their characteristics form marked point patterns with dependent marks.

Denote a marked point with a location $x$ and a mark $m$ as $y = [x;m]$. Assume that for marked points $y_1, y_2, \ldots, y_n$ we know their marks but not their locations. First, we need to sort the marks. If a negative association between marks is modeled, we sort marks in decreasing order such that a mark of the $i$-th point is larger than a mark of $(i+1)$-th point, while for a positive association marks are sorted in increasing order. Then we allocate the marked points sequentially by the mark value in a bounded region $A$ in the following fashion. The location $x_1$ of the first point $y_1$ has uniform distribution over $A$. The location of the $k$-th point depends on both the locations and the marks of the earlier points $y_1, y_2, \ldots, y_{k-1}$ and it is determined by a conditional density

$$f_k(y|y_{k-1},\ldots,y_1) \propto \exp(-U_k(y|y_{k-1},\ldots,y_1)) \tag{1}$$

where $U_k(y|\ldots)$ can be interpreted as the impact of all previously allocated points on the point $y$. The function $U_k(y|\ldots)$ is supposed to be an additive function of pairs of points, namely

$$U_k(y|y_{k-1},\ldots,y_1) = \sum_{i=1}^{k-1} I(y, y_i)) \tag{2}$$

where $I(y,y_i)$ is the influence of one of the earlier points on the point $y$. To make this model realistic, we assume that the influence of any older point decreases with increasing distance from the affected point. The simplest influence function can be represented in the following form

$$I(y,z) = \theta 1(d(y,z) < R) \tag{3}$$

where $d(y,z)$ is the distance between two points $y$ and $z$, $R$ is the interaction radius, which may depend on the points, and $\theta$ is a parameter characterizing the strength of the influence. For negative association of marks, the parameter $\theta$ should be positive, and if a positive association of marks is modeled, then $\theta$ should be negative. If $\theta = 0$, it corresponds to the case where there is no interaction between points, and therefore marks are independent. This model can be considered as a particular case of the multivariate point process with hierarchical interactions studied by Grabarnik and Särkkä (2009), who applied their model to an analysis of the spatial structure of a forest stand. In the current example (Fig. 20.7), we choose as the mark distribution the normal distribution with the expectation 24 and the variance 9, and the influence function $I(y,y_i)$ as

$$I(y,z) = \theta 1(d(y,z) < m(z)/8)\frac{m(z)}{4d(y,z)} \tag{4}$$

where $m(z)$ is the diameter of tree located at point $z$ and the interaction parameter $\theta = 0.3$. The chosen parameterization corresponds to an ecologically realistic situation.

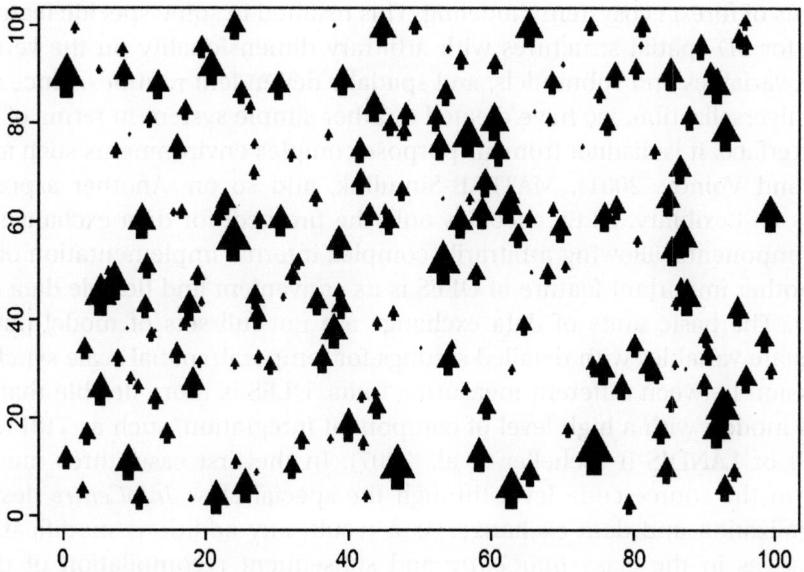


**FIGURE 20.7** Simulated spatial distribution of trees.

## 20.5 Conclusion

A new component-based framework is suggested. It is a scalable system, so it can be used to build large simulation models of terrestrial ecosystems in different climatic regions, taking into account various impacts (climate change, wildfires, windfalls, etc.) and silvicultural practices. The proposed framework has a number of merits:

- It allows easy combining of submodels and provides a range of routines to facilitate accounting of spatially explicit interactions within submodels.
- The system of models within the DLES framework has a standard description based on human-readable XML. This description includes the list of submodels and a definition of their interaction.
- DLES allows data exchange between submodels written in different programming languages and works with different spatial and temporal resolutions by means of a special system unit which automatically converts data with different measuring units. The parameters of these conversions are described in a special XML-based document.
- It is possible to run several simulation experiments simultaneously and combine and analyze the results of these simulations in one diagram.
- There is a set of statistical procedures for data processing, model comparison, and sensitivity analysis.

The main task was to develop a tool which facilitates combining of models with a discrete time step and 0, 1, 2, or 3 discrete spatial dimensions. Special attention was paid to the requirements of forest ecosystem modeling. This resulted in some specific features, such as support for 3D spatial structures with arbitrary dimensionality on the vertical axis, object-level variables and submodels, and spatially dependent routines. Since we didn't need full universalization, we have created a rather simple system in terms of structure and user interface. It is distinct from all-purpose complex environments such as STELLA (Costanza and Voinov, 2001), MATLAB-Simulink, and so on. Another aspect of this approach is its flexibility. DLES specifies only the protocol for data exchange between different components, allowing arbitrarily complex internal implementation of the submodels. Another important feature of DLES is its convenient and flexible data exchange mechanism. The basic units of data exchange are not full sets of model outputs but individual state variables with detailed settings for temporal/spatial scale synchronizing and conversion between different measuring units. DLES is more flexible than component-based models with a high level of component integration, such as TRIPLEX (Peng et al., 2002) or LANDIS-II (Scheller et al., 2007). In the first case, three models were combined on the source-code level through the special class *InfoCentre* designed for model initialization and data exchange. As a result, any additions/modifications may require changes in the class *InfoCentre* and subsequent recompilation of the whole project. LANDIS-II consists of the core, the shell, and extensions which are more specialized than those in DLES. The core is not only an extension manager. It also defines data structures for representing simulated ecosystems and implements many other functions. There are some strictly predefined types of extensions, e.g. succession, disturbance, and so on (Scheller et al., 2010).

The approach used in DLES is most similar to OpenMI (Gregersen et al., 2007), Capsis (de Coligny, 2005), and Eclpss (Wenderholm, 2005), because our framework is based on similar ideology, uses similar concepts, and differs mainly in terms of technical details. The most significant difference from the OpenMI is a unified user interface and unified management of input data, which are common for all models in the system. The orientation of OpenMI to watershed modeling has defined some other differences in supported data types and their management. Unlike Capsis, DLES works with models written in any programming language that supports COM. Since DLES links components in runtime, changes in the structure of the model system do not require recompilation of the entire system. Eclpss framework (Wenderholm, 2005) provides a wide set of tools for design and management of models, while DLES itself allows only data exchange between components. Shell and scheme editor are independent tools not included in the kernel. This resulted in simplification of user interface.

In contrast to many other frameworks, the kernel unit and user interface in DLES are separate. This allows various independent implementations of the GUI for different tasks.

Thus, DLES is the optimal tool to solve our problems because it provides the best (in our view) balance between flexibility in configuring the structure of the model system and possibilities in internal implementation of its individual components.

Future development of integrated modeling faces a number of problems. Once created, software should be further developed and maintained. When we have a deal with proprietary systems we can't extend their functional capabilities or change their features. Therefore, the framework for integrated modeling should be fully open-source.

Another question of principle is "a certain disparity between the software developer and the researcher views upon models and modules" (Voinov et al., 2004). So, for a software developer a module is a "black-box," in contrast to researchers, who use modules as tools for understanding system operation. Theoretically, models should be shared on the Web as executable modules with corresponding concept descriptions to avoid this contradiction. Therefore, standardization of the design process and unification of component interfaces are needed. We believe that these steps will make the models more transparent and likely to be reused. A researcher or decision maker can combine components to make the system meet his or her requirements. Some steps have already been taken, including working out standard requirements for the description of sub-models (Grimm et al., 2006; Gregersen et al., 2007). All steps listed above are possible only through an active community of modelers.

Nowadays, in the field of ecological modeling there is a transition from empirical models with local application to complex simulation models consisting of a large number of blocks. Such systems are usually focused on the application of models at landscape level or higher, up to global level. Models are not only research tools but also a means to assess possible development trends and to help in subsequent decision making. This requires the involvement not only of ecological models but also of economic and social models, and so on. Such processes can be greatly facilitated by special tools for the integration of models.

DLES is a registered software (certificate no. 2012610395).

# Acknowledgments

We are grateful to the team of the Laboratory of Ecosystem Modelling (Institute of Physico-Chemical and Biological Problems in Soil Science of the Russian Academy of Sciences) for help and discussion. This work was supported by RFBR projects 09-04-01209, 12-04-01269 and 12-04-01527, and Programme 16 of RAS Presidium "Changes of climate and environment: natural disasters".

# References

Argent, R.M., Voinov, A., Maxwell, T., Cuddy, S.M., Rahman, J.M., Seaton, S., et al., 2006. Comparing modelling frameworks: a workshop approach. Environ. Model. Software 21 (7), 895–910.

Chertov, O.G., Komarov, A.S., Nadporozhskaya, M.A., Bykhovets, S.S., Zudin, S.L., 2001. ROMUL: a model of forest soil organic matter dynamics as a substantial tool for forest ecosystem modelling. Ecol. Model. 138 (1–3), 289–308.

Costanza, R., Voinov, A., 2001. Modeling ecological and economic systems with STELLA: Part III. Ecol. Model. 143, 1–7.

de Coligny, F., 2005. Capsis: computer-aided projection for strategies in silviculture, a software platform for forestry modellers. In: Proceedings of Workshop on Information Science for Agriculture and Environment (ISAE) 3–4 June 2005. GuiZhou Normal University, GuiYang, P.R. China.

de Coligny, F., Ancelin, Ph., Cornu, G., Courbaud, B., Dreyfus, Ph., Goreaud, F., Saint-Andre, L., 2002. CAPSIS: Computer-Aided Projection for strategies in silviculture: open architecture for a shared forest-modelling platform. In: Proceedings of the IUFRO Working Party S5.01-04 Conference (September 2002). Harrison, British Columbia, Canada, pp. 371–380.

Diggle, P.J., 2003. Statistical Analysis of Spatial Point Patterns, second ed. Hodder Education Publishers, London. ISBN 978-0-3407-4070-5.

Grabarnik, P., Särkkä, A., 2009. Modelling the spatial structure of forest stands by multivariate point processes with hierarchical interactions. Ecol. Model. 220, 1232–1240.

Gray, M.A., 2011. Introduction to the Simulation of Dynamics using Simulink. Chapman and Hall/CRC, Boca Raton, FL. ISBN: 978-1-4398-1897-8.

Gregersen, J.B., Gijsbers, P.J.A., Westen, S.J.P., 2007. OpenMI: open modelling interface. J. Hydroinform. 9 (3), 175–191.

Grimm, V., Berger, U., Bastiansen, F., Eliassen, S., Ginot, V., Giske, J., et al., 2006. A standard protocol for describing individual-based and agent-based models. Ecol. Model. 198, 115–126.

Haefner, J.W., Poole, G.C., Dunn, P.V., Decker, R.T., 1991. Edge effects in computer models of spatial competition. Ecol. Model. 56, 221–244.

He, H., Mladenoff, D., Boeder, J., 1999. An object-oriented forest landscape model and its representation of tree species. Ecol. Model. 119 (1), 1–19.

He, H.S., Larsen, D.R., Mladenoff, D.J., 2002. Exploring component-based approaches in forest landscape modeling. Environ. Model. Software 17, 519–529.

Illian, J., Penttinen, A., Stoyan, H., Stoyan, D., 2008. Statistical Analysis and Modelling of Spatial Point Patterns. Wiley-Interscience, Chichester. ISBN 978-0-4700-1491-2.

Knox, R.G., Kalb, V.L., Levine, E.R., Kendig, D.J., 1997. A problem-solving workbench for interactive simulation of ecosystems. IEEE Comput. Sci. Eng. 4 (3), 52–60.

Komarov, A.S., Chertov, O.G., Zudin, S.L., Nadporozhskaya, M.A., Mikhailov, A.V., Bykhovets, S.S., et al., 2003. EFIMOD 2: a model of growth and cycling of elements in boreal forest ecosystems. Ecol. Model. 170, 373–392.

Konsynski, B.R., Kottemann, J.E., Nunamaker Jr., J.F., Stott, J.W., 1985. PLEXSYS84: an integrated development environment for information systems. J. Manage. Inform. Sys. 1 (3), 64–104.

Liu, J., Peng, C., Dang, Q., Apps, M.J., Jiang, H., 2002. A component object model strategy for reusing ecosystem models. Comput. Electron. Agr. 35, 17–33.

Lukina, N.V., Polyanskaya, L.M., Orlova, M.A., 2008. Soil Nutrient Condition in Northern Taiga Forest. Nauka, Moscow (in Russian). ISBN 978-5-0203-5585-9.

Lynch, S., 2010. Dynamical Systems with Applications Using MAPLE, second ed. Birkhäuser, Boston. ISBN 978-0-8176-4389-8.

May, H.K., Conery, J.S., 2003. FOREST: a system for developing and evaluating ecosystem simulation models. In: Proceedings from Summer Computer Simulation Conference. Montreal, QC, July 2003, pp. 119–127.

Moore, R.V., Tindall, C.I., 2005. An overview of the open modelling interface and environment (the OpenMI). Environ. Sci. Policy 8 (3), 279–286.

Peng, C., Liu, J., Dang, Q., Apps, M.J., Jiang, H., 2002. TRIPLEX: a generic hybrid model for predicting forest growth and carbon and nitrogen dynamics. Ecol. Model. 153, 109–130.

Robinson, A., Ek, A., St Paul, M., 1998. ACRONYM: a hierarchical tree and forest growth model framework. Staff Paper Series 127. Retrieved from. http://www.forestry.umn.edu/publications/staffpapers/Staffpaper127.pdf

Rogerson, D., 1997. Inside COM. Microsoft Press, Redmond, WA. ISBN 978-1-5723-1349-1.

Scheller, R.M., Domingo, J.B., Sturtevant, B.R., Williams, J.S., Rudy, A., Eric, J., et al., 2007. Design, development, and application of LANDIS-II, a spatial landscape simulation model with flexible temporal and spatial resolution. Ecol. Model. 201, 409–419.

Scheller, R.M., Sturtevant, B.R., Gustafson, E.J., Ward, B.C., Mladenoff, D.J., 2010. Increasing the reliability of ecological models using modern software engineering techniques. Front. Ecol. Environ. 8 (5), 253–260.

van Ittersum, M.K., Ewert, F., Heckelei, T., Wery, J., Olsson, J.A., Andersen, E., et al., 2008. Integrated assessment of agricultural systems: a component-based framework for the European Union (SEAMLESS). Agri. Syst. 96 (1–3), 150–165.

Vanguelova, E., Pitman, R., Benham, S., Wilkinson, M., Komarov, A., Khoraskina, Y., et al., 2011. Long term changes in soil carbon at Alice Holt forest modelled by ROMUL (forest soil organic matter process model). Poster presentation at COST 0803 meeting Carbon balance after disturbances and drought, 27 June–1 July, 2011, Barcelona, Spain.

Veryard, R., 2000. On determining the requirements for software components. SCIPIO. Retrieved from. http://www.users.globalnet.co.uk/~rxv/cbb/cbb-requirements.pdf

Voinov, A., Costanza, R., Wainger, L., Boumans, R., Villa, F., Maxwell, T., et al., 1999. Patuxent landscape model: integrated ecological economic modeling of a watershed. Environ. Model. Software 14 (5), 473–491.

Voinov, A., Fitz, C., Boumans, R., Costanza, R., 2004. Modular ecosystem modeling. Environ. Model. Software 19, 285–304.

Vorotyntsev, A.V., 2009. Ecological models as scenarios in net library of computable models. Proceedings of the 1st National Conference on Ecological Modelling, Pushchino, Russia, pp. 66–67 (in Russian).

Wenderholm, E., 2005. Eclpss: a Java-based framework for parallel ecosystem simulation and modeling. Environ. Model. Software 20, 1081–1100.

Wilson, G.V., 2006. Where's the real bottleneck in scientific computing? Am. Sci. 94, 5–6.

Wolfram, S., 2003. The Mathematica Book, fifth ed. Wolfram Media, Champaign, IL. ISBN 978-1-5795-5022-6.

PART 7

# Models of Landscapes

# 21

# Understanding Forest Changes to Support Planning: A Fine-Scale Markov Chain Approach

Marco Ciolli*, Clara Tattoni[†], Fabrizio Ferretti[‡]

*UNIVERSITÀ DI TRENTO, DIPARTIMENTO DI INGEGNERIA CIVILE E AMBIENTALE (DICA), VIA MESIANO 77, 38123 TRENTO, ITALY, [†]MUSEO DELLE SCIENZE, SEZIONE ZOOLOGIA DEI VERTEBRATI, VIA CALEPINA 14, 38122 TRENTO, ITALY, [‡]UNITÀ DI RICERCA PER LA GESTIONE DEI SISTEMI FORESTALI DELL'APPENNINO, CONSIGLIO PER LA RICERCA E LA SPERIMENTAZIONE IN AGRICOLTURA (CRA), ISERNIA, ITALY

## 21.1 Introduction

In land use planning, there is an increasing demand for the ability to predict future scenarios to guide decision making. Land use and cover have profound impacts on the environment, the economy, and society. Generally, there are major conflicts among the urban, agricultural, and natural uses of land (Araya and Cabral, 2010; Chen et al., 2010; Chuanga et al., 2011; Falahatkar et al., 2011; Muller and Middleton, 1994).

In different countries, researchers and planners are required to recognize several different processes that influence land use and cover. In rapidly developing areas, many studies show that urban areas have expanded during recent decades and that natural areas and woodland have been reduced (Fenglei et al., 2008; Tang et al., 2007). In contrast, in European mountain areas (Sitzia et al., 2010) and also in the Italian Alps, both national forest inventory (De Natale et al., 2005) and local (Del Favero, 2004; IPLA s.p.a., 2004; Piussi, 2006; Regione Veneto, 2006; Wolynski, 2005; Sitzia, 2009) studies have reported an increase of urban areas and the abandonment of marginal rural areas, the latter of which are progressively recolonized by species of shrubs and trees.

Forest land cover in the Alps has changed extensively in recent decades. Since the 1950s, the progressive urbanization of mountain and rural dwellers has intensified (Chauchard et al., 2007; Sitzia et al., 2010). Urban areas have been enlarging, invading large expanses of agricultural and natural land, while marginal agricultural areas and farmland have gradually been abandoned due to their distance from cities and their low productivity (Tattoni et al., 2011).

This shift in land use disrupted the age-long balance between human activity and the ecosystem, a balance that created a mosaic on mountain slopes in which forests, pastures,

Models of the Ecological Hierarchy. DOI: http://dx.doi.org/10.1016/B978-0-444-59396-2.00021-3

ISSN 0167-8892, 

and agricultural areas formed the distinctive alpine landscape. This human-maintained balance between forest, meadows, and settlements created a diversity of habitats well suited for a wide range of wildlife species, such as the capercaillie (*Tetrao urogallus*), the corncrake (*Crex crex*), and several plant species (e.g., *Nardus stricta*) (Brambilla et al., 2010). Large meadows are disappearing and causing dramatic changes in the landscape, affecting biodiversity and social and cultural dynamics (Gimmi et al., 2008; MacDonald et al., 2000; Sergio and Pedrini, 2007;Tasser and Tappeiner, 2002; Weiss, 2004).

To conserve mountain ecosystems and biodiversity and to develop new tools for planners and managers (Geneletti and van Duren, 2008), a detailed understanding of past forest dynamics, their relationship with human activities (e.g., forestry, agricultural, and grazing practices) and their possible future evolution is needed.

During the last decades, geographic information systems (GIS), historical maps, aerial imagery, and remotely sensed images have proven very effective in studying land change dynamics. These tools have been widely used (particularly in forestry) to assess changes over time and to predict future scenarios based on long-term sets of observations (Benfield et al., 2005; Thongmanivong et al., 2005; Jupiter and Marion, 2008; Geri et al., 2008; Corona et al., 2007; Calvo-Iglesias et al., 2006; Garbarino et al., 2006; Kozak, 2004; Ciolli et al., 2002; Poorzady and Bakhtiari, 2009; Gautam et al., 2004; Musaoglu et al., 2005; Liu et al., 2009).

Agarwal et al. (2002) presented a framework to compare models of land use change with respect to scale (spatial and temporal), complexity, and their ability to incorporate space, time, and human decision making. Several different approaches have been developed to predict future land use transformations. In particular, spatially explicit Markov chain (MC) models have proven effective in the forestry sector, incorporating remotely sensed and GIS data (Brown et al., 2000; Solow and Smith, 2006; Yuan, 2010; Lopez et al., 2001; Balzter, 2000; Del Rio et al., 2005; Cabral and Zamyatin, 2009). A feature common to many studies producing MC projections is the coarse scale (30 m to 1 km) (Cabral and Zamyatin, 2009; Yuan, 2010). However, a finer scale can represent the landscape in a more detailed way, thereby improving landscape management, planning resolution, and planning accuracy.

The purpose of this chapter is to present an approach to help forest and landscape managers forecast future forest coverage in the Alps. This approach examines coverage on a fine scale (5 m), focusing on habitats for which conservation status is at risk. From the knowledge of changes in forest coverage over the last 150 years, it is possible to create a fine-scale MC model to project the observed trends of lower human pressure and land abandonment.

The steps of the procedure can be summarized as follows:

- Quantify the change in past forest coverage;
- Analyze the changes at the landscape level;
- Analyze the changes in the timberline;
- Forecast future changes in the area under the current trend scenario; and
- Evaluate the effects of the changes on conservation priority areas.

## 21.2 Study Area

We selected a representative area of the eastern Italian Alps, the Paneveggio-Pale di S. Martino Natural Park (194 $km^2$) located in northeastern Italy (see Fig. 21.1), in the Autonomous Province of Trento as a study case. The Park belongs to the Natura 2000 network, as it overlaps with Special Areas of Conservation (SAC), the dolomitic Pale di San Martino, the Lagorai mountain range, and the Forest of Paneveggio. Six of the typical alpine habitats present in the park are prioritized for conservation according to the "Habitat" Directive (92/43/EEC), while two are classified as endangered at the regional level according to the local red list (Lasen, 2006).

The human population living in the study area has been monitored since 1921, and was in decline until the 1970s and concentrated in villages at the bottoms of valleys (Zanella et al., 2010; Sitzia, 2009). Although traditional livestock farming is still practiced both in and outside the Park, the quantity and quality of livestock has changed over time

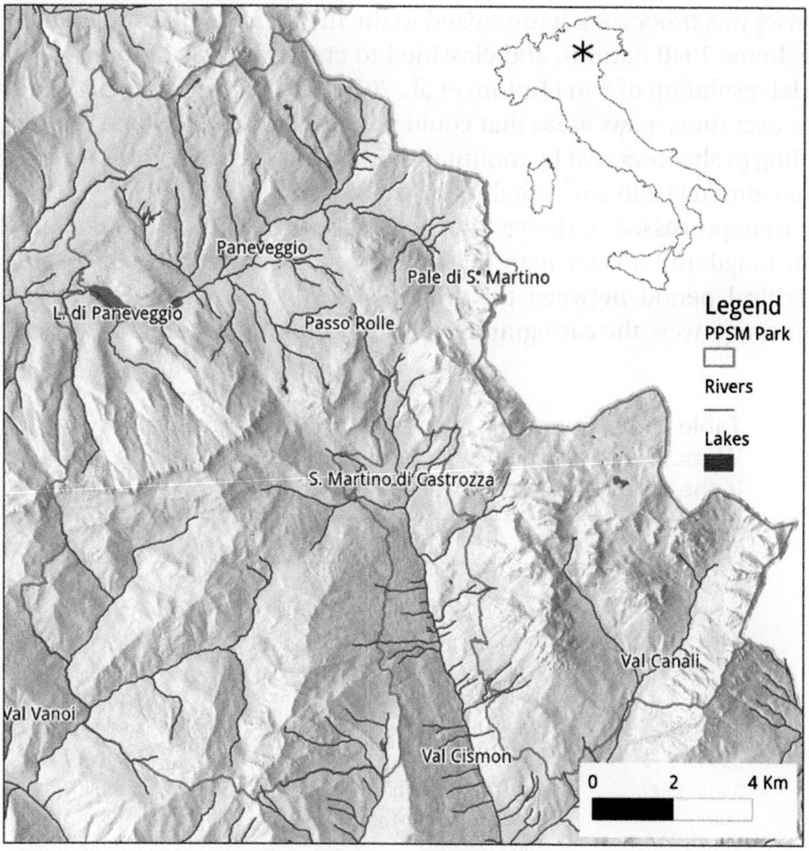


**FIGURE 21.1** Study area. The line denotes the boundary of the Park, top right is the location of the area within Italy. S. Martino di Castrozza and Paneveggio are the main villages in the area. For color version of this figure, the reader is referred to the online version of this book.

(Zanella et al., 2010). Many of the cottages of alpine herdsmen have been abandoned, and the surrounding pastures are being invaded by forests. Human and livestock pressure on the environment was very strong until the end of the 1800s, then decreased consistently until 1970s and holding constant since then. Sylviculture in the area has changed throughout history and was influenced by World War I; however, apart from a few integral reserves within the Park, the forest has been managed following close-to-nature sylviculture criteria (Motta, 2002), and no clear felling has occurred since the 1950s.

## 21.3 Materials and Methods

The sources of information for forest coverage in the past are varied and include aerial photos and historical maps covering a time span of 147 years (from 1859 to 2006) with a many time points represented (Table 21.1).

The dataset was processed, harmonized to the Italian reference system (Gauss-Boaga, West Zone, Rome 1940 datum), and classified to create a series of forest coverage maps with a spatial resolution of 5 m (Tattoni et al., 2010, 2011). To ensure spatial coverage will be constant over time, a few areas that could not be processed in some photos (i.e., areas corresponding to shadows cast by mountains or clouds) were removed from the analysis, leaving approximately 260 $km^2$ suitable for analysis.

All data were processed to derive forest coverage dynamics (Tattoni et al., 2011). The 1936 Italian Kingdom Forestry map fills a temporal gap often present in studies in the Alps: the critical period between the two world wars (Tappeiner et al., 2007). Other fundamental maps were the cartography of Natura 2000 habitats of the Park developed by

**Table 21.1** Summary of the Material Used for This Work: Survey Date is Reported in the Date Column; Scale is the Average Scale of Aerial Photos

| Year | Map | Scale |
|---|---|---|
| 1854–1859 | Habsburgic Cadastre Map | 1:2880 |
| 1936 | Italian Kingdom Forestry Map | 1:100000 |
| 1954 | G.A.I Flight | 1:35000 |
| 1973 | Flight Rossi s.r.l. | 1:10000 |
| 1983 | Flight CGR "Volo Alta Quota" | 1:10000 |
| 1987–1990 | Timberline Regional Cartography | 1:50000 |
| 1994–1995 | Flight CGR "Volo Italia" | 1:10000 |
| 1998–1999 | Flight CGR "Volo IT2000" | 1:10000 |
| 2000 | Digital Elevation Model | 1:10000 |
| 2003 | Land use map | 1:10000 |
| 2005–2006 | Flight CGR "TerraitalyTm" | 1:10000 |
| 2009 | NATURA 2000 habitat map | 1:10000 |

Lasen (2006), the Timberline Regional Cartography by Piussi (1992), and the Land Cover map of the Trento Province. These maps were used to refine the suitability maps for the MC cellular automaton model.

Two Free and Open Source Software packages, GRASS GIS (version 6.4; GRASS Development Team, 2010) and Quantum GIS (version 1.7.0; Quantum GIS Development Team, 2010), were selected to perform most of the data manipulation, storage, and analysis, and to produce figures and map layouts. Idrisi, the Andes edition (Eastman, 2006a) was selected to create MC models and simulations.

Because forest patterns are related to complex ecological processes (McGarigal, 2002), Landscape analysis was performed to quantify the changes in landscape structure from 1954 to 2006, focusing on the single class "forest" and the matrix that includes all of the areas classified as "nonforested." Landscape metrics were calculated using GRASS (Tattoni et al., 2010) selecting the landscape indices (mean patch size, patch number, and sum of edges) more frequently used to assess structural features in studies of natural reforestation (Sitzia et al., 2010). These metrics are regarded as optimal for explaining forest changes at the landscape level with two classes (McGarigal, 2002; Sundell-Turnera and Rodewald, 2008) and can be useful for understanding the spatial dynamics of the afforestation process (Vacchiano et al., 2006, Tattoni et al., 2011).

The Timberline Regional Cartography (Piussi, 1992) was used to compare the results of the image analysis. The timberline area (above 1700 m) was divided into different altimetric belts of variable heights to allow comparisons with Piussi's field surveys. One belt extended from 1700 m to 1900 m, another 11 belts of 20 m each extended up to 2120 m, and the remaining belt spanned from 2120 m to the highest altitude(s). Tree coverage was calculated according to two different thresholds: 30% corresponding with the threshold chosen by Piussi (1992) and 50% following the FAO official classification for the class "forest" (Tattoni et al., 2010). Piussi considered four coverage classes: "forest", "beyond limits" (areas above isolated trees), "potential" (areas that could potentially be colonized by trees in the future), and "excluded potential" (areas in which limiting factors prevent colonization by trees).

## 21.4 Markov Chain: Cellular Automata

Future scenarios can be predicted using spatially explicit MC models that allow one to model the state of a complex system based on its preceding state. A spatial process can be modeled using Markov chains when it is characterized by a finite number of states and known probabilities of moving from one state to another. A transition matrix containing the probabilities of an MC specifies the rates of transition between different states. Balzter (2000) gives a detailed description of the mathematical aspects of Markovian processes and their application to vegetation studies. Here, two states for land use are considered: covered by forest and not covered by forest. The probability matrix was computed from the analyses of the forest at different times and then used for future projections. Cellular

Automaton model (CA) (Cabral and Zamyatin, 2009) makes the Markovian process spatially explicit and overcomes a limit of the MC: the assumption of statistical independence of spatial units. This step is necessary because the changes in land cover are not spatially independent. The discrete CA models are based on a regular grid of cells, wherein each cell can have different statuses related to the statuses of neighboring cells; therefore, applying a CA to the MC results constrains the predictions according to landscape features. The transition probability matrix issued by the MC and the maps (created by GIS analysis) describing spatial transition rules are necessary inputs of an MC–CA model because they provide information on how many pixels are going to change and where the changes will occur.

The MC–CA inputs consist of the following:

- The land cover maps for two different time steps (for initial state definition);
- The transition probability matrix, which indicates for each cell the probability of changing from one class to another;
- The transition area matrix, which contains the number of changing cells, calculated by multiplying the probability matrix by the number of cells of each class;
- The rules of spatial transition, defined using a suitability map for each class. A suitability map represents the future probability that each cell has of changing to a given class.

In an iterative process, CA uses the suitability maps to select which cells are going to change and in which direction. At each step, a new map is created by CA, according to a multiobjective land allocation process. In this way, conflicts in land use transitions are resolved. Each new map becomes the input for the following iteration (Eastman, 2006a).

Model calibration and validation were possible due to the long time series available. To calibrate the model, we varied a number of parameters (Logofet and Korotko, 2002) to identify the best accordance between computed and observed state variables. Model validation is usually pursued by testing the selected parameters with an independent set of data; but in cases of land use change, this procedure is not possible because each state is affected by the previous state.

We used the technique of validation for spatially explicit land change models suggested by Pontius et al. (2004) and implemented in the Validate module for the Idrisi GIS. This technique assesses the predictive power of the model, making projections within the series.

The MC–CA implementation performed by Idrisi allows one to change two parameters: the proportional error and the contiguity filter. The former takes into account the error in the classification of the land use maps for each land cover class; the default setting is 0.15 because a common value of accuracy for a land use map is 85%. The latter parameter ensures spatial coherence so that land use change does not occur at random in space but occurs close to land of similar use classes; it is set by default to a Boolean filter of $5 \times 5$ pixels (Eastman, 2006a).

# 21.5 Results

## 21.5.1 Forest Coverage in the Past

A time series of high-resolution (5 m) raster maps showing forest coverage over time (Fig. 21.3) was created using image classification and error correction (Tattoni et al., 2010). Classification accuracy was always better than the 85% threshold required for land use classes (Eastman, 2006b); thus, all sets were considered suitable for future scenario projections. The area of forest coverage increased by 38% since 1859, starting from 97.9 $km^2$ of woodland in 1859 (35% of the study area) and reaching 133.9 $km^2$ (49% of the study area) in 2006 (Table 21.2). Afforestation accelerated along with the abandonment of traditional rural activities, leading to an increase in forest coverage of approximately 26% since the middle of the nineteenth century. These patterns mimic those observed in other mountain areas across Europe that have undergone similar

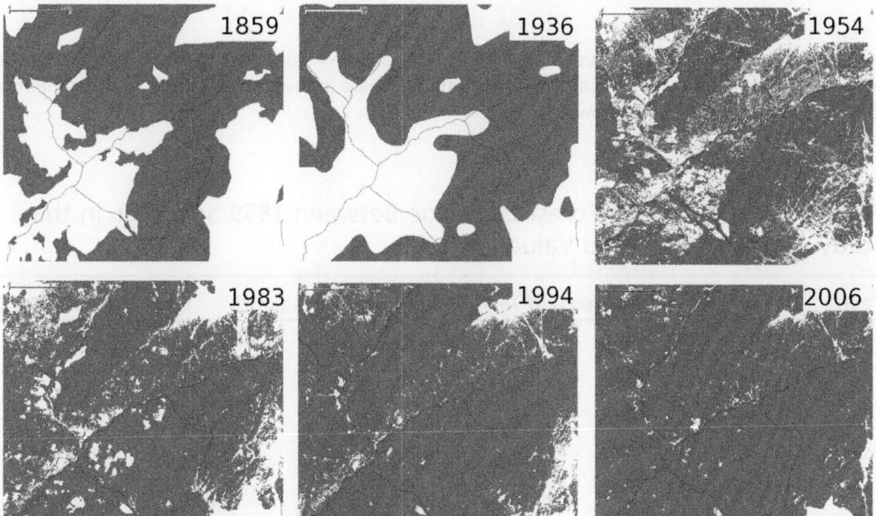

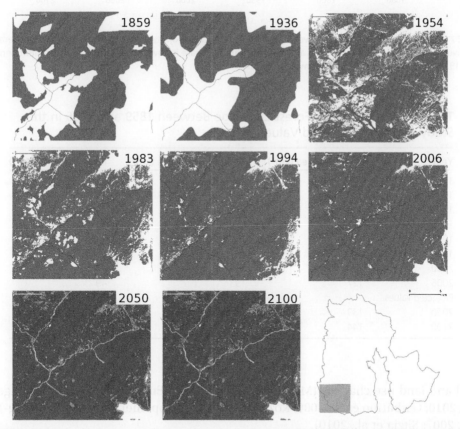


**FIGURE 21.2** Changes in forest coverage in the enlarged area (bottom left) of the study area. The first six images were derived from maps and photos, while the coverages tagged as 2050 and 2100 are the results of the MC–CA projections. For color version of this figure, the reader is referred to the online version of this book.

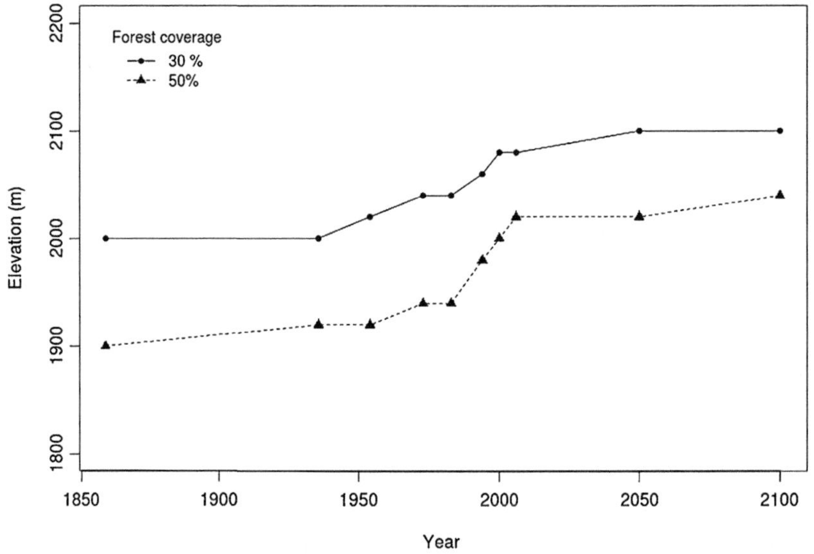


**FIGURE 21.3** Rise of the timberline, observed from 1859 to 2006 and projected after 2006. Two different coverage values were used to describe the forested areas. After a rapid increase in coverage after the abandonment period of the 1970s, the trend continues at a slower pace.

**Table 21.2** Changes in Forest Coverage Between 1859 and 2006 in the Study Area and Projected Values to 2100

| Year | Forested Area ($km^2$) | Non forested Area ($km^2$) | Forest Coverage (%) |
|---|---|---|---|
| 1859 | 92 | 168 | 35 |
| 1936 | 95 | 165 | 36 |
| 1954 | 102 | 158 | 39 |
| 1973 | 103 | 157 | 39 |
| 1983 | 111 | 149 | 43 |
| 1994 | 117 | 143 | 45 |
| 2000 | 125 | 135 | 48 |
| 2006 | 127 | 133 | 49 |
| Projected values | | | |
| 2050 | 139 | 121 | 53 |
| 2100 | 144 | 116 | 55 |

social and land use changes (MacDonald et al., 2000; European Environment Agency, 2004, 2010; Garbarino et al., 2006; Kozak et al., 2007; Tappeiner et al., 2007; Gehrig-Fasel et al., 2007; Sitzia et al., 2010).

Changes in the forest landscape between 1954 and 2006 were quantified via landscape analysis of forest maps derived from aerial photos (details in Tattoni et al., 2010).

The number of forested patches decreased, but both the average and maximum patch area increased over time. These changes reflect the merging of marginal patches into larger forests, creating a more continuous landscape (see Fig. 21.2). Areas of forest became progressively more compact and forest edges decreased in length and complexity; as a result, the amount of ecotones declined from 1954 to 2006, creating a more homogeneous landscape. Instead of forming new expansion kernels, forest colonization has followed a dynamic of enlargement and merging of existing patches, filling the open gaps abandoned by human activities. This pattern of colonization was also observed by Motta et al. (1999, 2002) in other studies of forest stands in the area.

The timberline contributed to forest expansion, rising at a rate of approximately 1.3 m/year from 1954 to the present and of approximately 0.6 m/year from 1859 to the present. The timberline estimated in 1983 and 1994 was compared with the 1992 field survey reported on the Timberline Regional Cartography: forested areas identified with photo interpretations coincided with field survey data or overlapped with land defined as "potential." Both the increase in forest surface area and the rise in the timberline observed in this work have similar trends in other areas of the Alps (Wallentin et al., 2008; Gehrig-Fasel et al., 2007). Measures of forest coverage as calculated from photo interpretations were always lower than those calculated from the Timberline Regional Cartography "potential" areas, suggesting that at these elevations, the colonization of forest colonization is far from the theoretical potential.

Field measurements from 2005 to 2007 (Ciolli et al., 2007) validated the use of image interpretation for estimating forest cover and timberline change (Tattoni et al., 2010); therefore, the obtained data are appropriate for MC projection.

Building the transition probability matrices (computed from 1954 onwards and used in the MC-CA modeling phase) would not have been possible without the solid base of the comparison of forest coverage in different time periods.

## 21.5.2 Markov Chain Settings for Landscape Analysis

To create reliable future projections, it is advisable to assess the accuracy and predictive power of the chosen model. When dealing with time series, model calibration and testing can be performed within the series itself, using two time steps as the input and a later one as comparison for the model output (Logofet and Korotko, 2002; Pontius et al., 2004).

MC–CA simulations require as input two initial states of the system: the transition probabilities (see Table 21.3) and the suitability maps for each land cover class (to guide the transition of pixels in a spatially explicit way). The development of the suitability maps is a crucial step of the simulation. In this study, the transition from open areas to forest was investigated via landscape analysis, the results of which demonstrated that new trees are likely to grow near the edges of existing forest

**Table 21.3** Probability that a Pixel of Forest Retained its Forest Status or Transitioned from Nonforest Status Over Time. The Probabilities of Forest Persistence (F-F, Forest to Forest Transition) or Increase (NF-F, Nonforested to Forest Transition) are Presented in Absolute Values and Decade Standardized. The Time Span of the Reference Period Is Reported in Years for the Reader's Convenience

| Period | Time Span | F-F | NF-F | F-F per Decade | NF-F per Decade |
|---|---|---|---|---|---|
| 1954–1973 | 19 | 0.75 | 0.17 | 0.40 | 0.09 |
| 1973–1983 | 10 | 0.79 | 0.19 | 0.79 | 0.19 |
| 1983–1994 | 11 | 0.85 | 0.17 | 0.77 | 0.16 |
| 1994–2000 | 6 | 0.92 | 0.14 | 1.00 | 0.23 |
| 2000–2006 | 6 | 0.90 | 0.11 | 1.00 | 0.18 |

rather than in the middle of open areas. Therefore, the suitability maps were created accordingly, with two states in the MC system. The following two maps were created:

- Forest suitability map: The probability of transitioning to the forest class was computed according to the proximity of existing forests, with the probability of growing new forest decreasing with increasing distance from the edge. Some ecological and man-made constraints were added to the map to give null suitability values to water bodies, glaciers and bare rocks as well as roads, urban areas, and ski slopes kept clear of trees.
- Nonforest suitability map: The future expansion of open areas is unlikely to occur in a forest expansion scenario, but this transition can take place as a consequence of natural events such as landslides, storms, the falling of old trees, or maintenance operations near roads, networks, and urban areas. A greater transitional probability was given to cells near these features.

Model calibration was performed by running the MC–CA simulation to model the forest coverage in 1994, performing simulations at different time intervals and varying the contiguity filter and proportional error parameters. The effects of different filter sizes and proportional errors were also tested (Tattoni et al., 2011). Model results were compared using the Kappa index of agreement (Rosenfield and Fitzpatrick-Lins, 1986) between the projected scenario and the "ground truth" of 1994 and by visually evaluating the spatial accuracy of the prediction as recommended by Pontius et al. (2004). The best calibration results were obtained from those simulations lacking down-weight filter and proportional error, with 91.3% of pixels correctly assigned. These parameters were used to predict forest expansion from 1994 to 2006, and the results were compared with the actual 2006 map using the Kappa index. The model correctly predicted 95.6% of the cells, with a satisfying spatial pattern visible in the enlarged area of Fig. 21.4 (first row of images). The predictive power of the model and the accuracy of the spatial

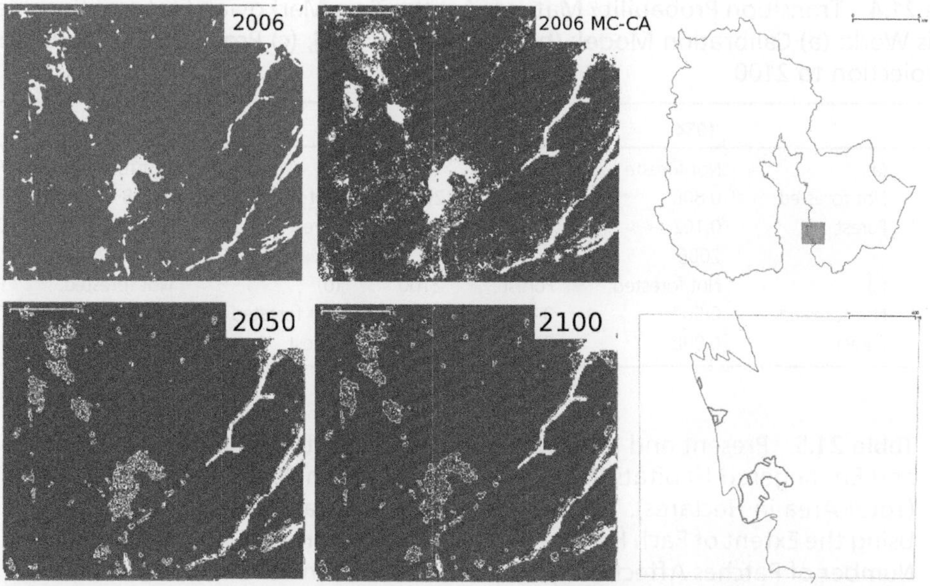


**FIGURE 21.4** MC–CA results. The image marked 2006 depicts the actual forest coverage that year from the orthophoto, and the next is the result of the inner validation of the MC–CA model. The bottom row shows the change in forest coverage according to the model in two open areas classified as priority habitat. The bottom right image presents the extent of this habitat in 2006 and the park boundary (thinner line). The enlarged area is presented in the upper right. For color version of this figure, the reader is referred to the online version of this book.

pattern within the series suggest that future projections can be made with high accuracy.

## 21.5.3 Future Scenarios of Forest Coverage

If current trends continue, forest expansion is likely to follow the same dynamics observed over the last decades. Simulations can be run under the assumptions that human pressures, the economic situation, forestry management, and climate will not change in the coming years. Future scenarios of forest coverage were obtained running the MC–CA simulation with the best parameters issued from the set up and validation process described above. The initial time steps selected were 1954 and 2006, which were used to calculate the transitional probabilities (Table 21.4) to apply the projection to the years 2050 and 2100. According to the simulations, afforestation will occur with a lower rate of expansion than that observed over the last 50 years, reaching a total coverage of 55% of the study area (Table 21.2). This trend can be considered realistic because the space available for forest expansion will become increasingly limited by elevation and residual open areas. Open habitats will likely decline or disappear in the future, as shown in Table 21.5 and Fig. 21.4 (bottom row of pictures). The 2050 and 2100 projections present more homogeneous

**Table 21.4** Transition Probability Matrices for the Two Markovian Statuses Considered in this Work: (a) Calibration Model; (b) Validation Model; (c) Projection to 2050; and (d) Projection to 2100

| | | 1954 | | | | 1954 | |
|---|---|---|---|---|---|---|---|
| 1994 | (a) | Not forested | Forest | | (b) | Not forested | Forest |
| | Not forested | 0.806 | 0.194 | 2006 | Not forested | 0.818 | 0.182 |
| | Forest | 0.102 | 0.898 | | Forest | 0.070 | 0.930 |
| | | 2006 | | | | 2006 | |
| 2050 | (c) | Not forested | Forest | 2100 | (d) | Not forested | Forest |
| | Not forested | 0.562 | 0.438 | | Not forested | 0.359 | 0.641 |
| | Forest | 0.098 | 0.902 | | Forest | 0.144 | 0.856 |

**Table 21.5** Present and Projected States of Nardetum and Other Priority and Endangered Habitats Within the Study Area in 2006, 2050, and 2100 (Total Area in Hectares). The Percentages of Habitat Loss Were Calculated Using the Extent of Each Habitat in 2006 as a Reference. $N > 50$ Indicates the Number of Patches Affected by a Loss of More Than 50% of the 2006 Area. The NATURA 2000 Nardetum Code Includes an * Indicating that Nardetum is a Priority Habitat

| | Nardetum *6230 | Other Endangered Grasslands and Meadows |
|---|---|---|
| *N* of patches | 20 | 73 |
| $N > 50$ (%), 2050 | 1 | 73 |
| $N > 50$ (%), 2100 | 19 | 73 |
| Area, 2006 | 21.8 | 75.47 |
| Area, 2050 | 13.2 | 25.56 |
| Area, 2100 | 5.93 | 17.88 |
| 2050 habitat loss (%) | 39.44 | 66.13 |
| 2100 habitat loss (%) | 72.81 | 76.31 |

landscapes (particularly the 2100 projection)—a scenario in which meadows and ecotones face dramatic reductions. In terms of ecological functions, open areas might be unable to support the same level of biodiversity before complete re-invasion, and consequences could arrive earlier than the moment of complete closure predicted by the model.

Regarding the timberline, forest coverage (calculated at either the 30% or 50% threshold, Fig. 21.3) increases in future projections. The colonization rate of the 2050 projection is in agreement with the medium tax rate of colonization calculated for the entire period (1859–2006) but smaller than the rate calculated for the period 1954–2006. The decreasing rate observed in the long-term projection is not unexpected, as the physical space available for forest expansion decreases yearly, and the occupation of forest at the highest altitudes is expected to take a long time.

#### *21.5.3.1 Priority Habitats*

According to European conservation laws, one of the main priority habitats for conservation in the Park in terms of surface is the Nardus grassland, a habitat typical of mountainous areas, which is at risk of being modified and reduced because of the ongoing invasion of trees.

Based on the simulations, 19 out of 20 *Nardeti* of the Park will undergo partial or total invasion by forests. In 2050, about 30% of this priority habitat is projected to be gone, with only one patch remaining intact and two reduced by 50%. In 2100, it is predicted that almost all Nardus grassland will vanish; eight patches will suffer a loss of 80% or more, such that they can no longer be labeled as grasslands, while the others will maintain a core opening that could preserve the habitat (Tattoni et al., 2011). The only one projected to persist intact is located at 2050 m, well above the tree line. The spatial trend predicted for one Nardus grassland area is presented in Fig. 21.4; the grassland will be completely covered by trees, the habitat will be converted to forest, and the Nardus patches that remain in the core will vanish.

Without intervention, many other European priority open areas of locally endangered habitats will share the same fate as the Nardus grasslands (Table 21.5). From a management perspective, it is important to address possible intervention in those areas most valuable for the conservation of biodiversity.

## 21.6 Discussion and Conclusions

The ability of forest planners and managers to predict future landscape changes is urgently needed. This need is particularly great in areas, such as the Alps, where human activity has undergone rapid change in recent decades. Creating reliable scenarios is a scientific challenge with important implications for managers, because land use planning can be informed by such projections. Nevertheless, research is seldom designed to provide answers that are directly applicable to management issues (MacDonald et al., 2000; Angelstam et al., 2005).

In an effort to improve projection ability, we chose a modeling scheme based on a fine spatial scale of 5 m. A common approach in similar studies (Brown et al., 2000; Cabral and Zamyatin, 2009; Yuan, 2010) is to use larger scales (30 m or 1 km); however, these scales are inadequate for capturing changes that occur in areas (such as the open areas of the present study) that are smaller than 1 ha. The coupling of a multitemporal study encompassing a 150-year data set with spatially explicit Markovian techniques demonstrated the ability of this technique to provide reliable information about past and future forest cover change.

The availability of historical data is of primary importance for understanding ecological processes and for accurate modeling (Becker et al., 2007; Schneeberger et al., 2007). Our use of historical maps revealed their usefulness in reconstructing long-term dynamics because they allowed for the calibration and validation of simulations within the series to predict the accuracy of future projections. However, the utility of historical maps will vary depending on their scale and resolution and the information contained within. Historical maps represent a potentially wealthy source of information still waiting to be exploited by researchers.

In land change projections, accuracy is expected to decrease over time (Pontius et al., 2004). The more distant a projection, the less accurate it is, and it should be interpreted more conservatively. In the present study, it is important to consider the length of the validation period. We consider the 2050 projection to be reliable because the model was validated on a time period of similar length; however, the 2100 projection encompasses greater uncertainty (Tattoni et al., 2011). Moreover, when evaluating a changing landscape, it is important to know at what point forest and human pressures will equilibrate. Change in land cover also affects the spatial relationships between habitats, such as the degree of isolation or connectivity among patches; these changes can have both positive and negative impacts on biodiversity and on the environment.

For example, an increase in forest coverage favors some species, such as tawny owls, brown bear, and red deer (Marchesi et al., 2006; Preatoni et al., 2005; Mattioli et al., 2001), and can have other positive effects on the environment, such as increases in soil stability and carbon storage (Veit et al., 2007). On the other hand, reductions of pastures, agricultural areas, and ecotones are linked to reductions in biodiversity (Tappeiner et al., 2007; Marriott et al., 2004; Sergio et al., 2005).

Generally, it is very difficult to estimate the effects of future changes on biodiversity (Chemini and Rizzoli, 2003; Sitzia et al., 2010). However, in the present study, we were able to quantify the future anticipated loss of high-biodiversity habitats by focusing on Natura 2000 sites and local red-listed habitats. Under the current trend, priority grasslands (especially the Nardetum) will vanish or be heavily altered over the next 50–100 years.

This study emphasizes the importance of predicting future trends of land cover. Such predictions can help planners and managers identify the major risks of losing biodiversity and choose feasible strategies to reduce such loss. In the present study, under the current trend, it is not realistic to actively manage all declining open areas, but it is very important to prioritize areas of need and to estimate the time left to act.

This study shows that combining a fine-scale, spatially explicit MC model with a longitudinal data series permits statistically accurate projections and likely spatial patterns. This type of procedure can be used to support management practices, but it is important to acknowledge that the suitability maps, which guide the spatial patterns of the simulations, play a fundamental role in MC–CA modeling. Therefore, it is essential to implement these maps with the same rigor as the other steps in the modeling process. An important consideration to take into account is the way suitability maps are built: good results can be obtained if maps are based on long-term observations instead of common-sense constraints (Houet and Hubert-Moy, 2006), heuristic rules (Brown et al., 2000), or expectations.

Taking into account the above considerations, and the factors unique to each case, the approach described in this study should be readily applicable elsewhere.

## Acknowledgments

The authors thank Paneveggio-Pale di S. Martino Natural Park, Prof. Bruno Zanon, CRA-MPF (Agricultural Research Council, Forest Monitoring and Planning Research Unit), and MIPAF-CFS (Ministero delle Politiche Agricole e Forestali and Corpo Forestale dello Stato).

## References

Agarwal, C., Green, G.M., Grove, J.M., Evans, T.P., Schweik, C.M., 2002. A Review and Assessment of Land-Use Change Models: Dynamics of Space, Time, and Human Choice. Gen. Technical Report NE-297. U.S. Department of Agriculture, Forest Service, Northeastern Research Station, Newtown Square, PA, 61 pp.

Angelstam, P., Boutin, S., Schmiegelow, F., Villard, M.-A., Drapeau, P., Host, G., Innes, J., Isachenko, G., Kuuluvainen, M., Moenkkoenen, M., Niemela, P., Niemi, G., Roberge, J., Spence, J., Stone, D., 2005. Targets for boreal forest biodiversity conservation – a rationale for macroecological research and adaptive management. Ecol. Bull. 51, 487–509.

Araya, Y.H., Cabral, P., 2010. Analysis and modeling of urban land cover change in Setúbal and Sesimbra, Portugal. Remote Sens. 2, 1549–1563.

Balzter, H., 2000. Markov chain models for vegetation dynamics. Ecol. Model. 126, 139–144.

Becker, A., Körner, C., Brun, J., Guisan, A., Tappeiner, U., 2007. Ecological and land use studies along elevational gradients. Mt. Res. Dev. 27 (1), 58–65.

Benfield, S., Guzman, H., Mair, J., 2005. Temporal mangrove dynamics in relation to coastal development in Pacific Panama. J. Environ. Manag. 76 (3), 263–276.

Brambilla, M., Casale, F., Bergero, V., Bogliani, G., Crovetto, G.M., Falco, R., Roati, M., Negri, I., 2010. Glorious past, uncertain present, bad future? Assessing effects of land-use changes on habitat suitability for a threatened farmland bird species. Biol. Conserv. 143 (11), 2770–2778.

Brown, D.G., Pijanowski, B.C., Duh, J.D., 2000. Modeling the relationships between land use and land cover on private lands in the upper Midwest, USA. J. Environ. Manag. 59, 247–263.

Cabral, P., Zamyatin, A., 2009. Markov processes in modeling land use and land cover changes in Sintra-Cascais, Portugal. Dyna 158, 191–198.

Calvo-Iglesias, M., Fra-Paleo, U., Crecente-Maseda, R., Díaz-Varela, R., 2006. Directions of change in land cover and landscape patterns from 1957 to 2000 in agricultural landscapes in NW Spain. Environ. Manage. 38 (6), 921–933.

Chauchard, S., Carcaillet, C., Guibal, F., 2007. Patterns of land-use abandonment control tree-recruitment and forest dynamics in Mediterranean mountains. Ecosystems 10, 936–948.

Chemini, C., Rizzoli, A., 2003. Land use change and biodiversity conservation in the Alps. J. Mt. Ecol. 7 (Suppl.), 1–7.

Chen, D., Liu, W., Tian, J., Luciani, P., 2010. Chapter 13 Evaluating the ecological and environmental impact of urbanization in the greater toronto area through multi-temporal remotely sensed data and landscape ecological measures. In: JiangB., YaoX. (Eds.), Geospatial Analysis and Modelling of Urban Structure and Dynamics, GeoJournal Library 99. Springer ScienceþBusiness Media B.V 2010.

Chuanga, C.-W., Lina, C.-Y., Chiena, C.-H., Chou, W.-C., 2011. Application of Markov-chain model for vegetation restoration assessment at landslide areas caused by a catastrophic earthquake in Central Taiwan. Ecol. Model. 222, 835–845.

Ciolli, M., Milesi, E., Zatelli, P., 2002. Digital analysis of multitemporal aerial images for forest and landscape change detection. In: Proceedings of the IUFRO (International Union of Forest Organizations) Palermo, Italy, December 2001, pp. 40–47.

Ciolli, M., Serafini, M., Tattoni, C., 2007. Storia della copertura vegetale nel Parco di Paneveggio Pale di S. Martino. Dendronatura n.1, 9–15.

Corona, P., Fattorini, L., Chirici, G., Valentini, R., Marchetti, M., 2007. Estimating forest area at the year 1990 by two-phase sampling on historical remotely sensed imagery in Italy. J. For. Res. 12, 8–13.

De Natale, F., Floris, A., Gasparini, P., Scrinzi, G., Tabacchi, G., Tosi, V., 2005. Inventario Nazionale delle Foreste e dei Serbatoi Forestali di Carbonio. Technical Report 39, CRA- Istituto Sperimentale per l'Assestamento Forestale e per l'Alpicoltura, Trento.

Del Favero, R., 2004. I boschi delle regioni alpine italiane. Tipologia, funzionamento, selvicoltura. Edizioni Cleup, pp 600.

Del Rio, S., Penas, A., Romero, R.P., 2005. Potential areas of deciduous forests in Spain (Castile and Leon) according to future climate change. Plant Biosyst. 139, 222–233.

Eastman, J.R., 2006a. Idrisi Andes. Clark University, Worcester, MA.

Eastman, J.R., 2006b. IDRISI Andes. Guide to GIS and image processing. Clark University, Worcester, MA. Accessed in IDRISI [15.00 Andes Edition].

European Environment Agency, 2004. Agriculture and the Environment in the EU accession Countries. Environmental Issue Report 37. Technical Report. European Environment Agency, Copenhagen.

European Environment Agency, 2010. 10 Messages for 2010 - Mountain Ecosystems. Technical Report 8. European Environment Agency, Denmark.

Falahatkar, S., Soffianian, A.R., Khajeddin, S.J., Ziaee, H.R., Nadoushan, M.A., 2011. Integration of remote sensing data and GIS for prediction of land cover map. Int. J. Geomatics Geosci. 1 (4), 847–864.

Fenglei, F., Yunpeng, W., Zhishi, W., 2008. Temporal and spatial change detecting (1998–2003) and predicting of land use and land cover in Core corridor of Pearl River Delta (China) by using TM and ETM+ images. Environ. Monit. Assess. 137, 127–147.

Garbarino, M., Lingua, E., Vacchiano, G., Motta, R., 2006. Scots pine forests in the NW Italian Alps. what has changed in the last 50 years? In: IUFRO Landscape Ecology Conference, Sept. 26–29, 2006 - Locorotondo, Bari (Italy).

Gautam, A.P., Shivakoti, G.P., Webb, E.L., 2004. Forest cover change, physiography, local economy, and institutions in a mountain watershed in Nepal. Environ. Manage. 33 (1), 48–61.

Gehrig-Fasel, J., Guisan, A., Zimmermann, N.E., 2007. Tree line shifts in the Swiss Alps: climate change or land abandonment? J. Veg. Sci. 18 (4), 571–582.

Geneletti, D., van Duren, I., 2008. Protected area zoning for conservation and use: a combination of spatial multicriteria and multiobjective evaluation. Landsc. Urban Plan. 85 (2), 97–110.

Geri, F., Giordano, M., Nucci, A., Rocchini, D., Chiarucci, A., 2008. Analisi multitemporale del paesaggio forestale della provincia di Siena mediante l'utilizzo di cartografie storiche. Forest@ 5(1) 82–91.

Gimmi, U., Burgi, M., Stuber, M., 2008. Reconstructing anthropogenic disturbance regimes in forest ecosystems: a case study from the Swiss Rhone Valley. Ecosystems 11, 113–124.

GRASS Development Team, 2010. Geographic Resources Analysis Support System (GRASS GIS) Software. http://grass.osgeo.org.

Houet, T., Hubert-Moy, L., 2006. Modelling and projecting land-use and land-cover changes with a cellular automaton in considering landscape trajectories: an improvement for simulation of plausible future states. EARSeL eProceedings 5 (1), 63–76.

IPLA s.p.a, 2004. Tipi Forestali del Piemonte – metodologia e guida per l'identificazione. Regione Piemonte, 204.

Jupiter, S., Marion, G., 2008. Changes in forest area along stream networks in an agricultural catchment of the great barrier reef lagoon. Environ. Manage. 42 (1), 66–79.

Kozak, J., 2004. Forest cover change in the western Carpathians in the past 180 years. a case study in the Orawa region in Poland. Mt. Res. Dev. 3 (4), 369–375.

Kozak, J., Estreguil, C., Vogt, P., 2007. Forest cover and pattern changes in the Carpathians over the last decades. Eur. J. For. Res. 126, 77–90.

Lasen, C., 2006. Habitat Natura 2000 in Trentino. Technical Report, Provincia Autonoma di Trento.

Liu, M., Hu, Y., Chang, Y., He, X., Zhang, W., 2009. Land use and land cover change analysis and prediction in the upper reaches of the Minjiang river, China. Environ. Manage. 43 (5), 899–907.

Logofet, D.O., Korotko, V.N., 2002. Hybrid optimisation: a heuristic solution to the Markov-chain calibration problem. Ecol. Model. 151, 51–61.

Lopez, E., Bocco, G., Mendoza, M., Duhau, E., 2001. Predicting land-cover and land-use change in the urban fringe a case in Morelia city, Mexico. Landsc. Urban Plan. 55, 271–285.

MacDonald, D., Crabtree, J., Wiesinger, G., Dax, T., Stamou, N., Fleury, P., Lazpita, J., Gibon, A., 2000. Agricultural abandonment in mountain areas of Europe: environmental consequences and policy response. J. Environ. Manage. 59, 47–69.

Marchesi, L., Sergio, F., Pedrini, P., 2006. Implications of temporal changes in forest dynamics on density, nest-site selection, diet and productivity of Tawny Owls Strix aluco in the Alps. Bird Study 53 (3), 310–318.

Marriott, C., Fothergill, M., Jeangros, B., Scotton, M., Lauault, F., 2004. Long-term impacts of extensification of grassland management on biodiversity and productivity in upland areas: a review. Agronomie 24, 447–461.

Mattioli, S., Meneguz, P.G., Brugnoli, A., Nicoloso, S., 2001. Red deer in Italy: recent changes in range and numbers. Hystrix It. J. Mamm. 12 (1), 27–35.

McGarigal, K., 2002. Encyclopedia of Environmentrics, volume 2, chapter Landscape pattern metrics, pp 1135–1142. John Wiley & Sons, Sussex, England.

Motta, R., 2002. Old-growth forests and sylviculture in the Italian Alps: the case-study of the strict reserve of Paneveggio (TN). Plant Biosyst. 136 (2), 223–232.

Motta, R., Nola, P., Piussi, P., 1999. Structure and stand development in three subalpine Norway spruce (*Picea abies* (l.) karst,) stands in Paneveggio (Trento, Italy). Glob. Ecol. Biogeogr. 8 (6), 455–471.

Motta, R., Nola, P., Piussi, P., 2002. Long-term investigations in a strict forest reserve in the eastern Italian Alps: spatio-temporal origin and development in two multi-layered subalpine stands. J. Ecol. 90, 495–507.

Muller, M.R., Middleton, J., 1994. A Markov model of land-use change dynamics in the Niagara Region, Ontario, Canada. Landsc. Ecol. 9 (2), 151–157.

Musaoglu, N., Tanik, A., Kocabas, V., 2005. Identification of land-cover changes through image processing and associated impacts on water reservoir conditions. Environ. Manage. 35 (2), 220–230.

Piussi, P., 1992. Carta del limite potenziale del bosco in Trentino. Servizio Foreste Caccia e Pesca della Provincia Autonoma di Trento.

Piussi, P., 2006. Nature-based Forestry in Central Europe: Alternatives to Industrial Forestry and Strict Preservation. Chapter Close to Nature Forestry Criteria and Coppice Management. pp 27–37.

Pontius, R.G.J., Huffaker, D., Denman, K., 2004. Useful techniques of validation for spatially explicit land-change models. Ecol. Modell. 179, 445–461.

Poorzady, M., Bakhtiari, F., 2009. Spatial and temporal changes of hyrcanian forest in Iran. iForest – Biogeosci. For. 2 (1), 198–206.

Preatoni, D.G., Mustoni, A., Martinoli, A., Carlini, E., Chiarenzi, B., Chiozzini, S., Van Dongenc, S., Wauters, L.A., Tosi, G., 2005. Conservation of brown bear in the Alps: space use and settlement behavior of reintroduced bears. Acta Oecol. 28 (3), 189–197.

Quantum GIS Development Team 2010. Quantum GIS Geographic Information System. http://www.qgis.org. (accessed 20.07.10).

Rosenfield, G.H., Fitzpatrick-Lins, K., 1986. A coefficient of agreement as a measure of thematic classification accuracy. Photogramm. Eng. Remote Sens. 52 (2), 223–227.

Schneeberger, N., Bürgi, M., Kienast, F., 2007. Rates of landscape change at the northern fringe of the Swiss Alps: historical and recent tendencies. Landsc. Urban Plan. 80, 127–136.

Sergio, F., Pedrini, P., 2007. Biodiversity gradients in the Alps: the overriding importance of elevation. Biodivers. Conserv. 16 (12), 3243–3254.

Sergio, F., Scandolara, C., Marchesi, L., Pedrini, P., Penteriani, V., 2005. Effect of agro-forestry and landscape changes on common buzzards (*Buteo buteo*) in the Alps: implications for conservation. Anim. Conserv. 8 (1), 17–25.

Sitzia, T., 2009. Ecologia e gestione dei boschi di neoformazione nel paesaggio Trentino. Provincia Autonoma di Trento, Servizio Foreste e Fauna.

Sitzia, T., Semenzato, P., Trentanovi, G., 2010. Natural reforestation is changing spatial patterns of rural mountain and hill landscapes: a global overview. For. Ecol. Manag. 259 (31), 1354–1362.

Solow, A.R., Smith, W.K., 2006. Using Markov chain successional models backwards. J. Appl. Ecol. 43, 185–188.

Sundell-Turnera, N.M., Rodewald, A.D., 2008. A comparison of landscape metrics for conservation planning. Landsc. Urban Plan. 86 (3–4), 219–225.

Tang, J., Wang, L., Yao, Z., 2007. Spatio-temporal urban landscape change analysis using the Markov chain model and a modified genetic algorithm. Int. J. Remote Sens. 28 (15), 3255–3271.

Tappeiner, U., Tasser, E., Leitinger, G., Tappeiner, G., 2007. Landnutzung in den Alpen: historische Entwicklung und zukünftige Szenarien. Alpine space - man & environment: Die Alpen im Jahr 2020. Innsbruck University Press 1, p. 23–39.

Tasser, E., Tappeiner, U., 2002. Impact of land use changes on mountain vegetation. Appl. Veg. Sci. 5, 173–184.

Tattoni, C., Ciolli, M., Ferretti, F., Cantiani, M.G., 2010. Monitoring spatial and temporal pattern of Paneveggio forest (Northern Italy) from 1859 to 2006. iForest – Biogeosci. For. 3 (1), 72–80.

Tattoni, C., Ciolli, M., Ferretti, F., 2011. The fate of priority areas for conservation in protected areas: a fine-scale Markov chain approach. Environ. Manage. 47 (2), 263–278.

Thongmanivong, S., Fujita, Y., Fox, J., 2005. Resource use dynamics and land-cover change in Ang Nhai village and Phou Phanang national reserve forest, Lao. Environ. Manage. 36 (3), 382–393.

Vacchiano, G., Garbarino, M., Lingua, E., Motta, R., 2006. Le pinete di Pino silvestre come testimoni delle trasformazioni del paesaggio montano in Piemonte e Val d' Aosta. In Atti della 10a Conferenza Nazionale ASITA, Bolzano 14–17/11/2006, volume 1. ASITA (Associazioni Scientifiche per le Informazioni Territoriali ed Ambientali).

Veneto, R., 2006. Carta Regionale dei tipi forestali: documento base. Venezia.

Veit, H., Scheurer, T., Köck, G., 2007. Landscape development in mountain regions. In: Proceedings of the ForumAlpinum 2007 18–21 April, Engelberg, Switzerland pp82. Austrian Academy of Sciences, Vienna.

Wallentin, G., Tappeiner, U., Strobla, J., Tasser, E., 2008. Understanding alpine tree line dynamics: an individual-based model. Ecol. Model. 218 (3–4), 235–246.

Weiss, G., 2004. The political practice of mountain forest restoration – comparing restoration concepts in four European countries. For. Ecol. Manag. 195, 1–13.

Wolynski, A., 2005. I frassineti un'occasione da non perdere. Terra trentina 7, 37–39.

Yuan, F., 2010. Urban growth monitoring and projection using remote sensing and geographic information systems: a case study in the twin cities metropolitan area, Minnesota. Geocarto Int. 25 (3), 213–230.

Zanella, A., Tattoni, C., Ciolli, M., 2010. Studio della variazione temporale della quantità e qualità del bestiame nel parco di Paneveggio Pale di S. Martino e influenza sui cambiamenti del paesaggio forestale. Dendronatura 1, 24–33.

# 22

# Development of a Program Tool for the Determination of the Landscape Visual Exposure Potential

**Imrich Jakab, Peter Petluš**

*DEPARTMENT OF ECOLOGY AND ENVIRONMENTAL SCIENCE, FACULTY OF NATURAL SCIENCES, CONSTANTINE THE PHILOSOPHER UNIVERSITY IN NITRA, NITRA, SLOVAKIA*

## 22.1 Introduction

The landscape visual connection assessment is not a new topic. It has been discussed in landscaping quite frequently. The landscape assessment has been used since the mid of the twentieth century as an object of sense-perception with scientific methods in the Anglo-American literature. Lynch (1960) and later also Zube et al. (1974), Litton and Tetlow (1978), Litton (1982), and Smardon et al. (1988) worked with the term landscape as with a visual type of information.

The authors of the above-mentioned as well as other works consider landscape as a scenic or visual source of information. We will sustain this definition of landscape in the following.

The importance of the landscape visual information assessment is associated with landscape planning activities (Clay and Smidt, 2004) and the visual impact on landscape regarding the environmental impact assessment.

The current assessments depend on expertise as well. It evaluates the landscape visual quality in a subjective way.

We realized that an exact and automatic landscape assessment as a visual information source is impossible to carry out with the help of software tools.

Since each landscape is unique, the visual landscape quality depends on the combination of unique landscape attributes. Zube et al. (1982), Jančura (1998), Štefunková (2000), Salašová (1996), Löw and Míchal (2003), Vorel et al. (2004), and other authors advocate the research expertise of landscape perception. In this regard, Coeterier (1996) mentioned examples of combined researches in cooperation with a respondent sample.

Our ambition is to create an objective assessment platform as a source of visual identifiable information. A software solution of landscape visual attributes is closely related to this issue in the works of Bishop and Hulse (1994), Shang and Bishop (2000), and Bishop et al. (2000).

**Models of the Ecological Hierarchy. DOI: http://dx.doi.org/10.1016/B978-0-444-59396-2.00022-5**

ISSN 0167-8892, 

Relief is limited for the visual landscape perception that defines the spatial elements of landscape to be perspective and visible at the same time. This limit is a basic starting point for solving issues of the landscape visual exposure.

We wish to create and test a model which will be used as an objectively measurable platform in the case of real or planned activities assessment with a visual impact on landscape. Such a tool may be considered to be complementary for landscape planning activities.

In our chapter, we will use the landscape visual exposure as a target concept and the visual exposure of the landscape potential as a source tool for finding the real landscape visual exposure.

The real landscape visual exposure predicts two fundamental factors: the shape of relief as a limit of the visual connection in landscape and the elements of the secondary landscape structure. Our software solution offers objective measurements of the potential relief connections without the elements of the secondary landscape structure.

The basic attribute entering the evaluation process is relief as characterized by the value of the visual exposure of landscape. The real exposure is affected by the quantitative abundance of the elements of the landscape structure and by their qualitative level through esthetic validation. An exact PVE is a hypothetical term in which the area is restricted by the potential of relief without the occurrence of the landscape elements. On the other hand, relief enters the process of the visual evaluation of landscape as a basic attribute.

Very subjective processes are frequently applied in the assessment of the landscape visual quality, which does not allow one to use the existing methodologies in different landscape types. Therefore, it is important to follow unbiased approaches based on quantitative data. Our chapter defines a value of each point on the relief surface. We start with the assumption that each point on the surface can be a potential viewpoint as well as a visible point. The limitation of the visual perception of landscape is the relief. It defines to what measure each spatial unit of landscape is a viewpoint and a visible point, respectively. The landscape visual exposure is thus expressed in terms of values. A given value is determined by the area from which we can identify a specific point in the landscape (the number of points in the landscape from which one point can be seen), or by the area which is identifiable from a specific point (the number of points in the landscape that can be seen from one point). The VEP is the main presumption for the visual exposure of the landscape reality—a real part of the visual landscape structure.

## 22.2 Viewshed Analysis

At present the most common GIS software does not possess a direct tool for the determination of the visual exposure of an area. It offers viewshed operations which are able to derive a new coverage from the DEM of the area. The newly created coverage shows those areas which are visible from one or more locations and which are coded as a binary image, with 1 indicating those areas which are visible and 0 those which are not (Fisher, 1995).

The basic algorithm for generating a viewshed from the DEM is based on the estimation of the elevation difference of intermediate pixels between the viewpoint and the target pixels. The determination whether the target pixels can be seen from the viewpoint is accomplished by examining each of the intermediate pixels between two cells to determine the corresponding "line-of-sight" (LOS). If the land surface rises above the LOS, the target is invisible. Otherwise, it is visible from the viewpoint. The LOS computation is repeated for all target pixels from a set of viewpoints, and the set of targets which are visible from the viewpoints form a viewshed (Burrough and McDonnel, 1998; Kima et al., 2004).

The viewshed depends on the elevation of the viewpoint. A change in the elevation results in a change in the viewshed. Different techniques have been developed concerning different ways by which (Achilleos and Tsouchlaraki, 2004):

1. The data are collected from the DEM for the viewshed calculation.
2. View Point and Target Point are defined horizontally and vertically.
3. The decision VISIBLE or NOT VISIBLE is made for a given point.

The viewshed analysis has been used in a wide range of applications, including the location of wind turbines (Miller et al., 1999; Bishop, 2002) and Observation Tower (Bao et al., 2010), assessment of landmarks visibility (Ogburn, 2006; Ohsawa and Kobayashi, 2005), analysis of archeological locations (Paliou et al., 2011), evaluation of urban design (Yang et al., 2007; Dean, Lizarraga-Blackard, 2007), and assessment of land qualities(Sevenant and Antrop, 2007).

The majority of the above-mentioned applications of the viewshed analysis uses a relatively small number of points of interest, resulting in a relatively small number of calculations. For instance, as Kima et al. (2004) mentioned in their work, the assessment of the visual intrusion of a development (e.g., a new wind turbine, new road, or new housing) is based on the evaluation of a single point, points along a line, or points on the perimeter of the investigated area. The LOS computations were in these cases repeated for all target pixels from one viewpoint or a set of viewpoints. This way of computation cannot be used for the cases in which the viewshed operations need to be used separately for all points on the terrain (e.g., the map of the visual exposure creation).

The solution of the computations in which all of the points ($n$) on the terrain are used as viewpoints and targets is partially described by Rana (2003). According to this work, it is the exhaustive but time-consuming Golden Case in which the visibility index computation time is of order $O(n^2)$.

The map of the visual exposure can also be considered to be the Golden Case solution. Therefore, each pixel of the input DEM carries a value of the visibility index. The map creation process of the potential visual exposure for a specific area requires a number of time-consuming computations.

The main factors that affect the total time of computation, besides the performance of the computer itself, are the numbers of viewpoints and target points in the viewshed computation. The number of points is influenced by the resolution of the input DEM, size of the investigated area, and maximal distance from the viewpoint inside of which the LOS analysis will be performed.

There are several possibilities for speeding up the Golden Case solution. One approach is described by Rana (2003) and Kima et al. (2004) as the reduction of the number of observers (viewpoints), targets, or both. One way to do so is to perform the analysis on a TIN rather than on a gridded DEM because there are far fewer points in a TIN than in a gridded DEM. An alternative approach is to use only a subset of the pixels as observers for each target.

Our approach to speed up the process of the visual exposure map creation is based on using parallel computations. The input map is first divided by a user into the required number of segments. The computation of the visibility in the individual segments can be performed by individual processors. This solution leads to an acceleration of the process, depending on the performance of the employed computer—from 2 or 4 processor-containing desktop computers, through computer clusters with dozens of processors, to supercomputers with several hundreds or thousands of processors.

The created segments present groups of viewpoints of approximately the same size. The number of the viewshed computations obtained by the lines-of-sight projecting within a digital model of relief remains the same, but the computations run simultaneously. Each processor can solve the computations in an individual segment, while the computation of the visibility goes beyond the borders of a segment. In other words, the target points, the parts of the computation, are not limited by the borders of segments; therefore, the computation of the visibility exceeds borders of these segments.

In the case of parallel computations, the source of viewpoints can be the created sectors, and the source of target points can be the whole input map or a map which was derived by including the border zone whose size is equal to the maximal distance from the viewpoint. Inside the newly derived border zone, the LOS analysis will be performed toward the borders of the individual sectors. Our parallel algorithm is designed for the distributed memory MIMD architecture. In order to maximize the utilization of the processors, the algorithm distributes its data among the processors and allows each processor to process the data asynchronously. This asynchronous operation ensures that each processor processes the data independently of the other processors and as fast as possible (Lanthier et al., 2003).

## 22.3 Development of a Toolkit for Modeling the Visual Exposure

As a basis for developing the toolkit, the GRASS GIS (Grass Development Team, 2011) was chosen because it is a leading open-source GIS Software.

The GRASS (Geographical Resources Analysis Support System) is a comprehensive GIS with the raster, topological vector, and image processing and with the graphics production functionality (GRASS Development Team, 2002). The GRASS GIS as a programming environment has a very stable and mature application programming interface (API) and provides a spectrum of functions for handling input/output subroutines and file management (Jasiewicz and Metz, 2011).

We developed a GRASS toolkit for solving the Viewshed analysis for a relatively large area, using the GRASS GIS modules (mainly the r.los) with the support of programming in the BASH language under OS GNU/Linux.

The module r.los generates a raster map output with cells that are visible from a user-specified observer location at a given height over the ground. The output map cell values represent the vertical angle (in degrees from the ground) required to see the cells from the observer location (Neteler and Mitasova, 2007).

The following input parameters of the r.los module (Grass Development Team, 2011) are important to create a map of the visual exposure:

- the raster map—A layer containing the elevation data used as a program input.
- the coordinates (*x*, *y*)—The Geographic coordinates identifying the desired location of the viewpoint.
- obs_elev—The value Height of the observer (in meters) above the viewpoint's elevation.
- max_dist—The maximal distance (in meters) from the viewpoint within which the LOS analysis will be performed.

The map of visual exposure as an output is a data matrix of a specified area where each cell of the area matrix possesses a value of the Visibility index transformed into selected spatial units ($m^2$, $km^2$, ares, or hectares) and where the value of each cell presents a quantitative expression of the visual exposure potential of the land (which is given by the territory visible from the cell).

Besides the automation of the r.los module performance for each pixel of the input, it is necessary to use other GRASS modules for the computation of the visibility index, selection of the region, import and export of the raster layers, etc.

The package of our toolkit consists of several modules offering the following functionalities (Grass Development Team, 2011):

- r.in.gdal—Allows a user to create a (binary) GRASS raster map layer, or an imagery group, from any GDAL-supported raster file format with an optional title.
- g.region—Allows the user to manage the settings of the current geographic region. This region defines the geographic area in which all GRASS displays and analyses will be carried out.
- r.buffer—Creates a raster map layer showing the buffer zones surrounding the cells that contain nonzero category values.
- r.stats—Calculates the area presented in each of the categories of the user-selected raster map layer(s).

- r.out.arc—Converts a raster map layer into an ASCII grid format
- r.patch —Creates a composite raster map layer by using a known category values from one (or more) map layer(s) to fill in the areas of "no data" in another map layer.
- Grass batch job—An alternative method of easily run jobs in the GRASS with a collection of commands in a shell script file. It is possible to apply it to launch the GRASS in the text mode and in a parallel solution of the GRASS jobs.

## 22.4 Visual Exposure Modeling using Parallel Computations

The created toolkit runs in a GRASS GIS terminal. An input raster layer is uploaded after starting the run (Fig. 22.1A). An input map containing the elevation data (e.g., digital model of relief) is transformed into the ASCII grid format using the GRASS GIS (the module r.in.gdal).

The size of the raster map which enters to the program algorithm must exceed the borders of the area of interest by the distance of the visibility. This condition is necessary for the visibility computation on the territory borders where calculations take into account the area lying outside the borders.

After running the toolkit, the GRASS terminal shows a simple guide for users in the form of questions and answers, leading to the specification of the input parameters, such as the units, computation region, and division of the input map into segments in the case of parallel computations.

The dialog window offers the following possibilities for the specification of the units in which the data are recorded:

- the number of pixels that are visible from a set point
- the area in square meters
- the area in square kilometers
- the area in ares
- the area in hectares

Two more questions in a new dialog window are used to provide the specification of the computational region in the GRASS (Fig. 22.1B and C). The first one is for the specification of the output map resolution. The second one is important for the specification of the borders of the computation (the border pixels). There are three possibilities to choose from. The first one defines the borders of the region for the calculation base on the MASK. The second one defines the borders according to the specification of the first and last pixels. This possibility allows one to divide the input map into smaller segments which can be used separately as the basic input maps. It can be used in parallel computing on computer clusters (or on a supercomputer with the cluster architecture) where each computer of the cluster is able to calculate an individual segment and divide this segment into the fragmental units according to the number of processors.

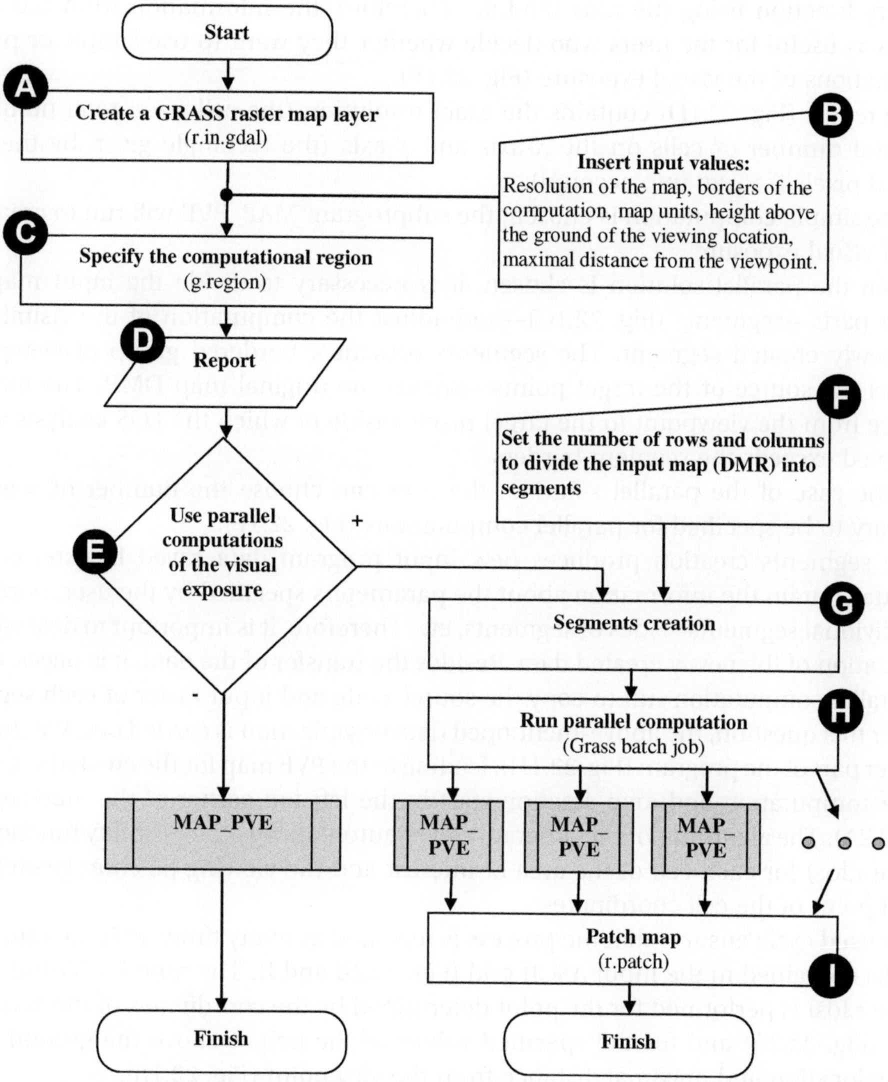


FIGURE 22.1 The flow-process diagram of the program algorithm.

The third possibility is the calculation of the visual exposure on the whole area of the input raster. In this case, we have to take into account that the results of the calculations in the area near the borders are distorted, and therefore the maximal distance from the viewpoint is beyond the borders of the input raster.

If we know the map resolution and the border of the computational region, the program will be able to assess how large area is used in the computation, which is reported as the final number of map pixels in the dialog window (Fig. 22.1D).

Whereas the Visibility function has to be performed individually for each pixel, the number of pixels in the computational region is equal to the number of runs of the

Visibility function using the r.los module. Therefore, the information from the dialog window is useful for the users who decide whether they want to use simple or parallel computations of the visual exposure (Fig. 22.1E).

The report (Fig. 22.1D) contains the exact resolution (the cell size), total number of cells, and number of cells on the *X*-axis and *Y*-axis (the rectangle given by the most outward pixels is taken into account).

If the simple computation is chosen, the subprogram MAP_PVE will run to create the map of visual exposure.

When the parallel solution is chosen, it is necessary to divide the input map into smaller parts—segments (Fig. 22.1G)—and adjust the computation of the visibility for each newly created segment. The segments present a bordered group of viewpoints, whereas the source of the target points remains the original map DMR. The maximal distance from the viewpoint to the target pixels inside of which the LOS analysis will be performed exceeds the segment borders.

In the case of the parallel solution, the user can choose the number of segments necessary to be specified for parallel computations (Fig. 22.1F).

The segments creation produces new input program data saved in external files. The data contain the information about the parameters specified by the user, borders of the individual segments, codes of segments, etc. Therefore, it is important to deal with the organization of the newly created data. Besides the transfer of the data, it is necessary for the parallel computation run to copy the source code and input raster of each segment.

After this question, the above-mentioned data organization is carried out. We then run a further part of the program (Fig. 22.1H), leading to the PVE map for the created segments.

The computation and map creation start in the left top corner of the map segment (Fig. 22.2A). The algorithm of the program solves automatically the Visibility function (the module r.los) for each cell of the area of interest and the viewing position given by the central point of the cell coordinates.

The used cycle ensures that the process is repeated as many times as is the number of the cells contained in the input ASCII grid (Fig. 22.2B and I). The function Visibility (the module r.los) is performed for the point determined by the coordinates of the actual cell center (Fig. 22.2C) and for the specified values of the height above the ground of the viewing location and maximal distance from the viewpoint (Fig. 22.1B).

The module r.los generates a raster output map in which the cells that are visible from the user-specified observer position are marked with a vertical angle (in degrees) required for the cells' visibility (the viewshed). The value of 0 is shown directly below the specified viewing position, 90 is to the right or left of the position, and 180 is directly above the observer. The angle of the cell containing the viewing position is undefined and set equal to 180 (Grass Development Team, 2011).

The creation of the visibility index for a given viewpoint is the next step of the algorithm. The visibility index is obtained by summarizing the results of the r.los module. We use the module r.stats for reporting the area statistics containing the number of the individual cells and the values of their vertical angles (Fig. 22.2D).

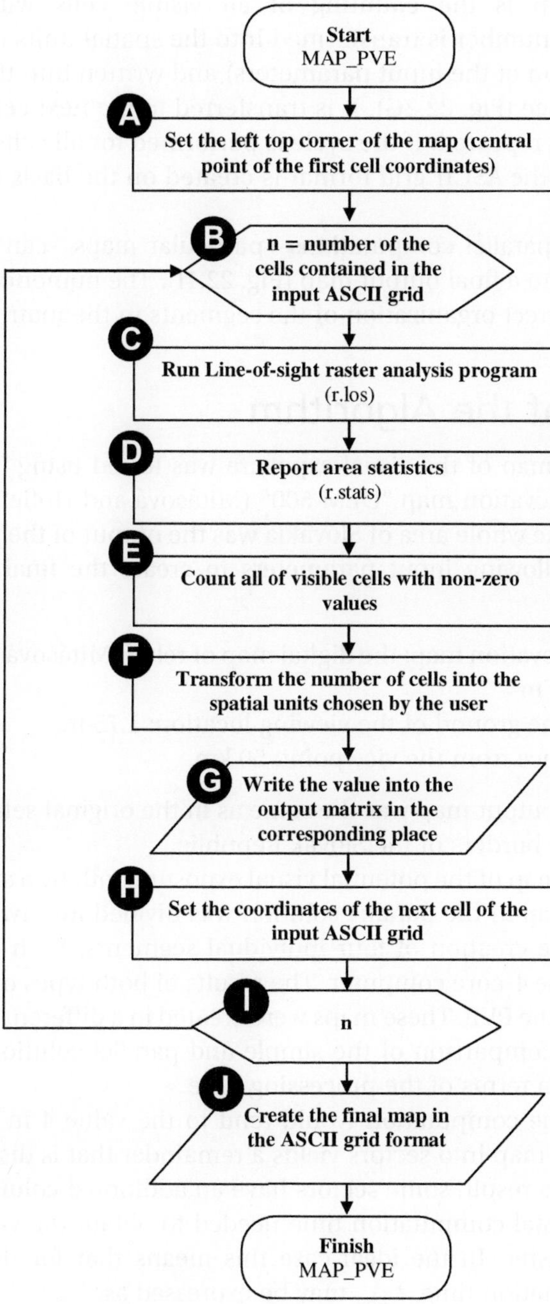


**FIGURE 22.2** The subprogram for the computation of the PVE map.

The following step is the counting of all visible cells with nonzero values (Fig. 22.2F). The final number is transformed into the spatial units (chosen by the user during the specification of the input parameters) and written into the output matrix in the corresponding place (Fig. 22.2G). It is transferred to the next cell later (Fig. 22.2H), and the whole cycle is repeated. If the cycle is performed for all cells in the input ASCII grid, the final map in the ASCII grid format is created on the basis of the values in the matrix.

The results of the parallel computations—particular maps—can be patched by last part of the program into a final output map (Fig. 22.1I). The numerical codes of the map segments enable a correct organization of the segments in the matrix.

## 22.5 Testing of the Algorithm

The algorithm of the map of the visual exposure was tested using the parallel computation on the digital elevation map "DEM 500" (Mitášová and Hofierka, 2004).

The PVE map for the whole area of Slovakia was the output of the process in our case. We have used the following input parameters to create the final map of the visual exposure potential:

- The raster digital elevation map: the digital map of relief (Mitášová and Hofierka, 2004) with resolution 500 m
- The height above the ground of the viewing location: 1.75 m.
- The maximal distance from the viewpoint: 50 km.

The resolution of the output map was the same as in the original setting. The borders of computation were the borders of the Slovak Republic.

We calculated the map of the potential visual exposure both by a simple and a parallel solution. The input map in the parallel solution was divided into two columns and two rows, which led to the creation of four individual segments. Both computations were performed on the same 4-core computer. The results of both types of computation were two identical maps of the PVE. These maps were created in a different time due to the way of computation. The comparison of the simple and parallel solutions showed that the latter was 3.78 faster in terms of the processing time.

The speed-up of the computation would tend to the value 4 in the ideal case. The integer division of the map into sectors yields a remainder that is distributed among the individual sectors. As a result, some sectors have an additional column or row of pixels, which increases the total computation time needed to obtain the visibility index for all cells of the output raster. In the ideal case this means that for the division with no remainder the computation time, $t_{calc}$, may be expressed as

$$t_{calc} = (P_{row} \text{ div } S_{row}) * (P_{col} \text{ div } S_{col}) * t_{rlos},$$

where $P_{col}$ and $P_{row}$ represent the number of pixels of the input map on the $X$-axis and $Y$-axis, respectively, $S_{row}$ a $S_{col}$ are the numbers of columns and rows in the division of the input map into the required segments as specified by a user, and $t_{\text{rlos}}$ is the computation time needed to calculate the visibility index for a single cell of the output map.

Our case corresponds to the worst scenario when the remainder occurs in the division of the input map into segments for columns as well as for rows. Then the computation time may be expressed as

$$t_{\text{calc}} = ((P_{\text{row}} \text{ div } S_{\text{row}}) * (P_{\text{col}} \text{ div } S_{\text{col}} + P_{\text{col}} \text{ div } S_{\text{col}} + P_{\text{row}} \text{ div } S_{\text{row}} + 1)) * t_{\text{rlos}}.$$

These relations for the computation time for the ideal and worst cases can be applied also to computations by computer clusters and supercomputers.

The division of the map into more segments and the application of more processors for the computation can lead to a proportional decrease of the computation time. Thus, the developed toolkit can be easily applied, using computer clusters and supercomputers for the computations of large areas in a high resolution.

## 22.6 Results and Discussion

The result of the computation process is a map of the potential visual exposure of the landscape in Slovakia. The computations were done for the borders of the potential visibility set equal to 50 km. The distance was verified in a real terrain and identified with the model distance.

The visual exposure of the Slovakia's landscape potential is surprising but logical. Ridges of hills, uplands, highlands, and mountains present visually the most exposed mountain parts. The altitude does not remarkably influence the values of the potential visual exposure of the landscape. We identified the mountains of Malé Karpaty, Považský Inovec, Tribeč, Vtáčnik, Kremnické vrchy, Štiavnické vrchy, Poľana, Vysoké Tatry, Čergov, Slánske vrchy, and Vihorlatské vrchy as compact areas with a high value of the visual exposure of the landscape potential.

The digital model of the visual exposure potential is a result of the software processing with the resulting scale of the visual exposure values from 1.25 to 4487.5 km$^2$. The maximal visually identified area takes 7850 km$^2$. The scale ranges from the minimal exposure cell (the light gray parts in the Fig. 22.3) with the initial value 1.25 km$^2$ of the possible visibility to the maximal exposure cells (the dark gray parts in the Fig. 22.3) with the maximal value 4487.5 km$^2$ of the possible visibility. The visibility of each point was derived from the area of the circle with the radius of 50 km, which presents the maximal range of visibility. The highest obtained value presents 57.16% of the total value of the visibility (Fig. 22.4). Our results showed that the highest value of the possible visibility was achieved for the Zobor Hill (586 m a.s.l.) in the Nitra region. The Zobor Hill belongs to the Tribeč Mountains and is situated in the contact zone of the Carpathian Mountains and Pannonia Basin.

Further investigations were focused on the identification of the localities with the relevant value of 25% of the possible maximum. We made a new digital model of the visual exposure potential with good readability of high visual quality (Fig. 22.5). This

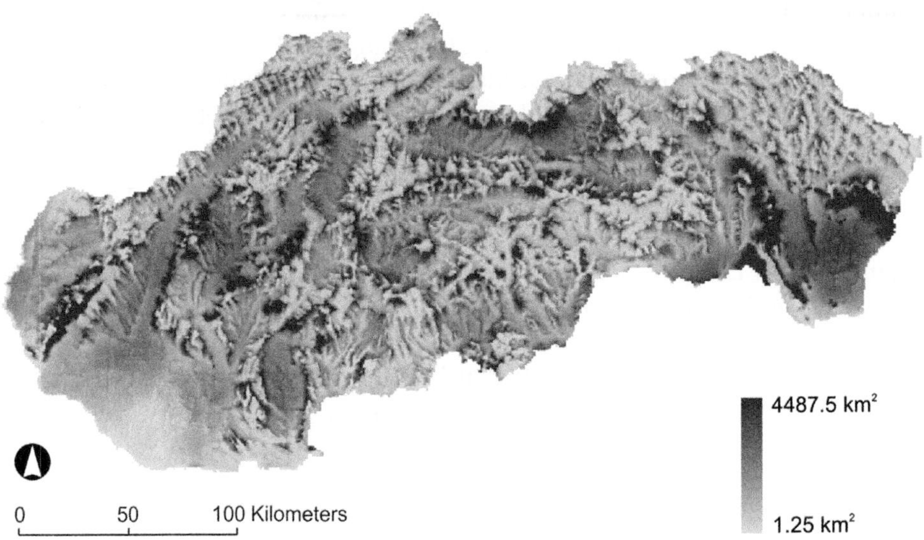


**FIGURE 22.3** The potential of the landscape visual exposure of Slovakia.

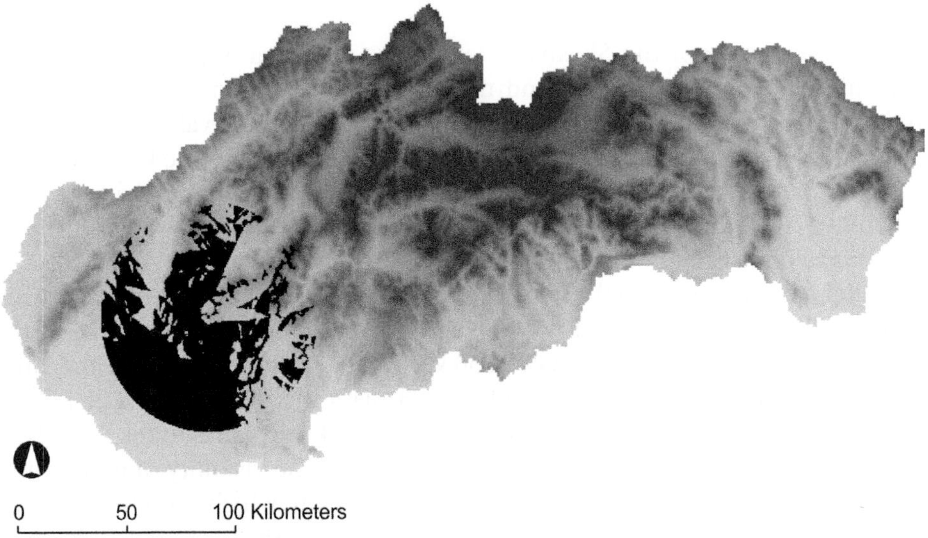


**FIGURE 22.4** The location of the highest value of the visual exposure obtained with the radius of 50 km presenting the maximal range of visibility. The Zobor Hill (586 m a.s.l.), Nitra.

digital model defines the areas with the visibility from 25 to 57.16% (the maximal value). A map with the most potentially visually exposed areas in the Slovakia's landscape was created by the data visualization. The highest visually exposed areas were the mountains of Malé Karpaty, Slánske vrchy, and Vihorlatské vrchy.

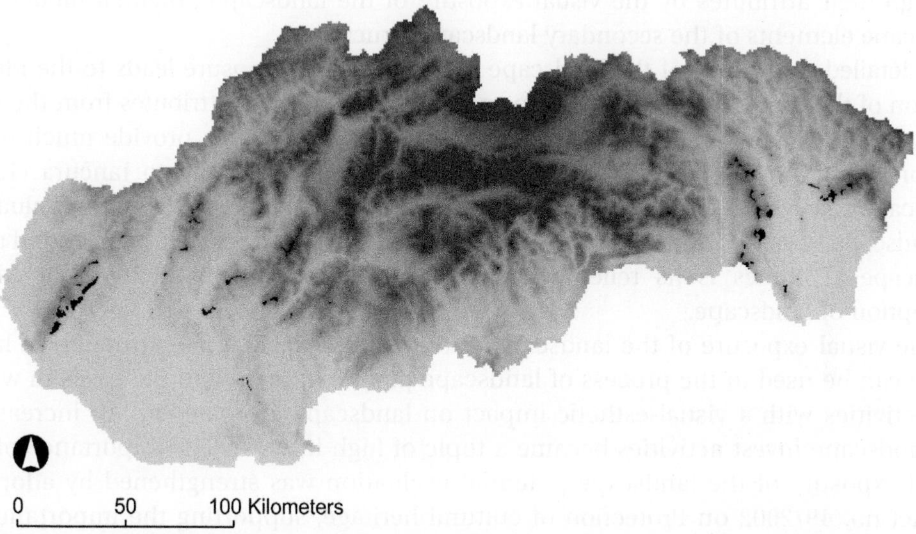


**FIGURE 22.5** The identification of the potential visible areas exceeding 25% of the possible maximal area.

These observations are in agreement with our assumptions because of the adjacent lowlands (the Podunajská and Záhorská nížina (a lowland) in contact with the mountains of Malé Karpaty on the west and the Východoslovenská nížina (a lowland) in contact with the mountains of Slanské vrchy and Vihorlatské vrchy on the east of Slovakia). Similarly, the mountains of Zoborské vrchy with the peaks Zobor (585 m a.s.l.) and Žibrica (617 m a.s.l.) are surrounded by the wide Podunajská pahorkatina highland, which enables one a wide view of the localities. Smaller areas with the relevant value of 25% of the possible maximum are scattered over the whole area of Slovakia with the culmination on hilly and mountainous parts of the Zoborské vrchy—Zobor (586 m a.s.l.), Štiavnické vrchy—Sitno (1009 m a.s.l.), Pohronský Inovec—Veľký Inovec (901 m a.s.l.), Považský Inovec—Marhát (748 m a.s.l.), Inovec (1042 m a.s.l.), Malá Fatra—Veľká lúka (1476 m a.s.l.), Veľký Kriváň (1709 m a.s.l.), Nízke Tatry—Chopok (2023 m a.s.l.), Kráľova Hoľa (1948 m a.s.l.), Vysoké Tatry—Gerlachovská štít (2664,4 m a.s.l.), Lomnický štít (2632 m a.s.l.), Poľana—Poľana (1458 m a.s.l.), Slánske vrchy—Šimonka (1092 m a.s.l.), Makovica (981 m a.s.l.), Bogota (855 m a.s.l.), and Vhorlatské vrchy—Vihorlat (1076 m a.s.l.). These localities are identical with the scenic and cultural–historical symbols of Slovakia.

Our results present a part of the testing in the tool development for a direct determination of the landscape potential visual exposure. The combination of the landscape potential visual exposure model and landscape elements of the secondary landscape structure leads directly to the definition of the landscape visual exposure. We are able to work at the level of the visual impact assessment as a part of the tool for the Environmental impact assessment which requires the identification of the basic landscape values through their attributes by the visual exposure of the landscape potential model with landscape elements of the secondary landscape structure.

A detailed evaluation of the landscape potential visual exposure leads to the identification of the areas that are highly important as the landscape attributes from the view point of the visual connections. The visual aspects of landscape provide much information about the landscape content and its character. According to Jančura (1998), landscape is a system of attributes. These attributes are used for the primary evaluation of landscape (e.g., we can evaluate the attributes that can be seen). The background of all landscape attributes is its relief which presents the basic limitation of the visual perception of landscape.

The visual exposure of the landscape potential as a basic visual structure of landscape can be used in the process of landscape planning, mainly in the cases in which the activities with a visual-esthetic impact on landscape are planned. An increase of the landscape invest activities became a topic of high interest. The importance of the visual exposure of the landscape potential evaluation was strengthened by adopting the Act no. 49/2002 on Protection of cultural heritage, supporting the importance of panoramas preservation. It is important to preserve the characteristic landscape panoramas and countryside views originating from the state of relief and area exposure. The problem is that we still do not have a system and regulation of objective assessment.

## 22.7 Conclusion

Our contribution was aimed at the use of the GIS software to develop a toolkit for modeling the visual exposure of landscape. The output of the process is a map of the visual exposure in which each pixel of the raster map carries information on the visibility index. The creation of the map requires the viewshed analysis applied to each pixel individually, resulting in a number of time-consuming computations. Therefore, parallel computations were applied to speed up the computation process.

The importance of the landscape visual information assessment is associated with the landscape planning activities and a visual impact on landscape. The current assessments depend on expertise which evaluates the landscape visual quality subjectively.

We wanted to create and test the model which would be used as an objectively measurable platform in the case of the assessment of real or planned activities with a visual impact on landscape. This would be a complementary tool for landscape planning activities.

## Acknowledgments

The work was supported by the grant KEGA no. 030UKF-4/2011.

## References


Achilleos, G., Tsouchlaraki, A., 2004. Visibility and viewshed algorithms in an information system for environmental management. In: Brebbia, C.A. (Ed.), Management Information Systems 2004: GIS and Remote Sensing. WIT Press, Ashurst, pp. 109–121.

Bao, S.T., Liu, S.B., Yao, J.R., Huang, S.L., 2010. 3D locating system of observation tower with multiple constraints. In: Tan, H. (Ed.), PIAGENG 2010: Photonics and Imaging for Agricultural Engineering. SPIE-INT Soc. Optical Engineering, Bellingham 77520K-77520K.

Bishop, I.D., 2002. Determination of thresholds of visual impact: the case of wind turbines. Environ. Plann. B. 29, 707–718. http://dx.doi.org/10.1068/b12854.

Bishop, I.D., Hulse, D.W., 1994. Prediction of scenic beauty using mapped data and geographic information systems. Landsc. Urban Plan 30, 59–70.

Bishop, I.D., Wherrett, J.R., Miller, D.R., 2000. Using image depth variables as predictors of visual quality. Environ. Plann. B Plann. Des 27, 865–875.

Burrough, P.A., McDonnel, R.A., 1998. Principles of Geographical Information Systems. Oxford University Press, London.

Clay, G.R., Smidt, R.K., 2004. Assessing the validity and reliability of descriptor variables used in scenic highway analysis. Landsc. Urban Plan 66, 239–255.

Coeterier, J.F., 1996. Dominant attributes in the perception and evaluation of the Dutch landscape. Landsc. Urban Plan 34, 27–44.

Dean, D.J., Lizarraga-Blackard, A.C., 2007. Modeling the magnitude and spatial distribution of aesthetic impacts. Environ. Plann. B 34, 121–138. http://dx.doi.org/10.1068/b30101.

Fisher, P.F., 1995. An exploration of probable viewsheds in landscape planning. Environ. Plann. B 22, 527–546. http://dx.doi.org/10.1068/b220527.

GRASS Development Team, 2011. Geographic Resources Analysis Support System (GRASS) software. Open Source Geospatial Foundation Project <. http://grass.osgeo.org .>.

Jančura, P., 1988. The present and historical landscape structures in landscape formation. Životné Prostredie. (Bratislava, Slovakia) 32, 236–240.

Jasiewicz, J., Metz, M., 2011. A new GRASS GIS toolkit for Hortonian analysis of drainage networks. Comput. Geosci. 37, 1162–1173. http://dx.doi.org/10.1016/j.cageo.2011.03.003.

Kima, Y.-H., Rana, S., Wise, S., 2004. Exploring multiple viewshed analysis using terrain features and optimisation techniques. Comput. Geosci. 30, 1019–1032.

Lanthier, M., Nussbaum, D., Sack, J.R., 2003. Parallel implementation of geometric shortest path algorithms. Parallel Comput. 29, 1445–1479.

Litton Jr., R.B., 1982. Visual assessment of natural landscapes. West Geogr. 20, 97–116.

Litton Jr., R.B., Tetlow, R.J., 1978. In: A landscape inventory framework: scenic analyses of the northern Great Plains. USDA Forest Service Research Paper RM, 135, p. 91.

Löw, J., Míchal, I., 2003. Krajinný ráz. (Kostelec nad Černými lesy, Czech republic) 552 str.

Lynch, K., 1960. Image of the City. The MIT Press, Massachusetts. ISBN 0 262 12004 6.

Miller, D.R., Wherrett, J.R., Morrice, J.G., Fisher, P.F., 1999. Geographic modelling of the visual impact of wind turbines. In: HinsonP. (Ed.), Wind Energy 1999: Wind Power Comes Of Age. Professional Engineering Publishing Ltd., Westminister, pp. 167–179.

Mitasova, H., Hofierka, J., 2004. Slovakia Precipitation Data. http://mpa.itc.it/grasstutor/data_menu2nd.php.

Neteler, M., Mitasova, H., 2007. Open Source GIS: A GRASS GIS Approach. Springer, New York.

Ogburn, D.E., 2006. Assessing the level of visibility of cultural objects in past landscapes. J. Archaeol. Sci. 33, 405–413. http://dx.doi.org/10.1016/j.jas.2005.08.005.

Ohsawa, Y., Kobayashi, T., 2005. An analytical model to assess the visibility of landmarks. Geogr. Anal. 37, 336–349. http://dx.doi.org/10.1111/j.1538-4632.2005.00586.x.

Paliou, E., Wheatley, D., Earl, G., 2011. Three-dimensional visibility analysis of architectural spaces: iconography and visibility of the wall paintings of Xeste 3 (Late Bronze Age Akrotiri). J. Archaeol. Sci. 38, 375–386. http://dx.doi.org/10.1016/j.jas.2010.09.016.

Rana, S., 2003. Fast approximation of visibility dominance using topographic features as targets and the associated uncertainty. Photogramm. Eng. Rem. Sens. 69, 881–888.

Salašová, A., 1996. Village restoration in the Czech Republic. Int. J. Herit. Stud 2, 160–171. http://dx.doi.org/10.1080/13527259608722169.

Sevenant, M., Antrop, M., 2007. Settlement models, land use and visibility in rural landscapes: two case studies in Greece. Landsc. Urban Plan. 80, 362–374. http://dx.doi.org/10.1016/j.landurbplan.2006.09.004.

Shang, H., Bishop, I.D., 2000. Visual thresholds for detection, recognition and visual impact in landscape settings. J. Environ. Psychol. 20, 125–140.

Smardon, R.C., 1988. Correspondence address perception and aesthetics of the urban environment: review of the role of vegetation. Landsc. Urban Plan. 15, 85–106.

Štefunková, D., 2000. The possibilities of implementation of landscape visual quality evaluation to the landscape-ecological planning. Ekologia 19, 199–206.

Vorel, I., Bukáček, R., Matějka, P., Sklenička, P., Culek, M., 2004. A Method for Assessing the Visual Impact on Landscape Character of Proposed Construction, Activities or Changes in Land Use (A Method for Spatial and Character Differentiation of an Area). Czech Technical University In Prague, Prague.

Yang, P.P.J., Putra, S.Y., Li, W.J., 2007. Viewsphere: a GIS-based 3D visibility analysis for urban design evaluation. Environ. Plann. B. 34, 971–992. http://dx.doi.org/10.1068/b32142.

Zube, E.H., Pitt, D.G., Anderson, T.W., 1974. Perception and measurement of scenic resources in the Southern Connecticut River Valley. Massachusetts. 191 R-74–1.

Zube, E.H., Sell, J.L., Taylor, J.G., 1982. Landscape perception, research, application and theory. Landsc. Plann 9, 1–35.

# 23

# Spatial Algorithms Applied to Landscape Diversity Estimate from Remote Sensing Data


Duccio Rocchini*, Niko Balkenhol[†], Luca Delucchi*, Anne Ghisla*, Heidi C. Hauffe*, Attila R. Imre**, Ferenc Jordán[‡], Harini Nagendra[$,¶], David Neale[‖], Carlo Ricotta[††], Claudio Varotto*, Cristiano Vernesi*, Martin Wegmann[‡‡], Thomas Wohlgemuth[$$], Markus Neteler*

*FONDAZIONE EDMUND MACH, RESEARCH AND INNOVATION CENTRE, DEPARTMENT OF BIODIVERSITY AND MOLECULAR ECOLOGY, VIA E. MACH 1, 38010 S. MICHELE ALL'ADIGE (TN), ITALY, [†]DEPARTMENT OF FOREST ZOOLOGY AND FOREST CONSERVATION, UNIVERSITY OF GOETTINGEN, BUESGENWEG 3, 37077 GOETTINGEN, GERMANY, **HAS CENTRE FOR ENERGY RESEARCH, H-1525, POB 49, BUDAPEST, HUNGARY, [‡]THE MICROSOFT RESEARCH - UNIVERSITY OF TRENTO CENTRE FOR COMPUTATIONAL AND SYSTEMS BIOLOGY, PIAZZA MANIFATTURA 1, 38068, ROVERETO, ITALY, [$]ASHOKA TRUST FOR RESEARCH IN ECOLOGY AND THE ENVIRONMENT, ROYAL ENCLAVE, SRIRAMPURA, JAKKUR P.O., BANGALORE 560064, INDIA, [¶]CENTER FOR THE STUDY OF INSTITUTIONS, POPULATION, AND ENVIRONMENTAL CHANGE, INDIANA UNIVERSITY, 408 N. INDIANA AVENUE, BLOOMINGTON, IN 47408, USA, [‖]DEPARTMENT OF PLANT SCIENCES, UNIVERSITY OF CALIFORNIA AT DAVIS, DAVIS, CA, USA, [††]DEPARTMENT OF ENVIRONMENTAL BIOLOGY, UNIVERSITY OF ROME ''LA SAPIENZA'', PIAZZALE ALDO MORO 5, 00185 ROME, ITALY, [‡‡]UNIVERSITY OF WÜRZBURG, INSTITUTE OF GEOGRAPHY, DEPARTMENT OF REMOTE SENSING, REMOTE SENSING AND BIODIVERSITY RESEARCH, AM HUBLAND, 97074 WÜRZBURG, GERMANY, [$$]WSL SWISS FEDERAL INSTITUTE FOR FOREST, SNOW AND LANDSCAPE RESEARCH, ZÜRCHERSTRASSE 111, CH-8903 BIRMENSDORF, SWITZERLAND


## 23.1 Introduction: Why Measuring Landscape Heterogeneity?

A recent assessment of progress toward achieving the 2010 targets of the Convention on Biological Diversity concluded that most indicators showed a decline in biodiversity, an increase in human pressure on natural and semi-natural ecosystems, and no significant changes in the rate of biodiversity loss (Butchart et al., 2010). A number of global organizations (e.g., WSSD of Johannesburg, Convention on Biological Diversity, International


Models of the Ecological Hierarchy. DOI: http://dx.doi.org/10.1016/B978-0-444-59396-2.00023-7
ISSN 0167-8892,

Panel on Biodiversity, and Ecosystem Services, UN Environment Programme, and UN Development Programme) are strongly advocating concrete efforts toward biodiversity conservation.

At present, the assessment of biodiversity at local and regional scales often relies on labor-intensive fieldwork for data collection (Ferretti and Chiarucci, 2003). An estimate of species numbers and distributions over large areas has always been a particularly challenging task for ecologists as a result of the intrinsic difficulty in judging the completeness of the species lists and in quantifying the sampling effort (e.g., Palmer, 1995). Time and economic constraints mean that field biologists are neither able to capture and/or observe every species nor estimate how species composition changes over time (e.g., Robinson et al., 1994; Palmer et al., 2002). In addition, these simple taxonomic approaches are susceptible to "taxonomic inflation," that is, the elevation of intraspecific taxa to the rank of species, which is one of the main reasons for the rapidly increasing number of species in certain groups (Isaac et al., 2004; Knapp et al., 2005). Bacaro et al. (2009) demonstrated that, at a local scale, the subjectivity of field biologists in acquiring species lists is expected to increase error variance instead of improving information on actual community diversity, resulting in large inaccuracies in the evaluation of species presence/absence. An alternative solution would be to describe ecosystems at a much lower resolution (e.g., at a functional group level, Lawton and Brown, 1994) to focus on redundancy and functional diversity (e.g., keystone species, Bond, 1994; Power et al., 1996).

Surprisingly, the more efficient and accurate ways for measuring biodiversity have largely been neglected. For example, in most cases, genetic variation is a prerequisite for improving the capacity of organisms to adapt to environmental changes with potential further speciation, thus possibly resulting in an increase of species diversity (Szathmáry et al., 2001). Appropriately collected and analyzed, genetic data can aid (sub)species identification, especially in the case of cryptic (morphologically identical) taxa (e.g., Roca et al. 2001; Wilting et al. 2011). With the advent of molecular methods for analyzing noninvasive samples such as faeces, urine, hair, and feathers, the presence of rare or elusive species in a landscape can often be established more reliably using genetic analyses (Waits and Paetkau 2005'Schwartz and Monfort 2008). Moreover, given (i) the fast progress in DNA barcoding applied to whole ecosystems (e.g., Forest et al., 2007; Hollingsworth et al., 2011) and (ii) the advent of next-generation sequencing technology and its forecasted cost reduction (Glenn, 2011), genetic data will prove invaluable for further testing of the methods for estimating species diversity described in this chapter.

However, while genetic approaches may be more accurate than traditional species inventories, and sample collection can be planned to avoid systematic bias between sample sites, collecting genetic data on large spatial scales remains slow, relatively expensive, and technically and logistically challenging.

Given the above limitations of field- and laboratory-based monitoring techniques, new innovative methods have been put forward for identifying environmental gradients that may explain the maximum change in species diversity (e.g., Hortal and Lobo, 2005). Among these, entropy (also referred to as "heterogeneity") measured by the spatial variation of remotely sensed spectral signals has been proposed as a proxy for species diversity (see Rocchini et al., 2010 for a review) since the entropy of the Earth's surface is closely related to physical and ecological diversity (Gillespie et al., 2008; Nagendra and Gadgil, 1999). This concept is encompassed in the Spectral Variation Hypothesis (Palmer et al., 2002; Rocchini, 2007), which predicts that the higher the habitat heterogeneity, the higher will be the species diversity therein. Depending on the scale and the habitat being considered, the Spectral Variation Hypothesis can be expected to hold true in many cases (e.g., Oindo and Skidmore, 2002; Rocchini et al., 2005; Levin et al., 2007; Bino et al., 2008; St-Louis et al., 2009).

The use of remote sensing for estimating environmental heterogeneity and (subsequently) species diversity represents a powerful tool since it creates a synoptic view of large areas with a high temporal resolution. For example, the availability of satellite-derived data with high spatial (IKONOS, Orbview-3, BGIS-2000 (Ball's Global Imaging System-2000), RapidEye) and spectral resolution (CHRIS (Compact High Resolution Imaging Spectrometer), Hyperion, GLI (Global Imager), MERIS (Medium Resolution Imaging Spectrometer), and MODIS (Moderate Resolution Imaging Spectrometer)) together with long-term programs like Landsat makes it possible to study any terrestrial region of the globe at a resolution of a few meters.

Importantly, freely available repositories like the Global Land Cover Facility of the University of Maryland (Fig. 23.1) and Open Source systems are also available

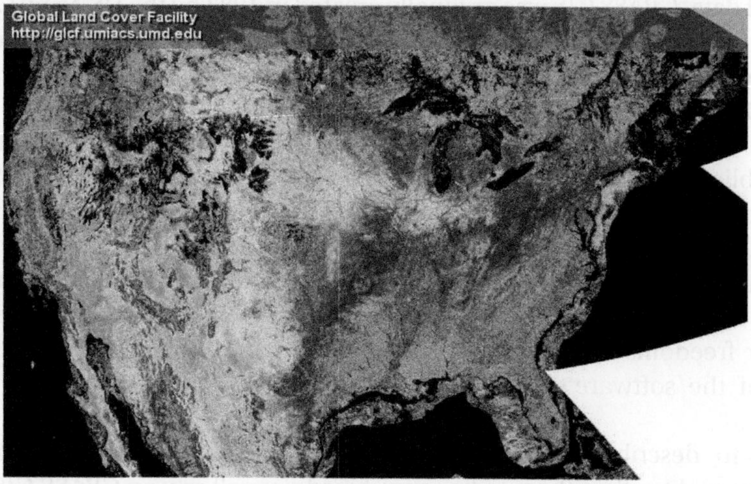


**FIGURE 23.1** Example of MODIS 16-days NDVI data of North America freely available from the global Land Cover Facility portal http://glcf.umiacs.umd.edu/. For color version of this figure, the reader is referred to the online version of this book.

for remotely sensed imagery analysis (Neteler et al., 2005; Neteler and Mitasova, 2008; Neteler et al., 2012). The aim of this book chapter is to review the main Free Open Source tools for analyzing landscape heterogeneity over different spatial scales.

## 23.2 Open Source Software Philosophy for a Free Calculation of Landscape Diversity

The idea of Free and Open Source (FOSS) software has been around for almost as long as software development (Neteler and Mitasova, 2008). The famous "four freedoms" paradigm, developed by Richard Stallman (1985, 1997) in his seminal work, proclaims that FOSS is defined as (i) the freedom to run the program for any purpose, (ii) the freedom to study how the program works and adapt it to one's own needs, (iii) the freedom to redistribute copies, and (iv) the freedom to improve the program and release such improvements to the public. This guarantees that the whole scientific community benefits from software development (also see Fogel, 2009; Rocchini and Neteler, 2012a).

In order to calculate landscape heterogeneity, full access to the source code is crucial to ensuring robust analysis output based on code peer-review, particularly where complex algorithms are concerned (Neteler and Mitasova, 2008). There are several well-known examples of FOSS in research fields such as Statistics (e.g., R Language and Environment for Statistical Computing, R Development Core Team, 2011), while GIS scientists and landscape ecologists benefit from a powerful GIS called "GRASS" (Geographical Resources Analysis Support System, http://grass.osgeo.org, see Neteler et al., 2012), which includes more than 350 modules for managing and analyzing geographical data. GRASS GIS was originally created in 1982 by the US Army Construction Engineering Research Laboratories, and it is currently one of the cutting-edge projects of the Open Source Geospatial Foundation (OSGeo.org, founded in 2006). As Neteler and Mitasova (2008) explain:

"The key development in recent GRASS history was the adoption of GNU GPL (General Public License, see http://www.gnu.org) in 1999. With this, GRASS embraced the Open Source philosophy, well known from the GNU/Linux development model, which stimulated its wide acceptance." Adoption of the FOSS license changed the development process of GRASS so that contributions to the source code became decentralized. The legal statements declared in the GPL are based on the aforementioned "four freedoms" paradigm (Stallman, 1985, 1997) and allow the user to take advantage of the software's full range of capabilities, and to distribute, study, and improve it.

In order to describe the following spatial algorithms for measuring landscape heterogeneity, and to allow them to be reproduced, we will rely on GRASS GIS and R with complete reference to the source code and/or commands to be applied to achieve estimates of landscape heterogeneity.

## 23.3 Spatial Algorithms Available for the Quantification of Landscape Heterogeneity from Remote Sensing Data

A satellite or more generally a remotely sensed image can be viewed as a matrix composed of

$$\text{Digital Numbers } \mathbf{M}_{\mathrm{DN}} = \begin{array}{ccccc} x1y1 & x2y1 & x3y1 & \cdots & xny1 \\ x1y2 & x2y2 & x3y2 & \cdots & xny2 \\ \vdots & \vdots & \vdots & \vdots & \vdots \\ x_1yn & x_2yn & x_3yn & \cdots & xnyn \end{array} \text{ distributed over spatial locations}$$

defined by $x$ and $y$ coordinates.

Specifically, let $U$ be a universe of entities $u$ (pixels or patches), each of which can be represented as a tuple $\{z(u)|s(u)\}$, where $z(u)$ = property of the $u$th entity related to its $s(u)$ spatial component (Goodchild et al., 1999). Notice that in this review we will explicitly consider $z(u)$ as either (i) the raw or preprocessed (e.g., with sensor calibration and atmospheric correction) Digital Number (DN) value of pixels $u$ composing a remotely sensed image $U$, or (ii) the class of each polygon $u$ composing a land use (or vegetation) map $U$. In the following paragraphs, we will rely on DNs in a remotely sensed image but the same reasoning holds even for classes in a land-use map.

### 23.3.1 Measures of Spectral Diversity Based on DN Variability

Rough measures of spectral variability include spectral variance and coefficient of variation into a certain local area, given each pixel DN value.

In addition, variability can be achieved by spectral unmixing or on fuzzy sets theory (Zadeh, 1965). A fuzzy set is defined as follows: let $U$ denote a universe of entities $u$, the fuzzy set $F$ turns out to be

$$F = \{(u, \mu_f(u))|u \in U\} \tag{1}$$

where the membership function $\mu_f$ associates the degree of membership into the set $F$ (see Zimmermann, 2001) for each entity $u \in U$. The degree of membership ranges in the interval [0.1], that is in the real range between 0 and 1.

Two major assumptions allow us to consider fuzzy sets as powerful tools for maintaining uncertainty information when aiming at mapping and analyzing landscape patterns: (i) membership of ecological entities to classes is not forced to occur within the integer range [0,1] as in Boolean logic; and (ii) considering different classes [A,B,…,N] the sum of the membership values $\sum(\mu_A, \mu_B, \ldots, \mu_N)$ does not necessarily equal 1 for each pixel or polygon.

When dealing with spectral variability information, once different land-use classes have been derived by fuzzy classification from a remotely sensed image, it is expected that the higher the variability in the fuzzy membership the higher will be the complexity/heterogeneity of the landscape under study and thus its diversity (Rocchini, 2010).

As stressed by Palmer et al. (2002), processing and classifying images can result in an important loss of information due to the degradation of continuous quantitative information into discrete classes. Therefore, in order to estimate local species diversity recorded by plot-based sampling design, these authors proposed the mean of the Euclidean distances from the spectral centroid of all the pixels' DN values composing the plot/area (Fig. 23.2) as a measure of spectral diversity.

#### *23.3.1.1 How Measures of DN Variability Relates to Biodiversity*

The Spectral Variation Hypothesis, based on simple measures of variability such as the distance from the spectral centroid, has been demonstrated to be a powerful tool in predicting species richness for several taxa such as vascular plants (e.g., Gould, 2000; Foody and Cutler, 2006; Levin et al., 2007), lichens (Waser et al., 2004), ants (Lassau et al., 2005), birds (Bino et al., 2008; St-Louis et al., 2009), and mammals (Oindo and Skidmore,

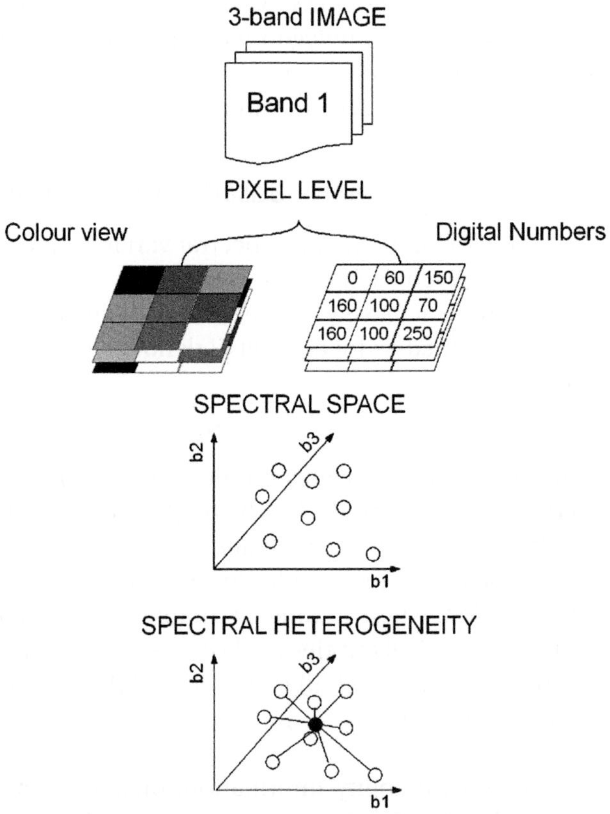


**FIGURE 23.2** Measure of spectral diversity by the distance from the spectral centroid. Image colors are derived from Digital Numbers, which can be viewed as coordinates in a multispectral space. The distance from the spectral centroid can thus represent the local variability of the image. Courtesy of Elsevier. *Source.* Rocchini et al. (2010).

2002), using a variety of models from simple univariate models (Gillespie, 2005) to multiscale univariate models (Rocchini, 2007), locally weighted methods such as geographically weighted regression (Foody, 2005), local smoothing surfaces (Nagendra et al., 2010), Generalized Additive Models (GAMs, Parviainen et al., 2009), and neural networks (Foody and Cutler, 2003).

Using spectral diversity relying on the above models, local species diversity has been mapped within different habitats and study areas, from the plants of the Arctic tundra region (Gould, 2000), savanna grasslands, and highland moors (Oindo and Skidmore, 2002), to the birds of the Mediterranean regions (Bacaro et al., 2011).

Such "direct" measures of remotely sensed variability are thus extremely powerful since in most cases they do not require image preprocessing, that is they are simply based on continuous variability. However, in some studies, maps or remotely sensed images are discrete rather than continuous; in such cases, information theory may prove more powerful in measuring heterogeneity as discussed in the following section.

## 23.3.2 Information Theory Applied to Measure Spectral Diversity

### *23.3.2.1 Point Descriptors of Landscape Diversity Based on Information Theory*

The standard measure of spectral diversity is mainly based on the Boltzmann formula (Boltzmann, 1872), further reworked by Shannon (1948) as $H = -\sum p \times \ln(p)$, where $p$ is the relative abundance (array of probabilities of occurrences $[p_1, p_2, p_3, \ldots, p_n]$) of each spectral reflectance value ( DN). It basically defines the level of uncertainty in achieving a determined DN value into a landscape (Bolliger, 2005).

According to Ricotta (2005), Shannon's entropy is the only measure that follows all the axioms of meaningful uncertainty:

**(i)** symmetry: invariance of uncertainty with respect to the permutation of the probability array;
**(ii)** expansibility: zero probabilities do not impact the final uncertainty result;
**(iii)** additivity: the uncertainty of the joint probability distribution is equal to the sum of the uncertainties of the individual probability distributions;
**(iv)** continuity: the function expressing uncertainty should be continuous in all its arguments (probabilities);
**(v)** range: the range of uncertainty is the array [0, log2($N$)] where $N$ = number of different reflectance values, that is the minimum value corresponds to a certain outcome (only one DN value is present) and the maximum value is related to all possible outcomes sharing the same probability to occur.

Calculating Shannon's entropy into GRASS GIS can be done relying on *r.li* functions such as:

```
r.li.shannon input=name config=string output=name
```

The config parameter is a configuration file basically referring to the grain and the extent to which it is being used (refer to http://grass.osgeo.org/manuals/html70_user/r.li.shannon.html).

Shannon's entropy can be incorporated into R by means of the vegan package (Oksanen et al., 2011) by:

**(i)** importing an image from GRASS GIS into R by spgrass6 library (Bivand, 2011):

```
library(spgrass6)
input.data <-readRAST6(c("image_band"), cat=c(TRUE))
```

the `cat` (category) argument can be put as `TRUE` in case of discrete data or `FALSE` in case of continuous data. In this case it has been put to `FALSE` imaging for a 8-bit coded image.

**(ii)** calculating Shannon index using the vegan package (Oksanen et al., 2011)

```
library(vegan)
H <- diversity(input.data$image_band, "shannon")
```

In addition to Shannon diversity, Simpson diversity also can be accounted for (Simpson, 1949).

Rocchini and Neteler (2012b) recently reported several problems with entropy-based metrics, that is a single index of diversity was not useful for either distinguishing different ecological situations or for discerning differences in richness *or* relative abundance. This is because areas differing in richness or relative abundance of reflectance values (DNs) may share similar Shannon index values. Instead, coupling such entropy- or reversed-dominance-based metrics with indices taking into account evenness would dramatically increase the information content of such metrics. Among these, the most widely used index is the Pielou evenness index $J = \frac{-\sum p \times \ln(p)}{\ln(N)}$ (Pielou, 1969) with $0 \leq J \leq 1$, which takes into account the maximum diversity given the same number of DNs $N$ and thus can be rewritten as $J = \frac{H}{H_{\max}}$ (see also Ricotta and Avena, 2003 for a review).

As stated for the Shannon diversity index, evenness can be calculated into GRASS by:

```
r.li.pielou input=name config=string output=name
```

or in R by:

```
N <- specnumber(input.data$image_band)
H <- diversity(input.data$image_band, "shannon")
J <- H/log(N)
```

This R-based code simply follows the aforementioned solution $J = \frac{-\sum p \times \ln(p)}{\ln(N)}$.

### *23.3.2.2 Generalized Entropy to Measure Spectral Diversity*

Generalized entropy-based algorithms for entropy calculation could allow us to solve the aforementioned problem since they encompass a continuum of diversity measures by varying one of the several parameters in their formula. As an example, Rényi (1970) proposed a generalized entropy:

$$H_\alpha = \frac{1}{1-\alpha}\ln\sum p^\alpha \tag{2}$$

which is extremely flexible and powerful since many popular diversity indices are simply special cases of $H_\alpha$. As an example, for $\alpha = 0$, $H_0 = \ln(N)$ namely the logarithm of richness ($N$ = number of DN values), maximum Shannon entropy index ($H_{\max}$), which is used as the denominator of the Pielou index, while for $\alpha = 2$, $H_2 = \ln(1/D)$, where $D$ is the Simpson Dominance index. For $\alpha = 1$, the Rényi entropy is not defined and its derivation, $H_1$ = Shannon's entropy $H$, is based on l'Hôpital's rule of calculus.

Hence, while traditional metrics *supply point descriptions of diversity*, in Rényi's framework there is a *continuum of possible diversity measures*, which differ in their sensitivity withrare and abundant DNs, becoming increasingly regulated by the commonest DNs while increasing the values of $\alpha$. That is why Rényi-generalized entropy has been referred to as a "continuum of diversity measures" (Ricotta et al., 2003).

Figure 23.3, based on a Landsat ETM+ image (left), shows how the variation of the parameter alpha into the Rényi formula may lead to very different results with higher or lower heterogeneity of the image being detected. This example demonstrates how using single metrics (point descriptors of diversity) instead of generalized formulas may lead to hidden heterogeneity of the landscape.

As far as we know, the first example in which Rényi entropy is provided in a Free environment is the LaDy software by Ricotta et al. (2003), a tool aimed at analyzing the diversity of classified images. A first attempt to allow researchers using generalized entropy in a Free and Open Source framework (e.g,. GRASS GIS) is the `r.diversity` function (Rocchini et al., under review); the code for this function is available from the GRASS GIS source code repository (http://svn.osgeo.org/grass/grass-addons/raster/r.diversity/) for further modifications, improvements, and if needed, bug fixing, and the reuse for a development of new indices based on new or still underused mathematical theory.

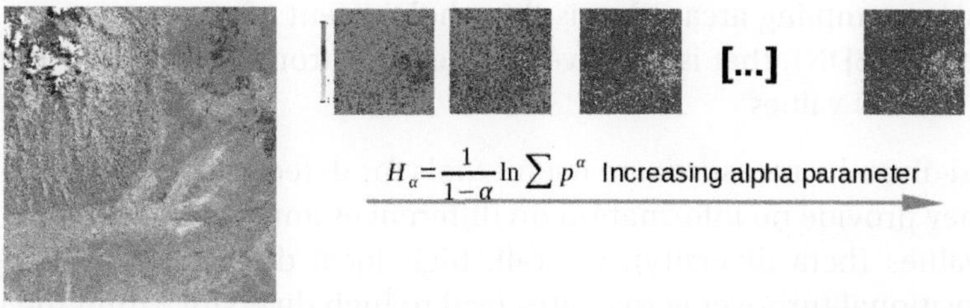


**FIGURE 23.3** Calculation of Rényi entropy with different alpha values on a remotely sensed image. Changing alpha will lead to different patterns of diversity over the landscape. Hence, using single-point descriptors of diversity (single alpha values) may hide diversity patterns. For color version of this figure, the reader is referred to the online version of this book.

#### 23.3.2.2.1 HOW INFORMATION THEORY APPLIED TO SPECTRAL DIVERSITY RELATES TO BIODIVERSITY

It has been demonstrated that the use of abundance information theory-based measures improves the ability of remotely sensed data to detect local species diversity (Oldeland et al., 2010) since they are less sensitive to the influence of rare incidental sets of values (although they are strictly dependent on the scale at which they are applied: refer to Ricotta et al. (2008) for an elegant explanation considering species-based measures of diversity based on the Grime's (1998) mass ratio hypothesis).

Indices based on relative abundances may reveal differences between sites with high spectral or species diversity. It has been shown using partial least square regression models that better results are achieved when mapping species diversity as measured by Shannon index instead of simple richness (Feilhauer and Schmidtlein, 2009). Although such tests were mainly performed by applying information theory to species data rather than spectral data, examples exist proving the power of information theory applied to land-use maps to estimate species diversity (Kumar et al., 2006).

### 23.3.3 Beta-Diversity and Spatial Turnover of Spectral Reflectances

Although most of the above methods allow the mapping of local species diversity in order to identify "hotspots" of high species diversity, given two or more hot spots, their spatial turnover may be more important, for example, for building efficient networks of protected areas or supporting policy decision-making (Chiarucci et al., 2008). This has inevitably led to the necessity of considering beta diversity (spectral and species turnover).

Notoriously, after Whittaker (1972), landscape as well as species diversity may be divided into different components each of which considers a different spatial extent. Considering the DNs of an image as values to calculate diversity, such components may be summarized following Whittaker (1972):

- **(i)** alpha or local diversity ($\alpha$DN), that is the number of different DNs within one plot and/or a local area/polygon; it can be measured also by the aforementioned information-theory-based indices;
- **(ii)** gamma or total diversity ($\gamma$DN), that is the number of different DNs considering N plots and/or sampling areas, that is the whole extent of an area;
- **(iii)** beta diversity ($\beta$DN), that is the diversity deriving from the turnover in the composition of the DN values.

While local diversity measures are useful tools for detecting diversity hot spots at the local scale, they provide no information on differences among sites in terms of turnover in spectral values (beta diversity). Overall, high local diversity values together with a high compositional turnover across sites lead to high diversity within the whole study area (regional diversity or gamma diversity, Chao et al., 2005; Rocchini et al., 2005; Tuomisto and Ruokolainen, 2006; Vellend, 2001; Bacaro and Ricotta, 2007; He and Zhang, 2009).

A straightforward method for summarizing beta diversity consists of looking at the differences between pairs of plots in terms of their species composition using one of the many possible (dis)similarity coefficients proposed in the ecological literature (Legendre and Legendre, 1998; Podani, 2000; Koleff et al., 2003). As a rule of thumb, if plot-to-plot similarity is computed using a normalized measure ranging between 0 and 1, such as the Jaccard coefficient $C_i = a/(a+b+c)$, for single pairs of plots beta diversity is directly related to dissimilarity and can be rewritten as $\beta = 1 - C_j$ . Here, the letters refer to the traditional 2 × 2 contingency table: *a* is the number of species present in both the plots, representing the intersection between plots, *b* is the number of species present solely in the first plot and absent from the second plot, and *c* is the number of species present solely in the second plot (Vellend, 2001; Koleff et al., 2003).

Another straightforward estimate of beta diversity is related to the additive formulation fostered by Lande (1996) $\beta = \gamma - \alpha$. This measure can be estimated using a rarefaction curve created either (i) by repeatedly resampling the pool of N plots or areas at random without replacement and plotting the average number of spectral values represented by 1, 2, ..., *N* plots (Gotelli and Colwell, 2001) or (ii) relying on the Shinozaki (1963) rarefaction analytical solution:

$$E(S_n) = S - \frac{\sum_{i=1}^{S} \binom{N - N_i}{n}}{\binom{N}{n}} \tag{3}$$

where $S$ = total number of DN values in the dataset, $N_i$ = number of plots/areas where species $i$ is found, and $n$ = number of randomly chosen plots/areas (Kobayashi, 1974; Shinozaki 1963).

Subsampling an image by means of $N$ plots/areas, that is spatial windows with a certain dimension, will lead to a presence-absence matrix $M_{DN}$ of $N$ plots per $S$ DNs.

Equation 3 provides a formal estimate of the number of DNs per number of windows where $N_i$ is the number of plots where a certain DN value $i$ is found resulting in a rarefaction curve (Fig. 23.4). It is expected that the higher the environmental heterogeneity of an area the higher will be the slope of such curve (higher beta diversity) since more values are added to the curve given the same number of plots (Fig. 23.5).

The R Software provides an elegant solution for calculating rarefaction based on both permutation and/or the analytical solution by the `specaccum` function as:

```
library(spgrass6)
input.data <-readRAST6(c("image_band"), cat=c(TRUE))
rarefaction <- specaccum(image_band, method= "exact") # using the
analytical solution
rarefaction <- specaccum(image_band, method= "random", permutatins
=10000) # using the permutation-based solution
```

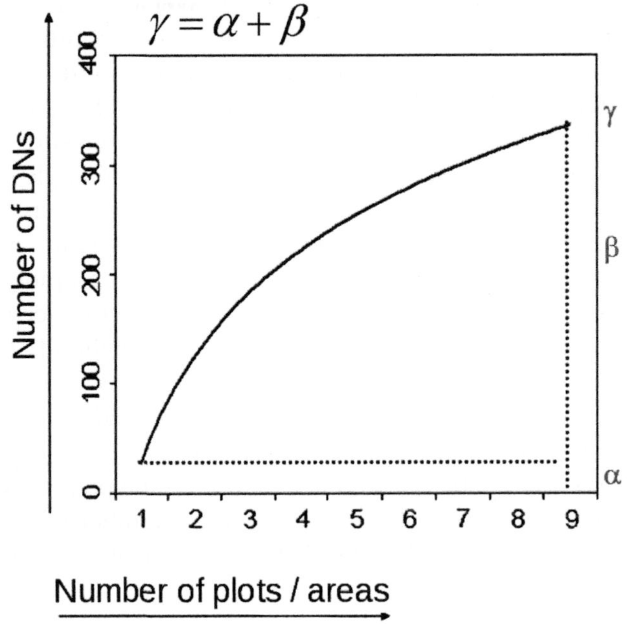


**FIGURE 23.4** γ-diversity is represented by the sum between α and β. This leads to considering β in the same unit of measurement (i.e., number of species) of α and γ.

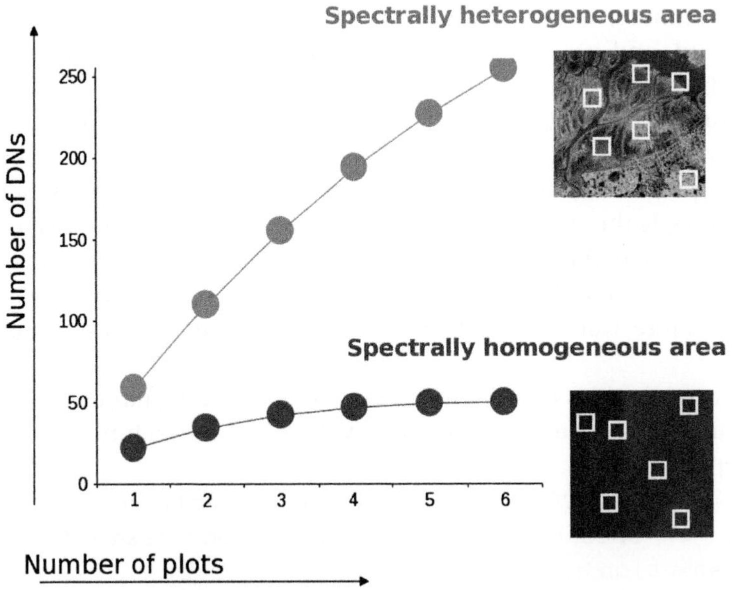


**FIGURE 23.5** Spectral rarefaction curves derived from two differently heterogeneous areas. For color version of this figure, the reader is referred to the online version of this book.

#### *23.3.3.1 How Spectral Turnover Relates to Biodiversity*

Matrices of pair-wise spectral distances versus species distances among plots have been shown to provide estimates of species turnover versus the variation of environmental properties over space (He and Zhang, 2009).

In addition, rarefaction curves are directly influenced by the environmental heterogeneity of the area sampled. Thus, it is expected that the greater the landscape heterogeneity, the greater the species diversity, including both fine-scale and coarse-scale species richness (i.e., alpha- and gamma-diversity, respectively), and compositional turnover, or beta diversity. As an example, Rocchini et al. (2011), studying the variability of alpine areas in Switzerland, demonstrated that areas with higher gradients like the Alps have higher spectral- and plant species-diversities compared to flat landscapes (Plateau area). However, this was strictly dependent on the scale at which the relationship was tested, with no differences among the areas at a local scale (alpha diversity) and significant differences at regional/global scales (beta diversity, Fig. 23.6).

### 23.3.4 Beyond Compositional Diversity: The Diversity of Shapes into the Landscape as Estimated by Fractal Geometry

The shapes of spatial entities, habitats, and ecosystems hold an important information for landscape ecologists since the shape carries key information about landscape structures and potentially, on the processes that generated them (O'Neill et al., 1988). As an example, shapes of particular landscape elements (patches and corridors) may be studied by topological indices, where quantifying the topology of the habitat network can provide indicators of vulnerability in either single-species (Baranyi et al., 2011) or multispecies situations (Jordán et al., 2007).

Fractals are statistically self-similar mathematical objects (Mandelbrot, 1982), that is most of their shape-related properties are scale invariant. Fractals have been used in landscape ecology since the foundation of fractal geometry (Mandelbrot, 1982). Even some of the forerunners of fractal geometry (like Richardson's plot for the length of Nile of the length of the coast of Norway or Korcak's plot for the area distribution of some islands) are coming from landscape analysis. There are several, profoundly different fractal dimensions, that is Hausdorff, box, Minkowski, Korcak, etc., (Mandelbrot, 1982; Feder, 1988; Russ, 1994; Imre and Bogaert, 2006; Imre et al., 2011); however, the one most often referred to as "fractal dimension" (D) is the Hausdorff dimension.

Imagine a single line filling a 2D space where its dimension can range from 1 to 2 once its complexity in the 2D space is so high that it completely fills it. The same reasoning can be applied to 2D spatial surfaces or objects whose fractal dimension ranges from 2 to theoretically 3 as their shape complexity increases (Fig. 23.7).

For a formal definition of the fractal Hausdorff dimension ($D$), let $M_{DN}$ be a binary matrix (lattice grid) 1/0 with pixel size $S$. The complexity of the shapes composing such

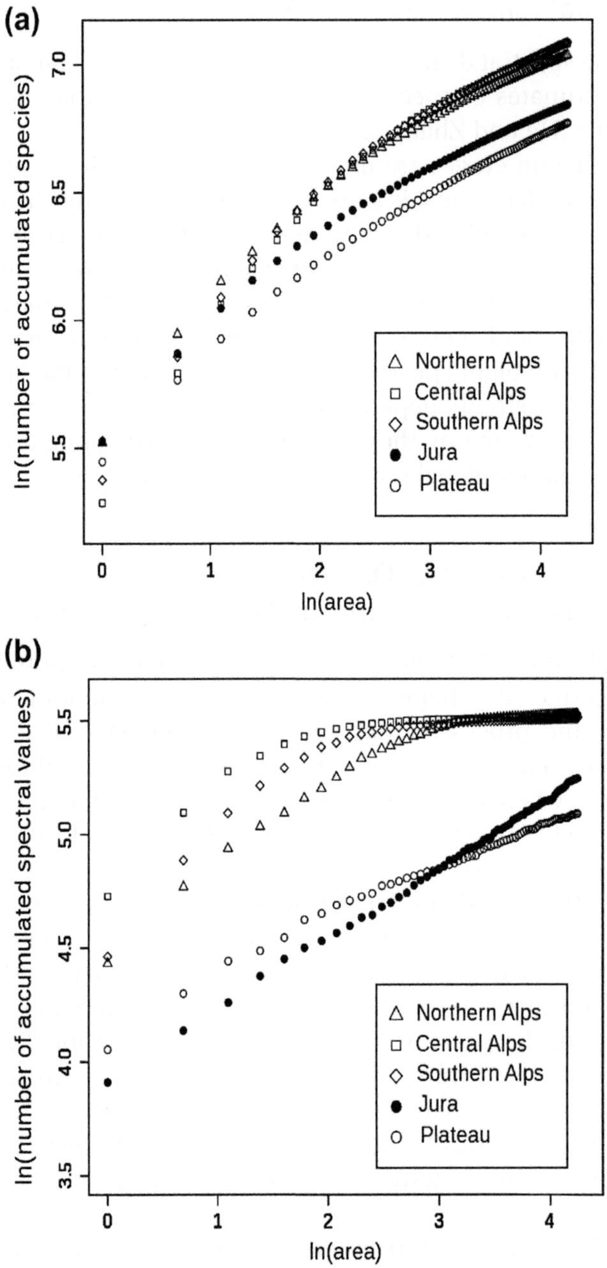


**FIGURE 23.6** Species (a) and spectral (b) rarefaction of Switzerland Biodiversity Monitoring data. Alpine regions with a higher spectral diversity show a higher spectral diversity of areas with a lower topographic complexity. Courtesy of Wiley Blackwell. *Source.* Rocchini et al. (2010).

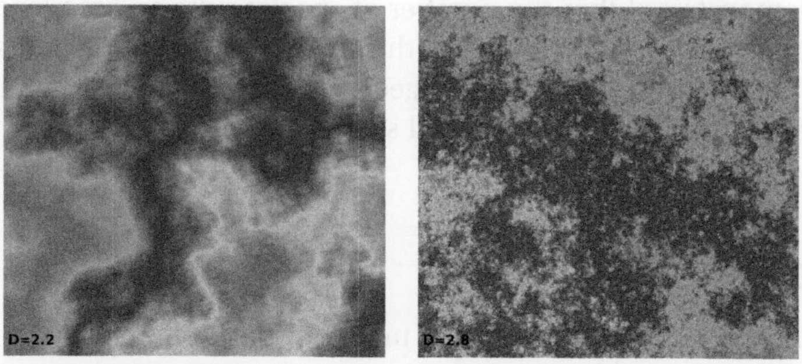


**FIGURE 23.7** Neutral lattice landscapes with different fractal (Hausdorff) dimensions. A higher fractal dimension is expected to host a higher species diversity on the strength of the high landscape complexity. For color version of this figure, the reader is referred to the online version of this book.

landscape is a function of the number of distinct fragments *N* into the landscape and the pixel dimension *r:*

$$D^m = \lim_{r \to 0} \frac{\ln(N)}{\ln\left(\frac{1}{r}\right)} \tag{4}$$

Figure 23.7 shows two lattice neutral models (maps) generated into GRASS GIS by:

```
r.surf.fractal output=D2.2 dimension=2.2
r.surf.fractal output=D2.8 dimension=2.8
```

with a different complexity (fractal dimension).

In addition to generating a neutral landscape with a predefined fractal dimension, GRASS GIS also allows the calculation of the landscape fractal dimension using the `r.boxcount` function (https://svn.osgeo.org/grass/grass-addons/raster/r.boxcount/description.html) based on the Minkowski–Bouligand box-counting method (refer to Feder, 1988) mainly applied to classified images (e.g., land-use maps).

This can be done into R also by packages such as *fdim* (Martinez de Pison Ascacibar et al., 2011) or *fractaldim* (Sevcikova et al., 2011).

#### *23.3.4.1 How Shape Complexity Relates to Biodiversity*

Evidence that there is an increase in species diversity with increasing shape complexity over the landscape has been shown for a number of different regions and habitat types such as Mediterranean semi-natural areas (Imre and Rocchini, 2009), prairies, and pine forests or Rocky Mountains, Colorado (Kumar et al., 2006), and Austrian meadows mixed with crops and vineyards (Moser et al., 2002),

Explicitly considering fractal dimension, Palmer (2007) proved that landscapes with a higher Haussdorf *D* dimension rapidly accumulate a higher number of species, for example in a species rarefaction curve since most environments are sampled at the finest scale. Hence, measuring the complexity of shapes over the landscape may represent an inexpensive proxy for the variation of species diversity in space. Furthermore, Šizling and

Storch (2004) demonstrated that the number of species over an area (species area relationship (SAR)) could be predicted using the fractal nature of their distribution over space, thus advocating fractal landscape geometry as one of the main determinants shaping species diversity at different spatial scales (Scheuring, 1991).

## 23.4 Coda

In this book chapter, we have discussed many of the available open source spatial algorithms for estimating landscape diversity.

Since landscape diversity often relates to species biodiversity at several ecological levels from species community diversity to genetic diversity, estimating landscape heterogeneity from remotely sensed data is an efficient and relatively cost-effective way of providing biodiversity estimates over large geographical areas. Depending on the study objectives, species diversity can also be modeled at appropriate scales in time and space.

In this book chapter we mainly relied on remote sensing and land use data, but the proposed tool chain can be applied to a variety of continuous fields (*sensu* Burrough and McDonnel, 1998) such as fractal surfaces (Lovejoy and Schertzer, 2010), fractional tree cover surfaces (Schwarz and Zimmermann, 2005), or elevation models (Hofer et al., 2008). The latter is particularly suitable on the strength of the known relation between topographic variability and species diversity considering both elevation and its derivatives (Hofer et al., 2008).

Despite the methodology being used for calculating heterogeneity/uncertainty over the landscape, final results can only be considered reliable once uncertainty has been shown to be consistent even when computed in different ways, according to the "branching principle" (Aczél and Daróczy, 1975).

With this book chapter we hope to stimulate further discussion about spatial algorithms applied to landscape diversity estimates and their use in a FOS environment.

## References

Aczél, J., Daróczy, Z., 1975. On Measures of Information and their Characterizations. Academic Press, New York.

Bacaro, G., Ricotta, C., 2007. A spatially explicit measure of beta diversity. Commun. Ecol. 8, 41–46.

Bacaro, G., Baragatti, E., Chiarucci, A., 2009. Using taxonomic data to assess and monitor biodiversity: are the tribes still fighting? J. Environ. Monitor. 11, 798–801.

Bacaro, G., Santi, E., Rocchini, D., Pezzo, F., Puglisi, L., Chiarucci, A., 2011. Geostatistical modelling of regional bird species richness: exploring environmental proxies for conservation purpose. Biodivers. Conserv. 20, 1677–1694.

Baranyi, G., Saura, S., Podani, J., Jordán, F., 2011. Contribution of habitat patches to network connectivity: redundancy and uniqueness of topological indices. Ecol. Indic. 11, 1301–1310.

Bino, G., Levin, N., Darawshi, S., van der Hal, N., Reich-Solomon, A., Kark, S., 2008. Landsat derived NDVI and spectral unmixing accurately predict bird species richness patterns in an urban landscape. Int. J. Remote Sens. 29, 3675–3700.

Bivand, R., 2011. spgrass6: Interface between GRASS 6+ Geographical Information System and R. R Package Version 1.0-7. http://cran.r-project.org/web/packages/spgrass6/index.html Accessed July 10th 2012.

Bolliger, J., 2005. Simulating complex landscapes with a generic model: sensitivity to qualitative and quantitative classifications. Ecol. Complex 2, 131–149.

Boltzmann, L., 1872. Weitere Studien über das Wärmegleichgewicht unter Gasmolekülen. Wien. Ber. 66, 275–370.

Bond, W.J., 1994. Keystone species. In: Schulze, E.D., Mooney, H.A. (Eds.), Biodiversity and Ecosystem Function. Springer, Berlin.

Burrough, P.A., McDonnell, R.A., 1998. Principles of Geographical Information Systems. Oxford University Press, New York.

Butchart, S.H.M., Walpole, M., Collen, B., van Strien, A., Scharlemann, J.P.W., Almond, R.E.A., Baillie, J.E.M., Scharlemann, J.P.W., Almond, R.E.A., Baillie, J.E.M., Bomhard, B., Brown, C., Bruno, J., Carpenter, K.E., Carr, G.M., Chanson, J., Chenery, A.M., Csirke, J., Davidson, N.C., Dentener, F., Foster, M., Galli, A., Galloway, J.N., Genovesi, P., Gregory, R.D., Hockings, M., Kapos, V., Lamarque, J.-F., Leverington, F., Loh, J., McGeoch, M.A., McRae, L., Minasyan, A., Hernández Morcillo, M., Oldfield, T.E.E., Pauly, D., Quader, S., Revenga, C., Sauer, J.R., Skolnik, B., Spear, D., Stanwell-Smith, D., Stuart, S.N., Symes, A., Tierney, M., Tyrrell, T.D., Vié, J.-C., Watson, R., 2010. Global biodiversity: indicators of recent declines. Science 328, 1164–1168.

Chao, A., Chazdon, R.L., Colwell, R.K., Shen, T.-J., 2005. A new statistical approach for assessing similarity of species composition with incidence and abundance data. Ecol. Lett. 8, 148–159.

Chiarucci, A., 2007. To sample or not to sample? That is the question for the vegetation scientist. Folia Geobot. 42, 209–216.

Chiarucci, A., Bacaro, G., Rocchini, D., 2008. Quantifying plant species diversity in a Natura 2000 Network: old ideas and new proposals. Biol. Conserv. 141, 2608–2618.

Feder, J., 1988. Fractals. Plenum Press, New York.

Feilhauer, H., Schmidtlein, S., 2009. Mapping continuous fields of forest alpha and beta diversity. Appl. Veg. Sci. 12, 429–439.

Ferretti, M., Chiarucci, A., 2003. Design concepts adopted in long-term forest monitoring programs in Europe: problems for the future? Sci. Total Environ. 310, 171–178.

Foody, G.M., 2005. Mapping the richness and composition of British breeding birds from coarse spatial resolution satellite sensor imagery. Int. J. Remote Sens. 26, 3943–3956.

Foody, G.M., Cutler, M.E.J., 2003. Tree biodiversity in protected and logged Bornean tropical rain forests and its measurement by satellite remote sensing. J. Biogeogr. 30, 1053–1066.

Foody, G.M., Cutler, M.E.J., 2006. Mapping the species richness and composition of tropical forests from remotely sensed data with neural networks. Ecol. Model. 195, 37–42.

Fogel, K., 2009. Producing Open source Software: How to Run a Successful Free Software Project. Creative Commons Attribution–ShareAlike (3.0) License. O'Reilly, Sebastopol.

Forest, F., Grenyer, R., Rouget, M., Davies, T.J., Cowling, R.M., Faith, D.P., Balmford, A., Manning, J.C., Proches, S., van der Bank, M., Reeves, G., Hedderson, T.A.J., Savolainen, V., 2007. Preserving the evolutionary potential of floras in biodiversity hotspots. Nature 445, 757–760.

Gillespie, T.W., 2005. Predicting woody-plant species richness in tropical dry forests: a case study from South Florida, USA. Ecol. Appl. 15, 27–37.

Gillespie, T.W., Foody, G.M., Rocchini, D., Giorgi, A.P., Saatchi, S., 2008. Measuring and modelling biodiversity from space. Prog. Phys. Geogr. 32, 203–221.

Glenn, T.C., 2011. Field guide to next-generation DNA sequencers. Mol. Ecol. Resour. 11, 759–769.

Goodchild, M.F., Egenhofer, M.J., Kemp, K.K., Mark, D.M., Sheppard, E.S., 1999. Introduction to the Varenius project. Int. J. Geogr. Inf. Sci. 13, 731–745.

Gotelli, N.J., Colwell, R.K., 2001. Quantifying biodiversity: procedures and pitfalls in the measurement and comparison of species richness. Ecol. Lett. 4, 379–391.

Gould, W., 2000. Remote sensing of vegetation, plant species richness, and regional biodiversity hot spots. Ecol. Appl. 10, 1861–1870.

Grime, J.P., 1998. Benefits of plant diversity to ecosystems: immediate, filter, and founder effects. J. Ecol. 86, 902–910.

He, K.S., Zhang, J., 2009. Testing the correlation between beta diversity and differences in productivity among global ecoregions, biomes, and biogeographical realms. Ecol. Inform. 4, 93–98.

Hofer, G., Wagner, H., Herzog, F., Edwards, P., 2008. Effects of topographic variability on the scaling of plant species richness in gradient dominated landscapes. Ecography 31, 131–139.

Hollingsworth, P.M., Graham, S.W., Little, D.P., 2011. Choosing and using a plant DNA barcode. PLoS One 6, e19254.

Hortal, J., Lobo, J.M., 2005. An ED-based protocol for the optimal sampling of biodiversity. Biodivers. Conserv. 14, 2913–2947.

Imre, A.R., Bogaert, J., 2006. The Minkowski–Bouligand dimension and the interior-to-edge ratio of habitats. Fractals 14, 49–53.

Imre, A., Rocchini, D., 2009. Explicitly accounting for pixel dimension in calculating classical and fractal landscape shape metrics. Acta Biotheor. 57, 349–360.

Imre, A.R., Cseh, D., Neteler, M., Rocchini, D., 2011. Characterization of landscape fragmentation and reforestation by the Korcak fractal dimension. Ecol. Indic. 11, 1134–1138.

Jordán, F., Magura, T., Tóthmérész, B., Vasas, V., Ködöböcz, V., 2007. Carabids (Coleoptera: Carabidae) in a forest patchwork: a connectivity analysis of the Bereg Plain landscape graph. Landsc. Ecol. 22, 1527–1539.

Knapp, S., Lughadha, E.N., Paton, A., 2005. Taxonomic inflation, species concepts and global species lists. Trends Ecol. Evol. 20, 7–8.

Kobayashi, S., 1974. The species-area relation I. A model for discrete sampling. Res. Popul. Ecol. 15, 223–237.

Koleff, P., Gaston, K.J., Lennon, J.J., 2003. Measuring beta diversity for presence–absence data. J. Anim. Ecol. 72, 367–382.

Kumar, S., Stohlgren, T.J., Chong, G.W., 2006. Spatial heterogeneity influences native and nonnative plant species richness. Ecology 87, 3186–3199.

Lande, R., 1996. Statistics and partitioning of species diversity, and similarity among multiple communities. Oikos 76, 5–13.

Lassau, S.A., Cassis, G., Flemons, P.K.J., Wilkie, L., Hochuli, D.F., 2005. Using high-resolution multispectral imagery to estimate habitat complexity in open-canopy forests: can we predict ant community patterns? Ecography 28, 495–504.

Lawton, J.H., Brown, V.K., 1994. Redundancy in ecosystems. In: Schulze, E.D., Mooney, H.A. (Eds.), Biodiversity and Ecosystem Function. Springer, Berlin, pp. 255–270.

Legendre, P., Legendre, L., 1998. Numerical Ecology. Elsevier, Amsterdam, The Netherlands.

Levin, N., Shmida, A., Levanoni, O., Tamari, H., Kark, S., 2007. Predicting mountain plant richness and rarity from space using satellite-derived vegetation indices. Divers. Distrib. 13, 692–703.

Lovejoy, S., Schertzer, D., 2010. On the simulation of continuous in scale universal multifractals, part I: spatially continuous processes. Comput. Geosci. 36, 1393–1403.

Mandelbrot, B.B., 1982. The Fractal Geometry of Nature. Freeman, New York.

Martinez de Pison Ascacibar, J., Ordieres Mere, J., Castejon Limas, M., de Cos Juez., J., 2011. fdim: Functions for Calculating Fractal Dimension. R package Version 1.0-7. http://CRAN.R-project.org/package=fdim.

Moser, D., Zechmeister, H., Plutzar, C., Sauberer, N., Wrbka, T., Grabherr, G., 2002. Landscape patch shape complexity as an effective measure for plant species richness in rural landscapes. Lands. Ecol. 17, 657–669.

Nagendra, H., Gadgil, M., 1999. Biodiversity assessment at multiple scales: linking remotely sensed data with field information. Proc. Natl. Acad. Sci. USA 96, 9154–9158.

Nagendra, H., Rocchini, D., Ghate, R., Sharma, B., Pareeth, S., 2010. Assessing plant diversity in a dry tropical forest: comparing the utility of Landsat and IKONOS satellite images. Remote Sens. 2, 478–496.

Neteler, M., Mitasova, H., 2008. Open source GIS: a GRASS GIS approach. In: The International Series in Engineering and Computer Science, third ed. Springer, New York.

Neteler, M., Grasso, D., Michelazzi, I., Miori, L., Merler, S., Furlanello, C., 2005. An integrated toolbox for image registration, fusion and classification. Int. J. Geoinf. 1, 51–61.

Neteler, M., Bowman, M.H., Landa, M., Metz, M., 2012. GRASS GIS: a multi-purpose open source GIS. Environ. Model. Softw. 31, 124–130.

Oksanen, J., Blanchet, F.G., Kindt, R., Legendre, P., O'Hara, R.B., Simpson, G.L., Solymos, P., Henry, M., Stevens, H., Wagner, H., 2011. vegan: Community Ecology Package. R Package Version 1.17-2. http://cran.r-project.org/web/packages/vegan/index.html.

Oindo, B.O., Skidmore, A.K., 2002. Interannual variability of NDVI and species richness in Kenya. Int. J. Remote Sens. 23, 285–298.

Oldeland, J., Wesuls, D., Rocchini, D., Schmidt, M., Jurgens, N., 2010. Does using species abundance data improve estimates of species diversity from remotely sensed spectral heterogeneity? Ecol. Indic. 10, 390–396.

O'Neill, R.V., Krummel, J.R., Gardner, R.H., Sugihara, G., Jackson, B., DeAngelis, D.L., Milne, B.T., Turner, M.G., Zygmunt, B., Christensen, S.W., Dale, V.H., Graham, R.L., 1988. Indices of landscape pattern. Landsc. Ecol. 1, 153–162.

Palmer, M.W., 1995. How should one count species? Nat. Areas J. 15, 124–135.

Palmer, M.W., 2007. Species-area curves and the geometry of nature. In: Storch, D., et al. (Eds.), Scaling Biodiversity. Cambridge University Press, Cambridge, UK, pp. 15–31.

Palmer, M.W., Earls, P., Hoagland, B.W., White, P.S., Wohlgemuth, T., 2002. Quantitative tools for perfecting species lists. Environmetrics 13, 121–137.

Parviainen, M., Luoto, M., Heikkinen, R.K., 2009. The role of local and landscape level measures of greenness in modelling boreal plant species richness. Ecol. Model. 220, 2690–2701.

Pielou, E.C., 1969. An Introduction to Mathematical Ecology. John Wiley, New York.

Podani, J., 2000. Introduction to the Exploration of Multivariate Biological Data. Backhuys, Leiden, The Netherlands.

Power, M.E., Tilman, D., Estes, J.A., Menge, B.A., Bond, W.J., Mills, L.S., Daily, G., Castilla, J.C., Lubchenco, J., Paine, R.T., 1996. Challenges in the quest for keystones. BioScience 46, 609–620.

R Development Core Team, 2011. R: A Language and Environment for Statistical Computing. R Foundation for Statistical Computing, Vienna, Austria. http://www.R-project.org.

Rényi, A., 1970. Probability Theory. North Holland Publishing Company, Amsterdam.

Ricotta, C., 2005. On possible measures for evaluating the degree of uncertainty of fuzzy thematic maps. Int. J. Remote Sens. 26, 5573–5583.

Ricotta, C., Avena, G., 2003. On the relationship between Pielou's evenness and landscape dominance within the context of Hill's diversity profiles. Ecol. Indic. 2, 361–365.

Ricotta, C., Corona, P., Marchetti, M., Chirici, G., Innamorati, S., 2003. LaDy: software for assessing local landscape diversity profiles of raster land cover maps using geographic windows. Environ. Model. Softw. 18, 373–378.

Ricotta, C., Godefroid, S., Celesti-Grapow, L., 2008. Common species have lower taxonomic diversity: evidence from the urban floras of Brussels and Rome. Divers. Distrib. 14, 530–537.

Robinson, G.R., Yurlina, M.E., Handel, S.N., 1994. A century of change in the Staten island flora: ecological correlates of species losses and invasions. Bull. Torrey Bot. Club 121, 119–129.

Roca, A.L., Georgiadis, N., Pecon-Slattery, J., O'Brien, S.J., 2001. Genetic evidence for two species of elephant in Africa. Science 293, 1473–1477.

Rocchini, D., 2007. Effects of spatial and spectral resolution in estimating ecosystem $\alpha$-diversity by satellite imagery. Remote Sens. Environ. 111, 423–434.

Rocchini, D., 2010. While Boolean sets non-gently rip: a theoretical framework on fuzzy sets for mapping landscape patterns. Ecol. Complex 7, 125–129.

Rocchini, D., Neteler, M., 2012a. Let the four freedoms paradigm apply to ecology. Trends in Ecology & Evolution 27, 310–311.

Rocchini, D., Neteler, M., 2012b. Spectral rank–abundance for measuring landscape diversity. Int. J. Remote Sens. 33, 4458–4470.

Rocchini, D., Andreini Butini, S., Chiarucci, A., 2005. Maximizing plant species inventory efficiency by means of remotely sensed spectral distances. Global Ecol. Biogeogr. 14, 431–437.

Rocchini, D., Balkenhol, N., Carter, G.A., Foody, G.M., Gillespie, T.W., He, K.S., Kark, S., Levin, N., Lucas, K., Luoto, M., Nagendra, H., Oldeland, J., Ricotta, C., Southworth, J., Neteler, M., 2010. Remotely sensed spectral heterogeneity as a proxy of species diversity: recent advances and open challenges. Ecol. Inform. 5, 318–329.

Rocchini, D., Delucchi, L., Bacaro, G., Cavallini, P., Feilhauer, H., Foody, G.M., He, K.S., Nagendra, H., Porta, C., Ricotta, C., Schmidtlein, S., Spano, L.D., Wegmann, M., Neteler, M., 2012. Calculating landscape diversity with information-theory based indices: A GRASS GIS solution. Ecol. Inform. http://dx.doi.org/doi:10.1016/j.ecoinf.2012.04.002.

Rocchini, D., McGlinn, D., Ricotta, C., Neteler, M., Wohlgemuth, T., 2011. Landscape complexity and spatial scale influence the relationship between remotely sensed spectral diversity and survey based plant species richness. J. Veg. Sci. 22, 688–698.

Russ, J.C., 1994. Fractal Surfaces. Plenum Press, New York.

Scheuring, J., 1991. The fractal nature of vegetation and the species-area relation. Theor. Popul. Biol. 39, 170–177.

Schwartz, M.K., Monfort, S.L., 2008. Genetic and endocrine tools for carnivore surveys. In: Long, R.A., MacKay, P., Ray, J.C., Zielinski, W.J. (Eds.), Noninvasive Survey Methods for North American Carnivores. Island Press, Washington D.C, pp. 228–250.

Schwarz, M., Zimmermann, N., 2005. A new GLM-based method for mapping tree cover continuous fields using regional MODIS reflectance data. Remote Sens. Environ. 95, 428–443.

Sevcikova, H., Gneiting, T., Percival, D., 2011. fractaldim: Estimation of Fractal Dimensions. R package Version 1.0-7. http://cran.r-project.org/web/packages/fractaldim/index.html.

Shannon, C., 1948. A mathematical theory of communication. Bell Syst. Techn. J. 27, 379–423.

Shinozaki,, K., 1963. Note on the species area curve. Proceedings of the 10th Annual Meeting of Ecological Society of Japan 50 (in Japanese). Ecological Society of Japan, Tokyo, Japan.

Simpson, E.H., 1949. Measurement of diversity. Nature 163, 688.

Šizling, A.L., Storch, D., 2004. Power-law species-area relationships and self-similar species distributions within finite areas. Ecol. Lett. 7, 60–68.

Stallman, R., 1985. The GNU Manifesto. Available at: http://www.gnu.org/gnu/manifesto.html

Stallman, R., 1997. The GNU Manifesto. In: Ermann, M.D., Williams, M.B., Shauf, M.S. (Eds.), Computers, Ethics and Society, second ed. Oxford University Press, Oxford, pp. 229–239.

St-Louis, V., Pidgeon, A.M., Clayton, M.K., Locke, B.A., Bash, D., Radeloff, V.C., 2009. Satellite image texture and a vegetation index predict avian biodiversity in the Chihuahuan Desert of New Mexico. Ecography 32, 468–480.

Stoerfer, A., Murphy, M.A., Spear, S.F., Holderegger, R., Waits, L.P., 2010. Landscape genetics: where are we now? Mol. Ecol. 19, 3496–3514.

Szathmáry, E., Jordán, F., Pál, C., 2001. Can genes explain biological complexity? Science 292, 1315–1316.

Tuomisto, H., Ruokolainen, K., 2006. Analyzing or explaining beta diversity? Understanding the targets of different methods of analysis. Ecology 87, 2697–2708.

Vellend, M., 2001. Do commonly used indices of β-diversity measure species turnover? J. Veg. Sci. 12, 545–552.

Waits, L.P., Paetkau, D., 2005. New non-invasive genetic sampling tools for wildlife biologists: a review of applications and recommendations for accurate data collection. J. Wildl. Manage. 69, 1419–1433.

Waser, L.T., Stofer, S., Schwarz, M., Küchler, M., Ivits, E., Scheidegger, C.H., 2004. Prediction of biodiversity: regression of lichen species richness on remote sensing data. Commun. Ecol. 5, 121–134.

Whittaker, R.H., 1972. Evolution and measurement of species diversity. Taxon 21, 213–251.

Wilting, A., Christiansen, P., Kitchener, A.C., Kemp, I.Y.M., Ambu, L., Fickel, J., 2011. Geographical variation in and evolutionary history of the Sunda clouded leopard (*Neofelis diardi*) (Mammalia: Carnivora: Felidae) with the description of a new subspecies from Borneo. Mol. Phylogenet. Evol. 58, 317–328.

Zadeh, L., 1965. Fuzzy sets. Inf. Control 8, 338–353.

Zimmermann, H.-J., 2001. Fuzzy Set Theory and its Applications, fourth ed. Springer, Berlin.

# 24

# Offsetting Policies for Biodiversity Conservation: The Need for Compensating Habitat Relocation

Karin Johst, Florian Hartig, Martin Drechsler
*HELMHOLTZ CENTRE FOR ENVIRONMENTAL RESEARCH-UFZ, DEPARTMENT OF ECOLOGICAL MODELLING, PERMOSERSTR. 15, 04318, LEIPZIG, GERMANY*

## 24.1 Introduction

Human activities alter environmental conditions on earth, with far-reaching consequences for biodiversity. In particular, land use change may lead to the transformation, fragmentation, or destruction of natural habitats. The last decades have shown that many species are not able to cope with these changes such that their existence is at serious risk. Species extinctions, in turn, may impact ecosystem dynamics, ecosystem functioning, and the resulting ecosystem services which are essential for human well-being (Loreau et al., 2001; Tilman, 1999). One potential policy response to halt biodiversity loss is the recent trend toward market-based conservation policies, known as conservation banking, biodiversity offsetting, or simply as tradable permit schemes (Salzman and Ruhl, 2000; Wissel and Wätzold, 2010; Ring et al., 2010).

The principle of these policies is that the development of land, for example to establish infrastructure projects or business parks in a landscape, requires adequate compensation in form of upgrading the ecological value of other sites. This upgrade can be performed by the developer, and also by other landowners, companies, or authorities. The successful reconstruction of ecologically valuable habitat is documented in the form of a permit issued by an authorized agency that ensures that quality standards are met (Bishop et al., 2009; Cuperus et al., 1999; Schippers et al., 2009; Wissel and Wätzold, 2010). Companies or private persons intending to carry out development projects have to buy or create such permits to satisfy the legal requirements for the compensation of any habitat destruction that is caused by the respective projects. This ensures that the overall amount of ecologically valuable habitat in the landscape remains constant.

One problem, however, is that such a market may lead to high rates of habitat destruction and habitat reconstruction (i.e., high habitat patch turnover) and therefore to

Models of the Ecological Hierarchy. DOI: http://dx.doi.org/10.1016/B978-0-444-59396-2.00024-9

ISSN 0167-8892, 

highly dynamic landscapes. These landscape dynamics may substantially affect the ecological effectiveness of offsetting policies (Hartig and Drechsler, 2009). The reason is that, in most cases, the goal of offsetting is not to preserve natural habitats for their own sake. Rather, one aims at ensuring the conservation and long-term viability of the species that depend on these habitats. Offsetting policies keep the total amount of habitat for these species constant, but they may lead to changes in the habitat allocation. Species are threatened by the resulting increased local extinction risks due to patch destruction and the temporally varying patch configurations (Johst et al., 2011; Hartig and Drechsler, 2009), and decision makers must explicitly take this threat into account in their policy design. A central question for the design of offsetting policies is therefore whether increased landscape dynamics pose a risk to the persistence of species, how this risk depends on species and landscape properties, and to which extent this risk can be compensated by additional safeguards and compensation measures. This question is not straightforward to answer because (1) there are trade-offs between different spatial and dynamic landscape attributes, (2) these trade-offs are generally nonlinear, and (3) they are different for different species.

Trade-offs between landscape attributes arise because a change in one landscape attribute (e.g., increased patch turnover) can often be compensated for by certain changes in another attribute (e.g., increased patch number or connectivity) in order to maintain a desired viability of a population. This option of compensation is important for management decisions. Trade-offs between spatial landscape attributes such as the number and size of habitat patches (e.g., Ovaskainen, 2002) or between the number and connectivity of patches (e.g., Moilanen and Hanski, 1998; Vuilleumier et al., 2007) have therefore frequently been discussed. It is also known that conservation management differs among species as species have different dispersal abilities or area requirements and/or differ in their sensitivity to environmental stochasticity (e.g., Vos et al., 2001; Jacquemyn et al., 2003). In the context of biodiversity offsetting it is therefore pivotal to know how habitat patch relocation and thus increased patch turnover impacts the species to be protected, whether this increased turnover can be compensated at all, and if so, which changes in landscape attributes can be used to achieve this compensation.

Here, we consider landscapes characterized by patchy habitats that are destroyed and created by anthropogenic influences resulting in stationary landscape dynamics (Hanski, 1999). Populations of species living in these patchy landscapes can be described as metapopulations. Metapopulation models assume that the local populations inhabiting the (habitat) patches may go extinct; however, extinct patches may be recolonized by dispersing individuals from other patches (e.g., Gyllenberg and Hanski, 1997; Keymer et al., 2000; Wilcox et al., 2006; Cornell and Ovaskainen, 2008). The survival of species living in such landscapes can be evaluated by the mean metapopulation lifetime, that is, the time during which on average one or more habitat patches in the landscape are occupied by the species (Frank and Wissel, 2002; Drechsler, 2009; Drechsler and Johst, 2010).

To describe biodiversity offsetting, we assume that for each patch destroyed, another patch has to be created elsewhere in the landscape. Thus, although the introduction of biodiversity offsetting into a landscape implies an increased habitat patch turnover, the total amount of habitat patches is maintained. Calculations of expected survival times of a metapopulation as a function of number, size, and connectivity of habitat patches and dynamic landscape attributes such as patch turnover usually require extensive computer simulations and a lengthy numerical analysis (e.g., Johst et al., 2002; Johst and Drechsler, 2003; Wintle et al., 2005; Wilcox et al., 2006; Vuilleumier et al., 2007; Hodgson et al., 2009). This makes a systematic exploration of trade-offs between landscape attributes cumbersome. Fortunately, there are approximation formulas that allow a rapid calculation of the mean metapopulation lifetime in static (Frank and Wissel, 2002; Drechsler, 2009) and dynamic landscapes (Drechsler and Johst, 2010).

Johst et al. (2011) have demonstrated the usefulness of such a formula for the rapid and convenient assessment of trade-offs between different landscape attributes for a variety of dynamic landscapes and species (for general applicability and validity of the formula, see Drechsler and Johst, 2010). We use this analytical formula here to investigate the necessity and the potential of compensation of increased patch turnover by improving spatial landscape attributes. For this, we go beyond considering two landscape attributes and resulting metapopulation lifetime (see Johst et al., 2011) and focus explicitly on trade-offs between three landscape attributes to maintain a given (desired) metapopulation lifetime as conservation aim: patch number, landscape radius (a landscape attribute related to the connectivity of patches), and patch turnover. These landscape attributes are important as they are relatively easy to monitor in real landscapes and it is feasible to manage them in the context of biodiversity offsetting (Hartig and Drechsler, 2009; Drechsler and Hartig, 2011).

With respect to these three landscape attributes, we investigate whether increased habitat patch turnover can be compensated at all under the constraint that metapopulation lifetime is to be maintained and if so, which changes in other landscape attributes are necessary to achieve this compensation. With respect to species attributes, we investigate which type of species can be protected effectively by biodiversity offsetting. This supports understanding for which species and in which landscapes biodiversity offsetting is an option for biodiversity conservation.

## 24.2 Methods

In the first part, we shortly present the formula for the mean metapopulation lifetime on which our analysis is based. This formula has been derived analytically and was validated by comparison of the output with that of a spatially explicit stochastic computer simulation model (for details see Drechsler and Johst, 2010). Mean metapopulation lifetime is calculated with this formula by a few macroscopic spatial and dynamic landscape attributes. In the following, we repeat the definition of these landscape attributes and how they are adapted to describe biodiversity offsetting policies. In the second part, we explain the landscape perspective of the formula, and in the third part how landscape attributes and species traits are related.

### 24.2.1 Analytical Formula for the Mean Metapopulation Lifetime

The mean metapopulation lifetime in a dynamic landscape characterized by patch destruction and creation is given by (Drechsler and Johst, 2010):

$$T_{\mathrm{dyn}} \approx \frac{1}{\tilde{e}_{\mathrm{dyn}}} \sum_{i=1}^{N_{\mathrm{dyn}}} \sum_{k=1}^{N_{\mathrm{dyn}}} \frac{1}{k} \frac{(N_{\mathrm{dyn}}-i)!}{(N_{\mathrm{dyn}}-k)!} \frac{1}{(N_{\mathrm{dyn}}-1)^{k-i}} {q_{\mathrm{dyn}}}^{k-i} \tag{1}$$

$N_{\mathrm{dyn}}$ in Eqn (1) is the number of patches in the dynamic landscape. As biodiversity offsetting allows destruction of a patch only if a new patch is created elsewhere in the landscape, in this analysis $N_{\mathrm{dyn}}$ is independent of the level of patch turnover.

$\tilde{e}_{\mathrm{dyn}}$ in Eqn (1) is the geometric mean over the local extinction rates $e_i$ including patch destruction at rate $\mu$:

$$\tilde{e}_{\mathrm{dyn}} = \prod_{i=1}^{N_{\mathrm{dyn}}} (\mu + e_i)^{1/N_{\mathrm{dyn}}} = \tilde{e} \prod_{i=1}^{N_{\mathrm{dyn}}} (1 + \mu/e_i)^{1/N_{\mathrm{dyn}}} \tag{2}$$

The extinction rates $e_i$ represent the rate of a local population on patch $i$ to go extinct. The patch destruction rate $\mu$ characterizes the level of habitat patch turnover caused by the relocation of patches under biodiversity offsetting policies.

The quantity $q_{\mathrm{dyn}}$ in Eqn (1) is the so-called aggregated colonization–extinction ratio:

$$q_{\mathrm{dyn}} = \frac{\overline{c}}{\tilde{e}_{\mathrm{dyn}}} H \tag{3}$$

It depends on $\tilde{e}_{\mathrm{dyn}}$ in Eqn (2) and $\overline{c}$ which is the power mean of the colonization rates $c_i$ of the patches:

$$\overline{c} = \left( \frac{1}{N_{\mathrm{dyn}}} \sum_{i=1}^{N_{\mathrm{dyn}}} c_i^{\eta/b} \right)^{b/\eta} \tag{4}$$

The colonization rates $c_i$ represent the rate by which individuals emigrate from a local populations on patch $i$ divided by the number of immigrants required for successful colonization of an empty patch.

Both colonization and extinction rates of local populations, $c_i$ and $e_i$ in Eqns (2) and (4), depend on patch sizes (scaled areas) $A_i$ in the following way (Drechsler, 2009):

$$e_i = \varepsilon A_i{}^{-\eta} \text{ and } c_i = m A_i^b. \tag{5}$$

The relations in Eqn (5) depend on the species-specific parameters $\eta$ and $b$ as well as $\varepsilon$ and $m$. The meaning of these parameters is discussed in detail in Section 24.2.3.

The quantity $H$ in Eqn (3) is a measure of patch connectivity relative to a landscape with perfectly connected patches. It resulted from the aggregation process during the derivation of the formula (Drechsler, 2009; Drechsler and Johst, 2010) and can take values between zero and one:

$$H = \min\left\{\frac{3\cdot 10^{-R/(15d)}}{(R/d)^2}\frac{(R/d)^{1.65}}{(R/d)^{1.65}+5}, 1\right\} \tag{6}$$

The parameter $d$ in Eqn (6) is the mean species' dispersal distance. It scales the quantity $R$ which is called "landscape radius" because it relates to the spatial extent of the landscape (see Section 24.2.2 for details). $H$ in Eqn (6) describes the ability of dispersers to actually reach the other patches in the landscape—if $R$ is small compared to $d$, species can easily reach all patches within $R$ and connectivity is high. In this case, the landscape consists of perfectly connected patches and $H$ takes its maximum of one. If $R$ is large compared to $d$, species can disperse only to adjacent patches such that connectivity among the patches is low. In this case the landscape consists of very poorly connected patches and $H$ is close to zero.

## 24.2.2 Landscape Perspective of the Formula

The formula (Eqn 1) assumes that the exact spatial structure of patches within a metapopulation network can, for the purpose of estimating mean metapopulation lifetime, be summarized by a few macroscopic landscape attributes. This assumption is a good approximation across a wide range of landscapes (Drechsler and Johst, 2010). The "landscape level" aggregation has the advantage that the metapopulation lifetime in Eqn (1) can directly be related to macroscopic landscape attributes, and those species characteristics which are important for conservation decisions. This is explained now in more detail.

Generally, metapopulation dynamics are based on the separation of local (patch) dynamics and regional dynamics (Hanski, 1999; Gyllenberg and Hanski, 1997). Metapopulation lifetime within the framework of the formula given in Eqn (1) depends on the following dynamic and spatial landscape attributes: patch destruction rate $\mu$ (as synonym for habitat patch turnover caused by biodiversity offsetting), habitat patch number $N_{dyn}$ in this dynamic landscape ($N_{dyn}$ is independent of the level of patch turnover in case of biodiversity offsetting), the landscape radius $R$ (affecting patch connectivity $H$ in Eqn (6)), and patch areas $A_i$ (which may differ among the patches leading from patch size homogeneity to patch size heterogeneity).

The landscape radius $R$ is the spatial extent of the dynamic landscape containing $N_{dyn}$ habitat patches. It can be regarded as:

$$R = \sqrt{\Delta x \Delta y/\pi} \tag{7}$$

where $\Delta x$ and $\Delta y$ are the distances between the most eastern and most western, and the most northern and most southern patches, respectively. Thus, $R$ represents the geographic range, within which patches can be destroyed and created again. In case of biodiversity offsetting, patches are created at other positions than those of the destroyed patches.

Differences in the patch areas $A_i$ reflect patch size heterogeneity. For the present analysis, we assume that the $A_i$ vary around their geometric mean $\tilde{A}$ according to $A_i = \delta_i \tilde{A}$. The largest and smallest patch have an area ratio of $Q = A_{max}/A_{min}$. Variation in patch size leads to different extinction and colonization rates among the patches according to Eqn (5). We set $\tilde{e} = 0.05$ assuming a reference patch area scaled to $\tilde{A} = 1$ and compare landscapes with patches of equal size ($Q = 1$) to landscapes in which patch areas differ by a factor $Q = 2$.

### 24.2.3 Relation of Species Traits to the Landscape Attributes

The analytical formula shows how single-species traits ($\varepsilon$, $m$, $\eta$, $b$ in Eqn (5), and $d$ in Eqn (6)) influence metapopulation lifetime. The parameters $\varepsilon$ and $m$, respectively, determine the general level of extinction and colonization rates related to $\tilde{A}$, whereas the parameters $\eta$ and $b$, respectively, determine how these rates respond to deviations $A_i \neq \tilde{A}$. These single parameters can be summarized in two aggregated parameters called $\gamma$ and $\beta$ (see Johst et al., 2011). Using these aggregated species-specific parameters instead of the variety of single parameters is advantageous as it allows a straightforward conceptual grouping of species along combinations of low/high values of these aggregated parameters (such as low/high $\gamma$ and $\beta$, respectively). This grouping simplifies both model analysis and parameter estimation in the field as information on the single parameters ($\varepsilon$, $m$, $\eta$, $b$) of many species is generally scarce.

The first aggregated species parameter $\gamma$ can be derived in the following way:

$$\gamma = \prod_{i=1}^{N} {\gamma_i}^{1/N} \quad \text{where} \quad \gamma_i = \frac{c_i}{e_i} = \frac{m}{\varepsilon} A_i^{b+\eta} \tag{8}$$

Here, $\gamma$ is the geometric mean over the $\gamma_i$ which represent the mean number of individuals emigrating from a local population $i$ of area $A_i$ during its expected lifetime $1/e_i$, divided by the number of immigrants required for the successful colonization of an empty patch. The aggregated quantity $\gamma$ can be interpreted as the mean number of emigrants from the reference area $\tilde{A} = 1$ during the expected lifetime $\tilde{e}^{-1} = 20$ years of this local population divided by the number of immigrants required for the successful colonization of an empty patch. Thus, species with high values of $\gamma$ are characterized by a high dispersal propensity (i.e., a high propensity to leave the habitat patch), whereas species with low values of $\gamma$ are characterized by a low dispersal propensity (i.e., a low propensity to leave a patch). The actual patch-specific values $\gamma_i$, however, are related to patch area according to Eqn (8).

The parameter $\gamma$ may be determined in (at least) three ways: (1) through direct measurement of the number of emigrants during the lifetime of the local populations plus an estimate of how many immigrants are required for the successful colonization of an empty patch, (2) through Eqn (5) (assuming the parameters $m$, $\varepsilon$, $b$, and $\eta$ are known), or (3) through a Bayesian Calibration with a Markov Chain Monte Carlo (MCMC) algorithm (Drechsler and Wätzold, 2003; O'Hara et al., 2002; Hartig et al., 2011) of a time series of patch occupancy data (which patch is occupied and which is not) which, among others, allows estimating $c_i$ and $e_i$.

The second aggregated species parameter $\beta$ is defined as:

$$\beta = b/\eta. \tag{9}$$

This definition follows from the fact that the power mean of colonization rates in Eqn (4) depends on the quotient of $b$ and $\eta$ only and not on the single values of $b$ and $\eta$ (Eqn (5)). The value of $b$ determines the dependence of the colonization rates on patch area (Eqn (5)). Typical values of $b$ range from 0.5 (emigration proportional to patch perimeter) to 1.0 (emigration proportional to patch area).

The value of $\eta$ determines the dependence of the local extinction risk on patch area (Eqn (5)). There are various reasons for such a "natural" extinction risk additionally to patch turnover, for example environmental variability such as temperature fluctuations or precipitation variability. The resulting extinction risk depends on both the strength of these environmental fluctuations and on the species sensitivity to it. Parameter $\eta$ typically ranges from 0.5 to 2, according to high and low sensitivity to and/or level of environmental stochasticity (e.g., Lande, 1993; Foley, 1997).

Considering these ranges of possible values of $b$ and $\eta$, the parameter $\beta$ in Eqn (9) may vary among species between about 0.25 and 2. $\beta$ combines with $b$ and $\eta$ different types of species traits, one related to species dispersal ($b$) and the other one ($\eta$) related to species sensitivity to environmental fluctuations aside from patch turnover, and is therefore not straightforward to interpret. Nevertheless, large values of $\beta$ (1, 2) can be associated with species experiencing high environmental stochasticity whereas small values of $\beta$ (0.25, 0.5) represent species experiencing low environmental stochasticity.

The parameter $\beta$ may be determined through Eqn (9) (assuming $b$ and $\eta$ are known), or from the rates $c_i$ and $e_i$ obtained from field data. In the latter approach, patch area $A_i$ in Eqn (5) is eliminated resulting in $-\beta$ as slope of the regression line plotting $c_i$ versus $e_i$:

$$c_i = m\varepsilon^{\beta} e_i^{-\beta} \quad \Rightarrow \quad \ln(c_i) = \ln(m\varepsilon^{\beta}) - \beta\ln(e_i)$$

Another species-specific parameter is the species' dispersal range $d$. However, as Eqn (6) shows, this parameter appears only as a ratio between $d$ and the landscape radius $R$ (Eqn (7)). As this ratio $R/d$ determines the level of connectivity $H$ in Eqn (6), it can be seen as a connectivity measure itself. High values $R/d$ correspond to low connectivity—species dispersal range $d$ is much lower than the landscape radius $R$, and only adjacent patches can be colonized. Low values $R/d$ correspond to high connectivity—species dispersal range exceeds the landscape radius and all patches can be easily reached by the dispersers (see also explanations to Eqn (6) above). The connectivity measure $R/d$ can be managed and modified in two ways. First, the landscape radius $R$ can be changed, either by implementing an offsetting policy where destruction is only allowed outside core areas, and creation must be undertaken inside core areas to decrease patch distances (zoning, see e.g., Wissel and Wätzold, 2010), or by including explicit incentives for increasing connectivity in the policy (Hartig and Drechsler, 2009). Second, the species mean dispersal distance $d$ can be influenced by establishing dispersal corridors or improving the matrix between the patches to decrease dispersal mortality.

### 24.2.4 Analysis

For the analysis, we group species according to the species-specific parameters $\gamma$ and $\beta$, using maximum and minimum values of both parameters in the following combinations:

$\beta = 0.25$ ($b = 0.5$, $\eta = 2$) combined with $\gamma = 4$ and $\gamma = 12$

$\beta = 0.5$ ($b = 1$, $\eta = 2$) combined with $\gamma = 4$ and $\gamma = 12$

$\beta = 1$ ($b = 0.5$, $\eta = 0.5$) combined with $\gamma = 4$ and $\gamma = 12$

$\beta = 2$ ($b = 1$, $\eta = 0.5$) combined with $\gamma = 4$ and $\gamma = 12$.

Hereby, we can systematically compare different species types. Roughly speaking, species with high values of $\gamma$ and $\beta$ (e.g., $\beta = 2$ and $\gamma = 12$) may be assigned to the so-called "r-strategists" (May, 1976; Chapman and Reiss, 1992), characterizing species highly influenced by environmental fluctuations but also exhibiting a high dispersal propensity. Accordingly, species with low values of $\gamma$ and $\beta$ (e.g., $\beta = 0.25$ and $\gamma = 4$) may be assigned to the so-called "K-strategists," characterizing species associated with low environmental fluctuations and low dispersal propensity. Although the concept of "r- versus K-strategists" has been discussed controversially, it is a well-known heuristics in ecological research for describing the variety of life-history strategies and facilitates the ecological interpretation of our species-specific parameters $\gamma$ and $\beta$.

We use the analytical formula in Eqn (1) to investigate to which extent increased patch turnover that originates from a biodiversity offsetting policy can be compensated by changes in spatial landscape attributes and how compensation requirements depend on species traits. The general scheme of our analysis is shown in Fig. 24.1. We calculate the mean metapopulation lifetime $T_{dyn}$ for a variety of combinations of the three landscape attributes: patch number $N_{dyn}$, landscape radius $R/d$,and patch turnover $\mu/\tilde{e}$. The landscape radius $R$ (scaled by the species mean dispersal distance $d$) is related to the connectivity of patches as explained above. The rate of patch destruction $\mu$ (scaled by the geometric mean of the local extinction rates $\tilde{e}$) describes the degree of patch turnover because each destroyed patch is replaced by a new patch created elsewhere in the landscape. We select four different patch turnover rates $\mu/\tilde{e}$ ranging from very low to high turnover. We plot those combinations of number of patches $N_{dyn}$, landscape radius $R/d$, and patch turnover $\mu/\tilde{e}$ that result in a given metapopulation lifetime of $T_{dyn} = (100 \pm 5)\tilde{e}^{-1}$ (also called iso-$T$ lines). In a biodiversity offsetting scheme, this lifetime would be the conservation target of the policy. The shape of the iso-$T$ lines for the three landscape attributes shows to which extent and in which ways compensation of

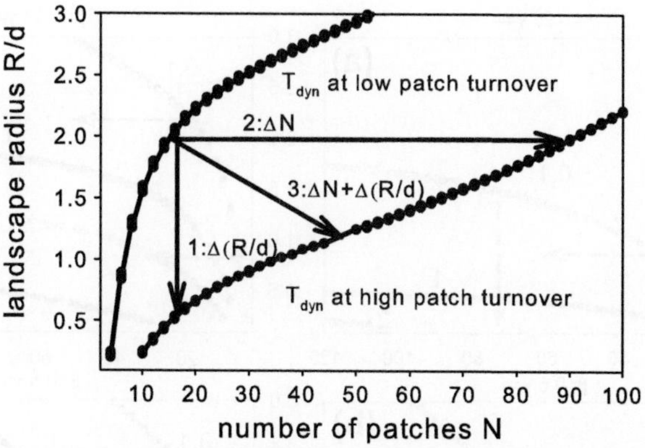


**FIGURE 24.1** Illustration of a three-dimensional trade-off between the number of patches $N_{dyn}$ (*x*-axis), the landscape radius in relation to the species mean dispersal distance *R*/*d* which is a measure of connectivity (*y*-axis), and patch turnover in relation to geometric mean local extinction rate $\mu/\tilde{e}$ (exemplified by trade-off curves for two different turnover rates). The trade-off curves describe combinations ($N_{dyn}$, *R*/*d*, $\mu/\tilde{e}$) of these three landscape attributes, resulting in the same mean metapopulation lifetime $T_{dyn}$. It can be seen that increased patch turnover (e.g., by the introduction of biodiversity offsetting) can be compensated in different ways, exemplified by the numbered arrows. Compensation option 1 is to increase connectivity. Compensation option 2 is to increase the number of patches. Compensation option 3 is to increase both number and connectivity of patches. Note that this combination requires smaller changes in both landscape attributes as compared to the previous two options.

increased patch turnover rates $\mu/\tilde{e}$ is possible by changing $N_{dyn}$ or *R*/*d*. We compare these trade-offs for the different species types.

## 24.3 Results

We analyze trade-offs between the three landscape attributes (Fig. 24.2): number of patches $N_{dyn}$ (*x*-axis), landscape radius *R*/*d* (*y*-axis), and the level of patch turnover $\mu/\tilde{e}$ (numbers at the curves) for different species types (different graphs). The general scheme of this analysis is shown in Fig. 24.1 and explained in Section 24.2.4.

Generally, increasing the landscape radius $R/d$ (i.e., decreasing patch connectivity) requires a higher number of patches $N_{dyn}$ to maintain metapopulation lifetime. Accordingly, decreasing the number of patches $N_{dyn}$ requires higher connectivity (i.e., a smaller landscape radius $R/d$) to maintain a given metapopulation lifetime (Fig. 24.2). Increasing patch turnover $\mu/\tilde{e}$ due to the introduction of biodiversity offsetting corresponds to going from iso-*T* line numbered with 0.1 (assuming a very low level of natural patch turnover) to iso-*T* lines numbered with 1, 2 up to 5 (depending on how strongly biodiversity offsetting increases patch turnover).

Maintenance of metapopulation lifetime requires increasing the number $N_{dyn}$ of patches (e.g., horizontal solid arrow in Fig. 24.2f) or decreasing the landscape radius $R/d$ (e.g., vertical solid arrow in Fig. 24.2f). Thus, compensation of increased patch turnover by improvements in other landscape attributes is generally possible.

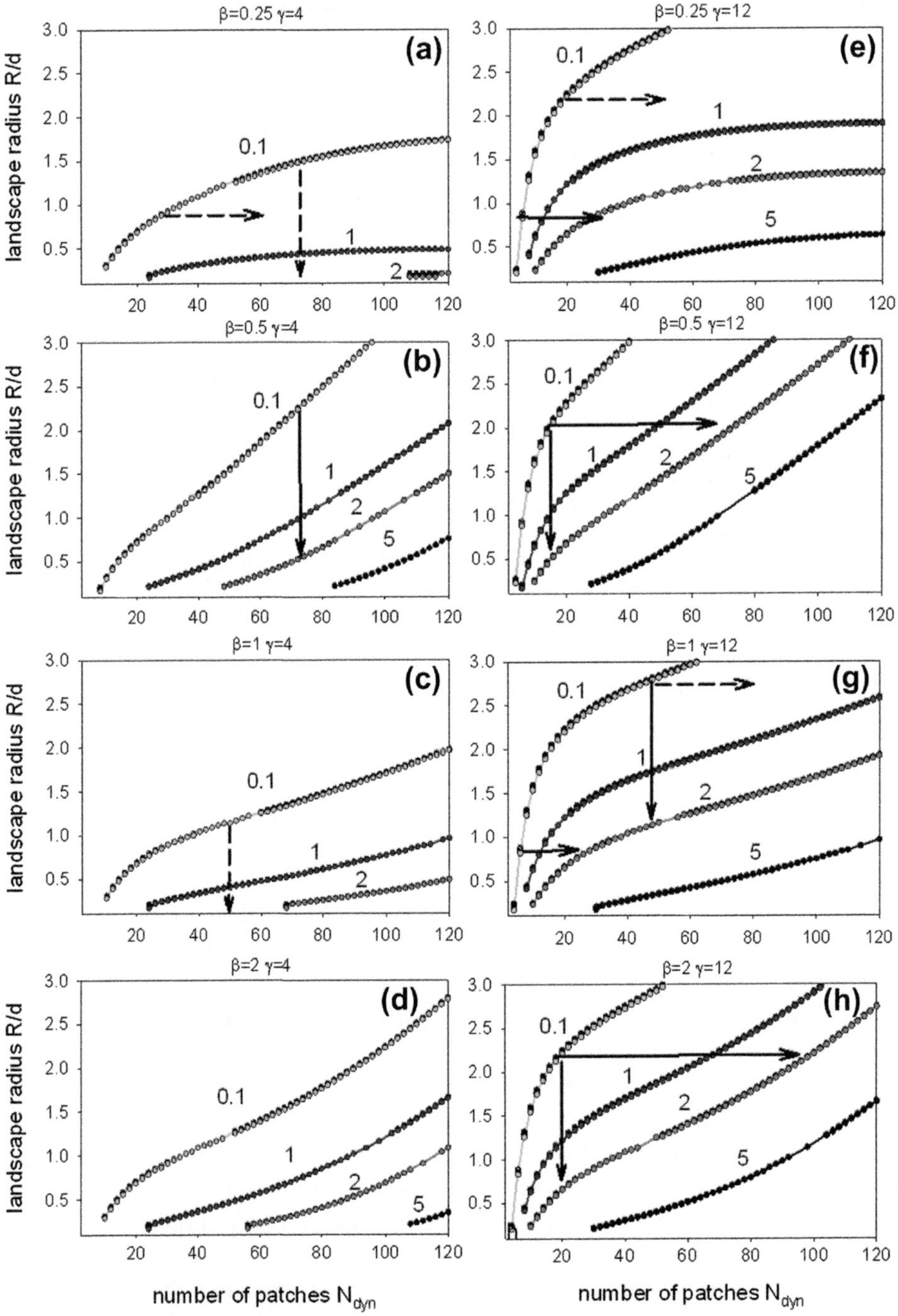


**FIGURE 24.2** Trade-offs between the number of patches $N_{dyn}$ (*x*-axis), the landscape radius *R* (*y*-axis), and the level of patch turnover $\mu/\tilde{e}$ which is given by the numbers at the curves, according to $\mu/\tilde{e} = 0.1,\ 1,\ 2,\ 5$. The landscape consists of patches differing in size by $Q = A_{max}/A_{min} = 2$. Combinations ($N_{dyn}$, *R*/*d*, $\mu/\tilde{e}$) that lead to the same mean

However, the height and shape of these trade-off curves considerably varies among the species types (different graphs in Fig. 24.2). The species-specific parameter $\gamma$ has an influence because it describes the relation between colonization and extinction rates of patches (see Eqn (8)). The species-specific parameter $\beta$ has an influence because the difference in the patch sizes $Q = A_{max}/A_{min}$ translates differently into differences in the colonization rates $c_{max}/c_{min}$ compared to that in the extinction rates $e_{max}/e_{min}$ (see Eqn (5)). How these heterogeneities differ depends on the parameters $\eta$ and $b$ combined in the aggregated parameter $\beta$ (Eqn (9)). Heterogeneity in the colonization rates also increases the power mean in Eqn (4) again depending on $\beta$.

Species with both low $\gamma$ and very low $\beta$ require high connectivity (i.e., extremely low $R/d$) and high $N_{dyn}$ to maintain metapopulation lifetime (Fig. 24.2a). They suffer from already low habitat patch turnover $\mu/\tilde{e}$ (i.e., iso-$T$ lines for higher turnover rates do not exist). Both increasing patch number $N_{dyn}$ and connectivity (i.e., lower $R/d$) cannot compensate for increasing patch turnover effectively (dashed arrows in Fig. 24.2a fail to reach iso-$T$ lines of turnover rates larger than $\mu/\tilde{e} = 1$). This species type is therefore not well suited for biodiversity offsetting.

Generally, species with low dispersal propensity $\gamma$ require higher connectivity (i.e., low $R/d$) and higher patch number $N_{dyn}$ to maintain metapopulation lifetime (Fig. 24.2a–d) than species with high $\gamma$ (Fig. 24.2e–h). For species with low $\gamma$, increased habitat patch turnover $\mu/\tilde{e}$ has to be compensated for by increasing both number and connectivity of patches. Compensating for increasing habitat patch turnover from $\mu/\tilde{e} = 0.1$ to $\mu/\tilde{e} = 2$ by increasing connectivity alone (without increasing patch number) is only effective if the patch number $N_{dyn}$ is already quite large (e.g., compare vertical solid arrow in Fig. 24.2b with vertical dashed arrow in Fig. 24.2c).

Species with high dispersal propensity $\gamma$ can better cope with patch turnover (Fig. 24.2e–h). For these species it is already effective to increase connectivity (i.e., decrease landscape radius $R/d$) without increasing $N_{dyn}$ to compensate for increasing turnover $\mu/\tilde{e}$ (compare, e.g., the vertical solid arrow in Fig. 24.2g with the vertical dashed arrow in Fig. 24.2c). Nevertheless, increasing both number and connectivity of patches relaxes the strong requirements if only one attribute alone would be increased.

Now we turn to the species-specific parameter $\beta$. For species experiencing very low environmental stochasticity ($\beta = 0.25$) increasing the number of patches alone does not effectively compensate for increasing patch turnover (see horizontal dashed arrows in Fig. 24.2a and e). Therefore, a parallel increase in connectivity is required. Exceptions are landscapes with very low number of patches and already high connectivity (e.g.,

metapopulation lifetime (called iso-$T$ lines) show how much alteration in one landscape attribute is required for compensating for the alteration in another one to maintain metapopulation lifetime. We calculate these iso-$T$ lines at $T_{dyn} = (100 \pm 5)\tilde{e}^{-1}$, which means that a metapopulation lifetime of about 100-times of local population lifetime is to be maintained. We set $\tilde{e} = 0.05$, assuming a reference patch area scaled to $\tilde{A} = 1$. The left panel shows the trade-offs for $\gamma = 4$, the right panel for $\gamma = 12$. Increasing values of parameter $\beta$ are shown top down ($\beta = 0.25, 0.5, 1, 2$). For color version of this figure, the reader is referred to the online version of this book.

horizontal solid arrow in Fig. 24.2e). For species with higher values of $\beta$ increased patch turnover due to the introduction of biodiversity offsetting can be compensated for by increasing patch number and/or connectivity (e.g., horizontal and/or vertical solid arrows in Fig. 24.2b and f).

In general, species with higher $\beta$ and high $\gamma$ (Fig. 24.2f to h) can cope well with patchy and dynamic landscapes, as the target metapopulation lifetime can be reached at even low connectivity (i.e., rather large $R/d$) and low patch number $N_{dyn}$. Increasing patch turnover $\mu/\tilde{e}$ can be compensated for by increasing patch number and/or connectivity. However, the effectiveness of the two types of compensation (patch number or connectivity) depends on the particular values of $\beta$ and $\gamma$. It also depends on the actual values of $N_{dyn}$, $R/d$, and $\mu/\tilde{e}$ because of the nonlinearity of the trade-offs (compare, e.g., dashed and solid horizontal arrow in Fig. 24.2g).

The form of the trade-offs also depends on patch size heterogeneity (i.e., landscape consists of patches of different size) or homogeneity (i.e., landscape consists of patches of equal size; Fig. 24.3 in comparison to Fig. 24.2). Note that in landscapes with equal patch sizes $\tilde{A} = 1$ trade-offs depend on the species-specific parameter $\gamma$ only and are independent of the species-specific parameter $\beta$ (as $\beta$ influences local colonization and extinction rates only if patch sizes deviate from $\tilde{A} = 1$). Similar to landscapes with patch size heterogeneity, species with low dispersal propensity $\gamma$ need high connectivity (i.e., low landscape radius $R/d$) and high number of patches $N_{dyn}$ to survive at all in patchy and dynamic landscapes (Fig. 24.3a). Only a very good connectivity can ensure that the low number of dispersers can actually reach and colonize other patches. Even an increase in the number of patches does not relax this strong demand for high connectivity (horizontal dashed arrow in Fig. 24.3a). Increasing patch turnover leads to the extinction of species with low $\gamma$ in these landscapes (no iso-$T$ lines exist for turnover rates $\mu/\tilde{e} > 1$). Thus, these species and landscapes are not well suited for biodiversity offsetting because adequate compensation is difficult.

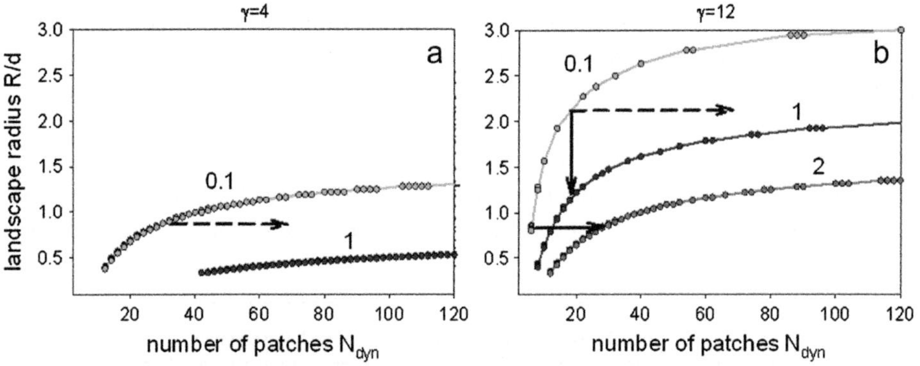


**FIGURE 24.3** The plot shows the same variables as in Fig. 24.2 but assuming homogeneous landscapes with patches of equal size scaled to $\tilde{A} = 1$ and $Q = A_{max}/A_{min} = 1$. The left panel shows the trade-offs for $\gamma = 4$, the right panel for $\gamma = 12$. In such landscapes, the parameter $\beta$ does not influence the trade-offs. For color version of this figure, the reader is referred to the online version of this book.

Species with large dispersal propensity $\gamma$ can better survive in dynamic landscapes with equal patch sizes (compare Fig. 24.3b and Fig. 24.3a). To compensate for increasing patch turnover, it is generally insufficient, however, to increase the number of patches alone (horizontal dashed arrows in Fig. 24.3a and b). Exceptions are landscapes with very low number of patches and already high connectivity (horizontal solid arrow in Fig. 24.3b). Increasing connectivity is therefore the most effective way to compensate for increasing turnover for species with large $\gamma$ in homogenous landscapes (vertical solid arrow in Fig. 24.3b). In contrast, in landscapes with patch size heterogeneity (but the same geometric mean patch size) increasing turnover can be effectively compensated by increasing the number of patches even without increasing connectivity (e.g., horizontal solid arrows in Fig. 24.2f and h, except for species with very low $\beta$, see horizontal dashed arrows in Fig. 24.2e).

We have also tested dynamic landscapes with larger patch sizes (results not shown). Although the iso-$T$ lines generally require lower patch number and connectivity if patches are larger, the general characteristics shown in Figs 24.2 and 24.3 are maintained.

## 24.4 Discussion

Market-based instruments such as biodiversity offsetting are gaining importance for biodiversity conservation. Different to more traditional conservation policies like landscape planning or designation of protected areas, offsetting policies allow certain amounts of habitat patch turnover, that is, the relocation of habitat in the landscape. This turnover can be problematic for some species, while being tolerable for others, particularly if adequate compensation is provided. Our analysis shows that offsetting schemes that do not take into account the induced patch turnover may fail to deliver their actual target, that is, ensuring the long-term viability of threatened species in these landscapes (see also Hartig and Drechsler, 2009). Therefore, decision makers must explicitly take into account patch turnover, and they must implement the responsibility for adequate compensation not only of habitat destruction, but also of habitat relocation into offsetting policies.

General guidelines for appropriate compensation of patch turnover have hitherto been missing. This is due to the large spatial scale (region, landscape) that has to be considered and the complex interplay between the spatial and dynamic landscape attributes and species population dynamics. Approaches to this problem must therefore consider environmental attributes at a landscape scale, the traits of the species to be protected, and the interacting impacts of landscape and species attributes on the species population dynamics and resulting species survival.

In this paper, we have used an analytical formula that is able to capture these challenges in a relatively straightforward way. This allows us to consider trade-offs between three landscape attributes: patch number $N_{\mathrm{dyn}}$ (corresponding to the total habitat area available for a species in a landscape), landscape radius $R$ (which, in relation to species dispersal distance, is a measure for connectivity), and habitat patch turnover $\mu$ (caused by the habitat relocation allowed in offsetting policies). Analyzing these trade-offs, we examined which landscapes are suitable for biodiversity offsetting, and which compensation would be necessary for the induced habitat patch turnover, depending on the targeted species types. In the following, we summarize our results in terms of robust guidelines for compensating increased habitat patch turnover. An illustration of these general guidelines is given in Fig. 24.4.

## 24.4.1 The Impact of Landscape Attributes in Relation to Species Attributes

Our results show that not only the landscape attributes per se are decisive for effective compensation but their relation to the species type or group which is to be protected (characterized by the species-specific parameters $\gamma$, $d$, and $\beta$).

Species with high values of $\gamma$, $d$, and $\beta$ ("r-strategist"-type, see Section 24.2.4) can generally cope well with patch turnover. These species send a large number of long-range dispersers into the landscape and can therefore follow the temporally varying patch configurations due to patch relocation. For species of the type "r-strategist" increased

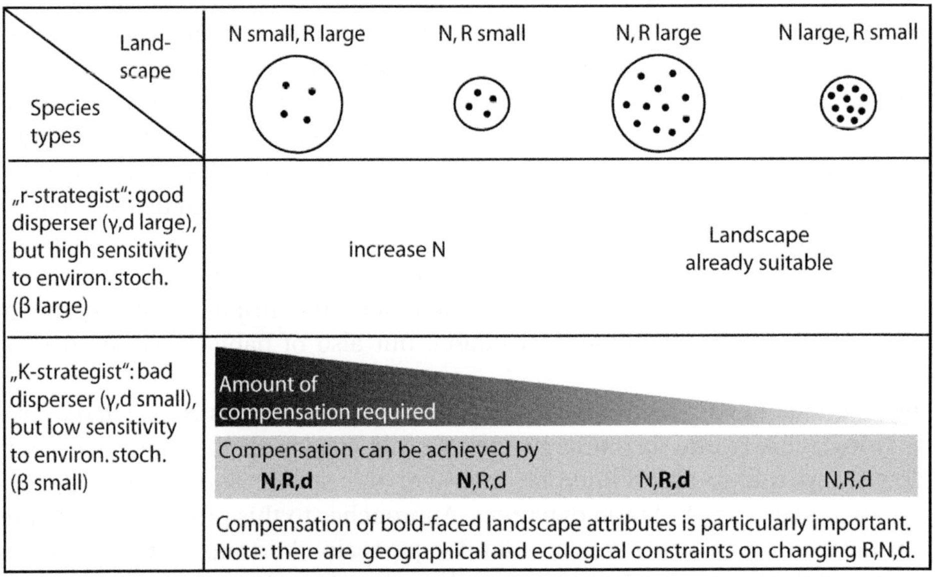


**FIGURE 24.4** Management recommendations for the compensation of increased habitat patch turnover induced by a biodiversity offsetting policy. Recommendations depend on species types (vertical axis) and landscape attributes (horizontal axis). *R* is the landscape radius within which patch turnover occurs and stands for the spatial extent of the landscape. *N* is the number of patches in the landscape, and *d* is the mean dispersal distance of the species. Increased patch turnover can be compensated by increasing patch number and connectivity. The number of patches *N* can be increased by establishing new habitat areas. Connectivity can be increased by decreasing patch distances (i.e., reducing *R*) or generating dispersal corridors (i.e., increasing *d*).

patch turnover can be compensated effectively by increasing patch number and (if necessary in specific situations) connectivity. They are therefore well suited for biodiversity offsetting (Fig. 24.4).

Species with low values of $\gamma$, $d$, and $\beta$ ("K-strategist"-type, see Section 24.2.4) suffer already from low habitat patch turnover. These species send a small number of short-range dispersers into the landscape and cannot therefore follow patch relocation effectively. Thus, this species type requires strong changes in the landscape attributes to compensate for increasing patch turnover and is less well suited for biodiversity offsetting (Fig. 24.4).

A robust guideline is, therefore, that species with high propensity to disperse and long dispersal distances can generally cope well with increased patch turnover, and compensation is usually feasible (preferably by increasing patch number as connectivity is not a limiting factor). However, it should be noted that even though deficits in the species dispersal propensity and dispersal distance can be overcome in principle by adapting patch number, connectivity, and patch size accordingly, the possibilities for compensation in practical situations may be constrained. This is explained in more detail in the following.

## 24.4.2 Uncertainty and Limitations of Compensation Options for Increased Patch Turnover

Our results show that only landscapes with well-connected and sufficiently many patches can compensate for the increased patch turnover induced by biodiversity offsetting such that metapopulation lifetime of the species living in them can be maintained or even increased (Figs 24.2 and 24.3; schematic summary in Fig. 24.4). At a first glance, this is not surprising because higher connectivity and/or more patches can counterbalance the increased local extinction risk $\tilde{e}_{dyn}$ (see Eqn (2)) that is caused by increased patch turnover (e.g., Keymer et al., 2000; Johst et al., 2002; Matlack and Leu, 2007). However, our results also show that the most effective type of compensation (higher number of patches or higher connectivity) depends on the species and landscape information and is affected by spatial and dynamical restrictions.

First, the effectiveness of different compensation types depends on the current state of the landscape, that is, on the level of already existing patch turnover, the number of patches involved, the degree of their connectivity, and their patch size heterogeneity. This dependence on the current level of spatial and dynamic landscape attributes is due to the general nonlinearity of the trade-offs which confirms former findings (discussed in detail in Johst et al., 2011). This implies that monitoring of the current state of the landscape is indispensable for deciding on the effectiveness of different compensation types in offsetting policies.

Second, the effectiveness of different compensation types depends on the species traits. In particular, if multiple species are to be protected the type that is most effective for one species may be ineffective for another species and may lead to conflicting recommendations. Moreover, interspecific interactions may impact the effectiveness of compensation types, and these types in turn may affect the resulting community response. Detailed monitoring of both landscape and species attributes is therefore necessary to make informed choices on the preferred type of compensation (higher number of patches or higher connectivity) in any particular case.

Third, there are spatial limits to the extent to which patch turnover can be compensated, posed by the landscape configuration. Although the number of patches $N_{dyn}$ can be increased by establishing additional habitat areas and connectivity can be increased by decreasing patch distances (i.e., reducing $R$) or aiding dispersal by dispersal corridors (i.e., increasing $d$), all these options have spatial restrictions. Thus, although we have shown in our figures a large range of patch number $N_{dyn}$ and landscape radius $R/d$ to visualize the shape of the trade-offs, in particular, combinations of low $R/d$ with high $N_{dyn}$ may not be feasible in nature.

Last but not least, our results show the existence of dynamic restrictions. At high patch turnover, the desired target metapopulation lifetime cannot be reached anymore. At which level of turnover this happens depends on the species traits (Fig. 24.2).

### 24.4.3 General Guidelines for the Design of Landscapes to Implement Biodiversity Offsetting

In conclusion, biodiversity offsetting schemes need, in many circumstances, compensation for increased habitat patch turnover by increasing the number of habitat patches and/or their connectivity. The effectiveness of the different compensation types depends on the traits of the species to be protected and the current level of spatial and dynamic landscape attributes. Therefore, detailed management recommendations depend on the specific situation. Each type of compensation (increasing the number of patches or increasing connectivity) has (1) uncertainties in its effect and (2) restrictions in its feasibility. Finally, one should be aware that too much patch turnover is always (in every landscape and for every species) detrimental. It might therefore be necessary to implement fixed limits to habitat turnover in real-world compensation policies.

## References


Bishop, K., Beven, K., Destouni, G., Abrahamsson, K., Andersson, L., Johnson, R.K., Rodhe, J., Hjerdt, N., 2009. Nature as the "natural" goal for water management: a conversation. Ambio 38, 209–214.

Chapman, J.L., Reiss, M.J., 1992. Ecology. Cambridge University Press, Cambridge, 294 pp.

Cornell, S.J., Ovaskainen, O., 2008. Exact asymptotic analysis for metapopulation dynamics on correlated dynamic landscapes. Theor. Popul. Biol. 74, 209–225.

Cuperus, R., Canters, K.J., de Haes, H.A.U., Friedman, D.S., 1999. Guidelines for ecological compensation associated with highways. Biol. Conserv. 90, 41–51.

Drechsler, M., 2009. Predicting metapopulation lifetime from macroscopic network properties. Math. Biosci. 218, 59–71.

Drechsler, M., Hartig, F., 2011. Conserving biodiversity with tradable permits under changing conservation costs and habitat restoration time lags. Ecol. Econ. 70, 533–541.

Drechsler, M., Johst, K., 2010. Rapid viability analysis for metapopulations in dynamic habitat networks. Proc. R. Soc. B 277, 1889–1897.

Drechsler, M., Wätzold, F., 2003. Species Conservation in the Face of Political Uncertainty. UFZ, Leipzig-Halle GmbH.

Foley, P., 1997. Extinction models for local populations. In: Hanski, I.A., Gilpin, M.E. (Eds.), Metapopulation Biology: Ecology, Genetics, and Evolution. Academic Press, San Diego, London, pp. 215–246.

Frank, K., Wissel, C., 2002. A formula for the mean lifetime of metapopulations in heterogeneous landscapes. Am. Nat. 159, 530–552.

Gyllenberg, M., Hanski, I., 1997. Habitat deterioration, habitat destruction, and metapopulation persistence in a heterogenous landscape. Theor. Popul. Biol. 52, 198–215.

Hanski, I., 1999. Habitat connectivity, habitat continuity, and metapopulations in dynamic landscapes. Oikos 87, 209–219.

Hartig, F., Calabrese, J.M., Reineking, B., Wiegand, T., Huth, A., 2011. Statistical inference for stochastic simulation models – theory and application. Ecol. Lett. 14, 816–827.

Hartig, F., Drechsler, M., 2009. Smart spatial incentives for market-based conservation. Biol. Conserv. 142, 779–788.

Hodgson, J.A., Moilanen, A., Thomas, C.D., 2009. Metapopulation responses to patch connectivity and quality are masked by successional habitat dynamics. Ecology 90, 1608–1619.

Jacquemyn, H., Butaye, J., Hermy, M., 2003. Influence of environmental and spatial variables on regional distribution of forest plant species in a fragmented and changing landscape. Ecography 26, 768–776.

Johst, K., Brandl, R., Eber, S., 2002. Metapopulation persistence in dynamic landscapes: the role of dispersal distance. Oikos 98, 263–270.

Johst, K., Drechsler, M., 2003. Are spatially correlated or uncorrelated disturbance regimes better for the survival of species? Oikos 103, 449–456.

Johst, K., Drechsler, M., van Teeffelen, A.J.A., Hartig, F., Vos, C.C., Wissel, S., Wätzold, F., Opdam, P., 2011. Biodiversity conservation in dynamic landscapes: trade-offs between number, connectivity and turnover of habitat patches. J. Appl. Ecol. 48, 1227–1235.

Keymer, J.E., Marquet, P.A., Velasco-Hernandez, J.X., Levin, S.A., 2000. Extinction thresholds and metapopulation persistence in dynamic landscapes. Am. Nat. 156, 478–494.

Lande, R., 1993. Risks of population extinction from demographic and environmental stochasticity and random catastrophes. Am. Nat. 142, 911–927.

Loreau, M., Naeem, S., Inchausti, P., Bengtsson, J., Grime, J.P., Hector, A., Hooper, D.U., Huston, M.A., Raffaelli, D., Schmid, B., Tilman, D., Wardle, D.A., 2001. Ecology – biodiversity and ecosystem functioning: current knowledge and future challenges. Science 294, 804–808.

Matlack, G.R., Leu, N.A., 2007. Persistence of dispersal-limited species in structured dynamic landscapes. Ecosystems 10, 1287–1298.

May, R.M., 1976. Theoretical ecology. In: Principles and Applications. Blackwell Scientific Publications, Oxford, London, p. 317.

Moilanen, A., Hanski, I., 1998. Metapopulation dynamics: effects of habitat quality and landscape structure. Ecology 79, 2503–2515.

O'Hara, R.B., Arjas, E., Toivonen, H., Hanski, I., 2002. Bayesian analysis of metapopulation data. Ecology 83, 2408–2415.

Ovaskainen, O., 2002. Long-term persistence of species and the SLOSS problem. J. Theor. Biol. 218, 419–433.

Ring, I., Drechsler, M., van Teeffelen, A.J.A., Irawan, S., Venter, O., 2010. Biodiversity conservation and climate mitigation: what role can economic instruments play? Curr. Opin. Environ. Sustain. 2, 50–58.

Salzman, J., Ruhl, J.B., 2000. Currencies and the commodification of environmental law. Stanford Law Rev. 53, 607–694.

Schippers, P., Snep, R.P.H., Schotman, A.G.M., Jochem, R., Stienen, E.W.M., Slim, P.A., 2009. Seabird metapopulations: searching for alternative breeding habitats. Popul. Ecol. 51, 459–470.

Tilman, D., 1999. The ecological consequences of changes in biodiversity: a search for general principles. Ecology 80, 1455–1474.

Vos, C.C., Verboom, J., Opdam, P.F.M., Ter Braak, C.J.F., 2001. Toward ecologically scaled landscape indices. Am. Nat. 157, 24–41.

Vuilleumier, S., Wilcox, C., Cairns, B.J., Possingham, H.P., 2007. How patch configuration affects the impact of disturbances on metapopulation persistence. Theor. Popul. Biol. 72, 77–85.

Wilcox, C., Cairns, B.J., Possingham, H.P., 2006. The role of habitat disturbance and recovery in metapopulation persistence. Ecology 87, 855–863.

Wintle, B.A., Bekessy, S.A., Venier, L.A., Pearce, J.L., Chisholm, R.A., 2005. Utility of dynamic-landscape metapopulation models for sustainable forest management. Conserv. Biol. 19, 1930–1943.

Wissel, S., Wätzold, F., 2010. A conceptual analysis of the application of tradable permits to biodiversity conservation. Conserv. Biol. 24, 404–411.

# 25

# Combining Habitat Suitability Models and Fluvial Functionality Data for a Multilayer Assessment of Riverine Vulnerability

Maria T. Carone*,[†], Tiziana Simoniello[†], Anna Loy*, Maria L. Carranza*

*ENVIX_LAB, STAT DEPARTMENT, UNIVERSITY OF MOLISE, C.DA FONTE LAPPONE, 86090, PESCHE (IS), ITALY, [†]IMAA, ITALIAN NATIONAL RESEARCH COUNCIL—CNR, C.DA S.TA LOJA, 85050, TITO (PZ), ITALY

## 25.1 Introduction

River/landscape systems are characterized by a complex multilayered structure (Brown et al., 2011; Poole, 2002) with many functional links and feedback processes, which make difficult the correct individuation of river criticalities to suggest management and/or restoration strategies. A detailed modeling that is able to take into account such links and processes can be very complicated, even for merely selecting the right variables for the whole system representation. A possible solution to overcome this problem, especially in operative contexts, can be represented by comprehensive indicators having links with all the system layers.

In order to obtain concise information on both landscape and fluvial ecosystem layers and to provide a tool for the assessment of the river vulnerability (VU) suitable for supporting management and/or restoration activities, we propose the integration of habitat suitability (HS) models for a wide-ranging key species of riverine habitats (the Eurasian otter, *Lutra lutra*) with fluvial functionality (FF) analyses based on indicators of riverine ecosystem equilibrium (index of FF).

HS models are widely considered as robust tools for the description of species' environmental niche and for a number of different ecological implications being that they are able to generate maps of suitable habitats for the studied species (see, e.g., Guisan and Zimmermann, 2000; Mladenoff et al., 1997). For our model implementation, we considered the otter for obtaining information at a broad scale (landscape layer) since the survival of such a species is strictly dependent on the environmental quality and the

**Models of the Ecological Hierarchy. DOI: http://dx.doi.org/10.1016/B978-0-444-59396-2.00025-0**
ISSN 0167-8892, 

equilibrium of fluvial habitats and on the good ecological functioning of their links with the surrounding territory.

Indices of FF (e.g., AUSRIVAS—Australian River Assessment System, ISC—Index of Stream Condition, RHS—River Habitat Survey, IFF—Index of Fluvial Functionality) represent the operational tools for the assessment of the well working of river metabolic functions, linked to the trophic relationships between the living organisms, and the nonmetabolic functions, such as fluvial contributions to the environmental biodiversity, solid transport loads, and so on (e.g., Ladson et al., 1999; Raven et al., 1997; Siligardi et al., 2003). Therefore, they can support the evaluation of river criticalities at the local ecosystem scale (river layer). In particular, for this purpose, we accounted the Italian standard protocol for FF (Siligardi et al., 2003) since it was specifically adapted for Italian river characteristics and widely tested and adopted at operational levels by the agencies for environmental protection. Moreover, it proved to be useful for other integrated approaches for fluvial assessment (e.g., Comiti et al., 2009; Munafò et al., 2005).

The effectiveness of the integrated information concerning fluvial VU was evaluated by performing a Landscape Ecology analysis focused on fragmentation. Such a process influences the ecological equilibrium at different scales (e.g., Lindenmayer and Fischer, 2006 and references therein), and landscape metrics analyses revealed a satisfactory reliability for providing information on landscape structure (e.g., Colson et al. 2011; Peng et al., 2010).

The approach, recently developed and tested in the Italian otter core area on a river subbasin showing a heavy anthropogenic influence, revealed a good capability to provide concise information for the fluvial VU prioritization (Carone, 2012, PhD thesis). Since such a prioritization still represents a crucial challenge for the environmental management, here we present a further test of the method in order to evaluate its exportability. The investigation was realized on a river subbasin presenting a lower anthropogenic presence, and so a more natural river/landscape structure than the subbasin analyzed for the setting up of the approach. All the procedural steps applied to both the subbasins were compared in order to assess the methodology response in systems having different levels of anthropogenic pressures.

# 25.2 Methods

## 25.2.1 Study Area

The study area is located in a portion of the otter core area, which hosts an important part of the Italian southern otter population (Basilicata region); it is represented by two river subbasin with different anthropogenic pressures: the High Agri River and the High Sinni River (Figure 25.1).

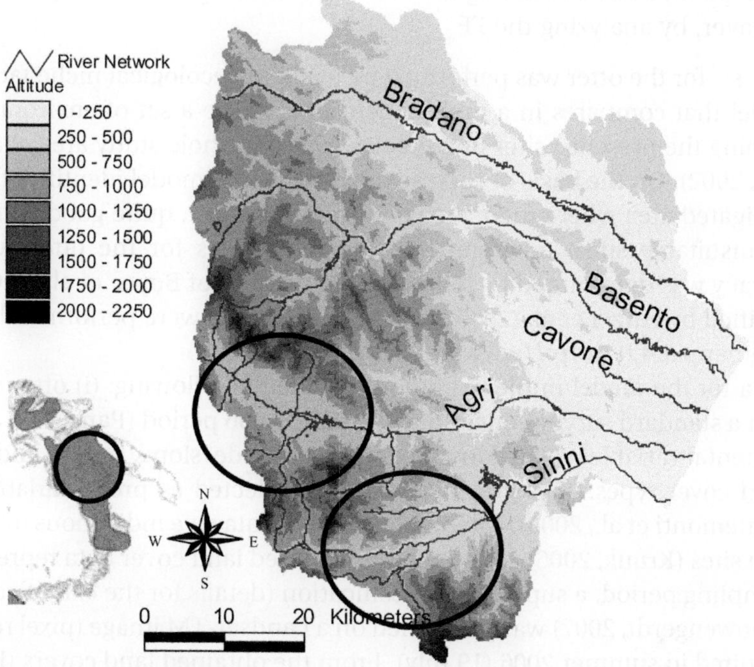


**FIGURE 25.1 Localization of the study area.** Bottom left, small picture: representation of the otter range with indication of Basilicata region (circle). Main picture: the Basilicata region, internal portion of the otter areal; in the circles the two river subbasins are utilized for the integration. For color version of this figure, the reader is referred to the online version of this book.

The studied river subbasins are similar for their morphology (mountainous areas) and hydraulic regimes, which are characterized by heavy winter water loads alternate by drought summers; conversely, they are highly different for the anthropic pressure.

The High Agri River shows a wide presence of agricultural covers mixed to natural portions interested by a National Park (Lucanian Apennines NP), long sections of the river's main axis are managed in concrete, and an important industrial site related to an oil-drilling activity is also present.

On the contrary, the High Sinni River shows a less important anthropogenic pressure than the High Agri River; larger sectors are interested by a National Park (Pollino NP), the urban areas are smaller and more scattered, and the land cover structure presents larger forested patches.

## 25.2.2 The Integrated Approach

In order to provide a tool to assess river VU and to classify such a VU in different priorities of intervention need, the proposed approach integrates information on the environmental criticalities coming from the river/landscape system layers. In particular, the different pieces of information take into account

- the landscape layer, by evaluating the HS for the endangered otter species
- the river layer, by analyzing the FF

The HS analysis for the otter was performed by using the ecological niche factor analysis (ENFA) model that compares in a multidimensional space a set of environmental variables describing the presence sites of a species with the whole study area characteristics (Hirzel et al., 2002). On the basis of such a comparison, the model identifies the fractions of the investigated area where there are good (optimal sites), quite good (suitable sites), and poor (unsuitable sites) environmental characteristics for the otter survival. The model accuracy was determined by following the method of Boyce et al. (2002) and was further modified by Hirzel et al. (2006). All the model steps were performed by using the software Biomapper 4.0 (http://www.unil.ch/biomapper).

Input data for the model implementation include the following: (i) otter occurrences derived from a standard survey concerning the 2002–2006 period (Panzacchi et al., 2010), (ii) environmental variables related to topography (altitude, slope, aspect, and convexity), and (iii) land cover types. Topographic data were selected as proxy variables for food availability (Remonti et al., 2009), whereas land cover data give indications on resting and reproductive sites (Kruuk, 2006). To obtain the required land cover data representative of the otter sampling period, a supervised classification (details for the classification can be found in Schowengerdt, 2007) was performed on a Landsat-TM image (pixel resolution 30 $\times$ 30 m) acquired in summer 2006 (19 July). From the obtained land covers (Figure 25.2), 11 classes were chosen as more significant for the otter's ecology (Loy et al., 2009). For each of the selected cover, to obtain continuous variables as required by the ENFA model, a frequency map was elaborated by evaluating the presence of the given cover within a circular window of 300 m radius. Topographic variables were originated from a digital elevation model having a pixel size of 20 $\times$ 20 m and resampled at the same spatial resolution of the satellite-derived land cover maps.

Finally, all the variables were clipped on a fluvial buffer 300 m wide since the studied species is rarely found far from the water. FF values were derived from field data collected in a 2003 summer campaign by following the Italian standard protocol (Siligardi et al., 2003), which is based on the evaluation of various river basin aspects (riparian vegetation, hydrology, fluvial morphology, etc.) at the operator scale. The protocol has been defined, starting from the Riparian Channel Environmental Inventory (RCE-I) (Petersen et al., 1987), through different applications and modifications (Siligardi and Maiolini, 1993; Siligardi et al., 2000), which made it specific and more reliable than other indices for the Italian river characteristics (Balestrini et al., 2004). The index is adopted as a standard tool for the analysis of FF by the Italian Environmental Protection Agency.

In order to integrate HS and FF data, both suitability and functionality values were categorized into three levels (low, medium, and high suitability/functionality) and combined in a GIS environment (ArcView 3.2) to obtain three levels of river VU as follows: (i) low ecological VU level, where the integration indicated that both HS and FF provide the highest values; (ii) medium ecological VU level, where the integration indicated that

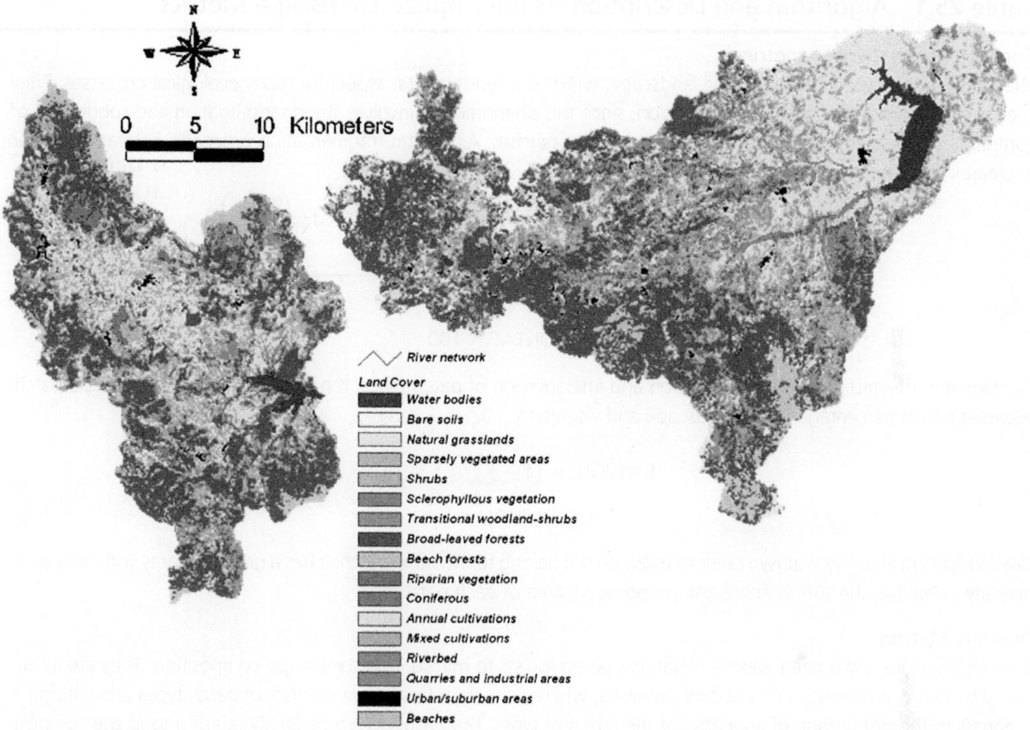


**FIGURE 25.2 Land cover maps of the studied subbasins, provided by a classification of a Landsat-TM summer image.** For color version of this figure, the reader is referred to the online version of this book.

both HS and FF provide a medium value, or at least one of them; (iii) high ecological VU level, where the integration indicated that both HS and FF provide the lowest values, or at least one of them. The integrated VU map is representative of environmental criticalities related to both landscape and fluvial ecosystem layers, and it can be read as map of priority in term of needed intervention since the higher the VU, the higher the urgency for intervention activities.

The provided prioritization was assessed by performing a further analysis based on landscape metrics (LM). A specific subset of landscape indices concerning fragmentation processes (Table 25.1) (McGarigal et al., 2002) was calculated at landscape level on land cover arrangements included within each sublandscape delineated by the functionality, suitability, and VU levels.

# 25.3 Results and Discussion

The evaluation of the predictive power for the implemented ENFA model provided good values of the Boyce index ($0.801 \pm 0.039$). In Figure 25.3 are shown the final HS maps of the two test areas, obtained by segmenting the suitability values (ranging from 0 to 1) into three levels (unsuitable, suitable, and optimal).

**Table 25.1** Algorithm and Description of the Applied Landscape Metrics

**Contagion/Interspersion metrics**

They refer to the overall texture of the landscape, which is a fundamental aspect for many ecological processes. They specifically describe the habitat fragmentation, since this phenomenon involves the disaggregation and subdivision of contiguous habitats into disaggregated and/or disjoint patches. As habitat fragmentation proceeds, habitat contagion decreases and habitat subdivision increases.

$$\mathrm{CONTAG} = \left( 1 + \frac{\sum_{i=1}^{m}\sum_{k=1}^{m}\left((P_i)\left(\frac{(g_{ik})}{\sum_{k=1}^{m} g_{ik}}\right)\right)\left(\ln(P_i)\frac{(g_{ik})}{\sum_{k=1}^{m} g_{ik}}\right)}{2\ln(m)} \right)(100)$$

$$0 \le \mathrm{CONTAG} \le 100$$

*Contagion* is affected by both the dispersion and interspersion of patch types. It has high values when a single patch occupies a high percentage of the landscape and vice-versa.

$$\mathrm{DIVISION} = (1 - \sum_{i=1}^{m}\sum_{j=1}^{n}(\frac{a_{ij}}{A})$$

$$0 \le \mathrm{DIVISION} < 1$$

*Division* is the probability that two random pixels do not belong to the same patch. It has a great similarity with Simpson's diversity index, but the sum is across the proportional area of each patch.

**Diversity Metrics**

Diversity measures have been applied by landscape ecologists to measure the landscape composition. They are influenced by two components: richness and evenness, where richness refers to the number of patch types present and evenness to the distribution of area among the different types. Both aspects are generally referred to as the compositional and structural components of diversity, respectively.

$$\mathrm{SHEI} = \frac{-S_{i=1}^{m}(P_i * \ln(P_i))}{\ln(m)}$$

$$0 \le \mathrm{SHEI} \le 1$$

*Shannon's Evenness Index* is expressed such that an even distribution of area among patch types results in maximum evenness.

$$\mathrm{SHDI} = -S_{i=1}^{m}(P_i * \ln(P_i))$$

$$\mathrm{SHDI} \ge 0, \text{ without limit}$$

*Shannon's Diversity Index* is a popular measure of diversity in community ecology, applied to landscapes.

**Area/Density Metrics**

The group represents all metrics dealing with the patches' number and size. Specifically, the number of patches can be considered to be the most basic aspect of landscape pattern that can affect a multitude of processes linked to fragmentation.

$$\mathrm{NP} = \mathrm{N}$$

$$\mathrm{NP} \ge 1, \text{ without limit}$$

*Number of Patches* is the simplest measure for expressing the subdivision of a class or a landscape

(from McGarigal et al., 2002)

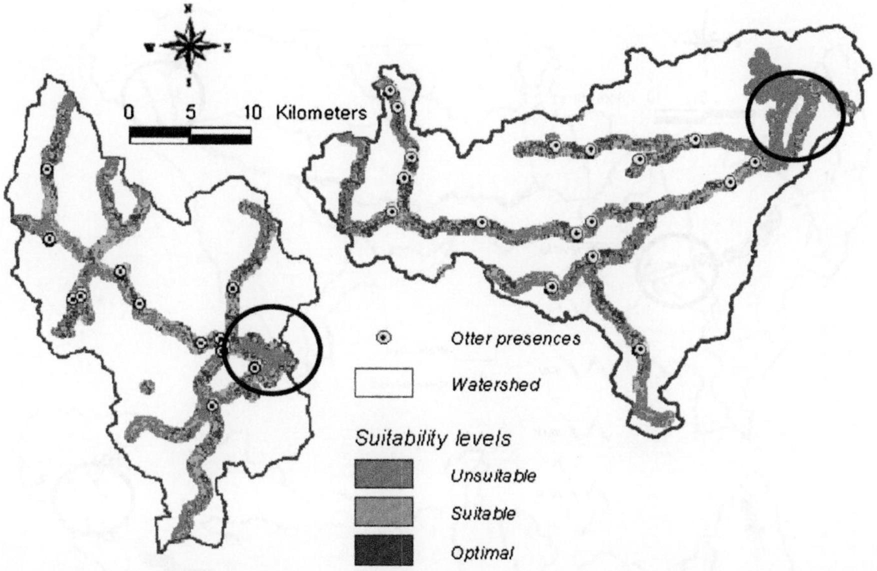


**FIGURE 25.3 Habitat suitability map for the two test areas.** On the left, the High Agri River; on the right, the High Sinni River; in the circles the Pertusillo Lake (left) and the Monte Cotugno Lake (right). For color version of this figure, the reader is referred to the online version of this book.

**Table 25.2** Relative Otter Distribution in the Different Levels of Suitability

| Suitability Levels | High Agri River (%) | High Sinni River (%) |
|---|---|---|
| Low | 47 | 41 |
| Medium | 20 | 30 |
| High | 33 | 30 |

In both the areas, it is possible to observe a mixed structure of optimal and suitable sites that cover half of the territory, with the same relative percentage of the optimal level (15%). Despite its structure more defined by the anthropogenic activities, the High Agri River shows a higher proportion of suitable areas (43%) compared to the High Sinni River (34%), and a lower relative presence of unsuitable clusters (42% for High Agri River and 51% for High Sinni River). Such percentages of suitability levels appear to be strongly linked to the wider incidence for High Agri River versus High Sinni River of covers with wooded vegetation (broad-leaved forests, riparian and transitional vegetation) that are present on 48.5% of the subbasin. The presence of vegetation cover jointly with its spatial distribution seems to characterize the noticeable high fragmentation of optimal and suitable sites, and consequently the otter occurrence per HS level (Table 25.2). A high percentage of presences in unsuitable sites was found; such sites have always small dimensions and are often included in bigger suitable clusters. This percentage is higher in the High Agri River, which shows also in the high suitable level a more consistent percentage of otter presences than in

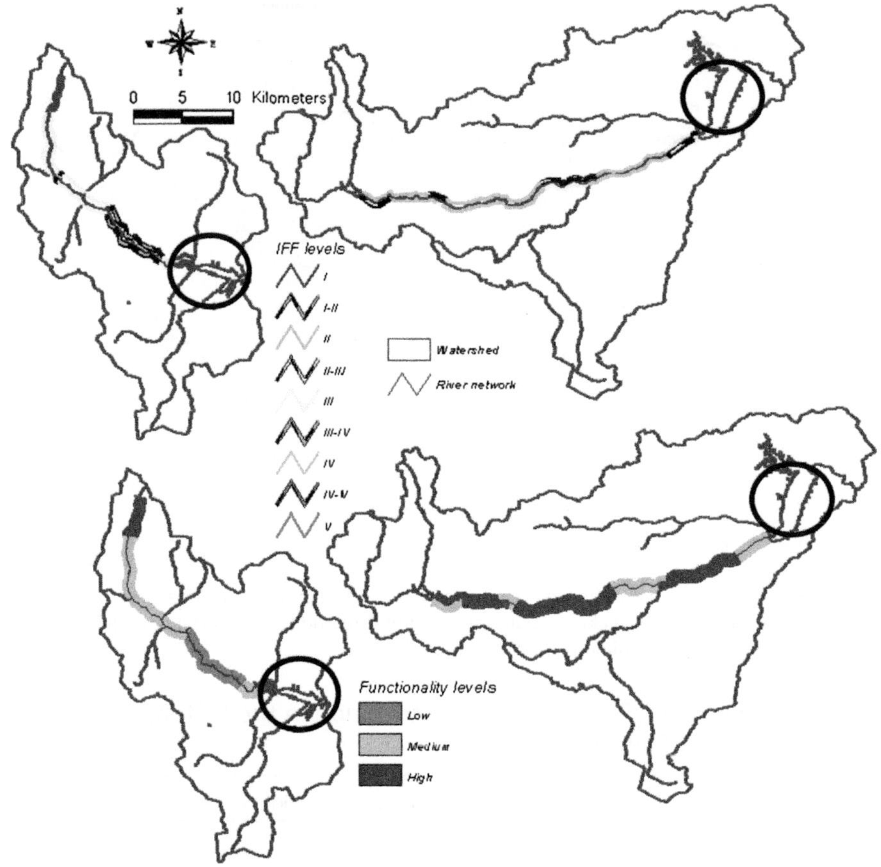


**FIGURE 25.4 FF map for the two test areas.** On the left, the High Agri River; on the right, the High Sinni River; in the circles the Pertusillo Lake (left) and the Monte Cotugno Lake (right). On the top of the figure are depicted all the classes provided by the application of the Italian standard protocol for FF; on the bottom are shown the relative maps reclassified into three levels of functionality from the lowest to the highest. For color version of this figure, the reader is referred to the online version of this book.

the medium level. Conversely, the High Sinni River evidences homogeneously distributed occurrences between medium and high suitable clusters.

The structure of such presence distribution can also be linked to the dynamics of the otter population in the study area. When a species is recovering, as the otter demonstrated to be in the last two decades in the studied territories (Loy and Racana, 1986; Loy et al., 2009), once saturated the most suitable sites it proceeds to colonize are also the less suitable ones (Begon et al., 1990). However, the unsuitable habitats colonized can act as ecological traps (Battin, 2004), leading to a decrease of the species population in future.

The FF analysis (Figure 25.4 top left) shows for the High Agri River a structure that clearly denotes an anthropogenic influence. The highest functionality levels are located along short segments in proximity of the river source and close to the Pertusillo Lake (areas with a lower agricultural and urban presence). On the contrary, a long river segment having low functionality is present in the territory portion with the highest urbanization and agricultural activity (see Figure 25.2). Such a segment is regulated with riverbanks in concrete; notwithstanding, due to a diffuse riparian recolonization inside the artificial riverbanks, it corresponds to a mixed structure in term of HS allowing for the otter presence. The High Sinni River (Figure 25.4 top right) shows a different structure, coherent with the more natural texture of its territory. The river axis belongs for the major part to the highest functionality classes. Segments of medium level (classes II–III and III) are located close at the extremities of the river sector, where are concentrated the bigger urbanized areas and along the river portions characterized by a large scarcely vegetated riverbed. The functionality maps reclassified for the integration step (Figure 25.4 bottom, left and right) better highlight the differences between the two subbasins.

The integrated VU analysis, taking into account the information coming from both the broad-scale HS analyses and the fine-scale FF analysis, summarizes the fluvial ecological criticalness of the two layers. The obtained VU levels are shown in Figure 25.5. The VU results underline for the High Agri River, in spite of the actual riparian vegetation presence, the crucial VU of the central part of the subbasin managed in concrete emphasizing the influence of the local aspects (see FF analysis) in this portion. In addition, even if the medium level is well spread in the remnant part of the river, it is possible to observe many highly vulnerable small portions inside it. In this case, the major influence in enhancing the VU is given by the presence of patches unsuitable for the otter habitat.

Such an aspect seems to be the most important in regulating the VU in the High Sinni River, since FF does not give evidence of criticalness (see Figure 25.4). In this territory, the percentage of unsuitable habitats represent a relevant weight in determining the overall

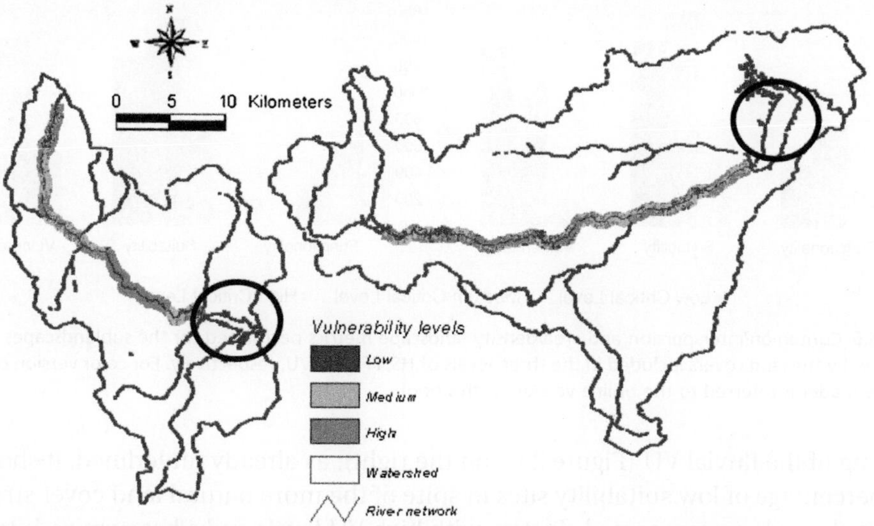


**FIGURE 25.5 Map of vulnerability (VU) for the two test areas.** On the left, the High Agri River; on the right, the High Sinni River; in the circles, the Pertusillo Lake (left) and Monte Cotugno Lake (right). Each level of VU corresponds to a different priority in term of intervention needs. For color version of this figure, the reader is referred to the online version of this book.

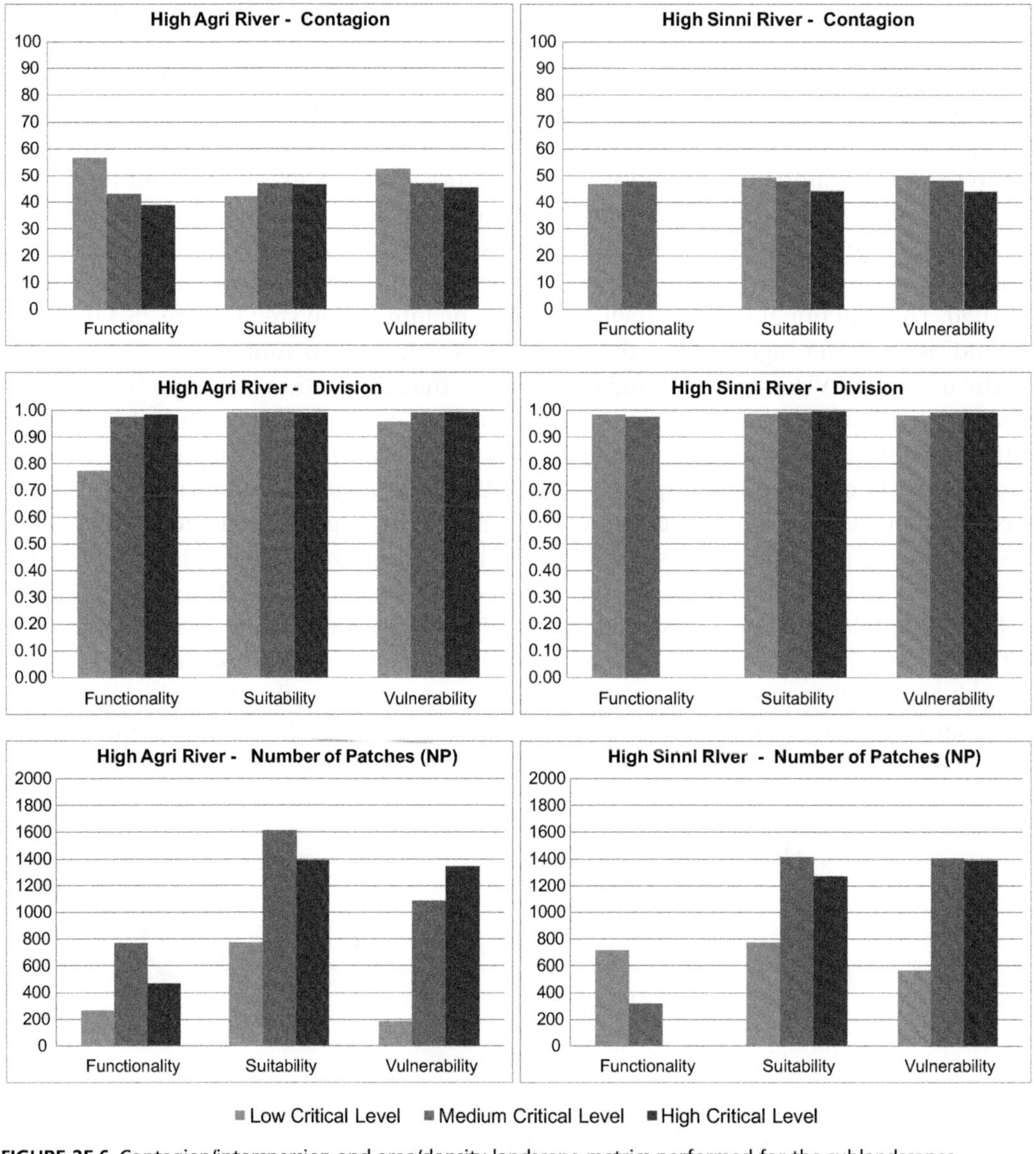


**FIGURE 25.6** Contagion/interspersion and area/density landscape metrics performed for the sublandscapes represented by the land covers included in the three levels of HS, FF, and VU, respectively. For color version of this figure, the reader is referred to the online version of this book.

increasing of the fluvial VU (Figure 25.5 on the right); as already underlined, it shows the major percentage of low suitability sites in spite of the more natural land cover structure. The map shows long compacted clusters with high VU levels and a heavy mixed structure of the three different levels in the remnant part of the river.

The obtained results allow for considering the information obtained by the proposed integrated VU analysis a useful tool in terms of prioritization for intervention needs. In

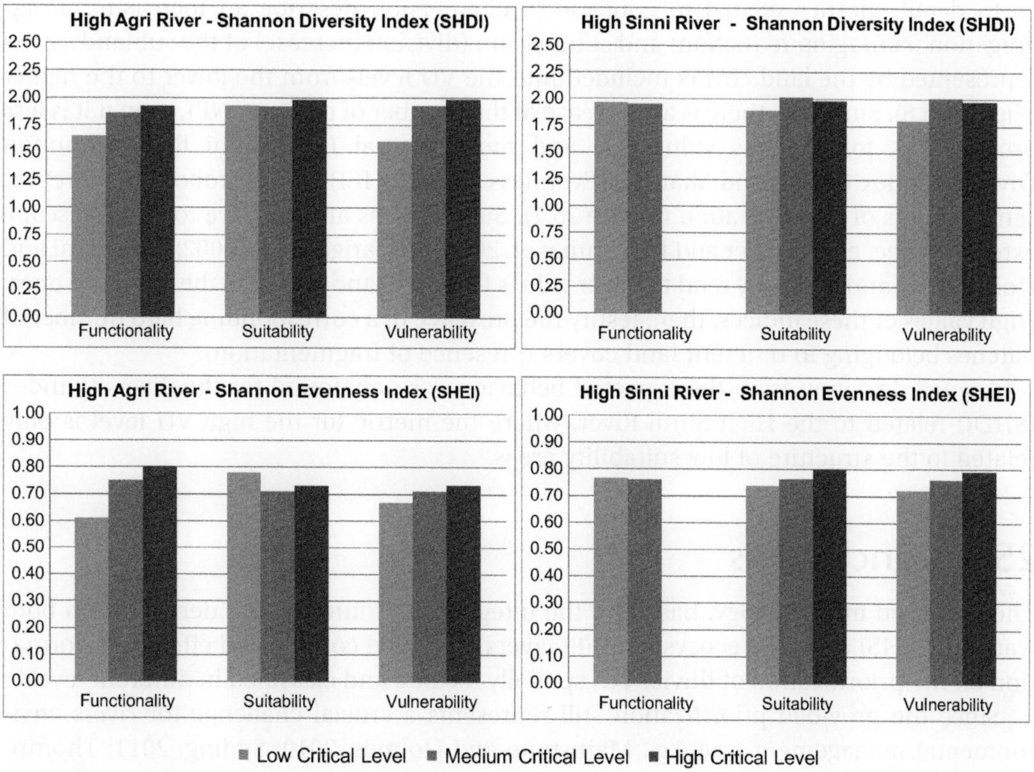


**FIGURE 25.7** Diversity landscape metrics performed for the sublandscapes represented by the land covers included into the three levels of HS, FF, and VU, respectively. For color version of this figure, the reader is referred to the online version of this book.

other words, where a high VU is found, the interested area must take precedence in term of intervention and/or management activities.

The assessment of this provided prioritization was also evaluated in terms of fragmentation representativeness. The results of the LM analysis are shown in Figures 25.6 and 25.7 where is represented the relative behavior of LM values moving from the low to the high level of VU.

Since a wide literature underlines that fragmentation is a major threats for the environmental equilibrium and may significantly alter ecological processes and biological communities (see, e.g., Collinge, 1996; Laurance, 2008; Marvier et al., 2004), we expected that the higher the VU, the higher the fragmentation suggested by the LM. For a comparison, the evaluation was also performed on the separated suitability and functionality components.

For both the subbasins, there is not a strict correspondence between the HS (Figure 25.3) and FF (Figure 25.4b) levels and the continuous increase or decrease of the corresponding LM values. This is particularly evident for HS levels, which clearly show a noncontinuity in the relative increase or decrease. On the contrary, a clear relationship between VU and fragmentation degree relative strength is observed.

In detail, metrics related to contagion/interspersion describe an increasing disaggregation (contagion reduction) and subdivision (division increase) of the sublandscapes represented by the land covers included into the VU levels from the lower to the higher (Figure 25.6). Similarly, there is an increase of the number of patches (NP), even if it is less pronounced, for the less vulnerable and more natural High Sinni River. Shannon Diversity Index (SHDI) and Shannon Evenness Index (SHEI) also corroborate the representativeness of fragmentation (Figure 25.7). Such indices are sensitive to the landscape extension (see, e.g., Fischer and Lindenmayer, 2007; McGarigal et al., 2002; Ramezani and Holm, 2011) and normally tend to lower values for small landscapes as the analyzed ones. High values of these indices, then, testify the presence of a corresponding high number of patches belonging to different land covers (presence of fragmentation).

A small deviation from the expected behavior was only found for the diversity index (SHDI) related to the High Sinni River, where the metric for the high VU level is only related to the structure of low suitability areas.

## 25.4 Conclusions

The proposed methodology, based on the integration of information derived from both watershed (HS) and river ecosystem (FF) layers, provided concise and effective information on the prioritization of fluvial VU, spatially located and classified in different levels.

Since the provided prioritization still represents a crucial challenge for rivers environmental management (see e.g., Mainstone and Holmes, 2010; Suding, 2011; Thompson, 2010), this method effectively contributes to such debate enhancing the calibration and the evaluation of the urgency of restoration activities.

The capability of VU to resume and emphasize the ecological criticalness confirms the hypothesis that for a multilayered system, such as a river-watershed system, an integrated approach is more efficient than a separated analyses, in agreement with the "Emergent Property Principle" (Odum and Barret, 1971). Therefore, the study adds helpful information to the field of river systems multilayered analysis, a field that is still little investigated (Bedoya et al., 2011).

The comparison of the results thus obtained on the two subbasins (one with more human presence and the other with a more natural structure) confirms the usefulness of the proposed integrated approach in different environmental conditions. Additional tests, also from other countries, using different FF indices, and on a wider set of LM, are needed to further prove the method's exportability and its usefulness for the final stakeholders.

## References

Balestrini, R., Cazzola, M., Buffagni, A., 2004. Characterizing hydromorphological features of selected Italian rivers: a comparative application of environmental indices. Hydrobiologia 516, 365–379.

Battin, J., 2004. When good animals love bad habitats: ecological traps and the conservation of animal populations. Conserv. Biol. 18 (6), 1482–1491.

Bedoya, D., Manolakos, E.S., Novotny, V., 2011. Characterization of biological responses under different environmental conditions: a hierarchical modeling approach. Ecol. Model 222, 532–545.

Begon, M., Harper, J.L., Townsend, C.R., 1990. Ecology, Individuals, Populations and Communities, second ed. Blackwell Scientific Publications, Boston.

Boyce, M.S., Vernier, P.R., Nielsen, S.E., Schmiegelow, F.K.A., 2002. Evaluating resource selection functions. Ecol. Model 157, 281–300.

Brown, B.L., Swan, C.M., Auerbach, D.A., Campbell Grant, E.H., Hitt, N.P., Maloney, K.O., Patrick, C., 2011. Metacommunity theory as a multispecies, multiscale framework for studying the influence of river network structure on riverine communities and ecosystems. J. N. Am. Benthol. Soc. 30 (1), 310–327.

Carone, 2012. Assessment of river ecosystems by a multitemporal and multilayered landscape approach. The semiaquatic Otter (Lutra lutra L.) as a paradigmatic species. PhD Thesis. University of Molise, Pesche (IS).

Collinge, S.K., 1996. Ecological consequences of habitat fragmentation: implications for landscape architecture and planning. Landscape Urban Plan. 36, 59–77.

Colson, F., Bogaert, J., Ceulemans, R., 2011. Fragmentation in the legal Amazon, Brazil: can landscape metrics indicate agricultural policy differences? Ecol. Indic. 11, 1467–1471.

Comiti, F., Mao, L., Lenzi, M.A., Siligardi, M., 2009. Artificial steps to stabilize mountain rivers: a post-project ecological assessment. River Res. Appl. 25, 639–659.

Fischer, J., Lindenmayer, D.B., 2007. Landscape modification and habitat fragmentation: a synthesis. Global Ecol. Biogeogr. 16 (3), 265–280.

Guisan, A., Zimmermann, N.E., 2000. Predictive habitat distribution models in ecology. Ecol. Model 135, 147–186.

Hirzel, A.H., Hausser, J., Chessel, D., Perrin, N., 2002. Ecological-niche factor analysis: how to compute habitat-suitability maps without absence data? Ecology 83 (7), 2027–2036.

Hirzel, A.H., Le Lay, G., Helfer, V., Randin, C., Guisan, A., 2006. Evaluating habitat suitability models with presence-only data. Ecol. Model 199 (2), 142–152.

Kruuk, H., 2006. Otters. Ecology, Behaviour and Conservation. Oxford University Press, Cambridge.

Ladson, A.R., White, L.J., Doolan, J.A., Finlayson, B.L., Hart, B.T., Lake, P.S., Tilleard, J.W., 1999. Development and testing of an index of stream condition for waterway management in Australia. Freshwater Biol. 41 (2), 453–468.

Laurance, W.F., 2008. Theory meets reality: how habitat fragmentation research has transcended island biogeography theory. Biol. Conserv. 141 (7), 1731–1744.

Lindenmayer, D.B., Fischer, J., 2006. Habitat Fragmentation and Landscape Change. An Ecological and Conservation Synthesis. Island Press Ed., Washington.

Loy, A., Racana A., 1986. La Lontra in Basilicata. WWF Italia. Serie Atti e Studi n. 5. (In Italian).

Loy, A., Carranza, M.L., Cianfrani, C., D'Alessandro, E., Bonesi, L., Di Marzio, P., 2009. Otter *Lutra lutra* population expansion: assessing habitat suitability and connectivity in south-central Italy. Folia Zool. 58 (3), 309–326.

Mainstone, C.P., Holmes, N.T.H., 2010. Embedding a strategic approach to river restoration in operational management processes-experiences in England. Aquat. Conserv. Mar. Freshw. Ecosyst. 20 (1), S82–S95.

Marvier, M., Kareiva, P., Neubert, M.G., 2004. Habitat destruction, fragmentation, and disturbance promote invasion by habitat generalist in a multispecies metapopulation. Risk Anal. 24 (4), 869–878.

McGarigal, K., Cushman, S.A., Neel, M.C., Ene, E., 2002. FRAGSTATS: Spatial Pattern Analysis Program for Categorical Maps. Computer Software Program Produced by the Authors at the University of Massachusetts, Amherst. Available at the following web site: www.umass.edu/landeco/research/fragstats/fragstats.html.

Mladenoff, D.J., Haight, R.C., Sickley, T.A., Wydeven, A.P., 1997. Causes and implications of species restoration in altered ecosystems: a spatial landscape projection of wolf population recovery. Bioscience 47, 21–31.

Munafò, M., Cecchi, G., Baiocco, F., Mancini, L., 2005. River pollution from non-point sources: a new simplified method of assessment. J. Environ. Manage. 77, 93–98.

Odum, E.P., Barret, G.W., 1971. Fundamentals of Ecology. Thomson Ed, Belmont, CA.

Panzacchi, M., Genovesi, P., Loy, A., 2010. Piano d'Azione Nazionale per la Conservazione della Lontra (Lutra lutra). Ministero dell'Ambiente e della Tutela del Territorio e del Mare – Istituto Superiore per la Protezione e la Ricerca Ambientale (in Italian), Roma.

Peng, J., Wang, Y., Zhang, Y., Wu, J., Li, W., Li, Y., 2010. Evaluating the effectiveness of landscape metrics in quantifying spatial patterns. Ecol. Indic. 10 (2), 217–223.

Petersen, R.C., Madsen, B.L., Wilzbach, M.A., Magadza, C.H., Parlberg, A., Kullberg, A., Kummins, K.M., 1987. Stream management: emerging global similarities. Ambio 16, 166–179.

Poole, G.C., 2002. Fluvial landscape ecology: addressing uniqueness within the river discontinuum. Freshwater Biol. 47, 641–660.

Ramezani, H., Holm, S., 2011. Sample based estimation of landscape metrics; accuracy of line intersect sampling for estimating edge density and Shannon's Diversity Index. Environ. Ecol. Stat. 18, 109–130.

Raven, P.J., Fox, O.P., Everard, M., Holmes, N.T.H., Dawson, F.H., 1997. River habitat survey: a new system for classifying rivers according to their habitat quality. In: Boon, P.J., Howell, D.J. (Eds.), Freshwater Quality – Defining the Indefinable. The Stationary Office, Edinburgh, pp. 215–234.

Remonti, L., Balestrieri, A., Prigioni, C., 2009. Altitudinal gradient of Eurasian otter (*Lutra lutra*) food niche in Mediterranean habitats. Can. J. Zool. 87 (4), 285–291.

Schowengerdt, R.A., 2007. Remote Sensing Models and Methods for Image Processing, third ed. Elsevier Academic Press, Burlington, MA 14. ISBN 10:0-12-369407-8.

Siligardi, M., Maiolini, B., 1993. L'inventario delle caratteristiche ambientali dei corsi d'acqua alpini. Guida all'uso della scheda R.C.E.-2. Biologia ambientale n° 2. (In Italian).

Siligardi, et al., 2000. I.F.F. Indice di Funzionalità Fluviale. Manuale ANPA. Agenzia Nazionale per la Protezione dell'Ambiente, Roma (In Italian).

Siligardi, et al., 2003. I.F.F. Indice di Funzionalità Fluviale. Manuale APAT. Agenzia per la Protezione dell'Ambiente e per i Servizi Tecnici, Roma (In Italian).

Suding, K.N., 2011. Toward an era of restoration in ecology: successes, failures and opportunities ahead. Ecol. Evol. Systemat. 42, 465–487.

Thompson, B.A., 2010. Planning for implementation: landscape-level restoration planning in an agricultural setting. Restor. Ecol. 19 (1), 5–13.

PART 8

# Global Models - Models of the Ecosphere

# 26

# Carbon Cycle Modeling and Principle of the Worst Scenario

Sergey I. Bartsev*,[1], Andrey G. Degermendzhi*, Pavel V. Belolipetsky*,[†]

*INSTITUTE OF BIOPHYSICS SB RAS, KRASNOYARSK, 660036, RUSSIA, [†]INSTITUTE OF COMPUTATIONAL MODELING SB RAS, KRASNOYARSK, 660036, RUSSIA

[1]EMAIL: BARTSEV@YANDEX.RU

## 26.1 Introduction

Available estimates suggest that increasing atmospheric carbon dioxide concentration is the major aggregate contributor to the rise in global temperature due to the greenhouse effect. The ocean contains 50 times more dissolved oxidized carbon than the atmosphere does, and 70% of the surface of the earth is covered by ocean. For these reasons, the prevalent opinion among scientists in the early twentieth century was that the ocean would prevent from increasing the $CO_2$ of the atmosphere due to industrial activity. This view prevailed until precise measurements of free-atmosphere $CO_2$ values showed an increasing trend of (at that time) 0.8 ppm/year (Keeling, 1961). During the last century, about half of the $CO_2$ released to the atmosphere by human activities was taken up by the ocean and the land biosphere (Prentice et al., 2001). Between 2000 and 2006, for instance, global anthropogenic carbon emissions were about 9.1 GtC/year, while carbon uptake by the ocean and the land was 2.2 and 2.8 GtC/year, respectively (Canadell et al., 2007).

Within the present century even larger emissions and a substantial global warming are expected. While there is considerable uncertainty about the continued uptake of $CO_2$ by both the land and the ocean, in fact, several model-intercomparison studies (Cramer et al., 2001; Friedlingstein et al., 2006) revealed that existing models describe somewhat inconsistent carbon cycle regimes and predict different responses to global change. And even confident knowledge of the most probable scenario of global change will not suffice to make practical decisions, because confidence intervals are very wide due to inevitable uncertainty of model parameters. It means that the probability of the scenario developing in the way that is most unfavorable to us does not equal to zero. Therefore, it seems obvious that for practical reasons we should primarily investigate the worst but possible variants, in which the contribution of the compensatory and alleviating mechanisms is the smallest. The vital question is "Can anthropogenic

Models of the Ecological Hierarchy. DOI: http://dx.doi.org/10.1016/B978-0-444-59396-2.00026-2
ISSN 0167-8892, 

influence lead to irreversible negative changes in the climate–biosphere system or a global ecological catastrophe?" The possibility of irreversible changes may be not very high, but one cannot ignore it.

## 26.2 Difficulties of Carbon Cycle Modeling

Friedlingstein et al. (2006) present a feedback analysis, which is based on globally averaged output of coupled and uncoupled simulations forced by historical and SRES A2 emissions of $CO_2$. They found that coupled climate–carbon cycle models simulated a negative sensitivity for both the land and the ocean carbon cycle to future climate. However, there is a large uncertainty on the magnitude of these sensitivities. Eight models attributed most of the changes to the land, while three attributed it to the ocean. Uncertainties for different parts of carbon cycle are even more contrast. The 2100 net land $CO_2$ flux ranges between an uptake of 11 GtC/year to a source of 6 GtC/year. The lowest uptake by the ocean reaches 3.8 GtC/year and the largest is 10 GtC/year.

If we consider terrestrial carbon cycle in more detail, we can see that these process-oriented models are complex combinations of scientific hypotheses; hence, their results depend on these embedded hypotheses. It is not yet clear how net primary production (NPP) and heterotrophic respiration (RH) will respond to rising $CO_2$ and to changing temperatures or rainfall. The effect of the changing climate on the carbon balance in terrestrial ecosystems is described mostly by relatively simple response functions and kinetic concepts of $CO_2$ uptake by photosynthesis and loss by respiration. The fundamental paradigm adopted by researchers over the past two decades was that photosynthetic uptake is stimulated both by increasing $CO_2$ and, in boreal and temperate regions, by rising temperature, although both the effects are expected to saturate at high levels of these variables. On the other hand, the biological processes underlying respiration are assumed to respond to temperature in an exponential way. But the temperature sensitivity of these functions differs between models. Some models use a Gaussian distributed temperature dependence of photosynthesis with a globally constant optimum temperature (Cox et al., 2001). Other models use an implicit temperature acclimation, with the consequence that photosynthesis has a broad optimum between 40 and 55 °C (Knorr, 2000). Matthews et al. (2005) showed that the value of this optimum temperature has a substantial impact on the simulated carbon–cycle climate feedback. This leads to the conclusion that the biosphere is able to provide negative feedback to rising $CO_2$ and temperature until the temperature climbs so high that the stimulating effect on respiration exceeds the $CO_2$ fertilization effect. This fundamental principle reflects the behavior of almost all the models in the comparative studies described earlier (Cramer et al., 2001; Friedlingstein et al., 2006).

Verification of models is hard because of several reasons. Differences in the behavior of the different coupled models regarding the simulated atmospheric $CO_2$ growth rate become important after 2025 (Friedlingstein et al., 2006). Other variables that are described by the models are NPP, RH, biomass, and soil organic pool. Unfortunately, it is impossible currently to verify the models according to these variables because both the modeling estimates—inventory and remote sensing—have large uncertainties. Let us consider, for example, NPP. Ito (2011) found that for estimates published between 2000 and 2010, the variability among estimates remains at 8–9 GtC/year. Many estimates derived using the remote-sensing approach were based on a few common global data of vegetation properties. Nevertheless, these estimates exhibit a substantial range of variability. Also different algorithms give very different values. For example, Nemani et al. (2003) derived quite different NPP values for the period 1982–1999 using two remote sensing products (GIMMS and PAL), but using the same NPP estimation algorithm, 49.98 GtC/year by GIMMS and 59.74 GtC/year by PAL. Also, Nemani et al. (2003) obtained linear trends in NPP at +0.23 (GtC/year)/year using GIMMS data and +0.12 (GtC/year)/year using PAL data during the period 1982–1999. Later Ito and Sasai (2006) obtained different linear trends in NPP using multiple models and climate data, ranging from +0.08 to +0.21 (GtC/year)/year during the period 1982–2001. The same problems exist with respect to RH, biomass, and soil organic pool estimates. As a result, model predictions are very uncertain and we have no means to eliminate these uncertainties at current time.

## 26.3. Principle of the Worst Scenario and Minimal Model of Biosphere

In order to deal with the described uncertainties, we formulated the principle of the worst scenario (Bartsev et al., 2008; Degermendzhi et al., 2008). Firstly, having identified the processes that can bring about irreversible negative changes quicker than others, one can leave aside interactions with longer characteristic times. Secondly, the worst scenario assumes the course of events, acceptable within the framework of the available data, when compensatory and alleviating mechanisms are for some reason ineffective and, therefore, need not be taken into account. We take into account catastrophic, threshold variants of global change, in which the magnitude of changes exceeds a certain threshold of stability of the "biosphere–climate" system, giving rise to avalanche-like and irreversible changes in global parameters. That is, in this study the term "catastrophic" corresponds rather to its meaning in the theory of catastrophes than to its everyday usage, when any serious natural cataclysm is perceived as a catastrophe. The minimal model of the biosphere–climate discussed in this study is an illustration of using the worst scenario principle. The main aim of constructing this minimal model is to illustrate the probability that the catastrophic variant of the development of the "biosphere–climate" system can take place and to determine the conditions that can bring about early formation of the catastrophic regime.

Following the worst scenario principle, let us construct a simplest zero-approximation model based on the reduced scheme of interactions in the system. One of the key mechanisms of the biosphere–climate interaction is the temperature—$CO_2$ concentration positive feedback. It is assumed that the positive feedback is affected through two loops of interactions. Elevation of atmospheric $CO_2$ causes, due to the greenhouse effect, a near-surface temperature rise, which, in turn, leads to an increase in soil microflora respiration and release of more $CO_2$. The second loop involves a decrease in the photosynthesis rate of land plants due to departure of the temperature from the optimum and, as a consequence, a decrease in the uptake of atmospheric $CO_2$, which amounts to the second positive feedback. We have chosen these two loops of interactions because they are high-rate processes that cause changes in the environment within an extremely short time.

Computer experiment with the previously published version of minimal model (Bartsev et al., 2008) showed that it can be simplified by excluding oceanic biota. This simplification is possible because it was shown that its contribution to the development of catastrophic biosphere regime is negligible.

Global model consists of five compartments between which carbon dioxide is exchanged: the atmosphere, land plants, the respective dead organic residues, surface and deep layers of the ocean. These compartments are interrelated through processes of growth, death, and decomposition of biomass and exchange of $CO_2$ between atmosphere and ocean. The model also contains the anthropogenic carbon source, which upsets the carbon balance of the system. The set of equations constituting the model has the following form:

$$\frac{dA}{dt} = S(y, T(A)) + C_{a\_up}BM_{out}(A) - P(x, A, T(A)) - C_{a\_down}AM_{in}(A) + \text{fuel}(t) \quad (1)$$

$$\frac{dx}{dt} = P(x, A, T(A)) - D(x) \quad (2)$$

$$\frac{dy}{dt} = D(x) - S(y, T(A)) \quad (3)$$

$$\frac{dB}{dt} = [C_{a\_down}AM_{in}(A) + C_{d\_up}U] - [C_{f\_down}B + C_{a\_up}BM_{out}(A)] \quad (4)$$

$$\frac{dU}{dt} = C_{f\_down}B - C_{d\_up}U \quad (5)$$

First equation describes the change of carbon amount in the atmosphere, second—in the vegetation biomass, third—in dead organic matter, fourth—in surface and fifth—in deep layers of ocean. Detailed description and assumptions, made for developing the model, were published earlier (Bartsev et al., 2008; Degermendzhi et al., 2008), so here we describe only the types of functions used in equations.

The function of the growth rate of plant biomass (GtC/year) has the following form:

$$P(x, A, T) = V_p \cdot x \cdot (x_{max} - x) \cdot V(A) \cdot f_p(T(A)) \quad (6)$$

where $x$ is the amount of carbon in the biomass of the plant compartment (GtC); $A$—atmospheric carbon (GtC); $T$—mean annual global surface temperature; $V_p$—the scale factor (1/(GtC × year)); $x_{max}$—the limited amount of biomass which depends on the limit of the density of vegetation cover (GtC) and is set in the model as $x_0G$, where $x_0$—the number of terrestrial plant biomass at present, $G$—a factor characterizing the opportunity to increase the number of plant biomass.

The function $V(A)$ describes the growth of biomass in relation to the atmospheric concentration of $CO_2$ in the form of well-known function of Monod:

$$V(A) = \frac{A}{K_A + A} \quad (7)$$

Monod equation usually considers concentrations, but since the volume of the reaction space (atmosphere) remains unchanged, the model uses the total amount of atmospheric carbon as a unit to simplify the customization of data. Empirical dependence of the rate of growth of plant biomass on temperature $T$ and the maximum temperature $T_{max}$ is the following:

$$f(T, T_{max}) = T^d(T_{max} - T), \quad \text{where } d = 1.5, \text{ and } 0 \le T \le T_{max} \quad (8)$$

Empirical dependence of the growth of average global surface temperature on the concentration of $CO_2$ was derived from the published data (Gifford, 1993) as follows:

$$T(A) = T_o + T_{del} \cdot \log_2\left(\frac{A}{A_0}\right) \quad (9)$$

where $A$ is the current amount of carbon in the atmosphere, $A_o$—the amount of carbon in the atmosphere at the time of measuring the average surface temperature $T_o$, which is equal to 15.5 at present; $T_{del}$—the climate sensitivity.

The view of function *fuel*($t$) depends on the external to biosphere source of $CO_2$ which brakes the closure of carbon cycle. In the case of future dynamics prediction this function is determined by possible scenario of burning fissile fuel.

Extinction rate of biomass (GtC/year) is written in a simple form as follows:

$$D(x) = V_d \cdot x \quad (10)$$

where $V_d$ is a scale factor; $x$—amount of carbon (Gt) in the biomass.

The rate of soil respiration (decomposition of dead organic matter) and $CO_2$ emissions in the atmosphere is described with the following function:

$$S(y, T) = V_S \cdot y \cdot f_M(T) \quad (11)$$

where $V_S$ is a scale factor; $y$—amount of carbon in dead biomass (Gt); $f_M(T)$—function of type (8) that expresses the temperature dependence of soil respiration, but for larger values of maximum temperature.

In Eqns (1) and (4), the term $C_{a\_down}AM_{in}(A)$ describes the absorption of carbon dioxide with the surface layer of the oceans, and the term $C_{a\_up}BM_{out}(A)$—emission of carbon dioxide from the ocean surface to the atmosphere. Variable coefficients $M_{in}(A) = e^{-0.03[T(A)-T_0]}$ and $M_{out}(A) = e^{0.03[T(A)-T_0]}$ describe the physical phenomenon of the declining of the gas solubility in the liquid at higher temperatures. These coefficients are

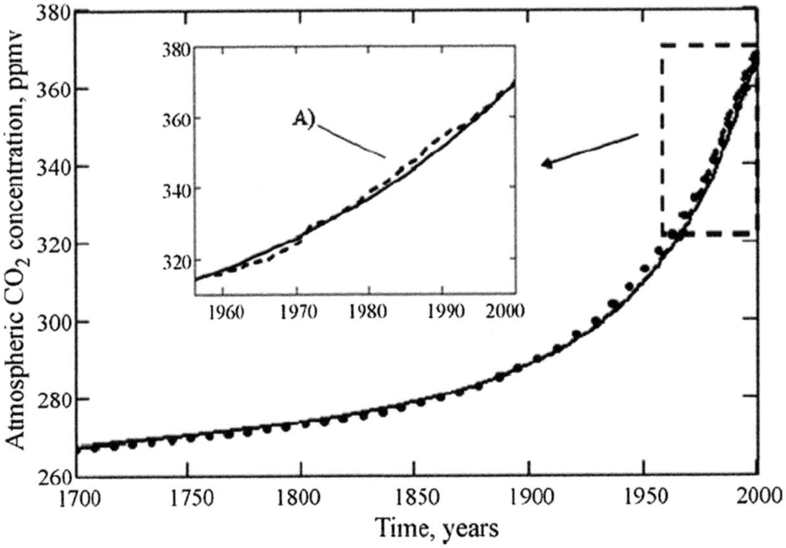


**FIGURE 26.1** A comparison of the observation data and computational experiments on carbon dioxide dynamics. (A) A fragment of comparison with Mauna Loa data.

obtained by a simple empirical approximation of data on the concentration of $CO_2$ in water at different temperatures (Kondratiev and Krapivin, 2004). In Eqn (5), the term $C_{d\_up}U$ describes upwelling and $C_{f\_down}B$—downwelling.

Various assumptions of carbon pool in the atmosphere and land compartments along with flows between them are represented in (Kondratiev and Krapivin, 2004; Semenov, 2004; IPCC 1996; IPCC Climate Change, 2001; Brovkin et al., 2002; Brovkin et al., 2004). In accordance with the worst-scenario principle the initial values of model's parameters were chosen as 850 GtC in vegetation biomass and 1100 GtC in sediment organic.

In Eqns (6), (10), and (11) scale factors were set just to make biomass growth and soil respiration equal to 55 GtC/year as estimated in the above mentioned papers. Without anthropogenic source of $CO_2$, the model stays in the stationary state which corresponds to biosphere and climate parameters at the end of the twentieth century.

Adjustment of the model parameters was performed based on the ice core data on atmospheric carbon dioxide concentration in the past, from 1700 to the present time (Fig. 26.1), the data on fossil fuel combustion, and the dynamics of the global temperature since 1860 (IPCC Climate Change, 2007; Houghton, 2003). Additionally, we used the data on atmospheric carbon dioxide concentration obtained at the Mauna Loa Observatory (Fig. 26.1A).

## 26.4 Minimal Model in Describing Past and Future Dynamics of Biosphere

To calculate the possible scenarios of the biosphere development, we used the variant of fossil fuel combustion labeled as A2 in the IPCC classification (IPCC Climate Change, 2007). In this case, the following dependences were considered as a function of fuel($t$): linear dependence where starting global emission of $CO_2$ is ~40 Gt in 2000 and increases annually by 1 Gt.

The purpose of computational experiments was to estimate the conditions under which a catastrophic scenario is realized and to determine the "irreversibility dates" corresponding to these conditions, that is, the moments after which the system collapses, even if human-induced carbon emissions have been stopped completely. The results of the model simulation at $T_{\text{del}} = 9\ ^{\circ}\text{C}$ are shown in Fig. 26.2.

The computer experiments demonstrated the presence of threshold of $CO_2$ emission dose. The prolongation of $CO_2$ emission by 1 year (cases 2 and 3 in Fig. 26.2) causes catastrophic changes in the system dynamics. In addition, the experiments showed significant compensatory role of the ocean—decreasing atmospheric $CO_2$ concentration after emission stopping (cases 1 and 2 in Fig. 26.2a). However, at high rate of $CO_2$ emission and its long duration, the ocean is not able to maintain the "biosphere–climate" system within the range favorable for the human.

To evaluate the possibility of catastrophical climate and biosphere changes described by the model, it is desirable to study the obtained historical samples when $CO_2$ changed rapidly or have large variations. Recent reconstructions of Earth's history have considerably improved our knowledge of episodes of rapid emissions of greenhouse gases and abrupt warming. The overall conditions and transient hyperthermals of the early Cenozoic represent an assortment of natural experiments that are showing the possibility of realizing the mechanisms described in the minimal model (Zachos et al., 2008, Schneider and Schneider, 2010, Pagani et al., 2010, Lunt et al., 2010). Using our model is possible to explain both physical and biogeochemical feedbacks causing increases in concentrations of greenhouse gases and abrupt warming. The transient warming events show characteristics that are indicative of short-term positive feedbacks, which accelerated and magnified the effects of initial carbon injection before weathering and other negative feedbacks restored the global carbon cycle to a steady state. Stable-isotope and other records suggest that the abrupt and massive carbon input followed an interval of gradual warming and preceded an interval of decreased carbon uptake.

For example, let us consider an abrupt warming about 55 million years ago—Paleocene–Eocene Thermal Maximum (PETM). PETM is one of the best known examples of a transient climate perturbation, associated with a brief, but intense, interval of global warming and a massive perturbation of the global carbon cycle from injection of isotopically light carbon into the ocean–atmosphere system (Pagani et al., 2006; Röhl et al.,

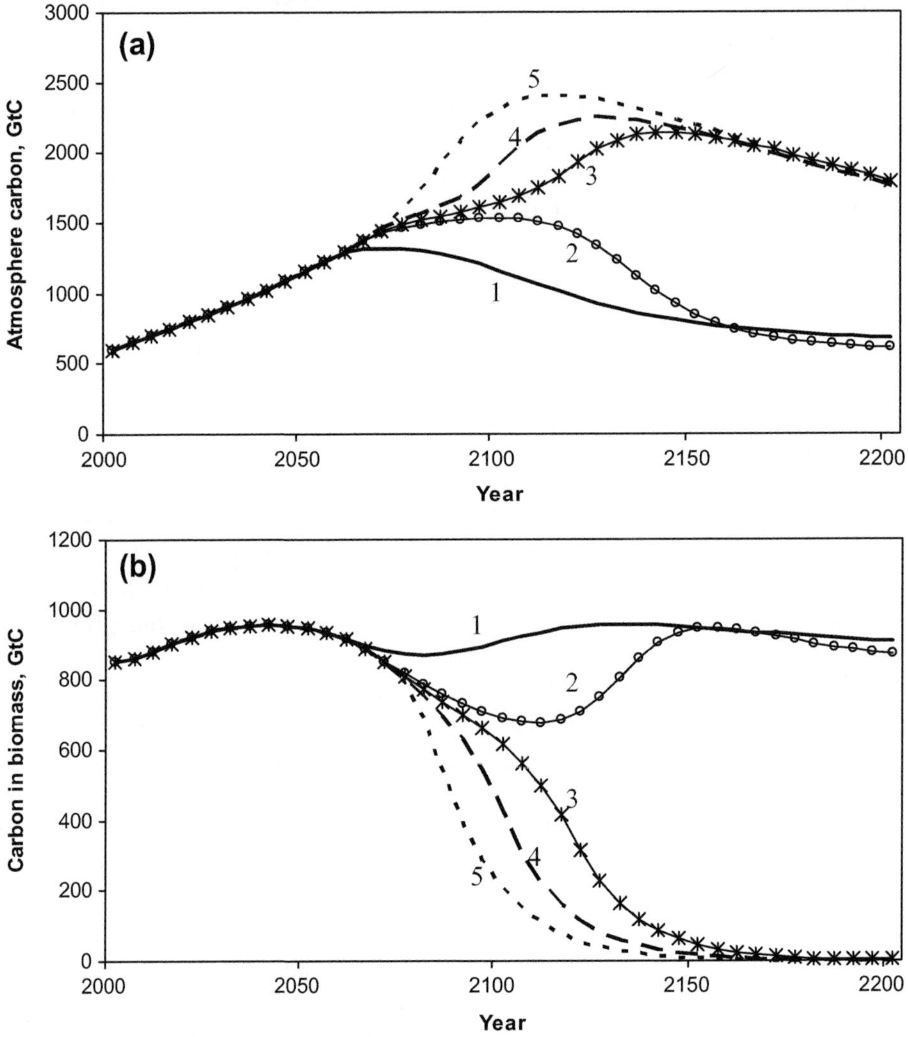


**FIGURE 26.2** Dynamics of carbon in atmosphere (a) and biomass (b) under different dates of completely stopping the emission: (1) 2060; (2) 2067; (3) 2068; (4) 2070; (5) 2080.

2007). Pagani et al. (2006) argue that it may be the best ancient analog for future increases in atmospheric $CO_2$. Kurtz et al. (2003) calculate pyrite sulfur ($S_{py}$) and organic carbon ($C_{org}$) burial rates from recently improved Cenozoic stable isotope records, and from these rates infer global changes in $C_{org}$ burial environments. They suggested that the Paleocene was a time of widespread terrestrial organic carbon burial, and the cause of PETM was the rapid oxidation of terrestrial organic carbon.

Using the described minimal model, it is possible to explain the release of terrestrial organic carbon. It was found that model is unstable with some parameters, that is, if biomass and dead organic matter reaches some values then small climate perturbation leads to self-sustaining release of organic carbon to atmosphere. Kurtz et al. (2003) argues that before PETM terrestrial ecosystems were a carbon sink for a long time. So there was a large carbon pool in terrestrial vegetation and soils. Also the last part of Paleocene was a time of gradual warming (Zachos et al., 2001). So before PETM terrestrial vegetation may be quiet near the temperature optimum for photosynthesis. In such conditions very small influence can lead to irreversible changes in our model. For example, it may be small increase of volcanic activity. Addition of some more geological cycle carbon to the atmosphere, something like 0.01 GtC/year, eventually lead to an event like PETM in our model (Fig. 26.3). To get such a dynamic, we changed the initial values of temperature and $CO_2$ concentration to 20 °C and 2000 ppm. It is argued that at that time both the parameters were higher than at present (Zachos et al., 2001). Another problem is that soil carbon pool is changing very quickly in the model that leads to quicker changes in atmospheric concentrations and temperature compared to observations. In order to avoid this disagreement, Eqn (3) can be written as follows:

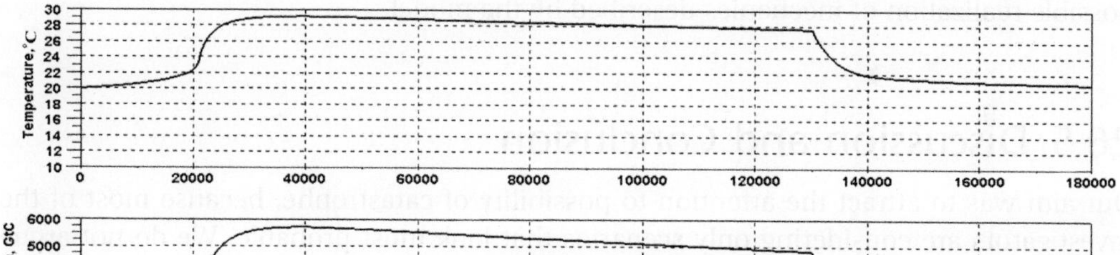

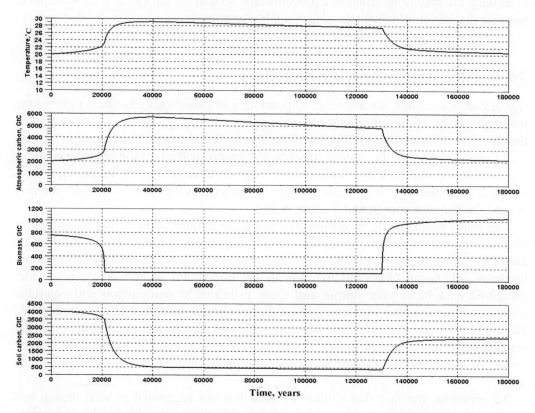


**FIGURE 26.3** Dynamics of global temperature, atmospheric carbon, vegetation carbon and soil organic carbon during PETM event simulated by the model.

$$\frac{dy}{dt} = D(x) - 0.95D(x) - 0.05S(y, T + \Delta T) \qquad (12)$$

In Eqn (12), we assume that most of the soil respiration is due to the oxidation of "fresh" dead biomass, and soil organic pool turnover times are longer than in Eqn (3). Model with Eqn (12) shows the same dynamics as initial model and does not eliminate the conclusion of possibility of irreversible changes.

It could be seen (Fig. 26.3) that after small temperature rise due to greenhouse warming, the system reaches the instability point that leads to the release of terrestrial organic carbon to atmosphere. The amount of accumulated organic carbon should be enough to explain the injection of isotopically light carbon into the ocean–atmosphere system (Kurtz et al., 2003). The $CO_2$ concentration rises to more then 4000 ppm accompanied by 5 °C rise in temperature. After that it takes a lot of time (about 100,000 years by estimates of Pagani et al., 2006) before a system starts to return to its prevent state. We can suppose that this time was spent for the adaptation of vegetation to new conditions. It was modeled via slowly increasing the maximal growth temperature $T_{max}$. Then the vegetation adopted biomass and soil organic carbon began to grow. And eventually the system returned to a prevent state. So may be the PETM is an example of possible realization of mechanics described by the model.

## 26.5 Discussion and Conclusion

Our aim was to attract the attention to possibility of catastrophe, because most of the investigators are considering only scenarios that look most probable. We do not argue that the worst-case scenario will be exactly realized. We only consider what might happen if all the parameters of the system "biosphere–climate" will be on the negative edges of the confidence intervals (but **inside** them!), which is unlikely, but in principle it is possible. The alternative scenario is plausible scenarios implemented in the vast majority of models that were mentioned in Section 26.1.

The main result of the investigation consists in confirmation of the possibility of negative and irreversible changes in the "biosphere–climate" system, caused by amplification of the positive feedback: "anthropogenic emission of $CO_2$—temperature increase—additional $CO_2$ emission due to decomposition of soil organics." Model described here looks very simple and quiet primitive in comparison with spatially distributed and coupled with atmosphere–ocean climate models (Friedlingstein et al., 2006). But we argue that this scenario is probable and due to large uncertainty of model predictions of carbon cycle feedbacks cannot be eliminated. Or in other words, by means of modeling one cannot prove that the probability of catastrophe is equal to zero.

Nevertheless, the fact that similar scenario has not happened at least during last million years is actually strong evidence against both positive feedbacks and catastrophe. However, described mechanisms could play a key role in some events in the more distant past. And then minimal mathematical model can be used to assess theoretical probability of the occurrence of certain hypothetic events in the history of the Earth's biosphere and climate. In this case, the worst scenario principle turns into the principle of the maximum favorability for the hypothetic event. Indeed, theoretical probability of the occurrence of the event implies that under a certain, quite realistic though unlikely, combination of circumstances this event can occur. According to this principle, it was shown that PETM, an event with large and rapid perturbations of carbon cycle and climate, can be explained by means of considered mechanism.

## Acknowledgments

The work was supported by the integration project SB RAS (No. 50-2009), RFBR (Grant No. 11-04-01116-a), and RFBR-KKFN (Grant No. 11-04-98089-a).

## References

Bartsev, S.I., Degermendzhi, A.G., Erokhin, D.V., 2008. Principle of the worst scenario in the modelling past and future of biosphere dynamics. Ecol. Model. 216, 160–171.

Brovkin, V., Bendsen, J., Claussen, M., Ganapolski, A., Kubatzki, C., Petoukhov, V., Andreev, A., 2002. Carbon cycle, vegetation, and climate dynamics in Holocene: experiments with the CLIMBER-2 model. Glob. Biogeochem. Cycles 16 (4), 1139–1143.

Brovkin, V., Sitch, S., Bloh von, W., Claussen, M., Bauer, E., Cramer, W., 2004. Role of land cover changes for atmospheric $CO_2$ increase and climate change during the last 150 years. Glob. Change Biol. 10, 1253–1266.

Canadell, J.G., Le Quere, C., Raupach, M.R., Field, C.B., Buitenhuis, E.T., Ciais, P., Conway, T.J., Gillett, N.P., Houghton, R.A., Marland, G., 2007. Contributions to accelerating atmospheric $CO_2$ growth from economic activity, carbon intensity, and efficiency of natural sinks. Proc. Natl. Acad. Sci. 104, 18866–18870.

Cox, P.M., Betts, R.A., Betts, A., Jones, C.D., Spall, S.A., Totterdell, I.J., 2002. Modelling vegetation and the carbon cycle as interactive elements of the climate system. Inte. Geophys. 83 (C), pp. 259–279.

Cramer, W., Bondeau, A., Woodward, F.I., et al., 2001. Global response of terrestrial ecosystem structure and function to $CO_2$ and climate change: results from six dynamic global vegetation models. Global Change Biol. 7, 357–373.

Degermendzhi, A.G., Bartsev, S.I., Gubanov, V.G., Erokhin, D.V., Shevirnogov, A.P., 2008. Forecast of biosphere dynamics using small-scale models (Hardcover). In: Global Climatology and Ecodynamics: Anthropogenic Changes to Planet Earth. Springer Praxis Books / Environmental Sciences, Berlin, pp. 241–300 (Chapter 10).

Friedlingstein, P., Cox, P.M., Betts, R., et al., 2006. Climate-carbon cycle feedback analysis, results from the C4MIP model intercomparison. J. Clim. 19, 3337–3353.

Gifford, R.M., 1993. Implications of $CO_2$ effects on vegetation for the global carbon budget. In: Heimann, M. (Ed.), The Global Carbon Cycle. Springer-Verlag, Berlin, pp. 159–199.

Houghton, R.A., 2003. Revised estimates of the annual net flux of carbon to the atmosphere from changes in land use and land management 1850–2000. Tellus 55B, 378–390.

IPCC, 1996. Climate Change. The Science of Climate Change, Intergovernmental Panel on Climate Change. Cambridge University Press, Cambridge (Chapter 2)65–131.

IPCC, 2001. Climate Change. In: Scientific Aspects. UNEP, p. 881. www.ipcc.ch.

IPCC Climate Change, 2007. The physical science basis. Contribution of working group I to the Fourth Assessment Report of the Intergovernmental Panel On Climate Change. In: Solomon, S., Qin, D., Manning, M., Chen, Z., Marquis, M., Averyt, K.B., M.Tignor, Miller, H.L. (Eds.). Cambridge University Press, Cambridge, United Kingdom and New York, NY, USA, p. 996. www.ipcc.ch.

Ito, A., 2011. A historical meta-analysis of global terrestrial net primary productivity: are estimates converging? Global Change Biol. 17, 3161–3175.

Ito, A., Sasai, T., 2006. A comparison of simulation results from two terrestrial carbon cycle models using three climate datasets. Tellus 58B, 513–522.

Keeling, C.D., 1961. The concentration and isotopic abundance of carbon dioxide in rural and marine air. Geochim. Cosmochim. Acta 24, 277–298.

Knorr, W., 2000. Annual and interannual $CO_2$ exchange of the terrestrial biosphere: process based simulations and uncertainties. Global Ecol. Biogeogr. 9, 225–252.

Kondratiev, K.Ya., Krapivin, V.F., 2004. Carbon Global Cycle Modeling, 336. FIZMATLIT, Moscow.

Kurtz, A.C., Kump, L.R., Arthur, M.A., Zachos, J.C., Paytan, A., 2003. Early Cenozoic decoupling of the global carbon and sulfur cycles. Paleoceanography 18.

Lunt, D.J., Haywood, A.M., Schmidt, G.A., Salzmann, U., Valdes, P.J., Dowsett, H.J., 2010. Earth system sensitivity inferred from Pliocene modelling and data. Nat. Geosci. 3, 60–64.

Matthews, H.D., Eby, M., Weaver, A.J., Hawkins, B.J., 2005. Primary productivity control of simulated carbon cycle-climate feedbacks. Geophys. Res. Lett. 32, L14708. doi:10.1029/2005GL022941.

Nemani, R.R., Keeling, C.D., Hashimoto, H., et al., 2003. Climate-driven increases in global terrestrial net primary production from 1982 to 1999. Science 300, 1560–1563.

Prentice, I.C., Farquhar, G.D., Fasham, M.J.R., et al., 2001. The carbon cycle and atmospheric carbon dioxide. In: Houghton JT, DingY, et al. (Eds.), Climate Change 2001: the Scientific Basis. Contribution of Working Group I to the Third Assessment Report of the Intergovernmental Panel on Climate Change. Cambridge University Press, Cambridge, pp. 183–237.

Pagani, M., Caldeira, K., Archer, D., Zachos, J.C., 2006. An ancient carbon mystery. Science 314, 1556–1557.

Pagani, M., Liu, Z., LaRiviere, J., Ravelo, A.C., 2010. High Earth-system climate sensitivity determined from Pliocene carbon dioxide concentrations. Nat. Geosci. 3 (1), 27–30.

Röhl, U., Westerhold, T., Bralower, T.J., Zachos, J.C., 2007. On the duration of the Paleocene–Eocene thermal maximum (PETM). Geochem. Geophys. Geosyst. 8.

Schneider, B., Schneider, R., 2010. Palaeoclimate: global warmth with little extra $CO_2$. Nature Geosci. 3, 6–7.

Semenov, S.M., 2004. Greenhouse gases and present climate of the Earth. M. (in Russian, extended summary in English). Meteorology and Hydrololgy Publishing Centre, Moscow. 175.

Zachos, J.C., Dickens, G.R., Zeebe, R.E., 2008. An early Cenozoic perspective on greenhouse warming and carbon-cycle dynamics. Nature 451, 279–283.

Zachos, J., Pagani, M., Sloan, L., Thomas, E., Billups, K., 2001. Trends, rhythms and aberrations in global climate 65 Ma to present. Science 292, 686–693.

# 27

# The Worst Scenario Principle and the Assessment of the Impact of Quality of Life for Biosphere Dynamics

Sergey I. Bartsev, Yuliya D. Ivanova, Anton L. Shchemel
*INSTITUTE OF BIOPHYSICS SB RAS, KRASNOYARSK, 660036, RUSSIA*

## 27.1 Introduction

Negative development trends of the system "biosphere–climate," at the first time global warming, caused by anthropogenic effects (IPCC, 2007) draw more and more concerns of public and politicians. Attempts to slow down these trends are being undertaken as agreements on limitations of greenhouse gases emission. However, implementations of these attempts are not so easy. In this way, besides natural questions of effectiveness and sufficiency of such steps, not all nations accept these agreements, which depress efficacy of them. It could be supposed that the reason is that volumes of greenhouse gases emission are formed by available power of those nations' citizens. Available power in turn determines their life quality (Alam et al., 1991; Dekker et al., 1995). In spite of the life quality being rather a multi-aspect showing, there are methods that make it possible to estimate it quite objectively and to provide comparison between nations considering the instant showing (Human Development Report, 2010).

It has been shown that the quality of life is dependent on the energy consumption up to an yearly energy consumption of ~4000 kWh electrical energy per capita (Dekker et al., 1995). This value can be obtained at 30% efficiency from burning about 1000 l of oil. According to World Resource Institute (http://www.wri.org/) data, about 80% of consumed energy is made from fossil fuels. So it is possible to estimate the yearly $CO_2$ emission per capita associated with providing worthy quality of life as ~900 kg of carbon per year. For comparison, the world average yearly carbon emission is ~1.8 ton per capita and the same in the developed countries is ~6 ton per capita. Further growth of human population (United Nations, 2004) and providing worthy quality of life to all human beings will cause further increasing of $CO_2$ emission at the present structure of energy production and consumption. For evaluating permissible level of $CO_2$ emission, mathematical models of climate and biosphere are used (IPCC, 2007; Stainforth et al., 2005; Tarko, 2005; C&C, 2010).

Models of the Ecological Hierarchy. DOI: http://dx.doi.org/10.1016/B978-0-444-59396-2.00027-4

However, traditional climate and biosphere models designated for predicting the most likely scenario consist of big number of equations and therefore it is very difficult to verify and analyze the key mechanisms of a possible global change using such models.

Furthermore, even reliable knowledge of the most likely scenario may not be used for practical solutions because the confidence intervals of that scenario will be very broad due to unavoidable uncertainty of model parameters (e.g., even estimations of global soil carbon amount are varying from 1000 to 2000 GtC) (Kondratiev and Krapivin, 2004; Semenov, 2004; IPCC Climate Change, 2001; Brovkin et al., 2004).

Therefore, it seems clear that for the practical purposes the worst but possible variant, in which the contribution of compensatory mechanisms and mitigation is the smallest and the values of model parameters lie on the adverse for humanity edges of confidence intervals, should be studied first. Based on the above, the "principle of the worst-case scenario" was formulated (Bartsev et al., 2008), which consists in selecting and studying only those processes that can quickly lead to negative changes in the biosphere, reasonably ignoring the possible compensatory mechanisms.

The additional reason to take into account the worst-case scenario is hardly probable but possible due to its extremely high cost of risk. Let us explain it by giving aviation accidents as an example. According to crash statistics that is presented on the website http://www.planecrashinfo.com/, probability of death for a passenger is about $10^{-7}$. People agree to use airplanes with such safety level. "Cost" of accident equals one human life. Thus VaR (value at risk which is equal to probability $\times$ cost production) is equal to $10^{-7}$. Taking equivalence of human lives as a postulate, and accepting VaR to be the same as for an individual, we can make a simple ratio to calculate the threshold probability of global ecological disaster. For example, considerable biosphere–climate changes will in varying degrees affect all of the humanity and could lead to mass deaths from starvation, diseases, climatic disasters, and wars over resources. Assume that an environmental disaster may lead to death of 1 billion people, and then its probability threshold, which should promote decision-makers to the activity, is about $10^{-16}$.

Mass estimates of possible climate change determined by greenhouse gas emissions, conducted by a group from Oxford (Stainforth et al., 2005), showed that from the whole ensemble of possible scenarios, about 1% of them demonstrate an extra-catastrophic temperature rise by 11 °C per doubling concentration of carbon dioxide in atmosphere. A possibility for climate sensitivity to reach 10 °C is also shown in Roe and Baker paper (2007) based on the analysis of existed uncertainties of known parameters and their contribution to climatic sensitivity. Such dramatic surface temperature raising no doubt will cause catastrophic consequences. Furthermore, because of the huge complexity of the biosphere and climate models and fundamentally unrecoverable error in the determination of the most model parameters, **no one** can give a guarantee that such a catastrophic rise in temperature has a negligible chance (about $10^{-16}$).

Currently, it is required to define what is the worth scenario. Conventional models of "biosphere–climate" system (e.g., Tarko, 2005; Bolin, 1981; Keeling and Whorf, 2001; Stainforth et al., 2005; Bolin, 1981; Keeling et al. 2001) are presented as smooth and reversible. These representations can be classified as optimistic, since in this case, it is theoretically possible to weaken the negative change by reducing human impacts and even to return biosphere to the initial state.

But this conception is not well-founded. On the contrary, nonlinearity of ecological systems makes it to allow the existence of thresholds of stability or resilience of local ecosystems and whole biosphere as a typical case. One of the possible mechanisms of irreversibility of changes in "biosphere–climate" system is discussed earlier (Bartsev et al., 2005, 2008).

Unlike the previous mechanism, the worst-case scenario suggests a catastrophic threshold for global changes when up to a certain threshold incremental changes are reversible, but in excess of this threshold in the biosphere–climate system, the global parameters keep changing in an avalanche and in an irreversible manner.

The aim of this work is to obtain estimates of the needed reduction of $CO_2$ emissions and thus reduction of the energy consumption level to prevent passing through the threshold of irreversibly accounting.

## 27.2 Results of the Simulation of the Minimal Model of Biosphere

For the purpose of this study, the generalized (including land and oceans) minimal model of the biosphere presented in Chapter 26 was used. Good correspondence with the observed data (Keeling and Whorf, 2001; Pritchard et al., 2001; Morgan et al., 2001) is achieved when parameter $G$ is equal to 1.6. According to the worst-case principle, let us chose the most inauspicious value of climate sensitivity which is 11 °C (Stainforth et al., 2005; Roe and Baker, 2007) and then estimate up to what level the emission of carbon should be reduced in comparison with A2 scenario (IPCC, 2007) to avoid irreversible catastrophic changes of biosphere described earlier (Bartsev et al., 2008). To make the representation of system dynamics better, the simulation period was increased by 100 years in comparison with IPCC predictions while preserving the same trend as in A2 scenario of increasing rates of fuel burning.

In Figs 27.1–27.4, the dynamics of annual mean surface temperature along with the amount of carbon of the main biosphere compartments are shown for various trends of increasing rates of burning fuels. Carbon enters the atmosphere from the main three sources: burning fossil fuels, mineralization of dying biomass and soil's organics, and degassing shallow ocean waters due to temperature increase.

Shocking increase of land surface temperature in the model corresponds to the extinction of complex life on the Earth. Unfortunately, there is such chance of extinction if climate sensitivity would be near the value calculated by the simulation (Stainforth et al., 2005) in 1% of all obtained scenarios. Even more dramatic variant of scenario was found earlier in simulations based on other assumptions (Karnaukhov, 2001). Due to the severity of the possible consequences, at least the obtained results should be considered.

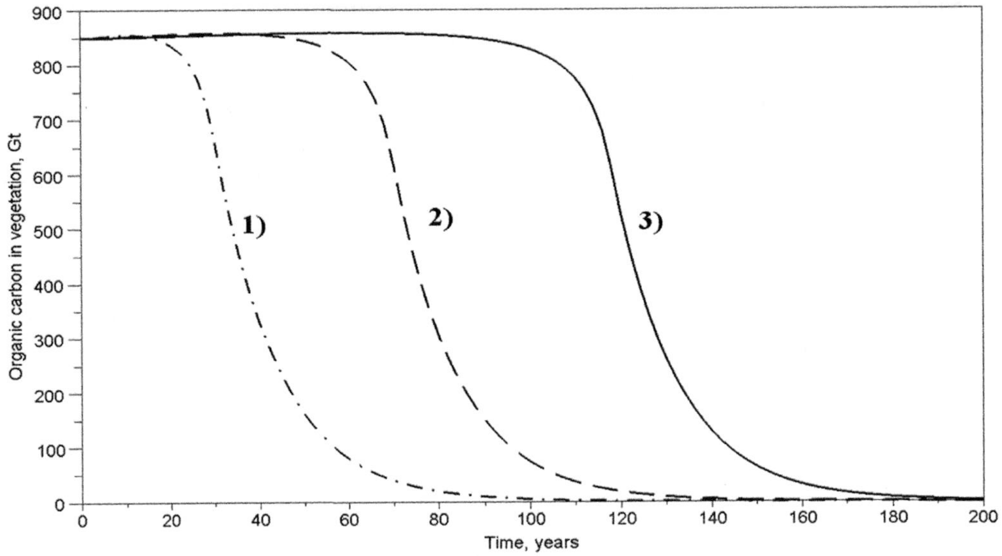


**FIGURE 27.1** The ensemble of trajectories of the amount of carbon in the biomass in the biosphere model for different growth rate of burning carbon-containing fuels. Curve (1) corresponds to the initial A2 scenario; curve (2) depicts 25% from the temps of increasing burning fuels as in the A2; and curve (3) depicts 10% correspondingly.

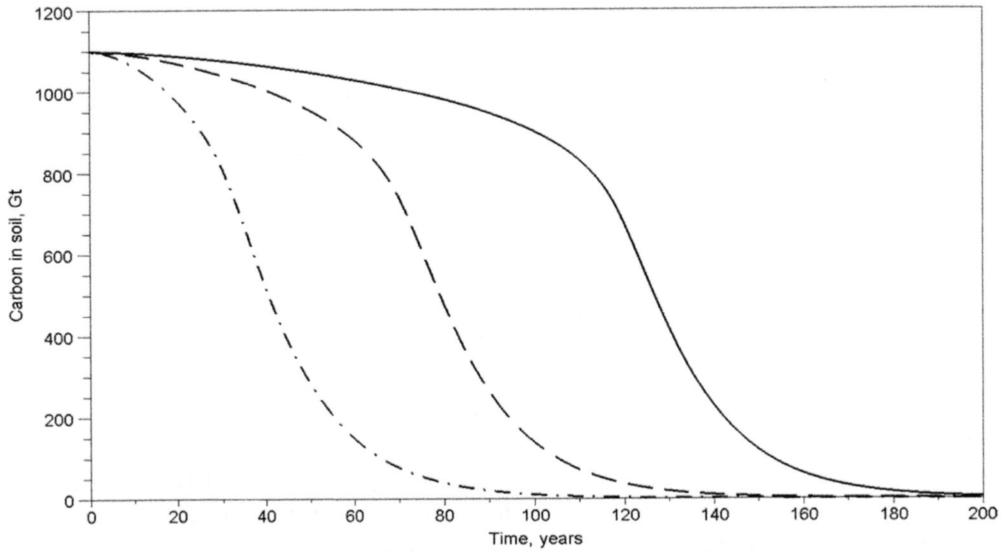


**FIGURE 27.2** The ensemble of trajectories of the amount of carbon in the soil in the biosphere model for different growth rate of burning carbon-containing fuels. Designations of the curves are the same as on the Fig. 27.1.

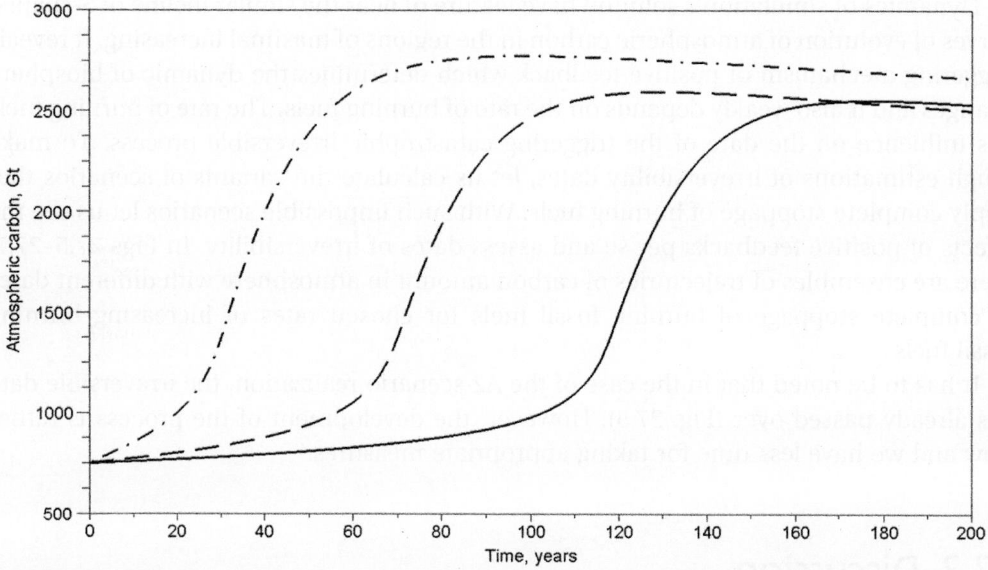


**FIGURE 27.3** The ensemble of trajectories of the amount of carbon in the atmosphere in the biosphere model for different growth rate of burning carbon-containing fuels. Designations of the curves are the same as in Fig. 27.1.

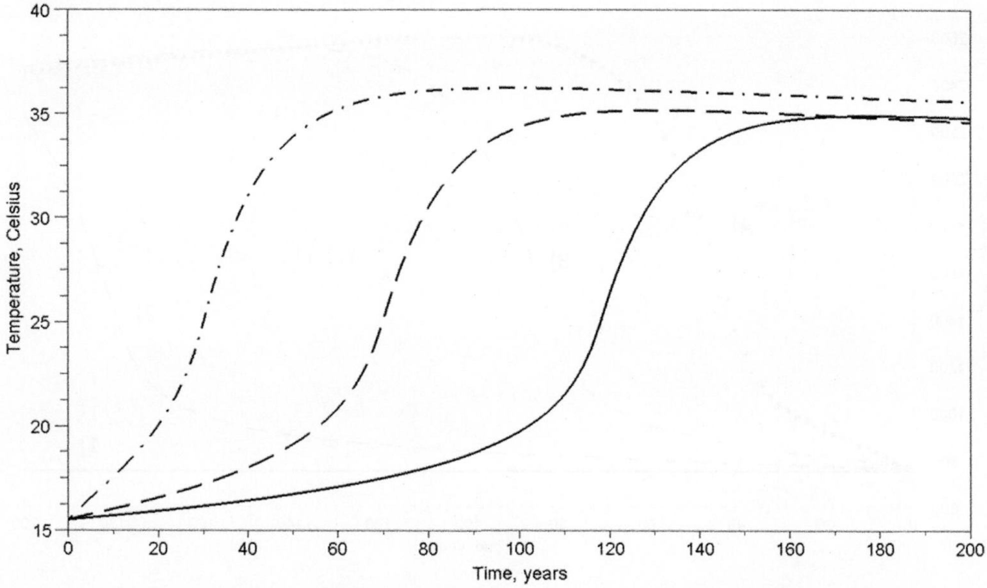


**FIGURE 27.4** The ensemble of trajectories of the average annual surface temperature in the biosphere model for different growth rate of burning carbon-containing fuels. Designations of the curves are the same as in Fig. 27.1.

Dynamics of simulation's solution have feature of near the similar incline of S-shaped curves of evolution of atmospheric carbon in the regions of maximal increasing. It reveals triggering mechanism of positive feedback which determines the dynamic of biosphere changes and it also weakly depends on the rate of burning fuels. The rate of burning fuels has influence on the date of the triggering catastrophic irreversible process. To make rough estimations of irreversibility dates, let us calculate the variants of scenarios that imply complete stoppage of burning fuels. With such impossible scenarios let us see the effects of positive feedbacks per se and assess dates of irreversibility. In Figs 27.5–27.7, there are ensembles of trajectories of carbon amount in atmosphere with different dates of complete stoppage of burning fossil fuels for chosen rates of increasing burning fossil fuels.

It has to be noted that in the case of the A2 scenario realization, the irreversible date has already passed over (Fig. 27.5). However, the development of the process is rather slow and we have less time for taking appropriate measures.

## 27.3 Discussion

The provided simulations show that if the climate sensitivity is close to 11 °C then we are already in the phase of irreversible changes. Further burning of fuels just makes the approach of the phase of accelerated emission of carbon to the atmosphere. In that situation, a dramatic reducing of carbon emission requires to get a period of time to

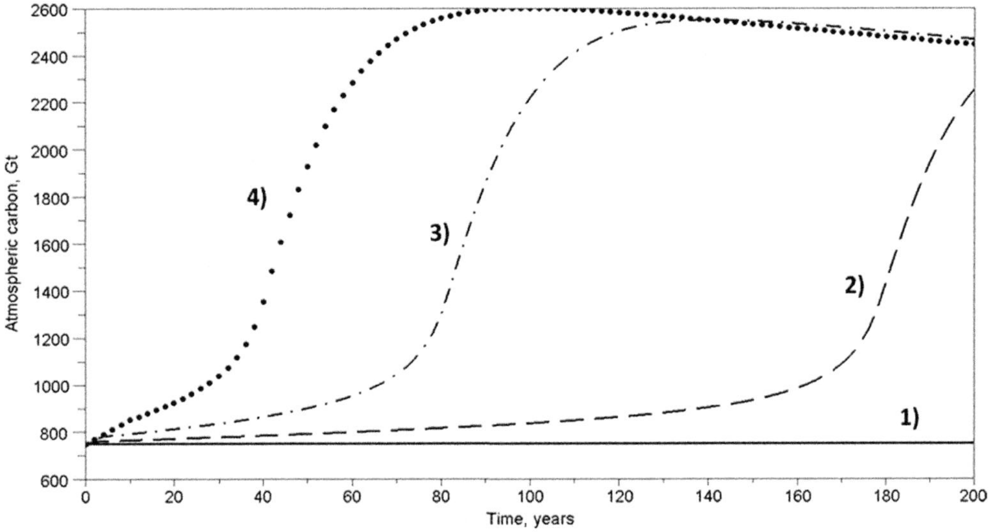


**FIGURE 27.5** The ensemble of trajectories of the amount of carbon in the atmosphere in the biosphere model for the A2 scenario and various dates of complete stoppage of burning fossil fuels: (1) stoppage in 2010; (2) stoppage in 2011; (3) stoppage in 2013; and (4) stoppage in 2020.

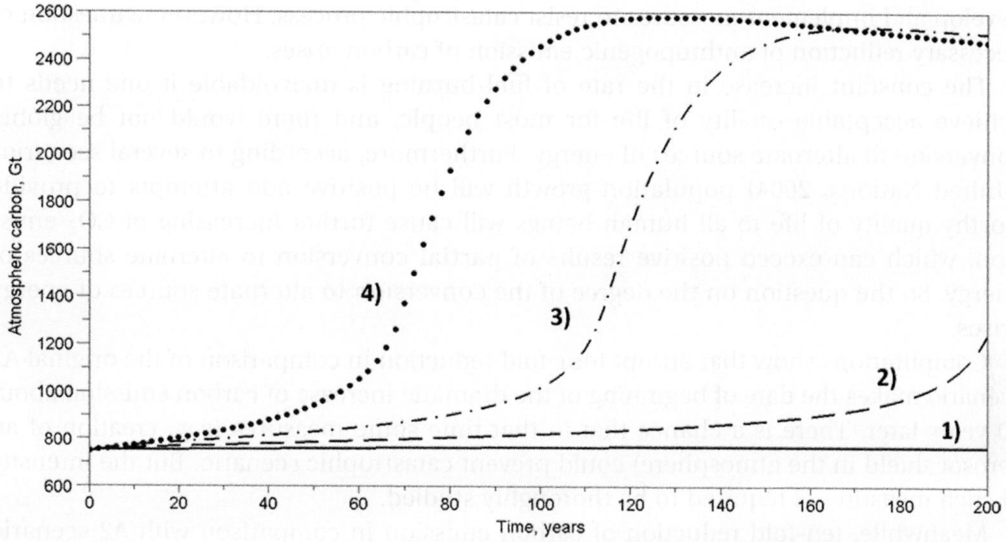


**FIGURE 27.6** The ensemble of trajectories of the amount of carbon in the atmosphere in the biosphere model for the 25% of A2 scenario's increasing of burning fuels rate and various dates of complete stoppage of burning fossil fuels: (1) stoppage in 2011; (2) stoppage in 2014; (3) stoppage in 2020; and (4) stoppage in 2080.

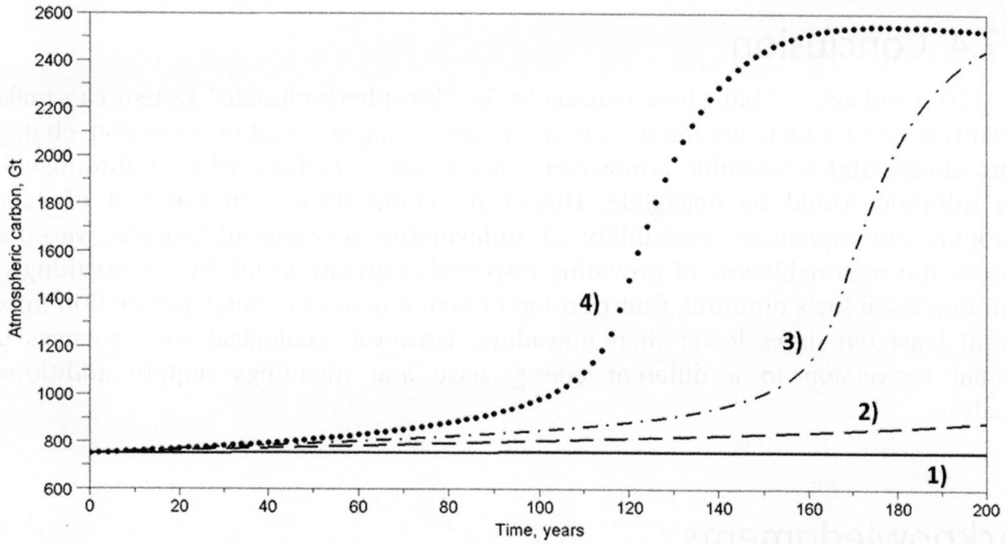


**FIGURE 27.7** The ensemble of trajectories of the amount of carbon in the atmosphere in the biosphere model for the 10% of A2 scenario's increasing of burning fuels rate and various dates of complete stoppage of burning fossil fuels: (1) stoppage in 2011; (2) stoppage in 2020; (3) stoppage in 2030; and (4) stoppage in 2110.

develop and implement measures to resist catastrophic process. However, a question of necessary reduction of anthropogenic emission of carbon arises.

The constant increase in the rate of fuel burning is unavoidable if one needs to achieve acceptable quality of life for most people, and there would not be global conversion to alternate sources of energy. Furthermore, according to several scenarios (United Nations, 2004) population growth will be positive and attempts to provide worthy quality of life to all human beings will cause further increasing of $CO_2$ emission, which can exceed positive results of **partial** conversion to alternate sources of energy. So the question on the degree of the conversion to alternate sources of energy arises.

Computations show that abrupt four-fold reduction in comparison of the original A2 scenario makes the date of beginning of the dramatic increase of carbon emission about 40 years later. There is a chance that in that time some measures (e.g., creation of an aerosol shield in the atmosphere) could prevent catastrophic scenario. But the intensity of such measures is required to be thoroughly studied.

Meanwhile, ten-fold reduction of carbon emission in comparison with A2 scenario makes the time of dramatic changes about 80 years later, which gives even more possibilities for taking measures. However, as we can see in Fig. 27.7, long-term burning fuels with even such a low rate anyway starts catastrophic events. This rate of burning fuels are characterized by emitting 180 kg per person of carbon annually, which is about five times less than emissions that are caused by energy supply (4000 kWh annually) for providing adequate life quality.

## 27.4 Conclusion

Positive feedbacks of biosphere genesis in the "biosphere–climate" system can make relatively slow global warming caused by anthropogenic carbon emission change into abrupt and irreversible atmosphere saturation by carbon, while anthropogenic contribution would be negligible. Higher (in comparison with the cost of catastrophic consequences) probability of unfavorable scenario of climate warming makes the reasonableness of providing respectable quality of life by the meaning of burning fossil fuels doubtful. Contribution of heat engines in energy production must be at least ten times lower than nowadays. However, ecological consequences of global conversion to a different energy base and meanings require additional analysis.

## Acknowledgments

The work was supported by the integration project of SB RAS No. 50 and RFBR Grant No. 11-04-01116-a.

# References

Alam, M.S., Bala, B.K., Huo, A.M.Z., Matin, M.A., 1991. A model for the quality of life as a function of electrical energy consumption. Energy 16 (4), 739–745.

Bartsev, S.I., Degermendzhi, A.G., Erokhin, D.V., 2005. Global minimal model of long-term carbon dynamics in the biosphere. Doklady Earth Sci. 401 (2), 326–329.

Bartsev, S.I., Degermendzhi, A.G., Erokhin, D.V., 2008. Principle of the worst scenario in the modelling past and future of biosphere dynamics. Ecol. Model. 216 (2), 160–171.

Bolin, B., 1981. Steady state and response characteristics of a simple model of the carbon cycle. In: SCOPE-16. Carbon Cycle Modeling. John Wiley and Sons, New York, pp. 315–331.

Brovkin, V., Sitch, S., Bloh von, W., Claussen, M., Bauer, E., Cramer, W., 2004. Role of land cover changes for atmospheric $CO_2$ increase and climate change during the last 150 years. Glob. Change Biol. 10, 1253–1266.

Dekker, P.-M., Meisen, P., Bruton, A.B., 1995. The GENI model: the interconnection of global power resources to obtain an optimal global sustainable energy solution. Simulation 64, 244–253.

C&C, Contraction & Convergence, Global Commons Institute, 2010. http://www.gci.org.uk/.

Human Development Report, 2010. 20th Anniversary, United Nations, 2010. http://hdr.undp.org/en/.

IPCC, 2001. Climate Change. 2001: Scientific aspects, UNEP, 881 P. www.ipcc.ch.

IPCC, 2007. Climate change 2007: the physical science basis. In: Solomon, S., Qin, D., Manning, M., Chen, Z., Marquis, M., Averyt, K.B., M.Tignor, Miller, H.L. (Eds.), Contribution of Working Group I to the Fourth Assessment Report of the Intergovernmental Panel on Climate Change. Cambridge University Press, Cambridge, NY, p. 996. www.ipcc.ch.

Karnaukhov, A.V., 2001. Role of the biosphere in the formation of the Earth's Climate: the greenhouse catastrophe. Biophysics 46 (6), 1078–1088.

Keeling C.D., Whorf T.P., the Carbon Dioxide Research Group, Scripps Institution of Oceanography (SIO). University of California, La Jolla, California, USA, 2001. August 13. (http://cdiac.ornl.gov/trends/co2/sio-mlo.htm).

Kondratiev, K.Ya., Krapivin, V.F., 2004. Carbon Global Cycle Modeling. FIZMATLIT, Moscow. 336.

Morgan, J.A., LeCain, D.R., Mosier, A.R., Milchunas, D.G., 2001. Elevated $CO_2$ enhances water relations and productivity and affects gas exchange in C3 and C4 grasses of the Colorado shortgrass steppe. Glob. Change Biol. 7, 451–466.

Pritchard, S.G., Davis, M.A., Mitchell, R.J., Prior, A.S., Boykin, D.L., Rogers, H.H., Runion, G.B., 2001. Root dynamics in an artificially constructed regenerating longleaf pine ecosystem are affected by atmospheric CO2 enrichment. Environ. Exp. Bot. 46, 35–69.

Roe, G., Baker, M., 2007. Why is climate sensitivity so unpredictable? Science 318, 629–632.

Semenov, S.M., 2004. Greenhouse Gases and Present Climate of the Earth. (in Russian, extended summary in English). Meteorology and Hydrololgy Publishing Centre, Moscow. 175.

Stainforth, D.A., Aina, T., Christensen, C., Collins, M., Faull, N., Frame, D.J., Kettleborough, J.A., Knight, S., Martin, A., Murphy, J.M., Piani, C., Sexton, D.L., Smith, A.R., Spicer, A.A., Thorpe, J., Allen, M.R., 2005. Uncertainty in predictions of the climate response to rising levels of greenhouse gases. Nature 433, 403–406.

Tarko, A.M., 2005. Anthropogenic changes of global biospheric processes. In: Mathematical Modeling. FIZMATLIT, Moscow, p. 232.

United Nations, 2004. World Population to 2300. Department of Economic and Social Affairs United Nations, New York. 240.

# 28

# Modeling and Evaluating the Global Energy Flow in Ecosystems and its Impacts on the Ecological Footprint

Safwat H. Shakir Hanna, Irvin W. Osborne-Lee

*TEXAS GULF COAST ENVIRONMENTAL DATA (TEXGED) CENTER, CHEMICAL ENGINEERING DEPARTMENT, PRAIRIE VIEW A&M UNIVERSITY, PRAIRIE VIEW, TX, 77446, USA*

## 28.1 Introduction

Energy is an essential element globally for living organisms. It also considered the dynamic of the global ecosystems for the production of goods and services (Shakir Hanna and Osborne-Lee, 2011). Considerable attention has been focused on the issue toward the sustainable energy for many reasons such as (1) increasing global demand on energy as the results of increasing human population (i.e., increasing the human ecological footprint on the Earth); (2) the scarcity of energy resources such as oil; (3) the costs of producing alternatives energy and fuel; and (4) further the need for faster research for production of new efficient energy with the suitable quantities, clean energy and economically feasible.

Materials may flow into or out of the ecosystem, be spatially transported, be stored or delayed, or be processed to change their biological, chemical, or technical attributes. The material flows are driven by energy, determined by an ecological potential, defined as an energy density by analogy of the ecosystem (Birket, 1994). This means that energy is recycled through the ecosystems to produce its function for recycling the materials.

Energy is the important key element for the ecosystems' pro ductivities for the production of materials, supplies, and transformation of these products in the ecosystems to meet the demands of human beings. Energy pathways occur due to the contribution of dead organic matter to detrital pools and those organisms that feed on them, reintroducing some of that energy back into the food web. Recognition of these cyclic energy pathways profoundly affects many aspects of ecology, such as trophic levels, control, and the importance of indirect effects (Fath and Halnes, 2007). Further, Hannon (1973) introduced a system for Material and energy flow analysis (MEFA) that included primary energy inputs such as solar energy as an intermediate part of the system, where

**Models of the Ecological Hierarchy. DOI: http://dx.doi.org/10.1016/B978-0-444-59396-2.00028-6**

ISSN 0167-8892, 

the total production of energy is calculated by the total energy, which is equal to the total consumption by ecosystem components plus the net energy loss by the system.

This chapter is a contemporary analysis of energy flow in global ecosystems and its impacts on human ecological footprint. The human ecological footprint is considered the most important factor in the energy flow in the ecosystems because of the unbalanced equilibrium between the ecological footprint growth (i.e., demands of human beings from the nature) against the productivity or regenerative biological capacity of nature or the ecosystems (Shakir and Osborne-Lee, 2011). Further, the chapter seeks to assess the dynamics of energy flow in the global Earth ecosystems for better understanding of how to conserve and save the lost energy by the use of modeling, to satisfy the needs of human beings now and in the future.

## 28.2 Modeling Energy Flow Theoretical Approach

In understanding the modeling of energy flow, we compose the model from several small compartments, each compartment itself composed of smaller parts, and finally compiling these parts together to form the bigger picture that describes the flow of energy flow. Therefore, the flow of energy starts at the source of energy, the Sun, to the Earth (Figs 28.1–28.3). The arrows in the diagrams explain the interactions and relationships

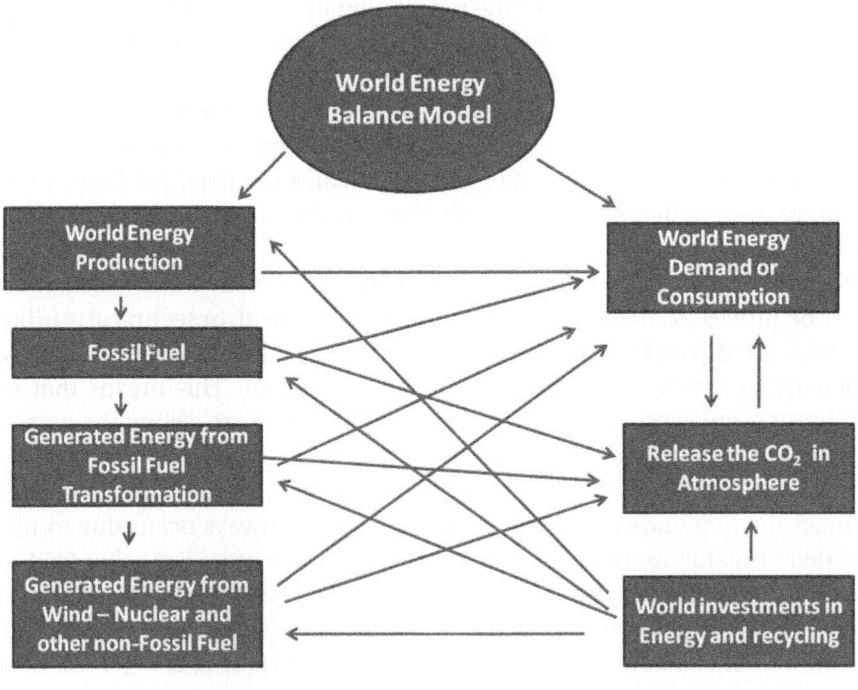


**FIGURE 28.1** World energy balance. For color version of this figure, the reader is referred to the online version of this book.

between the compartments. These figures show the world energy flow and transformation from the source of energy to be consumed and recycled in the globe, including the impact of energy use on the economy of the world and its consequences. In this respect, these figures show the world energy balance model between the world source of energy and energy consumption and demand; the energy relationship model that describes the interrelations between source of energy, energy forms, and energy transformation process; and the energy flows in the global ecosystems.

The energy flows from the Sun through green vegetation (producers) and then through the consumers. In this respect, the energy will flow from one trophic level to another. This process indicates that the energy has be.en stored from one level to another. However, the energy at the top of the food chain pyramid in the ecosystems will save only 0.1%. This means the bottom of the food chain stores most of the energy such as the plants and higher vegetative components of the ecosystems. In the present theoretical approach, we are studying the flow of energy from the Sun to primary producers and then to the consumer from different levels and finally to the level of human use. Therefore, the human use of products and services is a very important factor in determining the use of the energy availability.

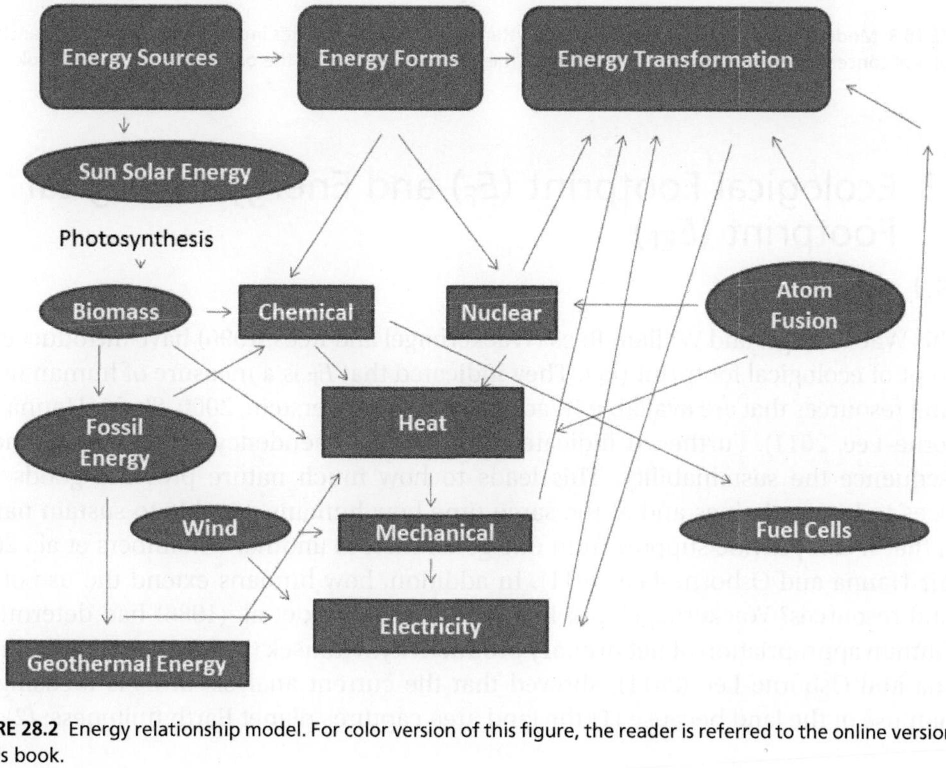


**FIGURE 28.2** Energy relationship model. For color version of this figure, the reader is referred to the online version of this book.

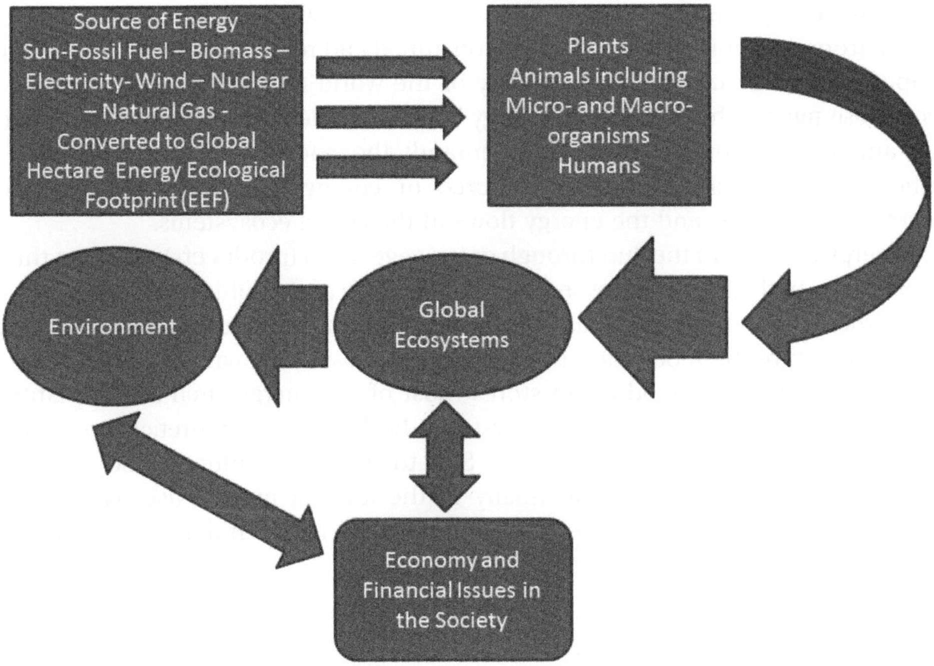


**FIGURE 28.3** Model of energy flow in the global ecosystems and environment, it's impact on society, and financial economical concerns. For color version of this figure, the reader is referred to the online version of this book.

# 28.3 Ecological Footprint ($E_F$) and Energy Ecological Footprint ($E_{EF}$)

## 28.3.1 Concept

Mathis Wackernagel and William Rees (Wackernagel and Rees, 1996) have introduced the concept of ecological footprint ($E_F$). They indicated that $E_F$ is a measure of human use of natural resources that are available (Wackernagel and Silverstein, 2000, Shakir Hanna and Osborne-Lee, 2011). Further, it indicates the human dependency on ecosphere and in consequence the sustainability. This leads to how much nature provides goods and services to human beings and at the same time how humans are able to sustain nature such that it can provide support from one generation to another (Chambers et al., 2000; Shakir Hanna and Osborne-Lee, 2011). In addition, how humans extend the use of the natural resources? Wackernagel and Rees (1996); Vitousek et al. (1986) had determined the human appropriation of net primary productivity. Vitousek (1986), Rees (2001); Shakir Hanna and Osborne-Lee (2011), showed that the current analysis of $E_F$ is focusing on human use of the land because (1) the land area captures planet Earth finiteness; (2) it is the function of the human and their interactions in the ecosystems at the global level of the ecosphere; (3) it supports photosynthesis, the energy conduit for the web of life, (Wackernagel and Rees, 1996); (4) the energy and its management as a part of $E_F$ has an economic values to the global ecosystems (Shakir Hanna and Osborne-, 2011); and (5) further, the ecological footprint is a function of all the parameters that interact between the power of ecosystem productivity, human interactions, and activities on a particular ecosystem or the demand from that ecosystem (Shakir Hanna and Osborne-Lee, 2011).

Ecological footprint accounting (Ewing et al., 2008) designed to represent actual human consumption of biological resources and the generation of wastes in terms of appropriated ecosystem area, which is compared to the biosphere's productive capacity in a given year. Since living renewable resources regenerate using solar energy, a population ecological footprint can be said to represent the area continuously required to generate a quantity of photosynthetic biomass energy and material equivalent to the amount used and dissipated by the population's consumption (Rees, 2003, 2006; Wackernagel and Galli, 2007). In other words, the accounting sheds the needed light on how much is being used and how much needs to be preserved that is essential to human well-being. Further, this method of accounting represents the human impact on the Earth in a clear manner (Moffatt, 2000).

The ecological footprint measures how much of nature's resources we have, how much of nature's resources we use, how we use it, and how the resources are distributed among the human population globally (Barrett et al., 2006). Additionally, Shakir Hanna and Osborne-Lee (2011) defined the ecological footprint ($E_F$) as a measure of the status of the global ecosystems' ability to provide service and goods to all human beings that are living on the Earth. Further, the global ecosystems should provide for the needs of all other organisms that exist on the planet and in consequence, which also play a part enabling nature to provide goods and services to human beings. Therefore, the global ecosystems provide for the needs of all living beings on the Earth, including the human beings (Shakir Hanna and Osborne-Lee, 2011).

In this way, the $E_F$ was born, as an indicator of the carrying capacity of regions, nations, and the globe, and sometimes extended as an indicator of sustainability (Chen et al., 2007; Penela and Villasante, 2008). Developed in the early 1990s, the $E_F$ was initially promoted as a planning tool (Wackernagel et al., 1999; Wackernagel and Silverstein, 2000) to estimate the magnitude of human consumption, which is currently know to exceed the biosphere's regenerative capacity (Wackernagel, 1999).

$E_F$ is defined as "the total area of productive land and water required on a continuous basis to produce the resources that the population consumes and to assimilate the wastes that the population produces, wherever on Earth, the relevant land and water are located" (Wackernagel and Rees, 1996; Rees, 2000). Its starting point is the assumption that both consumption of resources and waste generation can be converted into the biologically productive land that is necessary to maintain those consumption levels. According to

Wackernagel and Rees (1996, 1997), $E_F$ unsustainability is assessed by examining the available and the needed Earth surface areas, assuming that populations with a larger $E_F$ than their domestic land base are unsustainable (Lenzen et al., 2003). Nevertheless, along the time, the objectives present in the initial formulation of this indicator, which emphasized the efficiency of the $E_F$ as a sustainability indicator and planning tool (Wackernagel and Rees, 1996), have been focused and redirected toward more specific and may be fewer ambitious issues than the initial one. Probably influenced by the numerous criticisms received, the $E_F$ now stresses the accounting of natural capital and the documentation of the ecological overshoot (Wackernagel et al., 2004). Without changing the essence of the concept and the calculation method, the present formulation focuses on the study of how much the regenerative capacity of the biosphere is required to renew the resource throughout of a defined population in a given year with the prevailing technology and resource management of that year (Monfreda et al., 2004). Further, Wackernagel et al. (2004) indicated that the $E_F$ is a baseline measure of unsustainable overshoot.

### 28.3.2 Energy Ecological Footprint ($E_{EF}$) Importance

Energy ecological footprint (i.e., the demand for available energy in a specific space and time) is a part of the ecological footprint, which is a measure of sustainability. Consumed goods and activities are taken into account when calculating the footprint. Because energy is a part of produced goods and activities, any production of goods and services must on sustaining and conserving natural resources to the extent as to be preserved for the next generations.

Humankind depends on a variety of ecological goods and services, including energy, provided by nature. For sustainability, these must be available in increasing quantities from somewhere on the planet as population and mean per capita resource consumption increase (Overby, 1985). In general, energy represents a capacity that can be used in creating the goods and services that are needed for the global humanity. In this respect, the human carrying capacity in the world is considered as the maximum rates of resource harvesting and waste generation that can be sustained indefinitely without progressively impairing the productivity and functional integrity of relevant ecosystems wherever the latter may be located. The size of the corresponding population would be a function of technological sophistication and mean per capita material standards (Rees, 1995). Philosophically, energy is the most important part of the ecological footprint and as part of the generation and recycling the material and supplies in the global ecosystem, including the human race.

The energy ecological footprint as a component of the ecological footprint is very important because it is a major dynamic concept that allows for processing ecosystems at the global scale. Further, without energy, all the components of ecosystems of biotic and nonbiotic will cease to function and in consequence, it represents the dynamic force for the process of life.

## 28.4 Scientific Approach and Methodologies of Measuring Energy Ecological Footprint

We use the following definition of energy footprint: An energy footprint is a measure of land required to absorb the $CO_2$ emissions. This approach focuses on the outcome of energy use that is $CO_2$ emissions, to highlight the problem and pave the way for corrective action to be taken.

Data used in this chapter were collected from different data set of series available on the websites of World Research Institute (WRI)-Earth-Trends [(2000),(1960-2005)], World Bank ([1960-2008]), Food and Agricultural Organization (FAO) [(2008, 2008a)], United Nation Environmental Program (UNEP) [2009], United Nation Development Program (UNDP) [2010, 2010a], World Wildlife Fund (WWF) [(2002),(2004),(2006), and (Humphrey, 2008)] and Global Footprint Network [(Kitzes et al., 2008 and Kitzesa et al., 2009)]. The data were analyzed using the regression, correlation, and statistical methodologies using Sigma Plot Software (Version 8, Version 11), SAS, 2D software of SPSSSCIENCE [2002], and STELLA software.

The parameters that were synthesized from the data sets that were collected are (1) energy maintenance index (EMI), (2) energy global biological capacity ($E_{GBC}$) in billion global hectares, (3) energy global biological demand ($E_{GBD}$) in billion global hectares, (4) global biological capacity ($G_{BC}$) in billion global hectares, (5) global biological demand ($G_{BD}$) in billion global hectares, (6) croplands footprint ($C_F$) in billion global hectares, (7) grazing land footprint ($G_{LF}$) in billion global hectares, (8) forest ground footprint ($F_{GF}$) excluding fuel wood in billion global hectares, (9) fish ground footprint ($F_F$) in billion global hectares, (10) total energy footprint ($T_{EF}$) in billion global hectares, (11) built-up land ($B_L$) in billion global hectares, (12) global deficit capacity ($G_{DC}$) in billion global hectares, and (13) sensitivity index as % of intensity use of energy (SIIUE).

Energy maintenance index (EMI) or (energy sustainability index) is a percentage of the total energy global biological capacity ($E_{GBC}$) (i.e., total availability or supply of energy) in global hectares to the total energy global biological demand ($E_{GBD}$) (i.e., consumption or demand) in global hectares from the Earth. This index explains the ability of the Earth to regenerate energy from the prospective of natural resources availability. In other words, it explains the status of the Earth in providing energy for the production of goods and services to the human population needs. The ($E_{GBC}$) is defined as the ability of the Earth to produce renewable and natural resources energy and other sources of energy in term of global hectare per capita. $E_{GBD}$ is defined as the energy resources consumptions by human beings in terms of global hectare per capita. It is important to indicate the calculations of total $E_{GBC}$ and total $E_{GBD}$ are the most important factors in the calculations of indices that have been discussed in this chapter. Further, SIIUE is defined as the ratio of % of ratio of growth domestic product ($G_{DP}$)/total energy footprint ($T_{EF}$) produced to the ratio of growth domestic product ($G_{DP}$)/energy ecological footprint ($E_{EF}$) consumed (i.e., SIIUE). In this respect, the SIIUE covers the use and production of energy as a ratio of growth domestic product per capita, which explains the intensity of consumption of resources corresponding to the economic output. The formulas for calculating the following indices are as follows:

$$\text{EMI index} = E_{GBC}/E_{GBD} \times 100 \tag{1}$$

$$G_{BC} = C_F + G_{LF} + F_{GF} + F_F + T_{EF} + B_L \text{ (i.e., production)} \tag{2}$$

$$G_{BD} = C_F + G_{LF} + F_{GF} + F_F + E_{EF} + B_L \text{ (i.e., consumption)} \tag{3}$$

$$G_{DC} = G_{BC} - G_{BD} \tag{4}$$

$$E_{GBC} = G_{BC} \times \text{factor of energy production annually} \tag{5}$$

$$E_{GBD} = G_{BD} \times \text{factor of energy consumption annually} \tag{6}$$

$$\text{SIIUE} = \text{Ratio of } (\mathrm{G_{DP}/E_{EF}}) \text{ produced/ratio of } (\mathrm{G_{DP}/E_{EF}}) \text{ as } \% \tag{7}$$

These estimates are calculated according to several authors of ECOTEC-U.K group (2001) and Wackernagel et al. (2002) using the following: (1) net primary productivity (NPP); (2) growing crops for food, animal feed; (3) fiber and oil; (4) grazing animals; (5) woods from forest; (6) fishing from marine and freshwater; (7) infrastructure for housing and building transportation and electrical power; and (8) burning for fuel in $CO_2$ emissions. These calculations depended on the governmental, UN data series from the World Bank. These calculations were converted in billion hectares from the global-sized hectares. Further, the factor of energy production annually is a factor calculated according to the real energy produced that is converted to global hectare and estimated as a percentage of the total biological capacity of the Earth. Factor of energy consumption annually is a factor calculated according to the real energy consumed is converted to global hectare and estimated as a percentage of the total global footprint of the Earth. It was found that the energy produced and consumed is equal to 33–53% of the produced biological capacity of the Earth. This assumption is depending on that the percentage of energy produced or consumed are equal in percentage converted to global hectares. Additional consideration is that the energy flow and recycled in the ecological system is equal in percentage due to the first law of thermodynamics.

Predictions of total $E_{GBC}$, total $E_{GBD}$, and SIIUE for the Earth in the next decades were calculated to show the impact of the growth of the world population based on annual data from a series of years 1960–2008 for almost 48 years of published data. The predictions for the years 2009–2050 were analyzed using correlation and regression line models (Tables 28.1 and 28.2).

**Table 28.1** Global Population, Energy Global Biological Capacity ($E_{GBC}$), Energy Global Biological Demand ($E_{GBD}$), and Maintenance Index of the Energy from Our Planet Earth from Year 1961 to Year 2008 – Data are in 10 Years Intervals*

| Year | Global Population (billion) | Surface Area (billion ha) | Total Global Biological Demand ($G_{BD}$) (billion global ha) | Global Biological Demand (global ha/surface areas global ha as %) | Total Energy Global Biological Demand ($E_{GBD}$) (billion global ha) | Total Global Biological Capacity ($G_{BC}$) (billion global ha) | Total Energy Global Biological Capacity ($E_{GBC}$) (billion global ha) | Deficit in Energy (global billion ha) $= E_{GBC} - E_{GBD}$ | Maintenance Index $= E_{GBC}/E_{GBD}$ | Sensitivity Index of Intensity Use of Energy (SIIUE) as % | Energy Produced from the Sun According to Krenz (1976) (billon cal/capita) |
|---|---|---|---|---|---|---|---|---|---|---|---|
| 1961 | 3.08 | 13.39 | 7.47 | 55.80 | 2.51 | 10.90 | 3.66 | 1.15 | 1.46 | 72.10 | 8.80 |
| 1970 | 3.70 | 13.39 | 9.50 | 70.96 | 3.84 | 11.00 | 4.45 | 0.61 | 1.15 | 84.11 | 7.07 |
| 1980 | 4.44 | 13.39 | 11.25 | 84.03 | 5.08 | 11.13 | 5.03 | −0.05 | 1.01 | 98.44 | 5.89 |
| 1990 | 5.27 | 13.39 | 12.93 | 96.58 | 6.08 | 11.38 | 5.35 | −0.73 | 0.88 | 114.52 | 4.96 |
| 2000 | 6.10 | 13.39 | 15.1 | 112.73 | 7.30 | 12.00 | 5.80 | −1.50 | 0.79 | 130.60 | 4.29 |
| 2008 | 6.69 | 13.41 | 18.8 | 140.20 | 8.80 | 12.00 | 5.92 | −2.88 | 0.67 | 142.03 | 3.91 |

*Data Sources are World Bank, FAO, WWF, Ecological Footprint Network, WRI-Earth Trends—US Estimates.

**Table 28.2** Predicted Values Calculated for Global Population, Energy Global Biological Capacity ($E_{GBC}$), Energy Global Biological Demand ($E_{GBD}$), and Maintenance Index of the Energy from Our Planet Earth from Year 2009 to Year 2050. The Presented Data Are in 10 Years intervals on the Bases of Current Trend of Estimate of Population Growth Rate

| Year | Global Population (billion) | Surface Area (billion ha) | Total Global Biological Demand ($G_{BD}$) (billion global ha) | Global Biological Demand (global ha/surface areas global hectare as %) | Total Energy Global Biological Demand ($E_{GBD}$) (billion global ha) | Total Global Biological Capacity ($G_{BC}$) (billion global ha) | Total Energy Global Biological Capacity ($E_{GBC}$) (billion global ha) | Deficit in Energy (global billion ha) $= E_{GBC} - E_{GBD}$ | Maintenance Index $= E_{GBC}/E_{GBD}$ | Sensitivity Index of Intensity Use of Energy (SIIUE) as % | Energy Produced from the Sun According to Krenz (1976) (billon cal/capita) |
|---|---|---|---|---|---|---|---|---|---|---|---|
| 2009 | 6.74 | 13.41 | 16.98 | 126.61 | 8.42 | 11.94 | 5.92 | −2.50 | 0.70 | 143.02 | 3.89 |
| 2010 | 6.79 | 13.41 | 17.11 | 127.60 | 8.50 | 11.95 | 5.94 | −2.56 | 0.69 | 144.03 | 3.86 |
| 2020 | 7.34 | 13.41 | 18.50 | 137.94 | 9.34 | 12.13 | 6.12 | −3.22 | 0.66 | 154.52 | 3.57 |
| 2030 | 7.92 | 13.41 | 19.99 | 149.11 | 10.25 | 12.31 | 6.31 | −3.94 | 0.62 | 165.85 | 3.31 |
| 2040 | 8.55 | 13.41 | 21.61 | 161.16 | 11.22 | 12.52 | 6.50 | −4.72 | 0.58 | 178.08 | 3.06 |
| 2050 | 9.23 | 13.41 | 23.36 | 174.18 | 12.28 | 12.73 | 6.70 | −5.58 | 0.55 | 191.28 | 2.84 |

Accordingly, Krenz (1976) had estimated that reflectivity and emissivity constants are averaged to the values that depend on the cloud coverage and atmospheric composition at 227 W/m$^2$ of the Earth. From the conversions of the Earth watts of energy per square meter to calories, the Earth will receive and emit in the range of 26.1 28.0 million trillion calories annually. The availability of energy in calories for each person on this Earth is calculated by the author to be 8.49 billion calories/capita in the year 1961 and it reached to 3.91 billion calories in the year 2008. This is an indication of lower the per capita share in the global calories that can be produced by the Sun sources of energy as the main source of energy in this Earth as the result of the increasing human population in an alarming rate. However, this type of energy is not usable by human beings. In this respect, the most important part of this energy is the energy absorbed and converted to fossil fuel or converted to useful available energy to flow in the ecosystems or in the ecosphere and recycled between its components and not the lost heat energy. It is estimated that the energy used to form the net primary production (NPP), which is the main source of energy transformed, is about less than 1% of the Sun energy.

## 28.5 Analysis of Global Energy Ecological Footprint and Impacts of Human Population on the Energy Resources

The total size of the energy footprint in 2008 was 8.80 billion hectares. The Earth has about 12.00 billion hectares of biologically productive space (this includes the actually used cropland as well as the additional potential cropland), and the total global ecological footprint in 2008 was 18.80 billion hectares. So energy use comprises about half of the world ecological footprint on the Earth (i.e., 46.6%). The energy footprint increased from 2.5 billion hectares in 1961 to 8.8 billion hectares in 2008. However, it is predicted that the energy footprint in 2050 will reach 12.28 billion hectares. However, this predicted energy demands from Earth have a deficit in energy and it is about 45%. Figure 28.4 shows the model of energy calculations for the production and consumption, including energy ecological footprint ($E_{EF}$). This has been shown in (Tables 28.1 and 28.2) and (Figs 28.5 and 28.6). In Fig. 28.5, the energy footprint produced showed increased energy production with increasing global human population. However, Fig. 28.6 showed a sharp deficit of energy in global production of energy per capita against the increasing global population.

On the other hand, in Fig. 28.7, when the global population increases, there is a decrease in the total energy footprint in global hectares (i.e., consumption of total energy). Further, there is an increase in per capita consumption of the energy footprint as shown in (Fig. 28.8). As human population and the economic development continue to grow, additional energy resources are necessary to foster the continuation of life and production to support the growing population. Accordingly, the energy resources such as petroleum, coal, and natural gas will be needed to tap to, and other means of energy production, including hydroelectric, geothermal, solar, wind, and nuclear resources, will

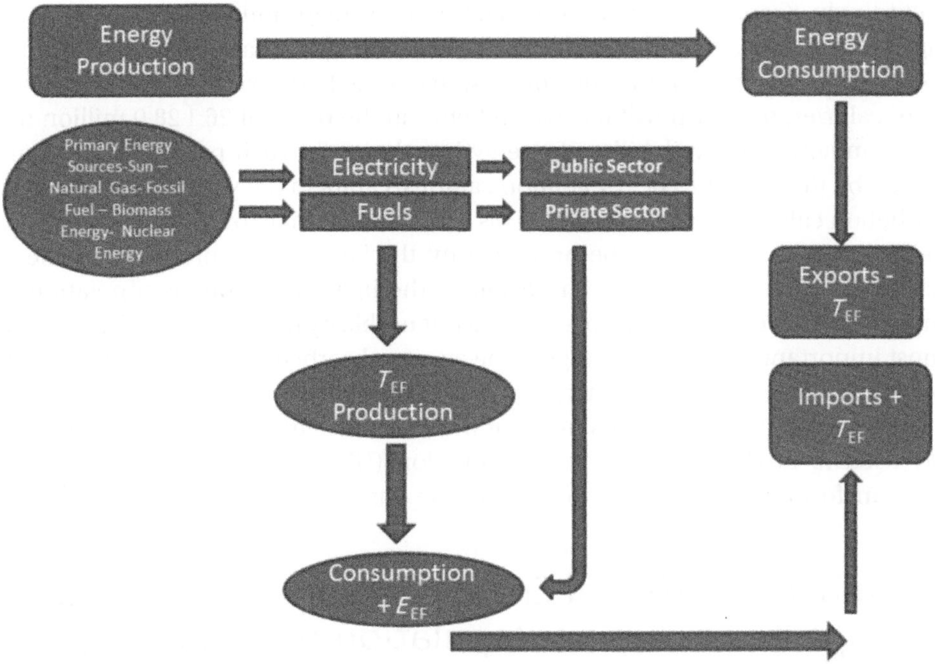


**FIGURE 28.4** Model for energy production and energy ecological footprint ($E_{EF}$). For color version of this figure, the reader is referred to the online version of this book.

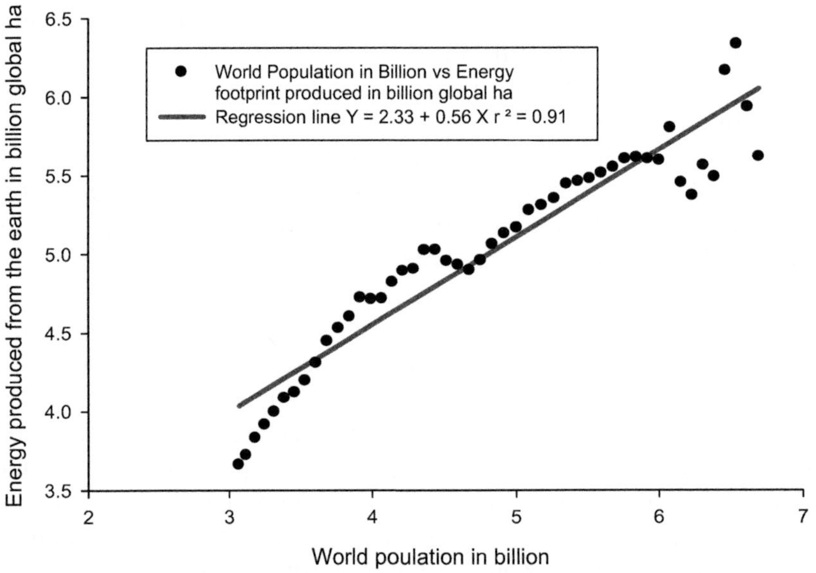


**FIGURE 28.5** Relationship between world population in billion and total energy footprint ($T_{EF}$) produced from Earth in billion global hectares. For color version of this figure, the reader is referred to the online version of this book.

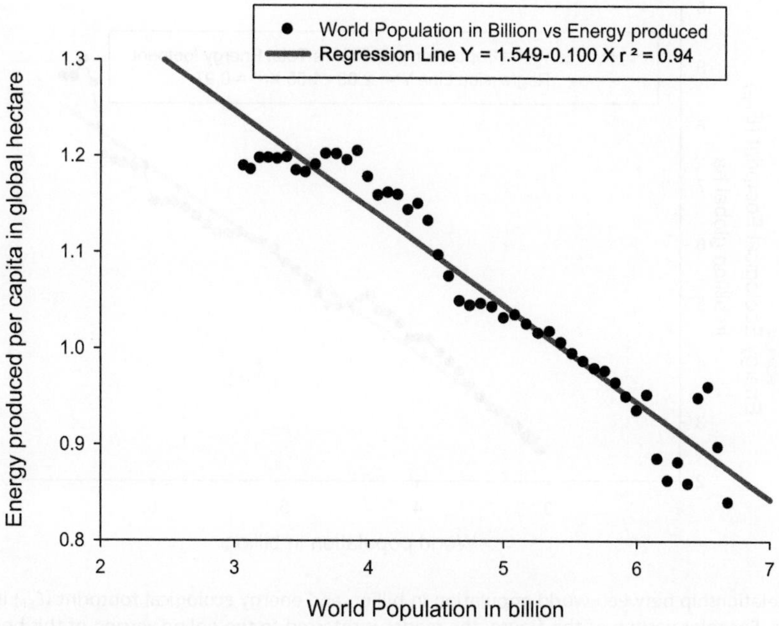


**FIGURE 28.6** Relationship between world population and total energy produced (global ha/capita). For color version of this figure, the reader is referred to the online version of this book.

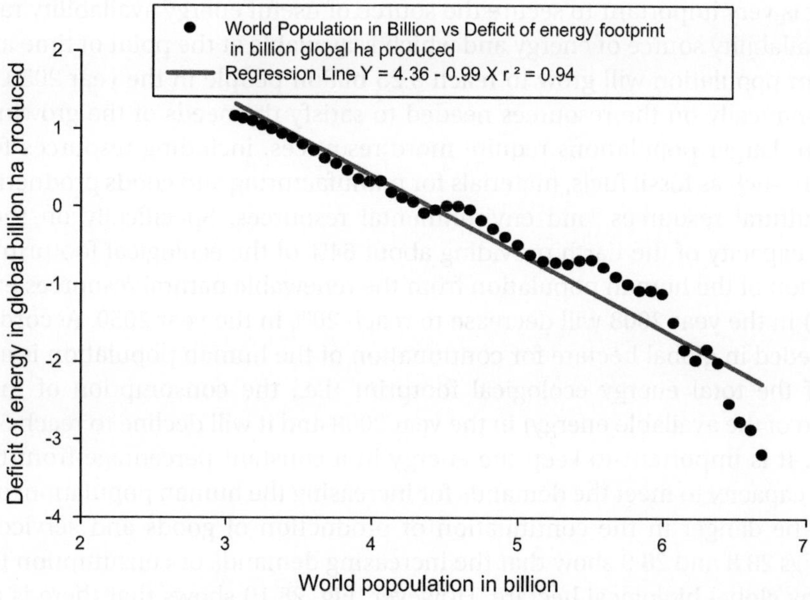


**FIGURE 28.7** Relationship between world population in billion and the deficit of energy footprint produced in global billion hectares. For color version of this figure, the reader is referred to the online version of this book.

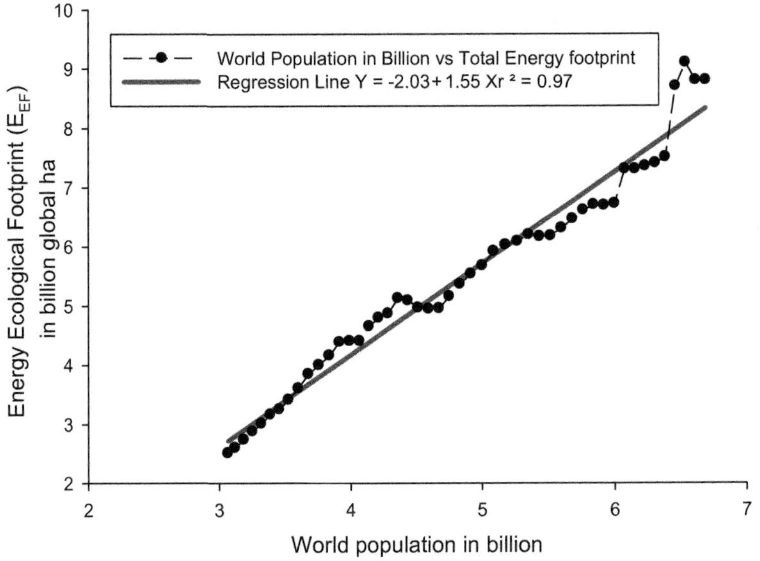


**FIGURE 28.8** Relationship between world population in billion and energy ecological footprint ($E_{EF}$) in billion global hectares. For color version of this figure, the reader is referred to the online version of this book.

need to be considered and implemented. Though the major energy source comes from the Sun, it is very important to secure the source of useful energy availability rather than the nonavailability source of energy and which is available at the point of time and space. The human population will grow to reach 9.23 billion people in the year 2050. This will affect dramatically on the resources needed to satisfy the needs of the growing human population. Larger populations require more resources, including resources for energy production, such as fossil fuels, materials for manufacturing and goods production, water and agricultural resources, and environmental resources. Specifically on the hectare biological capacity of the Earth providing about 64% of the ecological footprint (i.e., the consumption of the human population from the renewable natural resources existing on this Earth) in the year 2008 will decrease to reach 20% in the year 2050. Accordingly, the energy needed in global hectare for continuation of the human population is accounted to 67% of the total energy ecological footprint (i.e., the consumption of the human population of the available energy) in the year 2008 and it will decline to reach 55% in the year 2050. It is important to keep the energy in a constant percentage from the global biological capacity to meet the demands for increasing the human population; otherwise, there will be danger in the continuation of production of goods and services. In this respect, Figs 28.8 and 28.9 show that the increasing demands or consumption for energy as noted by global biological hectare. However, Fig. 28.10 shows that there is a positive relationship between energy consumption in calories produced per capita from the Earth biocapacity in billion global hectares. Additionally, Fig. 28.11 shows the relationship

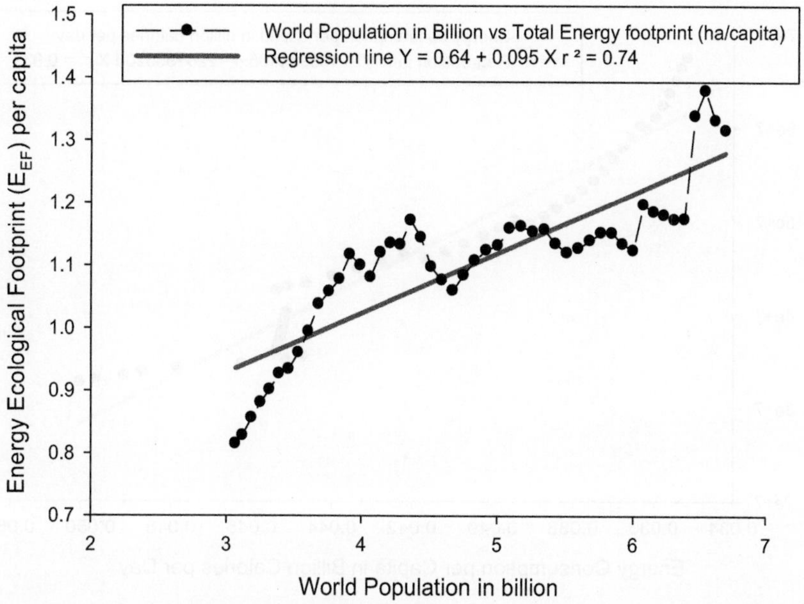


**FIGURE 28.9** Relationship between world population and energy ecological footprint ($E_{EF}$) in (ha/capita). For color version of this figure, the reader is referred to the online version of this book.

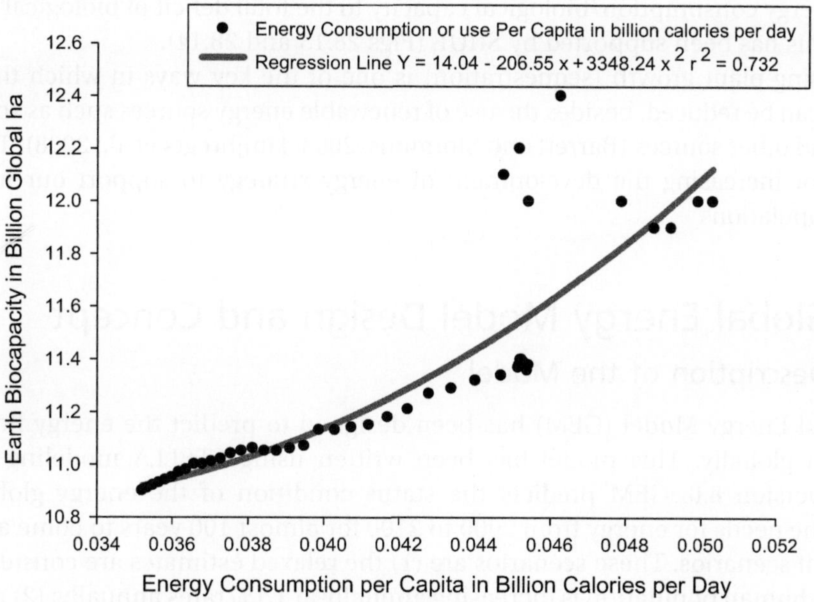


**FIGURE 28.10** Relationship between energy consumption per capita and bio-capacity of the Earth. For color version of this figure, the reader is referred to the online version of this book.

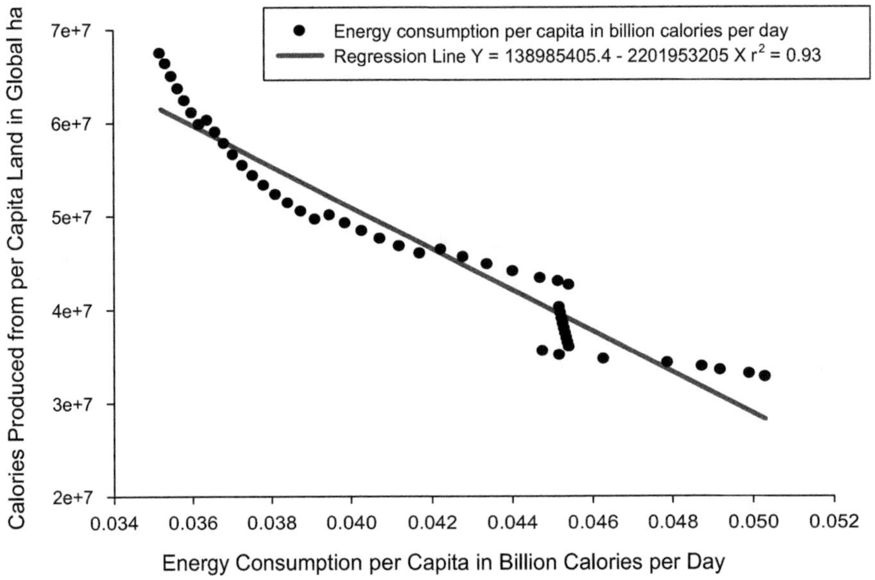


**FIGURE 28.11** Relationship between energy consumption in calories. Per capita and calories produced from global hectares per capita. For color version of this figure, the reader is referred to the online version of this book.

between energy consumption in calories per capita and calories produced from global hectare per capita. Accordingly, Fig. 28.12 indicates that the relationship between the ratio of energy consumption/biological capacity to the total deficit of biological capacity. Further, this has been supported by SIIUE (Figs 28.13 and 28.14).

Increasing plant growth (sequestration) is one of the key ways in which the energy footprint can be reduced, besides the use of renewable energy sources such as solar, wind energy, and other sources (Barrett and Simmons, 2003; Huijbregts et al., 2008). This is the concept for increasing the development of energy strategy to support our increasing human populations

# 28.6 Global Energy Model Design and Concept

## 28.6.1 Description of the Model

The Global Energy Model (GEM) has been designed to predict the energy flow in the ecosystem globally. This model has been written using STELLA modeling software package version 8.0. GEM predicts the status condition of the energy globally and predicts the needs for energy from 2000 to 2100 for almost 100 years to come according to different scenarios. These scenarios are (1) the relaxed estimates are considering the global net human population is increasing annually at 1.1% rates annually; (2) moderate estimates are considering the global net human population is increasing at 0.9% increase annually; (3) the optimum estimates are considering the global net human

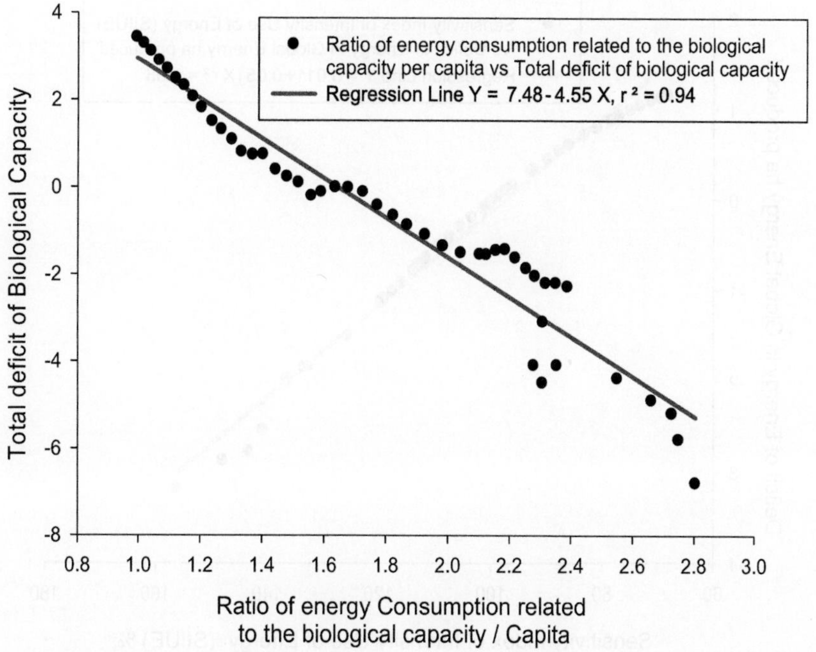


**FIGURE 28.12** Relationship between the ratio of energy consumption/biological capacity to total deficit of biological capacity. For color version of this figure, the reader is referred to the online version of this book.

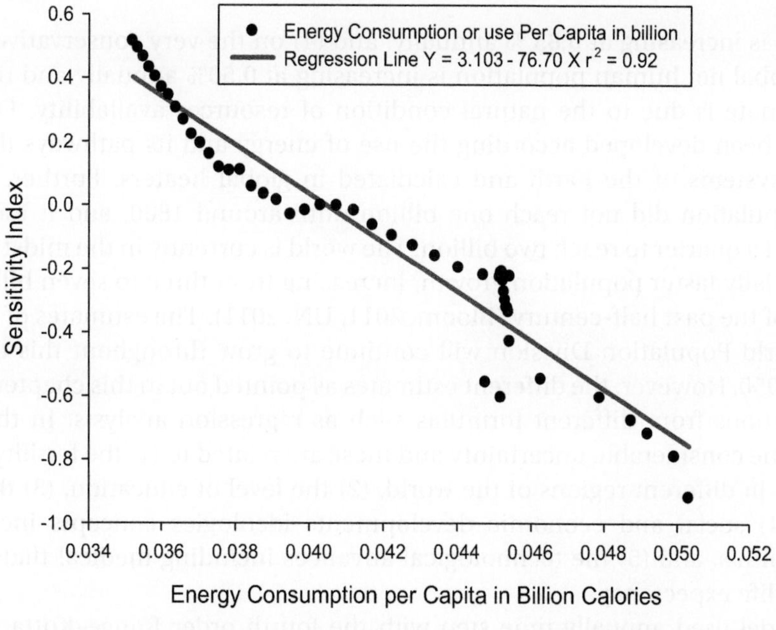


**FIGURE 28.13** Relationship between energy consumption per capita and sensitivity index. For color version of this figure, the reader is referred to the online version of this book.

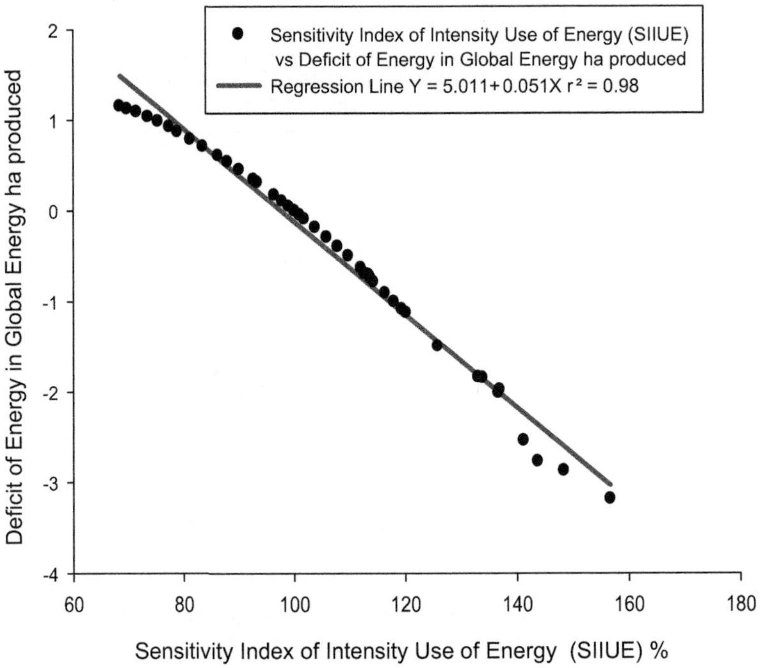


**FIGURE 28.14** Relationship between sensitivity index of intensity use of energy and deficit in energy produced in global hectare. For color version of this figure, the reader is referred to the online version of this book.

population is increasing at 0.85 % annually; and (4) on the very conservative estimates that the global net human population is increasing at 0.50% annually and this conservative estimate is due to the natural condition of resources availability. Further, the model has been developed according the use of energy and its pathways through the global ecosystems of the Earth and calculated in global heaters. Further, the global human population did not reach one billion until around 1800, and it took another century and a quarter to reach two billion. The world is currently in the midst of a period of substantially faster population growth, increasing from three to seven billion within the space of the past half-century (Bloom, 2011; UN, 2011). The estimates of the United Nation World Population Division will continue to grow throughout this century 9.3 billion in 2050. However, the different estimates as pointed out in this chapter are due to the calculations from different formulas such as regression analysis. In this respect, there is some considerable uncertainty and these are related to (1) the fertility of human population in different regions of the world, (2) the level of education, (3) the political systems, (4) social and economic developments ideologies concepts including the people wealth's, and (5) the technological advances including medical that allows the increasing life expectancy.

The model used annually time step with the fourth order Runge–Kutta integration method, STELLA Modeling System (Meadows, 2008). The simulation period could be from one year to several years and could be used for a short-time period of stimulation. Background data and literature parameters were used to initialize the model and a short-term data collected from different sources and data sets of series available on the websites of World Research Institute (WRI)-Earth-Trends, World Bank, Food and Agricultural Organization (FAO), United States Department of Agriculture (USDA), United Nation Development Program, World Wildlife Fund (WWF), and Global Footprint Network. Table 28.3 shows the list of variables and parameters in the Model and its interpretation.

**Table 28.3** List of Variables and Parameters That Are Used in the Global Energy Model (GEM) and Its Interpretation

| Variables | Interpretation | Unit Used in the Model |
|---|---|---|
| $W_{PS}$ | World population series | In billion individuals |
| $E_F$ | Total ecological footprint | In billion global hectares |
| EMI | Energy maintenance index | % |
| $E_{GBC}$ | Energy global biological capacity | In billion global hectares in the total energy |
| $E_{GBD}$ | Energy global biological demand | In billion global hectares of the total energy consumed from the Earth |
| $G_{BC}$ | Global biological capacity of the Earth | In billion global hectares that generate the biological capacity of the Earth |
| $G_{BD}$ | Global biological demand from the Earth | In billion global hectares that consumed from biological capacity by human beings from the Earth |
| $C_F$ | Cropland footprint | In billion global hectares of cropped lands |
| $G_{LF}$ | Grazing land footprint | In billion global hectares of grazed lands |
| $F_{GF}$ | Forest ground footprint | In billion global hectares of forest lands excluding fuel wood |
| $F_F$ | Fish ground footprint | In billion global hectares of fishing farming |
| $T_{EF}$ | Total energy footprint (produced) | In billion global hectares generated |
| $E_{EF}$ | Energy ecological footprint (consumed) | |
| $B_L$ | Built-up land | In billion global hectares |
| $G_{DC}$ | Global deficit capacity | In billion hectares |
| SIIUE | Sensitivity Index as % of intensity use of energy. This is % of ratio of growth domestic product ($G_{DP}$)/total energy footprint ($T_{EF}$) produced to ratio of growth domestic product ($G_{DP}$)/energy ecological footprint ($E_{EF}$) consumed | % |

The model used parameters such as world population series ($W_{PS}$) from year 1961 to 2009, ecological footprints, energy maintenance index (EMI) or (energy sustainability index) is a percentage of the total energy global biological capacity ($E_{GBC}$) (i.e., total availability or supply of energy) in global hectares to the total energy global biological demand ($E_{GBD}$) (i.e., consumption or demand) in global hectares from the Earth. Further, other terms such as the $G_{BC}$ which is the global biological capacity and $G_{BD}$ which is global biological demand, cropland footprint ($C_F$) in billion global hectares, grazing land footprint ($G_{LF}$) in billion global hectares, forest ground footprint ($F_{GF}$) excluding fuel wood in billion global hectares, fish ground footprint ($F_F$) in billion global hectares, Total energy footprint ($T_{EF}$) in billion global hectares, Built-up land ($B_L$) in billion global hectares and global deficit capacity ($G_{DC}$) in billion hectares, are used in calculations. Further, (SIIUE) is the sensitivity Index as % of Intensity Use of Energy and is defined as a ratio of % of ratio of growth domestic product ($G_{DP}$)/energy ecological footprint ($E_{EF}$) produced to ratio of growth domestic product ($G_{DP}$)/energy ecological footprint $E_{EF}$ consumed (i.e., sensitivity Index of the intensity of use of energy %).

## 28.6.2 Model Formulas Input in the Model

1.Energy footprint produced from Earth in billion global hectares = $2.33 + 0.56 \times W_{PS,}$- where $W_{PS}$ is the world population series.
2.Energy footprint produced per capita in global hectare = $1.549 - 0.100 \times W_{PS}$.
3.Deficit of the energy footprint produced from the Earth in billion global hectares = $4.36 - 0.99 \times W_{PS}$.
4.Total energy footprint (energy consumed) in billion global hectares = $-2.03 + 1.55 \times W_{PS}$.
5.Energy consumption per capita in global hectare = $0.64 + 0.095 \times W_{PS}$.
6.Global biological capacity ($G_{BC}$) of the Earth in billion global hectares as calculated in Wackernagel et al. (1999) and Shakir Hanna and Osborne-Lee (2011).
7.Global biological demand ($G_{BD}$) from the Earth in billion global hectares as calculated calculated in Wackernagel et al. (1999) and Shakir Hanna and Osborne-Lee (2011).
8.The energy production received from the Sun according to the assumption of Krenz (1976) is 227 $W/m^2$.

The above formulas were used because the code in STELLA software is very limited and consists mainly of establishing mathematical relationships between the various components. It is important to provide the STELLA software with the critical formulas that are needed to construct the model and to predict the future aspects of the problem discussed. Additionally, these are the rules of a system diagram in the STELLA software. In addition, the mathematical relationships in the form of equations and the factor parameters in the model, that one can input graphical relationships between variables, should be provided, (STELLA, 2001; Ouyang, 2008).

## 28.6.3 Simulation and Analysis

The simulation analysis output from the presented GEM model showed that the main interaction in the model is between energy availability at any point in time and estimates of average global per capita consumption. According to the model, the human population is expected to reach 9.3 billion people by 2050 on the Earth and might reach between 11.0 and 20.1 billion people in year 2100 at the 0.5% human population growth rate (i.e., the optimum estimates) and at 1.1% human population growth rate (i.e., the relaxed assumption estimates). However, the recent estimate of human population growth rate as indicated by Bloom (2011) is between 0.2% (i.e., conservative estimate) and 1.1% annually (i.e., relaxed estimates). Further, the current global consumption of energy is about 9.67 global hectares per person in year 2011 and will reach to 28.94 global hectares per person in year 2100. However, according to the very conservative assumption of human population estimates (0.2%), in the model, the decline in consumption evenly will reach to 15.01 global hectares per person by 2100. Additionally, the GEM model showed and predicted the deficit of energy per capita, the deficit of energy globally, the sensitivity index for maintaining the Earth energy, and sensitivity index of the intensity of energy use. The model outputs are shown in Figs 28.15–28.18. It is important to note from these figures, we can predict on the very relaxed estimates, that the energy global biological capacity will be 0.7 global hectares per capita and the energy maintenance index in year 2100 is 46.65%. This indicates a deficit in energy that is needed to be −15.5 global billion hectares by all human beings on this Earth in year 2100, and in spite of the increase in energy in the reservoir (i.e., the energy reservoir supplies globally), there is a deficit in energy supplies. On the moderate assumption (i.e., 0.85%), we can conclude that the human population will increase to reach about 15.65 billion people globally in 2100. This will lead the energy global biological capacity to reach 0.71 global hectares per capita, and the energy maintenance index to 49.91% in the year 2100. This also will indicate a deficit in the energy that is needed to be −11.1 global billion hectares by all human beings. On the optimum assumptions (i.e., 0.5% population growth rate), we can conclude that the human population will increase to reach in the year 2100 about 11.1 billion people globally. In consequence, the energy global biological capacity will be 0.77 global hectares per capita; the energy maintenance index in year 2100 is 56.46%. This indicates a deficit in energy needed in −6.65 global billion hectares by all human beings living on this Earth. Finally, on the very conservative estimates (i.e., 0.2%), we can conclude that the human population will increase to reach about 8.17 billion people globally in 2100. In consequence, the energy global biological capacity will be 0.85 global hectares per capita; the energy maintenance index in year 2100 is 64.93 %. This indicates a deficit in energy needed in −3.73 global billion hectares by all human beings living on this Earth.

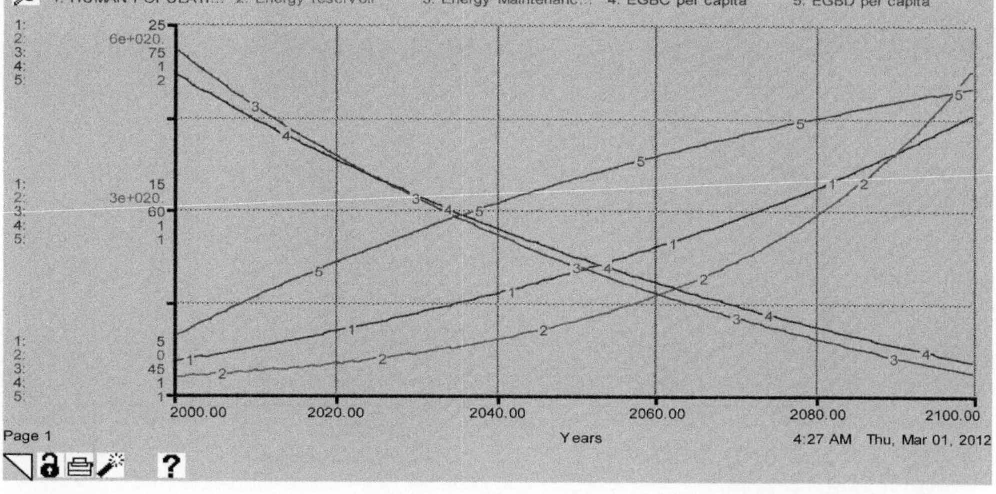


**FIGURE 28.15** The output of simulation of global energy model (GEM) including human population, energy reservoir (ER), energy maintenance index (EMI), energy global biological capacity ($E_{GBC}$) per capita, and energy global biological demands ($E_{GBD}$) per capita on the bases of relaxed assumptions. Symbols: (1) Human population in blue; (2) energy reservoir in red; (3) energy maintenance index in green; (4) energy global biological capacity ($E_{GBC}$) per capita in yellow; and (5) energy global biological demand ($E_{GBD}$) per capita in pale red. For interpretation of the references to color in this figure legend, the reader is referred to the online version of this book.

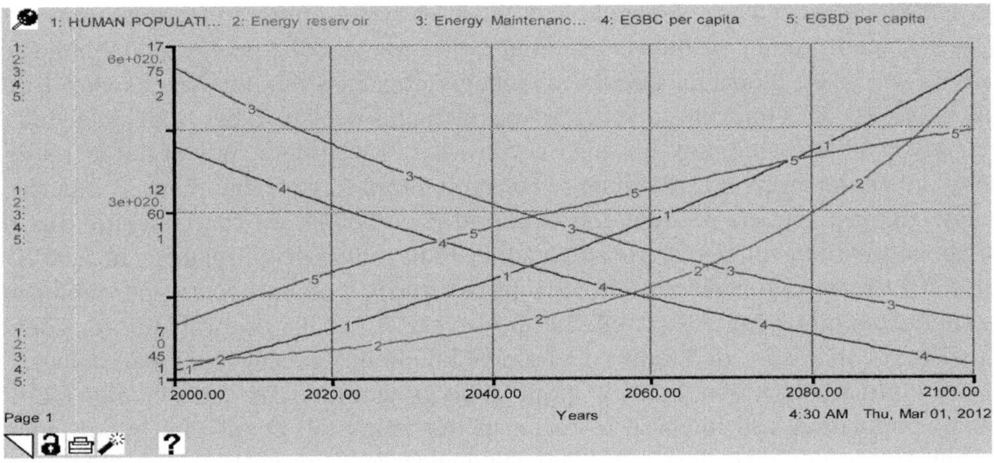


**FIGURE 28.16** The output of simulation of global energy model (GEM) including human population, energy reservoir (ER), energy maintenance index (EMI), energy global biological capacity ($E_{GBC}$) per capita, and energy global biological demands ($E_{GBD}$) per capita on the bases of moderate assumptions. Symbols: (1) Human population in blue; (2) energy reservoir in red; (3) energy maintenance index in green; (4) energy global biological capacity ($E_{GBC}$) per capita in yellow; and (5) energy global biological demand ($E_{GBD}$) per capita in pale red. For interpretation of the references to color in this figure legend, the reader is referred to the online version of this book.

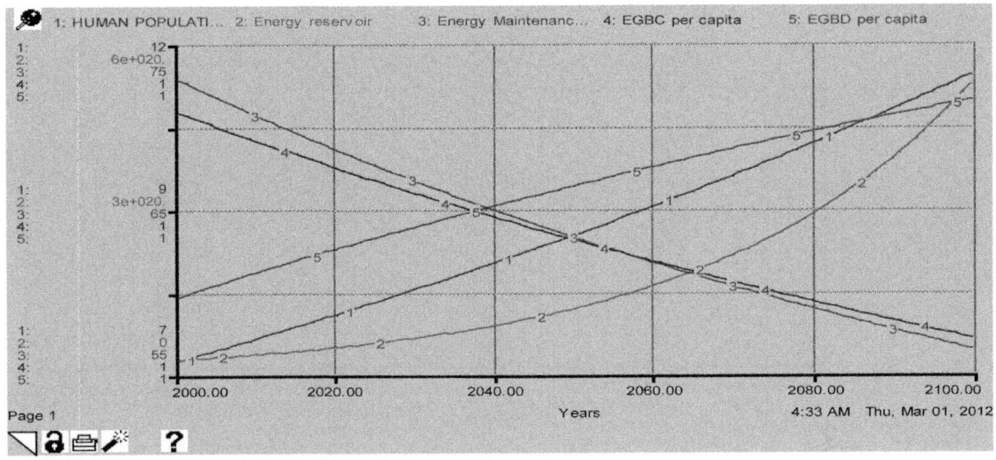


**FIGURE 28.17** The output of simulation of global energy model (GEM) including human population, energy reservoir (ER), energy maintenance index (EMI), energy global biological capacity ($E_{GBC}$) per capita, and energy global biological demands per capita on the bases of optimum assumption. Symbols: (1) Human population in blue; (2) energy reservoir in red; (3) energy maintenance index in green; (4) energy global biological capacity ($E_{GBC}$) per capita in yellow; and (5) energy global biological demand ($E_{GBD}$) per capita in pale red. For interpretation of the references to color in this figure legend, the reader is referred to the online version of this book.

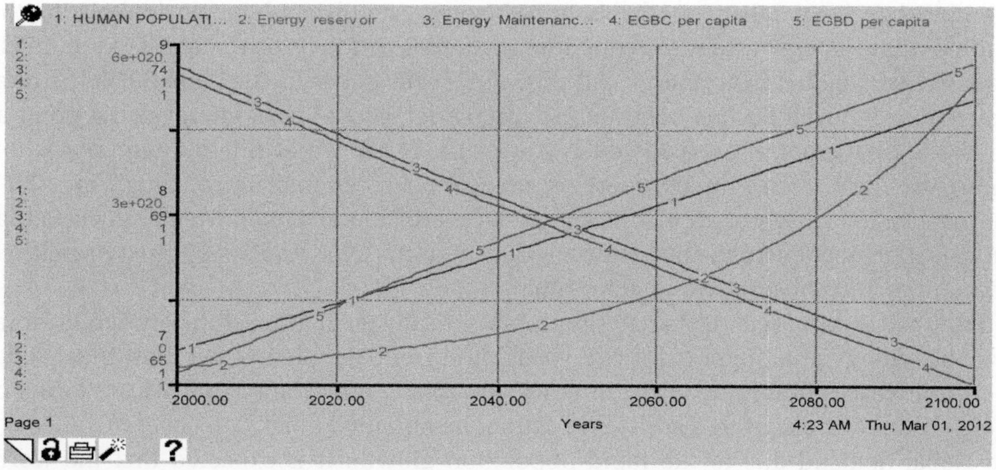


**FIGURE 28.18** The output of simulation of global energy model (GEM) including human population, energy reservoir (ER), energy maintenance index (EMI), energy global biological capacity ($E_{GBC}$) per capita, and energy global biological demands ($E_{GBD}$) per capita on the bases of very conservative assumptions. For color version of this figure, the reader is referred to the online version of this book.

From Figs 28.15–28.18, we can extract that energy global bological capacity ($E_{GBC}$) and energy global biological demand ($E_{GBD}$) will be in equilibrium in the years between 2035 and 2040. In addition, the EMI will be in equilibrium with increasing population between the years 2040 and 2060. This means that energy balance between the $E_{GBC}$ and $E_{GBD}$ will be in the equivalence between years 2040 and 2060 on the assumption of lower and upper estimates of global increasing population annually, and on the availability of efficient use of energy from the energy reservoir.

## 28.7 Discussion

As the biological capacity of the world is declining relative to the speed of growing population, this will provide lowering the per capita availability of food, energy supplies, croplands, and the supply for drinking water, in addition to the loss of world biodiversity from plants and animals due to human activities. Increasing the consumption of energy is associated with advanced economies, as well as with longer life expectancies, higher levels of education, and other indicators of social development. However, high-energy consumption can also fail to produce economic growth if it is misdirected, but there is no substantial economic growth without corresponding growth in energy consumption (UNFPA, 2001). Therefore, the energy is the most vital source of development.

The energy ecological footprint ($E_{FF}$) as one component of the ecological footprint ($E_F$) is defined as the areas required for sequestering the amount of $CO_2$ emissions from burning fossil fuels, for buffering the radiation from nuclear power, and for building dams to generate hydrological electricity (Wackernagel et al., 1999) (Fig. 28.4). Considerable attention has been taken on the issue of energy supply and in particularly to the nonrenewable energy which is constituent of 86% (NESS, 2004). Based on publicly available data, global peak energy will probably occur between 2025 and 2030; the total available energy will decline continuously thereafter. Global peak energy is the point at which the total amount of usable energy available to the worldwide human population from currently known primary energy sources reaches its maximum (Clugston, 2007). This conclusion closely supports the suggested model in this paper, which indicates that the global peak of energy will occur between the years 2030 and 2050 based on relaxed assumption and conservative assumption.

From the focus of scientific analysis, and specifically the relationship between the total available energy and support for the levels of current population, we find that global energy decreases due to decrease in nonrenewable energy such as oil. These forms of energy cannot be discovered easily and cannot be obtained quickly to cover the needs of the global demand due to the rapid population growth. It is expected that the world human population on the conservative assumption (at 0.5% growth rate) as indicated in the suggested model will reach 11.2 billion people by the year 2100, and this leads to energy maintenance in the global ecosystem by 56% only, and this means the shortage of energy will lead by 44%. However, on the current trend of population growth (i.e., 1.0% population growth rate), the model suggested that the global human population will reach about 18.2 billion people in year 2100, and this will lead to energy maintenance of global ecosystem by 47% only. This means a shortage of energy will lead by 53%. Decreasing the energy supply to human population will lead to the following :

1. Reduction of their abilities to produce the goods and services that the ecosystem can supply.
2. Reduction of the human beings' ability to move and support the transportation of goods and services.
3. Reduction of ability of the human population to cultivate land and to produce crops that are needed for food.
4. Reduction of the goods production in the factories.
5. Reduction of the productivity in service sectors.
6. Finally, the impact will be enormous in every aspects of life.

From the simulation model that is presented, the global peak of energy use will happen between the years 2030 and 2040, and then the available energy will be declining continuously until we will be able to find and be able to transfer the energy to useful energy or significantly break through the quantity and quality energy that be efficient and significant for use by all human populations. Further demographic projections (Bloom, 2011) suggest that the global human population growth will be 0.2% on a very conservative assumption and will stabilize at about 9–10 billion people by mid-century. These projections rest on two fundamental assumptions. The first is that the energy needed to fuel development, and the associated decline in fertility will keep pace with energy demand far into the future. The second is that the demographic transition is irreversible such that once countries start the path to lower fertility, they cannot reverse to higher fertility. Both of these assumptions are problematic and may have an effect on population projections.

DeLong et.al. (2010) examined the above assumptions, and they showed that by year 2050: (1) the human population growth rates are negatively related to per/capita energy consumption; (2) with zero growth occurring at ~13 kW, the global human population will stop growing; and (3) the current supply of energy is far below the supply needed to fuel the global human population. This conclusion has supported the present model that described above.

As populations grow and demand increases, the search for water, food, and energy resources and the resulting impact on the environment are calling sustainability into question. The limits of technologies and the wisdom of our use of them are growing challenges, and questions of governance, social organization, and human rights are increasingly important to a sustainable outcome (DeLong et al., 2010; Bloom, 2011).

Finally, in order to monitor the energy use and consumption, we need global efforts from governments, nongovernmental agencies, and all people to be concerned about their future and concern about the future generations for the optimum use of energy and natural resources. This will include the intention of supporting a wide range of education activities to educate the public about their importance of conserving energy and keep the human population at the level that can support the human needs. This conclusion will indicate an alarm to the leaders of the world in the United Nation with all its agencies to take appropriate actions to conserve world resources for the future generations. Otherwise we are facing severe competition on these natural resources (i.e., renewable and nonrenewable), and none can predict the consequences of this situation. It is important that leaders of the world concentrate on the education of the population about the disastrous consequence of human population growth against the regenerative speedy process of the energy, the biological capacity of the ecosystems to sustain the demand with production of these resources.

## 28.8 Conclusions

The energy is the most important factor globally for activates and processing the work and life maintenance on the Earth. Without the energy, the life will stop and the processing of the global ecosystem will not function. The ecosystems operate by virtue the flow of energy through the components of the systems and transformation. In this chapter, we presented a theoretical GEM which including in it the relationship between human global population and the energy as part of ecological footprint (i.e., produced and consumed from our environment). The simulation of the model has been used to predict the impact of use of energy in the ecosystems, the production and deficient of energy, and the breakeven point of the availability of energy that supports ecosystems and human population in the next century. Further, the GEM is based mainly on long-term aggregate effects of energy consumption against the energy global biological capacities, and human population (i.e., either increasing or declining) on the basic assumption of very conservative or relaxed assumption of population growth. The main interaction in the model is between energy availability at any point in time and estimates of average global per capita consumption. According to the model, the human population is expected to reach 9.3 billion people by 2050 on the Earth and might reach between 11.0 and 20.1 billion people in year 2100 at the very conservative estimates and at the relaxed assumption of human population estimates (i.e., growth of human population at 0.2% and 1.1% annually). Further, the current global consumption of energy is about 9.67 global hectares per person in year 2011 and will reach to 28.94 global hectares per person in year 2100. However, according to the very conservative assumption of human population estimates (0.2%), in the model, the declines in consumption will reach to 15.01 global hectares per person by 2100. According to the very conservative and relaxed assumption of population growth rate (i.e., at 0.2% and 1.1%) is that the energy model proposes the growth of the population goes through two phases, the first consisting of rising life expectancies and the second characterized by a drop in fertility. The global human population transits from the high birth to death rates through one of high birth and low death rates, to one of low birth and death rates; it is expected that the energy will be higher to use until the stabilization of population and we believe that the energy will be the source of the power of the global ecosystems for the continuation of existence. Therefore, the GEM is theorizing two scenarios: one is the growing human population to the extent that there will be unbalanced of the availability and consumption of energy against the human population, and the second is the declining of human population to the extent that will no overpopulation and in consequence the reduction of demands for energy. This will lead to allowing the number of people to become in line with our resources and energy availabilities. From these scenarios, the expected decline of population through different disasters will be enormous on the future human populations. It is very important to understand the magnitude of reduction of usable and available energy on the society and all human beings including our activities and the ecosystems processing the goods and services that will be affecting the progressing of human beings.

It is very important that we have the knowledge that any human being on this Earth and in consequence the global human population living are depending on energy/material flows to and from different places around the world is the major energy flow system. The production, consumption, and policy decisions in any given local or areas have the potential to create unseen unsustainable burdens on connected productive ecosystems in distant locals. Further, the ecological change in one region has the potential to create another problem for other regions and creates unstability of other regions ecologically. In consequence, the sustainability of these regions and society should be monitored. The sustainability of energy and other natural resources that the humanity is depending on may be well maintained through the years 2025–2035, according to the prediction from GEM model. This conclusion will affect the sustainability of our global ecosphere and in consequence the humanity.

Addressing the humanity issue is that humans facing unbalance between consumption and demand in a very near future will affect their level of welfare. Therefore, addressing sustainability increases tensions between intragenerational and intergenerational equity, because not every policy will benefit poor people today as well as future generations. The key policy questions relate to the transition to renewable energy, development links with the green economy and green growth, and other market mechanisms, such as green taxes, cap and trade schemes for the environment, and regulatory frameworks to prevent (UNDP 2010). The UN Development Program report 2010 indicated that "in order to preserve the full range of natural ecosystems required for human well-being, environmental and energy access objectives must be embedded in policies that influence human activities, including the key productive sectors of the economy, as well as national development planning frameworks and budgets, institutions, governance, and market-based mechanisms." This is the pathway to sustainable development globally. Finally, education is a valuable tool that must be promoted at every opportunity in order to avoid the worries about overpopulation. Some of the opinions suggested that population is declining naturally, and will soon stabilize at a manageable number. However, in order to hasten the fall of fertility rates, usually through the education and empowerment of women. Further, what we should not expect is that this approach will make a significant contribution to resolving the population problem in time to balance the resources with growing population. Education and empowerment take time, and there is far too little time remaining before the first wave of impacts is upon us. The dramatic changes that could happen as the result of lack of energy are the increase the possibilities of war, starvation, diseases and finally the death and this will lead to the reduction of human population. Therefore if we concentrate on educating and empowering women we will be able to reduce fertility rate more likely to happen and especially in the developing countries, where the population is growing at an alarming rate.

## Acknowledgments

The authors would like to thank Dr. Samir Ghabbour, Cairo University, Egypt and Dr. Magdy Khalil, Ain Shams University, Egypt for their valuable comments. Additionally, the Office of Research and Development and Roy G. Perry, College of Engineering, Prairie View A&M University, The Texas A&M University System, supported this research work.

## References

Barrett, J., Simmons, C., 2003. An Ecological Footprint of the UK: Providing a Tool to Measure the Sustainability of Local Authorities. Stockholm Environment Institute Stockholm, York, U.K.

Birket, S.H., 1994. On the interconnection of ecological system model. Ecol. Model. 72, 153–174.

Bloom, D.E., 2011. 7 Billion and counting. Science 333, 562–568.

Chambers, N., Simmons, C., Wackernagel, M., 2000. Sharing Nature's Interest: Ecological Footprint as an Indicator of Sustainability. Earthscan, London.

Chen, B., Chen, G.Q., Yang, Z.F., Jiang, M.M., 2007. Ecological footprint accounting for energy and resource in China. Energy Policy 35, 1599–1609.

Clugston, C., 2007. Global peak energy: implications for future human populations. Energy Bull. http://energybulletin.net/node/34120.

DeLong, J.P., Burger, O., Hamilton, M.J., 2010. Current demographics suggest future energy supplies will be inadequate to slow human population growth. PLoS ONE 5 (10), e13206. doi: 10.1371/journal.pone.0013206.

ECOTEC, U.K., 2001. Ecological Footprint, European Parliament, Directorate General for Research, Directorate A, The STOA Programme March, 2001.

Ewing, B., Reed, A., Rizk, S.M., Galli, A., Wackernagel, M., Kitzes, J., 2008. Calculation Methodology for the National Footprint Accounts, 2008 ed.ition. Global Footprint Network, Oakland, CA 94607-3510, USA. http://www.footprintnetwork.org.

FAO, 1960–2008. FAOSTAT FAO Statistical Database–Data Series 1960–2008. Food and Agriculture Organization, UN, Rome, Italy. www.fao.org.

FAO, 2008. FAOSTAT Data Series. Food and Agriculture Organization, UN. www.fao.org.

FAO, 2008a. FAOSTAT (Food and Agriculture Organization, United Nations, Rome) 2008.

Fath, B.D., Halnes, G., 2007. Cyclic energy pathways in ecological food webs. Ecol. Model. 208, 17–24.

Hannon, B., 1973. The structure of ecosystems. J. Theor. Biol. 41, 535–546.

Huijbregts, M.A.J., Hellweg, S., Frischknecht, R., Hungerbühler, K., Hendriks, A.J., 2008a. Analysis ecological footprint accounting in the life cycle assessment of products. Ecol. Econ. 64, 798–807.

Humphrey, S., Chapagain, A., Bourne, G., Mott, R., Oglethorpe, J., Gonzales, A., Atkin, M., Loh, J., Collen, B., McRae, L., Carranza, T.T., Pamplin, F.A., Amin, R., Baillie, J.E.M., Goldfinger, S., Wackernagel, M., Stechbart, M., Rizk, S., Reed, A., Kitzes, J., Peller, A., Niazi, S., Ewing, B., Galli, A., Wada, Y., Moran, D., Williams, R., Backer, W.D., Hoekstra, A.Y., Mekonnen, M., 2008b. Living Planet Report 2008. World Wide Fund (WWF), International Avenue du Mont-Blanc 1196 Gland, Switzerland. www.panda.org.

Jenkins, M., Jakubowska, J., Gaillard, V., Groombridge, B., Wackernagel, M., Monfreda, C., Deumling, D., Gurarie, E., Friedman, S., Linares, A.C., Sánchez, M.A.V., Falfán, I.S.L., Loh, J., Randers, S.J., Monfreda, C., 2002. Living Planet Report 2002. World Wildlife Fund (WWF), International. Avenue du Mont-Blanc, CH-1196 Gland, Switzerland. www.panda.org.

Kitzes, J., Galli, A., Rizk, S., Reed, A., Wackernagel, M., 2008. Guidebook to the National Footprint Accounts:, 2008 Edition Global Footprint Network Oakland, CA 94607-3510, USA.

Kitzesa, J., Gallib, A., Baglianic, M., Barrettd, J., Digee, G., Edef, S., Erg, K., Giljumh, S., Haberlg, H., Hailsi, S., Jolia-Ferrierj, L., Jungwirthk, S., Lenzenl, M., Lewis, K., Lohn, J., Marchettinib, N., Messingero, H., Milnek, K., Molesp, R., Monfredaq, C., Moranr, D., K.NakanosPyhälät, A., Reissue, W., C.Simmonsm, Wackernagela, M., Wadav, Y., Walshq, C., Wiedmann, T., 2009. A research agenda for improving national ecological footprint accounts. Ecol. Econ. 68, 1991–2007.

Krenz, J.H., 1976. Energy Conversion and Utilization. Allyn and Bacon, Inc., USA, 359 p.

Lenzen, M., Lundie, S., Bransgrove, G., Charet, L., Sack, F., 2003. Assessing the ecological footprint of a large metropolitan water supplier: lessons for water management and planning towards sustainability. J. Environ. Plann. Manage. 46, 113–141.

Meadows, D. 2008. Thinking in System: A primer. Edited by Diana Wright 242 pp. Chelsea Green Publishing.

Moffatt, I., 2000. Ecological footprints and sustainable development. Ecol. Econ. 32, 359–362.

Monfreda, C., Wackernagel, M., Deumling, D., 2004. Establishing national natural capital accounts based on detailed ecological footprint and biological capacity accounts. Land Use Policy 21, 231–246.

Ness, G.D., 2004. Population growth and energy. Encyclopedia Energy 5, 107–116.

Ouyang, Y., 2008. Modeling the mechanisms for uptake and translocation of dioxane in a soil-plant ecosystem with STELLA. J. Contam. Hydrol. 95, 17–29.

Overby, R., 1985. The Urban economic environmental challenge: Improvement of human welfare by building and managing urban ecosystems. Paper presented in Hong Kong to the POLMET '85 Urban Environmental Conference. Washington DC: The World Bank.

Penela, A.C., Villasante, C.S., 2008. Applying physical input–output tables of energy to estimate the energy ecological footprint (EEF) of Galicia (NW) Spain. Energy Policy 36, 1148–1163.

Rees, W.E., 1995. Cumulative environmental assessment and global change. Environ. Impact Assess. Rev. 15, 295–309.

Rees, W.E., 2000. Eco-footprint analysis: merits and brickbats. Ecol. Econ. 32, 371–374.

Rees, W.E., 2001. Ecological Footprint, Concept of., Encyclopedia of Biodiversity. Academic Press. 229–244.

Rees, W.E., 2003. Economic development and environmental protection: an ecological economics perspective. Environ. Monit. Assess. 86, 29–45.

Rees, W.E., 2006. Ecological footprints and bio-capacity: essential elements in sustainability assessment. In: Dewulf, J., Langenhove, H.V. (Eds.), Renewables-Based Technology: Sustainability Assessment. John Wiley and Sons, Chichester, pp. 143–158.

Shakir-Hanna, S.H., Osborne-Lee, I.W., 2011. Sustainable economy of the ecological footprint: economic analysis and impacts. In: Villacampa, Y., Brebbia, C.A. (Eds.), Ecosystems and Sustainable Development. VIII. WIT Press, South Hampton, U.K., pp. 313–326.

SPSSSCIENC, 2002. Sigmaplot Version 8. SPSSSCIENCE. www.spssscience.com/sigmaplot.

STELLA, 2001. An Introduction to System Thinking by Barry Richmond. High Performance Systems, Inc. The System Thinking Company, ISBN 0-9704921-1-1, 165 p.

UN, 2011. The State of World Population 2011.

UNDP, 2010. Accelerating Progress Towards the Millennium Development Goals. K. Office Supplies, Ltd, New York.

UNDP, 2010a. Human Development Report 2010. 20th Anniversary Edition. The Real Wealth of Nations: pathways to human development. United Nation Development Program. Palgrave Macmillan Publ. 228 p.

UNEP, 2009. A Planet in Ecological Debt, Arendal Maps and Graphics Library. UNEP. http://maps.grida.no/go/graphic/a-planet-in-ecological-debt <.

UNFPA, 2001. The State of World Population 2001.

Vitousek, P.M., Ehrlich, P.R., Erhlich, A.H., Matson, P.A., 1986. Human appropriation of the products of photosynthesis. Bioscience 36, 368–373.

Wackernagel, M., 1999. An evaluation of the ecological footprint. Ecol. Econ. 31, 317–318.

Wackernagel, M., Galli, A., 2007. An overview on ecological footprint and sustainable development: a chat with Mathis Wackernagel. Int. J. Ecodyn. 2, 1–9.

Wackernagel, M., Rees, W., 1996. Our Ecological Footprint: Reducing Human Impact on the Earth. New Society Publishers, Gabriola Island, British Columbia.

Wackernagel, M., Rees, W.E., 1997. Perceptual and structural barriers to investing in natural capital: economics from an ecological footprint perspective. Ecol. Econ. 20, 3–24.

Wackernagel, M., Silverstein, J., 2000. Big things first: focusing on the scale imperative with the ecological footprint. Ecol. Econ. 32, 391–394.

Wackernagel, M., Monfreda, C., Moran, D., Goldfinger, S., Deumling, D., Murray, M., 2004. National Footprint and Biocapacity Accounts. http://www.footprintnetwork.Org Oakland, CA.

Wackernagel, M., Onisto, L., Bello, P., Linares, A.C., Falfán, I.L.P., García, J.M.N., Guerrero, A.S.R., Guerrero, G.S.R., 1999. National natural capital accounting with the ecological footprint concept. Ecol. Econ. 29, 375–390.

Wackernagel, M., Schulz, N.B., Deumling, D., Linares, C.A., Jenkins, M., Kapos, V., Monfreda, C.LoJ., Myers, N., Norgaard, R., Randers, J., 2002. Tracking the ecological overshoot of the human economy. Proc. Nat. Acad. Sci. USA (PNAS), 9266–9271.

World Bank, 1960–2008. Data Series and Research. The World Bank Organization. www.worldbank.org1960-2008.

WRI, 2000. World Resources 2000–2001-People and Ecosystems: The Fraying Web of Life. World Resources Institute (WRI), Washington DC.

WRI, 1960–2005 Series. EarthTrends Environmental Information, World Resource Institute (WRI). World Resource Institute:WRI http://Earthtrends.wri.org/.

WWF, The Global Conservation Organization, living_planet_report/footprint.

WWF, 2002. Living Planet Report 2002. World Wide Fund for Nature (WWF), Geneva Switzerland. http://www.panda.org/livingplanet.

WWF, 2004. Living Planet Report 2004. World-Wide Fund for Nature International (WWF), Global Footprint Network, UNEP World Conservation Monitoring Centre. WWF, Gland, Switzerland. http://www.panda.org/livingplanet.

WWF, 2006. Living Planet Report 2006. World Wide Fund (WWF) for Nature, Gland, Switzerland. http://www.panda.org/livingplanet.

PART

9

# Models of Interactions Between Hierarchical Levels and Hierarchy Theory

# 29

# The Dynamics Linking Biological Hierarchies, Fish Stocks and Ecosystems: Implications for Fisheries Management

Francisco Arreguín-Sánchez

*CENTRO INTERDISCIPLINARIO DE CIENCIAS MARINAS DEL IPN, APARTADO POSTAL 592, LA PAZ, 23000, BAJA CALIFORNIA SUR, MEXICO, E-MAIL: FARREGUI@IPN.MX*

## 29.1 Resource Management, Stability and Change

Traditionally, fisheries management has been oriented to the target species. Even if we recognize the importance of the ecosystem, the institutional structure is organized into a single species-based management profile. In practice, it is not yet sufficiently clear how ecosystem-based management can be implemented. However, it is clear that in this approach to management, the dynamics of vertical processes linking the biological hierarchies or levels of biological organization, particularly populations and ecosystems, should be studied and understood.

The populational focus of fisheries science had solid foundations in the assumption of a reasonably stable environment, a concept that has certainly sustained the decision-making process and the implementation of management strategies. If we consider Fig. 29.1, which represents the global anomaly of temperature during the past century, and take this anomaly as an indicator of the state of the environment, we see from this trend that the period from 1920 through 1980 was reasonably stable. The remarkable development of fisheries science occurred during this period, as did the great concepts that lent scientific support to fisheries management. The theoretical bases (e.g., Baranov, 1918), the logistical equation (e.g., Graham, 1935, 1938), individual growth (Bertalanffy, 1938), the surplus-yield models (Schaefer, 1954, 1957; Pella and Tomlinson, 1969; Fox, 1970; Clark, 1976; Walter, 1978, 1981), recruitment models (e.g., Ricker, 1954; Beverton and Holt, 1957), virtual population analysis (VPA) (Gulland, 1965; Pope, 1972; Pope and Shepherd, 1985), multispecies surplus-yield models (Pope, 1979; Pauly, 1979a), multi-species VPA (Helgason and Gislason, 1979; Pope, 1979b), and catch-at-age analysis (e.g., Deriso et al., 1985) all included the common element of the assumption of

**Models of the Ecological Hierarchy. DOI: http://dx.doi.org/10.1016/B978-0-444-59396-2.00029-8**

ISSN 0167-8892, 

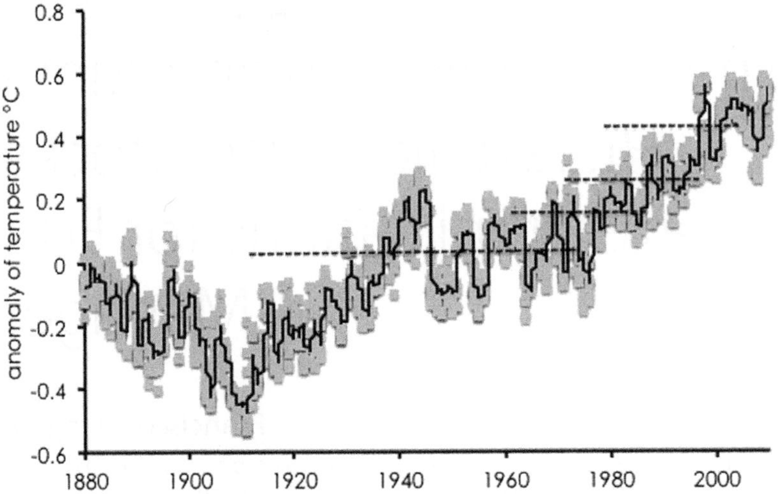


**FIGURE 29.1** Anomaly of global temperature. Note the stable period between mid-1930 and 1980 as well as the rise period after.

a reasonably stable carrying capacity. Consequently, the primary driver of stock variation was fishing, and it was therefore clear that effective management would result from the control of fishing mortality.

Since the 1980s, applied statistics has become relevant to population models because of the need for increased precision in fishing mortality estimates and especially to increase the probability of success of conservation of exploited populations (see Hilborn and Walters, 1992). Overfishing and the collapse of several of the world's fisheries (FAO, 2010) were noted, and the rationale that a lower uncertainty in the estimates would generate a higher probability of success in implementing management decisions was adopted. However, the pattern shown by the anomaly of temperature in recent decades suggests that natural systems are continually changing following an increasing trend that is present in addition to environmental fluctuations. This observation suggests that the assumption of stability is no longer valid and that the view that more accurate population estimates will increase the probability of successful management decisions is therefore uncertain. Indeed, the environment now appears to be a significative driver of the observed patterns of variation of exploited populations. In this context, the resources obey not only their own dynamics but also the dynamics of the ecosystem to adapt to a constantly changing environment. Under these conditions, it is expected that ecosystem approaches to fish resource management will reduce the uncertainty and increase the efficiency of management actions and, therefore, we must understand and characterize the dynamics among both the population and the ecosystem levels of organization and adapt management strategies to their joint dynamics under conditions of continuous change.

## 29.2 Population Level of Organization

Figure 29.2 shows examples of long-term changes in four target species of commercial fisheries in the Bank of Campeche, southern Gulf of Mexico. One of these fisheries, pink shrimp (*Farfantepenaeus duorarum*), is diagnosed as collapsed (DOF, 2010; Wakida-Kusunoki et al., 2006; Arreguín-Sánchez et al., 2008a,b; Arreguín-Sánchez, 2009). The current harvest is 5% of the catch taken in the 1950s and 1960s.The red grouper (*Epinephelus morio*) is diagnosed as overfished, with a current stock size of 35% of the estimated size at the beginning of the 1970s (Burgos-Rosas and Pérez-Pérez, 2006; DOF, 2010). The current yields of octopus (*Octopus maya*), an endemic species, are almost three times greater than those obtained in the late 1970s (DOF, 2010; Pérez-Pérez et al., 2006). The lobster (*Panulirus argus*) increases catches after 1980. These four cases suggest a relationship between commercially important stocks with long-term changes in environmental variables, such as sea-surface temperature, salinity, sea level, observed and estimated (Zetina-Rejón, 2004) primary production, and the index of the North Atlantic Oscillation (http://www.esrl.noaa.gov/psd/data/climateindices/list/). If we take the catch of pink shrimp, lobster and octopus (Arreguín-Sánchez and Arcos-Huitrón, 2007) as an index of relative abundance (because they are well-established fisheries) and also consider the estimated population size for red grouper (Arreguín-Sánchez et al., 1996, 1997; Giménez-Hurtado, 2005), we observe a clear response associated with long-term environmental change in all the four cases. In turn, these findings suggest an effect at the ecosystem level, with fishing representing a factor contributing to the decline in abundance. The meaning of these results is that an environmental effect at the ecosystem level affects populations, which respond differently.

## 29.3 Ecosystem-Level Organization

To explore the long-term effect of the joint dynamics of populations and the ecosystem and to understand the underlying processes, an example of a trophic model, built with the Ecopath program (Christensen and Pauly, 1992), of the coastal region known as Atasta on the southern coast of the Gulf of Mexico was used. The model was constructed from the data collected over a 15 months period (Ramos-Miranda et al., 2005) on more than 140 fish and 20 species of macroinvertebrates at over 35 collection sites (Fig. 29.3a) and arranged in 33 functional groups (Arreguín-Sánchez et al., submitted). The Ecosim model (Walters et al., 1997) was applied to incorporate the observed short-term variation in primary production as environmental forcing and to calibrate the model with relative abundance data from 12 functional groups (Fig. 29.3b).

To evaluate the effect of long-term environmental forcing on the ecosystem, we used the sustainability index proposed by Ulanowicz et al., (2009). This index is based on the information theory and represents the self-organizing capacity of the ecosystem, as expressed by the following relation:

$$F = -\left[\frac{e}{\log(e)}\right]\alpha^{\beta}\log(\alpha^{\beta}) \tag{1}$$

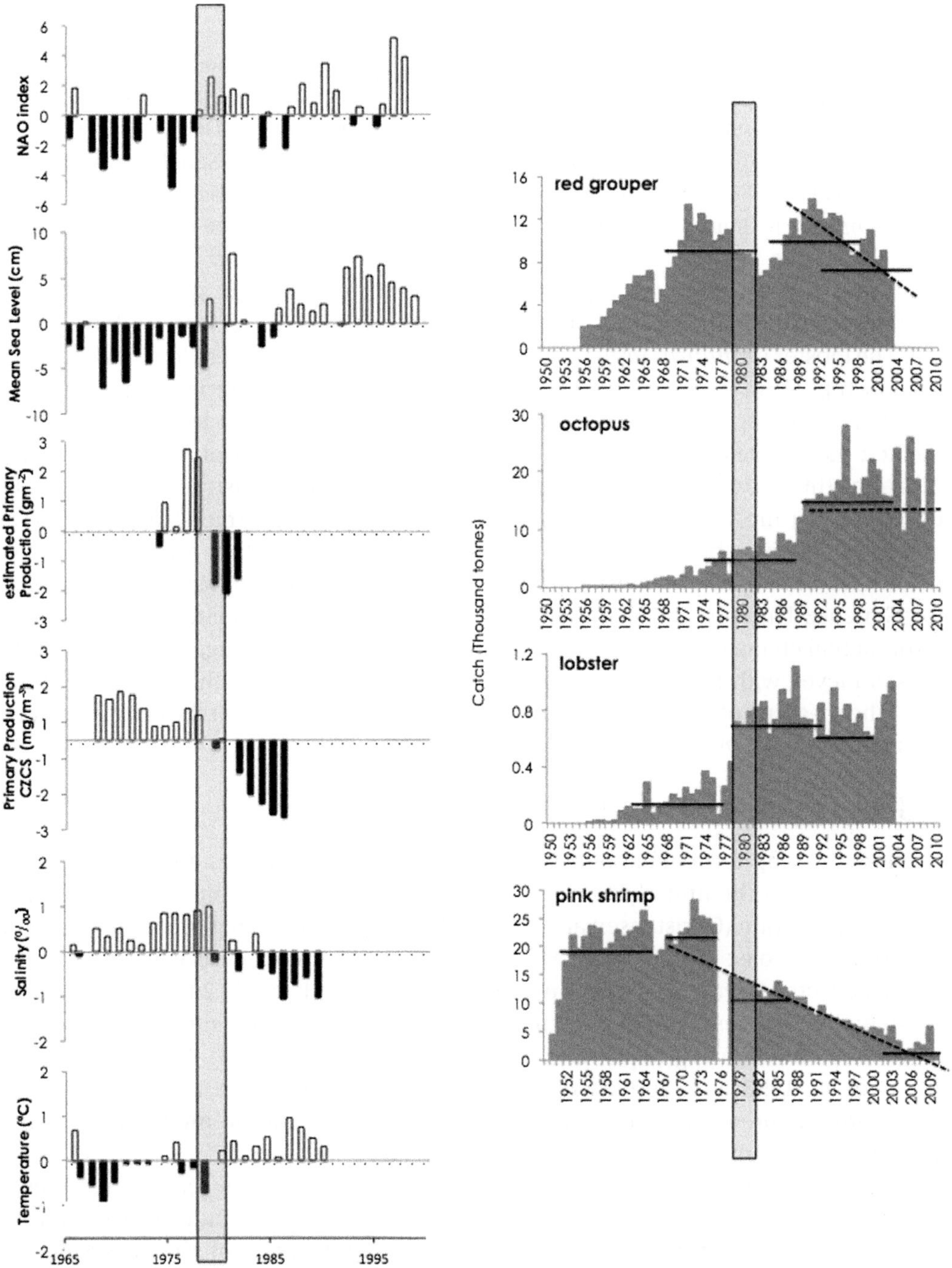


**FIGURE 29.2** Historical annual catches of pink shrimp (*F. duorarum*), octopus (*O. maya*), lobster (*Panuliru sargus*), and red grouper (*E. morio*), and their relationship with long-term environmental variables such as temperatura, salinity, observed and estimated primary poduction, mean sea level and the North Atlantic Oscillation (NOA) index. Catch tendencies change appears to be associated with the change of phase of environmental variables suggesting climate change effects on stocks.

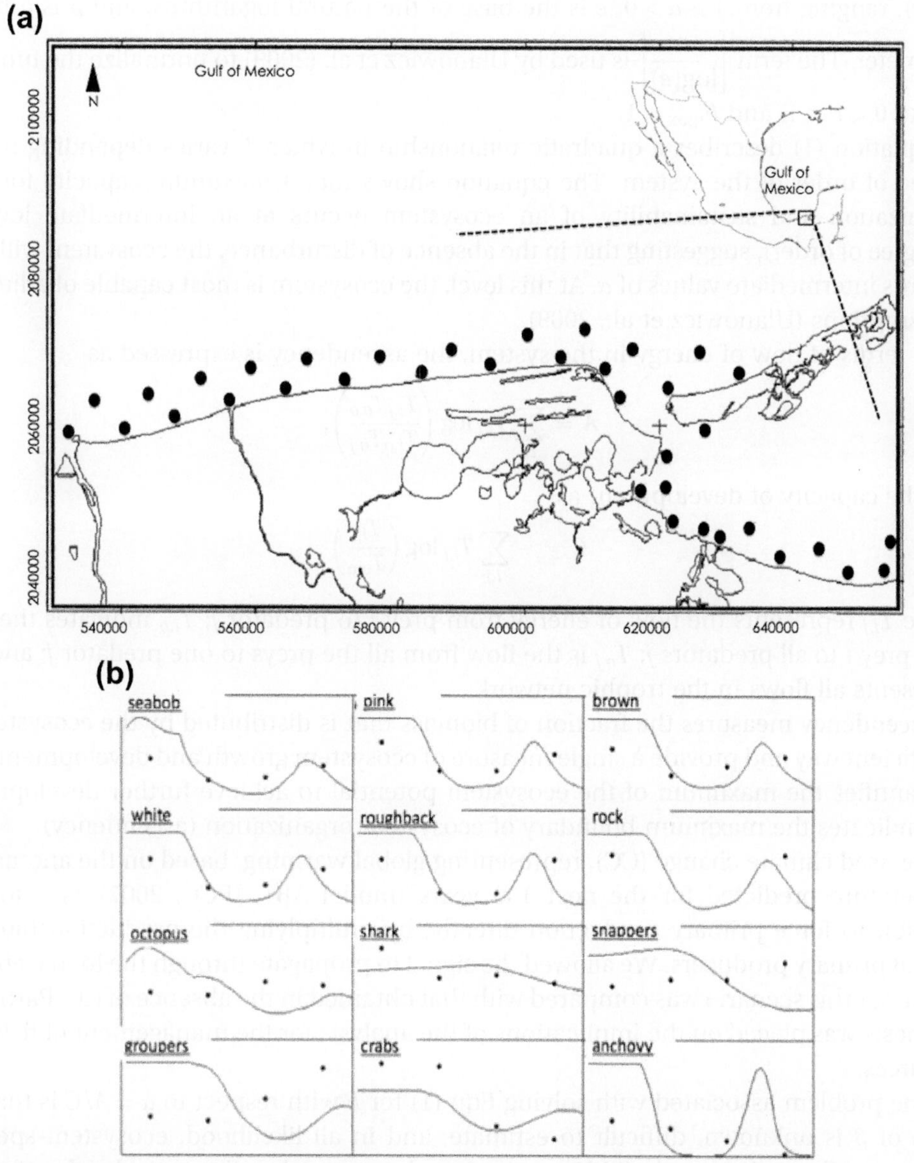


**FIGURE 29.3** (a) Coastal ecosystem model of Atasta in the southern Gulf of Mexico used to exemplify biological hierarchies dynamics. Dots indicate sampling sites. (b) Short-term fitting process to calibrate time dynamic Ecosim model. Dots are observed data, lines are simulated biomasses.

where $F$ is the ecosystem fitness (as a measure of sustainability), the potential of the ecosystem to evolve or self-organize, ranging from $1 > F > 0$; $\alpha = A/C$, where $A$ is the ascendency and $C$ is the development capacity of the ecosystem (Ulanowicz, 1986), so that $\alpha$ is a relative measure of the organized power flowing within the system (degree of order), ranging from $1 > \alpha > 0$; $e$ is the base of the natural logarithms; and $\beta$ is a fitting parameter. The term $\left[\frac{e}{\log(e)}\right]$ is used by Ulanowicz et al. (2009) to normalize the function so that $0 < F < 1$, and $F_{max} = 1$.

Equation (1) describes a quadratic relationship in which $F$ varies depending on the degree of order of the system. The equation shows that a maximum capacity for self-organization and sustainability of an ecosystem occurs at an intermediate level of $\alpha$ (degree of order), suggesting that in the absence of disturbance, the ecosystem will tend towards intermediate values of $\alpha$. At this level, the ecosystem is most capable of adjusting to disruptions (Ulanowicz et al., 2009).

In terms of flow of energy in the system, the ascendency is expressed as

$$A = \sum_{i,j} T_{i,j} \log\left(\frac{T_{i,j} T_{o,o}}{T_{i,o} T_{o,j}}\right), \tag{2}$$

and the capacity of development as

$$C = -\sum_{i,j} T_{i,j} \log\left(\frac{T_{i,j}}{T_{o,o}}\right) \tag{3}$$

where $T_{i,j}$ represents the flow of energy from prey $i$ to predator $j$; $T_{i,o}$ indicates the flow from prey $i$ to all predators $j$; $T_{o,j}$ is the flow from all the preys to one predator $j$; and $T_{o,o}$ represents all flows in the trophic network.

Ascendency measures the fraction of biomass that is distributed by the ecosystem in an efficient way and provide a single measure of ecosystem growth and development, and $C$ quantifies the maximum of the ecosystem potential to achieve further development, and indicates the maximum boundary of ecosystem organization (ascendency).

We used climate change (CC), representing global warming, based on the anomaly of temperature predicted for the next 100 years (model AB1, IPCC, 2007) as a forcing function to force primary production directly, by multiplying the production/biomass ratio of primary producers. We allowed the signal to propagate through the food web. The output for this scenario was compared with that obtained in the absence of CC. Particular emphasis was placed on the implications of the analysis for the management of fisheries resources.

One problem associated with solving Eqn (1) for $F$ with respect to $\alpha = A/C$ is that the value of $\beta$ is unknown, difficult to estimate, and in all likelihood, ecosystem-specific. However, Ulanowicz et al. (2009) argued and proposed a theoretical value for the constants in this equation. This value can be used as reference value. In the case of the coastal ecosystem of Atasta, we consider two options: (1) assume that the ecosystem is currently in a state of greater capacity for self-organization, $F_{max,act}$, with values of $\alpha = 0.29$ and $\beta = 0.84$; and (2) the theoretical value proposed by Ulanowicz et al. (2009) that yield the maximum self-organization, $F_{max,teo}$ with values for $\alpha = 0.46$ and $\beta = 1.29$. Figure 29.4 represents the state of the ecosystem without CC for the two alternative conditions assumed for the ecosystem state. The figure shows the limits of variation

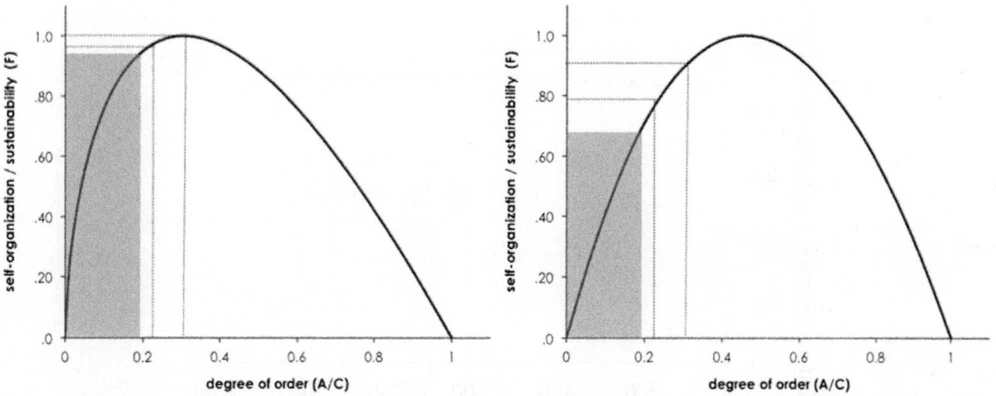


**FIGURE 29.4** Ranges of variation of the sustainability (self-organization) index proposed by Ulanowicz et al. (2009) for the coastal ecosystem of Atasta, southern Gulf of Mexico, simulated using Ecosim model. Left assuming ecosystem is under maximum sustainability at the present. Right, assuming theoretical values of sustainability proposed by Ulanowicz et al. (2009). Dashed lines represent the loss of sustainability when forcing primary production with anomaly of the sea surface temperature but without warming CC tendency. Gray area represents loss of sustainability in both scenarios forcing primary producers under warming (all forcing scenarios of simulation were over the next 100 years).

under these conditions. As expected, low variation in $F_{max,act}$ was observed despite the estimated change in $\alpha$. However, if we considered the theoretical values of the parameters of Eqn (1), the value varied in the range of $0.78 \leq F_{max,teo} \leq 0.93$. This variation suggested a loss of ecosystem sustainability (a loss of the ability to self-organize).

Figure 29.4 also represents the ecosystem state under the influence of CC. In both the cases, a significant loss of sustainability is observed. The values of sustainability ranged from $0 \leq F_{max,act} \leq 0.95$ and $0 \leq F_{max,teo} \leq 0.65$ for the hypothetical current state and for the theoretical values of the constants of Eqn (1), respectively. Figure 29.5 shows the annual trend for the same indicators. In the absence of CC, the average trend of $F$ represents stability around $F_{max,act}$ with some variation. In the presence of CC, $F$ decreases over time. Moreover, the variation increases with time. This trend has two aspects: (1) the decrease of sustainability and (2) in the growing scale of variability, we can observe similar values of $F$ (sustainability) at different points in the time series. This condition raises the following key question: Do two or more similar values of $F$ in the time series represent the same state of the ecosystem? The answer has important management implications. If the state is the same, management strategies could be maintained over the long term. However, if they represent different ecosystem states, management strategies should be tailored to the particularities of each state.

Figure 29.6 shows the structure of the ecosystem according to trophic level (TL) and the relative changes in the biomass of its components for three different time points on the trajectory of the ecosystem disturbed by CC at which $F$ takes similar values. It is clear that the structure of the ecosystem (particularly secondary consumers, TL > 2.5) differs at these points and they represent different states of the ecosystem.

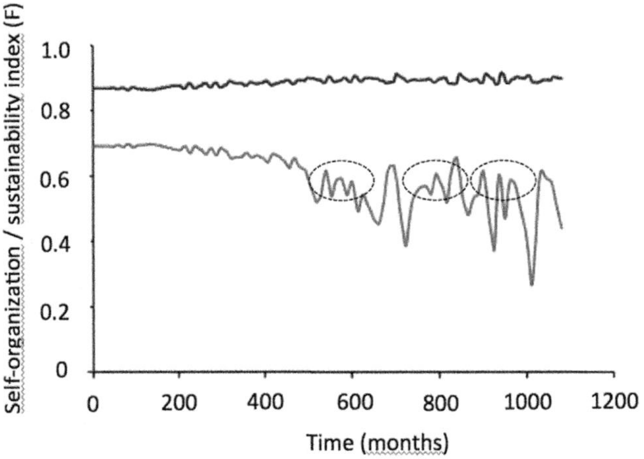


**FIGURE 29.5** Time series over 100 years of sustainability (auto-organization) index for the coastal ecosystem of Atasta, Southern Gulf of Mexico. Black line represents sea surface temperature effect without CC; gray line with effect of warming CC; and dashed circles show same levels of sustainability at different times.

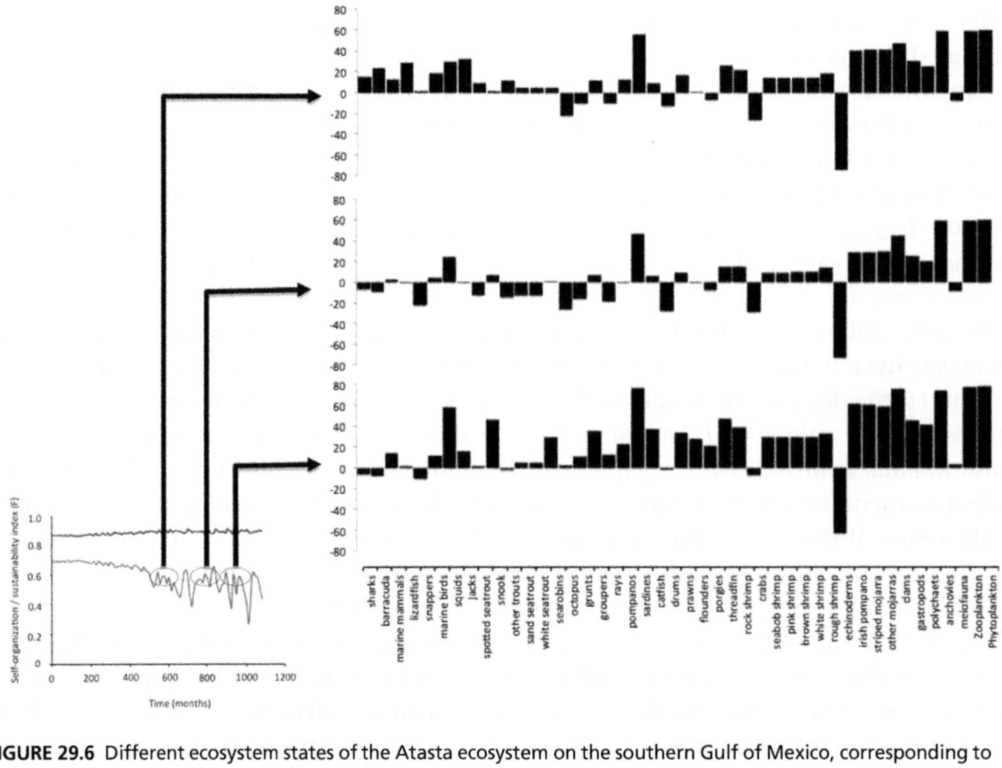


**FIGURE 29.6** Different ecosystem states of the Atasta ecosystem on the southern Gulf of Mexico, corresponding to a same level of sustainability at different times.

## 29.4 Implications for Management

At the level of individual stocks, Arreguín-Sánchez (2009) described in detail the environmental impact that caused the decrease in the size of the population of pink shrimp on the Campeche Bank. The population decreases showed a strong relationship with the decreasing recruitment rate over the past four decades (Fig. 29.2). In contrast, it is shown at the level of simulation that the long-term trend caused variations in the ecosystem sustainability expressed by *F* and thereby allowed different ecosystem states. This trend appears to coincide with the right-hand portion of the temperature anomaly shown in Fig. 29.1. In its upwards trend, this graph shows fluctuations that have short-term stable states with a duration of approximately one decade. These short-term stable states are separated by short periods with rapid changes in temperature. The observed pattern suggests possible successive stable states of the environment and then the ecosystem.

The trends of commercial stocks of pink shrimp, grouper, lobster and octopus in Fig. 29.2, viewed in the context of CC, may be considered together with the results of the simulations. These trends and results suggest that it is possible that the current state of the resources respond to different states of the ecosystem. If this view is reasonable, then it seems unlikely that current management strategies based on limiting fishing mortality will foster the recovery of pink shrimp and red grouper stocks, as are stated by the current management objectives that are displayed in the National Fishing Chart in Mexico (DOF, 2010, Arreguín-Sánchez et al., 2006). Interestingly, the significant increase in octopus abundance (expressed by the consistent increase in yield) appears to be related to the reduction in the abundance of grouper, the main predator of octopus. This effect was previously reported by Arreguín-Sánchez (2000).

In this context, it is important to note that the management of these resources is based on the concept of biological reference points (BRPs). This approach was also defined within a framework of stable ecosystems (e.g., our case history of the Campeche Bank). Three decades ago, a stable-ecosystem framework was represented by the state of the Campeche Bank ecosystem for nearly four decades. Through the control of fishing, it was expected that the abundance of the managed resources would return to a desirable state close to the BRP and referred to the maximum sustainable yields estimated from historical fishing data (DOF, 2010; Bugos-Rosas and Pérez-Pérez, 2006; Pérez-Pérez et al., 2006; Wakida-Kusunoki et al., 2006). However, it seems clear that currently, these BRP levels are not consistent with the current state of the ecosystem. Accordingly, the present basis of management should be adapted to incorporate BRP values estimated from each particular ecosystem state. These states, as suggested by the trend of the temperature anomaly (Fig. 29.1) and in agreement with the simulation results reported here, could presumably be changing on a decadal time scale.

An important aspect to consider in the light of CC is the need for adaptive and dynamic management associated to different sustainable ecosystem's states. Such an adaptive management would be near real-time and would reflect the state of the ecosystem. This approach is independent of the CC hypotheses that we adopt and would accommodate either global warming or global cooling. In either case, we would need to generate different baselines to define BRPs corresponding to different states of the ecosystem and thus applying to the resources. Of course, this condition would be expected to generate new situations, such as new institutional arrangements, laws and regulations that would contend with legitimate ecosystem-based management that considered the overall dynamics of the resource, and the ecosystem, the two levels of biological organization. In a highly dynamic decision-making process, the involvement of stakeholders becomes relevant for adopting, in a short time, management strategies (in a bottom-up management or mixed-management fashion) and to promote the need for continuous monitoring to generate information at the different levels of biological organization.

# 29.5 Population to Ecosystem Dynamics

## 29.5.1 Biological Organization Levels

Within the theoretical concepts of population dynamics, survival, expressed as $\exp^{-(P/B)_i}$, represents the performance of the population $i$. $P/B$ is the production/biomass ratio, which can be assumed to be equivalent to the total instantaneous mortality rate (Allen, 1971). Mortality can be represented by the sum of the total outflows of population $i$, considering predation mortality, export, fishing, and other causes including migration. In contrast, Zorach and Ulanowicz (2003) and Ulanowicz et al. (2009) refer to the number of roles $\varrho_i$, of a population in the ecosystem (the number of functions in the food web, the number of TLs expressed). This concept represents the role of a population in the dynamics of the ecosystem. Here, we refer to this concept as population performance within the ecosystem, expressed in terms of the amount of flow and connectivity of the population and represented by the following relation:

$$\varrho_i = \exp^{AMI_i} \tag{4}$$

where $\mathrm{AMI}_i$ is the average mutual information (limitation), expressed as

$$\mathrm{AMI}_i = \sum_{i,j} \left(\frac{T_{i,j}}{T_{o,o}}\right) \log\left(\frac{T_{i,j} T_{o,o}}{T_{i,o} T_{o,j}}\right), \tag{5}$$

where $T$ has the same meaning as that in Eqn (1).

Clearly, if both the indexes are associated through the flow of energy in the food web, their relationship defines the vertical dynamic linking the two levels of biological organization, namely the population and the ecosystem. Figure 29.7a shows the relationship between these two performance variables for the functional groups of the ecosystem for the coastal area of Atasta. Here, the simulation considered the effects of CC. These relationships were constructed to represent the effects of environmental forcing on primary producers using the Atlantic Multidecadal Oscillation (AMO) index (www.esrl.noaa.gov/psd/data/timeseries/AMO/). This index includes the Gulf of Mexico and

**(a)**

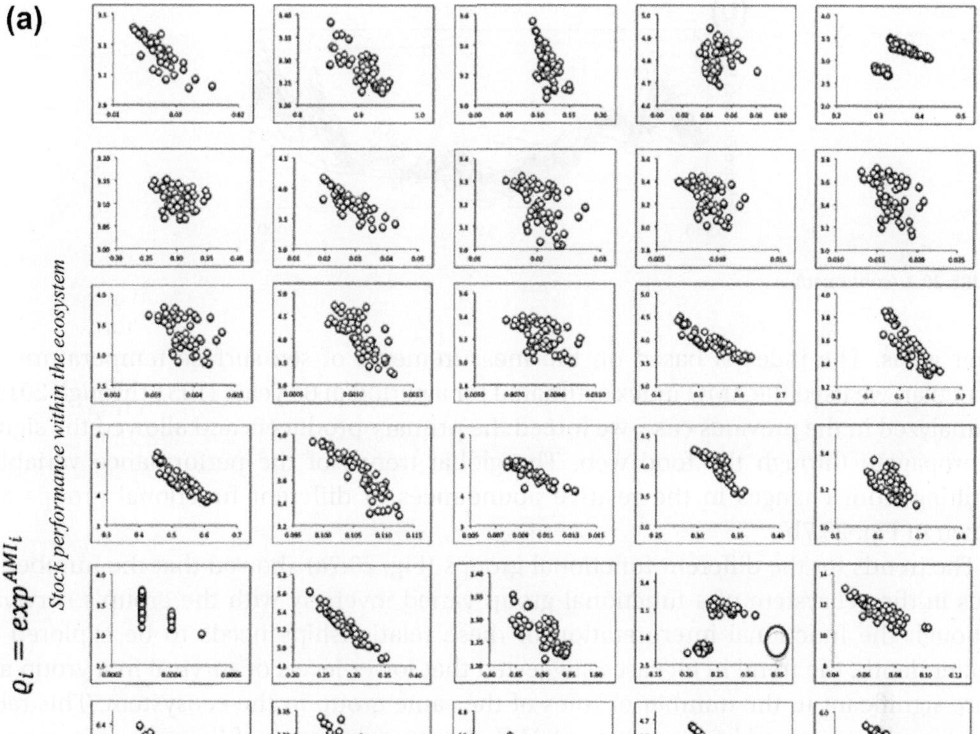

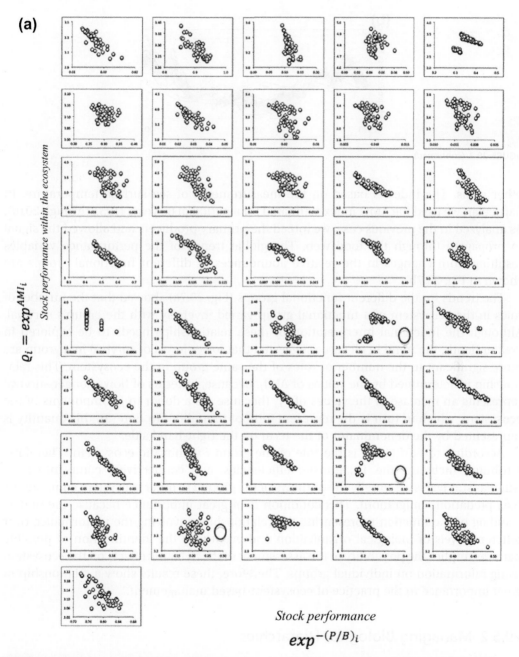


**FIGURE 29.7** (a) Plots showing stock-to-ecosystem performance for each species/functional groups of the Atasta coastal ecosystem of the Southern Gulf of Mexico, when forced by the Atlantic Multidecadal Oscillation (AMO) index over the period 1951–2010. Note the inverse relationship among survival (independent variable) and ecosystem roles (dependent variable) of each group when used as performance variables. (b) Simulated changes in relative biomasses for the Atasta coastal ecosystem of the Southern Gulf of Mexico, when forced by the Atlantic Multidecadal Oscillation (AMO) index over the period 1951–2010. For color version of this figure, the reader is referred to the online version of this book.

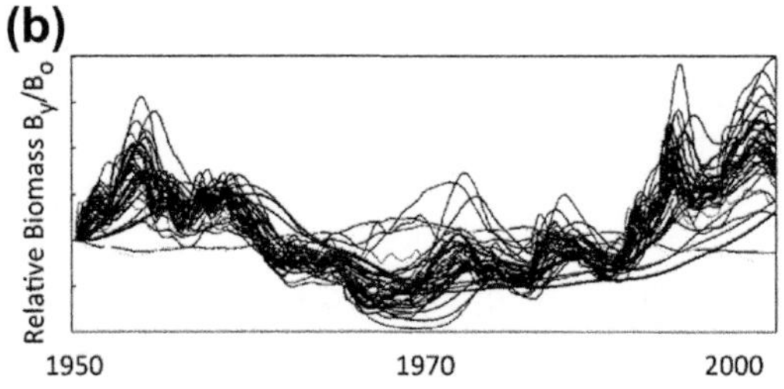


**FIGURE 29.7** *(continued)*.

other areas. The index is based on the measurements of sea surface temperature. In particular, we used the AMO index estimated for a period of 60 years (1951 through 2010). As analyzed in the previous case, we forced the primary producers and allowed the signal to propagate through the food web. The global trends of the performance variables resulting from changes in the relative abundances of different functional groups are shown in Fig. 29.7b.

The trends for the different functional groups (Fig. 29.7a) showed that the number of roles in the ecosystem of a functional group varied inversely with the group's survival. Although the functional interpretation of these relationships needs to be explored in greater depth, the trend in all cases suggested that lower levels of survival in a group are more significant to the number of roles of the same group in the ecosystem. This relationship is determined by the values of $AMI_i$, because, in terms of flows, longer survival represents an increase in the values of $T_{i,j}$ (because $i$ is a donor to $j$ components of the ecosystem, which in turn is the representation of mortality). In Eqn (5), this quantity is represented by the denominator of the term on the right-hand side.

According to Eqn (4), $\varrho_i$ is the role of the group $i$ within the ecosystem. Therefore, attributes, including the TL, transfer efficiencies, and connectivity, come into play, whereas survival is essentially the outflows of the group $i$ to other ecosystem components (i.e., predation, emigration). This condition is of great importance because the understanding of the function representing the relationships involving the performance over different levels of biological organization and eventually the identification of possible patterns implies the possibility of the assessment of the dynamic effects on the ecosystem using information on individual groups. Therefore, these results show a relationship of great importance to the practice of ecosystem-based management.

## 29.5.2 Managing Biological Hierarchies

According to current practice, strategies of ecosystem-based management are based on ecosystem information as a framework for use in decision making about target stocks. In the best case, for example, such criteria encompass a set of criteria based on the harvest rate of the target species relative to the population reference levels as a desired goal of management. These criteria also consider the amount of incidental catch and the physical impacts of the fishing gear, particularly, in trawl fisheries. The criteria reflect the goal of reducing or avoiding such impacts. Endangered species are usually considered in the light of the protection of biodiversity. Measures are taken for the protection of these species. A number of studies show the degree to which managers consider or apply this knowledge to the decision-making process (ITTC, 2010; DOF, 2010, www.msc.org). In general, these practices, even if we consider ecosystem-level criteria, are aimed at a single-species decision process and do not allow the effects on both levels of organization.

An aspect that we have not been able to assess properly is the issue of simultaneously managing multiple species when managing a target fish stock. In principle, institutional arrangements remain focused on criteria for the management of a single stock. In this context, decisions aimed simultaneously at multiple species appear relatively complex. However, strategies based on the type of relationships shown in Fig. 29.7a could motivate the design of criteria for monitoring other stocks at the ecosystem level in a management environment that involves the application of control measures to the target resource. The certification process of the Gulf of California sardine fishery (www.msc.org/track-a-fishery/certified/pacific/gulf-of_california-mexico-sardine) offers an example of such complexity. Lengthy discussions were required, in part, because of the need to ensure that the stock remaining after fishing was sufficient to ensure ecosystem functions. The central question in these considerations was the practical evaluation of these aspects of management. A similar situation occurs in the shrimp trawl fisheries. In these fisheries, it is generally accepted that the fishing gear damages the ecosystem through the large amount of bycatch and through the physical effect on the bottom (Eayrs, 2007; Gillett, 2008). In the first case, the obvious questions involve the effect on the maintenance of the ecosystem and the way to assess this effect. In both the cases, the basic element is the ability to assess the effects of a fishery and its management on the rest of the system. The formal evaluation of these effects will provide values of reference and monitoring criteria for managing at the ecosystem level. It is precisely clear that the vertical dynamics linking the two different levels of organization, population, and ecosystem, offer a significant possibility.

Finally, one of the relationships derived from Fig. 29.7a is that the performance variables are directly linked to relevant aspects of survival at each level of organization, to fishing mortality, and to the performance of stocks within the ecosystem, and in turn to the ability of the ecosystem to self-organize. According to Ulanowicz et al. (2009), this ability is a measure of ecosystem sustainability and resilience.

## Acknowledgments

The author expresses thanks for partial support through projects SEP-CONACYT (contract 104974), ANR-CONACYT (111465), GEF-LME Gulf of Mexico, SIP-IPN (2011785, 20121444), WWF-K88, and thanks the National Polytechnic Institute of Mexico for support through the EDI and COFAA programs.

# References


Allen, K.R., 1971. Relation between production and biomass. J. Fish. Res. Bd. Can. 28, 1573–1581.

Arreguín-Sánchez, F., 2000. Octopus – red grouper interaction in the exploited ecosystem of the northern continental shelf of Yucatan, Mexico. Ecol. Model. 129, 119–129.

Arreguín-Sánchez, F., 2009. Cambio climátio y el colpaso de la pesquería de camarón rosado (*Farfantepenaeus duorarum*) de la Sonda de Campeche. In: Rivera-Arriaga, E., Azuz-Adeath, I., Villalobos-Zapata, G.J., Alpuche-Gual, L. (Eds.), Cambio Climático en México un Enfoque Costero-Marino. Universidad Autónoma de Campeche, México, pp. 399–410.

Arreguín-Sánchez, F.M., Contreras, Moreno, V., Burgos, R., Valdes, R., . Population dynamics and stock assessment of red grouper (Epinephelus morio) fishery on Campeche Bank. Mexico. In: Arreguín-Sánchez, F., Munro, J.L., Balgos, M.C., Pauly, D. (Eds.), Biology, Fisheries and Culture of Tropical Groupers and Snappers, pp. 202–217. ICLARM Conf. Proc. 48, p. 449.

Arreguín-Sánchez, F., Contreras, M., Valdés, R., Moreno, V., Burgos, R., 1997. Biología y pesquería del mero (Epinephelus morio) del Banco de Campeche, México. In: Flores-Hernández, D., Sánchez-Gil, P., Seijo y, J.C., Arreguín-Sánchez, F. (Eds.), Análisis y diagnóstico de los recursos pesqueros críticos del Golfo de México EPOMEX Serie Científica 7. México.

Arreguín-Sánchez, F., Beléndez-Moreno, L., Méndez-Gómez-Humarán, I., Solana-Sansores, R., Rangel-Dávalos, C. (Eds.), 2006. Sustentabilidad y Pesca Responsable en México. Evaluación y Manejo. Instituto Nacional de Pesca, SAGARPA, México.

Arreguín-Sánchez, F., Arcos-Huitrón, E., 2007. Fisheries catch statistics for Mexico. In: Zeller, D., Pauly, D. (Eds.), Reconstruction of marine fisheries catches for key countries and regions (1950–2005). Fisheries Centre, University of British Columbia, Canada, pp. 81–103. Fisheries Centre Research Reports 15 (2).

Arreguín-Sánchez, F., Zetina-Rejón, M., Ramírez-Rodríguez, M., 2008a. Exploring ecosystem-based harvesting strategies to recover the collapsed pink shrimp (*Farfantepenaeus duorarum*) fishery in the southern Gulf of Mexico. Ecol. Model. 214, 83–94.

Arreguín-Sánchez, F., Ramírez-Rodríguez, M., Zetina-Rejón, M.J., Cruz-Escalona, V.H., 2008b. Natural hazards, stock depletion, and stock management in the Southern Gulf of Mexico pink shrimp fishery. Amer. Fish Soc. Sympos. 64, 419–428.

Arreguín-Sánchez, F., Ramos-Miranda, J., Flores-Hernández, D. Exploring effects of restricted fishing areas in a coastal small scale shrimp fishery of the Southern Gulf of Mexico, Ecol. Model. submitted for publication.

Baranov, F.I., 1918. On the question of the biological basis of fisheries. Nauchn. Issled. Ikthiologicheskii Inst. Zzv 1, 81–128 (in Russian).

Bertalanffy, L. Von, 1938. A quantitative theory of organic growth (inquiries on growth laws 11). Hum. Biol. 10 (Z), 181–213.

Beverton, R.J.H., Holt, S.J., 1957. On the dynamics of exploited fish populations. Fish. Invest. Ser. II 19, 533.

Burgos-Rosas, R., Pérez-Pérez, M., 2006. Mero (Epinephelus morio). In: Arreguín-Sánchez, F., Beléndez-Moreno, L., Méndez-Gómez-Humarán, I., Solana-Sansores, R., Rangel-Dávalos, C. (Eds.), Sustentabilidad y Pesca Responsable en México. Evaluación y Manejo. Instituto Nacional de Pesca, SAGARPA, México, pp. 505–521.

Christensen, V., Pauly, D., 1992. Ecopath II—a software for balancing steady-state ecosystem models and calculating network characteristics. Ecol. Model. 61, 169–185.

Clark, C.W., 1976. Mathematical Bioeconomics: The Optimal Management of Renewable Resources. Wiley, New York. 352p.

Deriso, R.B., Quinn, T.J., Neal, P.R., 1985. Catch-age analysis with auxiliary information. Can. J. Fish. Aq. Sci. 42, 815–824.

DOF, 2010. Actualización de la Carta Nacional Pesquera. Diario Oficial de la Federación, México. Diciembre 2.

Eayrs, S., 2007. A guide to bycatch reduction in tropical shrimp-trawl fisheries, Revised ed. FAO, Rome. 2007108.

FAO, 2010. Estado mundial de la pesca y la acuacultura. FAO, Roma. 210.

Fox, W., 1970. An exponential surplus-yield model for optimizing exploited fish populations. Trans. Am. Fish. Soc. 99 (1), 80–88.

Gillett, R., 2008. Global Study of Shrimp Fisheries. FAO Fisheries Technical Paper. No. 475. FAO, Rome. 2008. 331p.

Giménez-Hurtado, E., 2005. Análisis de la pesca del mero *Epinephelus morio.* (Serranidae: Pisces, Valenciennes 1928) en el Banco de Campeche. Tesis Doctorado. Centro de Investigaciones Biológicas del Noroeste. La Paz, Baja California Sur, México. 93.

Graham, M., 1935. Modern theory of exploiting a fishery and application to North Sea trawling. J. Cons. Intern. Expl. Mer. 10, 264–274.

Graham, M., 1938. Rates of fishing and natural mortality from the data of marking experiments. J. Cons. Intern. Expl. Mer. 13, 76–90.

Gulland, J.A., 1965. Estimation of mortality rates. Annex to rep. Arctic Fish. Working group, ICES, CM. 1965. (3) pp. 9.

Helgason, T., Gislason, H., 1979. VPA-Analysis with species interaction due to predation ICES CM 1979/G:52.

Hilbom, R., Walters, C.J., 1992. Quantitative Fisheries Stock Assessment. Choice, Dynamics and Uncertainty. Chapman and Hall, New York, 570 pp.

IPCC, 2007. Climate Change 2007: the Physical Science Basis. Contribution of Working Group I to the Fourth Assessment Report of the Intergovernmental Panel on Climate Change. In: Solomon, S., Qin, D., Manning, M., Chen, Z., Marquis, M., Averyt, K.B., Tignor, M., Miller, H.L. (Eds.), Climate Change 2007: the Physical Science Basis. Contribution of Working Group I to the Fourth Assessment Report of the Intergovernmental Panel on Climate Change, 996. Cambridge University Press, Cambridge, United Kingdom and New York, NY, USA.

ITTC, 2010. Tunas and billfishes in the Eastern Pacific ocean in 2008. Fishery Status Report 7, 143.

Pauly, D., 1979. Theory and management of tropical multispecies stocks: a review, with emphasis on the southeast Asia demersal fisheries. ICLARM Studies Review, 135.

Pella, J.J., Tomlinson, P.K., 1969. A generalized stock production model. Bull. IATTC 13 (3), 420–496.

Pérez-Pérez, M., Wakida-Kusunoki, A., Solana-Sansores, R., Burgos-Rosas, R., Santos, J., 2006. La Pesquería de pulpo. 523–543. In: Arreguín-Sánchez, F., Beléndez-Moreno, L., Méndez-Gómez-Humarán, I., Solana-Sansores, R., Rangel-Dávalos, C. (Eds.), Sustentabilidad y Pesca Responsable en Mexico. Evaluación y Manejo. Instituto Nacional de Pesca, SAGARPA, México.

Pope, J.G., 1972. An investigation of accuracy of Virtual Population Analysis. Znt. Comm. NW Atl. Fish. Res. Bull. 9, 65–74.

Pope, J., 1979a. Stock assessment in multi species fisheries with special reference to the trawl fishery in the Gulf of Thailand. South China Sea Fish. Development Coordination Programme, SCS/DEV/79. 10.

Pope, J.G., 1979b. A modified cohort analysis in which constant natural mortality is replaced by estimates of predation levels. ICES CM 1979/H:16.

Pope, J.G., Shepherd, J.G., 1985. A comparison of the performance of various methods for tuning VPA using effort data. J. Cons. Int. Explor. Mer. 42, 129–151.

Ramos-Miranda, J., Quiniou, L., Flores-Hernández, D., Do-Chi, T., Ayala-Pérez, L., Sosa-López, A., 2005. Spatial and Temporal Changes in the Nekton of the Terminos Lagoon, vol. 66. Campeche, México (2): pp. 513-530.

Ricker, W.E., 1954. Stock and recruitment. J. Fish. Res. Board Canada 11 (5), 560–623.

Schaefer, M.B., 1954. Some aspects of the dynamics of populations important to the management of the commercial marine fisheries. Inter-Am. Trop. Tuna Comm. Bull. 1, 27–56.

Shaefer, M.B., 1957. A study of the dynamics of the fishery for the yellow fin tuna in the eastern tropical Pacific Ocean. Inter-Am. Trop. Tuna Comm. Bull. 2, 247–268.

Ulanowicz, R., 1986. Growth and Development: Ecosystem Phenomenology. Springer-Verlag, New York. 203p.

Ulanowicz, R.E., Goerner, S.J., Lietaer, B., Gomez, R., 2009. Quantifying sustainability: resilience, efficiency and the return of information theory. Ecol. Complex. 6, 27–33.

Wakida-Kusunoki, A.T., Solana-Sansores, R., Sandoval-Quintero, M.E., Núñez-Márquez, G., Uribe-Martínez, J., González-Cruz, A., Medellín-Ávila, M., 2006. Camarón del Golfo de México y Caribe. In: Arreguín-Sánchez, F., Beléndez-Moreno, L., Méndez-Gómez-Humarán, I., Solana-Sansores, R., Rangel-Dávalos, C. (Eds.), Sustentabilidad y Pesca Responsable en Mexico. Evaluación y Manejo. Instituto Nacional de Pesca, SAGARPA, México, pp. 427–476.

Walter, G.G., 1981. Surplus models for fisheries management. In: Chapmanans, D.G., Gallucci, F. (Eds.), Quantitative Population Dynamics. International Cooperaion Publishing House, Fairland, Maryland, pp. 151–180.

Walter, G.G., 1978. A surplus yield model incorporating recruitment and applied to a stock of Atlantic mackerel (*Scomber scombrus*). J. Fish. Res. Board Canada 35 (2), 225–234.

Walters, C., Christensen, V., Pauly, D., 1997. Structuring dynamic models of exploited ecosystems from trophic mass-balance assessments. Rev. Fish Biol. Fish 7, 1–34.

Zetina-Rejón, M.J., 2004. Efectos de la pesca en ecosistemas interdependientes: Laguna de Términos y Sonda de Campeche, México. Tesis Doctorado. Centro Interdisciplinario de Ciencias Marinas del IPN, México.

Zorach, A.C., Ulanowicz, R.E., 2003. Quantifying the complexity of flow networks: how many roles are there? Complexity 8 (3), 68–76.

# 30

# A Network Model of the Hierarchical Organization of Supra-Individual Biosystems

Federica Ciocchetta*, Davide Prandi#, Michele Forlin#, Ferenc Jordán*

*THE MICROSOFT RESEARCH, UNIVERSITY OF TRENTO CENTRE FOR COMPUTATIONAL AND SYSTEMS BIOLOGY, PIAZZA MANIFATTURA 1, 38068 ROVERETO, ITALY, #CENTRE FOR INTEGRATIVE BIOLOGY, UNIVERSITY OF TRENTO, VIA DELLE REGOLE 101, 38123 TRENTO, ITALY

## 30.1 Introduction

Living systems evolve in a modular way. The hierarchical organization of modules ranges from microcells to the biosphere. These levels are more or less discrete (O'Neill et al. 1989). Although most of biological research focuses on particular organization levels, there is an old interest (Allen and Starr 1982) and an increasing need for connecting these levels. It is now generally recognized that hierarchical organization implies both the horizontal connections (interactions) between modules at the same level and the vertical connections between modules at different levels. One of the tools to study the horizontal interactions is network analysis. Analyzing vertical effects and integrating this with horizontal network models is a challenge (e.g., because of the different spatial and temporal scales; O'Neill et al. 1986, 1989). However, linking the two is essential, for example, vertical interactions can test horizontal innovations in the adaptive cycle model (Holling 2001). The loose vertical/horizontal coupling can ensure the decomposability of hierarchical systems (Wu 1999).

Studying ecological processes seem to be more helpful than static, structural analyses for this kind of integration (Allen and Starr 1982; Senft et al. 1987), especially because functioning is also modular in hierarchical systems (Wu and David 2002). The strength of interactions influences the discreteness of the organizational modules (Kolasa and Pickett 1989) and the effective complexity of the system (Kolasa 2005). It is important to note that the interaction system of hierarchical organizations is vertically asymmetrical but not unidirectional (Wu and David 2002), as in hierarchies of graph theory. Thus, both bottom-up (initial conditions) and top-down (constraints) forces play their roles (Wu and David 2002). There are several ways how the hierarchy has been conceptualized in

Models of the Ecological Hierarchy. DOI: http://dx.doi.org/10.1016/B978-0-444-59396-2.00030-4
ISSN 0167-8892, 

biology, for example, the hierarchy of feedbacks (Jörgensen et al. 1998), energy hierarchy (Brown et al. 2004), spatial hierarchy in large-scale systems (O'Neill et al. 1989), and hierarchy in community organization (generalists versus specialists: Waltho and Kolasa 1994; abundant versus rare species: Kolasa 2006). We focus on hierarchy in terms of organizational levels.

Since there are a variety of interconnected, parallel processes in hierarchical systems, there is a need for novel computational solutions. A process algebra-based approach seems to be promising here, as it can provide simple, process-based descriptions of ecosystems. We use the BlenX process algebra for building a stochastic simulation model and evaluate its results. The important question is whether social or landscape processes have larger effects on community dynamics under a particular set of conditions. We present an individual-based, stochastic, dynamical modeling framework, where horizontal and vertical interactions are described in the same network context for three levels: social networks (SNs) of different populations are linked in food webs (FWs), and the FWs of communities are linked in a metacommunity. Thus, processes characterizing the social group, the ecological community, and the landscape are integrated in the same model. We are interested in the dynamics of a particular metapopulation, how sensitive it is to demographical, local community ecological, and global landscape ecological processes. Since the model is very complicated with a large number of parameters, we deliberately narrow our interest to the local perturbation of one parameter set and focus on the conceptual and methodological novelties. We propose that in order to understand the decomposition of hierarchical systems, it may be useful to utilize the composability property of process algebra-based models (i.e., the complex systems can be relatively easily defined, given a small set of local rules).

## 30.2 A Network Approach: Networks at All Levels

Network models are useful tools for ecological research at several organizational levels (Jordán et al. 2010). Vertically linking these horizontal models is a new and great challenge, even if field evidence has been calling it for a long time. SN structure, representing the interactions among individuals, influences the behavior of the whole population (Lusseau, 2003; Gadagkar, 2001). Interspecific interactions affect spatial community dynamics (vanNouhuys and Hanski, 2005; Miller and Kneitel, 2005; Davies et al., 2005). Landscape level changes may have fundamental effects on local communities and their populations (Kruess and Tscharntke, 1994; Zabel and Tscharntke, 1998; Crooks and Soulé, 1999; Komonen et al., 2000; Polis et al., 1997). These vertical, interlevel effects can link different organizational levels and integrate horizontal interactions into a single multilayer ecosystem. The interplay of horizontal and vertical processes is a key property of living systems; understanding this may shed some light on better understanding the diversity and vulnerability of nature (Holling 2001). Trade-offs between competition ability and mobility (Thomas, 2000), local extinction—re-colonization dynamics (Spiller and Schoener, 1998)—and extinction following inbreeding (Saccheri et al., 1998) are well-known examples for hierarchical, multilevel processes.

# 30.3 Methods

## 30.3.1 The Model

We study FWs composed of five species: the E top predator consumes a "specialist" (C) and a "generalist" (D) prey. The former feeds on a single producer (A), while the latter one feeds on two primary producers (A and B) (Fig. 30.1). Each of the five habitat patches contains these identical FWs (but population size and interaction strength do differ between patches). Among habitat patches, individuals can migrate to different extent, specific to the species. Within each population, individual organisms may or may not be linked to each other, forming a dynamical SN (ties can be formed and broken). Structural properties of the SNs, the FWs, and the landscape graph (LG) influence the rates of processes at neighboring levels (i.e., SN on FW, FW on SN, FW on LG, and LG on FW). Direct effects between SN and LG are also possible but, for simplicity, we focus here only on their interactions mediated by the FW level. We focus on the population dynamics of species D and study its sensitivity to changing different kinds of parameters in the system.

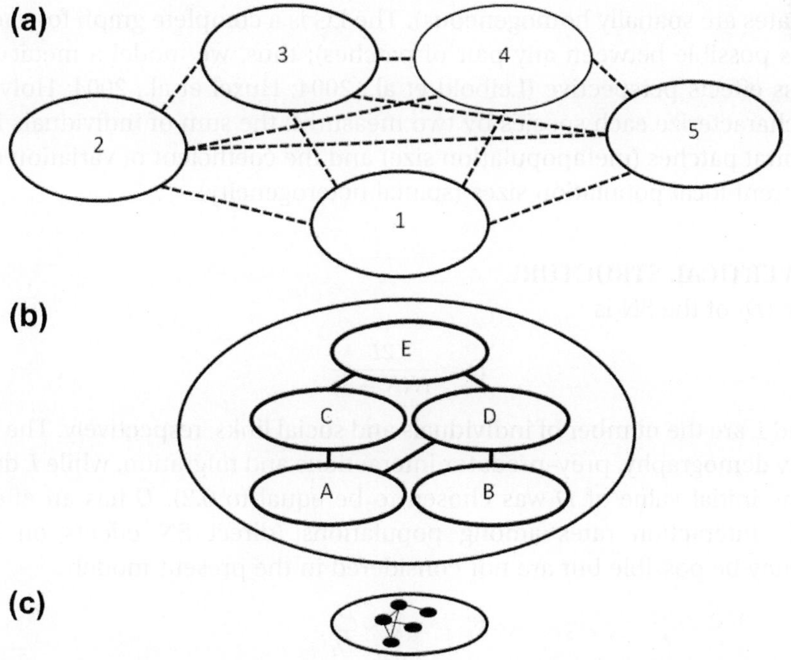


**FIGURE 30.1 The landscape is composed of five habitat patches.** In each patch, the local community is composed by five species and finally, every population is composed by a SN of individuals.

### 30.3.1.1 Structure

#### 30.3.1.1.1 HORIZONTAL STRUCTURE

Species are characterized by the initial number of individuals (*init*, in the simplest case, is constant in each patch and differs only for species). Population dynamics is governed by the reproduction rate $b_i$ (the difference between the birth and the death rate). For simplicity, we consider this to be spatially homogeneous. Also, conspecific individuals may be connected to each other: the initial SN density can be increased by $p$ (social tie formation: the probability of adding a nonexisting link) and decreased by $q$ (probability of removing an existing link) in each step. If $p = q$, SN density remains constant on the long term. The SN need not be a connected graph (i.e., two nodes $i$ and $j$ may become separated in different graph components).

We consider undirected FWs (population dynamical effects do spread in both directions between the prey and the predator). Since our dynamical model is quite complicated regarding the number of parameters, we keep FW topology simple and constant. This FW topology had already been structurally and dynamically analyzed (Jordán et al., 2002, 2003a). It is a highly reliable network topology (Jordán and Molnár, 1999), with low probability of disconnection. FW links represent prey–predator interactions with well-defined initial strength parameters ($k$). For simplicity, we sometimes denote $k$ values by the names of species (the rate of species D feeding on species A is denoted as DA).

Each species forms a metapopulation, where the migration rate is $m$ (for simplicity, migration rates are spatially homogeneous). The LG is a complete graph for each species (dispersal is possible between any pair of patches); thus, we model a metacommunity from a mass effects perspective (Leibold et al., 2004; Huxel et al., 2004; Holyoak et al., 2005). We characterize each species by two measures: the sum of individuals inhabiting the five habitat patches (metapopulation size) and the coefficient of variation calculated for the different local population sizes (spatial heterogeneity).

#### 30.3.1.1.2 VERTICAL STRUCTURE

The density ($D$) of the SN is

$$D = \frac{2L}{N(N-1)}, \tag{1}$$

where $N$ and $L$ are the number of individuals and social links, respectively. The value of $N$ is shaped by demography, prey–predator interactions and migration, while $L$ depends on $p$ and $q$ (the initial value of $D$ was chosen to be equal to 0.2). $D$ has an effect on the interspecific interaction rates among populations (direct SN effects on landscape processes may be possible but are not considered in the present model):

$$k_{t+1} = k_t \times \frac{D_j}{D_i}, \tag{2}$$

where the rate of interspecific interaction at time $(t+1)(k_{t+1})$ between consumer $j$ and food $i$ depends on the previous rate at time $t(k_t)$ as well as on the ratio of their SN densities. This effect can be understood as a dense SN resulting in a social group performing efficient information processing, leading to high consumption rate (e.g., by learning) and low rate of being consumed (e.g., by successful predator avoidance behavior).

The connectance of the FW influences both SN dynamics ($p$ and $q$) and landscape dynamics ($m$). For two hypothetical examples, in highly connected FWs, competition is generally more intense, and a more flexible SN (larger $p$) may better serve the population (FW on SN), while migration intensity (larger $m$) may also be increased (FW on LG). The interplay between SN and FW as well as between FW and LG is mutual. The connectance of a FW is as follows:

$$C = \frac{2R}{S(S-1)}, \tag{3}$$

where $S$ and $R$ are the number of species and interspecific interactions, respectively. The actual value of $p$ at time $(t+1)(p_{t+1})$ depends on $C$ according to

$$p_{t+1} = \frac{p_t}{1 - \frac{C}{10}}, \tag{4}$$

where $p_t$ is the previous value at time $t$. According to these assumed effects, in a dense FW, the SN dynamics of individual species is intense: both creation and deletion of links are likely.

Similarly, FW topology affects migration rates of all species. If the density of a FW is $C$, then the migration rates of all species will change as follows:

$$m_{t+1} = \frac{m_t}{1 - \frac{C}{10}}, \tag{5}$$

So, in a dense FW, species migrate frequently (because of increased competition, for example). The properties of the LG affect the interspecific interaction rates among populations (direct LG effects on social processes may be possible but are not considered in the present model):

$$k_{t+1} = k_t \times \frac{V_j}{V_i}, \tag{6}$$

where $V_i$ and $V_j$ are the variance of local population size in different habitat patches for prey $i$ and predator $j$, respectively. The distribution of individuals among habitat patches in the landscape is the higher-level property influencing the lower, community-level processes (interspecific interactions). The heterogeneous distribution of local population size is understood as the species becoming specialized in one patch but going to extinction in the other. Instead, homogeneous distribution indicates a habitat-generalist species. We consider the specialist species to be a better predator (higher capture rates) and a successful prey (lower rates of being captured). These bottom-up and top-down mechanisms unite all of the studied processes into a single, hierarchical, spatial ecosystem model (cf. Loreau et al., 2003b).

### *30.3.1.2 Dynamics*

We model the dynamics of this hierarchical system in a stochastic framework, written in the process algebra-based BlenX programming language (Dematté et al., 2008).

#### 30.3.1.2.1 THE BLENX LANGUAGE

BlenX is a stochastic programming language explicitly designed to model interactions of biological entities. BlenX represents a biological entity as a box, composed by a set of interfaces and a process. Interfaces represent the interaction capabilities of a box (e.g., social tie formation, predation, and migration). The internal program (P) codifies for the mechanism of the transformation of an interaction into an internal change (e.g., reproduction, change of interactive capabilities). All the possible interactions are associated with a rate, expressing their dynamics. These rates are used when analyzing the system in order to derive the actual rate in stochastic simulation (Gillespie 1977). The BlenX program can be executed within the Beta Workbench (BWB, Dematté et al. 2008), a set of tools to design, simulate, and analyze models written in BlenX. The BWB simulator implements an efficient variant of the Gillespie algorithm (Gillespie 1977), allowing for using the quantitative information of BlenX to run stochastic simulations. For a more accurate description of BlenX, see Dematté et al. (2008).

#### 30.3.1.2.2 THE BLENX MODEL

In the BlenX model of our hierarchical network, habitat patches are identified by names. Each box has a location interface with a specific type, representing the patch. Corridors are formed by migration actions from one patch to another. Each individual $i$ is represented by a box as illustrated in Fig. 30.2. They have four interfaces, representing the patch (location interface), interspecific interactions, migration, and an interface indicating whether the box is enabled to reproduce. The basal species (E, D, and C) have two additional interfaces, for intraspecific interactions and spontaneous reproduction (defined by $b$).

Migration of an individual of a given species in a given patch is described as changing its location (its rate is assumed not to depend on patch identity). In the present model, the interspecific interactions involve only the prey and the predator. If the predator eats the prey, there are two options (their relative frequency is defined by rates). One option is that the predator remains unchanged and the prey disappears. The alternative is that the predator reproduces and the prey disappears. These are clear simplifications, considering the variety of feeding behaviors and functional responses seen in nature. However, this kind of simplistic kinetics can be quite realistic in certain cases. The predator $i$ can interact with the prey $j$ by means of intercommunication on compatibles interfaces. The death of an individual $i$ is represented by the "death" process (and a predefined death rate). Birth rates are different for producers and consumers. Individuals of producers reproduce spontaneously (with a predefined birth rate), while the individuals of consumers are assumed to reproduce only after feeding as described earlier. Social interactions are also defined for conspecific individuals living in the same habitat patch.

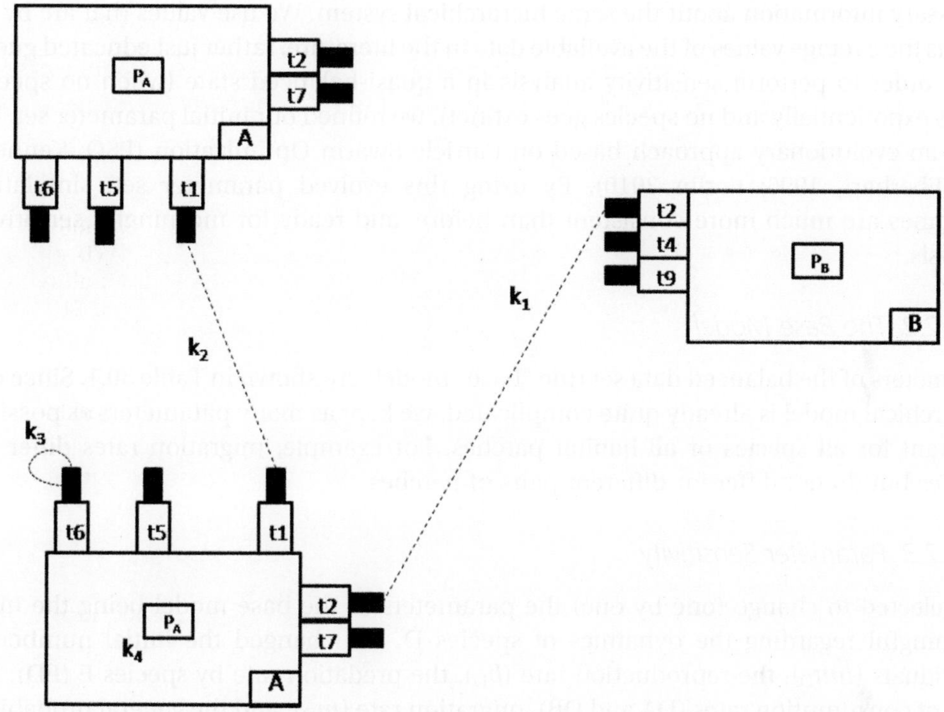


**FIGURE 30.2 The box representation of our process algebra-based ecosystem model.** An individual of species A can be involved in various processes, for example, with rate $k_1$, it can eat an individual of species B (compatibility coded by the same $t_2$ type of binder), with rate $k_2$ it can form a social tie to another conspecific A individual; with rate $k_3$, it can switch spatial location between habitat patches, and with rate $k_4$ it can reproduce. Some of these processes require two individuals, while others are performed by a single individual.

Fig. 30.2 provides a schematic representation of what can happen with an individual. This individual-based description (Grimm and Railsback, 2005; Grimm et al., 2005, 2006) allows for efficient stochastic simulations of this richly parameterized hierarchical system, based on simple kinetic laws between the boxes representing the organisms (see also Livi et al., 2011).

## 30.3.2 Simulation

We run 100 simulations for each parameter set and sampled the time-series after 60 steps.

### *30.3.2.1 Parameter Search*

All of the parameters we used vary massively in nature; thus, there is no way to build a "universal" model of typical parameters. Apart from the variability and lack of data, we also mention that the available data describe different systems; there is no hope to find all necessary information about the same hierarchical system. We use values that are by no means the average values of the available data in the literature, rather just educated guess.

In order to perform sensitivity analysis in a quasi-balanced state (when no species grows exponentially and no species goes extinct), we refined our initial parameter set. We used an evolutionary approach based on Particle Swarm Optimization (PSO; Kennedy and Eberhart, 1995; Forlin 2010). By using this evolved parameter set, simulation outcomes are much more consistent than before, and ready for meaningful sensitivity analysis.

#### *30.3.2.2 The Base Model*

Parameters of the balanced data set (the "base" model) are shown in Table 30.1. Since our hierarchical model is already quite complicated, we kept as many parameters as possible constant for all species or all habitat patches. For example, migration rates differ for species but do not differ for different pairs of patches.

#### *30.3.2.3 Parameter Sensitivity*

We selected to change (one by one) the parameters of the base model being the most meaningful regarding the dynamics of species D. We changed the initial number of individuals ($init_D$), the reproduction rate ($b_D$), the predation rate by species E (ED), the ratio of consumption rates (DA and DB), migration rate ($m_D$), and the general probability of social tie formation ($p$). We always tried ten alternative values for each parameter.

For each parameter set, species are characterized in four ways. For each local population, the average (av) and the coefficient of variation (CV) of its local population size were measured (based on 100 simulations). Calculating the CV (dynamical variability) is a more valuable information, regarding the stochastic nature of our model: it is increasingly recognized that variability itself has a great conservation value (Feest et al.,

**Table 30.1** The Parameters of the Base Model. The First Panel Shows the Migration Rate (*m*), the Reproduction Rate (*b*), and the Initial Population Size (*init*) for the Five Species. The Second Panel Shows the Feeding Rates (*k*)

| | A | B | C | D | E |
|---|---|---|---|---|---|
| *m* | 0.001 | 0.001 | 0.03 | 0.01 | 0.08 |
| *b* | 0.272199 | 0.119943 | 0.044891 | 0.236937 | 0.080983 |
| *init* | 30 | 30 | 12 | 16 | 4 |
| ***k*** | **A** | **B** | **C** | **D** | **E** |
| A | 0 | 0 | 0.000335 | 5.60E-05 | 0 |
| B | 0 | 0 | 0 | 2.90E-06 | 0 |
| C | 0 | 0 | 0 | 0 | 0.000348 |
| D | 0 | 0 | 0 | 0 | 1.73E-05 |
| E | 0 | 0 | 0 | 0 | 0 |

2010), as variability is the key to adaptability and evolvability. Especially in the case of small populations, by focusing on average values, one may lose essential information on variability. Based on both the average population size and the dynamical variability, the sum (metapopulation size) and the CV (spatial variability) of the local values were calculated for the total landscape.

Besides the series of the other parameters, a slightly different gradient was studied for the feeding habits of species D. In the original model, its feeding rate on species A was more intense (DB < DA). We calculated the sum of the feeding coefficients on species A (DA) and species B (DB), and distributed this constant amount of food intake in five ways: D feeding only on A, more on A than B (reversing the original strengths), on both A and B equally, more on B (original values), and only on B. These cases form a gradient from specialization on A versus B.

**Table 30.2** The Effects of Changing the Social Tie Formation (*p*) Parameter on Species D. The First Panel Shows the Average Population Size for Each Patch (Based on 100 Simulations). Boldface Numbers Show the Parameter Value and Results of the Base Model. Metapopulation Size (Sum) and Spatial Heterogeneity (CV) Are Given for Each *p* Value. The Second Panel Shows the Same Information Calculated for Dynamical Variability Instead of the Average

| | | | | av | | | |
|---|---|---|---|---|---|---|---|
| ***p*** | **Patch 1** | **Patch 2** | **Patch 3** | **Patch 4** | **Patch 5** | **Sum** | **CV** |
| 0.001 | 12.70 | 12.43 | 13.00 | 13.78 | 12.98 | 64.89 | 0.04 |
| 0.01 | 12.14 | 12.31 | 12.53 | 12.61 | 12.57 | 62.16 | 0.02 |
| 0.05 | 7.38 | 7.70 | 7.40 | 8.00 | 7.72 | 38.20 | 0.03 |
| 0.09 | 4.19 | 4.18 | 4.30 | 4.37 | 4.40 | 21.44 | 0.02 |
| **0.1** | **3.90** | **3.61** | **3.87** | **4.06** | **3.88** | **19.32** | **0.04** |
| 0.11 | 3.38 | 3.39 | 3.49 | 3.53 | 3.66 | 17.45 | 0.03 |
| 0.15 | 3.00 | 2.95 | 2.97 | 3.15 | 3.26 | 15.33 | 0.04 |
| 0.2 | 3.14 | 3.05 | 2.93 | 3.05 | 3.06 | 15.23 | 0.02 |
| 0.5 | 3.16 | 3.11 | 2.94 | 3.09 | 3.05 | 15.35 | 0.03 |
| 1 | 2.95 | 3.17 | 2.94 | 3.12 | 3.25 | 15.43 | 0.04 |
| | | | | **CV** | | | |
| ***p*** | **Patch 1** | **Patch 2** | **Patch 3** | **Patch 4** | **Patch 5** | **Sum** | **CV** |
| 0.001 | 0.16 | 0.19 | 0.18 | 0.16 | 0.18 | 0.87 | 0.07 |
| 0.01 | 0.20 | 0.18 | 0.17 | 0.17 | 0.18 | 0.89 | 0.07 |
| 0.05 | 0.22 | 0.21 | 0.21 | 0.22 | 0.21 | 1.07 | 0.03 |
| 0.09 | 0.29 | 0.28 | 0.30 | 0.29 | 0.31 | 1.47 | 0.04 |
| **0.1** | **0.31** | **0.30** | **0.29** | **0.31** | **0.30** | **1.52** | **0.03** |
| 0.11 | 0.30 | 0.30 | 0.32 | 0.32 | 0.28 | 1.52 | 0.05 |
| 0.15 | 0.33 | 0.29 | 0.33 | 0.30 | 0.31 | 1.57 | 0.05 |
| 0.2 | 0.32 | 0.32 | 0.35 | 0.31 | 0.27 | 1.57 | 0.08 |
| 0.5 | 0.32 | 0.32 | 0.29 | 0.28 | 0.27 | 1.48 | 0.07 |
| 1 | 0.32 | 0.30 | 0.33 | 0.30 | 0.29 | 1.55 | 0.06 |

## 30.4 Results

Table 30.2 shows patch-specific results for how $p$ influences the average population size and dynamical variability of species D. Metapopulation size and spatial heterogeneity are shown for both. Table 30.3 shows only the four measures (no patch-specific data), for all other studied parameters.

Social effects on species D are strong (Fig. 30.3a and g): more constant social structure (small $p$) is associated with larger metapopulation size and more or less constant spatial heterogeneity. Community effects on D have been quantified in both top-down and bottom-up direction. As the ED feeding link becomes stronger, the metapopulation size of D will decrease, with no clear trend in spatial heterogeneity (Fig. 30.3d and j). Specializing on either prey A or prey B seems to be more advantageous than a mixed diet

**Table 30.3** We Show All Parameter Values Used for Sensitivity Analysis ($p$: Social Tie Formation Probability; $m_D$: Migration Rate; $b_D$: Reproduction Rate; $init_D$: Initial Population Size; ED: Feeding Rate of Species E on Species D; DA/DB: the Ratio of the DA and DB Feeding Rates). Note that in the Latter Case We Have only Five Values (See Explanation in Text). Simulations Are Quantified Based on Both Average (AV) and dynamical Variability (CV) but only Metapopulation Size Is Shown. Boldface Numbers Always Mark the Base Model

| $p$ | av | CV | $m_D$ | av | CV | $b_D$ | av | CV |
|---|---|---|---|---|---|---|---|---|
| 0.001 | 64.89 | 0.865716 | 0.0001 | 19.4 | 1.388127 | 0.01 | 358.85 | 2.519488 |
| 0.01 | 62.16 | 0.891539 | 0.001 | 19.8 | 1.549011 | 0.1 | 25.79 | 1.610623 |
| 0.05 | 38.2 | 1.071461 | 0.005 | 21.07 | 1.884041 | 0.15 | 22.43 | 1.527775 |
| 0.09 | 21.44 | 1.465215 | 0.009 | 22.85 | 2.230705 | 0.2 | 20.49 | 1.466361 |
| **0.1** | **19.32** | **1.516331** | **0.01** | **19.32** | **1.516331** | **0.23** | **19.32** | **1.516331** |
| 0.11 | 17.45 | 1.519911 | 0.011 | 24.07 | 2.412564 | 0.3 | 18.41 | 1.463097 |
| 0.15 | 15.33 | 1.570017 | 0.015 | 25.69 | 2.51086 | 0.5 | 15.62 | 1.470573 |
| 0.2 | 15.23 | 1.572709 | 0.02 | 30.61 | 3.22823 | 1 | 13.6 | 1.452073 |
| 0.5 | 15.35 | 1.483091 | 0.05 | 5.98 | 11.48031 | 10 | 7.07 | 3.563893 |
| 1 | 15.43 | 1.547393 | 0.1 | 4.38 | 11.02667 | 100 | 0 | |

| *ED* | av | CV | $init_D$ | av | CV | *DA / DB* | av | CV |
|---|---|---|---|---|---|---|---|---|
| 0.000001 | 19.3 | 1.465919 | 1 | 19.49 | 1.46853 | all A | 19.84 | 1.486231 |
| 0.000005 | 19.99 | 1.470503 | 5 | 19.83 | 1.499592 | more A | 19.11 | 1.500539 |
| 0.00001 | 19.42 | 1.52149 | 10 | 19.4 | 1.574386 | equal | 19.55 | 1.407909 |
| 0.000017 | 19.31 | 1.605641 | 15 | 19.45 | 1.468933 | **more B** | **19.32** | **1.516331** |
| **1.73E-05** | **19.32** | **1.516331** | **16** | **19.32** | **1.516331** | all B | 19.6 | 1.491042 |
| 0.00002 | 19.42 | 1.504596 | 17 | 19.46 | 1.506806 | | | |
| 0.00005 | 19.35 | 1.492578 | 20 | 19.57 | 1.506419 | | | |
| 0.0001 | 19.04 | 1.519475 | 25 | 19.44 | 1.572372 | | | |
| 0.001 | 18.73 | 1.544358 | 50 | 19.73 | 1.513188 | | | |
| 0.01 | 18.4 | 1.692906 | 100 | 19.5 | 1.533017 | | | |

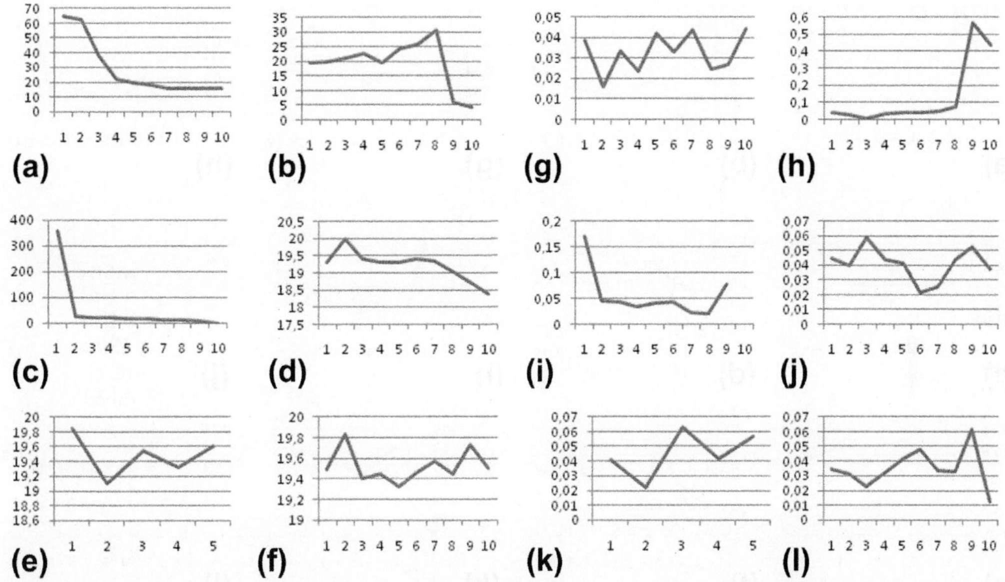


**FIGURE 30.3** The metapopulation size (a–f) and spatial heterogeneity (g–l) of species D, as a function of *p* (a and g), *m* (b and h), *b* (c and i), ED (d and j), DA/DB (e and k), and *init* (f and l). Simulations are quantified by average population size values. The sum of local population sizes measures metapopulation size and their coefficient of variation measures spatial heterogeneity. Parameter values are shown in Table 30.3 (the *x*-axis shows their series, with a reasonable value disturbed in both directions, see the text for more details). For color version of this figure, the reader is referred to the online version of this book.

(Fig. 30.3e and k). Landscape effects on D are weak for a range of parameters. However, when D migrates frequently (larger *m*D), its metapopulation size will drop suddenly and its spatial variability will increase (Fig. 30.3b and h). The effects on dynamical variability have also been measured: larger *p* and larger *m* resulted in higher dynamical variability (Fig. 30.4).

It is a question how important different kinds of disturbances are. In the range of the checked parameters, landscape (*m*) changes had largest effects on the mean population size of species D (Fig. 30.5a and c). The reproduction rate (*b*) is always a dominant factor, while the initial population size (*init*) matters mostly for the spatial heterogeneity of dynamical variability (Fig. 30.5d). Community effects (*prey*, *pred*) were weaker for the average behavior; however, these parameters could strongly influence the dynamical variability of species D.

## 30.5 Conclusions

Analyzing hierarchical biological systems is problematic, because of the large number of parameters needed and the lack of data describing the same system. We focused on conceptual and technical issues instead of trying to provide a very realistic case study.

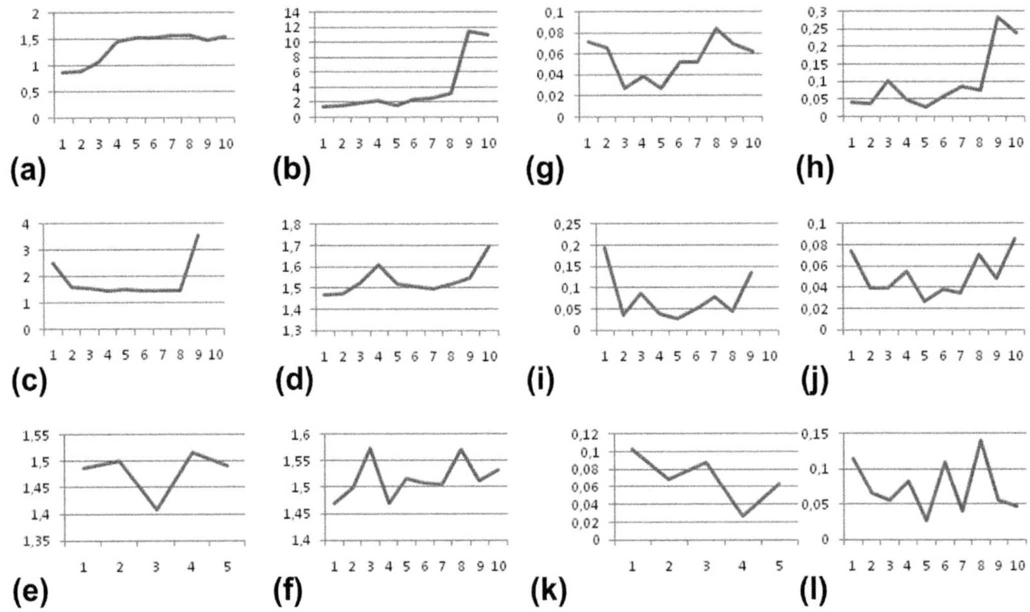


**FIGURE 30.4** The metapopulation size (a–f) and spatial heterogeneity (g–l) of species D, as a function of *p* (a and g), *m* (b and h), *b* (c and i), ED (d and j), DA/DB (e and k), and *init* (f and l). Simulations are quantified by the coefficient of variation (dynamical variability). The sum of local population sizes measures metapopulation size and their coefficient of variation measures spatial heterogeneity. Parameter values are shown in Table 30.3 (the *x*-axis shows their series, with a reasonable value disturbed in both directions, see the text for more details). For color version of this figure, the reader is referred to the online version of this book.

Our model was explicitly designed to study the joint dynamics of three levels of hierarchical organization in nature. Our results present the sensitivity of a single evolved parameter set. Systematic mapping of the parameter space is impossible but the model can be filled with realistic parameters of actual interest and the neighborhood of this parameter space can be mapped in the high-dimensional parameter space.

We quantified the relative importance of various kinds of parameters in influencing the metapopulation dynamics of species D. The effect of migration on the dynamical variability of species D (Fig. 30.4b) seems to support field observations that low dispersal results in more stable communities (Davies et al., 2005). Some of our findings may support earlier results on a metacommunity (Melián et al., 2005) showing higher interpatch variance for omnivores (see Fig. 30.3h). Since the effects of changing the migration rate result in curves with intermediate maxima (Fig. 30.3b), we can also support the hypothesis of highest community insurance at an intermediate dispersal (Loreau et al., 2003a).

Potential future extensions include exploring a larger region of the parameter space, the potential effects of landscape topology (Jordán et al., 2003b, 2007; Pascual-Hortal and Saura, 2006; Baranyi et al., 2011) and adding direct links between SN and LG (see Holt et al., 2005).

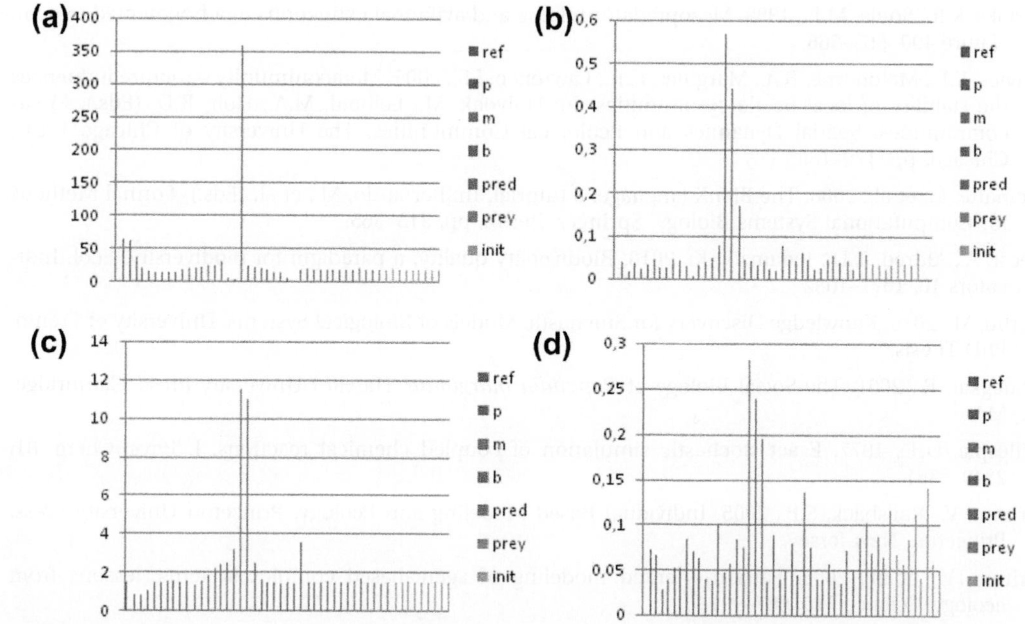


**FIGURE 30.5** Comparing the sensitivity of species D to all parameters changed. Metapopulation size and spatial heterogeneity based on the average are shown on the *y*-axis in (a) and (b), while the same based on dynamical variability are shown in (c) and (d). "ref" shows the result of the base model, "prey" corresponds to DA/DB, and "predator" corresponds to ED. Different parameter sets correspond to different columns (the *x*-axis has no particular meaning). For color version of this figure, the reader is referred to the online version of this book.

The main features of our approach are that they are highly integrative (the model integrates processes and data from several disciplines). Our focus was on the hierarchical perspective (vertical processes linking different organizational levels) and provided information based on stochastic simulations (measuring dynamical variability). We suggest that this kind of integrative model may contribute to better understand the role of indirect determination in ecosystems (Patten, 1981, 1991; László, 2004).

# Acknowledgments

We are grateful to Marco Scotti and Wei-chung Liu for technical help and collaboration.

# References

Allen, T.F.H., Starr, T.B., 1982. Hierarchy: Perspectives for Ecological Complexity. The University of Chicago Press, Chicago.

Baranyi, G., Saura, S., Podani, J., Jordán, F., 2011. Contribution of habitat patches to network connectivity: redundancy and uniqueness of topological indices. Ecol. Indicators 11, 1301–1310.

Brown, M.T., Odum, H.T., Jörgensen, S.E., 2004. Energy hierarchy and transformity in the universe. Ecol. Model. 178, 17–28.

Crooks, K.R., Soulé, M.E., 1999. Mesopredator release and avifaunal extinctions in a fragmented system. Nature 400, 563–566.

Davies, K.F., Melbourne, B.A., Margules, C.R., Lawrence, J.F., 2005. Metacommunity structure influences the stability of local beetle communities. In: Holyoak, M., Leibold, M.A., Holt, R.D. (Eds.), Metacommunities. Spatial Dynamics and Ecological Communities. The University of Chicago Press, Chicago, pp. 170–188.

Dematté, L., et al., 2008. The BlenX language: a tutorial. In: Bernardo, M., et al. (Eds.), Formal Methods for Computational Systems Biology. Springer, Berlin, pp. 313–365.

Feest, A., Aldred, T.D., Jedamzik, K., 2010. Biodiversity quality: a paradigm for biodiversity. Ecol. Indicators 10, 1077–1082.

Forlin, M., 2010. Knowledge Discovery for Stochastic Models of Biological Systems. University of Trento. PhD Thesis.

Gadagkar, R., 2001. The Social Biology of *Ropalidia marginata.* Harvard University Press, Cambridge, MA.

Gillespie, D.T., 1977. Exact stochastic simulation of coupled chemical reactions. J. Phys. Chem. 81, 2340–2361.

Grimm, V., Railsback, S.F., 2005. Individual-Based Modeling and Ecology. Princeton University Press, Princeton, New Jersey.

Grimm, V., et al., 2005. Pattern-oriented modeling of agent-based complex systems: lessons from ecology. Science 310, 987–991.

Grimm, V., et al., 2006. A standard protocol for describing individual-based and agent-based models. Ecol. Model. 198, 115–126.

Holling, C.S., 2001. Understanding the complexity of economic, ecological, and social systems. Ecosystems 4, 390–405.

Holt, R.D., Holyoak, M., Leibold, M.A., 2005. Future directions in metacommunity ecology. In: Holyoak, M., Leibold, M.A., Holt, R.D. (Eds.), Metacommunities. Spatial Dynamics and Ecological Communities. The University of Chicago Press, Chicago, pp. 465–490.

Holyoak, M., Leibold, M.A., Mouquet, N.M., Holt, R.D., Hoopes, M.F., 2005. Metacommunities: a framework for large-scale community ecology. In: Holyoak, M., Leibold, M.A., Holt, R.D. (Eds.), Metacommunities. Spatial Dynamics and Ecological Communities. The University of Chicago Press, Chicago, pp. 1–32.

Huxel, G.R., Polis, G.A., Holt, R.D., 2004. At the frontier of the integration of food web ecology and landscape ecology. In: Polis, G.A., Power, M.E., Huxel, G.R. (Eds.), Food Webs at the Landscape Level. The University of Chicago Press, Chicago, pp. 434–451.

Jörgensen, S.E., Mejer, H., Nielsen, S.N., 1998. Ecosystem as self-organizing critical systems. Ecol. Model. 111, 261–268.

Jordán, F., Molnár, I., 1999. Reliable flows and preferred patterns in food webs. Evol. Ecol. Res. 1, 591–609.

Jordán, F., Scheuring, I., Vida, G., 2002. Species positions and extinction dynamics in simple food webs. J. Theor. Biol. 215, 441–448.

Jordán, F., Scheuring, I., Molnár, I., 2003a. Persistence and flow reliability in simple food webs. Ecol. Model. 161, 117–124.

Jordán, F., Báldi, A., Orci, K.M., Rácz, I., Varga, Z., 2003b. Characterizing the importance of habitat patches and corridors in maintaining the landscape connectivity of a *Pholidoptera transsylvanica* (Orthoptera) metapopulation. Landscape Ecol. 18, 83–92.

Jordán, F., Magura, T., Tóthmérész, B., Vasas, V., Ködöböcz, V., 2007. Carabids (Coleoptera: Carabidae) in a forest patchwork: a connectivity analysis of the Bereg Plain landscape graph. Landscape Ecol. 22, 1527–1539.

Jordán, F., Baranyi, G., Ciocchetta, F., 2010. A hierarchy of networks spanning from individual organisms to ecological landscapes. In: Estrada, E., et al. (Eds.), Network Science: Complexity in Nature and Technology. Springer Verlag, Berlin, pp. 165–184.

Kennedy, J., Eberhart, R., 1995. Particle swarm optimization. Proceedings of IEEE International Conference on Neural Networks. vol. IV, 1942–1948.

Kolasa, J., 2005. Complexity, system integration, and susceptibility to change: biodiversity connection. Ecol. Complex. 2, 431–442.

Kolasa, J., 2006. A community ecology perspective on variability in complex systems: the effects of hierarchy and integration. Ecol. Complex. 3, 71–79.

Kolasa, J., Pickett, S.T.A., 1989. Ecological systems and the concept of biological organization. PNAS 86, 8837–8841.

Komonen, A., et al., 2000. Forest fragmentation truncates a food chain based on an old growth forest bracket fungus. Oikos 90, 119–126.

Kruess, A., Tscharntke, T., 1994. Habitat fragmentation, species loss, and biological control. Science 264, 1581–1584.

László, E., 2004. Nonlocal coherence in the living world. Ecol. Complex. 1, 7–15.

Leibold, M.A., Holyoak, M., Mouquet, N., Amarasekare, P., Chase, J.M., Hoopes, M.F., Holt, R.D., Shurin, J.B., Law, R., Tilman, D., Loreau, M., Gonzalez, A., 2004. The metacommunity concept: a framework for multi-scale community ecology. Ecol. Lett. 7, 601–613.

Livi, C.M., Jordán, F., Lecca, P., Okey, T.A., 2011. Identifying key species in ecosystems with stochastic sensitivity analysis. Ecol. Model. 222, 2542–2551.

Loreau, M., Mouquet, N., Gonzalez, A., 2003a. Biodiversity as spatial insurance in heterogeneous landscapes. Proc. Natl. Acad. Sci. USA 100, 12765–12770.

Loreau, M., Mouquet, N., Holt, R.D., 2003b. Meta-ecosystems: a theoretical framework for a spatial ecosystem ecology. Ecol. Lett. 6, 673–679.

Lusseau, D., 2003. The emergent properties of a dolphin social network. Proc. R. Soc. Lond. Ser. B. Biol. Sci. 270, S186–S188.

Melián, C.J., Bascompte, J., Jordano, P., 2005. Spatial structure and dynamics in a marine food web. In: Belgrano, A., Scharler, U.M., Dunne, J.D., Ulanowicz, R.E. (Eds.), Aquatic Food Webs: an Ecosystem Approach. Oxford University Press, Oxford, pp. 19–24.

Miller, T.E., Kneitel, J.M., 2005. Inquiline communities in pitcher plants as a prototypical metacommunity. In: Holyoak, M., Leibold, M.A., Holt, R.D. (Eds.), Metacommunities. Spatial Dynamics and Ecological Communities. The University of Chicago Press, Chicago, pp. 122–145.

O'Neill, R.V., DeAngelis, D.L., Waide, J.B., Allen, T.F.H., 1986. A Hierarchical Concept of Ecosystems. Princeton University Press, Princeton.

O'Neill, R.V., Johnson, A.R., King, A.W., 1989. A hierarchical framework for the analysis of scale. Landscape Ecol. 3, 193–205.

Pascual-Hortal, L., Saura, S., 2006. Comparison and development of new graph-based landscape connectivity indices: towards the prioritization of habitat patches for conservation. Landscape Ecol. 21, 959–967.

Patten, B.C., 1981. Environs: the superniches of ecosystems. Am. Zool. 21, 845–852.

Patten, B.C., 1991. Concluding remarks. Network ecology: indirect determination of the life-environment relationship in ecosystems. In: Higashi, M., Burns, T.P. (Eds.), Theoretical Studies of Ecosystems—the Network Perspective. Cambridge University Press, Cambridge, pp. 288–351.

Polis, G.A., Anderson, W.B., Holt, R.D., 1997. Toward an integration of landscape and food web ecology: the dynamics of spatially subsidized food webs. Annu. Rev. Ecol. Syst. 28, 289–316.

Saccheri, I., Kuussaari, M., Kankare, M., Vikman, P., Fortelius, W., Hanski, I., 1998. Inbreeding and extinction in a butterfly metapopulation. Nature 392, 491–494.

Senft, R.L., Coughenour, M.B., Bailey, D.W., et al., 1987. Large herbivore foraging and ecological hierarchies. Bioscience 37, 789–799.

Spiller, D.A., Schoener, T.W., 1998. Lizards reduce spider species richness by excluding rare species. Ecology 79, 503–516.

Thomas, C.D., 2000. Dispersal and extinction in fragmented landscapes. Proc. R. Soc. Lond. B 267, 139–145.

vanNouhuys, S., Hanski, I., 2005. Metacommunities of butterflies, their host plants, and their parasitoids. In: Holyoak, M., Leibold, M.A., Holt, R.D. (Eds.), Metacommunities. Spatial Dynamics and Ecological Communities. The University of Chicago Press, Chicago, pp. 99–121.

Waltho, N., Kolasa, J., 1994. Organization of instabilities in multispecies systems, a test of hierarchy theory. PNAS 91, 1682–1685.

Wu, J., 1999. Hierarchy and scaling: extrapolating information along a scaling ladder. 1999. Can. J. Remote Sens. 25, 367–380. 1999.

Wu, J., David, J.L., 2002. A spatially explicit hierarchical approach to modeling complex ecological systems: theory and applications. Ecol. Model. 153, 7–26.

Zabel, J., Tscharntke, T., 1998. Does fragmentation of Urtica habitats affect phytophagous and predatory insects differntially? Oecologia 116, 419–425.

# 31

# Hierarchical Energy Dissipation in Populations

Fernando R. Momo*,**, Santiago R. Doyle*, José E. Ure*
*INSTITUTO DE CIENCIAS, UNIVERSIDAD NACIONAL DE GENERAL SARMIENTO, J.M. GUTIÉRREZ 1150, LOS POLVORINES, ARGENTINA; CONICET, ARGENTINA, **INEDES, UNIVERSIDAD NACIONAL DE LUJÁN, ARGENTINA

*This chapter is devoted to three great ecologists: Leonardo Malacalza, Eduardo Rapoport and Ramón Margalef (in memoriam). They all have been a guide, an inspiration and an example.*

## 31.1 Introduction to Populations as Thermodynamic Systems

Biological populations are characterized by structural and demographic magnitudes such as density, age or stage structure, and growth and mortality rates. These are the typical magnitudes that are studied by ecologists, who analyze the relationships among them in terms of population dynamics. There is, however, another point of view to study biological populations, which consider them as open thermodynamic systems, i.e., systems that interchange matter and energy with their environment, and are, moreover, the units of entropy production and entropy exchange (Michaelian, 2005).

Populations are constituted by individuals, which are complex dissipative systems that remain far from thermodynamic equilibrium (Ulanowicz and Hannon, 1987). But the population is a system of a higher order of complexity: in a population, individuals born and die, also grow, and, eventually, suffer metamorphosis or changes in their reproductive stage.

The set of all interacting individuals constitute the system that we call population. It is a complex dissipative system, which is capable to regulate certain emergent properties.

If we adopt this point of view, it is clear that populations are also far from its thermodynamic equilibrium. Populations are, in general, in some stationary state characterized by a given biomass, density, age (or stage) structure, and a specific regime of energy and entropy fluxes.

In order to formalize these concepts, we must carefully define the different terms in stake.

We will focus animal populations, and start with a very imperfect analogy, considering the population like a sort of "turbine" activated by a "fall" of energy, which is sourced by

Models of the Ecological Hierarchy. DOI: http://dx.doi.org/10.1016/B978-0-444-59396-2.00031-6 

food. Because populations are open systems, they dissipate low-quality energy (heat) and pour degraded matter to the environment. Of course, food is not the sole energy that population receives; however, we assume here that it is at thermal equilibrium and other forms of energy such as heat or light are in a stationary regime. Then, the biological work made by individuals (that is translated as demographic changes at the population level) will be supported by the chemical energy in the food.

The energetic gradient under which populations work is given by the difference of free energy between source and sink (i.e., between foods and waste). We will call this gradient $\Delta E_{SourSink}$ (see Fig. 31.1), that is, the free energy contained in the food ($E_{So}$) minus the energy contained in the detritus ($E_{Si}$). The flux of energy running throughout a population is directly proportional to this gradient.

According to Prigogine (1978), in nonequilibrium stationary states, the perturbation in the produced entropy is given by

$$\frac{1}{2}\frac{\mathrm{d}}{\mathrm{d}t}\delta^2 S = \sum(\delta J_p \delta X_p). \tag{1}$$

Here, $\delta S$ is the perturbation in entropy production, the terms $\delta J_p$are the deviation in the rates of various irreversible processes, and $\delta X_p$ are the deviations in the generalized forces (affinities, gradients of temperature, etc.).

The question is what happens when the system (i.e., the population) is far from equilibrium and, moreover, far from the stationary regime?

In order to tackle this problem, we propos to follow the previous analogy and consider that, like in another physical systems, when a flux that is proportional to a given gradient exists, entropy dissipation is proportional to the square of that gradient (it is non linear). We can synthesize these as follows:

$$\begin{aligned} &\frac{\mathrm{d}E}{\mathrm{d}t} \propto k_1 \Delta E_{\mathrm{SourSink}} \\ &\frac{\mathrm{d}_e S}{\mathrm{d}t} \propto k_2 (\Delta E_{\mathrm{SourSink}})^2. \end{aligned} \tag{2}$$

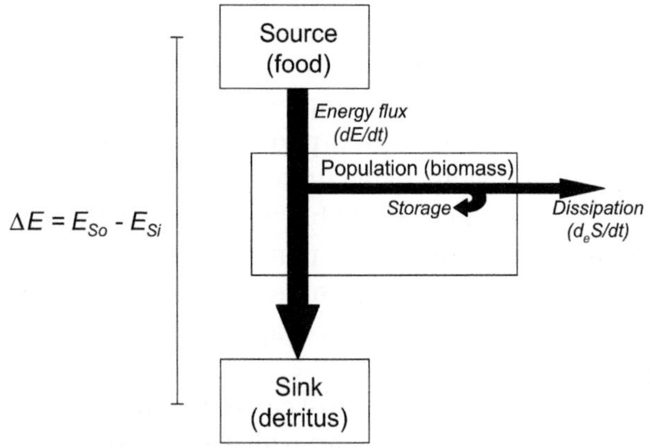

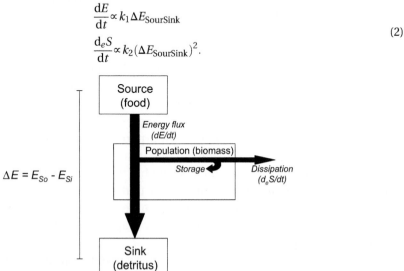

**FIGURE 31.1** Conceptual representation of biological populations as dissipative systems.

In Eqn (2), $\frac{d_eS}{dt}$ represents the amount of entropy dissipated by the population to its environment per unit of time.

Energy flows (Fig. 31.1) can be associated to population dynamics and to different demographic phenomena: inputs or energy are transformed basically in work and biomass. The biological work is measurable throughout the metabolic rate of the individuals in the population and, due to the second law of thermodynamics it is in a direct correlation with a fraction of the entropy exchange between organisms and environment. The other fraction of the entropy exportation is measured by mortality.

The amount of energy that is not spent is stored as new biomass.

Finally, egestion, excretion, and other matter exchange phenomena are responsible of the remainder flux of energy (and entropy) to the detritus (the sink).

When $\Delta E$ varies abruptly, the regime of the system becomes far from the stationary state; the rapid (nonlinear) increase of the dissipated entropy allows the system to recovery the stationary regime but in a new state. In a population, this new state will be defined by a given age and size structure.

## 31.2 Demography and Thermodynamics

In a recent work, we have shown that several modifications in the population demography can be interpreted in a thermodynamic way (Momo et al., 2010). In fact, we can say that populations are dissipative systems because they maintain their age structure and biomass by pumping entropy to their environment.

Assuming that if populations make some kind of mechanical work, this work will be transformed finally to heat, and by analogy with other thermodynamic systems, we can postulate the following relationship:

$$\frac{dS}{dU} = \frac{1}{T} \tag{3}$$

where $S$ is the entropy, $U$ is the internal energy, and $T$ is temperature. However, this relationship is incomplete for ecological purposes because dissipative systems maintain their organization by mean of fluxes of entropy and energy, and it is more accurate to write:

$$\frac{dS/dt}{dU/dt} = \frac{1}{T}. \tag{4}$$

Equation (4) represents the relationship between entropy expulsion, and the energy exchange between the system and its environment. If both fluxes are calculated, the magnitude $\tau = \frac{1}{T}$ may be interpreted as a demographic sensitivity.

Two demographic tools can be used to this approach: the first are the Leslie matrices, which represent populations divided into age classes of the same length as the time step, and that can be used to simulate the population dynamics; the second are the survival curves, which show the percent of a cohort that survives until a given age, and can be associated to different demographic traits suggesting how population allocate the available energy.

In Momo et al. (2010), we have also shown that Leslie matrices contain information about the way in which a population dissipates energy and pumps entropy outside the system, and have also shown that *r*-selected populations have higher entropic costs than *K*-selected populations, whereas populations having Type I survival curves (high survival of immature and low survival of adults) have lower entropic costs than populations with Type IV (high larvae mortality) survival curves.

Consider now a Leslie matrix (*A*) with three age classes, where only the last age class reproduces

$$A = \begin{bmatrix} 0 & 0 & f \\ s_1 & 0 & 0 \\ 0 & s_2 & s_3 \end{bmatrix}, \tag{5}$$

where $f$ is the fertility and $s_i$ are survival rates.

In this demographic model, we can represent different trade-off, as we will show in a moment. In that work, authors assumed that $f = 1/(s_1 s_2)$ in order to have a $\lambda_1 > 1$ (growing populations); then, numerical simulations were performed considering different cases having more or less fertility combined with more or less survivals (i.e., the *r–K* trade-off). In addition, the relationship between $s_1$ and $s_2$ was changed in order to simulate the shift between different survival curves (from type I to IV).

Computing the demographic entropy ($H = \Sigma pi \log(pi)$) for each gradient, authors observed that both trade-offs are not equivalent and *K*-selected populations maximize their demographic entropy (Fig. 31.2).

Despite the interest of this approximation, it is strongly limited because the relationship shown in Eqn (3) is only true under linear and near to equilibrium conditions, and the demographic entropy considered in the model is only a measurement of the relative abundance of individuals of different ages; so, it is not a true entropy in a thermodynamic sense, and it does not represent a flux. In consequence, it is necessary to deepen in the problem in order to catch the complete picture including nonlinear and far from equilibrium regimes and energy and entropy fluxes.

## 31.3 Body Size and Metabolic Rate

The metabolic rate is the fundamental biological rate, because it is a measure of energy uptake, transformation, and allocation (Brown et al., 2004). Metabolic rate is linked to the rates of many other biological activities at various hierarchical levels of organization (Brown et al., 2004; Glazier, 2005). Moreover, studies of metabolism are useful tools for understanding the patterns of energy flow in populations and ecosystems (Glazier, 1991; Doyle and Momo, 2009; Doyle et al., 2012).

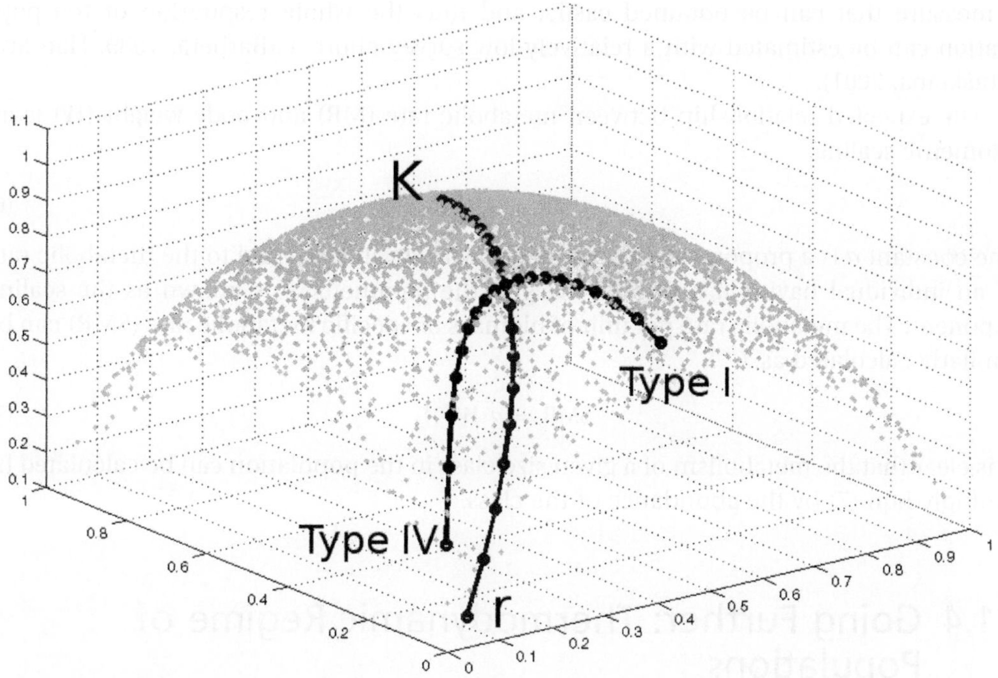


**FIGURE 31.2** Phase space of populations following *r–K* or Type IV—Type I trade-offs. Horizontal axis are percent of ages 2 and 3; the vertical axis indicates the entropy. Points indicate the level of demographic entropy.

Several variables influence the metabolic rate of organisms; among these, temperature and body size (generally understood as body weight) are the most important (Gillooly et al., 2001).

Temperature is the dominant abiotic factor that affects metabolic rate in ectothermic organisms, and its effect on metabolic rate is due to the thermodynamic nature of the kinetics of chemical reactions and enzyme conformational changes (Hill et al., 2004). In ectothermic animals, metabolic rate increases exponentially with temperature within a certain range up to a maximum value, and further increase in temperature produces a decrease in metabolic rate (Schmidt-Nielsen, 1997). The acute response of metabolic rate to temperature is due to the thermodynamic nature of the kinetics of chemical reactions and enzyme conformational changes (Hill et al., 2004).

Body size is the most biotic factor that regulates metabolic rate, and it has been recognized like that since the seminal work of Kleiber (1932). More recently, the books by Peters (1983) and Schmidt-Nielsen (1984) have summarized some regularities relating the body size with the physiological and population parameters, including metabolic rate. Knowledge of the relationship between body size and metabolic rate has multiple applications in ecology; for instance, it is possible to estimate the whole respiration of a population from its body size distribution. In an animal population, size distribution is a measure that can be obtained easily, and thus the whole respiration of the population can be estimated with a relatively low survey effort (LaBarbera, 1989; Han and Straškraba, 2001).

The expected relationship between metabolic rate (MR) and body weight ($W$) is an allometric scaling:

$$\mathrm{MR} = a \cdot W^b. \tag{6}$$

The constant $a$ is a proportionality parameter whose value is equal to the metabolic rate of an individual having a unit body weight; the parameter $b$ is known as the scaling exponent. The metabolic rate per unit of biomass or specific metabolic rate (SMR) can be similarly calculated as:

$$\mathrm{SMR} = a \cdot W^{b-1}. \tag{7}$$

It is clear that the metabolism of a given size class in the population can be calculated by multiply Eqn (7) by the abundance of the class.

## 31.4 Going Further: Thermodynamic Regime of Populations

Let us go back to the Prigogine (1978) approximation: if a population is in a stationary state, its amount of internal (structural) entropy remains roughly constant. If it is assumed that the system is near the equilibrium, we can expand the function of entropy as follows:

$$S = S^* + \delta S + \frac{1}{2}\delta^2 S, \tag{8}$$

were $S^*$ denotes the entropy in the stationary state, and $\delta S$ is the fluctuation in entropy around that stationary value. Because $\delta^2 S$ can be considered a Lyapunov function of the system (Jørgensen and Svirezhev, 2004), a given state will be stable (in both the thermodynamics sense and the Lyapunov sense) if it is verified that $\delta S = 0$, $\delta^2 S < 0$, and $\frac{\mathrm{d}}{\mathrm{d}t}(\delta^2 S) > 0$. In this condition, biomass, density, and age structure of the population remain constant, while $\frac{\mathrm{d}_i S}{\mathrm{d}t} = \frac{\mathrm{d}_e S}{\mathrm{d}t}$, and the total amount of internal entropy of the population remains constant. Hence, the mortality rate, integrated through all ages, equals the birthrate of the population.

However, populations rarely are near to the equilibrium and therefore these relationships ought to be different, because the entropy pumping to the environment grows nonlinearly with the energy gradient as is shown in Eqn (2). The distance from equilibrium can be measured by mean of one of the classic magnitudes used in the literature, among which *exergy* is one which has been most widely used (Jørgensen and Svirezhev,

2004). Meysman and Bruers (2007) have proposed the use of the disequilibrium between resource and waste products as a meaningful measure of the distance from equilibrium of an ecological system. In the context of this chapter, this last approximation is a good way to relate the forcing variable (the energetic cascade) with the response variable (the rate of entropy exportation) and, moreover, the state variable of the population (its structural entropy). We will assume that, under realistic conditions, the energetic distances between food and detritus are large, and the population has a very low probability to be near to the thermodynamic equilibrium.

In a far from equilibrium condition, populations fight against the Second Law in order to maintain not only their biomass, but their organization. In this regime, higher-order interaction terms in the relationship between forces and fluxes are not negligible.

## 31.5 Symmetry Breaking and Hierarchical Responses of Population

We postulate that, when the difference of free energy between the source (food) and the sink (detritus) starts to grow, the exportation of entropy to the environment ($d_eS/dt$) starts to vary according to the power relation shown in Eqn (2). One might ask, however, what is exactly $d_eS/dt$? We consider here that this rate is composed by two terms: the first represents the metabolic dissipation of all individuals in the population, and the second represents the mortality of every age or stage class.

The metabolic rate of a given individual, as previously exposed, is mainly determined by its body size and environmental temperature. As a consequence, the amount of entropy exchanged by each size class in the population will be given by the product between the mean body size of the class, the SMR for that size, and the relative abundance of the stage; all these magnitudes are multiplyed by a constant to adjust units and divided by *T*, the absolute temperature.

Mortality rates are also determined by the body size, but affect the relative abundance of each stage.

Briefly, we can write the following:

$$\frac{d_eS}{dt} = \frac{k_3 n}{T} \sum_{i=1}^{m} \mathrm{SMR}_i p_i \overline{B}_i + k_4 n \sum_{i=1}^{m} \mu_i p_i, \tag{9}$$

where $p_i = n_i/n$ is the relative abundance of the $i$-stage, $\overline{B}_i$ is the mean body biomass of the $i$-stage, and $\mu_i$ is its mortality rate.

Considering that $\Delta E$ is growing, the flux of entropy must vary over time and we can write:

$$\frac{d}{dt}\left(\frac{d_eS}{dt}\right) = \frac{d}{dt}\left(\frac{k_3 n}{T} \sum_{i=1}^{m} \mathrm{SMR}_i p_i \overline{B}_i + k_4 n \sum_{i=1}^{m} \mu_i p_i\right). \tag{10}$$

If we expand Eqn (10), under isothermal conditions and assuming that $\overline{B_i}$ are constant, we obtain the following expression:

$$\frac{d_e^2 S}{dt^2} = \frac{dn}{dt}\left(\frac{k_3}{T}\sum_{i=1}^{m}\text{SMR}_i p_i \overline{B}_i + k_4\sum_{i=1}^{m}\mu_i p_i\right) + \\ +n\left\{\frac{k_3}{T}\left[\sum_{i=1}^{m}\left(\frac{dp_i}{dt}\text{SMR}_i \overline{B}_i\right) + \sum_{i=1}^{m}\left(p_i\overline{B}_i\frac{d\text{SMR}_i}{dt}\right)\right] + k_4\left(\sum_{i=1}^{m}\left(\frac{dp_i}{dt}\mu_i\right) + \sum_{i=1}^{m}\left(p_1\frac{d\mu_i}{dt}\right)\right)\right\} \tag{11}$$

We can see that the first term in Eqn (11) represents the effect of the population density variation. In the second term, we have the effect of the whole population metabolism given by the variation in the relative abundances of stages: $\frac{k_3}{T}\left[\sum_{i=1}^{m}\left(\frac{dp_i}{dt}\text{SMR}_i \overline{B}_i\right)\right]$. Similarly, the whole population metabolism can be modified by changes in the specific metabolic rate of each size class: $\frac{k_3}{T}\left[\sum_{i=1}^{m}\left(p_i\overline{B}_i\frac{d\text{SMR}_i}{dt}\right)\right]$. The remaining terms indicate the variations in the flux of entropy that depend on total mortality, influenced by relative abundances and group mortalities variations. All these effects are summarized in Table 31.1.

As the difference between source and sink is increasing ($d\Delta E/dt > 0$), the flux of entropy to the environment also increases. When the ratio between this rate and the energy flow rises the critical values, symmetry can break, in consequence the population rises a new stationary state (i.e., it becomes a new dissipative structure) characterized by a particular amount of structural entropy.

However, the different terms of Eqn (11) change with a particular timing. The first effect in be apparent is the physiological one (term C of Table 31.1) because individuals

**Table 31.1** Mean of the Components of the Eqn (11) (Changes in the Rate of Entropy Exchange Between Population and Environment)

| | | |
|---|---|---|
| A | $\frac{dn}{dt}\left(\frac{k_3}{T}\sum_{i=1}^{m}\text{SMR}_i p_i\overline{B}_i + k_4\sum_{i=1}^{m}\mu_i p_i\right)$ | Is the global effect produced by the change in the population density |
| B | $n\frac{k_3}{T}\left[\sum_{i=1}^{m}\left(\frac{dp_i}{dt}\text{SMR}_i \overline{B}_i\right)\right]$ | Represents the changes in the total metabolism of the population given by changes in the relative abundances of the different size classes |
| C | $n\frac{k_3}{T}\left[\sum_{i=1}^{m}\left(p_i\overline{B}_i\frac{d\text{SMR}_i}{dt}\right)\right]$ | Is the change in metabolism produced by the variation in the SMR of each size class. It represents the physiological effect of modifying the energetic quality and quantity of food |
| D | $nk_4\left[\sum_{i=1}^{m}\left(\frac{dp_i}{dt}\mu_i\right)\right]$ | Is the change in total mortality given by the modification of the relative abundances of size classes |
| E | $nk_4\left[\sum_{i=1}^{m}\left(p_1\frac{d\mu_i}{dt}\right)\right]$ | Is the change in the mortality effect due to the modification in stage-specific rates of mortality |

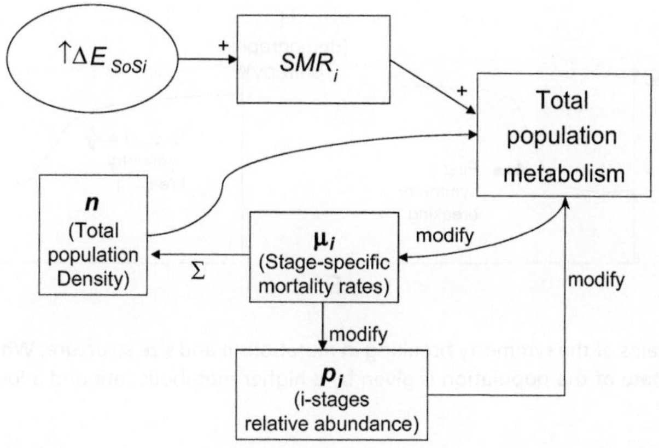


**FIGURE 31.3** Conceptual relationships between the different components of the entropy flux variation.

respond to a better input of food. The well-known effect of food on metabolic rates is the reason by which standard measurement of metabolism are made in fasted animals, and has a characteristic time scales of hours to days. Moreover, a trend to higher metabolic rate with increasing potential food availability, a related but different phenomenon operating at larger time scales, has been observed in several invertebrates (Brockington, 2001; Brockington and Peck, 2001; Fraser et al., 2002; Doyle et al., in press).

This change produces in a second instance a modification in the survival probability for each size class; if the food is better, all stages have lower mortality rates but the effect is mediated by the allometric dependence of the size. In consequence, term E is the second in be modified and causes a shift in relative abundances of stages (term D). Changes in relative abundances produce modifications in the total metabolism of the population (term B).

Finally, the population density (term A) changes until the biomass rises a new state stage. These phenomena are summarized in Fig. 31.3.

In this way, a sort of "cascade" of effects is caused by a change in energy flow throughout the population. This succession of changes can be detected experimentally by the measurement of the metabolic rate of the population and the Shannon entropy of size classes (stages or ages), that is, the demographic entropy *sensu* (Demetrius et al., 2004, 2009). If we plot these magnitudes versus the energetic difference between food and detritus, the phase transitions will be apparent by abrupt changes in the slope of each curve (Fig. 31.4).

## 31.6 Concluding Remarks

Ecological populations involve a set of nested biological phenomena, from physiological to demographic ones. This hierarchical structure constraints and determinates the thermodynamic response to changes in the flux of energy. Each biological process has

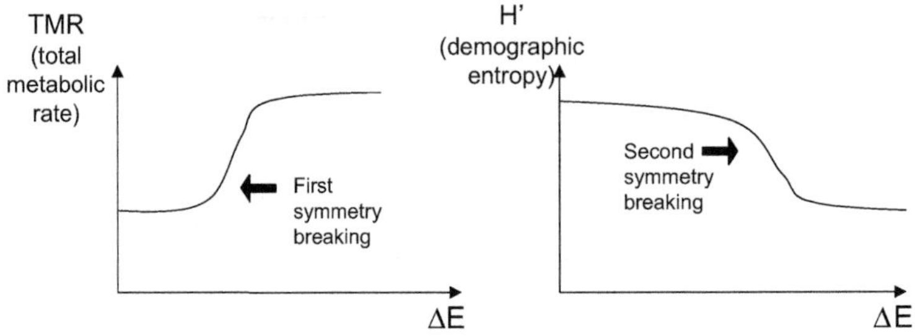


**FIGURE 31.4** Phase diagrams of the symmetry breaking in metabolism and size structure. When the energetic skip is high, the final stable state of the population is given by a higher metabolic rate and a lower diversity of sizes.

a characteristic velocity of response to environmental changes, so there is a successive group of responses in the population.

The consideration of nonlinear effects in thermodynamic processes allows us to predict some general timing in the responses and, in this way, it is possible to test experimentally the validity of our model and hypotheses. First of all, the metabolic rate of individuals should show changes under an increment of energy availability. In a second phase, we should see differential changes in the mortality rates of the different size classes. Finally, the size structure of the population should change together with its density.

Experiments that investigate these predictions will be very important in order to continue building a strong thermodynamic theory of ecological systems.

# References

Brockington, S., 2001. The seasonal energetics of the Antarctic bivalve *Laternula elliptica* (King and Broderip) at Rothera Point, Adelaide Island. Polar Biol. 24, 523–530.

Brockington, S., Peck, L.S., 2001. Seasonality of respiration and ammonium excretion in the Antarctic echinoid *Sterechinus neumayeri*. Mar. Ecol. Prog. Ser. 219, 159–168.

Brown, J.H., Gillooly, J.F., Allen, A.P., Savage, V.M., West, G.B., 2004. Toward a metabolic theory of ecology. Ecology 85 (7), 1771–1789.

Demetrius, L., Gundlach, V.M., Ochs, G., 2004. Complexity and demographic stability in population models. Theor. Popul. Biol. 65, 211–225.

Demetrius, L., Legendre, S., Harremöes, P., 2009. Evolutionary entropy: a predictor of body size, metabolic rate and maximal life span. Bull. Math. Biol. 71, 800–818.

Doyle, S.R., Momo, F.R., 2009. Effect of body size and temperature on the metabolic rate of *Hyalella cusvispina* (Amphipoda). Crustaceana 82 (11), 1423–1439.

Doyle, S.R., Momo, F.R., Brêthes, J.C., Ferreyra, G.A., 2012. Metabolic rate and food availability of the Antarctic amphipod *Gondogeneia antarctica* (Chevreux 1906): seasonal variation in allometric scaling and temperature dependence. Polar Biol. 35 (3): 413-424.

Fraser, K.P.P., Clarke, A., Peck, L.S., 2002. Feast and famine in Antarctica: seasonal physiology in the limpet *Nacella concinna*. Mar. Ecol. Prog. Ser. 242, 169–177.

Gillooly, J.F., Brown, J.H., West, G.B., Savage, V.M., Charnov, E.L., 2001. Effects of size and temperature on metabolic rate. Science 293 (5538), 2248.

Glazier, D.S., 1991. Separating the respiration rates of embryos and brooding females of *Daphnia magna*: implications for the cost of brooding and the allometry of metabolic rate. Limnol. Oceanogr. 36 (2), 354–362.

Glazier, D.S., 2005. Beyond the "3/4-power law": variation in the intra-and interspecific scaling of metabolic rate in animals. Biol. Rev. Camb. Philos. Soc. 80 (4), 611–662.

Han, B.P., Straškraba, M., 2001. Size dependence of biomass spectra and abundance spectra: the optimal distributions. Ecol. Model. 145 (2–3), 175–187.

Hill, R.W., Wyse, G.A., Anderson, M., 2004. Animal Physiology. Sinauer Associates, Sunderland, Massachusetts, 770 pp.

Jørgensen, S., Svirezhev, Y., 2004. Towards a Thermodynamic Theory for Ecological Systems. Elsevier, 366 pp.

Kleiber, M., 1932. Body size and metabolism. Hilgardia 6, 315–353.

LaBarbera, M., 1989. Analyzing body size as a factor in ecology and evolution. Annu. Rev. Ecol. Syst. 20, 97–117.

Meysman, F.J.R., Bruers, S., 2007. A thermodynamic perspective on food webs: quantifying entropy production within detrital-based ecosystems. J. Theor. Biol. 249, 124–139.

Michaelian, K., 2005. Thermodynamics stability of ecosystems. J. Theor. Biol. 237, 323–335.

Momo, F.R., Doyle, S., Ure, J.E., 2010. Leslie matrices and survival curves contain thermodynamical information. In: Mondaini (Ed.), BIOMAT 2009—International Symposium on Mathematical and Computational Biology. World Scientific Publishing, pp. 243–249.

Peters, R.H., 1983. The Ecological Implications of Body Size. Cambridge University Press, Cambridge, U.K., 345 pp.

Prigogine, I., 1978. Time, structure, and fluctuations. Science 201 (4358), 777–785.

Schmidt-Nielsen, 1984. Scaling: Why is Animal Size So Important? Cambridge University Press, Cambridge, U.K., 256 pp.

Schmidt-Nielsen, 1997. Animal Physiology: Adaptation and Environment, fifth ed. Cambridge University Press, Cambridge, U.K., 607 pp.

Ulanowicz, R., Hannon, B., 1987. Life and the production of entropy. Proc. R. Soc. B 232, 181–192.

PART 10

# How to Model the Different Levels of the Ecological Hierarchy?

# 32

# Conclusive Remarks About Hierarchy Theory and Multilevels Models

Sven Erik Jørgensen*,$, Ferenc Jordán†

*COPENHAGEN UNIVERSITY, UNIVERSITY PARK 2, 2100 COPENHAGEN Ø, DENMARK, †THE MICROSOFT RESEARCH UNIVERSITY OF TRENTO, CENTRE FOR COMPUTATIONAL AND SYSTEMS BIOLOGY, PIAZZA MANIFATTURA 1, 38068, ROVERETO, TN, ITALY, $CORRESPONDING AUTHOR, E-MAIL: MSIJAPAN@HOTMAIL.COM

In the Introduction chapter, a short overview of the hierarchical theory was presented. The theory presented is complete as it gives clearly the differences of the basic properties on the different levels of the ecological hierarchy; see Table 1 in the Introduction. There is no doubt that nature is organized in hierarchical levels and it has, as it has been demonstrated in the Introduction, very clear advantages, which can be summarized in the following points:

1. The variations are damped when we go up through the hierarchical levels. Variations and disturbances of the lower levels have dynamics that may affect the lower levels in the hierarchy, but due to the hierarchical organization, the higher levels are much less affected because the variations caused by the disturbances are leveled off.
2. A malfunction of one level can easily be eliminated by replacing a few components on the lower level.
3. The process rates are determined by the openness, which is higher than the lower level. By cooperation of the components in the lower level, the higher levels can obtain the results of these dynamic properties.
4. The levels have synergistic interactions.

If we want to model one level of the ecological hierarchy, it is necessary at least in some situations to include the lower and upper levels as they influence the processes on the focal level. Furthermore, the properties of the hierarchical organization have to be reflected in the ecological models of various hierarchical levels. These considerations lead to the following questions:

1. How can we model these interactions?
2. How can we develop multilevel ecological models?
3. Is our present hierarchical theory sufficient?

Models of the Ecological Hierarchy. DOI: http://dx.doi.org/10.1016/B978-0-444-59396-2.00032-8
ISSN 0167-8892, 

An overview of the theory has been presented briefly in the last chapters of the book, but are additions to the existing theory needed? Chapter 29 shows how ascendency can be used to link two levels in an effort to make a better environmental management of the fish stock.

In order to better understand the ecology of a population, its internal social structure and its interspecific interactions are equally important. While we know a lot about these factors, separately, their integration is still relatively poor. Chapter 30 presents a simple, dynamical model of a hierarchical, three-level ecological system. Individuals of populations form social networks, populations of communities form food webs, and communities in the landscape are linked in a landscape graph. Several parameters are used and sensitivity analysis quantifies the behavior of the system.

Chapters 29 and 30 give an answer to the first two questions in two specific cases. The presented methods may be applicable more generally, but there is no doubt that we need far more experience in the development of multilevel hierarchical models in ecology. It is our hope that this book may inspire young modelers to gain more experience in this type of ecological models, because we need models that can capture this part of the characteristic properties of ecological systems. This book has demonstrated clearly that we can model the various levels, but we have insufficient experience in models combining two or more ecological levels to account properly for the important interactions between levels.

The last chapter, Chapter 31, has the title "Hierarchical Energy Dissipation in Populations" and it may hopefully inspire to a further development of hierarchical theory. The chapter is very convincing in its use of thermodynamics to gain more insight in how the ecological hierarchical levels are working together. We need certainly more approaches that attempt to unite thermodynamics and hierarchical theory. Such approaches will most probably facilitate the use of hierarchical theory in the development of ecological models or in the environmental management. Wider applications of entropy, exergy, power, and ascendency of several hierarchical levels should be tested simultaneously, and it will probably inevitably lead to better models of two or more hierarchical levels and to a deeper insight into how ecosystems are working.

# Subject Index

Page numbers followed by "f" indicates figures and "t" indicates tables

**T**

Printed and bound by CPI Group (UK) Ltd, Croydon, CR0 4YY
08/05/2025
01864822-0002